ASCLEPIUS

ASCLEPIUS

COLLECTION AND INTERPRETATION OF THE TESTIMONIES

BY

EMMA J. EDELSTEIN

AND

LUDWIG EDELSTEIN

VOLUMES I AND II

With a New Introduction by
GARY B. FERNGREN

THE JOHNS HOPKINS UNIVERSITY PRESS
BALTIMORE AND LONDON

The Johns Hopkins University Press
2715 North Charles Street
Baltimore, Maryland 21218-4363
The Johns Hopkins Press Ltd., London

Library of Congress Cataloging-in-Publication Data

Edelstein, Emma Jeannette Levy, 1904–
   Asclepius : collection and interpretation of the testimonies / by Emma J. Edelstein and
Ludwig Edelstein ; with a new introduction by Gary B. Ferngren. — Johns Hopkins
paperbacks ed.
     p.   cm.
   Originally published: Baltimore : Johns Hopkins Press, 1945.
   Includes bibliographical references and indexes.
   ISBN 0-8018-5769-4 (pbk. : alk. paper)
   1. Aesculapius (Greek deity)—Cult. I. Edelstein, Ludwig, 1902–1965. II. Title.
BL820.A4E37 1998
291.2′11—dc21   97-44543
                                                         CIP

A catalog record for this book is available from the British Library.

HENRICO E. SIGERIST

GRATI ANIMI TESTIMONIUM

TABLE OF CONTENTS

VOLUME ONE: COLLECTION OF THE TESTIMONIES

I. LEGEND

II. DESCENDANTS

III. DEIFICATION AND DIVINE NATURE

VI. IMAGES

VII. SANCTUARIES

VOLUME TWO: INTERPRETATION OF THE TESTIMONIES

INTRODUCTION, 1998

Ludwig Edelstein was born in 1902 into a wealthy Jewish family in Berlin.[1] Because of his delicate health, he was educated by private tutors before attending a humanistic gymnasium. He studied classics at the University of Berlin under Werner Jaeger and philosophy and sociology with Eduard Spranger. In 1929 he was awarded the Ph.D. degree in Greek, Latin, and philosophy from the University of Heidelberg, graduating summa cum laude, with a dissertation that analyzed the Hippocratic treatise *On Airs, Waters, and Places*. During his years at Heidelberg, Edelstein was introduced into the circle of Marianne Weber, widow of Max Weber. Here he established lifelong friendships with several leading German intellectuals, among whom were the philosopher Erich Frank and the Indologist Heinrich Zimmer. It was in Heidelberg, too, that Edelstein met a fellow graduate student of classics and archaeology, Emma J. Levy (b. 1904), who was also completing her doctorate. They were married in 1928.

The Edelsteins settled in Berlin, where Ludwig was forced by his father's illness and the Great Depression to take over the family business temporarily. In 1930 he was appointed assistant at the Institute of the History of Medicine and in 1932 lecturer in the history of the exact sciences in classical antiquity in the University of Berlin. But in the following year Adolf Hitler came to power and Edelstein's position was terminated. The Edelsteins traveled to Italy for a year's study, hoping thereafter to return to Germany so that he could resume his academic career. It became increasingly clear, however, that a return was not possible as long as the Nazi regime held power, and Edelstein accepted the invitation of Henry Sigerist to join the staff of the newly founded Institute for the History of Medicine at Johns Hopkins in Baltimore. Sigerist had recently been lured away from the Institute of the History of Medicine in Leipzig to become director of the Institute at Hopkins. Here Edelstein found another young historian of medicine who had been recruited from

---

[1] More complete biographical information may be found in Harold Cherniss, "Ludwig Edelstein," *Yearbook of the American Philosophical Society*, 1965, 130–38 ; George Boas, "Ludwig Edelstein," *Rockefeller University Review* 1 (Nov.-Dec. 1965) : 17–19 ; Fridolf Kudlien, Detlev W. Bronk, and Lloyd G. Stevenson, "In Memoriam Ludwig Edelstein, 1902–1965," *Journal of the History of Medicine and Allied Sciences* 21 (1966) : 173–83 ; Owsei Temkin, "In Memory of Ludwig Edelstein," *Bulletin of the History of Medicine* 40 (1966) : 1–13 ; and "Editors' Introduction," in *Ancient Medicine: Selected Papers of Ludwig Edelstein*, eds. Owsei Temkin and C. Lilian Temkin (Baltimore: Johns Hopkins Press, 1967), vii–xiv.

Germany by Sigerist, Owsei Temkin. Baltimore became the Edelsteins' new home and there they made many new friends, including Arthur O. Lovejoy and other members of the History of Ideas Club, of which Edelstein became an active member. From 1934 until 1947 Edelstein worked closely with his two colleagues to make the Institute the leading center of medical history in the United States.[2]

In 1947 Sigerist left Hopkins to return to his native Switzerland and Edelstein accepted an invitation to join the Classics Department of the University of Washington. Edelstein had always regarded himself primarily as a philologist and a philosopher and he looked forward to finding a home in a department of classics. He had not yet completed his first year in Seattle when he was offered the position of professor of Greek at Berkeley, which he accepted. Soon after his arrival he found the campus divided over the requirement by the Board of Regents that members of the faculty submit to a loyalty oath. Edelstein's memories of the Nazification of German universities led to his strenuous opposition to the oath, and he became one of only eighteen faculty members (of a total of eight hundred) to refuse to take it. Although the Supreme Court of California eventually upheld the stand taken by those who would not sign the oath, Edelstein left Berkeley in 1951 to return to Hopkins, where, one year later, he became the first holder of a newly inaugurated chair of Humanistic Studies.

The years of Edelstein's second period at Hopkins were some of his happiest, but in 1957 Emma was diagnosed with cancer, and she died in the following year. Edelstein's stoicism prevented him from displaying his grief openly, but such was the intellectual and spiritual bond between them that he never fully reconciled himself to her death. In 1960 he accepted a position at the Rockefeller Institute of New York (now Rockefeller University), which he held concurrently with a part-time position at Hopkins until his death in 1965.

As a young man Edelstein had high ambitions as a scholar and he began early to make a name for himself in the study of ancient medicine. Owsei Temkin described the occasion on which he first heard Ludwig Edelstein lecture at the Institute of the History of Medicine in Leipzig in 1930:

---

[2] For an excellent account of these early years of the Institute and its role in the formation of medical history as a professional discipline in the United States, see Elizabeth Fee and Theodore M. Brown, "'Anything but Amabilis': Henry Sigerist's Impact on the History of Medicine in America," in *Making Medical History : The Life and Times of Henry E. Sigerist*, ed. Elizabeth Fee and Theodore M. Brown (Baltimore : Johns Hopkins University Press, 1997), 333–70.

I had heard of this young classical scholar who was working on Hippocrates, and who had remarked that the historians of medicine knew nothing about Hippocrates. Yet I at least had read the entire Littré edition of the Hippocratic works and was curious to hear and to meet the man who could so belittle us! Edelstein spoke for two hours, without notes, slowly, but without hesitation or any slip of the tongue, without rhetorical embellishments or emotional appeal, with the simple forcefulness of logic based on interpretive mastery of every single Hippocratic work and close familiarity with Greek civilization. The Hippocratic physician, the meaning of Hippocratic prognosis, Hippocrates himself, and the history of the collection ascribed to him, all emerged from this lecture in an entirely new light. All of us, including some well-known classicists, listened spellbound.[3]

Edelstein's early work was marked by an iconoclastic spirit. " One sometimes has the impression," wrote Fridolf Kudlien, " that young Edelstein had taken a fancy to questioning everything and to upsetting the commonest of opinions."[4] While younger colleagues like Temkin found him provocative, older colleagues were offended. Edelstein proposed novel interpretations that challenged some of the central assumptions that undergirded the study of Greek medicine. Among the subjects he treated were the social status of the Greek physician, the authorship of the Hippocratic treatises, and the origin and role of the Hippocratic oath.

In 1931 his revised dissertation, *Peri aerön und die Sammlung der hippokratischen Schriften,* was published.[5] The book attracted attention in great part because Edelstein rejected the idealized portrayal of the Greek physician that was then popular. When Edelstein entered the field, the study of Greek medicine was dominated by a kind of presentism. Medical historians regarded Greek medicine as scientific in its outlook and they portrayed Greek physicians as the earliest medical professionals.[6] Edelstein drew attention to the historical and philosophical context of Greek medicine, demonstrating the way in which it was informed by the

---

[3] Owsei Temkin, " The Double Face of Janus," in *The Double Face of Janus and Other Essays in the History of Medicine* (Baltimore : Johns Hopkins University Press, 1977), 18.

[4] Kudlien, et al., " In Memoriam," 174.

[5] Problemata, Heft 4 (Berlin : Weidmannsche Buchhandlung, 1931). Chapters 2 and 3 have been translated into English as " Hippocratic Prognosis " and " The Hippocratic Physician " in Temkin and Temkin, *Ancient Medicine,* 65–85 and 87–110.

[6] On the romantic idealization of Greece that was a part of the older German philological tradition, see Heinrich von Staden, "' Hard Realism ' and ' A Few Romantic Moves ' : Henry Sigerist's Versions of Ancient Greece," in Fee and Brown, *Making Medical History,* 136–61.

particular time and intellectual milieu in which it developed. He argued
that the physician in the classical world was an itinerant craftsman who
learned his trade by apprenticeship. Because the Greek doctor had to
compete with his rivals, he devised prognosis as a means of securing his
reputation, on which his livelihood depended, by accurately predicting
the course of his patients' illnesses. Since there was no medical licensure
or specific training required, anyone could practice medicine. Hence the
typical Greek physician always retained the relatively humble position of
a craftsman. This picture has become, since Edelstein first propounded it,
a widely accepted image of the ancient medical practitioner, but at the
time it was both novel and shocking to those who had come to believe
that the Greek physician was a scientifically trained professional. Edelstein
also expressed skepticism regarding what could be known about the
historical Hippocrates and he doubted that any of the treatises in the
Hippocratic Corpus could rightly be attributed to him. He developed his
views regarding Hippocrates at greater length in two articles that came to
exercise such wide influence that the burden of proof since then has fallen
on those who defend the Hippocratic authorship of any particular
treatise.[7]

Even in his earliest contributions Edelstein demonstrated those virtues
that were to characterize all his published work : a desire to reach independent conclusions based solely on the evidence and regardless of the
current *status quaestionis* ; a willingness to doubt received opinion, irrespective of the authority that lay behind it; and a deeply rooted historical
consciousness that understood that medicine, like all the arts, is culturally
and temporally bound. In 1943 Edelstein turned his attention to the
Hippocratic oath.[8] Medical historians had long accepted the oath as the
foundational document of Greek medical ethics whose timeless principles
had dominated ancient medical practice. Edelstein sought to understand
it as a document that had its origin in a particular time and place. He
argued that it was a late and esoteric document whose precepts were not

---

[7] " Hippokrates von Kos " in A. Pauly, G. Wissowa, and W. Kroll, *Real-Enzyklopädie der
klassischen Altertumswissenschaft*, Supplementband 6 (Stuttgart : J. B. Metzlersche, 1935),
cols. 1290–1345 ; and " The Genuine Works of Hippocrates," *Bulletin of the History of
Medicine* 7 (1939) : 236–48, reprinted in Temkin and Temkin, *Ancient Medicine*, 133–44.
For a more recent assessment that supports Edelstein's skepticism regarding Hippocrates's
authorship of treatises in the Hippocratic Corpus, see G. E. R. Lloyd, " The Hippocratic
Question," *Classical Quarterly* N. S. 25–26 (1975) : 177–92 (reprinted in his *Methods and
Problems in Greek Science : Selected Papers* [Cambridge : Cambridge University Press,
1991], 194–223).

[8] *The Hippocratic Oath : Text, Translation, and Interpretation*. Supplements to the
Bulletin of the History of Medicine, no. 1 (Baltimore : Johns Hopkins Press, 1943).

consonant with much that was characteristic of medical practice as reflected in other treatises of the Hippocratic Corpus. He attempted to prove that it had originated among a small group of physicians who followed the principles of the Pythagorean school.[9]

In 1945 there appeared the only work of scholarship on which the Edelsteins collaborated—their study of the Greek healing god Asclepius.[10] The first volume, which was the work of Emma Edelstein, contained the most important primary sources for Asclepius : 861 numbered literary and epigraphic testimonies in Greek or Latin text with English translation. The second volume, which was written by Ludwig Edelstein, provided a narrative account of the cult of Asclepius, in which the Edelsteins proposed what was in many ways a radical new interpretation. First, they rejected the modern theory that Asclepius had begun as a chthonic deity in favor of the view that he was a culture hero who was later deified. They argued that he was originally a physician from Tricca in Thessaly, a craftsman who was ennobled by Homer in the Iliad. Having thereafter gained legendary status, he became the patron of physicians, who named themselves Asclepiads, or " descendants of Asclepius." In the late sixth century B.C., he was raised to the status of a god in Epidaurus, which became the center of a cult that spread throughout the Mediterranean world.

Secondly, the Edelsteins rejected the view that Greek rational medicine had originated in the temples of Asclepius. According to one tradition, Hippocrates had learned medicine from reading the accounts of cures that were posted in Asclepius' temple on the island of Cos. The Edelsteins maintained that Greek physicians called themselves Asclepiads, but not because they learned medicine from the temples of Asclepius. Rather, as itinerant craftsmen who had lost the ancestral protection of their clans when they left their homes, they formed guilds that enabled them to enjoy protection, as descendants of Asclepius, in the foreign communities in which they practiced medicine.

The Edelsteins argued that there was no conflict in the Greek world between religious healing and healing by physicians. As a physician

---

[9] Edelstein's thesis of a Pythagorean origin of the Hippocratic oath has not found general acceptance among medical historians. For a comprehensive discussion of the oath see Heinrich von Staden, "' In a pure and holy way ' : Personal and Professional Conduct in the Hippocratic Oath," *Journal of the History of Medicine and Allied Sciences* 51 (1996) : 404–37.

[10] *Asclepius : A Collection and Interpretation of the Testimonies.* Texts and Documents, Second Series, no. 2. Publications of the Institute of the History of Medicine (Baltimore : Johns Hopkins Press, 1945).

himself, Asclepius was the patron of physicians and the guardian of their craft. Physicians made offerings to him and supported his cult, but they practiced rational, not religious, medicine. When they could not heal (which was common in an age in which medicine could do little), their patients were free to seek the care of Asclepius, where they hoped to experience healing through dreams and visions, which pilgrims experienced while spending the night in his sanctuary (a practice called incubation). The Edelsteins also rejected the theory of earlier scholars that Asclepius's ability to heal by means of incubation was attributable to fraud and deception by the priests who administered the sanctuaries. They maintained that unlike other healing deities of the ancient world, Asclepius's healings were not the result of pious superstition or magical practices, but that the god healed by rational means. But how does one account for the dreams, the visions, and the cures themselves? The Edelsteins explained the dreams and visions of Asclepius as the products of wish-fulfillment or the incubants' memories of their everyday experiences, including their recollection of previous treatment by physicians. They invoked a multiplicity of explanations to account for Asclepius's healings : the spontaneous recovery of suppliants afflicted with psychosomatic conditions, the natural recovery of incubants who suffered from mild disorders they thought to be more serious than they actually were, and the likelihood that pilgrims were more faithful in following the god's " prescriptions " than they were those of their own physicians. Finally, the Edelsteins contended that Asclepius was the philanthropic god par excellence of the classical world, a deity who evoked personal devotion because his worshippers saw in him a savior-god who (unlike the Olympians) demonstrated concern for all, irrespective of class or status. As a result, his cult spread widely and he became by late antiquity the chief pagan competitor of Christ.

Some of the Edelsteins' interpretations of the evidence were controversial, and not all of them found widespread acceptance. As a philologist, Ludwig Edelstein was concerned with the interpretion of classical texts. Owsei Temkin has remarked that he often based far-reaching conclusions on quite fragmentary texts.[11] Slender evidence might demand cautious conclusions ; yet the manner in which Edelstein built his case obscured its sometimes speculative nature. Readers of his essays are familiar with what Temkin described as the " presentation of his arguments as cogent demonstrations with inescapable results."[12] It is his relentless logic, as

---

[11] Temkin, " In Memory," 5.
[12] " Editors' Introduction," in Temkin and Temkin, eds., *Ancient Medicine*, ix.

much as the evidence on which it is based, that often compels the reader's assent. The Edelsteins' analysis of the primary sources for Asclepius is tightly constructed and carries great authority. But it is at times over speculative, while at crucial points it goes beyond the evidence.

The Edelsteins limited their testimonies to the literary and epigraphic evidence. They included no archaeological evidence, such as temples, statues, or coins, although they occasionally made reference to it. This exclusion necessarily weakens the basis on which the interpretive portion of the study rests (as the Edelsteins themselves admitted) and it imparts a narrowly philological perspective to the work, which perhaps appears more old-fashioned today than it did when the work first appeared. An analysis of the sanctuaries of Asclepius at Cos, Epidaurus, and Pergamum, and of the numerous statues of Asclepius, as well as coin types and inscriptions, would have greatly enriched their study.[13] Moreover, the Edelsteins eschewed an approach that compared Asclepius with other healing deities. Healing was one of the most typical functions of Greek and Roman gods and heroes. The Greeks sought healing from nearly all the gods, but several deities and heroes became known specifically for

[13] Alessandra Semeria has compiled a census of all known Asclepieia in mainland Greece and the islands, together with a bibliography of literary, numismatic, and epigraphic evidence as well as topographical and excavation reports, for each (" Per un censimento degli *Asklepieia* della Grecia continentale e delle isole," *Annali della Scuola Normale Superiore di Pisa, Classe di Lettere e Filosofia*, series 3, 16 [1986] : 931–58). Comprehensive discussions of the evidence of the archaeological sites may be found in Roland Martin and Henri Metzger, " La personnalité d'Asclépios," in *La religion grecque* (Presses Universitaires de France, 1976), 69–109 ; and Fritz Graf, "Heiligtum und Ritual das Beispiel der Griechisch-Römischen Asklepieia," in *Le Sanctuaire grec* (Entretiens sur l'Antiquité classique, vol. 37), ed. Albert Schachter and Jean Bingen (Geneva : Fondation Hardt, 1992), 159–99. For a short description of the archaeological remains of the Asclepieion at Cos see Susan M. Sherwin-White, *Ancient Cos : An Historical Study from the Dorian Settlement to the Imperial Period*, Hypomnemata 51 (Göttingen : Vandenhoeck & Ruprecht, 1978), 340–46. On the Asclepieion at Epidaurus see Nicolaos Yalouris, " Epidauros," in *The Princeton Encyclopedia of Classical Sites*, ed. Richard Stillwell (Princeton : Princeton University Press, 1976), 311–14. On the Asclepieion at Pergamum see Otfried Deubner, *Das Asklepieion von Pergamon : Ein kurz vorläufige Beschreibung* (Berlin : Verlag für Kunstwissenschaft, 1938) ; and on that at Corinth see Carl Roebuck, *The Asklepieion and Lerna : Corinth*, vol. 14 (Princeton : Princeton University Press, 1951) and Mabel Lang, *Cure and Cult in Ancient Corinth : American Excavations at Old Corinth*, Corinth Notes no. 1 (Princeton : Princeton University Press, 1977). On the Asclepieion at Athens see S. B. Aleshire, *The Athenian Asklepieion : The People, Their Dedications, and the Inventories* (Amsterdam : J. C. Gieben, 1989). For descriptions of the most recently discovered of the major Asclepieia, that at Messene, see Anastasios K. Orlandos, " Νεώτεραι έρευναί έν Μεσσήνηι (1957–1973)," in *Neue Forschungen in griechischen Heiligtümern*, ed. Ulf Jantzen (Tübingen : Wasmuth, 1976), 9–38 ; and Christian Habicht, *Pausanias' Guide to Ancient Greece* (Berkeley : University of California Press, 1985), 37–63. On the iconography of Asclepius see B. Holtzmann, " Asklepios," in *Lexicon Iconographicum Mythologicae Classicae*, ed. J. C. Balty, J. Boardman, Ph. Bruneau, et al, II. 1 (Zurich: Artemis Verlag, 1984), 863–97.

their healing ability. They included (in addition to Asclepius) Apollo, Amphiaraus, Trophonius, Isis, Serapis, and others.[14] The Edelsteins drew few parallels between Asclepius and his rivals. It is true that their method was an inductive one in which they attempted to reconstruct the cult of a single god by assembling every available scrap of evidence, a method that does not lend itself to a broadly comparative study. Nevertheless, by isolating one Greek healing cult (albeit the most important) from all others, the Edelsteins created a picture of the role of Asclepius in the classical world that lacked a wider perspective. Nor did they place Asclepius in the historical context of the pre-Greek healing cults, including those that had developed in the ancient Near East long before the advent of the worship of Asclepius.[15]

If we move from methodological considerations to the interpretation that the Edelsteins placed on the evidence, we find that a number of their conclusions are open to challenge. The thesis that Asclepius was a culture hero who was deified relatively late has not gained universal acceptance, with several scholars continuing to argue that he was regarded as a god from the beginning.[16] It is probable, too, that the Edelsteins viewed the cult of Asclepius too benignly. They underrated the elements of priestly manipulation, superstition, and propaganda that were found in his cult. Pace the Edelsteins, the records of the miracle-cures (*iamata*) from Epidaurus reveal the priests' willingness to exaggerate accounts of healing, which they shrewdly exploited for propaganda purposes.[17] Their thesis that the temple healing of Asclepius was *sui generis* is open to question. They maintained that whereas other healing cults resorted to irrational means of healing (e.g., magic and witchcraft), that of Asclepius employed rational means that imitated methods used by secular healers. But, in fact, Asclepius also used irrational and fantastic means of healing not very

[14] Two older but still valuable treatments are W. A. Jayne, *The Healing Gods of Ancient Civilizations* (New Haven : Yale University Press, 1925) ; and E. Thrämer, " Health and Gods of Healing (Greek)," in *Encyclopedia of Religion and Ethics*, ed. James Hastings, vol. 6 (New York : Charles Scribner's Sons, 1914), 540–53.

[15] It has long been argued that elements of Greek healing cults (e.g., incubation) were borrowed from ancient near-eastern practices. For a discussion of these elements in the cult of Asclepius, see Walter Burkert, *The Orientalizing Revolution : Near Eastern Influence on Greek Culture in the Early Archaic Age*, trans. Margaret E. Pinder and Walter Burkert (Cambridge : Harvard University Press, 1992), 75–79.

[16] See, for example, Gregory Vlastos, " Religion and Medicine in the Cult of Asclepius : A Review Article," *Review of Religion* 13 (1949) : 269–90, especially 270–76 ; C. [Karl] Kerényi, *Asklepios : Archetypal Image of the Physician's Existence*, trans. Ralph Manheim (New York : Pantheon/Bollingen Foundation, 1959), xiii–xviii ; and Christa Benedum, " Asklepiosmythos und archäologischer Befund," *Medizinhistorisches Journal* 22 (1987) : 48–61.

[17] Vlastos, " Religion and Medicine," 276–80.

different from those of other deities.[18] In emphasizing what they describe as the unique philanthropic spirit and benevolence of the god, the Edelsteins' picture tends toward idealization. The philanthropy of the god was by no means unique; nor is the evidence that he had a special concern for the poor unambiguous. The Edelsteins argued that the hostels attached to the Asclepieia were the first hospitals, where those who were too poor to afford the attention of physicians were cared for. If they were correct, the Edelsteins would have had a valid basis for comparing the philanthropy of Asclepius with that of early Christianity. In fact, however, the evidence for their thesis that the Asclepieia were the first hospitals is very meager.[19] In stressing the importance of Asclepius as a pagan rival of Christ, the Edelsteins misrepresented the early Christians' ministry to the sick, which did not compete with Asclepius in claiming to offer supernatural healing. Rather it created a previously unavailable role of charitable concern in caring for the sick, which ultimately led to the creation of the first hospitals.[20]

It is an indication of the Edelsteins' erudition and thoroughness that although their study has not gone unchallenged on the grounds of both historical method and the construction they placed on the evidence, it remains after fifty years the fundamental treatment of its subject. They have put all subsequent students of Asclepius in their debt by assembling the most important literary and epigraphic documents related to his cult.[21] Other studies have reached different conclusions from theirs, sometimes reflecting not merely alternative interpretations of the evidence but different presuppositions (e.g., the application of anthropological, Jungian, or psychoanalytical models).[22] Historical scholarship by its very nature faces the likelihood of revision in every new generation, and the Edelsteins' attempt to reconstruct the cult of Asclepius will sooner or later be

---

[18] Vlastos, "Religion and Medicine," 280–88.

[19] Vlastos, "Religion and Medicine," 288–90.

[20] See Gary B. Ferngren, "Early Christianity as a Religion of Healing," *Bulletin of the History of Medicine* 66 (1992) : 1–15.

[21] The Edelsteins omitted some important documents from the Testimonies, such as the third and fourth tablets of the *iamata* from Epidaurus [T423], which they excluded because of their damaged condition. The texts and translation of these tablets (C and D) may be found in Lynn R. LiDonnici, *The Epidaurian Miracle Inscriptions : Text, Translation, and Commentary* (Atlanta : Scholars Press, 1995), 116–31. Perhaps the arrangement of the Testimonies in chronological order rather than by subject matter would have provided a better sense of the historical development of the cult. It is, moreover, confusing to the reader to find several texts (e.g., the Hymn of Isyllus) divided up and listed as separate Testimonies.

[22] See, for example, E. R. Dodds, *The Greeks and the Irrational* (Berkeley : University of California Press, 1951), 110–16 ; Carl Alfred Meier, *Ancient Incubation and Modern Psychotherapy*, trans. Monica Curtis (Evanston, Ill. : Northwestern University Press,

modified.[23] But their painstaking collection of the primary sources will never be outdated by later studies. In that respect it belongs to the relatively small number of works of scholarship that have enduring value. Given the marked growth of interest in the history of ancient medicine in the past generation, it is likely that this study will find more readers today than it did when it first appeared. It is appropriate that this should be the case, since Ludwig Edelstein did so much to deepen our understanding of ancient medicine by effecting what Lloyd Stevenson called " the quiet revolution in the study of Greek medicine."[24] That revolution remains his abiding legacy.[25]

Gary B. Ferngren

1967) (originally published as *Antike Inkubation und moderne Psychotherapie*, Studien aus dem C. G. Jung Institut 1 [Zurich : Rascher, 1949]) ; and Robert Rousselle, " Healing Cults in Antiquity : The Dream Cures of Asclepius of Epidaurus," *Journal of Psychohistory* 12, no. 3 (1985) : 339–52.

[23] Subsequent studies have already done much to add to our understanding of some primary sources on which the Edelsteins' study was based. For example, on the Sacred Tales of Aelius Aristides see C. A. Behr, *Aelius Aristides and the Sacred Tales* (Amsterdam: Adolf M. Hakkert, 1968). A number of subjects that the Edelsteins treated have received further elucidation. For example, on sacred dreams see Steven M. Oberhelman, " The Interpretation of Prescriptive Dreams in Ancient Greek Medicine," *Journal of the History of Medicine and Allied Sciences* 36 (1981) : 416–24 ; " The Diagnostic Dream in Ancient Medical Theory and Practice," *Bulletin of the History of Medicine* 61 (1987) : 47–60 ; and " Dreams in Graeco-Roman Medicine," in *Aufstieg und Niedergang der Römischen Welt* II. 37. 1, ed. Wolfgang Haase (Berlin : Walter de Gruyter, 1993), 121–56.

[24] Kudlien, et al., " In Memoriam," 183.

[25] In the preparation of this essay I am indebted, as always, to the helpful suggestions of Darrel W. Amundsen of Western Washington University.

FOREWORD

The interest in Asclepius and his cult centers in two considerations: Asclepius, almost throughout antiquity, was the main representative of divine healing, a highly important form of ancient medical treatment, and as such never opposed by ancient physicians; moreover, the worship of Asclepius, beyond its medical significance, came to play such a rôle in the religious life of later centuries that in the final stage of paganism, of all the genuinely Greek gods, Asclepius was judged the foremost antagonist of Christ. It is with regard to both aspects of the subject that this book has been written; it is devoted to Asclepius, the god of medicine and the Savior.

Aiming thus at an understanding of the Asclepius problem in its widest sense, the first volume contains a collection of the ancient references to Asclepius' life and deeds, his cult, his temples, and his images. The second volume is concerned with an analysis of the material assembled. The way in which the book has been composed, therefore, differs considerably from that usually followed in interpreting ancient religious phenomena. To be sure, the treatises dealing with Greek and Roman deities adduce the testimonies on which their interpretation is based, but seldom do they more than name the writer and the book from which the inferences are drawn; it rarely happens that quotations are put down *verbatim*. This is also true of the many studies concerning Asclepius. No one, however, is able to look up references to hundreds of different authors, and often very obscure ones at that. On the other hand, an intimate acquaintance with the scattered statements preserved is the more necessary because no continuous exposition of religious beliefs is extant. Consequently, when the testimonies are not at the disposal of the reader, it is almost impossible for him to come to a considered judgment about the theories proposed by the author. There is then, it seems, justification, even need, for an interpretation which makes the sources accessible while evaluating their meaning. Nor is this the first time that such an undertaking has been proposed. Welcker wrote to Wilhelm v. Humboldt on January 13, 1823, that he had often wondered if it would not be advisable to collect the ancient texts concerning all the Greek gods. For only in this manner, he thought, could a solid foundation be laid for an understanding of Greek religion.

One thing, however, must be emphasized: the material here taken into consideration is restricted to the written evidence. Certainly, artists contributed to the evolution of religious concepts. The Asclepius ideal as it was gradually developed by sculptors and painters expressed and determined the sentiments and emotions which inspired the worshippers of the god. Many data of religious significance, therefore, can be gleaned from a study of the Asclepius statues. The architectural finds yield considerable information in regard to the details of the rites performed and of the cures prescribed in the temples of Asclepius. The coins reflect certain local variations of the saga or of the worship. Yet, all such material has on principle been excluded from this book; it has only occasionally been mentioned to elucidate important points which otherwise could not be fully discussed. The study of the monuments requires an approach very different from that to be applied to the study of the literary remains; to combine both methods within the compass of one inquiry seemed impossible. However, there can hardly be any doubt that the interpretation of the written evidence can safely precede the interpretation of the archaeological evidence; the former is more complete and less ambiguous than the latter. One might even claim that from the words there must first be reconstructed a setting, so to say, against which the productions of art can be placed in order to receive full appreciation. At any rate, though by the restriction of the material the picture arrived at necessarily remains defective, a representation of the Asclepius religion based only on the written documents should nevertheless contain all essential features.

[Henry Sigerist]

PREFACE

Literary references and inscriptions, the material to which this collection is limited, are two kinds of testimony of very different value. The statements contained in books are mainly indicative of the religious beliefs of the upper classes; the utterances inscribed on stones primarily re-echo the attitude of the average people. True, these two sides have to be taken into consideration if one wishes to understand religion not only as it was conceived but also as it was practiced. When the Epidaurian tablets became known, the data revealed through them completely revolutionized the views commonly held as to the character of divine healing, views which before the excavations started had of necessity been based on literary evidence alone. Yet, of such startling discoveries there have been few indeed. It would be impossible to reprint all the thousands of inscriptions without swelling the book to an unwieldy size; it also seems unnecessary and even superfluous. A hundred dedications to Asclepius the Savior, giving the name of the god and the devotee, do not teach more than does one. A selection of inscriptions can and must be made. Cures, rites, paeans, all such documents by which light is thrown upon problems of general interest, have of course been reproduced; but facts of merely individual significance or reports of only local interest have been discarded. Even the temptation of tracing the dissemination of the cult according to inscriptions has been resisted. Such a study, interesting and necessary as it is for certain special purposes, from a general point of view gives only cumulative evidence for what follows from the other testimonies, namely that the Asclepius cult was universally accepted.

The literary testimonies, on the other hand, it was necessary to assemble in their entirety. For the poet, the philosopher, the historian, the scholar gave voice to the silent thoughts of the people, to their fears and hopes, to their presumptuous demands on the god, and to their reverent submission to the deity. More than any one else these writers, representing the higher aspects of religious thought, shaped that concept of Asclepius which came to be accepted by the ancients. Of course, even the great authors sometimes repeat the same story. Such duplications have not been printed in full, but are referred to in footnotes. Nor has every mention of Asclepius' name been included. Otherwise, however, completeness is here at least intended and, it is hoped, achieved, with due allowance for unavoidable oversight and for omissions caused by unreliable indices.

The classification of the collected material has been planned to bring out the full impact of the story to be told. First, those texts are given which present as complete a picture as can be gathered from any single document and which, besides, illustrate different views held by the ancients in regard to Asclepius. Then, the Asclepius legend as such is illuminated in its various features; Asclepius' place of origin, his parentage, his education, his deeds, his death are related (I). There follows the evidence concerning Asclepius' descendants, heroic and human, their relationship to the god, their own accomplishments and fates (II). After that, the ancient theories concerning Asclepius' elevation to godhead, the interpretation of his divine nature, are united with the proposed etymologies of his name and with the description of his divine retinue. The philosophical and mystical aspects of his divinity are expounded; he is portrayed as the Anti-Christ (III). Then the god is shown " in action "; what is known about his relation to physicians and patients, about his contributions to medicine, and about his cures, this is all reported (IV). Next, the rites performed by patient and worshipper, the sacrifices and festivals, the prayers and hymns are attested (V). There follow an enumeration and description of the statues and pictures as well as of the attributes associated with Asclepius according to the literary evidence (VI). Finally an account is given of the temples mentioned in literature, of the sagas connected with these sanctuaries, of the dedications found in them (VII).

Within this framework the individual passages have been grouped in subdivisions according to convenience rather than according to any objective standard, if indeed objective standards can be found for arranging a multitude of chance statements which are united only by their being related to Asclepius. Each testimony has been placed under the rubric for which it seemed to furnish the most precious information. If it contains many data of equal importance it has, in some instances, been reprinted in part under various headings. Otherwise, repetitions have been avoided as far as possible, and through extensive cross-references attention has been drawn to all passages which may have any bearing on one particular problem. Since the material is classified according to subjects, no subject index has been provided, only an *index locorum* has been compiled.

The passages themselves are given in the original and in translations: this in the expectation that the book may be consulted not only by specialists but also by those who are interested in the subject from a more general point of view and who might find it difficult to understand the Greek and Latin texts. Besides, these translations make specific the sense in which the various statements are employed in the interpretation of the testimonies. Use has been made of the available translations as far as they seemed satisfactory. They are listed in the *index locorum* together with the editions from which the texts have been taken.[1] Where no adequate translations or none at all could be found, the passages have been translated by Dr. Evelyn H. Clift in collaboration with the authors.

Throughout the final revision of this volume we have had the good fortune to enjoy the sound and expert advice of Dr. Paul A. Clement. In numerous instances his suggestions enabled us to amend our work; his familiarity with the body of inscriptions proved especially valuable to us. Dr. Clement also read the galley proofs. The same laborious task was undertaken by Professor Paul Friedländer whose judicious criticism and unfailing knowledge, which he liberally put at our disposal, saved us from many an error. Moreover, comments which he made on difficult texts brought about a fuller understanding of their content. To both our friends we feel deeply grateful for their ever ready help over a period of long months and for the generosity with which they gave us of their time. It is with pleasure that in addition we acknowledge the cooperation of the printers, Mr. F. Furst and his associates, and the pains which they took in setting up the Greek and Latin passages.

Baltimore, Md.,   E. J. E.
    December, 1944   L. E.

---

[1] Slight deviations from these versions are indicated by *; considerable alterations by **.

I

COLLECTION OF THE TESTIMONIES

I. LEGEND

BIOGRAPHICAL ACCOUNTS

1. Pindarus, Pythiae, III, 1-58.

ΙΕΡΩΝΙ ΣΤΡΑΚΟΣΙΩΙ
ΚΕΛΗΤΙ

στρ. α΄

   Ἤθελον Χείρωνά κε Φιλλυρίδαν,
   εἰ χρεὼν τοῦθ᾽ ἀμετέρας ἀπὸ γλώσσας κοινὸν εὔξασθαι ἔπος,
   ζώειν τὸν ἀποιχόμενον,
   Οὐρανίδα γόνον εὐρυμέδοντα Κρόνου, βάσσαισί τ᾽ ἄρχειν Παλίου
      φῆρ᾽ ἀγρότερον,
5 νοῦν ἔχοντ᾽ ἀνδρῶν φίλον· οἷος ἐὼν θρέψεν ποτὲ
   τέκτονα νωδυνίας ἄμερον γυιαρκέος Ἀσκληπιόν,
   ἥρωα παντοδαπᾶν ἀλκτῆρα νούσων.

ἀντ. α΄

   τὸν μὲν εὐΐππου Φλεγύα θυγάτηρ
   πρὶν τελέσσαι ματροπόλῳ σὺν Ἐλειθυίᾳ, δαμεῖσα χρυσέοις
10 τόξοισιν ὕπ᾽ Ἀρτέμιδος,
   εἰς Ἀΐδα δόμον ἐν θαλάμῳ κατέβα τέχναις Ἀπόλλωνος. χόλος δ᾽
      οὐκ ἀλίθιος
   γίνεται παίδων Διός. ἁ δ᾽ ἀποφλαυρίξαισά νιν
   ἀμπλακίαισι φρενῶν, ἄλλον αἴνησεν γάμον κρύβδαν πατρός,
   πρόσθεν ἀκειρεκόμᾳ μιχθεῖσα Φοίβῳ,

ἐπ. α΄

15 καὶ φέροισα σπέρμα θεοῦ καθαρόν·
   οὐδ᾽ ἔμειν᾽ ἐλθεῖν τράπεζαν νυμφίαν
   οὐδὲ παμφώνων ἰαχὰν ὑμεναίων, ἅλικες
   οἷα παρθένοι φιλέοισιν ἑταῖραι
   ἑσπερίαις ὑποκουρίζεσθ᾽ ἀοιδαῖς· ἀλλά τοι
20 ἤρατο τῶν ἀπεόντων· οἷα καὶ πολλοὶ πάθον.
   ἔστι δὲ φῦλον ἐν ἀνθρώποισι ματαιότατον,
   ὅστις αἰσχύνων ἐπιχώρια παπταίνει τὰ πόρσω,
   μεταμώνια θηρεύων ἀκράντοις ἐλπίσιν.

στρ. β΄

   ἔσχε τοιαύταν μεγάλαν ἀάταν
25 καλλιπέπλου λῆμα Κορωνίδος. ἐλθόντος γὰρ εὐνάσθη ξένου

1

λέκτροισιν ἀπ᾽ Ἀρκαδίας.
οὐδ᾽ ἔλαθε σκοπόν· ἐν δ᾽ ἄρα μηλοδόκῳ Πυθῶνι τόσσαις ἄϊεν
 ναοῦ βασιλεὺς
Λοξίας, κοινᾶνι παρ᾽ εὐθυτάτῳ γνώμαν πιθών,
πάντα ἴσαντι νόῳ· ψευδέων δ᾽ οὐχ ἅπτεται· κλέπτει τέ νιν
30 οὐ θεὸς οὐ βροτὸς ἔργοις οὔτε βουλαῖς.

 ἀντ. β´
 καὶ τότε γνοὺς Ἴσχυος Εἰλατίδα
ξεινίαν κοίταν ἀθέμιν τε δόλον, πέμψεν κασιγνήταν μένει
 θύοισαν ἀμαιμακέτῳ
ἐς Λακέρειαν, ἐπεὶ παρὰ Βοιβιάδος κρημνοῖσιν ᾤκει παρθένος.
 δαίμων δ᾽ ἕτερος
35 ἐς κακὸν τρέψαις ἐδαμάσσατό νιν· καὶ γειτόνων
πολλοὶ ἐπαῦρον, ἁμᾷ δ᾽ ἔφθαρεν· πολλὰν δ᾽ ὄρει πῦρ ἐξ ἑνὸς·
σπέρματος ἐνθορὸν ἀΐστωσεν ὕλαν.

 ἐπ. β´
 ἀλλ᾽ ἐπεὶ τείχει θέσαν ἐν ξυλίνῳ
σύγγονοι κούραν, σέλας δ᾽ ἀμφέδραμεν
40 λάβρον Ἁφαίστου, τότ᾽ ἔειπεν Ἀπόλλων· "Οὐκέτι
τλάσομαι ψυχᾷ γένος ἀμὸν ὀλέσσαι
οἰκτροτάτῳ θανάτῳ ματρὸς βαρείᾳ σὺν πάθᾳ."
ὣς φάτο· βάματι δ᾽ ἐν πρώτῳ κιχὼν παῖδ᾽ ἐκ νεκροῦ
ἅρπασε· καιομένα δ᾽ αὐτῷ διέφαινε πυρά·
45 καί ῥά νιν Μάγνητι φέρων πόρε Κενταύρῳ διδάξαι
πολυπήμονας ἀνθρώποισιν ἰᾶσθαι νόσους.

 στρ. γ´
 τοὺς μὲν ὦν, ὅσσοι μόλον αὐτοφύτων
ἑλκέων ξυνάονες, ἢ πολιῷ χαλκῷ μέλη τετρωμένοι
ἢ χερμάδι τηλεβόλῳ,
50 ἢ θερινῷ πυρὶ περθόμενοι δέμας ἢ χειμῶνι, λύσαις ἄλλον ἀλλοίων
 ἀχέων
ἔξαγεν, τοὺς μὲν μαλακαῖς ἐπαοιδαῖς ἀμφέπων,
τοὺς δὲ προσανέα πίνοντας, ἢ γυίοις περάπτων πάντοθεν
φάρμακα, τοὺς δὲ τομαῖς ἔστασεν ὀρθούς.

 ἀντ. γ´
 ἀλλὰ κέρδει καὶ σοφία δέδεται.
55 ἔτραπεν καὶ κεῖνον ἀγάνορι μισθῷ χρυσὸς ἐν χερσὶν φανεὶς
ἄνδρ᾽ ἐκ θανάτου κομίσαι
ἤδη ἁλωκότα· χερσὶ δ᾽ ἄρα Κρονίων ῥίψαις δι᾽ ἀμφοῖν ἀμπνοὰν
 στέρνων καθέλεν
ὠκέως, αἴθων δὲ κεραυνὸς ἐνέσκιμψεν μόρον.

For Hieron of Syracuse, Winner in the Horse-race

If the poet's tongue might breathe the prayer that is on the lips of all, I would pray that Chiron, Philyra's son, who is dead and gone, were still alive,—he who once ruled far and wide as the offspring of Cronus, who was the son of Heaven. Would that that rugged monster with spirit kindly unto men were reigning still in Pelion's glens, even such as when, in olden days, he reared Asclepius,[1] that gentle craftsman who drove pain from the limbs,—that hero who gave aid in all manner of maladies.

Ere the daughter of Phlegyas famed for his horses could bear him, in the fulness of time, with the aid of Eleithyia, the goddess of child-birth, she was stricken in her chamber by the golden arrows of Artemis, and thus descended to the home of Hades by the counsels of Apollo. Not in vain is the wrath of the sons of Zeus. For she, in the errors of her heart, had lightly regarded that wrath; and, although she had aforetime consorted with Phoebus of the unshorn hair, and bare within her the pure seed of the god, yet without her father's knowledge she consented to be wedded to another. She waited not for the coming of the marriage feast, nor for the music of the full-voiced hymenaeal chorus, even the playful strains that maiden-mates love to utter in evening songs. No! she was enamoured of things otherwhere,—that passion, which many, ere now. have felt. For, among men, there is a foolish company of those, who, putting shame on their home, cast their glances afar, and pursue idle dreams in hopes that shall not be fulfilled.

Such was the strong infatuation that the spirit of the fair-robed Coronis had caught. For she slept in the couch of a stranger who came from Arcadia; but she escaped not the ken of the watchful god; for, although he was then at the sacrificial shrine of Pytho, yet Loxias, the king of the temple, perceived it in his mind that knoweth all things, with his thought convinced by an unerring prompter. He never deceiveth others; and he is not himself deceived by god or man, in deed or counsel.[2] Even so, at that time, he knew of her consorting with the stranger, Ischys, son of Elatus, and of her lawless deceit. Thereupon did he send his sister, Artemis, speeding with resist-less might, even to Lacereia, for the young maiden was dwelling by the banks of the Boebian lake; and an evil demon perverted her heart and laid her low, and many of her neighbors suffered the same fate, and perished with her; [3] even as, on a mountain, the fire that hath been sped by a single spark layeth low a mighty forest.

But, when the kinsmen had placed the girl in the midst of the wooden walls of the pyre, and the wild flame of the fire-god was playing around it, then spake Apollo: "No longer can I endure in my heart to slay my own child by a death most piteous, at the selfsame time as its mother's grievous doom." Thus he spake. He stepped forward but once, and anon he reached his child, and snatched it from the corse,[4] while the kindled pyre parting its flame opened for him a path; and he bare the babe away, and gave it to the Magnesian Centaur to teach it how to heal mortal men of painful maladies.

And those whosoever came suffering from the sores of nature, or with their limbs wounded either by gray bronze or by far-hurled stone, or with bodies wasting away with summer's heat or winter's cold, he loosed and delivered divers of them from diverse pains, tending some of them with kindly incantations, giving to others a soothing potion, or, haply, swathing their limbs with simples, or restoring others by the knife. But, alas! even the lore of leechcraft is enthralled by the love of gain;[5] even he was seduced, by a splendid fee of gold displayed upon the palm, to bring back from death one who was already its lawful prey. Therefore the son of Cronus with his hands hurled his shaft through both of them, and swiftly reft the breath from out their breasts, for they were stricken with sudden doom by the gleaming thunderbolt.[6] (*)

[1] Cf. T. 50-62.   [4] Cf. T. 2, v. 629; cf. however T. 26.
[2] Cf. T. 42.   [5] Cf. T. 99-103.
[3] Cf. T. 24.   [6] Cf. T. 105-15.

2. Ovidius, Metamorphoses, II, 542-648.

> Pulchrior in tota quam Larisaea Coronis
> non fuit Haemonia: placuit tibi, Delphice, certe,
> dum vel casta fuit vel inobservata; sed ales
> 545 sensit adulterium Phoebeius, utque latentem
> detegeret culpam, non exorabilis index,
> ad dominum tendebat iter. . . . .

. . . .

nec coeptum dimittit iter dominoque iacentem
cum iuvene Haemonio vidisse Coronida narrat.
600 laurea delapsa est audito crimine amantis,
et pariter vultusque deo plectrumque colorque
excidit, utque animus tumida fervebat ab ira,
arma adsueta capit flexumque a cornibus arcum
tendit et illa suo totiens cum pectore iuncta
605 indevitato traiecit pectora telo.
icta dedit gemitum tractoque a corpore ferro
candida puniceo perfudit membra cruore
et dixit: "Potui poenas tibi, Phoebe, dedisse,
sed peperisse prius; duo nunc moriemur in una."
610 hactenus, et pariter vitam cum sanguine fudit;
corpus inane animae frigus letale secutum est.
    Paenitet heu! sero poenae crudelis amantem,
seque, quod audierit, quod sic exarserit, odit;
odit avem, per quam crimen causamque dolendi
615 scire coactus erat, nec non arcumque manumque
odit cumque manu temeraria tela sagittas
conlapsamque fovet seraque ope vincere fata
nititur et medicas exercet inaniter artes.
quae postquam frustra temptata rogumque parari
620 vidit et arsuros supremis ignibus artus,
tum vero gemitus (neque enim caelestia tingui
ora licet lacrimis) alto de corde petitos
edidit, haud aliter quam cum spectante iuvenca
lactentis vituli dextra libratus ab aure
625 tempora discussit claro cava malleus ictu.
ut tamen ingratos in pectora fudit odores
et dedit amplexus iniustaque iusta peregit,
non tulit in cineres labi sua Phoebus eosdem
semina, sed natum flammis uteroque parentis
630 eripuit geminique tulit Chironis in antrum,
sperantemque sibi non falsae praemia linguae
inter aves albas vetuit consistere corvum.
    Semifer interea divinae stirpis alumno
laetus erat mixtoque oneri gaudebat honore;
635 ecce venit rutilis umeros protecta capillis
filia centauri, quam quondam nympha Chariclo

fluminis in rapidi ripis enisa vocavit
Ocyroen: non haec artes contenta paternas
edidicisse fuit, fatorum arcana canebat.
640 ergo ubi vaticinos concepit mente furores
incaluitque deo, quem clausum pectore habebat,
adspicit infantem " toto " que " salutifer orbi
cresce, puer! " dixit; " tibi se mortalia saepe
corpora debebunt, animas tibi reddere ademptas
645 fas erit, idque semel dis indignantibus ausus
posse dare hoc iterum flamma prohibebere avita,
eque deo corpus fies exsangue deusque,
qui modo corpus eras, et bis tua fata novabis."

In all Thessaly there was no fairer maid than Coronis of
Larissa. She surely found favor in thy eyes, O Delphic god,
so long as she was chaste—or undetected. But the bird of
Phoebus discovered her unchastity, and was posting with all
speed, hardhearted tell-tale, to his master to disclose the sin
he had spied out. . . .

He [sc., the raven] continued on his way to his master, and
then told him that he had seen Coronis lying beside the youth
of Thessaly. When that charge was heard the laurel glided
from the lover's head; together countenance and color changed,
and the quill dropped from the hand of the god. And as his
heart became hot with swelling anger he seized his accustomed
arms, strung his bent bow from the horns, and transfixed with
unerring shaft the bosom which had been so often pressed to
his own. The smitten maid groaned in agony, and, as she
drew out the arrow, her white limbs were drenched with her
red blood. " 'Twas right, O Phoebus," she said, " that I should
suffer thus from you, but first I should have borne my child.
But now two of us shall die in one." And while she spoke her
life ebbed out with her streaming blood, and soon her body, its
life all spent, lay cold in death.

The lover, alas! too late repents his cruel act; he hates himself because he listened to the tale and was so quick to break
out in wrath. He hates the bird by which he has been compelled to know the offence that brought his grief, bow and hand
he hates, and with that hand the hasty arrows too. He fondles
the fallen girl, and too late tries to bring help and to conquer

fate; but his healing arts are exercised in vain. When his efforts were of no avail, and he saw the pyre made ready with the funeral fires which were to consume her limbs, then indeed—for the cheeks of the heavenly gods may not be wet with tears—from his deep heart he uttered piteous groans; such groans as the young cow utters when before her eyes the hammer high poised from beside the right ear crashes with its resounding blow through the hollow temples of her suckling calf. The god pours fragrant incense on her unconscious breast, gives her the last embrace, and performs all the fit offices unfitly for the dead. But that his own son should perish in the same funeral fires he cannot brook. He snatched the unborn child from his mother's womb and from the devouring flames,[1] and bore him for safe keeping to the cave of two-formed Chiron. But the raven, which had hoped only for reward from his truth-telling, he forbad to take his place among white birds.

Meantime the Centaur was rejoicing in his foster-child of heavenly stock, glad at the honor which the task brought with it, when lo! there comes his daughter, her shoulders over-mantled with red-gold locks, whom once the nymph, Chariclo, bearing her to him upon the banks of the swift stream, had called thereafter Ocyrhoë. She was not satisfied to have learnt her father's art, but she sang prophecy. So when she felt in her soul the prophetic madness, and was warmed by the divine fire prisoned in her breast, she looked upon the child and cried: " O child, health-bringer to the whole world, speed thy growth. Often shall mortal bodies owe their lives to thee, and to thee shall it be counted right to restore the spirits of the departed. But having dared this once in scorn of the gods, from power to give life a second time thou shalt be stayed by thy grandsire's lightning. So, from a god shalt thou become but a lifeless corpse; but from this corpse shalt thou again become a god and twice renew thy fates." (*)

[1] Cf. T. 1, vv. 43-44; cf. however T. 26.

3. Apollodorus, Bibliotheca, III, 10, 3, 5-4, 1.

Πρὸς δὲ ταύταις Ἀρσινόην ἐγέννησε· ταύτῃ μίγνυται Ἀπόλλων, ἡ δὲ Ἀσκληπιὸν γεννᾷ. τινὲς δὲ Ἀσκληπιὸν οὐκ ἐξ Ἀρσινόης τῆς Λευκίππου λέγουσιν, ἀλλ' ἐκ Κορωνίδος τῆς Φλεγύου ἐν

Θεσσαλίᾳ. καί φασιν ἐρασθῆναι ταύτης Ἀπόλλωνα καὶ εὐθέως
συνελθεῖν· τὴν δὲ παρὰ τὴν τοῦ πατρὸς γνώμην ἑλομένην Ἰσχύϊ
τῷ Καινέως ἀδελφῷ συνοικεῖν. Ἀπόλλων δὲ τὸν μὲν ἀπαγ-
γείλαντα κόρακα καταρᾶται, ὃν τέως λευκὸν ὄντα ἐποίησε μέλανα,
αὐτὴν δὲ ἀπέκτεινε. καιομένης δὲ ταύτης ἁρπάσας τὸ βρέφος
ἐκ τῆς πυρᾶς πρὸς Χείρωνα τὸν Κένταυρον ἤνεγκε, παρ᾽ οὗ καὶ
τὴν ἰατρικὴν καὶ τὴν κυνηγετικὴν τρεφόμενος ἐδιδάχθη. καὶ
γενόμενος χειρουργικὸς καὶ τὴν τέχνην ἀσκήσας ἐπὶ πολὺ οὐ
μόνον ἐκώλυέ τινας ἀποθνήσκειν, ἀλλ᾽ ἀνήγειρε καὶ τοὺς ἀπο-
θανόντας· παρὰ γὰρ Ἀθηνᾶς λαβὼν τὸ ἐκ τῶν φλεβῶν τῆς
Γοργόνος ῥυὲν αἷμα, τῷ μὲν ἐκ τῶν ἀριστερῶν ῥυέντι πρὸς φθορὰν
ἀνθρώπων ἐχρῆτο, τῷ δὲ ἐκ τῶν δεξιῶν πρὸς σωτηρίαν, καὶ διὰ
τοῦτο τοὺς τεθνηκότας ἀνήγειρεν. [εὗρον δέ τινας λεγομένους
ἀναστῆναι ὑπ᾽ αὐτοῦ, Καπανέα καὶ Λυκοῦργον, ὡς Στησίχορός
φησιν ⟨ἐν⟩ Ἐριφύλῃ, Ἱππόλυτον, ὡς ὁ τὰ Ναυπακτικὰ συγγράψας λέγει, Τυνδάρεων, ὥς φησι Πανύασσις, Ὑμέναιον, ὡς
οἱ Ὀρφικοὶ λέγουσι, Γλαῦκον τὸν Μίνωος, ὡς Ἀμελησαγόρας
λέγει]. Ζεὺς δὲ φοβηθεὶς μὴ λαβόντες ἄνθρωποι θεραπείαν παρ᾽
αὐτοῦ βοηθῶσιν ἀλλήλοις, ἐκεραύνωσεν αὐτόν. καὶ διὰ τοῦτο
ὀργισθεὶς Ἀπόλλων κτείνει Κύκλωπας τοὺς τὸν κεραυνὸν Διὶ
κατασκευάσαντας. Ζεὺς δὲ ἐμέλλησε ῥίπτειν αὐτὸν εἰς Τάρταρον,
δεηθείσης δὲ Λητοῦς ἐκέλευσεν αὐτὸν ἐνιαυτὸν ἀνδρὶ θητεῦσαι.

Besides them [sc., Hilaira and Phoebe] he [sc., Leucippus]
begat Arsinoë: with her Apollo had intercourse, and she bore
Asclepius. But some affirm that Asclepius was not a son of
Arsinoë, daughter of Leucippus, but that he was a son of Coronis, daughter of Phlegyas in Thessaly.[1] And they say that
Apollo loved her and at once consorted with her, but that she,
in accordance with her father's judgment, chose Ischys and
married him, who was a brother of Caeneus. Apollo cursed the
raven that brought the tidings and made him black instead of
white, as he had been before;[2] but Coronis he killed. As she
was burning, he snatched the babe from the pyre and brought
it to Chiron, the Centaur,[3] by whom he was brought up and
taught the arts of healing and hunting. And having become a
surgeon, and carried the art to a great pitch, he not only prevented some from dying, but even raised up the dead; for he
received from Athena the blood that flowed from the veins of the

Gorgon,[4] and while he used the blood that flowed from her left side for the bane of mankind, he used the blood that flowed from her right side for salvation, and by that means he raised the dead. [I found some who are reported to have been raised by him, to wit, Capaneus and Lycurgus, as Stesichorus says in the *Eriphyle*; Hippolytus, as the author of the *Naupactica* reports; Tyndareus, as Panyassis says; Hymenaeus, as the Orphics report; and Glaucus, son of Minos, as Amelesagoras relates.] [5] But Zeus, fearing that men might acquire the healing art from him and so come to the rescue of each other, smote him with a thunderbolt.[6] Angry on that account, Apollo slew the Cyclops who had fashioned the thunderbolt for Zeus. But Zeus intended to hurl him to Tartarus; however, at the intercession of Leto he ordered him to serve as a thrall to a man for a year.(**)

[1] Cf. T. 34; 35; 37.　　　　　　　[2] Cf. T. 41-45.
[3] Cf. T. 1, v. 45; 50-62.
[4] Cf. T. 97; 3a. Tatianus, *Adv. Graecos*, 8, 2-3: Καὶ μετὰ τὴν Γοργοῦς καρατομίαν, . . . τὰς σταγόνας τῶν αἱμάτων ἡ Ἀθηνᾶ καὶ ὁ Ἀσκληπιὸς διενείμαντο· καὶ ὁ μὲν ἀπ' αὐτῶν ἔσωζεν, ἡ δὲ ἀπὸ τῶν ὁμοίων λύθρων ἀνθρωποκτόνος [ἢ πολεμοποιὸς] ἐγίνετο. Concerning the representation of Gorgon at Epidaurus, cf. T. 630; 739.
[5] Concerning this list, cf. T. 69.　　　　[6] Cf. T. 105-15.

4. Diodorus, Bibliotheca Historica, IV, 71, 1-4.

Τούτων δ' ἡμῖν διευκρινημένων πειρασόμεθα διελθεῖν περὶ Ἀσκληπιοῦ καὶ τῶν ἀπογόνων αὐτοῦ. μυθολογοῦσι τοίνυν Ἀσκληπιὸν Ἀπόλλωνος υἱὸν ὑπάρχειν καὶ Κορωνίδος, φύσει δὲ καὶ ἀγχινοίᾳ διενεγκόντα ζηλῶσαι τὴν ἰατρικὴν ἐπιστήμην, καὶ πολλὰ τῶν συντεινόντων πρὸς ὑγίειαν ἀνθρώπων ἐξευρεῖν. ἐπὶ τοσοῦτο δὲ προβῆναι τῇ δόξῃ ὥστε πολλοὺς τῶν ἀπεγνωσμένων ἀρρώστων παραδόξως θεραπεύειν, καὶ διὰ τοῦτο πολλοὺς δοκεῖν τῶν τετελευτηκότων ποιεῖν πάλιν ζῶντας. διὸ καὶ τὸν μὲν Ἅιδην μυθολογοῦσιν ἐγκαλοῦντα τῷ Ἀσκληπιῷ κατηγορίαν αὐτοῦ ποιήσασθαι πρὸς τὸν Δία ὡς τῆς ἐπαρχίας αὐτοῦ ταπεινουμένης· ἐλάττους γὰρ ἀεὶ γίνεσθαι τοὺς τετελευτηκότας, θεραπευομένους ὑπὸ τοῦ Ἀσκληπιοῦ. καὶ τὸν μὲν Δία παροξυνθέντα [καὶ] κεραυνώσαντα τὸν Ἀσκληπιὸν διαφθεῖραι, τὸν δ' Ἀπόλλωνα διὰ τὴν ἀναίρεσιν τούτου παροξυνθέντα φονεῦσαι τοὺς τὸν κεραυνὸν τῷ Διὶ κατασκευάσαντας Κύκλωπας. ἐπὶ δὲ τῇ τούτων τελευτῇ παροξυνθέντα τὸν Δία προστάξαι τῷ Ἀπόλλωνι θητεῦσαι παρ' ἀνθρώπῳ καὶ ταύτην τὴν τιμωρίαν λαβεῖν παρ' αὐτοῦ τῶν ἐγκλημάτων. Ἀσκληπιοῦ δέ φασι γενομένους υἱοὺς Μαχάονα καὶ Ποδαλείριον . . . περὶ μὲν οὖν Ἀσκληπιοῦ καὶ τῶν υἱῶν αὐτοῦ τοῖς ῥηθεῖσιν ἀρκεσθησόμεθα . . .

Now that we have discussed these [sc., Lapiths and Centaurs] thoroughly, we shall try to deal with Asclepius and his descendants. This is the legend they tell: Asclepius is the son of Apollo and Coronis;[1] and excelling in nature and keenness of mind, he strove after medical knowledge and discovered many of the things that tend towards the health of mankind.[2] He advanced so far in reputation as to cure many of the sick who had been despaired of, contrary to all expectations, and for this reason he seemed to revive again many who had died. Wherefore the story is current that even Hades, bringing a charge against Asclepius, accused him before Zeus on the ground that the scope of his power was diminishing, for the dead were continually growing fewer because they were healed by Asclepius. Zeus then, being roused to anger, struck Asclepius with a thunderbolt and killed him;[3] whereupon Apollo, provoked by the destruction of Asclepius, slew the Cyclops who fashioned the thunderbolt for Zeus. Zeus then, enraged at their death, commanded Apollo to serve as a thrall to a mortal and exacted from him this punishment for the charges against him. Furthermore, they say that Machaon and Podalirius were sons of Asclepius[4] . . . This will suffice in regard to Asclepius and his sons. . . .

[1] Cf. T. 29; T. 355; cf. also T. 21 ff.   [3] Cf. T. 1, vv. 55-58; T. 105-15.
[2] Cf. T. 348 ff.   [4] Cf. T. 135 ff., especially 147.

5. Theodoretus, Graecarum Affectionum Curatio, VIII, 19-23.*

Καὶ τὸν Ἀσκληπιὸν δέ φησιν Ἀπολλόδωρος κατὰ μέν τινας Ἀρσινόης εἶναι υἱόν, κατὰ δὲ ἄλλους Κορωνίδος, λάθρα μὲν Ἀπόλλωνι ξυνελθεῖν βιασθείσης, κυησάσης δὲ καὶ τεκούσης καὶ ἐκθεμένης τὸ βρέφος· τοῦτο δὲ κυνηγέτας τινὰς εὑρηκότας ὑπὸ κυνὸς τρεφόμενον λαβεῖν καὶ κομίσαι λέγει Χείρωνι τῷ Κενταύρῳ. εἶτα ἐκεῖ τραφῆναί τε καὶ ἀσκηθῆναι τὴν ἰατρικὴν ἐπιστήμην, ἐν Τρίκκῃ δὲ πρῶτον καὶ Ἐπιδαύρῳ δοῦναι πεῖραν τῆς τέχνης. οὕτω δὲ ἄκρως, φησίν, ἐπαιδεύθη καὶ μάλα γε σπουδαίως, ὡς μὴ μόνον τοὺς ἀρρωστοῦντας ἰᾶσθαι, ἀλλὰ καί τινας τῶν τετελευτηκότων ἐγείρειν· διὸ δὴ τὸν τερπικέραυνον χαλεπήναντα πρηστῆρσι βαλεῖν καὶ ἐξαγαγεῖν τῆς ζωῆς. τοιγαροῦν καὶ ἄνθρωπος ἦν,

οὐδὲ τοῖς ἄλλοις ἀνθρώποις παραπλησίως τραφείς, ἀλλὰ καὶ
παρὰ κυνὸς τῆς πρώτης ἀπολαύσας τροφῆς καὶ τῷ τῶν θηρευ-
όντων οἴκτῳ διασωθείς, καὶ οὐκ ἐκ θείας σοφίας καὶ ἐπιστήμης
τὸ ἰατρεύειν ἔχων, ἀλλὰ παρὰ Χείρωνος τήνδε τὴν τέχνην
ἐκπαιδευθείς. οὗτος δὲ ὁ Χείρων καὶ Ἀχιλλέως ἐγένετο παιδευτής·
ὅθεν δῆλον ὅτι κατὰ τὸν χρόνον τὸν αὐτὸν ἐγενέσθην, ἢ μικρῷ
τινι οὗτος Ἀχιλλέως πρεσβύτερος, ἐπειδήπερ καὶ τοῖς Ἀργο-
ναύταις ξυνέπλευσε, καὶ ὁ τούτου γε παῖς ὁ Μαχάων ξυνε-
πολέμησεν Ἀχιλλεῖ. καὶ ὁ κεραυνὸς δέ γε, καὶ ὁ γενόμενος
ἐμπρησμός, δηλοῖ τῆς φύσεως τὸ ἐπίκηρον. ἀλλὰ καὶ οὕτω γε
φύντα, ὡς εἴρηται, καὶ σωθέντα γε καὶ τραφέντα καὶ ἐμπρησ-
θέντα τοῖς ἄλλοις θεοῖς κατέλεξαν καὶ τεμένη γε καθωσίωσαν
καὶ καθιέρωσαν βωμούς· καὶ λοιβῇ καὶ κνίσῃ, πάλαι μὲν
προφανῶς, νῦν δὲ ἴσως ἐν παραβύστῳ γεραίρετε, καὶ τὰ τούτου
ἀγάλματα θείας ἀξιοῦτε τιμῆς καὶ τὸν ἐνειλημένον αὐτῷ θαυμάζετε
δράκοντα καὶ ξύμβολον εἶναι τῆς ἰατρικῆς φατε, ὅτι καθάπερ
ἐκεῖνος ἀποδύεται τὸ γῆρας, οὕτως ἡ ἰατρικὴ τῶν νόσων ἐλευθεροῖ.
ὅτι δὲ ἐπὶ τῶν Ὁμήρου χρόνων οὐδέπω οὗτος τῆς θεοποιίας
ἐτετυχήκει, μάρτυς ὁ ποιητής, οὐ τὸν Ἀσκληπιὸν ἀλλὰ τὸν
Παίηονα δείξας τὰ τοῦ Ἄρεως θεραπεύσαντα τραύματα. δεῖ γὰρ
τουτουσὶ τοὺς θεοὺς καὶ τραύματα ἔχειν καὶ ἰατρούς. καὶ τὸν
Μαχάονα δὲ οὐ θεοῦ λέγει υἱὸν οὐδέ γε ἡμιθέου, ἀλλ᾽ ἰατροῦ·
ἔφη γάρ·

         φῶτ᾽ Ἀσκληπιοῦ υἱὸν ἀμύμονος ἰητῆρος.

* = Apollodorus Atheniensis, Fr. 138 [Jacoby].

And about Asclepius Apollodorus says that according to some
authorities he was the son of Arsinoë, according to others, of
Coronis, who, forced secretly to consort with Apollo, became
pregnant, bore a child, and exposed it. Some hunters found
the child being nourished by a dog,[1] rescued it and, he says,
conveyed it to Chiron, the Centaur; there, then, the child was
reared and learned the art of medicine.[2] In Tricca first and in
Epidaurus he gave proof of his art. So excellently and
thoroughly was he trained, says Apollodorus, that not only did
he heal the sick but even dared to revive some of those who had
died.[3] Wherefore the Thunderer, in his anger, struck him with
a bolt of lightning and deprived him of life. Therefore he too
was a mortal—not, however, reared like other men, but getting
his first nourishment even from a dog, and saved by the com-

passion of hunters, he had medical science not from divine
wisdom and knowledge but rather was educated in this art by
Chiron. This same Chiron was also the teacher of Achilles.
Wherefrom it is apparent that they were both born about the
same time or Asclepius was even a little older than Achilles,
especially since he also sailed with the Argonauts [4] and his son,
Machaon, was a fellow-soldier of Achilles.[5] The thunderbolt
as well as the conflagration that occurred indicate the mortality
of his nature.

But this Asclepius, who was born, rescued, reared, and burnt
in the afore-mentioned way, they enrolled among the other gods,
made sacred precincts for him, and consecrated altars to him.[6]
You honor him with libations and sacrifices, once openly, but
now probably in a secret corner; you deem his statues worthy
of divine honor; you marvel at the serpent curling around him
and say that it is the symbol of the healing art, because just as
the serpent sloughs the skin of old age, so the medical art
releases from illness.[7] But that at the time of Homer, Asclepius
had not yet been deified, the poet testifies in showing that
Paeon, not Asclepius, healed the wounds of Ares.[8] For gods
of this kind must have wounds and physicians. And he [sc.,
Homer] calls Machaon the son neither of a god, nor of a demigod, but of a physician, for he says, " [Call] the man, [the]
son of Asclepius, the blameless physician." [9]

[1] Cf. T. 7 ; 9 ; 9b ; cf. also T. 630.
[2] Cf. T. 1, v. 45 ; T. 50-62.
[3] Cf. e. g. T. 97.
[4] Cf. T. 63 ; 64.
[5] Cf. T. 166.
[6] Cf. T. 232 ff.
[7] Cf. T. 6.
[8] Homer, Iliad, V, 899 ; cf. T. 257.
[9] T. 164, v. 194 ; cf. however T. 255.

6. Cornutus, Theologiae Graecae Compendium, Cp. 33.

Κατ' ἀκόλουθον πάλιν τὸν Ἀσκληπιὸν υἱὸν αὐτοῦ ἔφασαν
γενέσθαι, τὸν δοκοῦντα τοῖς ἀνθρώποις ὑποδεδειχέναι τὴν
ἰατρικήν· ἐχρῆν γὰρ καὶ τούτῳ τῷ τόπῳ θεῖόν τι ἐπιστῆσαι.
ὠνομάσθη δὲ ὁ Ἀσκληπιὸς ἀπὸ τοῦ ἠπίως ἰᾶσθαι καὶ ἀναβάλλεσθαι τὴν κατὰ τὸν θάνατον γινομένην ἀπόσκλησιν. διὰ τοῦτο
γὰρ δράκοντα αὐτῷ παριστᾶσιν, ἐμφαίνοντες ὅτι ὅμοιόν τι τούτῳ
πάσχουσιν οἱ χρώμενοι τῇ ἰατρικῇ κατὰ τὸ οἱονεὶ ἀνανεάζειν ἐκ
τῶν νόσων καὶ ἐκδύεσθαι τὸ γῆρας, ἅμα δ' ἐπεὶ προσοχῆς ὁ
δράκων σημεῖον, ἧς πολλῆς δεῖ πρὸς τὰς θεραπείας. καὶ τὸ
βάκτρον δὲ τοιούτου τινὸς ἔοικεν εἶναι σύμβολον· παρίσταται

γὰρ δι' αὐτοῦ ὅτι, εἰ μὴ ταύταις ταῖς ἐπινοίαις ἐπεστηριζόμεθα,
ὅσον ἐπὶ τὸ συνεχῶς εἰς ἀρρωστίαν ἐμπίπτειν, κἂν θᾶττον τοῦ
δέοντος σφαλλόμενοι κατεπίπτομεν. λέγεται δὲ ὁ Χείρων τετρο-
φέναι τὸν Ἀσκληπιὸν κἂν τοῖς τῆς ἰατρικῆς θεωρήμασιν ἠσκη-
κέναι, τὴν διὰ τῶν χειρῶν ἐνέργειαν τῆς τέχνης ἐμφαίνειν αὐτῶν
βουλομένων. παραδέδοται δὲ καὶ γυνὴ τοῦ Ἀσκληπιοῦ, Ἠπιόνη,
τοῦ ὀνόματος οὐκ ἀργῶς εἰς τὸν μῦθον παρειλημμένου, δηλοῦντος
δὲ τὸ πραϋντικὸν τῶν ὀχλήσεων διὰ τῆς ἠπίου φαρμακείας.

Consequently, then, they said that Asclepius was his [sc.,
Apollo's] son, reputed to have taught the medical art to mankind. For even this branch of learning had to be under the
tutelage of something divine.[1] Asclepius derived his name from
healing soothingly and from deferring the withering that comes
with death.[2] For this reason, therefore, they give him a serpent
as an attribute, indicating that those who avail themselves of
medical science undergo a process similar to the serpent in that
they, as it were, grow young again after illnesses and slough off
old age;[3] also because the serpent is a sign of attention, much
of which is required in medical treatments. The staff also seems
to be a symbol of some similar thing. For by means of this it
is set before our minds that unless we are supported by such
inventions as these, in so far as falling continually into sickness
is concerned, stumbling along we would fall even sooner than
necessary.[4] Chiron is said to have reared Asclepius and to have
trained him in the principles of medicine. They say this because
they want to emphasize the operation of the medical art by
means of the hands. There is mentioned in legend also the
wife of Asclepius, Epione, whose name is not idly introduced
into the legend, for it signifies the alleviation of troubles through
the agency of soothing simples.

[1] Cf. T. 342.　　　　　　　　　　　[3] Cf. T. 5.
[2] Cf. T. 268 ff.　　　　　　　　　　[4] Cf. T. 705.

7. Pausanias, Descriptio Graeciae, II, 26, 3-5.

Ἀσκληπιοῦ δὲ ἱερὰν μάλιστα εἶναι τὴν γῆν ἐπὶ λόγῳ
συμβέβηκε τοιῷδε. Φλεγύαν Ἐπιδαύριοί φασιν ἐλθεῖν ἐς Πελοπόννησον πρόφασιν μὲν ἐπὶ θέᾳ τῆς χώρας, ἔργῳ δὲ κατάσκοπον
πλήθους τῶν ἐνοικούντων καὶ εἰ τὸ πολὺ μάχιμον εἴη τῶν
ἀνθρώπων. ἦν γὰρ δὴ Φλεγύας πολεμικώτατος τῶν τότε καὶ
ἐπιὼν ἑκάστοτε ἐφ' οὓς τύχοι τοὺς καρποὺς ἔφερε καὶ ἤλαυνε τὴν

λείαν. ὅτε δὲ παρεγένετο ἐς Πελοπόννησον, εἵπετο ἡ θυγάτηρ
αὐτῷ, λεληθυῖα ἔτι τὸν πατέρα ὅτι ἐξ Ἀπόλλωνος εἶχεν ἐν γαστρί.
ὡς δὲ ἐν τῇ γῇ τῇ Ἐπιδαυρίων ἔτεκεν, ἐκτίθησι τὸν παῖδα ἐς τὸ
ὄρος τοῦτο ὃ δὴ Τίτθιον ὀνομάζουσιν ἐφ' ἡμῶν, τηνικαῦτα δὲ
ἐκαλεῖτο Μύρτιον. ἐκκειμένῳ δὲ ἐδίδου μέν οἱ γάλα μία τῶν
περὶ τὸ ὄρος ποιμαινομένων αἰγῶν, ἐφύλασσε δὲ ὁ κύων ὁ τοῦ
αἰπολίου φρουρός. Ἀρεσθάνας δὲ—ὄνομα γὰρ τῷ ποιμένι τοῦτο
ἦν—ὡς τὸν ἀριθμὸν οὐχ εὕρισκεν ὁμολογοῦντα τῶν αἰγῶν καὶ
ὁ κύων ἅμα ἀπεστάτει τῆς ποίμνης, οὕτω τὸν Ἀρεσθάναν ἐς πᾶν
φασιν ἀφικνεῖσθαι ζητήσεως, εὑρόντα δὲ ἐπιθυμῆσαι τὸν παῖδα
ἀνελέσθαι. καὶ ὡς ἐγγὺς ἐγίνετο, ἀστραπὴν ἰδεῖν ἐκλάμψαν
ἀπὸ τοῦ παιδός, νομίσαντα δὲ εἶναι θεῖόν τι, ὥσπερ ἦν, ἀπο-
τραπέσθαι. ὁ δὲ αὐτίκα ἐπὶ γῆν καὶ θάλλασσαν πᾶσαν ἠγγέλλετο
τά τε ἄλλα ὁπόσα βούλοιτο εὑρίσκειν ἐπὶ τοῖς κάμνουσι καὶ ὅτι
ἀνίστησι τεθνεῶτας.

That the [Epidaurian] land is especially sacred to Asclepius
is due to the following reason. The Epidaurians say that
Phlegyas came to the Peloponnesus, ostensibly to see the land,
but really to spy out the number of the inhabitants, and whether
the greater part of them was warlike. For Phlegyas was the
greatest soldier of his time, and, making forays on every occasion against whatever people he chanced upon, he carried off
the crops and lifted the cattle. When he went to the Peloponnesus, he was accompanied by his daughter, who all along had
kept hidden from her father that she was with child by Apollo.
In the country of the Epidaurians [1] she bore a child and exposed
him on the mountain called Nipple at the present day, but then
named Myrtium. As the child lay exposed one of the goats that
pastured about the mountain gave him milk, and the watch-dog
of the herd guarded him. And when Aresthanas—for this
was the herdsman's name—discovered that the tale of the goats
was not full, and that the watch-dog also was absent from the
herd, he left, they say, no stone unturned in his search, and on
finding the child desired to take him up. As he drew near, he
saw lightning that flashed from the child, and, thinking that
it was something divine, as in fact it was, he turned away.
Presently it was reported over every land and sea that the child
was discovering everything he wished to heal the sick, and that
he was raising dead men to life.(*)

[1] Cf. T. 18; 19.

8. Cyrillus, Contra Julianum, VI, 805 B -808 B.

Σοβαρεύεται μὲν οὖν Διὸς ἐκφῦναι λέγων τὸν Ἀσκληπιὸν ἐν τοῖς νοητοῖς, ὧφθαί γε μὴν ἐν ἀνθρώπου μορφῇ. ὁ γάρ τοι παρὰ πᾶσι λόγος ἐκ Κορωνίδος αὐτὸν καὶ Ἀπόλλωνος γενέσθαι βούλεται, μεμοιχευμένης κατὰ τὸ εἰκὸς παρά του τῶν ἱερ⟨έων τ⟩ῶν ἐν Ἀπόλλωνος ἱερῷ. οὑτοσὶ δὲ σεμνόν τι καὶ μέγα περὶ αὐτοῦ καὶ φρονεῖ καὶ λέγει ψευδοεπὴς ὢν ἀεί. . . . τουτονὶ καὶ πρῶτόν φασι τὴν ἰατρικὴν εὐτεχνίαν μεταχειρίσασθαί τε καὶ ἀνευρεῖν ἄμεινον ἢ οἱ πρὸ αὐτοῦ γεγονότες· παραδοῦναι δὲ καὶ Ἀσκληπιῷ, ὃς ἐπείπερ τῆς τέχνης γέγονεν ἐν καλῷ οὐκέτι ταῖς Αἰγυπτίων πόλεσιν ἐμφιλοχωρεῖν ἠνέσχετο, ἀλλὰ γὰρ ἐρασι-χρήματός τε ὑπάρχων καὶ λημμάτων αἰσχρῶν ἐκτόπως ἡττώμενος πᾶσαν ὡς ἔπος εἰπεῖν περιεφοίτησε χώραν, ἐρανιζόμενος τὰ παρὰ πολλῶν ὤνιόν τε προθεὶς τοῖς ἀλγοῦσι τὸ ἐπιτήδευμα. ἀρθεὶς δὲ ὑπὸ τῆς τέχνης καὶ μέγα φρονήσας ἐπ᾽ αὐτῇ καὶ θεὸν ἑαυτὸν ὠνόμαζε καὶ ἐκ νεκρῶν ἀναστῆσαί τινας δύνασθαι παραληρῶν ἐφαντάζετο. προσβαλὼν δὲ τοῖς Ἐπιδαυρίοις καὶ δόξαν ἁρπάσας οὐκ ἀγεννῆ παρ᾽ αὐτοῖς, τοῦ καὶ θεὸς εἶναί τις, βάλλεται κεραυνῷ, ποινὴν αὐτῷ τὴν πρεπωδεστάτην ἐπαφέντος Θεοῦ, καὶ διὰ τοῦ παθεῖν ἐξελέγχοντος οὐδὲν ὄντα παντελῶς, πλὴν ὅτι γῆν καὶ σποδόν. οἷα δὲ περὶ αὐτοῦ φησιν ὁ μελοποιὸς Πίνδαρος πῶς οὐκ ἄξιον ἰδεῖν; . . . καὶ μὴν καὶ ὁ τραγῳδὸς Εὐριπίδης εἰπεῖν τὸν Ἀπόλλω διατείνεται . . . εἰ μὲν οὖν εἴργασταί τι τῶν θεοπρεπῶν ὁ Κορωνίδος υἱός· εἰ τυφλοῖς ὀμμάτων ἐρρυηκότων ἐνῆκε τὸ φῶς· εἰ παλινδρομεῖν ἐκέλευσε τοῖς τεθνεῶσιν εἰς ζωήν· εἰ τοὺς ἐν πηρώσει σκελῶν, ἀρτίποδας ἀποφήνας, οἴκαδε χαίροντας ἱέναι προστέταχε· καταπλουτείτω καὶ πρὸς ἡμῶν τὸ χρῆναι θαυμάζεσθαι. εἰ δὲ γέγονεν εὐδόκιμος περί τινας τῶν ἠρρωστηκότων, ἐπαινείσθω μὲν καὶ αὐτὸς ὡς ἄριστα μετελθὼν τῆς ἰατρικῆς ἐμπειρίας τὸ χρῆμα. μὴ μὴν ἔτι καὶ ἀτιμαζέτω τὸ ‘Γνῶθι σαυτόν.’

He [sc., Julian] carries himself pompously, saying that Ascle-pius was born of Zeus among the intelligible gods, but was seen in the form of man.[1] For the current story wants to make him born of Apollo and Coronis, when she was seduced, in all probability by one of the priests in the temple of Apollo. But this man thinks there is something holy and noble about him and says so in his usual lying manner. . . . He [sc., Apis], they say, first pursued the medical art and discovered something better than his predecessors. And he also imparted this knowledge to Asclepius, [2] who, when he had become proficient in the art, did not content himself any longer with dwelling in the cities of the Egyptians, but, being fond of money, and being extraordinarily a slave to shameful gain,[3] traversed practically all the land, collecting money from many, offering his treatments for sale to the sick. Exalted by his art and becoming pompous because of it, he even called himself a god and foolishly boasted that he could raise people from the dead. Going to the Epidaurians, and acquiring among them a not ignoble reputation even of being a god, he is struck with a thunderbolt,[4] God letting loose upon him the punishment most suitable to him; and, through his suffering, proving that he is nothing at all except dust and ashes. What Pindar, the lyric poet, tells about him, how should it not be fitting to look at this [5] . . . and the tragic poet Euripides strains himself to make Apollo say [6] . . . If, then, the son of Coronis accomplished anything meet for a god; if he restored to the blind the sight which had slipped away from their eyes; if he bade the dead return to life; if, making the lame swift of foot, he commanded them to go home rejoicing, then let him be enriched with our due admiration, too; if he was in high repute among some of the most feeble, let him, too, be praised as most nobly going about the task of his medical skill. Yet let him not dishonor the " Understand thyself."

[1] Cf. T. 307.
[2] Cf. T. 358.
[3] Cf. T. 99-103.

[4] Cf. T. 105-15.
[5] T. 1, vv. 55-58.
[6] T. 107.

9. Lactantius, Divinae Institutiones, I, 10, 1-2.

Aesculapius et ipse non sine flagitio Apollinis natus quid fecit aliud divinis honoribus dignum nisi quod sanavit Hippolytum? mortem sane habuit clariorem quod a deo meruit fulminari. hunc Tarquitius de illustribus viris disserens ait incertis parentibus natum, expositum et a venatoribus inventum, canino lacte nutritum, Chironi traditum didicisse medicinam; fuisse autem Messenium, sed Epidauri moratum. Tullius etiam Cynosuris ait sepultum.

Asclepius who himself was not born without Apollo's disgrace, what else did he accomplish worthy of divine honors, except that he healed Hippolytus?[1] His death was certainly more renowned, because he earned the distinction of being struck with lightning by a god. Tarquitius [sc., Priscus, a contemporary of Cicero], talking about illustrious men, says that he was born of uncertain parents,[2] exposed, and being nourished by a dog[3] was found by some hunters; then, he was turned over to Chiron and learned the art of medicine; he says, moreover, that he was a Messenian, but lived at Epidaurus. Cicero says also that he was buried at Cynosura.[4]

[1] Cf. T. 74–82.

[2] Cf. 9a. Tertullianus, *Ad Nationes*, II, 14: Is [sc. Aesculapius] Apollinis filius . . . vel potius spurius ut incerto patre . . . ; cf. also T. 233; T. 9b; cf. T. 8. Concerning various other parentages cf. T. 21 ff.

[3] Cf. T. 5; 7; cf. also 9b. Lactantius, *Institutionum Epitoma*, 8, 1-2: Aesculapium Tarquitius tradit ex incertis parentibus natum et ob id expositum atque a venatoribus collectum, caninis uberibus educatum, Chironi in disciplinam datum. hic Epidauri moratus est, Cynosuris, ut Cicero ait [T. 116] sepultus, cum esset ictu fulminis interemptus; cf. T. 691.　　　　　　[4] T. 116; cf. also T. 101.

PLACE OF ORIGIN

TRICCA

10. Homerus, Ilias, II, 729-31.

Οἳ δ' εἶχον Τρίκκην καὶ Ἰθώμην κλωμακόεσσαν,
οἵ τ' ἔχον Οἰχαλίην, πόλιν Εὐρύτου Οἰχαλιῆος,
τῶν αὖθ' ἡγείσθην Ἀσκληπιοῦ δύο παῖδε,

. . . .

And they that held Tricca[1] and Ithome of the crags, and they who had Oechalia, city of Oechalian Eurytus, these again were led by the two sons of Asclepius . . .[2]

[1] Cf. 10a. Homerus, *Ilias*, IV, 201-2: . . . κρατεραὶ στίχες . . . λαῶν, οἵ οἱ [sc., τῷ Μαχάονι] ἕποντο Τρίκης ἐξ ἱπποβότοιο; cf. T. 164.　　　[2] Cf. T. 135.

11. Strabo, Geographica, XIV, 1, 39.

. . . ἕτερος δ' ἐστὶ Ληθαῖος ὁ ἐν Γορτύνῃ καὶ ὁ περὶ Τρίκκην,
ἐφ' ᾧ ὁ Ἀσκληπιὸς γεννηθῆναι λέγεται . . .

. . . there is another Lethaeus (river) in Gortyna, and another near Tricca, where Asclepius is said to have been born . . .

12. Hyginus, Fabulae, XIV, 21.

Argonautae convocati

. . . Asclepius Apollinis et Coronidis filius, a Tricca.

The assembly of the Argonauts [1]

. . . Asclepius, the son of Apollo and Coronis, from Tricca.[2]

[1] Cf. T. 63; 64.   [2] Cf. T. 137a.

13. Eusebius, Praeparatio Evangelica, III, 14, 6.

Καὶ ὁ Ἀσκληπιὸς πάλιν περὶ ἑαυτοῦ·
Τρίκκης ἐξ ἱερῆς ἥκω θεός, ὅν ποτε μήτηρ
Φοίβῳ ὑπευνασθεῖσα κυεῖ σοφίης βασιλῆα
ἴδριν ἰητορίης Ἀσκληπιόν.

And Asclepius again, about himself:
I come from holy Tricca, a god whom his mother bore in
wedlock with Apollo, Asclepius, skilful king of medical wisdom.

VARIOUS PLACES IN THESSALY

14. Homerus, Hymni, XVI, 1-3.[1]

. . . Ἀσκληπιὸν . . .
. . . τὸν ἐγείνατο δῖα Κορωνὶς
Δωτίῳ ἐν πεδίῳ . . .

. . . Asclepius . . . in the Dotian plain[2] fair Coronis bare
him . . .

[1] Cf. T. 31.   [2] Cf. T. 21.

15. Apollonius Rhodius, Argonautica, IV, 616-17.[1]

. . . τὸν ἐν λιπαρῇ Λακερείῃ
δῖα Κορωνὶς ἔτικτεν ἐπὶ προχοῇς Ἀμύροιο.

. . . whom divine Coronis bare in bright Lacereia at the mouth
of Amyrus.[2]

[1] Cf. T. 28.   [2] Cf. T. 24.

MESSENIA

16. Pausanias, Descriptio Graeciae, II, 26, 7.[1]

Ἀπολλοφάνει γὰρ τῷ Ἀρκάδι ἐς Δελφοὺς ἐλθόντι καὶ ἐρομένῳ
τὸν θεὸν εἰ . . . Μεσσηνίοις πολίτης εἴη, ἔχρησεν ἡ Πυθία . . .

οὗτος ὁ χρησμὸς δηλοῖ μάλιστα . . . Ἡσίοδον ἢ τῶν τινα ἐμπεποιηκότων ἐς τὰ Ἡσιόδου τὰ ἔπη συνθέντα ἐς τὴν Μεσσηνίων
χάριν.

For when Apollophanes, the Arcadian, came to Delphi and
asked the god if [Asclepius] was . . . a citizen of Messene, the
Pythian priestess gave this response. . . . This oracle makes
it quite clear that . . . the story was a fiction invented by
Hesiod,[2] or by one of Hesiod's interpolators, just to please the
Messenians.[3] (*)

[1] Cf. T. 36.   [2] = Hesiod, Fr. 87 [Rzach].
[3] Cf. also T. 39; 39a; 9.

ARCADIA

17. Pausanias, Descriptio Graeciae, VIII, 25, 11.

Ὁ δὲ Λάδων . . . παρέξεισιν . . . τὰ δὲ ἐν δεξιᾷ παρὰ Ἀσκληπιοῦ Παιδὸς ἱερόν, ἔνθα Τρυγόνος μνῆμά ἐστι [τροφοῦ]. τροφὸν
δὲ Ἀσκληπιοῦ τὴν Τρυγόνα εἶναι λέγουσιν· ἐν γὰρ τῇ Θελπούσῃ
τῷ Ἀσκληπιῷ παιδὶ ἐκκειμένῳ φασὶν ἐπιτυχόντα Αὐτόλαον
Ἀρκάδος υἱὸν νόθον ἀνελέσθαι τὸ παιδίον, καὶ ἐπὶ τούτῳ παῖδα
Ἀσκληπιὸν * * * εἰκότα εἶναι μᾶλλον ἡγούμην, ὃ καὶ ἐδήλωσα
ἐν τοῖς Ἐπιδαυρίων.

The Ladon . . . passes . . . on the right a sanctuary of the
Child Asclepius, where is the tomb of Trygon, who is said to
have been the nurse of Asclepius. For the story is that Asclepius, when little, was exposed in Thelpusa, but was found by
Autolaüs, the illegitimate son of Arcas, who reared the baby,
and for this reason the child Asclepius[1] * * * I considered [a
different version] more likely as also I set forth in my account
of Epidaurus.[2]

[1] Cf. T. 773; 679.
[2] Cf. T. 739; cf. also T. 7.

EPIDAURUS

18. Inscriptiones Graecae, IV², 1, no. 128, iv, 48-50 [Isyllus; *ca.* 300
B. C.].[1]

Ἐν δὲ θυώδει τεμένει τέ-
κε|τό νιν Αἴγλα, γονίμαν δ' ἔλυσεν ὠδῖ|να Διὸς
παῖς μετὰ Μοιρᾶν Λάχεσίς τε μαῖα ἀγανά.

Then in the perfumed temple Aigle bore him [sc., Asclepius], and the son of Zeus together with the Fates and Lachesis the noble midwife eased her birth pains.

[1] T. 32; 594.

19. Pausanias, Descriptio Graeciae, II, 26, 4.[1]

Ὡς δὲ ἐν τῇ γῇ τῇ Ἐπιδαυρίων ἔτεκεν, ἐκτίθησι τὸν παῖδα . . .

In the country of the Epidaurians she [sc., Phlegyas' daughter] bore a son, and exposed him. . . .[2]

[1] Cf. T. 7.
[2] Cf. 19a. Pausanias, Descriptio Graeciae, II, 26, 8: Μαρτυρεῖ δέ μοι καὶ τόδε ἐν Ἐπιδαύρῳ τὸν θεὸν γενέσθαι; cf. T. 564; 709.

20. Lactantius Placidus, Commentarii in Statium, Ad Thebaidem, III, 398.

Epidaurius: civitas in Graecia, unde Aesculapius fuisse dicitur . . .

Epidaurius: a city in Greece, whence Asclepius is said to have come . . .[1]

[1] Cf. T. 352.

CORONIS

21. Hesiodus, Fr. 122.

Ἤ οἵη Διδύμους ἱεροὺς ναίουσα κολωνοὺς
Δωτίῳ ἐν πεδίῳ πολυβότρυος ἄντ᾽ Ἀμύροιο
νίψατο Βοιβιάδος λίμνης πόδα παρθένος ἀδμής.

Or like her [sc., Coronis] who lived by the holy Twin Hills in the plain of Dotium[1] over against Amyrus[2] rich in grapes, and washed her feet in the Boebian lake, a maid unwed.

[1] Cf. T. 14.   [2] Cf. T. 24; 28.

22. Hesiodus, Fr. 123.

Τῆμος ἄρ᾽ ἄγγελος ἦλθε κόραξ ἱερῆς ἀπὸ δαιτὸς
Πυθὼ ἐς ἠγαθέην, φράσσεν δ᾽ ἄρα ἔργ᾽ ἀίδηλα
Φοίβῳ ἀκερσεκόμῃ, ὅτι Ἴσχυς γῆμε Κόρωνιν
Εἰλατίδης, Φλεγύαο διογνήτοιο θύγατρα.

There came, then, a messenger from the sacred feast to holy Pytho, a raven,[1] and he told unshorn Phoebus of secret deeds,

that Ischys son of Elatus had wedded Coronis the daughter of
Phlegyas of birth divine. (*)

[1] Cf. T. 41 ff.

23. Scholia in Pindarum, Ad Pythias, III, 25 c.

Εἰ γὰρ ὁ πατὴρ αὐτὴν ἐβεβίαστο συγγενέσθαι τῷ Ἰσχΰϊ,
συνέγνω ἂν αὐτῇ ὁ Ἀπόλλων. διατί δὲ προὐτίμησε τὸν Ἰσχυν
τοῦ Ἀπόλλωνος; Ἀκουσίλαός φησιν, ὡς κατὰ δέος ὑπεροψίας
θνητῷ βουληθεῖσα συνεῖναι.

For if her father had compelled her to marry Ischys, Apollo
would have forgiven her. But for what reason, then, did she
prefer Ischys to Apollo? Acusilaus [early 5th century] [1] says
that for fear of contempt she rather wanted to marry a mortal.

[1] = Fr. 17 [Jacoby].

24. Scholia in Pindarum, Ad Pythias, III, 59.

Ὅτι ἐν Λακερείᾳ ᾦκει Κορωνίς, Φερεκύδης ἐν ᾱ ἱστορεῖ, πρὸς
ταῖς πηγαῖς τοῦ Ἀμύρου· καὶ περὶ τοῦ κόρακος διηγεῖται, καὶ
ὅτι Ἄρτεμιν ἔπεμψεν ὁ Ἀπόλλων, ἣ πολλὰς ἅμα γυναῖκας
ἀπέκτεινεν, Ἰσχυν δὲ Ἀπόλλων ἀποκτείνει, τὸν δὲ Ἀσκληπιὸν
δίδωσι Χείρωνι.

That Coronis lived in Lacereia, near the source of the Amyrus,[1]
Pherecydes relates in book one,[2] and he also tells about the
raven;[3] and that Apollo sent Artemis, who at one stroke slew
many women. But Ischys, Apollo kills; while Asclepius he
gives to Chiron.

[1] Cf. T. 15; 21.
[2] = 24a. Pherecydes, Fr. 3a [Jacoby].
[3] Cf. however T. 1; cf. T. 43, and in general T. 41 ff.

25. Hyginus, Astronomica, II, 40.

Ister et complures dixerunt, Coronida Phlegyae filiam fuisse:
hanc autem ex Apolline Aesculapium procreasse, sed postea
Ischyn, Elati filium, cum ea concubuisse.

Ister [ca. 250 B. C.] [1] and several others have said that Coronis
was the daughter of Phlegyas; and that she bore Asclepius to
Apollo, but that later Ischys, son of Elatus, married her.

[1] = 25a. Ister, Fr. 36 [Müller].

26. Pausanias, Descriptio Graeciae, II, 26, 6.

> Λέγεται δὲ καὶ ἄλλος ἐπ' αὐτῷ λόγος, Κορωνίδα κύουσαν
> Ἀσκληπιὸν Ἴσχυϊ τῷ Ἐλάτου συγγενέσθαι, καὶ τὴν μὲν ἀπο-
> θανεῖν ὑπὸ Ἀρτέμιδος ἀμυνομένης τῆς ἐς τὸν Ἀπόλλωνα ὕβρεως,
> ἐξημμένης δὲ ἤδη τῆς πυρᾶς ἁρπάσαι λέγεται τὸν παῖδα Ἑρμῆς
> ἀπὸ τῆς φλογός.

> There is also another tradition concerning him [sc., Asclepius]:
> Coronis, when with child with Asclepius, had intercourse with
> Ischys, son of Elatus. She was killed by Artemis to punish her
> for her insolence against Apollo, but when the pyre was already
> lighted Hermes is said to have snatched the child from the
> flames.[1] (*)

[1] Cf. however T. 1, vv. 43-44.

27. Servius, Commentarii in Aeneidem, VI, 618.

> Phlegyas autem, Ixionis pater, habuit Coronidem filiam, quam
> Apollo vitiavit, unde suscepit Aesculapium. quod pater dolens,
> incendit Apollinis templum et eius sagittis est ad inferos
> trusus . . .

> Now Phlegyas, the father of Ixion, had a daughter Coronis,
> whom Apollo violated; whereupon she became pregnant with
> Asclepius. Her father, being distraught at this, set fire to the
> temple of Apollo and was driven to the underworld by the god's
> arrows . . .

28. Apollonius Rhodius, Argonautica, IV, 611-17.

> . . . . Κελτοὶ δ' ἐπὶ βάξιν ἔθεντο,
> ὡς ἄρ' Ἀπόλλωνος τάδε δάκρυα Λητοΐδαο
> συμφέρεται δίναις, ἅ τε μυρία χεῦε πάροιθεν,
> ἦμος Ὑπερβορέων ἱερὸν γένος εἰσαφίκανεν,
> οὐρανὸν αἰγλήεντα λιπὼν ἐκ πατρὸς ἐνιπῆς,
> χωόμενος περὶ παιδί, τὸν ἐν λιπαρῇ Λακερείῃ
> δῖα Κορωνὶς ἔτικτεν ἐπὶ προχοῆς Ἀμύροιο.

> But the Celts have attached this story to them [sc., the drops of
> amber],[1] that these are the tears of Leto's son, Apollo, that
> are borne along by the eddies, the countless tears that he shed
> aforetime when he came to the sacred race of the Hyperboreans

and, at the chiding of his father, left shining heaven, being in wrath concerning his son whom divine Coronis bare in bright Lacereia [2] at the mouth of Amyrus.[3] (*)

[1] Cf. T. 110.　　　　　[2] Cf. T. 15; 24.　　　　　[3] Cf. T. 21.

29. Diodorus, Bibliotheca Historica, V, 74, 6.[1]

Ἀπόλλωνος δὲ καὶ Κορωνίδος Ἀσκληπιὸν γενηθέντα . . .

Of Apollo and Coronis Asclepius was the son [2] . . .

[1] Cf. T. 355.
[2] Cf. T. 4; cf. **29a.** Hyginus, *Fabulae,* CLXI: Apollinis filii: Asclepius ex Coronide Phlegyae filia; cf. also T. 237; 238; cf. **29b.** Lucianus, *Alexander,* 38: Εἶτα Λητοῦς ἐγίγνετο λοχεία καὶ Ἀπόλλωνος γοναὶ καὶ Κορωνίδος γάμος καὶ Ἀσκληπιὸς ἐτίκτετο; cf. **29c.** Arnobius, *Adv. Nationes,* I, 36: Aesculapius . . . Coronide . . . natus.

30. Eustathius, Commentarii ad Homeri Iliadem, II, 729.

Πολλαχοῦ δὲ τῆς ποιήσεως ἐν τοῖς ἑξῆς μνεία γίνεται τῶν τοιούτων δύο Ἀσκληπιαδῶν, οὓς τοῦ Λαπιθίου γένους φασὶ κατάγεσθαι. Στίλβης γὰρ καὶ Ἀπόλλωνος ὁ Λαπίθης, οὗ ἀπόγονος Ἀσκληπιός, υἱὸς καὶ αὐτὸς Ἀπόλλωνος καὶ νύμφης Κορωνίδος· Ἀσκληπιοῦ δὲ οἱ ῥηθέντες δύο ἰητῆρες.

Subsequently in many parts of the poem there is reference to two such sons of Asclepius, who, they say, are descended from the Lapithian race. For Lapith was the son of Stilbe and Apollo, and his descendant was Asclepius, he too the son of Apollo, and of the maiden Coronis. Of Asclepius the aforementioned two physicians were sons.[1]

[1] Cf. T. 135 ff.

31. Homerus, Hymni, XVI, 1-5.

Εἰς Ἀσκληπιόν
Ἰητῆρα νόσων Ἀσκληπιὸν ἄρχομ᾽ ἀείδειν,
υἱὸν Ἀπόλλωνος τὸν ἐγείνατο δῖα Κορωνὶς
Δωτίῳ ἐν πεδίῳ κούρη Φλεγύου βασιλῆος,
χάρμα μέγ᾽ ἀνθρώποισι, κακῶν θελκτῆρ᾽ ὀδυνάων.
Καὶ σὺ μὲν οὕτω χαῖρε, ἄναξ· λίτομαι δέ σ᾽ ἀοιδῇ.

To Asclepius [1]

I begin to sing of Asclepius, son of Apollo and healer of sicknesses. In the Dotian plain [2] fair Coronis, daughter of King

Phlegyas, bare him, a great joy to men, a soother of cruel pangs. And so hail to you, lord: in my song I make my prayer to thee!

[1] Cf. T. 587 ff.   [2] Cf. T. 14.

32. Inscriptiones Graecae, IV², 1, no. 128, iv, 40-50 [Isyllus; *ca.* 300 B. C.].

> Φλεγύας δ᾽, [ὃς]
> πατρίδ᾽ Ἐπίδαυρον ἔναιεν, | θυγατέρα Μάλου γαμ[[ε]]-
> εῖ, τὰν Ἐρατὼ γεί|νατο μάτηρ, Κλεοφήμα δ᾽ ὀνομάσθη. | ἐγ
> δὲ Φλεγύα γένετο, Αἴγλα δ᾽ ὀνομάσθη· | τόδ᾽ ἐπώνυμον·
> τὸ κάλλος δὲ Κορωνὶς ἐπεκλήθη. | κατιδὼν δὲ ὁ χρυ-
> σότοξος Φοῖβος ἐμ Μά|λου δόμοις παρθενίαν ὥραν
> ἔλυσε· | λεχέων δ᾽ ἱμεροέντων ἐπέβας, Λα|τώιε κόρε
> χρυσοκόμα. | σέβομαί σε· ἐν δὲ θυώδει τεμένει τέ-
> κε|τό νιν Αἴγλα, γονίμαν δ᾽ ἔλυσεν ὠδῖ|να Διὸς
> παῖς μετὰ Μοιρᾶν Λάχεσίς τε μαῖα ἀγανά.

Phlegyas, who dwelt in Epidaurus, his fatherland, married the daughter of Malos, whom Erato bore, and her name was Kleophema. By Phlegyas then a child was begotten and she was named Aigle; this was her name, but because of her beauty she was also called Coronis. Then Phoebus of the golden bow, beholding her in the palace of Malos, ended her maidenhood. You went into her lovely bed, O golden haired son of Leto. I revere you. Then in the perfumed temple Aigle bore the child, and the son of Zeus together with the Fates and Lachesis the noble midwife eased her birth pains.[1]

[1] Cf. T. 18; cf. also T. 594.

33. Homerus, Hymni, III, 207-13.

> Πῶς τ᾽ ἄρ σ᾽ ὑμνήσω πάντως εὔυμνον ἐόντα;
> ἠέ σ᾽ ἐνὶ μνηστῆσιν ἀείδω καὶ φιλότητι,
> ὅππως μνωόμενος ἔκιες Ἀζαντίδα κούρην
> Ἴσχυ᾽ ἅμ᾽ ἀντιθέῳ, Ἐλατιονίδῃ εὐΐππῳ;
> ἢ ἅμα Φόρβαντι Τριοπέῳ γένος, ἢ ἅμ᾽ Ἐρευθεῖ;
> ἢ ἅμα Λευκίππῳ καὶ Λευκίπποιο δάμαρτι
> πεζός, ὁ δ᾽ ἵπποισιν;

How then shall I sing of you [*sc.*, Apollo]—though in all ways you are a worthy theme for song? Shall I sing of you as a wooer and in the fields of love, how you went wooing the daughter of Azan along with godlike Ischys, the son of well-horsed Elatius, or with Phorbas sprung from Triops, or with Ereutheus, or with Leucippus and the wife of Leucippus, you on foot, he with his chariot?

34. Scholia in Pindarum, Ad Pythias, III, 14.

Ἀριστείδης δὲ ἐν τῷ περὶ Κνίδου κτίσεως συγγράμματί φησιν οὕτως· Ἀσκληπιὸς Ἀπόλλωνος παῖς καὶ Ἀρσινόης. αὕτη δὲ παρθένυς οὖσα ὠνομάζετο Κορωνίς, Λευκίππου δὲ θυγάτηρ ἦν τοῦ Ἀμύκλα τοῦ Λακεδαίμονος. Ἀσκληπιοῦ δὲ καὶ Ἠπιόνης Ποδαλείριος καὶ Μαχάων.

Aristides [time unknown] in his book on the founding of Cnidus[1] speaks thus: Asclepius, the son of Apollo and Arsinoë. But when a young maiden, she was called Coronis; she was the daughter of Leucippus, the son of Amyclas, the Lacedaemonian. Children of Asclepius and Epione[2] were Podalirius and Machaon.[3]

[1] = Aristides, Fr. 22 [Müller].   [2] Cf. T. 162; cf. also e. g. T. 280.   [3] Cf. T. 135 ff.

35. Scholia in Pindarum, Ad Pythias, III, 15.

Καὶ Σωκράτης γόνον Ἀρσινόης τὸν Ἀσκληπιὸν ἀποφαίνει, παῖδα δὲ Κορωνίδος εἰσποίητον.

And Socrates [sc., the historian, time unknown][1] points out that Asclepius was the son of Arsinoë, but the adopted child of Coronis.

[1] = Socrates, Fr. 1 [Müller]; cf. T. 52.

ARSINOË

36. Pausanias, Descriptio Graeciae, II, 26, 7.

Ὁ δὲ τρίτος τῶν λόγων ἥκιστα ἐμοὶ δοκεῖν ἀληθής ἐστιν, Ἀρσινόης ποιήσας εἶναι τῆς Λευκίππου παῖδα Ἀσκληπιόν. Ἀπολλο-φάνει γὰρ τῷ Ἀρκάδι ἐς Δελφοὺς ἐλθόντι καὶ ἐρομένῳ τὸν θεὸν εἰ γένοιτο ἐξ Ἀρσινόης Ἀσκληπιὸς καὶ Μεσσηνίοις πολίτης εἴη, ἔχρησεν ἡ Πυθία·

᾽Ω μέγα χάρμα βροτοῖς βλαστὼν ᾽Ασκληπιὲ πᾶσιν,
ὃν Φλεγυηὶς ἔτικτεν ἐμοὶ φιλότητι μιγεῖσα
ἱμερόεσσα Κορωνὶς ἐνὶ κραναῇ ᾽Επιδαύρῳ.

οὗτος ὁ χρησμὸς δηλοῖ μάλιστα οὐκ ὄντα ᾽Ασκληπιὸν ᾽Αρσινόης,
ἀλλὰ ῾Ησίοδον ἢ τῶν τινα ἐμπεποιηκότων ἐς τὰ ῾Ησιόδου τὰ
ἔπη συνθέντα ἐς τὴν Μεσσηνίων χάριν.

The third account is, in my opinion, the farthest from the truth,
making Asclepius to be the son of Arsinoë, the daughter of
Leucippus. For when Apollophanes, the Arcadian, came to
Delphi and asked the god if Asclepius was the son of Arsinoë
and therefore a citizen of Messene,[1] the Pythian priestess gave
this response:

" O Asclepius, born to bestow great joy upon all mortals,
whom the daughter of Phlegyas, lovely Coronis, bore,
united with me in love, in rugged land, Epidaurus."[2]

This oracle makes it quite clear that Asclepius was not a son
of Arsinoë, and that the story was a fiction invented by
Hesiod,[3] or by one of Hesiod's interpolators, just to please the
Messenians. (*)

[1] Cf. T. 16.   [2] Cf. T. 19a.   [3] = 36a. Hesiodus, Fr. 87 [Rzach].

37.   Scholia in Pindarum, Ad Pythias, III, 14.[1]

Τὸν ᾽Ασκληπιὸν οἱ μὲν ᾽Αρσινόης, οἱ δὲ Κορωνίδος φασὶν εἶναι.
᾽Ασκληπιάδης δέ φησι τὴν ᾽Αρσινόην Λευκίππου εἶναι τοῦ
Περιήρους, ἧς καὶ ᾽Απόλλωνος ᾽Ασκληπιὸς καὶ θυγάτηρ ᾽Εριῶπις·

ἡ δ᾽ ἔτεκ᾽ ἐν μεγάροις ᾽Ασκληπιόν, ὄρχαμον ἀνδρῶν,
Φοίβῳ ὑποδμηθεῖσα, ἐυπλόκαμόν τ᾽ ᾽Εριῶπιν.
καὶ ᾽Αρσινόης ὁμοίως·

᾽Αρσινόη δὲ μιγεῖσα Διὸς καὶ Λητοῦς υἱῷ
τίκτ᾽ ᾽Ασκληπιὸν υἱὸν ἀμύμονά τε κρατερόν τε.

Some say Asclepius was the son of Arsinoë, others of Coronis.[2]
But Asclepiades says [3] that Arsinoë was the daughter of Leucippus, Perieres' son, and that to her and Apollo Asclepius and
a daughter, Eriopis, were born:

"And she bare in the palace Asclepius, leader of men, and
Eriopis with the lovely hair, being subject in love to Phoebus."

And of Arsinoë likewise:

"And Arsinoë was joined with the son of Zeus and Leto
and bare
a son Asclepius, blameless and strong."(*)

[1] = Hesiod, Fr. 87 [Rzach]; cf. T. 16.
[2] Cf. T. 3.
[3] = Asclepiades, Fr. 32 [Jacoby].

38. Pausanias, Descriptio Graeciae, IV, 3, 1-2.

. . . ἐς δὲ Νέστορα τὸν Νηλέως περιῆλθε Μεσσηνίων ἡ ἀρχὴ
τῶν τε ἄλλων καὶ ὅσων πρότερον ἐβασίλευεν Ἴδας, πλὴν ὅσοι
τοῖς Ἀσκληπιοῦ παισὶν αὐτῶν ὑπήκουον. καὶ γὰρ τοὺς Ἀσκληπιοῦ παῖδας στρατεῦσαί φασιν ἐπ᾽ Ἴλιον Μεσσηνίους ὄντας,
Ἀρσινόης γὰρ Ἀσκληπιὸν τῆς Λευκίππου καὶ οὐ Κορωνίδος
παῖδα εἶναι· καὶ Τρίκκαν τε καλοῦσιν ἔρημον ἐν τῇ Μεσσηνίᾳ
χωρίον καὶ ἔπη τῶν Ὁμήρου καταλέγουσιν, ἐν οἷς τὸν Μαχάονα
ὁ Νέστωρ τῷ ὀϊστῷ βεβλημένον περιέπων ἐστὶν εὐνοϊκῶς· οὐκ
ἂν οὖν αὐτὸν ⟨εἰ⟩ μὴ ἐς γείτονα καὶ ἀνθρώπων βασιλέα ὁμοφύλων
προθυμίαν τοσήνδε γε ἐπιδείξασθαι.

. . . and the kingdom of Messenia passed to Nestor, the son of
Neleus, including among the other parts also that ruled formerly
by Idas, but not that subject to the sons of Asclepius. For they
say that the sons of Asclepius who went to Troy were Messenians,[1] Asclepius being the son of Arsinoë, daughter of Leucippus, not the son of Coronis, and they call a desolate spot in
Messenia by the name of Tricca and quote the lines of Homer,[2]
in which Nestor tends Machaon kindly, when he has been
wounded by the arrow. He would not have shown such readi-
ness except to a neighbor and king of a kindred people.(*)

[1] Cf. T. 190.　　　　　　　　　[2] Cf. T. 166.

39. Pausanias, Descriptio Graeciae, III, 26, 4.

Ἐφ᾽ ὅτῳ μὲν δή ἐστιν ὄνομα τῇ πόλει Λεύκτρα, οὐκ οἶδα· εἰ δ᾽
ἄρα ἀπὸ Λευκίππου τοῦ Περιήρους, ὡς οἱ Μεσσήνιοί φασι, τούτου
μοι δοκοῦσιν ἕνεκα οἱ ταύτῃ θεῶν μάλιστα Ἀσκληπιὸν τιμᾶν,
ἅτε Ἀρσινόης παῖδα εἶναι τῆς Λευκίππου νομίζοντες.

Why the city has this name, Leuctra, I do not know. If
indeed it is derived from Leucippus, the son of Perieres, as

the Messenians say, it is for this reason, I think, that the inhabitants honor Asclepius most of the gods, supposing him to be the son of Arsinoë, the daughter of Leucippus.[1] (*)

[1] Cf. 39a. Pausanias, *Descriptio Graeciae*, IV, 31, 12: Ἀσκληπιός, Ἀρσινόης ὢν λόγῳ τῷ Μεσσηνίων; cf. T. 657.

40. Cicero, De Natura Deorum, III, 22, 57.[1]

. . . tertius Arsippi et Arsinoae . . .

. . . the third [sc., Asclepius] is the son of Arsippus and Arsinoë[2] . . .

[1] Cf. T. 116; 118; 379.   [2] Cf. T. 380; 380a.

RAVEN

41. Scholia in Pindarum, Ad Pythias, III, 52 b.

Ὁ δὲ Ἀρτέμων τὸν Πίνδαρον ἐπαινεῖ, ὅτι παρακρουσάμενος τὴν περὶ τὸν κόρακα ἱστορίαν [αὐτὸν δι᾽ ἑαυτοῦ ἐγνωκέναι φησὶ τὸν Ἀπόλλωνα] · ἱστορεῖται γάρ, ὅτι τὴν Ἴσχυος μίξιν ἐδήλωσεν αὐτῷ ὁ κόραξ, παρὸ καὶ δυσχεράναντα ἐπὶ τῇ ἀγγελίᾳ ἀντὶ λευκοῦ μέλανα αὐτὸν ποιῆσαι· τοῦτον δὴ τὸν μῦθον διωσάμενόν φησι τὸν Πίνδαρον τῷ ἑαυτοῦ νῷ καταλαβεῖν τὸν Ἀπόλλωνα τὰ πεπραγμένα τῇ Κορωνίδι· παράλογον γὰρ τὸν ἄλλοις μαντευόμενον αὐτὸν μὴ συμβαλεῖν τὰ κατ᾽ αὐτοῦ δρώμενα. χαίρειν οὖν φράσας τῷ μύθῳ τέλεον ὄντι ληρώδει αὐτόν φησι τὸν Ἀπόλλωνα παρὰ τοῦ νοῦ πυθόμενον ἐπιπέμψαι τὴν Ἄρτεμιν τῇ Κορωνίδι. τὸν δὲ περὶ τὸν κόρακα μῦθόν φησι καὶ Ἡσίοδον μνημονεύοντα λέγειν οὕτως . . .

Artemon [Hellenistic writer(?)][1] commends Pindar because he avoids the story about the raven (of his own accord, he says, Apollo learned of the affair). For there is a legend that the raven revealed to Apollo the intercourse of Ischys [sc., with Coronis]. Wherefore Apollo, vexed by the raven's disclosure, made him black instead of white. Rejecting this myth, he says Pindar has Apollo himself discover what Coronis had done,[2] for it is unreasonable that the same one who gives prophecies to others should not surmise what was being done against himself. Discarding the myth, then, as absolutely absurd, he says Apollo himself, learning of the deed from his own mind,

sent Artemis against Coronis.[8] He says Hesiod, too, recalling
the myth about the raven, said the following . . .[4]

[1] = Artemon Pergamenus, Fr. 7 [Müller].
[2] Cf. T. 1, vv. 27 ff.; T. 42.
[3] Cf. T. 1, v. 32.
[4] T. 22; cf. T. 24.

42. Scholia in Pindarum, Ad Pythias, III, 48 c-d.

Ἡνίκα οὖν ἡ Κορωνὶς τῷ Ἴσχυϊ συνῆλθεν, ἐν Πυθῶνι τυχὼν ὁ
Ἀπόλλων ᾔσθετο.
καλῶς δὲ ὁ Πίνδαρος τὴν περὶ τὸν κόρακα παρακρουσάμενος
ἱστορίαν δι᾿ ἑαυτοῦ φησι τὸν Ἀπόλλωνα κατειληφέναι τὴν
Κορωνίδα.

When, then, Coronis had intercourse with Ischys, Apollo, who
happened to be in Pytho, apprehended it.
Rightly, however, Pindar, avoiding the tale about the raven,
says that Apollo detected Coronis by himself.[1]

[1] Cf. T. 41.

43. Scholia in Pindarum, Ad Pythias, III, 59.

. . . καὶ περὶ τοῦ κόρακος διηγεῖται . . .

. . . and he [sc., Pherecydes] [1] also tells about the raven . . .
[1] Cf. T. 24.

44. Callimachus, Hecale, Fr. 34, Col. IV.

" [δεί]ελος ἀλλ᾿ ἢ νὺξ ἢ ἔνδιος ἢ ἔσετ᾿ ἠώς,
εὖτε κόραξ, [ὅ]ς νῦν γε καὶ ἂν κύκνοισιν ἐρίζοι
καὶ γάλακι χροιὴν καὶ κύματος ἄκρῳ ἀώτῳ,
κυάνεον φὴ πίσσαν ἐπὶ πτερὸν οὐλοὸν ἕξει,
ἀγγελίης ἐπίχε[ι]ρα τά οἱ ποτε Φοῖβος ὀπάσσει,
ὁππότε [κ]εν Φλεγύαο Κορωνίδος ἀμφὶ θυγατρὸς
Ἴσχυϊ πληξίππῳ σπομένης μιερόν τ[ι] πύθηται."

" But evening it shall be or night or noon or morn when the
raven, which now might vie for color with swans, or milk, or
the foam that tips the wave, shall put on a sad plumage black
as pitch, the guerdon that Phoebus shall one day give him for
his news, when he learns terrible tidings of Coronis, daughter
of Phlegyas, even that she has gone with knightly Ischys."

45. Ovidius, Metamorphoses, II, 534-41.

. . . .
quam tu nuper eras, cum candidus ante fuisses,
535 corve loquax, subito nigrantes versus in alas.
nam fuit haec quondam niveis argentea pennis·
ales, ut aequaret totas sine labe columbas,
nec servaturis vigili Capitolia voce
cederet anseribus nec amanti flumina cygno.
540 lingua fuit damno: lingua faciente loquaci,
qui color albus erat, nunc est contrarius albo.

. . . at the same time that thy plumage, talking raven, though white before, had been suddenly changed to black. For he had once been a bird of silvery-white plumage, so that he rivalled the spotless doves, nor yielded to the geese which one day were to save the Capitol with their watchful cries, nor to the river-loving swan. But his tongue was his undoing. Through his tattling tongue's fault the talking bird, which once was white, is now the opposite of white.[1]

[1] Cf. T. 2.

BIRTH

46. Servius, Commentarii in Aeneidem, VII, 761.

Tunc Diana eius castitate commota revocavit eum in vitam per Aesculapium, filium Apollinis et Coronidis, qui natus erat exsecto matris ventre, ideo quod, cum Apollo audisset a corvo, eius custode, eam adulterium committere, iratus Coronidem maturo iam partu confixit sagittis—corvum vero nigrum fecit ex albo—et exsecto ventre Coronidis produxit ita Aesculapium, qui factus est medicinae peritus. hunc postea Iuppiter propter revocatum Hippolytum interemit.

Then Diana, moved to pity by his [sc., Hippolytus'] chastity, recalled him to life through Asclepius, the son of Apollo and Coronis, who was born when his mother's womb was cut open. The reason for this was that when Apollo had heard from the raven, her guardian, that she had committed adultery, in a burst of anger he transfixed Coronis with his arrows, just as she was about to give birth to the child—the raven, moreover, he made

black instead of white—and cutting open her womb he thus
brought forth Asclepius who became skilled in medicine. Afterwards, Jupiter slew him because he had restored Hippolytus
to life.[1]

[1] Cf. T. 74 ff.; 100; 111.

47. Servius, Commentarii in Aeneidem, X, 316.

Omnes qui secto matris ventre procreantur, ideo sunt Apollini
consecrati, quia deus est medicinae, per quam lucem sortiuntur:
unde Aesculapius eius fingitur filius; ita enim eum procreatum
supra diximus.

All who are born through cutting open the mother's womb for
this reason are consecrated to Apollo because he is the god of
medicine by means of which they come forth to the light: therefore Asclepius is reputed to be the son of Apollo. That he was
thus born, we said previously.[1]

[1] Cf. T. 46.

48. Hyginus, Fabulae, CCII. 1-2.

Coronis

Apollo cum Coronida Phlegyae filiam gravidam fecisset, corvum custodem ei dedit, ne quis eam violaret. cum ea Ischys
Elati filius concubuit; ob id ab Iove fulmine est interfectus.
Apollo Coronidem gravidam percussit et interfecit; cuius ex
utero exsectum Asclepium educavit, at corvum qui custodiam
praebuerat ex albo in nigrum commutavit.

Coronis

When Apollo had made pregnant Coronis, the daughter of
Phlegyas, he gave her a raven as a guardian that no one might
violate her. Ischys, the son of Elatus, had intercourse with
her; for this reason he was killed with a thunderbolt by Jupiter.
Apollo struck the pregnant Coronis and killed her; by cutting
open her womb, he extracted Asclepius from it, but the raven
who had provided the guard [for Coronis] he changed from
white to black.

49. Lactantius Placidus, Commentarii in Statium, Ad Thebaidem, III, 506.

Apollo, cum Coronidem gravidam fecisset, corvum ei custodem apposuit, ne quis ad eam occulte temerator accederet. cum hac Lycus occulte concubuit, quem fulmine Iuppiter exstinxit. ipsam Coronidem Apollo sagittis occidit, cuius mortuae exsecto utero Aesculapium produxit in lucem. unde Vergilius: 'fulmine poenigenam Stygias detrusit ad undas' id est per poenam matris natum.

When Apollo had made Coronis pregnant, he assigned a raven as a guardian over her lest any rash person should secretly approach her. With her by stealth lay Lycus whom Jupiter destroyed with a thunderbolt. Coronis herself Apollo slew with his arrows; from her womb, cut open when she was dead, he brought forth Asclepius into the light of day. Wherefore Vergil says: 'with his thunderbolt he [sc., Jupiter] hurled down to the Stygian waters 'poenigenam,'[1] that means him who was born through the punishment of his mother.[2]

[1] Cf. however T. 77, v. 773: Phoebigenam.
[2] Cf. 49a. Servius, *Commentarii in Aeneidem*, VII, 773: Poenigenam, matris poena genitum: alii ' Phoebigenam ' legunt, ut Probus.

CHIRON

50. Homerus, Ilias, IV, 218-19.[1]

. . . . ἐπ' ἄρ' ἤπια φάρμακα εἰδὼς
πάσσε, τά οἵ ποτε πατρὶ φίλα φρονέων πόρε Χείρων.

. . . he [sc., Machaon] skilfully spread thereon soothing simples, which of old Chiron had given to his father [sc., Asclepius] with kindly thought.(*)

[1] Cf. T. 164; cf. also T. 174.

51. Pindarus, Nemeae, III, 54-56.

Βαθυμῆτα Χίρων τράφε λιθίνῳ
Ἰάσον' ἔνδον τέγει, καὶ ἔπειτεν Ἀσκλαπιόν,
τὸν φαρμάκων δίδαξε μαλακόχειρα νόμον·

The sage Chiron, dwelling under a rocky roof, nurtured Jason, and after him Asclepius, whom he taught the gentle-handed lore of simples.[1](*)

[1] Cf. T. 1, vv. 1-7; 45-53.

52. Scholia in Pindarum, Ad Nemeas, III, 92.

> . . . ὅτι δὲ ἐτράφη παρὰ τῷ Χείρωνι ὁ Ἰάσων, Ἡσίοδός φησιν
> . . . ἀλλὰ καὶ ὁ Ἀσκληπιὸς παρὰ Χείρωνι, φησίν, ἐτράφη, ὡς
> καὶ Σωκράτης ὁ Ἀργεῖός φησι.

> . . . that Jason was educated by Chiron Hesiod [1] says . . . but
> Asclepius too, he [sc., Pindar] says, was educated by Chiron,
> as Socrates the Argive [2] [time unknown] also asserts.

[1] = Fr. 19 [Rzach].　　　　　　　　[2] = Fr. 2 [Müller]; cf. T. 35.

53. Scholia in Pindarum, Ad Pythias, III, 79.

> . . . τὸν ἑαυτοῦ παῖδα Ἀπόλλων ἐνεχείρισεν τῷ Κενταύρῳ τῷ
> Θεσσαλῷ ἀναδιδάσκειν τὰς πολλῶν πημάτων αἰτίας τοῖς ἀν-
> θρώποις νόσους θεραπεύειν.

> . . . his own child Apollo entrusted to the Thessalian Centaur,
> to teach him to cure diseases, the causes of many of the miseries
> of mankind.[1]

[1] Cf. T. 1, vv. 45 ff.

54. Scholia in Pindarum, Ad Pythias, III, 9.

> . . . ὁποῖος ὢν τήν τε φύσιν καὶ τὸν νοῦν ὁ Χείρων ἀνέθρεψε
> τὸν τῆς ἀνωδυνίας καὶ ὑγείας πρᾳότατον κατασκευαστὴν ἥρωα
> Ἀσκληπιόν . . .

> . . . being of such nature and mind, Chiron educated the most
> gentle bestower of painlessness and health, the hero [1] Ascle-
> pius . . .

[1] Cf. T. 255 ff.

55. Scholia in Pindarum, Ad Pythias, III, 102 b.

> Ῥίψαις δι' ἀμφοῖν· δι' αὐτοῦ τε τοῦ Ἀσκληπιοῦ καὶ τοῦ
> ἰατρευθέντος, ὥστε καὶ τὸν Ἀσκληπιὸν ἀνῃρῆσθαι καὶ τὸν
> ἀναβεβιωκότα. ὁ δὲ Σωκράτης τὸν Ἀσκληπιόν φησι καὶ τὸν
> διδάξαντα αὐτὸν Χείρωνα. οὐδεὶς δὲ τοῦτο μαρτυρεῖ.

> Piercing through both of them: [1] through both Asclepius him-
> self and the one who was cured, so that Asclepius as well as
> the man who had returned to life were destroyed. Socrates

[the Argive],[2] however, says Asclepius and Chiron who taught him [were struck by the thunderbolt]. But no one substantiates this.

[1] Cf. T. 1, v. 57.　　　　　　　　[2] = Fr. 3 [Müller]; cf. T. 35; 52.

56. Xenophon, Cynegeticus, I, 1-6.

Τὸ μὲν εὕρημα θεῶν, Ἀπόλλωνος καὶ Ἀρτέμιδος, ἄγραι καὶ
κύνες· ἔδοσαν δὲ καὶ ἐτίμησαν τούτῳ Χείρωνα διὰ δικαιότητα.
ὁ δὲ λαβὼν ἐχάρη τῷ δώρῳ καὶ ἐχρῆτο· καὶ ἐγένοντο αὐτῷ
μαθηταὶ κυνηγεσίων τε καὶ ἑτέρων καλῶν Κέφαλος, Ἀσκληπιός,
. . . Μαχάων, Ποδαλείριος, . . . ὧν κατὰ χρόνον ἕκαστος ὑπὸ
θεῶν ἐτιμήθη. . . .
Κέφαλος μὲν καὶ ὑπὸ θεᾶς ἡρπάσθη, Ἀσκληπιὸς δὲ μειζόνων
ἔτυχεν, ἀνιστάναι μὲν τεθνεῶτας, νοσοῦντας δὲ ἰᾶσθαι· διὰ δὲ
ταῦτα θεὸς ὢν * παρ᾽ ἀνθρώποις ἀείμνηστον κλέος ἔχει.

* θεὸς ὢν M s omisit Ruehl　θεὸς ὡς Valckenaer.

Game and hounds are the invention of gods, of Apollo and Artemis. They bestowed it on Chiron and honored him therewith for his righteousness. And he, receiving it, rejoiced in the gift, and used it. And he had for pupils in venery and in other noble pursuits: Cephalus, Asclepius . . . Machaon, Podalirius, . . . of whom each in his time was honored by gods. . . . Cephalus was carried away by a goddess. Asclepius won yet greater preferment—to raise the dead and to heal the sick;[1] and for these things being a god he has everlasting fame among men.(*)

[1] Cf. T. 67; 243.

57. Philostratus, Heroicus, 9.

Χείρωνα δὲ τὸν ἐν Πηλίῳ γενέσθαι μέν φησιν ἀνθρώπῳ ὅμοιον,
σοφὸν δὲ καὶ λόγους καὶ ἔργα—θήρας τε γὰρ ποικίλης ἥπτετο
καὶ τὰ πολεμικὰ ἐπαίδευε καὶ ἰατροὺς ἀπέφαινε καὶ μουσικοὺς
ἥρμοττε καὶ δικαίους ἐποίει— . . . φοιτῆσαι δὲ αὐτῷ Ἀσκληπιὸν Τελαμῶνά τε καὶ Πηλέα καὶ Θησέα . . .

Chiron, he [sc., Protesilaus, the hero] says, who lived on Pelion, was like a man and clever in word and deed—for he engaged in various kinds of hunting and also taught the science of war; he trained physicians; guided musicians; and made righteous men . . . Asclepius and Telamon, as well as Peleus and Theseus resorted to him for instruction . . .[1]

[1] Cf. **57a**. Philostratus, *Heroicus*, 10, 1: Βουλομένου δὲ Χείρωνος ἰατρικὴν διδάσκειν αὐτὸν [*sc.*, τὸν Παλαμήδην] ʽ ἐγώ,ʼ ἔφη, ʽ ὦ Χείρων, ἰατρικὴν μὲν ἡδέως οὐκ οὖσαν ἂν εὗρον, εὑρημένην δὲ οὐκ ἀξιῶ μανθάνειν, καὶ ἄλλως τὸ ὑπέρσοφόν σου τῆς τέχνης ἀπήχθηται μὲν Διί, ἀπήχθηται δὲ Μοίραις, καὶ διήειν ἂν τὰ ᾿Ασκληπιοῦ, εἰ μὴ ἐνταῦθα ἐβέβλητο.ʼ

58. Anonymus, Vita Sophoclis, 11.

> Ἔσχε δὲ καὶ τὴν τοῦ Ἄλκωνος ἱερωσύνην, ὃς ἥρως μετὰ ᾿Ασκληπιοῦ παρὰ Χείρωνι . . .

He [*sc.*, Sophocles] [1] also held the priesthood of Alcon who, a hero, together with Asclepius, [was] . . . with Chiron . . .

[1] Cf. T. **587** ff.

59. Ps. Eratosthenes, Catasterismi, I, 40.

> Οὗτος δοκεῖ Χείρων εἶναι ὁ ἐν τῷ Πηλίῳ οἰκήσας, δικαιοσύνῃ δὲ ὑπερενέγκας πάντας ἀνθρώπους καὶ παιδεύσας ᾿Ασκληπιόν τε καὶ ᾿Αχιλλέα.

This [*sc.*, the Centaur] seems to be Chiron who dwelt on Pelion, surpassing all men in righteousness, the teacher of both Asclepius and Achilles.[1]

[1] Cf. **59a**. Commentaria in Aratum, *Scholia*, 436: Οὗτος δὲ ὁ Χείρων, ὁ ἐν τῷ σπηλαίῳ κατοικήσας, δικαιοσύνῃ πάντας ἀνθρώπους ὑπερβάλλων καὶ παιδεύσας ᾿Ασκληπιόν, τὸν υἱὸν ᾿Απόλλωνος . . . ; cf. **59b**. Commentaria in Aratum, *Aratus Latinus*, 436-42: Hic videtur Chiron esse qui in Pelio habitavit iustitia quidem superans homines omnes et ipse correxit Asclepium . . . ; cf. **59c**. *Scholia in Caesaris Germanici Aratea*, 417: [Chiron] . . . magister Aesculapii et Achillis aliorumque heroum; cf. **59d**. Hyginus, *Astronomica*, II, 38: Hic [*sc.*, Centaurus] . . . Aesculapium . . . nutrisse existimatur.

60. Heraclitus, Quaestiones Homericae, 15.

> Χείρων γὰρ αὐτὸν ἐδίδαξε, δικαιότατος Κενταύρων, ὃς πάσῃ μὲν ἐκέκαστο σοφίᾳ, περιττῶς δὲ τῇ ἰατρικῇ, ὅπου γνώριμον αὐτῷ φασιν εἶναι καὶ ᾿Ασκληπιόν.

For Chiron educated him [*sc.*, Achilles], Chiron, the most righteous of the Centaurs, who excelled in every branch of knowledge, and surpassingly so in the art of medicine, in which even Asclepius is said to have been his pupil.[1]

[1] Cf. **60a**. Servius, *Commentarii in Georgica*, III, 550: Chiron . . . adeo medicinae peritus, ut etiam Aesculapium docuisse dicatur; cf. **60b**. Eustathius, *Commentarii ad Iliadem*, IV, 202: ᾿Ασκληπιὸς δὲ Χείρωνος μαθητής, Χείρων δὲ εὑρετὴς ἰατρικῆς καὶ λυρικῆς; cf. T. **6**.

61. Scholia in Caesaris Germanici Aratea, 291.

> Alii eum Chironem esse dixerunt Saturni et Philyrae filium, quod iustissimus maximeque pius atque hospitalis esset, a quo Aesculapius medicina, Achilles cithara, in astrologia Hercules . . . * litteras didicisse dicantur.

*Lacunam indicavit Eyssenhardt.

> Some said that [Sagittarius] was Chiron, the son of Saturn and Philyra, because he was most righteous, and particularly reverent and hospitable; from him Asclepius is said to have received instruction in medicine, Achilles in the lyre, Heracles in astronomy . . .[1]

[1] Cf. **61a.** Lucius Ampelius, *Liber Memorialis*, 2, 9: Alii Chironem dicunt [*sc.*, Sagittarium] quod iustus et pius [doctus] ⟨atque⟩ hospitalis fuerit. ab eo Aesculapius ⟨doctus⟩ medicinam, Achilles citharam et alia multa.

62. Justinus, De Monarchia, 6, 23.

> Ἀσκληπιὸς καὶ Ἀπόλλων παρὰ Χείρωνι τῷ Κενταύρῳ ἰᾶσθαι διδάσκονται, τὸ καινότατον παρὰ ἀνθρώπῳ θεοί.

> Asclepius and Apollo were taught the art of healing by Chiron, the Centaur—the strangest thing, gods taught by a man!

ACCOMPLISHMENTS

HEROIC DEEDS

63. Clemens Alexandrinus, Stromateis, I, 21, 105.

> Ἀσκληπιός τε καὶ Διόσκουροι συνέπλεον αὐτοῖς, ὡς μαρτυρεῖ ὁ Ῥόδιος Ἀπολλώνιος ἐν τοῖς Ἀργοναυτικοῖς.

> Both Asclepius and the Dioscuri sailed with them [*sc.*, the Argonauts], as Apollonius of Rhodes testifies in the *Argonautica*.[1]

[1] The passage is not preserved; cf. T. 129; cf. T. 12; 5.

64. Eusebii Caesariensis Chronici Canones Latine vertit Eusebius Hieronymus, p. 89, 11.

> Ea quae de . . . Argonautis dicuntur, in quibus fuerunt Hercules Asclepius Castor et Pollux . . .

> What is said of . . . the Argonauts, among whom were Heracles, Asclepius, Castor and Pollux . . .

65. Hyginus, Fabulae, CLXXIII, 1.

> Qui ad aprum Calydonium ierunt
> Castor et Pollux Iovis filii . . . Aesculapius Apollinis.

> Those who went to hunt the Calydonian boar
> Castor and Pollux, sons of Jupiter . . . Asclepius, Apollo's son.

MYTHICAL REVIVING AND HEALING

66. Aeschylus, Agamemnon, 1019-24.

> Τὸ δ' ἐπὶ γᾶν πεσὸν ἅπαξ θανάσιμον
> προπάροιθ' ἀνδρὸς μέλαν αἷμα τίς ἂν
> πάλιν ἀγκαλέσαιτ' ἐπαείδων;
> οὐδὲ τὸν ὀρθοδαῆ
> τῶν φθιμένων ἀνάγειν
> Ζεὺς ἀπέπαυσεν ἐπ' εὐλαβείᾳ;

> But man's dark blood, once it hath flowed to earth in death, who by chanting spells shall call it back? Even him who possessed the skill to raise from the dead—did not Zeus put a stop to him as a precaution? [1](*)

[1] Cf. T. 1, vv. 55-8; T. 105.

67. Euripides, Alcestis, 122-29.

> . . . εἰ φῶς τόδ' ἦν
> ὄμμασιν δεδορκὼς
> Φοίβου παῖς . . . . . .
> . . . δμαθέντας γὰρ ἀνίστη,
> πρὶν αὐτὸν εἷλε διόβολον
> πλῆκτρον πυρὸς κεραυνίου.

> . . . were life's light in the eyes of Phoebus' son . . . for he was wont to raise the dead,[1] until there smote him the fiery bolt of thunder hurled by Zeus.[2]

[1] Cf. T. 56; 243.   [2] Cf. T. 105 ff.; esp. 107.

68. Libanius, Orationes, XX, 8.

> Ὁ μὲν γὰρ Ἀσκληπιὸς ἐνί τινι λέγεται λῦσαι θάνατον καὶ
> Ἡρακλῆς μιᾷ τινι γυναικί . . .

Asclepius is said to have released one man from death and Heracles one woman . . .

69. Sextus Empiricus, Adversus Mathematicos, I, 260-62.

Αἱ δέ γε κατὰ μέρος ἱστορίαι ἄπειροί τε διὰ τὸ πλῆθός εἰσι καὶ οὐχ ἑστῶσαι διὰ τὸ μὴ τὰ αὐτὰ περὶ τοῦ αὐτοῦ παρὰ πᾶσιν ἱστορεῖσθαι. οἷον (οὐκ ἄτοπον γὰρ ἵνα συμφυέσι τε καὶ οἰκείοις χρησώμεθα τῶν πραγμάτων παραδείγμασιν) ὑπόθεσιν γὰρ ἑαυτοῖς ψευδῆ λαμβάνοντες οἱ ἱστορικοὶ τὸν ἀρχηγὸν ἡμῶν τῆς ἐπιστήμης Ἀσκληπιὸν κεκεραυνῶσθαι λέγουσιν, οὐκ ἀρκούμενοι τῷ ψεύσματι, ἐν ᾧ καὶ ποικίλως αὐτὸ μεταπλάττουσι, Στησίχορος μὲν ἐν Ἐριφύλῃ εἰπὼν ὅτι τινὰς τῶν ἐπὶ Θήβαις πεσόντων ἀνιστᾷ, Πολύανθος δὲ ὁ Κυρηναῖος ἐν τῷ περὶ τῆς Ἀσκληπιαδῶν γενέσεως ὅτι τὰς Προίτου θυγατέρας κατὰ χόλον Ἥρας ἐμμανεῖς γενομένας ἰάσατο, Πανύασις δὲ διὰ τὸ νεκρὸν Τυνδάρεω ἀναστῆσαι, Στάφυλος δὲ ἐν τῷ περὶ Ἀρκάδων ὅτι Ἱππόλυτον ἐθεράπευσε φεύγοντα ἐκ Τροιζῆνος κατὰ τὰς παραδεδομένας κατ᾽ αὐτοῦ ἐν τοῖς τραγῳδουμένοις φήμας, Φύλαρχος δὲ ἐν τῇ ἐννάτῃ διὰ τὸ τοὺς Φινέως υἱοὺς τυφλωθέντας ἀποκαταστῆσαι, χαριζόμενον αὐτῶν τῇ μητρὶ Κλεοπάτρᾳ τῇ Ἐρεχθέως, Τελέσαρχος δὲ ἐν τῷ Ἀργολικῷ ὅτι τὸν Ὠρίωνα ἐπεβάλετο ἀναστῆσαι. οὐ τοίνυν τῆς οὕτως ἀπὸ ψευδοῦς ὑποθέσεως ἀρχομένης καὶ ἀδιεξιτήτου κατὰ πλῆθος καὶ πρὸς τὴν ἑκάστου προαίρεσιν μεταπλαττομένης γένοιτ᾽ ἄν τις τεχνικὴ θεωρία.

At all events, the special histories are countless in number and do not hold their ground because the same information is not given about the same subject by all of them. For instance (it is of course not inappropriate that we use familiar and fitting examples of these matters under discussion), making a false assumption on their own authority, the historians say that Asclepius, the founder of our science,[1] was transfixed with a thunderbolt;[2] not content with the false statement, they even indulge in inventing many variations, Stesichorus [640-550 B. C.(?)] saying in his *Eriphyle*[3] that Asclepius revived some of those who fell at Thebes; Polyanthus, the Cyrenaean [time unknown], again relating in his book *On the Origin of the Asclepiads*[4] that he healed the daughters of Proetus who had been driven mad through the wrath of Hera; Panyassis [*ca.* 500 B. C.] saying it was because he resuscitated the corpse of

Tyndareus;[5] Staphylus [Hellenistic author], on the other hand, writing in his book *About the Arcadians*,[6] that he healed Hippolytus when he fled from Troezen, according to the legends handed down about him in the tragedians; or Phylarchus [3rd century B. C.] saying in his *ninth book*[7] that it was because he restored the sight of the sons of Phineus when they were blinded, doing a favor to their mother, Cleopatra, the daughter of Erechtheus; whereas Telesarchus [Hellenistic historian] says in his *History of Argolis* that he undertook to restore Orion to life. When a hypothesis originates in falsehood in such a fashion and is for the most part incapable of proof and remodelled according to the whim of each historian, then no scientific theory can be formulated.

[1] Cf. T. 341.
[2] Cf. T. 105 ff.
[3] Cf. T. 70; 73.
[4] = Fr. 1 [Jacoby] (= Polyarchus?).
[5] = Fr. 19 [Kinkel]; cf. T. 70; 71; cf. also T. 84; 85; 104.
[6] = Fr. 8 [Müller].
[7] = Fr. 18 [Jacoby]; cf. T. 71; 72.

70. Apollodorus, Bibliotheca, III, 10, 3, 9-10.[1]

Εὗρον δέ τινας λεγομένους ἀναστῆναι ὑπ' αὐτοῦ, Καπανέα καὶ Λυκοῦργον, ὡς Στησίχορός φησιν ⟨ἐν⟩ Ἐριφύλῃ, Ἱππόλυτον, ὡς ὁ τὰ Ναυπακτικὰ συγγράψας λέγει, Τυνδάρεων, ὥς φησι Πανύασσις, Ὑμέναιον, ὡς οἱ Ὀρφικοὶ λέγουσι, Γλαῦκον τὸν Μίνωος, ὡς Ἀμελησαγόρας λέγει.

I found some who are reported to have been raised by him, to wit, Capaneus and Lycurgus, as Stesichorus [640-555 B. C. (?)] says in the *Eriphyle*;[2] Hippolytus, as the author of the *Naupactica* [6th century B. C.(?)] reports;[3] Tyndareus, as Panyassis [*ca.* 500 B. C.] says;[4] Hymenaeus, as the Orphics report;[5] and Glaucus, son of Minos, as Amelesagoras [5th century B. C.] relates.[6]

[1] Cf. T. 3.
[2] Cf. T. 69; 73.
[3] = Fr. 11 [Kinkel].
[4] = Fr. 19 [Kinkel]; cf. T. 69; 71; 84; 85; 104.
[5] = Fr. 40 [Kern]; cf. T. 71; 72.
[6] = Fr. 2 [Müller]; cf. T. 71.

71. Scholia in Euripidem, Ad Alcestim, 1.

. . . Φερεκύδης δὲ οὔ φησι τοὺς Κύκλωπας ὑπὸ Ἀπόλλωνος ἀνῃρῆσθαι, ἀλλὰ τοὺς υἱοὺς αὐτῶν, γράφων οὕτως. '. . . κτείνει δὲ αὐτοὺς Ἀπόλλων Διὶ μεμφθείς, ὅτι κτείνει Ζεὺς Ἀσκληπιὸν

τὸν παῖδα αὐτοῦ κεραυνῷ ἐν Πυθῶνι. ἀνίστη γὰρ ἰώμενος τοὺς
τεθνεῶτας.᾽ . . . ᾽Απολλόδωρος δέ φησι κεραυνωθῆναι τὸν
᾽Ασκληπιὸν ἐπὶ τῷ τὸν ᾽Ιππόλυτον ἀναστῆσαι, ᾽Αμελησαγόρας
δὲ ὅτι Γλαῦκον, Πανύασσις ⟨δὲ⟩ ὅτι Τυνδάρεων, οἱ δὲ ᾽Ορφικοὶ
ὅτι ῾Υμέναιον· Στησίχορος δὲ ἐπὶ Καπανεῖ καὶ Λυκούργῳ,
Φερεκύδης δὲ ἐν τῇ ā * τῶν ᾽Ιστοριῶν τοὺς ἐν Δελφοῖς φησι
θνῄσκοντας αὐτὸν ἀναβιώσκειν, Φύλαρχος δὲ διὰ τοὺς Φινείδας,
Τελέσαρχος δὲ δι᾽ ᾽Ωρίωνα, Πολύαρχος δὲ ὁ Κυρηναῖος διὰ τὸ
τὰς Προίτου θιγατέρας αὐτὸν ἰάσασθαι κεραυνωθῆναί φησιν.

* ā coni. Jacoby η MSS (Schwartz Kern).

. . . Pherecydes [5th century B. C.], however, does not say
that the Cyclops were destroyed by Apollo, but rather their
sons, for he writes [1] ". . . Apollo killed them, bearing a grudge
against Zeus, because Zeus slew Asclepius, his son, with his
thunderbolt in Pytho.[2] The reason was that Asclepius healed
the dead and returned them to life." Apollodorus [2nd century B. C.] says [3] that Asclepius was struck by lightning because he restored Hippolytus to life. But Amelesagoras [5th
century B. C.] says [4] it was Glaucus, and Panyassis [ca. 500
B. C.],[5] Tyndareus. The Orphics claim [6] it was Hymenaeus.
Stesichorus [640-555 B. C. (?)] declares it was because of Capaneus and Lycurgus. Pherecydes [5th century B. C.] in the
first book of his *Histories* says [7] he resurrected the dead in
Delphi.[8] Phylarchus [3rd century B. C.] says [9] it was because
of the Phinidae; Telesarchus [Hellenistic historian], because
of Orion; Polyarchus, the Cyrenaean [time unknown], asserts [10]
that he was struck with the thunderbolt for healing the
daughters of Proetus.

[1] = Fr. 35 a [Jacoby].
[2] Cf. T. 108.
[3] = Fr. 139 [Jacoby].
[4] = Fr. 2 [Müller]; cf. T. 70.
[9] = Fr. 18 [Jacoby]; cf. T. 69; 72.
[5] = Fr. 19 [Kinkel]; cf. T. 69; 70; 84; 104.
[6] = Fr. 40 [Kern]; cf. T. 70; 72.
[7] = Fr. 35 a [Jacoby].
[8] Cf. T. 98.
[10] = Fr. 1 [Jacoby] (= Polyanthus?); cf. T. 69.

72. Scholia in Pindarum, Ad Pythias, III, 96.

᾽Αλλὰ καὶ ἡ σοφία κέρδει δέδεται καὶ ἥττηται. μετέστρεψε γὰρ
καὶ τὸν θεόν, φησὶ δὲ τὸν ᾽Ασκληπιόν, τῷ πλείονι μισθῷ ὁ χρυσὸς
ἐν ταῖς χερσὶ φανείς, ὥστε ἄνδρα ἐκ θανάτου ἀναγαγεῖν ἤδη τῷ
μοιριδίῳ ληφθέντα. λέγεται δὲ ὁ ᾽Ασκληπιὸς χρυσῷ δελεασθεὶς

ἀναστῆσαι Ἱππόλυτον τεθνηκότα· οἱ δὲ Τυνδάρεων, ἔτεροι
Καπανέα, οἱ δὲ Γλαῦκον, οἱ δὲ Ὀρφικοὶ Ὑμέναιον, Στησίχορος
δὲ ἐπὶ Καπανεῖ καὶ Λυκούργῳ· οἱ δὲ διὰ τὸ τὰς Προιτίδας ἰάσασθαι, οἱ δὲ διὰ τὸ τὸν Ὠρίωνα· Φύλαρχος ὅτι τοὺς Φινείδας
ἰάσατο, Φερεκύδης δὲ ὅτι τοὺς ἐν Δελφοῖς θνήσκοντας ἀναβιοῦν
ἐποίησεν.

But even wisdom has been shackled by gain and overcome by
it. For the gold displayed upon the palm corrupted even the
god—he means Asclepius—so that for greater gain he resurrected from death a man who had already been overtaken by the
day of destiny. Enticed by gold,[1] Asclepius is reported to have
revived Hippolytus[2] after he had died. Some say it was Tyndareus, others Capaneus, others Glaucus; the Orphics say
Hymenaeus,[3] Stesichorus [ca. 640-555 B. C. (?)] connects this
legend with Capaneus and Lycurgus.[4] Some say it was because he healed the Proetides, others because [he healed] Orion.
Phylarchus [3rd century B. C.] says that he healed the Phinidae,[5] Pherecydes [5th century B. C.] that he made those who
died at Delphi live again.[6]

[1] Cf. T. 1, v. 55; 99; 100.　　　　[4] Cf. T. 71.
[2] Cf. T. 74 ff.; cf. also T. 121.　　[5] = Fr. 18 [Jacoby]; cf. T. 69; 71.
[3] = Fr. 40 [Kern]; cf. T. 70; 71.　　[6] = Fr. 35 b [Jacoby].

73. Philodemus, De Pietate, 52.

Ἀσκληπι[ὸν δὲ Ζε]ὺς ἐκεραύνωσ[εν, ὡς μ]ὲν ὁ τὰ Ναυπα[κτι]κὰ
συνγράψας [κὰ]ν Ἀσκληπιῶ[ι Τελ]έστης καὶ Κεινη[σίας] ὁ
μελοποιός, ὅ[τι τὸ]ν Ἱππόλυτον [παρα]κληθεὶς ὑπ' Ἀρ[τέμι]δος
ἀνέστ[η]σε[ν, ὡς δ' ἐ]ν Ἐριφύλῃ Σ[τησίχορ]ος, ὅτι Κα[πανέα
καὶ Λυ]κοῦρ[γον].

Zeus struck Asclepius by lightning,[1] as the writer of the *Naupactica* [6th century B. C. (?)] says[2] and Telestes [4th century B. C.] in the *Asclepius*[3] and the lyric poet Cinesias [4th
century B. C.], because he raised Hippolytus[4] from the dead
at the instance of Artemis; but, as Stesichorus [ca. 640-555
B. C. (?)] says in the *Eriphyle*, it was because he raised
Capaneus and Lycurgus.(*)

[1] Cf. T. 105 ff.　　　　[3] Cf. T. 106.
[2] = Fr. 12 [Kinkel]; cf. T. 106.　[4] Cf. T. 74 ff.

74. Scholia in Pindarum, Ad Pythias, III, 96.[1]

Λέγεται δὲ ὁ Ἀσκληπιὸς . . . ἀναστῆσαι Ἱππόλυτον . . .

Asclepius is said . . . to have revived Hippolytus . . .[2]

[1] Cf. T. **72**.   [2] Cf. also T. **69**; **100**; **9**.

75. Ovidius, Fasti, VI, 743-62.

Exciderat curru lorisque morantibus artus
    Hippolytus lacero corpore raptus erat
745 reddideratque animam, multum indignante **Diana**.
    ' nulla ' Coronides ' causa doloris ' ait;
  ' namque pio iuveni vitam sine volnere reddam
    et cedent arti tristia fata meae.'
gramina continuo loculis depromit eburnis
750   (profuerant Glauci manibus illa prius,
tunc cum observatas augur descendit in herbas,
    usus et auxilio est anguis ab angue dato),
pectora ter tetigit, ter verba salubria dixit:
    depositum terra sustulit ille caput.
755 lucus eum nemorisque sui Dictynna recessu
    celat: Aricino Virbius ille lacu.
at Clymenus Clothoque dolent; haec, fila reneri,
    hic, fieri regni iura minora sui.
Iuppiter exemplum veritus direxit in ipsum
760   fulmina, qui nimiae noverat artis opem.
Phoebe, querebaris: deus est; placare parenti:
    propter te, fieri quod vetat, ipse facit.

Hippolytus fell from the car, and, his limbs entangled by the
reins, his mangled body was whirled along, till he gave up the
ghost, much to Diana's rage. " There is no need for grief,"
said the son of Coronis, " for I will restore the pious youth
to life all unscathed, and to my leech-craft gloomy fate shall
yield." Straightway he drew from an ivory casket simples
that before had stood Glaucus' ghost in good stead, what time
the seer went down to pluck the herbs he had remarked, and
the snake was succoured by a snake. Thrice he touched the
youth's breast, thrice he spoke healing words; then Hippolytus
lifted his head, low laid upon the ground. He found a hiding-
place in a sacred grove and in the depths of Dictynna's own

woodland; he became Virbius of the Arician Lake.[1] But
Clymenus and Clotho grieved, she that life's broken thread
should be respun, he that his kingdom's rights should be
infringed. Fearing the example thus set, Jupiter aimed a
thunderbolt at him[2] who knew the resources of a too potent
art. Phoebus, thou didst complain: but he [sc., Asclepius]
is a god now, be reconciled to thy parent: he does himself for
thy sake what he forbids to be done.

[1] Cf. T. 739, 4; cf. also T. 77a.　　　　　　　[2] Cf. T. 105 ff.

76. Ovidius, Metamorphoses, XV, 531-36.

Vidi quoque luce carentia regna
et lacerum fovi Phlegethontide corpus in unda,
nec nisi Apollineae valido medicamine prolis
reddita vita foret; quam postquam fortibus herbis
atque ope Paeonia Dite indignante recepi,
tum mihi . . .

Further, I [sc., Hippolytus] saw the rayless world of death
and bathed my torn body in the waves of Phlegethon. And
there should I still be had not my life been restored by the
potent remedies of Apollo's son. And when I had regained it
by the help of strong herbs and Paeonian aid, though it was
against the will of Dis, then . . . (*)

77. Vergilius, Aeneis, VII, 765-73.

Namque ferunt fama Hippolytum, postquam arte novercae
occiderit patriasque explerit sanguine poenas
turbatis distractus equis, ad sidera rursus
aetheria et superas caeli venisse sub auras,
Paeoniis revocatum herbis et amore Dianae.
tum pater omnipotens, aliquem indignatus ab umbris
mortalem infernis ad lumina surgere vitae,
ipse repertorem medicinae talis et artis
fulmine Phoebigenam Stygias detrusit ad undas.

For they tell how that Hippolytus, when he fell by his step-
dame's craft, and slaked a sire's vengeance in blood, torn
asunder by frightened steeds—came again to the starry firma-

ment and heaven's upper air, recalled by the Paeonian Healer's herbs[1] and Diana's love. Then the Father omnipotent, wroth that any mortal should rise from the nether shades to the light of life, himself with his thunder[2] hurled down to the Stygian waters the finder of such healing-craft,[3] the Phoebus-born.[4] (*)

[1] Cf. **77a**. Marcellus, *Epistula ad Filios*, 1: Nam si quid umquam congruum sanitati curationique hominum vel ab aliis comperi vel ipse usu adprobavi vel legendo cognovi, id sparsum inconditumque collegi et in unum corpus quasi disiecta et lacerata Asclepius Virbii [*i.e.* Hippolyti] membra conposui; cf. T. **96**.
[2] Cf. T. **105** ff.   [3] Cf. T. **349** ff.   [4] Cf. T. **49**.

78. Lactantius, Divinae Institutiones, I, 17, 15.

> Altera cum paene amatorem suum perdidisset, qui erat turbatis distractus equis, praestantissimum medicum Asclepium curando iuveni advocavit eumque sanatum . . .

> Another [*sc.*, Diana], when she had almost lost her lover [*sc.*, Hippolytus], who was 'torn asunder by frightened steeds,'[1] called in the most excellent physician Asclepius for the treatment of the youth; and when he was healed . . .

[1] Cf. T. **77**,v. 767.

79. Lactantius Placidus, Commentarii in Statium, Ad Thebaidem, V, 434.

> . . . quod fabricatores fulminis sagitta percusserat quia Asclepius, Apollinis filius, fuerat fulminatus. rogatu enim Dianae Hippolytum ad auras superas revocarat. ut Virgilius ' Paeoniis revocatum herbis et amore Dianae.'

> . . . because he [*sc.*, Apollo] had pierced with his arrow the forgers of the thunderbolt, since Asclepius, the son of Apollo, had been struck with a thunderbolt. For at Diana's request, he had recalled Hippolytus to the upper air. As Vergil says[1] ' recalled by the Paeonian Healer's herbs and Diana's love.'

[1] Cf. T. **77**, v. 769.

80. Lactantius Placidus, Commentarii in Statium, Ad Thebaidem, VI, 353 (375).

> Fulminato Aesculapio, quod revocare ad vitam ausus fuisset Hippolytum, pater Apollo, ubi se vidit orbatum, sagittis Cyclopas occidit, qui Iovis fulmina fabricare consueverant.

When Asclepius was struck with a thunderbolt because he had
dared to restore Hippolytus to life, his father Apollo, seeing
himself bereft of his son, with his darts slew the Cyclops who
had been wont to fashion Jove's thunderbolts.

81. Hyginus, Fabulae, XLIX, 1.

Aesculapius Apollinis filius Glauco Minois filio vitam reddidisse
sive Hippolyto dicitur, quem Iuppiter ob id fulmine percussit.

Asclepius, the son of Apollo, is said to have restored life to
Glaucus, the son of Minos,[1] or else to Hippolytus; and for this
act Jupiter struck him with a thunderbolt.

[1] Cf. T. 70; 71.

82. Libanius, Orationes, XIII, 42.

Ὅπερ γὰρ Ἀσκληπιόν φασιν Ἱππολύτῳ γενέσθαι, τοῦτ᾽ αὐτὸς
ἐγένου τῷ τῆς οἰκουμένης σώματι. τεθνεῶτάς τε ἀνέστησας καὶ
βασιλείας ὄνομα νῦν, εἴπερ ποτέ, προσέλαβεν ἔργον.

For what they say Asclepius was to Hippolytus, this you [sc.,
Julian] were to the body of the inhabited world; you raised up
the dead and the name of majesty now if ever has assumed
actuality.

83. Clementina Homilia, V, 15.

. . . αὐτὸς ὁ Ζεὺς Γανυμήδους ἐρᾷ . . . ὁ δὲ Ἀσκληπιὸς
Ἱππολύτου.

. . . Zeus himself loved Ganymede . . . and Asclepius loved
Hippolytus.

84. Lucianus, De Saltatione, 45.

Οὐκ ὀλίγα δὲ καὶ ἡ Λακεδαίμων τοιαῦτα παρέχεται . . . καὶ
τὴν Τυνδάρεω ἀνάστασιν καὶ τὴν Διὸς ἐπὶ τούτῳ κατ᾽ Ἀσκληπιοῦ
ὀργήν.

Lacedaemon, too, furnishes not a few such examples . . . also the
resurrection of Tyndareus, and Zeus' anger at Asclepius over it.[1]

[1] Cf. T. 69; 70; 71; 85; 104.

85. Plinius, Naturalis Historia, XXIX, 1 (1), 3.

Auxit deinde famam etiam crimine, ictum fulmine Aesculapium fabulata, quoniam Tyndareum revocavisset ad vitam.

Then it [sc., the art of medicine] enhanced its reputation even by a crime, circulating the legend that Asclepius was struck with a thunderbolt because he had recalled Tyndareus [1] to life.

[1] Cf. T. 69; 70; 71; 84; 104.

86. Propertius, Elegiae, II, 1, 57-62.

Omnes humanos sanat medicina dolores:
  solus amor morbi non amat artificem.
tarda Philoctetae sanavit crura Machaon,
  Phoenicis Chiron lumina Phillyrides,
et deus extinctum Cressis Epidaurius herbis
  restituit patriis Androgeona focis,

. . . .

Medicine cures all the anguish of mankind; love alone loves no physician of its ill. Machaon healed Philoctetes' limping feet,[1] Chiron the son of Philyra opened the eyes of Phoenix, the Epidaurian god restored the dead Androgeon to his father's hearth by power of Cretan herbs . . .

[1] Cf. T. 152; 175; cf. also T. 174; 176-176b; 201; 202.

87. Pausanias, Descriptio Graeciae, III, 19, 7.

Ἡρακλῆς . . . ἀκεσθεὶς τὸ τραῦμα τὸ ἐς τὴν κοτύλην οἱ γενόμενον ἐν τῇ πρὸς Ἱπποκόωντα καὶ τοὺς παῖδας προτέρᾳ μάχῃ.

Heracles . . . was cured [sc., by Asclepius] of the wound in the hip joint that he received in the former fight with Hippocoon and his sons.[1] (*)

[1] Cf. T. 763.

88. Pausanias, Descriptio Graeciae, III, 20, 5.

Δήμητρος ἐπίκλησιν Ἐλευσινίας ἐστὶν ἱερόν · ἐνταῦθα Ἡρακλέα Λακεδαιμόνιοι κρυφθῆναί φασιν ὑπὸ Ἀσκληπιοῦ τὸ τραῦμα ἰώμενον.

There [*sc.*, between Taletum and Euoras] is a sanctuary of
Demeter surnamed Eleusinian. Here according to the Lacedaemonian story Heracles was hidden while his wound was
being treated by Asclepius.[1] (*)

[1] For the connection with Heracles cf. also T. 265.

89. Nicander, Theriaca, 685-88.

> Ἄγρει καὶ πάνακες Φλεγυήιον, ὅ ῥρα τε πρῶτος
> Παιήων Μέλανος ποταμοῦ παρὰ χεῖλος ἄμερξεν,
> Ἀμφιτρυωνιάδαο θέρων Ἰφικλέος ἕλκος,
> εὖτε σὺν Ἡρακλῆϊ κακὴν ἐπυράκτεεν ὕδρην.

Take also the Phlegyeian all-heal which Paeeon first plucked
from along the edge of the Black River, curing the wound of
Iphicles, the son of Amphitryon, at the time when he burned
the evil Hydra in company with Heracles.

90. Scholia in Nicandrum, Ad Theriaca, 687.

> Θέρων δὲ ἀντὶ τοῦ ἰώμενος τὸ τοῦ Ἰφικλέος ἕλκος, ὅπερ πέπονθε
> χρανθεὶς τῷ αἵματι τῆς ὕδρας, ὅτε αὐτὴν σὺν τῷ ἀδελφῷ Ἡρακλεῖ
> ἔκαιεν. Ἡρακλέους γὰρ τὴν ὕδραν ἀναιροῦντος ὁ ἀδελφὸς αὐτοῦ
> ὁ Ἰφικλῆς πληγεὶς ὑπ' αὐτῆς ἐθεραπεύθη ὑπ' Ἀσκληπιοῦ. ἢ
> Ἰφικλέος ἔρνος, ἵν' ᾖ λέγων τὸν Ἰόλαον πεπλῆχθαι ὑπ' αὐτῆς,
> ὅπερ ἄμεινον. ἰδίως δὲ ταῦτα ἱστορεῖ Νίκανδρος. οὐ γὰρ
> Ἰφικλῆς ἀλλ' ὁ τούτου παῖς ὁ Ἰόλαος συνεστράτευσεν Ἡρακλεῖ
> ἐπὶ τὴν ὕδραν.

Θέρων (warming) instead of ἰώμενος (healing) the wound of
Iphicles, which he sustained when he was smeared with the
blood of the Hydra at the time that he slew it with his brother
Heracles. For when Heracles destroyed the Hydra, his brother
Iphicles, being struck by it, was cured by Asclepius. Or Ἰφικλέος
ἔρνος (the child of Iphicles), so that he may be saying that
Iolaus was struck by the Hydra, which is better. Nicander tells
this story in a unique fashion, for not Iphicles but his son Iolaus
fought with Heracles against the Hydra.

91. Eutecnius, Metaphrasis Theriacorum Nicandri, 685-88.

> Ἄθρει δὴ καὶ ταύτην τὴν βοτάνην, ἧς ἐστιν εὑρετὴς Ἀσκληπιός,
> ὄνομα δὲ ἐπιλέγεται τὸ ταύτης πάνακες. ἐν γὰρ δὴ τῇ χώρᾳ τῇ

Φλεγυῶν, πλησίον Μέλανος τοῦ ποταμοῦ, καὶ παρὰ τὸ χεῖλος διαγωνιζομένου τοῦ τῆς Ἀλκμήνης πρὸς τὴν ὕδραν παιδός, παρὼν τιτρώσκεται ἅμ' αὐτῷ ὑπὸ τῆς ὕδρας ὁ Ἴφικλος, καὶ ὁ Ἀσκληπιὸς αὐτόθεν λαβὼν τὴν βοτάνην ἐπαμύνει αὐτῷ βεβλημένῳ, καὶ οὕτως ἰᾶται.

Now observe this herb, the discoverer of which is Asclepius. The name of this herb is given as panacea.[1] For when the son of Alcmene was struggling with the Hydra in the land of the Phlegyae, near the Black River, and along its edge, Iphicles, being there with him, was wounded by the Hydra; and Asclepius, taking from there the herb, came to his aid when he had been struck, and in this way healed him.

[1] Cf. T. 98; 370 ff.

92. Sophocles, Philoctetes, 1437-38.

> Ἐγὼ δ' Ἀσκληπιὸν
> παυστῆρα πέμψω σῆς νόσου πρὸς Ἴλιον.

And I [sc., Heracles] will send Asclepius to Troy[1] to heal thy [sc., Philoctetes'] illness. (*)

[1] Cf. however T. 135 ff.; 152.

93. Scholia in Lycophronem, Ad Alexandram, 1054.

> . . . θεραπεύσας δὲ Ἀσκλην τὸν Ἐπιδαύρου τύραννον ὀφθαλ-
> μιῶντα σφοδρῶς . . .

. . . but after he [sc., Asclepius] had cured Ascles, the tyrant of Epidaurus, who suffered seriously from ophthalmia . . .[1]

[1] Cf. T. 271 ff.

EARTHLY LIFE AND CHARACTER

94. Justinus, Apologia, 22, 6.

> Ὧι δὲ λέγομεν χωλοὺς καὶ παραλυτικοὺς καὶ ἐκ γενετῆς πονηροὺς ὑγιεῖς πεποιηκέναι αὐτὸν καὶ νεκροὺς ἀνεγεῖραι, ὅμοια τοῖς ὑπὸ Ἀσκληπιοῦ γεγενῆσθαι λεγομένοις καὶ ταῦτα φάσκειν δόξομεν.

When we say that He [sc., Jesus] made well the lame and the paralytic and those who were feeble from birth and that he resurrected the dead, we shall seem to be mentioning deeds similar to and even identical with those which were said to have been performed by Asclepius.[1] (**)

[1] Cf. T. 423, 36 and 37; 67.

95. Justinus, Dialogus, 69, 3.

Ὅταν δὲ τὸν Ἀσκληπιὸν νεκροὺς ἀνεγείραντα καὶ τὰ ἄλλα πάθη θεραπεύσαντα παραφέρῃ, οὐχὶ τὰς περὶ Χριστοῦ ὁμοίως προφητείας μεμιμῆσθαι τοῦτον καὶ ἐπὶ τούτῳ φημί;

And when he [sc., the devil] brings forward Asclepius as the raiser of the dead and healer of the other diseases, may I not say that in this matter likewise he has imitated the prophecies [1] about Christ? (*)

[1] Cf. T. 332.

96. Ausonius, Opuscula, XVI, p. 197, 17-19.

Quem tu aut ut Aesculapius redintegrabis ad vitam, aut ut Plato iuvante Vulcano liberabis infamia.

This [sc., book] you [sc., Symmachus] will either restore to life, like Asclepius, or like Plato [who burnt his juvenile poetry], you will free it from shame with Vulcan's assistance.[1]

[1] Cf. T. 77a.

97. Apollodorus, Bibliotheca, III, 10, 3, 8-9.[1]

Καὶ γενόμενος χειρουργικὸς καὶ τὴν τέχνην ἀσκήσας ἐπὶ πολὺ οὐ μόνον ἐκώλυέ τινας ἀποθνήσκειν, ἀλλ' ἐνήγειρε καὶ τοὺς ἀποθανόντας· παρὰ γὰρ Ἀθηνᾶς λαβὼν τὸ ἐκ τῶν φλεβῶν τῆς Γοργόνος ῥυὲν αἷμα, τῷ μὲν ἐκ τῶν ἀριστερῶν ῥυέντι πρὸς φθορὰν ἀνθρώπων ἐχρῆτο, τῷ δὲ ἐκ τῶν δεξιῶν πρὸς σωτηρίαν, καὶ διὰ τοῦτο τοὺς τεθνηκότας ἀνήγειρεν.

And having become a surgeon, and carried the art to a great pitch, he not only prevented some from dying, but even raised up the dead; for he had from Athena the blood that flowed from the veins of the Gorgon, and while he used the blood that flowed from her left side for the bane of mankind, he used the blood that flowed from her right side for salvation, and by that means he raised the dead.

[1] Cf. T. 3.

98. Scholia in Nicandrum, Ad Theriaca, 685.

Πάνακες· καὶ γὰρ τὸ πάνακες Φλεγνήιον τὸ παιανικὸν εἶδος τοῦ φυτοῦ. Φλεγναῖον δὲ τὸ ἐν Φλεγύαις, ἔθνει τῆς Φωκίδος, φυόμενον. Φλεγύαι γὰρ ἔθνος Φωκίδος παρὰ Δελφοῖς ᾤκησαν. ἢ ἐπεὶ οἱ Δελφοὶ Ἀπόλλωνος ἱερὸν Φλεγνήιον ἱδρύσαντο· ἢ ἐπεὶ ὁ Ἀσκληπιὸς ἰατρεῦσαι λέγεται ἐν Δελφοῖς. ὁ γὰρ Ἀσκληπιὸς Κορωνίδος ἦν υἱὸς τῆς Φλεγύου θυγατρός.

Πάνακες [a plant] : for the Phlegyeian plant is the healing form of the plant; and Φλεγναῖον means that which grows among the Phlegyae, a people of Phocis. For the Phlegyae, a people of Phocis, lived near Delphi. Or because Delphi established the Phlegyeian shrine of Apollo. Or because Asclepius is said to have practiced medicine at Delphi.[1] For Asclepius was the son of Coronis, daughter of Phlegyas.

[1] Cf. T. 71.

99. Plato, Res Publica, III, 408 B-C.

Πρέπει, ἦν δ᾽ ἐγώ· καίτοι ἀπειθοῦντές γε ἡμῖν οἱ τραγῳδοποιοί τε καὶ Πίνδαρος Ἀπόλλωνος μέν φασιν Ἀσκληπιὸν εἶναι, ὑπὸ δὲ χρυσοῦ πεισθῆναι πλούσιον ἄνδρα θανάσιμον ἤδη ὄντα ἰάσασθαι, ὅθεν δὴ καὶ κεραυνωθῆναι αὐτόν· ἡμεῖς δὲ κατὰ τὰ προειρημένα οὐ πειθόμεθα αὐτοῖς ἀμφότερα, ἀλλ᾽ εἰ μὲν θεοῦ ἦν, οὐκ ἦν, φήσομεν, αἰσχροκερδής, εἰ δ᾽ αἰσχροκερδής, οὐκ ἦν θεοῦ.

" 'Tis fitting," said I; " and yet in disregard of our principles the tragedians and Pindar affirm that Asclepius, though he was the son of Apollo, was bribed by gold to heal a man already at the point of death, and that for this cause he was struck by the lightning.[1] But we in accordance with the aforesaid principles refuse to believe both statements, but if he was the son of a god he was not avaricious,[2] we will insist, and if he was avaricious, he was not the son of a god." (*)

[1] Cf. T. 1, vv. 55-58; T. 105 ff.   [2] Cf. T. 104; 320; cf. however T. 455.

100. Scholia in Pindarum, Ad Pythias, III, 96.

Λέγεται δὲ ὁ Ἀσκληπιὸς χρυσῷ δελεασθεὶς ἀναστῆσαι Ἱππόλυτον τεθνηκότα.

Enticed by gold, Asclepius is said to have revived Hippolytus [1]
after he had died.

[1] Cf. T. 72; 74; 121.

101. Clemens Alexandrinus, Protrepticus, II, 30, 1.

Ἔχεις καὶ ἰατρόν, οὐχὶ χαλκέα μόνον ἐν θεοῖς· ὁ δὲ ἰατρὸς
φιλάργυρος ἦν, Ἀσκληπιὸς ὄνομα αὐτῷ. καί σοι τὸν σὸν παρα-
θήσομαι ποιητήν, τὸν Βοιώτιον Πίνδαρον . . . καὶ Εὐριπίδης
. . . οὗτος μὲν οὖν κεῖται κεραυνωθεὶς ἐν τοῖς Κυνοσούριδος
ὁρίοις.

You have also a physician, not only a blacksmith, among the
gods. The physician was covetous; Asclepius was his name.
I shall quote your poet, the Boeotian Pindar [1] . . . and Euripides [2] . . . ; he lies buried, struck with a thunderbolt, in the
territory of Cynosura. [3]

[1] Follows *Pythiae*, III, 96 [T. 1]; cf. also **101a**. Arnobius, *Adv. Nationes*, IV, 24:
Numquid cupidinis atque avaritiae causa, sicut canit Boeotius Pindarus, Aesculapium
fulminis transfixum esse telo?
[2] Follows T. 107, v. 3.　　　　　　　　　　[3] Cf. T. 9; 116.

102. Tertullianus, Apologeticus, XIV, 5-6.

Est et ille de lyricis, Pindarum dico, qui Aesculapium canit
avaritiae merito, quia medicinam nocenter exercebat, fulmine
iudicatum. malus Iuppiter, si fulmen illius est, impius in nepotem, invidus in artificem! haec neque vera prodi neque falsa
confingi apud religiosissimos oportebat.

There is also the lyric poet, I mean Pindar, who sings that
Asclepius by reason of his greed, because he practiced medicine
injuriously, was punished with a thunderbolt. Jupiter is evil
if the thunderbolt is his, without feeling for his grandson,
jealous of the skilled practitioner. [1] Such stories, if true, ought
never to have been revealed; if false, ought never to have been
invented among truly religious people. (*)

[1] Cf. T. 233.

103. Tertullianus, Ad Nationes, II, 14.

Sedenim Pindarus meritum eius non occultavit: cupiditatem et
avaritiam lucri in eo dicit vindicatam, qua quidem ille vivos ad

mortem, non mortuos autem ad vitam praevaricatione venalis
medicinae agebat. dicitur etiam mater eius eodem casu obisse,
meritoque quae tam periculosam mundo bestiam ediderat isdem
quasi scalis ad caelum erupisse. et tamen Athenienses scient
eiusmodi deis sacrificare. nam Aesculapio et matri inter
mortuos parentant.

Pindar, indeed, has not concealed his [*sc.*, Asclepius'] true
desert; according to him, he [*sc.*, Asclepius] was punished for
his avarice and love of gain, influenced by which he would bring
the living to their death, rather than the dead to life, by the
unlawful use of his medical art which he put up for sale. It is
said that his mother was killed by the same fate, and it was
only right that she, who had bestowed so dangerous a beast on
the world, should escape to heaven by the same ladder, as it
were. And yet, the Athenians will not be at a loss how to
sacrifice to gods of such a fashion, for they pay honors to
Asclepius and his mother amongst their dead [worthies].

104. Marcianus Aristides, Apologia, 10, 5-6.

Τὸν δὲ Ἀσκληπιὸν παρεισάγουσι θεὸν εἶναι, ἰατρὸν ὄντα καὶ
κατασκευάζοντα φάρμακα καὶ σύνθεσιν ἐμπλάστρων χάριν
τροφῆς· ἐπενδεὴς γὰρ ἦν· ὕστερον δὲ κεραυνοῦσθαι αὐτὸν ὑπὸ
τοῦ Διὸς διὰ Τυνδάρεων Lacedaemonium καὶ ἀποθανεῖν. εἰ δὲ
Ἀσκληπιὸς θεὸς ὢν κεραυνωθεὶς οὐκ ἠδυνήθη ἑαυτῷ βοηθῆσαι,
πῶς ἄλλοις βοηθήσει; ⟨Ut natura divina egens aut attonita sit
fieri non potest.⟩

They introduce Asclepius as a god—a physician who devises
remedies and compounds of salves to earn his livelihood; for
he was indigent.[1] But later he was struck with a thunderbolt
by Zeus because of Spartan Tyndareus,[2] and he died. Now if
Asclepius, although he was a god, was unable to render aid to
himself when struck with the thunderbolt, how shall he succor
others? ⟨It is impossible that divine nature is indigent or
thundered at.⟩

[1] Cf. T. 99.
[2] Cf. T. 69; 70; 71; 84; 85.

DEATH

105. Hesiodus, Fr. 125.

> Πατὴρ δ' ἀνδρῶν τε θεῶν τε
> χώσατ', ἀπ' Οὐλύμπου δὲ βαλὼν ψολόεντι κεραυνῷ
> ἔκτανε Λητοΐδην, Φοίβῳ σὺν θυμὸν ὀρίνων.

And the father of men and gods was wrath, and from Olympus he smote the descendant of Leto [1] with a smouldering thunderbolt and killed him,[2] arousing the anger of Phoebus.(*)

[1] Cf. 105a. Pindarus, *Pythiae*, III, 67: . . . ἤ τινα Λατοΐδα κεκλημένον ἢ πατέρος.
[2] Cf. T. 1, v. 57; 66; 99.

106. Philodemus, De Pietate, 17.[1]

> Τὸν Ἀσκλ[ηπιὸν δ' ὑ]πὸ Διὸς κα[τακτα]θῆναι γέγρ[αφεν Ἡ-]
> σίοδος κα[ὶ Πίνδ]αρος καὶ Φε[ρεκύδης] ὁ Ἀθηναῖος [καὶ Πανύ-]
> ασσις καὶ Ἄ[νδρων] καὶ Ἀκουσ[ίλαος. καὶ] Εὐριπίδ[ης οὕτω]
> λέγει . . . καὶ ὁ τ[ὰ Ναυ]πάκτια ποι[ήσας] καὶ Τελέστ[ης Ἀσ-]
> κληπιῷ.

That Asclepius was killed by Zeus Hesiod has written,[2] and Pindar[3] and Pherecydes the Athenian[4] and Panyassis[5] and Andron[6] and Acusilaus.[1] And Euripides says this.[7] . . . And the author of the *Naupactia*[8] and Telestes in his *Asclepius* [say that Asclepius was killed by Zeus].[9]

[1] = Acusilaus, Fr. 18 [Jacoby].
[2] = Fr. 125 [Rzach] = T. 105.
[3] T. 1, v. 57.
[4] = Fr. 35c [Jacoby]; cf. T. 71.
[5] = Fr. 19 [Kinkel]; cf. T. 69; 70; 71; 84.
[6] = Fr. 17 [Jacoby].
[7] T. 107.
[8] = Fr. 12 [Kinkel]; cf. T. 73.
[9] = II, 156 [Diehl]; cf. T. 73; 613.

107. Euripides, Alcestis, 1-7.

> Ὦ δώματ' Ἀδμήτει', ἐν οἷς ἔτλην ἐγὼ
> θῆσσαν τράπεζαν αἰνέσαι θεός περ ὤν.
> Ζεὺς γὰρ κατακτὰς παῖδα τὸν ἐμὸν αἴτιος
> Ἀσκληπιόν, στέρνοισιν ἐμβαλὼν φλόγα·
> οὗ δὴ χολωθεὶς τέκτονας Δίου πυρὸς
> κτείνω Κύκλωπας· καί με θητεύειν πατὴρ
> θνητῷ παρ' ἀνδρὶ τῶνδ' ἄποιν' ἠνάγκασεν.

Hail, house of Admetus, in which I stooped, god though I am, to endure the fare of a slave. For Zeus constrained me, who slew my son, Asclepius, hurling his bolt upon his breast;[1]

and I, in wrath thereat, slew the Cyclops, the fashioners of the heavenly fire; and in requital for this the Father forced me to serve as a menial in the house of a mortal man.[2]

[1] Cf. T. 67.

[2] Cf. 107a. Hesiodus, Fr. 88: Πῶς γὰρ τοὺς αὐτοὺς (τοὺς Κύκλωπας) θεοῖς ἐναλιγκίους λέγει [sc., ὁ Ἡσίοδος] καὶ ἐν τῷ τῶν Λευκιππίδων καταλόγῳ ὑπὸ Ἀπόλλωνος ἀνῃρῆσθαι ποιεῖ; 107b. Hesiodus, Fr. 126: Ἀνδρῶν δ' ἐν τοῖς Συγγενικοῖς Ἀ[δμή]τῳ λέγει τὸν Ἀπόλλω θητεῦσαι Δ[ιὸς] ἐπιτάξαντος· Ἡσίοδος δὲ καὶ Ἀκο[υσ]ίλαος μέλλειν [μὲν] εἰς τὸν Τάρταρον ὑπὸ τοῦ Διὸς ἐ[μβλη]θῆναι, τῆς δ[ὲ Λητοῦς] ἱκετευσά[σης ἀν]δρὶ θη[τεῦσαι]; 107c. Hesiodus, Fr. 127: Ἡ διὰ στόματος καὶ δημώδης ἱστορία περὶ τῆς Ἀπόλλωνος θητείας παρ' Ἀδμήτῳ αὕτη ἐστίν, ᾗ κέχρηται νῦν Εὐριπίδης. οὕτως δέ φησι καὶ Ἡσίοδος καὶ Ἀσκληπιάδης ἐν Τραγῳδουμένοις.

108. Scholia in Euripidem, Ad Alcestim, 1.

Φερεκύδης δὲ οὔ φησι τοὺς Κύκλωπας ὑπὸ Ἀπόλλωνος ἀνῃρῆσθαι, ἀλλὰ τοὺς υἱοὺς αὐτῶν, γράφων οὕτως· "παρ' αὐτὸν [τὸν Ἄδμητον] ἔρχεται Ἀπόλλων θητεύσων ἐνιαυτόν, Διὸς κελεύσαντος, ὅτι κτείνει τοὺς Βρόντεω καὶ Στερόπεω καὶ Ἄργεω παῖδας. κτείνει δὲ αὐτοὺς Ἀπόλλων Διὶ μεμφθείς, ὅτι κτείνει Ζεὺς Ἀσκληπιὸν τὸν παῖδα αὐτοῦ κεραυνῷ ἐν Πυθῶνι. ἀνίστη γὰρ ἰώμενος τοὺς τεθνεῶτας."

Pherecydes,[1] however, does not say that the Cyclops were destroyed by Apollo, but rather their sons, writing thus: "At his house [that is, in the palace of Admetus] Apollo comes to serve as a thrall for a year at the bidding of Zeus, because he slays the sons of Bronteus and Steropeus and Argeus. Apollo slays them in despite of Zeus, because Zeus kills Asclepius,[2] his son, with a thunderbolt in Pytho. For by his cures he restored the dead to life."

[1] = Fr. 35 A [Jacoby]; cf. T. 71.

[2] Cf. 108a. Ps. Eratosthenes, Catasterismi, I, 6: Τοῦτο τὸ βέλος ἐστὶ τοξικόν, ὅ φασι εἶναι Ἀπόλλωνος, ᾧ τε δὴ τοὺς Κύκλωπας τῷ Διὶ κεραυνὸν ἐργασαμένους ἀπέκτεινε δι' Ἀσκληπιόν. Cf. 108b. Commentaria in Aratum, Aratus Latinus, 311-313: Hoc est Iaculum, quod per arcum mittitur, quem dicunt Apollinis, quando interfecit omnes cycnos, qui Iovis fulmen furaverant. quos interemit per Asclepium. quem et abdidit ad aquilonem. 108c. Hyginus, Astronomica, II, 15: Ut Eratosthenes autem de Sagitta demonstrat, hac Apollo Cyclopas interfecit, qui fulmen Iovi fecerunt, quo Aesculapium interfectum complures dixerunt; cf. 108d. Scholia in Caesaris Germanici Aratea, 318: Haec esse dicitur sagitta Apollinis, qua Cyclopes interfecit eos, qui Iovis fulmen fecerunt, quod eo telo Aesculapius filius eius a Iove esset interfectus.

109. Scholia in Lucianum, Ad Jovem Confutatum, 8.

Ὁ Ἀσκληπιὸς υἱὸς ἦν τοῦ Ἀπόλλωνος. οὗτος τῇ ἰδίᾳ τέχνῃ τῇ ἰατρικῇ χρώμενος, ὡς ματαιολογοῦσι, τοὺς ἀποθνῄσκοντας ἤγειρεν, καὶ ὀργισθεὶς ὁ Ζεὺς ἐκεραύνωσεν αὐτόν. ἐπὶ τούτῳ λυπηθεὶς ὁ Ἀπόλλων τοὺς Κύκλωπας τοὺς εἰωθότας κατα-σκευάζειν τοὺς κεραυνοὺς κατετόξευσεν, καὶ θυμωθεὶς ὁ Ζεὺς ἐποίησε τὸν Ἀπόλλωνα τῷ Ἀδμήτῳ δουλεῦσαι.

Asclepius was the son of Apollo. Using his peculiar medical art, so they absurdly claim, he restored the dead to life, and Zeus, in his wrath, struck him with a bolt of lightning.[1] Grieved at this, Apollo struck down with his arrows the Cyclops who were wont to forge the thunderbolts, and Zeus, enraged, made Apollo serve as a thrall to Admetus.

[1] Cf. 109a. Lucianus, *De Morte Peregrini*, 4: Ἀλλ' ὅτι διὰ πυρὸς ἐξάγειν τοῦ βίου διέγνωκεν ἑαυτόν [*sc.*, ὁ Περεγρῖνος], εἰς κενοδοξίαν τινὲς τοῦτο ἀναφέρουσιν. οὐ γὰρ Ἡρακλῆς οὕτως; οὐ γὰρ Ἀσκληπιὸς καὶ Διόνυσος κεραυνῷ; cf. 109b. *Scholia in Lucianum, Ad Dialogos Deorum*, XV, 1: Ὁ Ἀσκληπιός, ὡς λέγουσιν, ἄριστος ὢν ἰατρὸς τοὺς ἀποθνῄσκοντας ἤγειρε, καὶ ἐπὶ τούτῳ ὁ Ζεὺς ὀργισθεὶς ἐκεραύνωσεν αὐτόν.

110.　Scholia in Apollonium Rhodium, Ad Argonautica, IV, 611–17.

Οἱ Κελτοὶ ἐφήμισαν τὸ ἤλεκτρον δάκρυον εἶναι μὴ τῶν Ἡλιάδων, ἀλλὰ Ἀπόλλωνος, ἡνίκα ὑπὸ Διὸς ἐκελεύσθη θητεύειν ἐπὶ τῷ φόνῳ τοῦ Ἀσκληπιοῦ καὶ τῷ τῶν Κυκλώπων.

The Celts said that amber was not the tears of the Heliades,[1] but of Apollo when he had been enjoined by Zeus to serve as a thrall in consequence of the murder of Asclepius and that of the Cyclops.

[1] Cf. T. **28.**

111.　Servius, Commentarii in Aeneidem, VI, 398.

. . . Amphrysius fluvius est Thessaliae, circa quem Apollo spoliatus divinitate a Iove irato Admeti regis pavit armenta ideo, quia occiderat Cyclopas, fabricatores fulminum, quibus Aesculapius extinctus est, Apollinis filius, quia Hippolytum ab inferis herbarum potentia revocaverat.

. . . the Amphrysius is a river in Thessaly near which Apollo, when deprived of his divinity by the enraged Zeus, pastured the herds of King Admetus. This happened because he had slain the Cyclops, the forgers of the thunderbolts with which Asclepius, Apollo's son, had been destroyed for the reason that he had recalled Hippolytus from the underworld by the power of his simples.[1]

[1] Cf. T. **46.**

112. Origenes, Contra Celsum, III, 23.

> . . . ἀλλὰ καὶ ὁ θάνατος αὐτοῦ ἐξ ἐπιβουλῆς ἀνθρώπων γέγονε, καὶ οὐδὲν ὅμοιον ἔσχε τῷ πρὸς τὸν Ἀσκληπιὸν κεραυνῷ.

> . . . but His death was indeed the result of a conspiracy of men and bore no resemblance to the death of Asclepius by lightning.[1] (*)

[1] Cf. T. 250.

113. Ambrosius, De Virginibus, III, 176, 7.

> Dent Aesculapio quod mortuum reformaverit, dummodo profiteantur quod fulminatus ipse non evaserit.

> To Asclepius let them grant that he revived the dead, provided only that they admit that he himself did not escape the lightning.[1]

[1] Cf. **113a.** Theophilus Antiochenus, Ad Autolycum, I, 9: Τί δέ μοι λέγειν Ἄττιν ἀποκοπτόμενον . . . ἢ Ἀσκληπιὸν κεραυνούμενον; **113b.** Anonymus, S. Ignatii Martyrium per Symeonem Metaphrastem conscriptum, VII, 5: Ἀμέλει καὶ Ζεὺς μὲν ὁ πρῶτος ὑμῖν καὶ μέγιστος τῶν θεῶν ἐν Κρήτῃ τέθαπται, Ἀσκληπιὸς δὲ κεραυνῷ βέβληται. . . .

114. Firmicus Maternus, De Errore Profanarum Religionum, XII, 8.

> Castores sepelit Sparta, ardet aput Oetam Hercules, et Aesculapius alibi fulminatur.

> Sparta provided a burial spot for Castor and Pollux, Heracles burned on the funeral pyre on Oeta, and Asclepius was destroyed by lightning somewhere else.

115. Heraclitus, De Incredibilibus, XXVI.

> Περὶ Ἀσκληπιοῦ

> Λέγουσιν αὐτὸν κεκεραυνῶσθαι. εἴη δὲ ἂν πιθανώτερον οὕτω· ἰατρικὴν κινήσας καὶ ὑψώσας αὐτὸς ὑπὸ πυρετοῦ φλεχθεὶς ὤλετο. ὅθεν διὰ τὴν φλεγμονὴν αὐτὸν κεραυνωθῆναι λέγουσιν.

> Concerning Asclepius

> They say that he was struck by a bolt of lightning. This would be more credible if expressed thus: having furthered and exalted the profession of medicine, he broke out in a burning fever and died. Wherefore, because of the inflammation, they say he was struck by lightning.

TOMB

116. Cicero, De Natura Deorum, III, 22, 57.

> Aesculapiorum . . . secundus . . . is fulmine percussus dicitur humatus esse Cynosuris.

> Of the Asclepii . . . the second . . . is said to have been struck by lightning and buried at Cynosura.[1]

[1] Cf. T. 9; 9b; 101; 379; 380.

117. Anonymus, S. Ignatii Martyrium Romanum, III, 2.

> Αὐτίκα γοῦν Ζεὺς μὲν ἐν Κρήτῃ τέθαπται, Ἀσκληπιὸς δὲ κεραυνοβοληθεὶς ἐν Κυνοσούρῃ, Ἀφροδίτη . . . Ἡρακλῆς . . .

> At any rate, Zeus, for example, was buried in Crete, Asclepius, after having been struck with a thunderbolt, [was buried] in Cynosura, . . . Aphrodite . . . Heracles . . .

118. Cicero, De Natura Deorum, III, 22, 57.

> Aesculapiorum . . . tertius Arsippi et Arsinoae . . . cuius in Arcadia non longe a Lusio flumine sepulcrum et lucus ostenditur.

> Of the Asclepii . . . the third, the son of Arsippus and Arsinoë . . . whose tomb and grove are shown in Arcadia, not far from the river Lusius.[1]

[1] Cf. T. 380.

119. Clementina Homilia, VI, 21.

> . . . καὶ οὕτως τελευτήσαντος τὸν τάφον Κρῆτες ἐπιδείκνυσιν . . . Ἀσκληπιὸς ἐν Ἐπιδαύρῳ· καὶ ἄλλων πολλῶν τοιούτων φαίνονται τάφοι.

> . . . of [sc., Zeus] who died in this manner the Cretans show the tomb . . . Asclepius [sc., is buried] in Epidaurus; and the tombs of many more such gods are to be seen.

120. Ps. Clementinae Recognitiones, X, 24.

> Sed et filiorum eius, qui apud eos dii putantur, sepulcra singulis quibusque in locis manifestissime demonstrantur: Mercurii

apud Hermopolim . . . in Epidauro Aesculapii; quique omnes non solum ut homines defuncti, sed ut nequam homines puniti pro sceleribus docentur, et tamen ab stultis hominibus ut dii adorantur.

But also the sepulchres of his [sc., Jupiter's] sons, who are regarded amongst these [sc., the Gentiles] as gods, are openly pointed out, one in one place and another in another: that of Mercury at Hermopolis . . . that of Asclepius in Epidaurus. And all these are spoken of, not only as men who have died, but as wicked men who have been punished for their crimes; and yet they are adored as gods by foolish men.

STAR

121. Ps. Eratosthenes, Catasterismi, I, 6.

⟨Π⟩ερὶ τοῦ Ὀφιούχου.

Οὗτός ἐστιν ὁ ἐπὶ τοῦ Σκορπίου συνεστηκώς, ἔχων ἐν ταῖς χερσὶν ἀμφοτέραις ὄφιν· λέγεται δὲ εἶναι Ἀσκληπιὸς ὑπό τινων ἀστρολόγων· ὃν ὁ Ζεὺς χαριζόμενος Ἀπόλλωνι [διὰ τὴν κεραυνοβολίαν] δοκεῖ αὐτῷ τιμὴν ἀπονεῖμαι ταύτην· ἐν γὰρ τοῖς ἀνθρώποις ὤν, τῇ τέχνῃ ἰατρικῇ χρώμενος τοὺς ἤδη τεθνηκότας ἤγειρεν· ὀργισθέντα δὲ κεραυνοβολῆσαι τὴν οἰκίαν αὐτοῦ· τοῦτον δὲ εἰς τὰ ἄστρα ἀναγαγεῖν διὰ τὸν Ἀπόλλωνα.

Ὀφιούχου.

Οὗτός ἐστιν ὁ ἐπὶ Σκορπίου ἑστηκώς, ἔχων ἐν ἀμφοτέραις χερσὶν [τὸν] ὄφιν· λέγεται δὲ εἶναι Ἀσκληπιός, ὃν Ζεὺς χαριζόμενος Ἀπόλλωνι εἰς τὰ ἄστρα ἀνήγαγεν· τούτου τέχνῃ ἰατρικῇ χρωμένου, ὡς καὶ τοὺς ἤδη τεθνηκότας ἐγείρειν, ἐν οἷς καὶ ἔσχατον Ἱππόλυτον τὸν Θησέως, καὶ τῶν θεῶν δυσχερῶς τοῦτο φερόντων, εἰ αἱ τιμαὶ καταλυθήσονται αὐτῶν τηλικαῦτα ἔργα Ἀσκληπιοῦ ἐπιτελοῦντος, λέγεται τὸν Δία ὀργισθέντα κεραυνοβολῆσαι τὴν οἰκίαν αὐτοῦ, εἶτα διὰ τὸν Ἀπόλλωνα τοῦτον εἰς τὰ ἄστρα ἀναγαγεῖν· ἔχει δὲ ἐπιφάνειαν ἱκανὴν ἐπὶ τοῦ μεγίστου ἄστρου ὤν, λέγω δὴ τοῦ Σκορπίου, εὐσήμῳ τῷ τύπῳ φαινόμενος.

Concerning Ophiuchus.

This is the constellation which is in Scorpion, holding in both hands a serpent. By some astronomers it is reputed to be Asclepius. Zeus seems to have accorded him this honor as a favor to Apollo [because of the hurling of the thunderbolt]. For when Asclepius dwelt among men, by using his medical skill, he revived those already dead. In a rage Zeus hurled a thunderbolt at his dwelling, but him he raised up among the stars for Apollo's sake.

Ophiuchus.[1]

This is the constellation which is in Scorpion, having in both hands a serpent. It is said, indeed, to be Asclepius whom Zeus, as a favor to Apollo, raised up among the stars. He employed his medical skill to revive even those already dead, among whom the last was Hippolytus,[2] the son of Theseus, and the gods took this unfavorably lest their honors be curtailed if Asclepius were able to accomplish such deeds; therefore it is said that Zeus in his wrath hurled a thunderbolt at his dwelling; then for Apollo's sake he raised Asclepius to the stars. It has a sufficient brilliance although it is in the largest star, that is, Scorpion, and it shines forth in clear outline.

[1] Cf. 121a. Hyginus, *Astronomica*, II, 14: . . . [Ophiuchum] Aesculapium finxerunt, quem Juppiter Apollinis causa inter astra collocavit; 121b. *Scholia in Caesaris Germanici Aratea*, 71: Hic Ophiuchus erit: hic est qui supra Scorpionem stat tenens duabus manibus anguem qui esse dicitur. Aesclapius rogatu Apollinis patris ab Jove astris inlatus, qui ab eo fulmine erat interfectus, quod esset indignatus Iuppiter eum arte sua rursus post obitum defunctis animas restituere ad vitam.

[2] Cf. T. 74 ff.; 100.

122. Servius, Commentarii in Aeneidem, XI, 259.

Haec autem numina quae inter sidera non videmus, licet sua signa propria non habeant, cum aliis potestate sunt permixta. ut Ophiuchus ipse est Aesculapius, Gemini Apollinis et Herculis esse dicuntur, sic Minervae Aries esse dinoscitur.

These divinities, which we do not see among the stars, inasmuch as they do not have their own characteristic signs, share their powers with others. Just as Ophiuchus is Asclepius himself, and the Gemini are said to be Apollo's and Heracles' stars, so the Ram is recognized as the star of Minerva.

II. DESCENDANTS

THE HOMERIC ASCLEPIUS

123. Plato, Res Publica, X, 599 C.

> . . . ἐρωτῶντες . . . τίνας ὑγιεῖς ποιητής τις τῶν παλαιῶν ἢ
> τῶν νέων λέγεται πεποιηκέναι, ὥσπερ Ἀσκληπιός, ἢ τίνας
> μαθητὰς ἰατρικῆς κατελίπετο, ὥσπερ ἐκεῖνος τοὺς ἐκγόνους, . . .

> . . . asking. . . what men any poet, old or new, is reported
> to have restored to health as Asclepius did, or what disciples
> of the medical art he left after him as Asclepius did his
> descendants . . .

124. Plato, Res Publica, III, 407 C-D.

> Οὐκοῦν ταῦτα γιγνώσκοντα φῶμεν καὶ Ἀσκληπιὸν τοὺς μὲν
> φύσει τε καὶ διαίτῃ ὑγιεινῶς ἔχοντας τὰ σώματα, νόσημα δέ τι
> ἀποκεκριμένον ἴσχοντας ἐν αὑτοῖς, τούτοις μὲν καὶ ταύτῃ τῇ
> ἕξει καταδεῖξαι ἰατρικήν, φαρμάκοις τε καὶ τομαῖς τὰ νοσήματα
> ἐκβάλλοντα αὐτῶν τὴν εἰωθυῖαν προστάττειν δίαιταν, ἵνα μὴ τὰ
> πολιτικὰ βλάπτοι, τὰ δ' εἴσω διὰ παντὸς νενοσηκότα σώματα
> οὐκ ἐπιχειρεῖν διαίταις κατὰ σμικρὸν ἀπαντλοῦντα καὶ ἐπι-
> χέοντα μακρὸν καὶ κακὸν βίον ἀνθρώπῳ ποιεῖν, καὶ ἔκγονα αὐτῶν,
> ὡς τὸ εἰκός, ἕτερα τοιαῦτα φυτεύειν, ἀλλὰ τὸν μὴ δυνάμενον ἐν
> τῇ καθεστηκυίᾳ περιόδῳ ζῆν μὴ οἴεσθαι δεῖν θεραπεύειν, ὡς
> οὔτε αὑτῷ οὔτε πόλει λυσιτελῆ; Πολιτικόν, ἔφη, λέγεις Ἀσκληπιόν.
> Δῆλον, ἦν δ' ἐγώ.

"Then shall we not say that it was because Asclepius knew
this—that for those who were by nature and course of life
sound of body but had some localized disease, that for such,
I say, and for this habit he revealed the art of medicine, and,
driving out their disease by drugs and surgery, prescribed for
them their customary regimen in order not to interfere with
their civic duties, but that, when bodies were diseased inwardly
and throughout, he did not attempt by diet and by gradual
evacuations and infusions to prolong a wretched existence for

the man and have him beget in all likelihood similar wretched offspring? But if a man was incapable of living in the established round and order of life, he did not think it worth while to treat him, since such a fellow is of no use either to himself or to the state." "A most politic[1] Asclepius you're telling us of," he said. "Obviously," said I.

[1] Cf. T. **421**, v. 726: φιλόπολις. Cf. **124a**. Plato, *Res Publica*, III, 406 A-C: . . . τῇ παιδαγωγικῇ τῶν νοσημάτων ταύτῃ τῇ νῦν ἰατρικῇ πρὸ τοῦ Ἀσκληπιάδαι οὐκ ἐχρῶντο, ὥς φασι, πρὶν Ἡρόδικον γενέσθαι· . . . Ἀσκληπιὸς οὐκ ἀγνοίᾳ οὐδὲ ἀπειρίᾳ τούτου τοῦ εἴδους τῆς ἰατρικῆς τοῖς ἐκγόνοις οὐ κατέδειξεν αὐτό, ἀλλ' εἰδὼς ὅτι πᾶσι τοῖς εὐνομουμένοις ἔργον τι ἑκάστῳ ἐν τῇ πόλει προστέτακται, ὃ ἀναγκαῖον ἐργάζεσθαι, καὶ οὐδενὶ σχολὴ διὰ βίου κάμνειν ἰατρευομένῳ. ὃ ἡμεῖς γελοίως ἐπὶ μὲν τῶν δημιουργῶν αἰσθανόμεθα, ἐπὶ δὲ τῶν πλουσίων τε καὶ εὐδαιμόνων δοκούντων εἶναι οὐκ αἰσθανόμεθα.

125. Lucianus, Philopseudes, 10.

. . . ὁ γοῦν Ἀσκληπιὸς αὐτὸς καὶ οἱ παῖδες αὐτοῦ ἤπια φάρμακα πάσσοντες ἐθεράπευον τοὺς νοσοῦντας . . .

. . . in fact, Asclepius himself and his sons ministered to the sick by laying on healing drugs . . .[1]

[1] Cf. T. **164**, v. 218; **143**.

126. Plutarchus, De Curiositate, 7, 518 D.

Οὕτω δ' ἑκάστῳ λυπηρόν ἐστιν ἡ τῶν περὶ αὐτὸν κακῶν ἀνακάλυψις, ὥστε πολλοὺς ἀποθανεῖν πρότερον ἢ δεῖξαί τι τῶν ἀπορρήτων νοσημάτων ἰατροῖς. φέρε γὰρ Ἡρόφιλον ἢ Ἐρασίστρατον ἢ τὸν Ἀσκληπιὸν αὐτόν, ὅτ' ἦν ἄνθρωπος, ἔχοντα τὰ φάρμακα καὶ τὰ ὄργανα κατ' οἰκίαν προσιστάμενον ἀνακρίνειν, μή τις ἔχει σύριγγα παρὰ δακτύλιον ἢ γυνὴ καρκίνον ἐν ὑστέρᾳ . . .

But so unpleasant is it to everybody to have his private ills brought to light, that many have died rather than acquaint the doctors with their secret ailments. For imagine Herophilus, or Erasistratus, or even Asclepius himself, when he was a mortal, going with their drugs and surgical instruments from house to house, to inquire what man had a fistula *in ano*, or what woman had a cancer in her womb . . .(*)

DATE OF HIS LIFETIME

127. Celsus, De Medicina, Prooemium, 3.

> . . . Aesculapius . . . huius . . . duo filii Podalirius et Machaon bello Troiano ducem Agamemnonem secuti . . .

> . . . Asclepius . . . whose two sons Podalirius and Machaon[1] in the Trojan War followed the leader Agamemnon . . .

[1] Cf. T. 135 ff.

128. Arnobius, Adversus Nationes, II, 74.

> Immo illud exquirimus potius, cur, si Herculem oportuit nasci, si Aesculapium Mercurium Liberum aliosque nonnullos, qui et conciliis adiungerentur deorum et mortalibus aliquid utilitatis adferrent, tam sero a Jove sint proditi, ut sola illos posteritas sciret, superiorum vero ignoraret antiquitas?

> Nay, this we rather ask, why, if it was fitting that Heracles should be born, and Asclepius, Mercury, Liber, and some others that they might be both added to the assemblies of the gods and might do men some service,—why they were produced so late by Jupiter that only later ages should know them, while the past ages of earlier men knew them not?(*)

129. Clemens Alexandrinus, Stromateis, I, 21, 105.

> Προτερεῖν ἄρα Μωυσῆς ἀποδείκνυται τῆς μὲν Διονύσου ἀπο-θεώσεως ἔτη ἑξακόσια τέσσαρα . . . ἀπὸ δὲ Διονύσου ἐπὶ Ἡρακλέα καὶ τοὺς περὶ Ἰάσονα ἀριστεῖς τοὺς ἐν τῇ Ἀργοῖ πλεύσαντας συνάγεται ἔτη ἑξήκοντα τρία. Ἀσκληπιός τε καὶ Διόσκουροι συνέπλεον αὐτοῖς, ὡς μαρτυρεῖ ὁ Ῥόδιος Ἀπολλώνιος ἐν τοῖς Ἀργοναυτικοῖς. ἀπὸ δὲ τῆς Ἡρακλέους ἐν Ἄργει βασι-λείας ἐπὶ τὴν Ἡρακλέους αὐτοῦ καὶ Ἀσκληπιοῦ ἀποθέωσιν ἔτη συνάγεται τριάκοντα ὀκτὼ κατὰ τὸν χρονογράφον Ἀπολλόδωρον.

> Moses, then, is shown to have preceded the deification of Dionysus by 604 years . . . From Dionysus, then, to Heracles and the noble comrades of Jason who sailed in the Argo 63 years are reckoned. Asclepius and the Dioscuri sailed with them, as Apollonius of Rhodes recounts in the *Argonauts*.[1] From the rule of Heracles in Argos to the deification of Hera-

cles himself and Asclepius 38 years elapsed,[2] according to
Apollodorus the Chronographer [2nd century B. C.].[3]

[1] Cf. T. 63.
[2] Cf. 129a. Isidorus, *Etymologiae*, IV, 3, 1-2: . . . et ars [*sc.*, medicinae] simul
cum auctore defecit latuitque per annos pene quingentos usque ad tempus Artaxerxis
regis Persarum; cf. T. 357.
[3] = Fr. 87 [Jacoby].

130. Eusebii Caesariensis Chronici Canones Latine vertit Eusebius
Hieronymus, 7a 22-b 12.

> . . . nam Moyses licet iunior supra dictis sit, ab omnibus
> tamen quos Graeci antiquissimos putant senior depraehenditur,
> Homero scilicet et Hesiodo Troianoque bello ac multo superius
> Hercule Museo Lino Chirone Orfeo Castore Polluce Aescu-
> lapio Libero Mercurio Apolline et ceteris diis gentium sacrisque
> vel vatibus ipsius quoque Jovis gestis quem Graecia in arce
> divinitatis conlocavit.

> . . . for although Moses is younger than the events and people
> [*sc.*, Abraham, etc.] mentioned above, he is shown to be older
> than all whom the Greeks considered most ancient, namely
> Homer and Hesiod and the Trojan War and much earlier than
> Heracles, Musaeus, Linus, Chiron, Orpheus, Castor and Pol-
> lux, Asclepius, Liber, Mercury, Apollo and the rest of the gods,
> as well as the sacred rites and priests of the Gentiles, and also
> older than the deeds of Jupiter himself whom Greece placed at
> the summit of the hierarchy of gods.

131. Eusebii Caesariensis Chronici Canones Latine vertit Eusebius
Hieronymus, 10a 25-b 7.

> . . . Liber et reliqui quos mox inferemus post CC annum
> Cecropis fuerunt, Linus scilicet et Zetus et Amphion Museus
> Orfeus Minos Perseus Aescolapius gemini Castores Hercules,
> cum quo Apollo servivit Admeto. . . .

> . . . Liber and the others whom we shall soon mention lived
> two hundred years after Cecrops, [the others including] Linus
> and Zetus and Amphion, Musaeus, Orpheus, Minos, Perseus,
> Asclepius, the twins Castor and Pollux, Heracles with whom
> Apollo served as a thrall to Admetus. . . .

132. Georgius Hamartolus, Chronicon, I, 30.

Κατὰ γὰρ τὴν Σεμίραμιν Ἀβραὰμ ἐγνωρίζετο, καὶ Ἰακὼβ κατὰ
Ἴναχον, καὶ Μωυσῆς κατὰ Κέκροπα τὸν διφυῆ, ὃν πρῶτον
βασιλέα τῆς Ἀττικῆς Ἕλληνες ἱστόρησαν. ὡς εἶναι τῶν παρ᾽
Ἕλλησιν ἀρχαίων σοφῶν Μωυσέα πρεσβύτερον, Ὁμήρου λέγω
καὶ Ἡσιόδου, καὶ τῶν Τρωικῶν, Ἡρακλέους τε καὶ Μουσαίου,
Λίνου τε καὶ Ὀρφέως καὶ Διοσκούρων, Ἀσκληπιοῦ καὶ Διονύσου
καὶ Ἑρμοῦ, Ἀπόλλωνός τε καὶ τῶν Ἑλληνικῶν μυστηρίων καὶ
τελετῶν, καὶ αὐτῶν τῶν τοῦ Διὸς πράξεων.

At the time of Semiramis Abraham flourished, and Jacob in
the days of Inachus, and Moses in the time of Cecrops, the
two-formed, whom the Greeks recount to have been the first
king of Attica. Thus Moses was older than the ancient wise
men among the Greeks, I mean older than Homer and Hesiod
and the Trojan War and Heracles and Musaeus, Linus and
Orpheus and the Dioscuri, Asclepius and Dionysus and
Hermes, Apollo and the Greek mysteries and mystic rites and
even the deeds of Zeus.

133. Georgius Hamartolus, Chronicon, I, 104.[1]

Ἐφ᾽ οὗ Προμηθεύς, καὶ Ὀρφεὺς ὁ Θρᾷξ, οἱ σοφώτατοι παρ᾽
Ἕλλησιν, καὶ Ἀσκληπιὸς ὁ ἰατρός, καὶ Λυκοῦργος ὁ Σπαρτιάτης
καὶ νομοθέτης τῶν Ἑλλήνων, ἐγνωρίζοντο.

In these days [sc., of Barach] both Prometheus and Orpheus
the Thracian, the wisest men among the Greeks, and Asclepius
the physician and Lycurgus the Spartan, lawgiver of the
Hellenes, flourished.

[1] Cf. **133a.** Georgius Cedrenus, *Historiae Compendium*, I, 95-96; **133b.** Joël, *Chrono-graphia*, 11, same text.

134. Chronicon Paschale, 79.

Τούτῳ τῷ ἔτει τὸ Ἴλιον ὑπὸ Ἰλίου ἐκτίσθη, καὶ Ἀσκληπιὸς τὴν
ἰατρικὴν μετῄει.

In this year [1405 B. C.] Ilium was founded by Ilius, and
Asclepius entered the profession of medicine.[1]

[1] Cf. **134a.** Georgius Cedrenus, *Historiae Compendium*, I, 147: Κατὰ τούτους τοὺς
χρόνους [sc., τοῦ Ἐσσεβῶνος] τὸ Ἴλιον ἐκτίσθη καὶ ἡ Τροία, καὶ Ἀσκληπιὸς τὴν ἰατρικὴν
ἐπιστήμην μετῄει; cf. T. 348-51.

MACHAON AND PODALIRIUS

135. Homerus, Ilias, II, 729-33.[1]

> Οἳ δ' εἶχον Τρίκκην καὶ Ἰθώμην κλωμακόεσσαν,
> οἵ τ' ἔχον Οἰχαλίην, πόλιν Εὐρύτου Οἰχαλιῆος,
> τῶν αὖθ' ἡγείσθην Ἀσκληπιοῦ δύο παῖδε,
> ἰητῆρ' ἀγαθώ, Ποδαλείριος ἠδὲ Μαχάων.
> τοῖς δὲ τριήκοντα γλαφυραὶ νέες ἐστιχόωντο.

And they that held Tricca and Ithome of the crags, and they who had Oechalia, city of Oechalian Eurytus, these again were led by the two sons of Asclepius, the skilled leeches Podalirius and Machaon. And with these were ranged thirty hollow ships. (*)

[1] Cf. T. 10.

136. Homerus, Ilias, XI, 833-36.

> " Ἰητροὶ μὲν γὰρ Ποδαλείριος ἠδὲ Μαχάων,
> τὸν μὲν ἐνὶ κλισίῃσιν ὀΐομαι ἕλκος ἔχοντα,
> χρηΐζοντα καὶ αὐτὸν ἀμύμονος ἰητῆρος,
> κεῖσθαι· ὁ δ' ἐν πεδίῳ Τρώων μένει ὀξὺν Ἄρηα."

" For the leeches, Podalirius and Machaon, the one methinks lieth wounded amid the huts, having need himself of a goodly leech, and the other in the plain abideth the sharp battle of the Trojans " [Eurypylus to Patroclus].[1]

[1] Cf. however T. 146.

137. Eustathius, Commentarii ad Homeri Iliadem, II, 729.

> Ποδαλείριος καὶ Μαχάων, Ἀσκληπιοῦ παῖδε, . . . ἰατροὶ ἀγαθοί,
> τριάκοντα νηῶν ἦρχον καὶ ἀνδρῶν, οἳ εἶχον Τρίκκην καὶ Ἰθώμην
> κλωμακόεσσαν καὶ Οἰχαλίαν, πόλιν Εὐρύτου Οἰχαλιέως.

Podalirius and Machaon, both sons of Asclepius . . . good physicians, commanded thirty ships and the men who held Tricca and Ithome of the crags and Oechalia, city of Oechalian Eurytus.[1]

[1] Cf. **137a.** Hyginus, *Fabulae*, XCVII, 6: Qui ad Troiam expugnatum ierunt et quot naves: . . . Machaon Asclepii et Coronidis filius a Tricca, navibus XX. Podalirius frater eius, navibus IX; cf. **137b.** Dictys Cretensis, *Ephemeris Belli Troiani*, I, 17: XXX [sc., naves] Podalirius et Machaon; cf. **137c.** Dares, *De Excidio Troiae Historia*, XIV: Podalirius et Machaon Aesculapii filii ex Tricca cum navibus numero XXXII; **137d.** Apollodorus, *Epitoma*, III, 14: Τρικκαίων Ποδαλείριος ⟨καὶ Μαχάων Ἀσκληπιοῦ⟩ λ'.

138. Eustathius, Commentarii ad Homeri Iliadem, XI, 833.

> . . . εἰπὼν δὲ ἐν τούτοις ὁ ποιητὴς τὸ ‘ ἰητροὶ Ποδαλείριος ἠδὲ
> Μαχάων ’ τοὺς ἐξόχους εἶπεν. ἄλλως γὰρ πολλοὶ ἐν Ἀχαιοῖς
> ἰατροί, ὡς δηλοῖ τὸ ‘ τοὺς μὲν ἰητροὶ πολυφάρμακοι ἀμφεπένοντο.’

> . . . saying here "the leeches Podalirius and Machaon" the
> poet means "the foremost." For besides there were many
> physicians among the Achaeans, as is clear from the words
> "of these took care physicians who knew many drugs." [1]

¹ Homer, *Iliad*, XVI, 28.

139. Eustathius, Commentarii ad Homeri Iliadem, IV, 202.

> Δύω ὄντων ἐπιφανῶν Ἕλλησιν ἰατρῶν, Ποδαλείριος μέν, οὗ ἡ
> ἐτυμολογία προγέγραπται, περὶ δίαιταν ἐπονεῖτο, Μαχάων δὲ
> περὶ τραύματα εἶχε, Τρίκκης καὶ αὐτὸς ἄρχων, πόλεως Θετταλικῆς
> . . . ἦν δὲ στρατιωτικώτερος ὁ Μαχάων, ὡς καὶ ἡ κλῆσις ὑπο-
> δηλοῖ. ἄμφω δὲ Ἀσκληπιάδαι ἦσαν, ἤτοι παῖδες Ἀσκληπιοῦ
> ἀμύμονος ἰητῆρος, ὥς φησιν ὁ ποιητής.

> Of the two notable physicians among the Greeks, one, Podali-
> rius, the etymology of whose name was previously given,[1]
> worked with the subject of diet; the other, Machaon, engaged
> in the treatment of wounds, he too being the ruler of Tricca,
> a Thessalian city . . . Machaon, however, was of a more war-
> like disposition, as even his name discloses. Both were
> Asclepiads, in fact sons of Asclepius the blameless physician,
> as the poet says.[2]

¹ Cf. T. 197a.   ² Cf. T. 164, v. 194.

140. Scholia in Homerum, Ad Iliadem, XI, 515.

> Οἱ μὲν οὖν φασὶν ὅτι τὸ χειρουργικὸν καὶ τὸ φαρμακευτικὸν
> εὕρητο παρὰ τοῖς παλαιοῖς. τοῦ γὰρ διαιτητικοῦ Ἡρόδικος μὲν
> ἤρξατο, συνετέλεσε δὲ καὶ Ἱπποκράτης, Πραξαγόρας, Χρύσιπ-
> πος. ἔνιοι δέ φασιν ὡς οὐδὲ ἐπὶ πάντας τοὺς ἰατροὺς ὁ ἔπαινος
> οὗτός ἐστι κοινός, ἀλλὰ τὸν Μαχάονα μόνον χειρουργεῖν θέλουσι·
> τὸν γὰρ Ποδαλείριον διαιτᾶσθαί φασι τὰς νόσους. καὶ τεκμήριον,
> ὅτι Ἀγαμέμνων τρωθέντος Μενελάου οὐ Ποδαλείριον καλεῖ, ἀλλὰ
> τὸν Μαχάονα. εἰ δὲ μὴ παράγει τινὰ διαιτώμενον, οὐ θαῦμα·
> διὰ γὰρ τὸ ἀπρεπὲς παρῆκε τὴν δίαιταν.

Some, then, say that the surgical and pharmaceutical branches of medicine were invented by the ancients. For Herodicus began dietetics, while Hippocrates, Praxagoras, and Chrysippus completed it. Some, however, say that this panegyric is not applied to all physicians, but rather they believe that Machaon practiced surgery only. For Podalirius, they claim, treated diseases by diet. The proof for this is that Agamemnon did not summon Podalirius, when Menelaus was wounded, but rather Machaon. But if he [sc., Homer] did not introduce someone being treated by diet,[1] it is not strange, for he omits treatment by diet because of its unseemliness.

[1] Cf. T. 143; cf. also T. 142.

141. Scholia in Homerum, Ad Iliadem, XI, 515 (BT).

Ἔνιοι δέ φασιν ὡς οὐδὲ ἐπὶ πάντας τοὺς ἰατροὺς ὁ ἔπαινος οὗτός ἐστι κοινός, ἀλλὰ ἐπὶ τὸν Μαχάονα, ὃν μόνον χειρουργεῖν τινες λέγουσι· τὸν γὰρ Ποδαλείριον διαιτᾶσθαι νόσους· καὶ τεκμήριον τούτου Ἀγαμέμνων τρωθέντος Μενελάου οὐκ ἄμφω ἐπὶ τὴν θεραπείαν καλεῖ, ἀλλὰ τὸν Μαχάονα. τοῦτο ἔοικε καὶ Ἀρκτῖνος ἐν Ἰλίου Πορθήσει νομίζειν ἐν οἷς φησι·

> αὐτὸς γάρ σφιν ἔδωκε πατὴρ κλυτὸς Ἐννοσίγαιος *
> ἀμφοτέροις, ἕτερον δ' ἑτέρου κυδίον' ἔθηκε·
> τῷ μὲν κουφοτέρας χεῖρας πόρεν ἔκ τε βέλεμνα
> σαρκὸς ἑλεῖν τμῆξαί τε καὶ ἕλκεα πάντ' ἀκέσασθαι,
> τῷ δ' ἄρ' ἀκριβέα πάντα ἐνὶ στήθεσσιν ἔθηκεν
> ἄσκοπά τε γνῶναι καὶ ἀναλθέα ἰήσασθαι·
> ὅς ῥα καὶ Αἴαντος πρῶτος μάθε χωομένοιο
> ὄμματά τ' ἀστράπτοντα βαρυνόμενόν τε νόημα.

* κλυτὸς Ἐννοσίγαιος Heyne Ἐννοσίγαιος πεσεῖν BT.

Some say that this panegyric is not meant to apply to all physicians, but to Machaon of whom certain people say that he practiced surgery only. For Podalirius, they claim, treated diseases by dietetics. And proof thereof is the fact that when Menelaus was wounded Agamemnon summoned for the treatment not both of them, but Machaon.[1] This even Arctinus [7th-6th c. B. C.] seems to think in the *Sack of Troy*, in which he says:

"For their father himself, the glorious Earthshaker,[2] gave them both honors, but one he made more renowned than the

other. To the one he gave more agile hands to draw the darts
from the flesh and to heal all wounds; to the other he gave the
power to know accurately in his heart all matters that are
unseen, and to heal things incapable of healing. He first understood the eyes flashing forth like lightning and the depressed
mind of the mad Ajax." [3]

[1] Cf. T. 164.   [2] Cf. T. 142.   [3] = Fr. 5 [Allen].

142. Eustathius, Commentarii ad Homeri Iliadem, XI, 514.

Ὅρα δὲ ὅτι περὶ τραύματα εἶχεν ὁ καλὸς οὗτος Ἑλληνικὸς ἰατρός,
διαίτης δὲ οὐ μέμνηται ὁ ποιητής. φασὶ γὰρ τὸ χειρουργικὸν
καὶ φαρμακευτικὸν μόνον εὑρῆσθαι παρὰ τοῖς παλαιοῖς, τοῦ δὲ
διαιτητικοῦ Ἱπποκράτην μὲν κατάρξαι, Ἡρόδικον δὲ συντελέσαι
καὶ Πραξαγόραν καὶ Χρύσιππον. τινὲς δὲ τὸν Μαχάονα μὲν
χειρουργεῖν ἐθέλουσι, Ποδαλείριον δέ, στρατιώτην ὄντα καὶ
αὐτόν, ὡς ἀλλαχοῦ δηλώσει ὁ ποιητής, ἀσκεῖν τὰ περὶ δίαιταν.
τεκμήριον δὲ ὁ βασιλεὺς εἰς θεραπείαν τοῦ Μενελάου βληθέντος
καλέσας τὸν Μαχάονα, οὐ μὴν τὸν Ποδαλείριον· μαρτυρεῖ δὲ
καὶ τὰ ἱστορούμενα ἔπη τὰ ἐπὶ τῇ Τρωικῇ Πορθήσει, ἐν οἷς
φέρεται περὶ Ποδαλειρίου καὶ Μαχάονος, ὡς ἄμφω μὲν Ποσειδῶνος ἦσαν, ἕτερον δ' ἑτέρου κυδίον' ἔθηκεν, ὁ Ποσειδῶν δηλαδή,
'τῷ μὲν κουφοτέρας — νόημα.'

Notice that this noble Greek physician [sc., Machaon] gave
his attention to the treatment of wounds, but the poet does not
mention diet;[1] for they say only surgery and pharmacy were
invented by the ancients, while Hippocrates began dietetics
and Herodicus, and Praxagoras, and Chrysippus completed it.
Some claim that Machaon practiced surgery, but that Podalirius, he, too, being a warrior, as the poet indicates in another
passage,[2] practiced the science of dietetics. Proof of this is that
the king summoned Machaon to cure the wounded Menelaus,
rather than Podalirius. Evidence for this is also given in the
epic accounts of the *Sack of Troy* in which it is related of
Podalirius and Machaon that both were sons of Poseidon, but
one he, that is Poseidon, made more renowned than the other.
For " to the one he gave more agile hands . . . "[3]

[1] Cf. T. 140; 143.
[2] Cf. T. 136.   [3] For the lines quoted cf. T. 141.

143. Plato, Res Publica, III, 407 E–408 A.

Πολιτικόν, ἔφη, λέγεις Ἀσκληπιόν. Δῆλον, ἦν δ᾽ ἐγώ· καὶ οἱ
παῖδες αὐτοῦ, ὅτι τοιοῦτος ἦν, οὐχ ὁρᾷς ὡς καὶ ἐν Τροίᾳ ἀγαθοὶ
πρὸς τὸν πόλεμον ἐφάνησαν, καὶ τῇ ἰατρικῇ, ὡς ἐγὼ λέγω,
ἐχρῶντο; ἢ οὐ μέμνησαι, ὅτι καὶ τῷ Μενέλεῳ ἐκ τοῦ τραύματος
οὗ ὁ Πάνδαρος ἔβαλεν

αἷμ᾽ ἐκμυζήσαντ᾽ ἐπί τε ἤπια φάρμακ᾽ ἔπασσον,

ὅ τι δ᾽ ἐχρῆν μετὰ τοῦτο ἢ πιεῖν ἢ φαγεῖν οὐδὲν μᾶλλον ἢ τῷ
Εὐρυπύλῳ προσέταττον, ὡς ἱκανῶν ὄντων τῶν φαρμάκων ἰάσασθαι ἄνδρας πρὸ τῶν τραυμάτων ὑγιεινούς τε καὶ κοσμίους ἐν
διαίτῃ, κἂν εἰ τύχοιεν ἐν τῷ παραχρῆμα κυκεῶνα πιόντες, νοσώδη
δὲ φύσει τε καὶ ἀκόλαστον οὔτε αὐτοῖς οὔτε τοῖς ἄλλοις ᾤοντο
λυσιτελεῖν ζῆν, οὐδ᾽ ἐπὶ τούτοις τὴν τέχνην δεῖν εἶναι, οὐδὲ
θεραπευτέον αὐτούς, οὐδ᾽ εἰ Μίδου πλουσιώτεροι εἶεν. Πάνυ
κομψούς, ἔφη, λέγεις Ἀσκληπιοῦ παῖδας.

"A most politic Asclepius you're telling us of," he [sc., Adei-
mantus] said. "Obviously," said I [sc., Socrates], "that was
his character. And his sons too, don't you see that at Troy
they approved themselves good fighting-men, and practiced
medicine as I described it? Don't you remember that in the
case of Menelaus too from the wound that Pandarus inflicted

They sucked the blood, and soothing simples sprinkled?[1]

But what he was to eat or drink thereafter they no more prescribed[2] than for Eurypylus, taking it for granted that the
remedies sufficed to heal men who before their wounds were
healthy and temperate in diet even if they did happen for the
nonce to drink a posset; but they thought that the life of a man
constitutionally sickly and intemperate was of no use to himself or others, and that the art of medicine should not be for
such nor should they be given treatment even if they were
richer than Midas." "Very ingenious fellows," he said, "you
make out these sons of Asclepius to be."

[1] Cf. however T. 164, vv. 218 f.　　　　　[2] Cf. T. 140; 141; 144.

144. Galenus, Utrum Medicinae sit an Gymnastices Hygieine, Cp. 33
[V, pp. 869-70 K.].

Εἰ δ᾽ ἔτι καὶ τρίτον ἄλλο μόριον ἰάσεως ὑπῆρχεν τὸ διαιτητικὸν
ἐν τοῖς καθ᾽ Ὅμηρον χρόνοις, ἐγὼ μὲν οὐκ ἔχω συμβαλεῖν,

ὁ δ' ἐμοῦ πρεσβύτερός τε ἅμα καὶ τὰ τῶν Ἑλλήνων πράγματα
πιθανώτερος ἐπίστασθαι, Πλάτων ὁ φιλόσοφος, οὐ πάνυ τι
χρῆσθαί φησι τοὺς παλαιοὺς Ἀσκληπιάδας τούτῳ τῷ μέρει τῆς
τέχνης.

Whether dietetics was known as a third part of medicine even
in Homer's time, I have no way of conjecturing; at any rate,
one who is older than I am and more trustworthy in his
knowledge of Greek affairs, Plato, the philosopher, says [1] that
the ancient Asclepiads did not particularly practice that part
of the art.

[1] T. 143.

145. Maximus Tyrius, Philosophumena, IV, 2a–3a.

Ἢ τὸ σκέμμα τουτὶ ἐοικέναι φῶμεν τοιῷδε, οἷον εἴ τις καὶ ἰατρικὴν
ἐνθυμηθεὶς τὴν πρώτην ἐκείνην πρὸς τὴν νέαν δὴ καὶ τοῖς νῦν
σώμασιν ἐπιτεταγμένην, σκοποῖ τὸ ἐν ἑκατέρᾳ βέλτιον καὶ
χεῖρον; ἀποκρίναιτο γὰρ ἂν αὐτῷ ὁ Ἀσκληπιός, ὅτι τὰς μὲν
ἄλλας τέχνας οὐ μεταποιοῦσιν οἱ χρόνοι· ὧν γὰρ ἡ αὐτὴ χρεία
ἀεί, τούτων παραπλήσια καὶ τὰ ἔργα· ἰατρικὴν δὲ ἀνάγκη
ἑπομένην τῇ κρατήσει τῶν σωμάτων, πράγματι οὐχ ἑστῶτι οὐδὲ
ὡμολογημένῳ, ἀλλὰ ταῖς κατὰ τὴν δίαιταν τροφαῖς ἀλλοιουμένῳ
καὶ μεταπίπτοντι, ἰάματα καὶ διαίτας αὐτῷ ἐξευρίσκειν ἄλλοτε
ἄλλας, προσφόρους τῇ παρούσῃ τροφῇ. μηδὲν οὖν ἡγοῦ τοὺς
υἱέας τοὺς ἐμούς, τὸν Μαχάονα ἐκεῖνον καὶ τὸν Ποδαλείριον,
ἧττόν τι εἶναι δεξιωτέρους ἰᾶσθαι τῶν αὖθις ἐπιτιθεμένων τῇ
τέχνῃ, καὶ τὰ σοφὰ ταῦτα καὶ παντοδαπὰ ἰάματα ἐξευρηκότων·
ἀλλὰ τότε μὲν ἡ τέχνη σώμασιν ὁμιλοῦσα οὐ θρυπτικοῖς,* οὐδὲ
ποικίλοις, οὐδὲ ἐκλελυμένοις παντάπασιν, ῥᾳδίως αὐτὰ μετε-
χειρίζετο, καὶ ἦν αὐτῆς ἔργόν τι ἁπλοῦν

ἰούς τ' ἐκτάμνειν, ἐπί τ' ἤπια φάρμακα πάσσειν·

τελευτῶσα δὲ νῦν, ὑπολισθαινόντων αὐτῇ τῶν σωμάτων εἰς
δίαιταν ποικιλωτέραν καὶ κρᾶσιν πονηράν, ἐξεποικίλθη τε αὐτὴ
καὶ μετέβαλλεν ἐκ τῆς πρόσθεν ἁπλότητος εἰς παντοδαπὸν σχῆμα.

Φέρε καὶ ὁ ποιητικὸς ὁμοῦ καὶ ὁ φιλόσοφος ἀποκρινάσθω
κατὰ τὸν Ἀσκληπιὸν ὑπὲρ τῶν ἐπιτηδευμάτων.

* θρυπτικοῖς Dübner θρεπτικοῖς MSS (except. τρεπτικοῖς a) Hobein.

Or shall we say the question is like something of this sort, as
if some one, pondering that first medical art in comparison with

medicine of today, which takes care of the bodies of today, should seek in each one the better and the worse? For to him Asclepius would reply that the times do not alter the other arts, for of what there is always the same use, of that there is similar performance also; but for medicine which must needs keep pace with the strength of the body—a thing which does not remain stationary nor fixed, but changes and alters together with the variations of daily diet—it is necessary to find different treatments and different modes of life for the body at one time and at another, befitting its state of nourishment at the time. Do not, then, think my sons Machaon and Podalirius are in any way less skilful in healing than those later in charge of the art who discovered such wise and varied treatments as these. In their time, however, the art of medicine, adapting itself to bodies that were not effeminate nor complex nor altogether enfeebled, administered to them with ease and its task was a simple thing:

> "to cut out the arrows and to spread thereon soothing simples "; [1]

but now that the art is at its peak, when bodies under its care are slipping into a more varied mode of life and dangerous mixture, the art itself became more varied and changed from its former simplicity to a manifold system.

Come, then, let the philosopher and the poet also make answer in the manner of Asclepius about their pursuits.

[1] T. **165**, v. 515.

---

146. Ennius, Hectoris Lytra, IV, 161-66 [Vahlen].[1]

Eurypylus

O Patricoles, ad vos adveniens, auxilium et vestras manus
peto, priusquam oppeto malam pestem mandatam hostili manu,
neque sanguis ullo potis est pacto profluens consistere,
si qui sapientia magis vestra mors devitari potest.
namque Aesculapi liberorum saucii opplent porticus;
non potest accedi.

Eurypylus

To you for aid I come, Patroclus, and your helping hands I beg
Before a cruel death encountering by foeman's hand bestowed,

And by no shift is 't possible the stream of flowing blood to
staunch,
To see if some way by your wisdom death can better be escaped,
For wounded crowd the entrance ways of the sons of Asclepius,
There is no access.[2]

[1] = Cicero, *Tusculanae Disputationes*, II, 16. 38.   [2] Cf. however T. 136.

147. Diodorus, Bibliotheca Historica, IV, 71.[1]

Ἀσκληπιοῦ δέ φασι γενομένους υἱοὺς Μαχάονα καὶ Ποδαλείριον,
καὶ τὴν τέχνην ἐκπονήσαντας, ἐπὶ Τροίαν συστρατεῦσαι τοῖς περὶ
τὸν Ἀγαμέμνονα· κατὰ δὲ τὸν πόλεμον μεγάλας χρείας αὐτοὺς
παρασχέσθαι τοῖς Ἕλλησι, θεραπεύοντας ἐμπειρότατα τοὺς
τιτρωσκομένους, καὶ διὰ τὰς εὐεργεσίας ταύτας ὑπὸ τῶν Ἑλλήνων
μεγάλης τυχεῖν δόξης· ἀτελεῖς δ' αὐτοὺς ἀφεῖναι τῶν κατὰ τὰς
μάχας κινδύνων καὶ τῶν ἄλλων λειτουργιῶν διὰ τὴν ὑπερβολὴν
τῆς ἐν τῷ θεραπεύειν εὐχρηστίας.

Furthermore, they say that Machaon and Podalirius were sons
of Asclepius, and that, having practiced the art [*sc.*, of medi-
cine],[2] they joined Agamemnon in the expedition against Troy.[3]
During the war they performed great service to the Greeks in
curing the wounded most skillfully, and because of these services, they won high repute from the Greeks, and they were
exempted from the dangers of battle and from other duties
because of their exceeding usefulness in the art of healing.

[1] Cf. T. 4.
[2] Cf. 147a. *Anthologia Latina*, I, 2, 719e, 2-3: . . . quod didicere olim Podalirius
atque Machaon a genitore suo; cf. T. 356a; 614.
[3] Cf. 147b. Pausanias, *Descriptio Graeciae*, IV, 31, 12: . . . καὶ Μαχάων καὶ Ποδαλείριος,
ὅτι ἔργου τοῦ πρὸς Ἰλίῳ καὶ τούτοις μέτεστι.

148. Xenophon, Cynegeticus, I, 14.

Μαχάων δὲ καὶ Ποδαλείριος παιδευθέντες τὰ αὐτὰ πάντα ἐγένοντο
καὶ τέχνας καὶ λόγους καὶ πολέμους ἀγαθοί.

Machaon and Podalirius, schooled in all the selfsame arts [*sc.*,
of hunting] proved in crafts and reasonings and wars good
men.

149. Dares, De Excidio Troiae Historia, XIII.

Podalirium crassum valentem superbum tristem. Machaonem
fortem magnum certum prudentem patientem misericordem.

Podalirius, stout, strong, haughty, stern. Machaon brave,
great, sure, wise, patient, compassionate.

150. Apollodorus, Bibliotheca, III, 10, 8, 1-3.

Παρεγένοντο δὲ εἰς Σπάρτην ἐπὶ τὸν Ἑλένης γάμον οἱ βασι-
λεύοντες Ἑλλάδος. ἦσαν δὲ οἱ μνηστευόμενοι οἵδε . . . Ποδαλείριος καὶ Μαχάων Ἀσκληπιοῦ . . .

Now the kings of Greece repaired to Sparta to win the hand
of Helen. The wooers were these . . . Podalirius and Machaon,
sons of Asclepius[1] . . .

[1] Cf. **150a.** Hyginus, *Fabulae*, LXXXI: Proci Helenae: . . . Podalirius . . . Machaon.

151. Dictys Cretensis, Ephemeris Belli Troiani, I, 14.

. . . Podalirius et Machaon Triccenses, Aesculapio geniti,
adsciti ad id bellum ob sollertiam medicinae artis . . .

. . . Podalirius and Machaon of Tricca, sons of Asclepius,
called to that war because of their skill in the art of medicine . . .

152. Sophocles, Philoctetes, 1329-34.

Καὶ παῦλαν ἴσθι τῆσδε μή ποτ᾽ ἂν τυχεῖν·
νόσου βαρείας, ἕως ἂν αὐτὸς ἥλιος
ταύτῃ μὲν αἴρῃ, τῇδε δ᾽ αὖ δύνῃ πάλιν,
πρὶν ἂν τὰ Τροίας πεδί᾽ ἑκὼν αὐτὸς μόλῃς,
καὶ τοῖν παρ᾽ ἡμῖν ἐντυχὼν Ἀσκληπίδαιν
νόσου μαλαχθῇς τῆσδε . . .

And from this sore disease shalt thou [*sc.*, Philoctetes] win
        rest
never, be sure, long as the selfsame sun
to eastward riseth, westward sets anon,
ere to Troy's plain thyself consenting come,
and meet Asclepius' sons[1] who bide with us,
be healed of thy disease . . .

[1] Cf. **152a.** Philostratus, *Heroicus*, V, 1: . . . ἰαθῆναι δὲ ὑπὸ τῶν Ἀσκληπιαδῶν αὐτός [*sc.*.
ὁ Φιλοκτήτης]; **152b.** Baebius Italicus, *Ilias Latina*, 218: Quem [*sc.*, Philocteten]
sequitur iuxta Podalirius atque Machaon; cf. however, T. 86; 92; 175; 201; 202.

153. Dictys Cretensis, Ephemeris Belli Troiani, II, 6.

> Hi itaque ad Telephum veniunt ac more regio invicem acceptis datisque donis Machaonem et Podalirium Aesculapii filios venire ac vulneri mederi iubent, qui inspecto crure * propere apta dolori medicamina inponunt.

* inspecto crure Vindingius   inspecta cura MSS Meister.

> Accordingly they [*sc.*, Agamemnon and Menelaus] come to Telephus and, having exchanged gifts in the royal manner, they bid Machaon and Podalirius, sons of Asclepius, come and cure the wound. After an examination of the leg they quickly apply medicines appropriate to the injury.

154. Dictys Cretensis, Ephemeris Belli Troiani, II, 10.

> Per idem tempus Telephus dolore vulneris eius, quod in proelio adversum Graecos acceperat, diu afflictatus, cum nullo remedio mederi posset, ad postremum Apollinis oraculo monitus, uti Achillem atque Aesculapii filios adhiberet, propere Argos navigat. dein cunctis ducibus causam adventus eius admirantibus oraculum refert atque ita orat, ne sibi praedictum remedium ab amicis negaretur. quae ubi accepere Achilles cum Machaone et Podalirio adhibentes curam vulneri brevi fidem oraculi firmavere.

> At the same time Telephus, who had long been suffering from the pain of that wound which he had received in the battle against the Greeks, being unable to heal it with any remedy, finally on the advice of Apollo's oracle that he employ Achilles and the sons of Asclepius, set sail quickly to Argus. Then to the assembled leaders who wondered at the reason for his coming he related the oracle and begged that the remedy vouchsafed to him be not denied by his friends. Achilles, together with Machaon and Podalirius, upon hearing this, treated the wound, and in a short time confirmed the truth of the oracle.

155. Dictys Cretensis, Ephemeris Belli Troiani, III, 19.

> Cursu longo certantibus Oilei Aiax victor excipitur, post quem secundus Polypoetes. duplici campo Machaon, singulari Eurypylus. . . . victores abeunt . . . dein ubi praemia certaminis

persoluta sunt, Achilles primum omnium Agamemnoni donum
. . . offert, secundo Nestori, Idomeneo tertio, post quos Poda-
lirio et Machaoni, dein reliquis pro merito ducibus . . .

Among those contending in a long race-course Ajax, son of
Oeleus, was announced as victor, with Polypoetes second after
him. In a double field Machaon was victor; in a single, Eurypylus. . . . Then, when the rewards for the contest were distributed, Achilles gave a gift to Agamemnon first of all . . . ,
to Nestor second, to Idomeneus third, and after them to Podalirius and Machaon, then to the other leaders according to
their due . . .

156. Pausanias, Descriptio Graeciae, III, 26, 10.

Ἀνασώσασθαι δὲ Νέστορα λέγεται τοῦ Μαχάονος τὰ ὀστᾶ·
Ποδαλείριον δέ, ὡς ὀπίσω πορθήσαντες Ἴλιον ἐκομίζοντο,
ἁμαρτεῖν τοῦ πλοῦ καὶ ἐς Σύρνον τῆς Καρικῆς ἠπείρου φασὶν
ἀποσωθέντα οἰκῆσαι.

It is said that the bones of Machaon were rescued by Nestor,
but that Podalirius, as they were returning after the sack of
Troy, was carried out of his course and reaching Syrnus on
the Carian mainland in safety, settled there.[1](*)

[1] Cf. T. **208**; **209**.

157. Hippocratis Vita Bruxellensis, 2-8.[1]

Yppocrates fuit genere Cous a Eraclide filius ex Finerata ortus
ab Asclepia stirpe. Asclepio enim ex Epiona Herculis filia duo
sunt creati successus, Podalirius et Macaon. quorum Macaon,
ut plurimi tradunt, Troiae excidio vitam finivit nulla subole
derelicta, Podalirius vero Sirnae consistens Rodi defecit, ut
Antimachus memorat in Thenito, filios nactus duos, Rodonem
et Ippolochon, ex Ifianassa, Ucalegontis filia.

Hippocrates was a Coan, a son of Heraclides by Phine-
rata and of the family of Asclepius.[2] For to Asclepius by
Epione, daughter of Heracles, were born two heirs, Podalirius
and Machaon, of whom Machaon, so most sources say, lost
his life in the sack of Troy,[3] leaving behind no offspring, while
Podalirius lived at Syrna and died on the island of Rhodes,

as Antimachus relates in his *Thenitus*, having begotten two sons, Rhodo and Hippolochus,[4] by Iphianassa, daughter of Ucalegon.

[1] = Antimachus, Fr. 150 [Wyss].   [3] Cf. T. 178.
[2] Cf. T. 213 ff.   [4] Cf. T. 213.

158. Lycophron, Alexandra, 1047-55.

Ὁ δ᾽ Αὐσονείων ἄγχι Κάλχαντος τάφων
δυοῖν ἀδελφοῖν ἅτερος ψευδηρίων
ξένην ἐπ᾽ ὀστέοισιν ὀγχήσει κόνιν.
δοραῖς δὲ μήλων τύμβον ἐγκοιμωμένοις
χρήσει καθ᾽ ὕπνον πᾶσι νημερτῆ φάτιν.
νόσων δ᾽ ἀκεστὴς Δαυνίοις κληθήσεται,
ὅταν κατικμαίνοντες Ἀλθαίνου ῥοαῖς
ἀρωγὸν αὐδήσωσιν Ἠπίου γόνον
ἀστοῖσι καὶ ποίμναισι πρευμενῆ μολεῖν.

And near the Ausonian cenotaph of Calchas one of two brothers [*sc.*, Machaon and Podalirius] shall have an alien soil over his bones and to men sleeping in sheepskins on his tomb[1] he shall declare in dreams his unerring message for all. And healer of diseases shall he be called by the Daunians, when they wash the sick with the waters of Althaenus and invoke the son of Epius[2] to their aid, that he may come gracious unto men and flocks. (*)

[1] Cf. T. 206; 294.   [2] Cf. T. 271.

159. Scholia in Lycophronem, Ad Alexandram, 1047.

Περὶ Ποδαλειρίου τοῦ Ἀσκληπιάδου ὁ λόγος· φησὶν οὖν ὅτι τεθνήξεται ἐν Ἰταλίᾳ πλησίον τῶν κενοταφίων τοῦ Κάλχαντος τοῦ ὑφ᾽ Ἡρακλέους ἐν Ἄργει ἀναιρεθέντος κονδύλῳ. τέθαπται οὖν ὁ Κάλχας ἐν Ἄργει, κενοτάφιον δὲ αὐτῷ ἐν Ἰταλίᾳ. ἐνταῦθα οὖν ὁ Ποδαλείριος τέθαπται.

The words concern Podalirius, son of Asclepius. The poet says that he will die in Italy, near the cenotaph of Calchas who was killed with a blow by Heracles in Argos. Calchas was then buried in Argos, but there is a cenotaph for him in Italy. There, then, Podalirius was buried.[1]

[1] Cf. T. 160; 205.

160. Scholia in Lycophronem, Ad Alexandram, 1048.

Τῶν β΄ ἀδελφῶν ὁ ἕτερος ἤγουν Μαχάονος καὶ Ποδαλειρίου· ὁ
μὲν γὰρ Μαχάων ἀνήρηται ἐν τῷ πολέμῳ ὑπὸ Εὐρυπύλου τοῦ
Τηλεφίδου, ὁ δὲ Ποδαλείριος ἐλθὼν ἐν Ἰταλίᾳ τελευτᾷ ἐκεῖσε.
τὸ δὲ ἑξῆς οὕτως· ὁ δὲ ἕτερος τῶν β΄ ἀδελφῶν πλησίον τῶν
Ἰταλικῶν ψευδηρίων τοῦ Κάλχαντος ταφήσεται.

One of the two brothers, that is to say, of Machaon and
Podalirius. One, Machaon, was killed during the war by Eury-
pylus, the son of Telephus;[1] the other, Podalirius, went to
Italy and died there.[2] The order of the words is the following:
One of the brothers will be buried near the Italian false tomb
of Calchas.

[1] Cf. T. 180 ff.　　　　　　　　　　[2] Cf. T. 159; 205.

161. Aristoteles, Peplus. Fr. 20.

Ἐν Τρίκκῃ ἐπὶ κενοταφίου Ποδαλειρίου καὶ Μαχάονος
οἵδ᾽ Ἀσκληπιάδαι Ποδαλείριος ἠδὲ Μαχάων
πρόσθεν μὲν θνητοί, νῦν δὲ θεῶν μέτοχοι.

In Tricca on the cenotaph of Podalirius and Machaon

These are the sons of Asclepius, Podalirius and Machaon,
mortals formerly, now partners of the gods.

162. Hippocrates, Epistulae, 10 [IX, p. 324, 16 L.].

Ἴθι οὖν μετὰ Ἀσκληπιοῦ πατρός, ἴθι μετὰ Ἡρακλέους θυγατρὸς
Ἠπιόνης, ἴθι μετὰ παίδων τῶν ἐπὶ Ἴλιον στρατευσαμένων . . .

Come then [sc., Hippocrates] with your father Asclepius,
come with Epione, the daughter of Heracles,[1] come with the
children[2] who went to fight against Ilium . . .

[1] Cf., however, T. 169.
[2] Cf. 162a. *Scholia in Pindarum, Ad Pythias,* III, 14: Ἀσκληπιοῦ δὲ καὶ Ἠπιόνης
Ποδαλείριος καὶ Μαχάων; cf. T. 280-280d; cf. also T. 34.

163. Marinus, Vita Procli, Cp. 32.

Καὶ μὴν ὁ ἐν Ἀδρόττοις θεὸς ἐναργῶς ἔδειξεν τοῦ θεοφιλοῦς
ἀνδρὸς τὴν πρὸς αὐτὸν οἰκειότητα . . . καὶ ἀποροῦντι αὐτῷ
καὶ εὐχομένῳ μαθεῖν τίς ἢ τίνες οἱ ἐπιφοιτῶντες καὶ τιμηθέντες

ἐν τῷ τόπῳ θεοί, διὰ τὸ μηδὲ τοὺς αὐτοὺς κρατεῖν παρὰ τοῖς
ἐπιχωρίοις λόγους—ἐνίων μὲν δοξαζόντων Ἀσκληπιοῦ εἶναι τὸ
ἱερὸν καὶ τοῦτο ἐκ πολλῶν τεκμηρίων πιστουμένων· καὶ γὰρ
ὄντως ἀκοαὶ λέγονταί που εἶναι τῷ τόπῳ, καὶ τράπεζά τις τῷ
θεῷ ἀνειμένη, καὶ χρησμοὶ δίδονται ἑκάστοτε ὑγιαστικοί, καὶ
ἐκ τῶν μεγίστων κινδύνων σώζονται παραδόξως οἱ προσιόντες·
ἑτέρων δὲ τινῶν οἰομένων τοὺς Διοσκούρους ἐπιφοιτᾶν τῷ τόπῳ
. . .—ἐκ τούτωῦ οὖν, ὡς εἴρηται, ἀποροῦντι τῷ φιλοσόφῳ καὶ
οὐκ ἔχοντι ἀπιστεῖν τοῖς ἱστορημένοις . . . ἐδόκει οἱ ὁ θεὸς
ὄναρ ἐπιφοιτᾶν καὶ ἐναργῶς ταῦτα ὑφηγεῖσθαι· "Τί δαί; Ἰαμ-
βλίχου οὐκ ἀκήκοας λέγοντος τίνες οἱ δύο, καὶ ὑμνοῦντος
Μαχάονά τε καὶ Ποδαλείριον;"

Indeed the god in Adrotta [in Lydia] clearly revealed how
familiar the man beloved of heaven [*sc.*, Proclus] was with
him. . . . For when he [*sc.*, Proclus] was pondering and
praying to learn which god or which gods regularly frequented
the district and were honored there, since the same stories
prevailed not even among the natives—for some imagine it is
the shrine of Asclepius and this they believe because of many
signs, for voices are actually said to be heard sometimes in the
place, and there is a table consecrated to the god, and curative
oracles are given out on various occasions, and from the
greatest dangers those who approach the place are unexpectedly
saved; others, however, think the Dioscuri frequent the
place . . . — when for this reason, then, as has been said,
the philosopher was in a quandary and could not distrust the
tales that were related, the god seemed to him to approach in
a dream and clearly to instruct him in these matters: " Well,
then, have you not heard Iamblichus explaining who the two
are in his hymn to Machaon and Podalirius? "

MACHAON

164. Homerus, Ilias, IV, 192–219.

Ἦ, καὶ Ταλθύβιον, θεῖον κήρυκα, προσηύδα·
"Ταλθύβι', ὅττι τάχιστα Μαχάονα δεῦρο κάλεσσον,
φῶτ' Ἀσκληπιοῦ υἱόν, ἀμύμονος ἰητῆρος,
195   ὄφρα ἴδῃ Μενέλαον ἀρήιον Ἀτρέος υἱόν,
ὅν τις ὀϊστεύσας ἔβαλεν, τόξων ἐὺ εἰδώς,

Τρώων ἢ Λυκίων, τῷ μὲν κλέος, ἄμμι δὲ πένθος."
Ὣς ἔφατ᾽, οὐδ᾽ ἄρα οἱ κῆρυξ ἀπίθησεν ἀκούσας,
βῆ δ᾽ ἰέναι κατὰ λαὸν Ἀχαιῶν χαλκοχιτώνων
200 παπταίνων ἥρωα Μαχάονα· τὸν δ᾽ ἐνόησεν
ἑσταότ᾽· ἀμφὶ δέ μιν κρατεραὶ στίχες ἀσπιστάων
λαῶν, οἵ οἱ ἕποντο Τρίκης ἐξ ἱπποβότοιο.
ἀγχοῦ δ᾽ ἱστάμενος ἔπεα πτερόεντα προσηύδα·
"ὄρσ᾽, Ἀσκληπιάδη, καλέει κρείων Ἀγαμέμνων,
205 ὄφρα ἴδῃς Μενέλαον ἀρήιον ἀρχὸν Ἀχαιῶν,
ὅν τις ὀιστεύσας ἔβαλεν, τόξων ἐὺ εἰδώς,
Τρώων ἢ Λυκίων, τῷ μὲν κλέος, ἄμμι δὲ πένθος."
Ὣς φάτο, τῷ δ᾽ ἄρα θυμὸν ἐνὶ στήθεσσιν ὄρινε·
βὰν δ᾽ ἰέναι καθ᾽ ὅμιλον ἀνὰ στρατὸν εὐρὺν Ἀχαιῶν.
210 ἀλλ᾽ ὅτε δή ῥ᾽ ἵκανον ὅθι ξανθὸς Μενέλαος
βλήμενος ἦν, περὶ δ᾽ αὐτὸν ἀγηγέραθ᾽ ὅσσοι ἄριστοι
κυκλόσ᾽, ὁ δ᾽ ἐν μέσσοισι παρίστατο ἰσόθεος φώς,
αὐτίκα δ᾽ ἐκ ζωστῆρος ἀρηρότος ἕλκεν ὀιστόν·
τοῦ δ᾽ ἐξελκομένοιο πάλιν ἄγεν ὀξέες ὄγκοι.
215 λῦσε δέ οἱ ζωστῆρα παναίολον ἠδ᾽ ὑπένερθε
ζῶμά τε καὶ μίτρην, τὴν χαλκῆες κάμον ἄνδρες.
αὐτὰρ ἐπεὶ ἴδεν ἕλκος, ὅθ᾽ ἔμπεσε πικρὸς ὀιστός,
αἷμ᾽ ἐκμυζήσας ἐπ᾽ ἄρ᾽ ἤπια φάρμακα εἰδὼς
πάσσε, τά οἵ ποτε πατρὶ φίλα φρονέων πόρε Χείρων.

Therewith he [*sc.*, Agamemnon] spake to Talthybius, the god-like herald: "Talthybius, make haste to call hither the man Machaon, son of Asclepius, the blameless physician, to see warlike Menelaus, son of Atreus, whom some man well skilled in archery hath smitten with an arrow, some Trojan or Lycian, compassing glory for himself but for us sorrow."

So he spake, and the herald failed not to hearken, as he heard, but went his way throughout the host of the brazen-coated Achaeans, glancing this way and that for the warrior Machaon; and he marked him as he stood, and round about him were the stalwart ranks of the shield-bearing hosts that followed him from Trica,[1] the pastureland of horses. And he came up to him, and spake winged words, saying: "Rouse thee, son of Asclepius; lord Agamemnon calleth thee to see warlike Menelaus, captain of the Achaeans, whom some man, well skilled in archery, hath smitten with an arrow, some Trojan or Lycian, compassing glory for himself but for us sorrow."

So he spake, and roused the heart in his breast, and they went
their way in the throng throughout the broad host of the
Achaeans. And when they were come where was fair-haired
Menelaus, wounded, and around him were gathered in a circle
all they that were chieftains, the godlike man came and stood
in their midst, and straightway drew forth the arrow from the
clasped belt; and as it was drawn forth the sharp barbs were
broken backwards. And he loosed the flashing belt and the
kilt beneath and the taslet that the copper-smiths fashioned.
But when he saw the wound where the bitter arrow had
lighted, he sucked out the blood, and with sure knowledge
spread thereon soothing simples,[2] which of old Chiron had
given to his father with kindly thought. (*)

[1] Cf. T. 10a.
[2] Cf. T. 143; cf. 164a. Ioannes Tzetzes, *Homerica*, 21: ... τὸν μὲν γὰρ [*sc.*, Μενέλαον]
Μαχάων ἰήσατο, φάρμακα πάσσων.

165. Homerus, Ilias, XI, 504-20.

Οὐδ᾽ ἄν πω χάζοντο κελεύθου δῖοι Ἀχαιοί,
505 εἰ μὴ Ἀλέξανδρος, Ἑλένης πόσις ἠϋκόμοιο,
παῦσεν ἀριστεύοντα Μαχάονα, ποιμένα λαῶν,
ἰῷ τριγλώχινι βαλὼν κατὰ δεξιὸν ὦμον.
τῷ ῥα περίδεισαν μένεα πνείοντες Ἀχαιοί,
μή πώς μιν πολέμοιο μετακλινθέντος ἕλοιεν.
510 αὐτίκα δ᾽ Ἰδομενεὺς προσεφώνεε Νέστορα δῖον·
"ὦ Νέστορ Νηληϊάδη, μέγα κῦδος Ἀχαιῶν,
ἄγρει, σῶν ὀχέων ἐπιβήσεο, πὰρ δὲ Μαχάων
βαινέτω, ἐς νῆας δὲ τάχιστ᾽ ἔχε μώνυχας ἵππους·
ἰητρὸς γὰρ ἀνὴρ πολλῶν ἀντάξιος ἄλλων
515 ἰούς τ᾽ ἐκτάμνειν ἐπί τ᾽ ἤπια φάρμακα πάσσειν."
Ὣς ἔφατ᾽, οὐδ᾽ ἀπίθησε Γερήνιος ἱππότα Νέστωρ.
αὐτίκα δ᾽ ὧν ὀχέων ἐπιβήσετο, πὰρ δὲ Μαχάων
βαῖν᾽, Ἀσκληπιοῦ υἱὸς ἀμύμονος ἰητῆρος·
μάστιξεν δ᾽ ἵππους, τὼ δ᾽ οὐκ ἀέκοντε πετέσθην
520 νῆας ἔπι γλαφυράς.

Yet would the goodly Achaeans in no wise have given ground
from their course, had not Alexander, the lord of fair-haired
Helen, stayed Machaon, shepherd of the host, in the midst of
his valorous deeds,[1] and smitten him on the right shoulder with

a three-barbed arrow. Then sorely did the Achaeans breathing
might fear for him, lest haply men should slay him in the turning of the fight. And forthwith Idomeneus spake to goodly
Nestor: " Nestor, son of Neleus, great glory of the Achaeans,
come, get thee upon thy chariot, and let Machaon mount beside
thee, and swiftly do thou drive to the ships thy single-hooved
horses. For a leech is of the worth of many other men for the
cutting out of arrows and the spreading of soothing simples."
So he spake, and the horseman, Nestor of Gerenia, failed not
to hearken. Forthwith he got him upon his chariot, and beside
him mounted Machaon, the son of Asclepius the blameless
physician; and he touched the horses with the lash, and nothing
loath the pair sped on to the hollow ships.

[1] Cf. **165a**. *Scholia in Homerum, Ad Iliadem*, XI, 506 (T): Λεληθότως ἐνέφηνεν, ὡς
διετίθετό τι ὁ Μαχάων·　τίς δὲ ἡ ἀριστεία, ἐπὶ κεφαλαίου εἰπὼν οὐ διέξεισι κατὰ μέρος ὡς
ἐπὶ τῶν ἄλλων.

166. Homerus, Ilias, XI, 596-615.

　　　　Ὡς οἱ μὲν μάρναντο δέμας πυρὸς αἰθομένοιο·
　　　　Νέστορα δ' ἐκ πολέμοιο φέρον Νηλήϊαι ἵπποι
　　　　ἱδρῶσαι, ἦγον δὲ Μαχάονα, ποιμένα λαῶν.
　　　　τὸν δὲ ἰδὼν ἐνόησε ποδάρκης δῖος Ἀχιλλεύς·
　600　ἑστήκει γὰρ ἐπὶ πρυμνῇ μεγακήτεϊ νηΐ,
　　　　εἰσορόων πόνον αἰπὺν ἰωκά τε δακρυόεσσαν.
　　　　αἶψα δ' ἑταῖρον ἑὸν Πατροκλῆα προσέειπε,
　　　　φθεγξάμενος παρὰ νηός· ὁ δὲ κλισίηθεν ἀκούσας
　　　　ἔκμολεν ἶσος Ἄρηϊ, κακοῦ δ' ἄρα οἱ πέλεν ἀρχή.
　605　τὸν πρότερος προσέειπε Μενοιτίου ἄλκιμος υἱός·
　　　　" τίπτέ με κικλήσκεις, Ἀχιλεῦ; τί δέ σε χρεὼ ἐμεῖο; "
　　　　τὸν δ' ἀπαμειβόμενος προσέφη πόδας ὠκὺς Ἀχιλλεύς·
　　　　" δῖε Μενοιτιάδη, τῷ ἐμῷ κεχαρισμένε θυμῷ,
　　　　νῦν ὀΐω περὶ γούνατ' ἐμὰ στήσεσθαι Ἀχαιοὺς
　610　λισσομένους· χρειὼ γὰρ ἱκάνεται οὐκέτ' ἀνεκτός.
　　　　ἀλλ' ἴθι νῦν, Πάτροκλε, Διῒ φίλε, Νέστορ' ἔρειο
　　　　ὅν τινα τοῦτον ἄγει βεβλημένον ἐκ πολέμοιο·
　　　　ἤτοι μὲν τά γ' ὄπισθε Μαχάονι πάντα ἔοικε
　　　　τῷ Ἀσκληπιάδῃ, ἀτὰρ οὐκ ἴδον ὄμματα φωτός·
　615　ἵπποι γάρ με παρήϊξαν πρόσσω μεμαυῖαι."

So fought they like unto blazing fire; but the mares of Neleus,
all bathed in sweat, bare Nestor forth from the battle, and bare

also Machaon, shepherd of the host. And swift-footed goodly Achilles beheld and marked him, for Achilles was standing by the stern of his ship, huge of hull, gazing upon the utter toil of battle and the tearful rout. And forthwith he spake to his comrade Patroclus, calling to him from beside the ship; and he heard, and came forth from the hut like unto Ares; and this to him was the beginning of evil. Then the valiant son of Menoetius addressed him first: " Wherefore dost thou call me, Achilles? What need hast thou of me?" And in answer to him spake Achilles, swift of foot: "Goodly son of Menoetius, dear to this heart of mine, now methinks will the Achaeans be standing about my knees in prayer, for need has come upon them that may no longer be borne. Yet go now, Patroclus, dear to Zeus, and ask Nestor who it is that he bringeth wounded from out the war. Of a truth from behind he seemeth in all things like Machaon, son of Asclepius, but I saw not the eyes of the man, for the horses darted by me, speeding eagerly onward." [1]

[1] Cf. **166a.** Ioannes Tzetzes, *Homerica*, 196-202: Αὐτὰρ Ἀλέξανδρος κεραελκέα τόξα ἐρύων/ Τυδείδην βάλεν ἠδὲ Μαχάονα Εὐρύπυλόν τε./ Νέστωρ δ' ὡς φορέεσκε Μαχάονα ἰητῆρα,/ Πάτροκλον ἧκεν Ἀχιλλεὺς ἐξερέοντα, τίς εἴη·/ ὃς δὴ Νέστορος ἐκ κλισίης παλίνορσος ὀρούων/ Εὐρυπύλου βεβολημένου ἄγριον ἕλκος ἀκεῖτο. **166b.** *Scholia in Homerum, Ad Iliadem*, XI, 614 (T): Ἢ ὡς πλησιοχώρου φροντίζει, ἢ ἐπεὶ Ἀσκληπιὸς καὶ Ἀχιλλεὺς συμφευτηταὶ παρὰ Χείρωνι.

167. Homerus, Ilias, XI, 648-52.

" Οὐχ ἕδος ἐστί, γεραιὲ διοτρεφές, οὐδέ με πείσεις.
αἰδοῖος νεμεσητὸς ὅ με προέηκε πυθέσθαι
ὅν τινα τοῦτον ἄγεις βεβλημένον. ἀλλὰ καὶ αὐτὸς
γιγνώσκω, ὁρόω δὲ Μαχάονα, ποιμένα λαῶν.
νῦν δὲ ἔπος ἐρέων πάλιν ἄγγελος εἴμ' Ἀχιλῆϊ
· · · · · · · · · · · · · · · · "

" I [*sc.*, Patroclus] may not sit, old sir [*sc.*, Nestor], fostered of Zeus, nor wilt thou persuade me. Revered and to be dreaded is he who sent me forth [*sc.*, Achilles] to learn who it is that thou bringest home wounded. But even of myself I know, and behold Machaon, shepherd of the host. And now will I go back again a messenger, to bear word to Achilles . . . ."

168. Homerus, Ilias, XIV, 1-8.

> Νέστορα δ' οὐκ ἔλαθεν ἰαχὴ πίνοντά περ ἔμπης,
> ἀλλ' Ἀσκληπιάδην ἔπεα πτερόεντα προσηύδα·
> " φράζεο, δῖε Μαχᾶον, ὅπως ἔσται τάδε ἔργα·
> μείζων δὴ παρὰ νηυσὶ βοὴ θαλερῶν αἰζηῶν.
> ἀλλὰ σὺ μὲν νῦν πῖνε καθήμενος αἴθοπα οἶνον,
> εἰς ὅ κε θερμὰ λοετρὰ ἐϋπλόκαμος Ἑκαμήδη
> θερμήνῃ καὶ λούσῃ ἄπο βρότον αἱματόεντα·
> αὐτὰρ ἐγὼν ἐλθὼν τάχα εἴσομαι ἐς περιωπήν."

And the cry of battle was not unmarked of Nestor, albeit at his wine, but he spake winged words to the son of Asclepius: "Bethink thee, goodly Machaon, how these things are to be; louder in sooth by the ships waxes the cry of lusty youths. Howbeit do thou now sit where thou art and quaff the flaming wine, until fair-tressed Hecamede shall heat for thee a warm bath, and wash from thee the clotted blood, but I will go straightway to a place of outlook and see what is toward."

169. Scholia in Homerum, Ad Iliadem, IV, 195.

> Μαχάων δὲ οὗτος υἱὸς Ἀσκληπιοῦ καὶ . . .* Ἀρσινόης ἢ Κορω-
> νίδος, κατὰ δέ τινας Ἠπιόνης τῆς Μέροπος, κατὰ δὲ Ἡσίοδον
> Ξάνθης.

* Lacunam indicavit Wilamowitz, *Isyllos*, p. 49, n. 12.

This Machaon was the son of Asclepius and . . . of Arsinoë or Coronis, but according to some authorities, the son of Epione, daughter of Merops,[1] according to Hesiod,[2] however, the son of Xanthe.

[1] Cf., however, T. **162.**　　　　　　　[2] = Fr. 89 [Rzach]; cf. T. **281d.**

170. Pausanias, Descriptio Graeciae, IV, 30, 3.[1]

> . . . Ἀντίκλειαν . . . τῆς δὲ Νικόμαχόν τε εἶναι καὶ Γόργασον,
> πατρὸς δὲ Μαχάονος τοῦ Ἀσκληπιοῦ.

. . . Anticleia . . . her children were Nicomachus and Gorgasus, their father being Machaon, the son of Asclepius.

[1] Cf. T. **191.**

171. Eustathius, Commentarii ad Homeri Iliadem, XI, 517.[1]

Ὅτι δὲ καὶ πατρῷα τέχνη τῷ Μαχάονι τὸ ἰατρεύειν, καὶ ὅτι
εὐγενὴς ὁ ἀνήρ, ἢ καὶ κατὰ μῦθον θεῖον γένος ὡς Ἀσκληπιάδης,
δηλοῖ μετ' ὀλίγα ὁ ποιητής, εἰπὼν αὐτὸν Ἀσκληπιοῦ υἱόν,
ἀμύμονος ἰητῆρος.

That medicine was for Machaon an art inherited from his
father, and that he was a well-born man or even, according to
myth, of divine race, as an Asclepiad, the poet indicates shortly
afterwards, calling him son of Asclepius, the blameless physician.[2]

[1] Cf. T. 256.   [2] Cf. T. 272a.

172. Statius, Silvae, I, 4, 112-14.

Citius non arte refectus
Telephus Haemonia, nec quae metuentis Atridae
saeva Machaonio coierunt vulnera suco.

Not more swiftly was Telephus restored by Haemonian skill,
nor the cruel wounds of which the Atride stood in terror
stanched by Machaon's healing balm. (*)

173. Proclus, Chrestomathia, p. 106, 26 [Allen].

Ἰαθεὶς δὲ οὗτος ὑπὸ Μαχάονος καὶ μονομαχήσας Ἀλεξάνδρῳ
κτείνει.

Having been healed by Machaon he [sc., Philoctetes] fought
in single combat with Alexander and slew him [according to
the Little Iliad].

174. Scholia in Pindarum, Ad Pythias, I, 109 a.

Φησὶ γὰρ Διονύσιος χρησμοῖς Ἀπόλλωνος ἀπολουσάμενον τὸν
Φιλοκτήτην ἀφυπνῶσαι, τὸν δὲ Μαχάονα ἀφελόντα τοῦ ἕλκους
τὰς διασαπείσας σάρκας καὶ ἐπικλύσαντα οἴνῳ τὸ τραῦμα ἐπι-
πάσαι βοτάνην, ἣν Ἀσκληπιὸς εἴληφε παρὰ Χείρωνος, καὶ οὕτως
ὑγιασθῆναι τὸν ἥρωα.

For Dionysius [Hellenistic author] says[1] that Philoctetes,
washed clean by the oracles of Apollo, fell asleep; then
Machaon, removing the gangrenous flesh from the festering
ulcer and deluging it with wine, sprinkled over the wound an

herb which Asclepius got from Chiron,[2] and in this way the hero was cured.[3]

[1] = Fr. 6 [Müller].　　[2] Cf. T. 50.　　[3] Cf. T. 86; 175 ff.; cf., however, T. 201; 202.

175. Propertius, Elegiae, II, 1, 59.

Tarda Philoctetae sanavit crura Machaon

. . . .

Machaon healed Philoctetes' limping feet[1] . . .

[1] Cf. T. 152; 201; 202; cf. also T. 86; 92.

176. Orphei Lithica, 346–54.

Καὶ σέο, δαιμονίη, μεμνήσομαι αὐτίκα, πέτρη,
αὐτοκασιγνήτη πολιῆς καὶ ὁμώνυμ᾽ ἐχίδνης·
ἥ ῥά ποτ᾽ εἰναετῆ λώβην ἀμενηνὸν ἔθηκας
ῥίμφα Φιλοκτήταο Μαχαονίης ὑπὸ τέχνης.
οὐ μέν πως κείνην Ποιάντιος ἔλπετο θυμῷ
νοῦσον ὑπεκφεύξεσθαι ἀταρτηρήν, ποθέων περ·
ἀλλ᾽ ὅγ᾽ ἀλεξικάκοιο μαθὼν παρὰ πατρὸς ἀρωγήν,
λᾶαν, ὅτις κ᾽ ἀκέοιτο, βαλὼν ἔπι φάρμακα μηρῷ,
σεῦεν ἐπὶ Τρώεσσιν Ἀλεξάνδροιο φονῆα.

At once I shall remember you, O divine rock,[1] own sister and name-sake of the hoary snake, you who once made the nine-year mutilation of Philoctetes swiftly vanish, through the skill of Machaon. The son of Poeas [sc., Philoctetes] did not hope in his heart to escape that baneful suffering, although he longed to. But Machaon, who learned succor from his father who wards off harm, applying as a remedy to the thigh the rock which might heal, let loose against the Trojans the murderer of Alexander.

[1] Cf. **176a.** Ioannes Tzetzes, *Posthomerica*, 583: Πέτρῃ δ᾽ ἀρτεμέα ἐχιήτιδι τεῦξε Μαχάων; cf. **176b.** *Scholia in Lycophronem, Ad Alexandram,* 1048: Ὁ μὲν γὰρ Μαχάων κατὰ Κόϊντον ἀνήρηται ἐν τῷ πολέμῳ ὑπ᾽ Εὐρυπύλου τοῦ υἱοῦ Τηλέφου, ὁ δὲ παλαιὸς Ὀρφεὺς καὶ μετὰ θάνατον Εὐρυπύλου ζῶντα παρεισάγει τὸν Μαχάονα ἰώμενον τὸν Φιλοκτήτην.

177. Vergilius, Aeneis, II, 259–64.

Illos patefactus ad auras
reddit equus, laetique cavo se robore promunt
Thessandrus Sthenelusque duces et dirus Ulixes,

demissum lapsi per funem, Acamasque Thoasque
Pelidesque Neoptolemus primusque Machaon
et Menelaus et ipse doli fabricator Epeos.

The opened horse restores them to the air, and there joyfully
come forth from the hollow wood Thessandrus and Sthenelus
the captains, and dread Ulysses, sliding down the lowered rope;
Acamas and Thoas and Neoptolemus son of Peleus, the leader
Machaon, Menelaus, and Epeus himself, who devised the fraud.[1]

[1] Cf. 177a. Hyginus, *Fabulae*, CVIII: . . . Epeus . . . equum . . . ligneum fecit,
eoque sunt collecti Menelaus, Ulixes, . . . Machaon. . . .

178. Hippocrates, Epistulae, 27 [IX, p. 426, 2-5 L.].

Οὕτως δὲ καὶ Ἀσκληπιοῦ παῖδες οὐ τέχνῃ μόνον, ἀλλὰ καὶ ὅπλοις
ἐπήρκεσαν Ἕλλησι· Μαχάων γέ τοι ψυχὴν κατέθετο ἐν τῇ
Τρωάδι, ὅτε, ὡς οἱ ταῦτα γράφοντες λέγουσι, ἐξ ἵππου ἐς πόλιν
τὴν Πριάμου εἰσῆλθε.

In this way too, then, the sons of Asclepius assisted the Greeks
not only by their skill but also by their arms. At any rate
Machaon lost his life in the Troad[1] when, as those who write
of this say, he climbed out of the Horse and entered the city of
Priam.

[1] Cf. T. 157.

179. Apollodorus, Epitoma, V, 1.

. . . Πενθεσίλεια . . . μάχης γενομένης πολλοὺς κτείνει, ἐν οἷς
καὶ Μαχάονα.

. . . Penthesileia . . . in the battle slew many, amongst them
also Machaon.

180. Quintus Smyrnaeus, Posthomerica, VI, 390-413.

390 Ὡς εἰπὼν κταμένοιο περικλυτὰ τεύχε' ἑλέσθαι
μήδετ' ἐπεσσύμενος· τοῦ δ' ἀντίος ἦλθε Μαχάων
χωόμενος Νιρῆος, ὅ οἱ σχεδὸν αἶσαν ἀνέτλη·
δουρὶ δέ μιν στονόεντι κατ' εὐρέος ἤλασεν ὤμου
δεξιτεροῦ, σύτο δ' αἷμα πολυσθενέος περ ἐόντος·
395 ἀλλ' οὐδ' ὣς ἀπόρουσεν ἀταρτηροῖο κυδοιμοῦ,

ἀλλ᾽, ὥς τίς τε λέων ἢ ἄγριος οὔρεσι κάπρος
μαίνετ᾽ ἐνὶ μέσσοισιν, ὅπως κ᾽ ἐπιόντα δαμάσσῃ,
ὅς ῥά μιν οὔτασε πρῶτος ὑποφθάμενος δι᾽ ὁμίλου·
τὰ φρονέων ἐπόρουσε Μαχάονι, καί ῥά μιν ὦκα
400 οὔτασεν ἐγχείῃ περιμήκεί τε στιβαρῇ τε
δεξιτερὸν κατὰ γλουτόν· ὁ δ᾽ οὐκ ἀνεχάζετ᾽ ὀπίσσω,
οὐδ᾽ ἐπιόντ᾽ ἀλέεινε, καὶ αἵματος ἐσσυμένοιο·
ἀλλ᾽ ἄρα καρπαλίμως περιμήκεα λᾶαν ἀείρας
κάββαλε κὰκ κεφαλῆς μεγαθύμου Τηλεφίδαο·
405 τοῦ δὲ κόρυς στονόεντα φόνον καὶ πῆμ᾽ ἀπάλαλκεν
ἐσσυμένως· ὁ δ᾽ ἔπειτα κραταιῷ χώσατο φωτὶ
Εὐρύπυλος μᾶλλον, μέγα δ᾽ ἀσχαλόων ἐνὶ θυμῷ
ὠκὺ διὰ στέρνοιο Μαχάονος ἤλασεν ἔγχος.
αἰχμὴ δ᾽ αἱματόεσσα μετάφρενον ἄχρις ἵκανεν·
410 ἤριπε δ᾽ ὡς ὅτε ταῦρος ὑπὸ γναθμοῖσι λέοντος·
ἀμφὶ δέ οἱ μελέεσσι μέγ᾽ ἔβραχεν αἰόλα τεύχη.
Εὐρύπυλος δέ οἱ αἶψα πολύστονον εἰρύσατ᾽ αἰχμὴν
ἐκ χροὸς οὐταμένοιο, καὶ εὐχόμενος μέγ᾽ ἀΰτει·

Thus speaking, he [sc., Eurypylus] planned swiftly to seize
the famed arms from the slain [sc., Nireus]. But against him
came Machaon, angered for Nireus who suffered doom at his
side. With death-dealing spear he drove against Eurypylus'
broad right shoulder. The blood gushed forth, mighty as he
was; but not so did he leap back from baneful combat; rather,
as in the mountains a lion or fierce boar rages in the midst of
the hunters, that he may kill the one who, hastening forward
first among the crowd, inflicted a wound upon him; in this
mood Eurypylus rushed against Machaon and instantly
wounded him in the right hip with his long stout lance. But
Machaon did not draw back nor did he avoid him as he advanced, even though the blood flowed fast. Swiftly heaving a
huge stone, he dashed it against the head of the boasting son
of Telephus. But the helmet straightway saved him from bale-
ful death and woe. Then Eurypylus grew even more enraged
at the mighty warrior and greatly wrathful in heart he launched
his swift spear through Machaon's breast. Through to his
midriff the bloody spearpoint reached. He fell, as when a bull
falls beneath a lion's jaws. And round his limbs his gleaming
armor clattered loudly. Whereupon Eurypylus plucked the

death-dealing spear from his wounded body, and boastfully cried . . .¹ (**)

¹ Cf. T. **160**; cf. **180a.** Ioannes Tzetzes, *Posthomerica*, 520-2: Κτεῖνε δὲ [*sc.*, ὁ Εὐρύπυλος] πολλοὺς Ἀργείων Ἀσκληπιάδην τε, ἤρω ἰητῆρα, Μαχάονα, κάλλιμον ἄνδρα, ὥς ῥα Κόϊντος ἔφη. ὁ δ᾽ ἄρ᾽ Ὀρφεὺς ἀλλ᾽ ἐπαείδει; cf. **180b.** Tzetzes, *ibid.*, 584: Τὸν δ᾽ [*sc.*, Μαχάονα] ὁ Κόϊντος ἔπεφνεν ὑπ᾽ Εὐρυπύλοιο βολῆσιν.

181. Quintus Smyrnaeus, Posthomerica, VI, 420-24.

Ἐσσὶ μὲν ἰητήρ, μάλα δ᾽ ἤπια φάρμακα οἶδας,
τοῖς πίσυνος τάχ᾽ ἔολπας ὑπεκφυγέειν κακὸν ἦμαρ.
ἀλλ᾽ οὐ μὰν οὐδ᾽ αὐτὸς ἀπ᾽ ἠνεμόεντος Ὀλύμπου
σεῖο πατὴρ τεὸν ἦτορ ἔτ᾽ ἐκ θανάτοιο σαώσει,
οὐδ᾽ εἴ τοι νέκταρ τε καὶ ἀμβροσίην καταχεύῃ.

You [*sc.*, Machaon] are a physician, you know soothing salves; relying on them perhaps you hoped to escape the fatal day. Yet not even your father himself from lofty Olympus shall any longer save your heart from death, not even if he pours nectar and ambrosia upon you.(**)

182. Quintus Smyrnaeus, Posthomerica, VI, 455-59.

Ὀψὲ δ᾽ ἀδελφειοῖο φόνον στονόεντα νόησε
βλημένου ἐν κονίῃ Ποδαλείριος, οὕνεκα νηυσὶν
ἧστο παρ᾽ ὠκυπόροισι τετυμμένα δούρασι φωτῶν
ἕλκε᾽ ἀκειόμενος. περὶ δ᾽ ἔντεα δύσατο πάντα
θυμὸν ἀδελφειοῖο χολούμενος· . . .

Late Podalirius learned of the wretched death of his brother, struck down in the dust, because he sat beside the swift ships, healing the wounds inflicted on men by spears. Angered in his heart for his brother's sake, he clothed himself in all his armor; . . . (**)

183. Quintus Smyrnaeus, Posthomerica, VII, 1-47.

Ἦμος δ᾽ οὐρανὸς ἄστρα κατέκρυφεν, ἔγρετο δ᾽ Ἠὼς
λαμπρὸν παμφανόωσα, κνέφας δ᾽ ἀνεχάσσατο νυκτός,
δὴ τότ᾽ ἀρήιοι υἷες ἐϋσθενέων Ἀργείων,
οἱ μὲν ἔβαν προπάροιθε νεῶν κρατερὴν ἐπὶ δῆριν
5   ἀντίον Εὐρυπύλοιο μεμαότες, οἱ δ᾽ ἀπάτερθεν
αὐτοῦ πὰρ νήεσσι Μαχάονα ταρχύσαντο

Νιρέα θ', ὃς μακάρεσσιν ἀειγενέεσσιν ἐῴκει
κάλλεΐ τ' ἀγλαΐῃ τε· βίη δ' οὐκ ἄλκιμος ἦεν·
οὐ γὰρ ἅμ' ἀνθρώποισι θεοὶ τελέουσιν ἅπαντα·
10   ἀλλ' ἐσθλῷ κακὸν ἄγχι παρίσταται ἔκ τινος αἴσης·
ὡς Νιρῆι ἄνακτι παρ' ἀγλαΐῃ ἐρατεινῇ
κεῖτ' ἀλαπαδνοσύνη· Δαναοὶ δέ οἱ οὐκ ἀμέλησαν,
ἀλλά ἑ ταρχύσαντο καὶ ὠδύραντ' ἐπὶ τύμβῳ,
ὅσσα Μαχάονα δῖον, ὃν ἀθανάτοισι θεοῖσιν
15   ἶσον ἀεὶ τίεσκον, ἐπεὶ πυκνὰ μήδεα ᾔδη·
αἶψα δ' ἄρ' ἀμφοτέροισι χυτὸν * περὶ σῆμα βάλοντο.
        Καὶ τότ' ἄρ' ἐν πεδίῳ ἔτι μαίνετο λοίγιος Ἄρης·
ὦρτο δ' ἄρ' ἀμφοτέρωθε μέγας κόναβος καὶ ἀϋτὴ
ῥηγνυμένων λάεσσι καὶ ἐγχείῃσι βοειῶν·
20   καί ῥ' οἱ μὲν πονέοντο πολυκμήτῳ ὑπ' Ἄρηι·
νωλεμέως δ' ἄρ' ἄπαστος ἐδητύος ἐν κονίῃσι
κεῖτο μέγα στενάχων Ποδαλείριος· οὐδ' ὅ γε σῆμα
λεῖπε κασιγνήτοιο· νόος δέ οἱ ὁρμαίνεσκε
χερσὶν ὑπὸ σφετέρῃσιν ἀνηλεγέως ἀπολέσθαι.
25   καί ῥ' ὁτὲ μὲν βάλε χεῖρας ἐπὶ ξίφος, ἄλλοτε δ' αὖτε
δίζετο φάρμακον αἰνόν· ἑοὶ δέ μιν εἶργον ἑταῖροι
πολλὰ παρηγορέοντες· ὁ δ' οὐκ ἀπέληγεν ἀνίης.
καί νύ κε θυμὸν ἑῇσιν ὑπαὶ παλάμῃσιν ὄλεσσεν
ἐσθλοῦ ἀδελφειοῖο νεοκμήτῳ ἐπὶ τύμβῳ,
30   εἰ μὴ Νηλέος υἱὸς ἐπέκλυεν, οὐδ' ἀμέλησεν
αἰνῶς τειρομένοιο· κίχεν δέ μιν ἄλλοτε μέν που
ἐκχύμενον περὶ σῆμα πολύστονον, ἄλλοτε δ' αὖτε
ἀμφὶ κάρη χεύοντα κόνιν καὶ στήθεα χερσὶ
θεινόμενον κρατερῇσι καὶ οὔνομα κικλήσκοντα
35   οἷο κασιγνήτοιο· περιστενάχοντο δ' ἄνακτα
δμῶες ὁμῶς ἑτάροισι· κακὴ δ' ἔχε πάντας ὀιζύς.
καί ῥ' ὅ γε μειλιχίοισι μέγ' ἀχνύμενον προσέειπεν·
" ἴσχεο λευγαλέοιο γόου καὶ πένθεος αἰνοῦ,
ὦ τέκος· οὐ γὰρ ἔοικε περίφρονα φῶτα γεγῶτα
40   μύρεσθ' οἷα γυναῖκα παρ' οὐκέτ' ἐόντι πεσόντα·
οὐ γὰρ ἀναστήσεις μιν ἔτ' ἐς φάος, οὕνεκ' ἄιστος
ψυχή οἱ πεπότηται ἐς ἠέρα, σῶμα δ' ἄνευθεν
πῦρ ὀλοὸν κατέδαψε καὶ ὀστέα δέξατο γαῖα·
αὔτως δ', ὡς ἀνέθηλε, καὶ ἔφθιτο. τέτλαθι δ' ἄλγος
45   ἄσπετον, ὥς περ ἔγωγε Μαχάονος οὔτι χερείω

παῖδ' ὀλέσας δηίοισιν ὑπ' ἀνδράσιν εὖ μὲν ἄκοντι
εὖ δὲ σαοφροσύνῃσι κεκασμένον. . . ."

* ἀμφοτέροισι χυτὸν coni. Zimmermann ἀμφοτέροις ταὐτὸν MSS ἀμφοτέροις αὐτὸν
Koechly Way.

When heaven hid the stars and dawn awoke, shining
brightly everywhere, and the darkness of night withdrew, then
some of the warlike sons of the valiant Argives marched from
the ships to mighty battle, pressing forward against Eurypylus;
some remained aloof from him, beside the ships, to bury
Machaon, and Nireus who resembled the eternal gods in beauty
and grace. But in strength he was weak, for the gods do not
grant all things at once to men. By some destiny evil always
attends good. So feebleness lay in the character of Nireus,
along with his lovely grace. Yet the Greeks did not neglect
him, but buried him and mourned for him at the tomb as much
as for godlike Machaon, whom they always regarded as the
equal of the deathless gods, for he knew skilful crafts. Swiftly,
then, for both of them they heaped up a mound. Meanwhile
in the plain deadly Ares still raged. From both sides arose a
mighty clamor and shouting as shields were shattered by stones
and spears. And while these men toiled in battle that claims
many spoils—all this time Podalirius lay in the dust, fasting
and groaning sorely, and he did not leave the tomb of his
brother. His heart longed to destroy himself recklessly by his
own hands. At times he put hands upon his sword; then again
he sought a deadly drug. But his comrades restrained him,
offering him consolation. Yet he did not give up his grief.
And he would have taken his life at his own hands over the
new-made tomb of his noble brother, had not Neleus' son heard
him and taken care of him in his grievous suffering. He found
him now stretched out beside the mournful tomb, now pouring
dust upon his head, beating his breast with mighty hands, and
calling on his brother's name. Around their lord slaves and
comrades mourned and bitter unhappiness claimed all. Then
with gentle words Nestor addressed the sorrowing brother:
"Put an end to these unhappy groans, this mournful lamen-
tation, my child. For it is not seemly for a prudent man to
weep like a woman, falling down beside one who no longer
lives. You will not bring him back again to the light once his
invisible soul has fled into the air, once the greedy fire has

consumed his body and earth has received his bones. As he grew, so he died. Bear then your unspeakable grief, even as I, when I lost at the hands of hostile men a son no less worthy than Machaon, excelling in war and in counsel. . . ." (**)

184. Quintus Smyrnaeus, Posthomerica, VII, 58-65.

"῏Ω πάτερ, ἄσχετον ἄλγος ἐμὸν καταδάμναται ἦτορ
ἀμφὶ κασιγνήτοιο περίφρονος, ὅς μ᾽ ἀτίταλλεν
οἰχομένοιο τοκῆος ἐς οὐρανὸν ὡς ἐὸν υἷα
σφῆσιν ἐν ἀγκοίνῃσι καὶ ἰητήρια νούσων
ἐκ θυμοῖο δίδαξε· μιῇ δ᾽ ἐνὶ δαιτὶ καὶ εὐνῇ
τερπόμεθα ξυνοῖσιν ἰαινόμενοι κτεάτεσσι·
τῷ μοι πένθος ἄλαστον ἐποίχεται· οὐδ᾽ ἔτι κείνου
τεθναότος φάος ἐσθλὸν ἐέλδομαι εἰσοράασθαι."

" O father, irrepressible grief masters my heart because of my prudent brother who, when our father departed to heaven, reared me as his son in his own arms, and gladly taught me the methods of healing diseases. We rejoiced in one table and one bed, taking delight in our common possessions. Pain that cannot be forgotten comes upon me; no longer do I desire to look upon the kindly light of day, now that he is dead " [Podalirius to Nestor].(**)

185. Quintus Smyrnaeus, Posthomerica, VII, 87-92.

". . . πέλει φάτις ἀνθρώποισιν
ἐσθλῶν μὲν νίσσεσθαι ἐς οὐρανὸν ἄφθιτον αἰεὶ
ψυχάς, ἀργαλέων δὲ ποτὶ ζόφον· ἔπλετο δ᾽ ἄμφω
σεῖο κασιγνήτῳ· καὶ μείλιχος ἔσκε βροτοῖσι,
καὶ πάϊς ἀθανάτοιο· θεῶν δ᾽ ἐς φῦλον ὀΐω
κεῖνον ἀνελθέμεναι σφετέρου πατρὸς ἐννεσίῃσιν."

". . . there is a saying among men that the souls of the good always go to imperishable heaven, while those of the wicked descend to the nether world. Both were true of your brother: he was gentle to men and the child of an immortal. To the company of the gods I believe he has risen through the counsels of your father " [Nestor to Podalirius].(**)

186. Pausanias, Descriptio Graeciae, III, 26, 9.

Ἐνταῦθα ἐν τῇ Γερηνίᾳ Μαχάονος τοῦ Ἀσκληπιοῦ μνῆμα καὶ
ἱερόν ἐστιν ἅγιον, καὶ ἀνθρώποις νόσων ἰάματα παρὰ τῷ Μαχάονι
ἔστιν εὑρέσθαι. καὶ Ῥόδον μὲν τὸ χωρίον τὸ ἱερὸν ὀνομάζουσιν,
ἄγαλμα δὲ τοῦ Μαχάονος χαλκοῦν ἐστιν ὀρθόν· ἐπίκειται δέ οἱ
τῇ κεφαλῇ στέφανος, ὃν οἱ Μεσσήνιοι κίφος καλοῦσι τῇ ἐπιχωρίῳ
φωνῇ. Μαχάονα δὲ ὑπὸ Εὐρυπύλου τοῦ Τηλέφου τελευτῆσαί
φησιν ὁ τὰ ἔπη ποιήσας τὴν μικρὰν Ἰλιάδα.

Here in Gerenia is a tomb of Machaon, son of Asclepius, and
a holy sanctuary.[1] With Machaon men may find cures for
diseases. They call the holy spot Rhodus; there is a standing
bronze statuette of Machaon; there is a crown on his head
which the Messenians in the local speech call "kiphos." The
author of the epic *The Little Iliad* says[2] that Machaon was
killed by Eurypylus,[3] son of Telephus. (*)

[1] Cf. **186a.** Pausanias, *ibid.*, IV, 3, 9: Γλαῦκος . . . καὶ Μαχάονι τῷ Ἀσκληπιοῦ πρῶτος
ἔθυσεν ἐν Γερηνίᾳ. . . .
[2] = Fr. 7 [Allen].
[3] Cf. **186b.** Hyginus, *Fabulae*, CXIII, 3: Nobilem quem quis occidit: . . . Eurypylus
Nireum, idem Machaonem; cf. T. 160; 607.

187. Pausanias, Descriptio Graeciae, II, 11, 6.

Ὕστερον δὲ Ἀλεξάνωρ ὁ Μαχάονος τοῦ Ἀσκληπιοῦ παραγενόμενος εἰς Σικυωνίαν ἐν Τιτάνῃ τὸ Ἀσκληπιεῖον ἐποίησε.

Afterwards Alexanor, the son of Machaon,[1] the son of Asclepius, came to Sicyonia and built the sanctuary of Asclepius at
Titane.[2]

[1] Cf. T. 188; 189; cf., however, T. 157.   [2] Cf. T. 749; 667.

188. Pausanias, Descriptio Graeciae, II, 23, 4.

Ἐξ ἀρχῆς δὲ ἱδρύσατο Σφῦρος τὸ ἱερόν, Μαχάονος μὲν υἱός,
ἀδελφὸς δὲ Ἀλεξάνορος τοῦ παρὰ Σικυωνίοις ἐν Τιτάνῃ τιμὰς
ἔχοντος.

The original founder of the sanctuary [at Argos] was
Sphyrus,[1] son of Machaon and brother of the Alexanor who
is honored among the Sicyonians in Titane.[2]

[1] Cf. T. 752.   [2] Cf. T. 749.

189. Pausanias, Descriptio Graeciae, II, 38, 6.

> . . . καὶ ἱερὸν τοῦ Πολεμοκράτους ἐστὶν ἐν ταύτῃ. ὁ δὲ Πολεμο-
> κράτης ἐστὶ καὶ οὗτος Μαχάονος υἱός, ἀδελφὸς δὲ Ἀλεξάνορος,
> καὶ ἰᾶται τοὺς ταύτῃ καὶ τιμὰς παρὰ τῶν προσοίκων ἔχει.

> . . . in which [sc., the village, Eua] there is a sanctuary of
> Polemocrates. This Polemocrates, too, is one of the sons of
> Machaon, and the brother of Alexanor;[1] he cures the people
> of the district, and receives honors from the neighbors.

[1] Cf. T. 187; 188.

190. Pausanias, Descriptio Graeciae, IV, 3, 2.[1]

> Οἳ δὲ καὶ μάλιστα ἤδη βεβαιοῦνται τὸν ἐς τοὺς Ἀσκληπιάδας
> λόγον, ἀποφαίνοντες ἐν Γερηνίᾳ Μαχάονος μνῆμα καὶ τὸ ἐν
> Φαραῖς τῶν Μαχάονος παίδων ἱερόν.

> But the surest warrant of their [sc., the Messenians'] account
> of the Asclepiads[1] is that they show the tomb of Machaon in
> Gerenia and the sanctuary of his sons at Pharae.[2] (*)

[1] Cf. T. 38.
[2] Cf. 190a. Pausanias, ibid., IV, 3, 10: . . . καὶ ἱερὸν τῷ Γοργάσῳ καὶ Νικομάχῳ τὸ ἐν
Φαραῖς ἐποίησεν [sc., ὁ Γλαῦκος].

191. Pausanias, Descriptio Graeciae, IV, 30, 3.

> Καὶ τάδε ἄλλα ἤκουσα ἐν Φαραῖς, Διοκλεῖ θυγατέρα . . .
> Ἀντίκλειαν γενέσθαι, τῆς δὲ Νικόμαχόν τε εἶναι καὶ Γόργασον,
> πατρὸς δὲ Μαχάονος τοῦ Ἀσκληπιοῦ· τούτους καταμεῖναί τε
> αὐτοῦ καὶ ὡς ὁ Διοκλῆς ἐτελεύτησε τὴν βασιλείαν ἐκδέξασθαι.
> διαμεμένηκε δὲ αὐτοῖς καὶ ἐς τόδε ἔτι νοσήματά τε καὶ τοὺς
> πεπηρωμένους τῶν ἀνθρώπων ἰᾶσθαι· καί σφισιν ἀντὶ τούτων
> θυσίας ἐς τὸ ἱερὸν καὶ ἀναθήματα ἄγουσιν.

> I heard also at Pharae that . . . a daughter Anticleia was born
> to Diocles, and that her children were Nicomachus[1] and Gor-
> gasus, their father being Machaon, the son of Asclepius. They
> remained at Pharae and succeeded to the kingdom on the death
> of Diocles. The power of healing diseases and curing the
> maimed has remained with them to this day, and in return
> for this, they bring sacrifices and votive offerings to their
> sanctuary.[2] (*)

[1] Cf. T. 192.　　　　　　　　　　　　　[2] Cf. T. 190a.

192. Diogenes Laertius, Vitae Philosophorum, V, 1.

> Ἀριστοτέλης Νικομάχου καὶ Φαιστίδος Σταγειρίτης. ὁ δὲ
> Νικόμαχος ἦν ἀπὸ Νικομάχου τοῦ Μαχάονος τοῦ Ἀσκληπιοῦ,
> καθά φησιν Ἕρμιππος ἐν τῷ Περὶ Ἀριστοτέλους·

Aristotle, son of Nicomachus and Phaestis, was a native of
Stagira. Nicomachus, as Hermippus [3rd c. B. C.] relates in
his book *On Aristotle*, traced his descent from Nicomachus,
the son of Machaon, who was the son of Asclepius.[1](*)

> [1] Cf. **192a.** Dionysius Halicarnasensis, *Epistula ad Ammaeum*, 5: Ἀριστοτέλης πατρὸς
> μὲν ἦν Νικομάχου τὸ γένος καὶ τὴν τέχνην ἀναφέροντος εἰς Μαχάονα τὸν Ἀσκληπιοῦ; cf.
> **192b.** Anonymus, *Vita Aristotelis*, p. 12, 24-6 [Westermann]: Ἀριστοτέλης, υἱὸς Νικο-
> μάχου καὶ Φαιστίδος (ὁ δὲ Νικόμαχος ἰατρὸς ἦν τοῦ τῶν Ἀσκληπιαδῶν γένους ἀπὸ Νικο-
> μάχου τοῦ Μαχάονος). . . .

193. Ps. Ammonius, Vita Aristotelis, p. 10, 6-11 [Westermann].

> Ἀμφότεροι δ᾿ οὗτοι, ὅ τε Νικόμαχος καὶ ἡ Φαιστίς, τὸ γένος
> εἶχον ἀπ᾿ Ἀσκληπιοῦ. μαρτυρεῖ δὲ τούτοις τὸ εἰς τὸν Ἀρισ-
> τοτέλην εἰρημένον ἐπίγραμμα τοῦτον ἔχον τὸν τρόπον·
>
> > Φαιστίδος ἦν μητρὸς καὶ Νικομάχου γενετῆρος
> > τῶν Ἀσκληπιαδῶν δῖος Ἀριστοτέλης.

Both of them, Nicomachus and Phaestis, traced their lineage
from Asclepius. Evidence for this is the epigram on Aristotle,
which runs as follows: "Of godlike Aristotle the mother was
Phaestis, the father Nicomachus, both Asclepiads."

194. Martialis, Epigrammata, II, 16, 5.

> Quid tibi cum medicis? dimitte Machaonas omnis.

What do you want with doctors? Dismiss all your Machaons.

195. Maximus Tyrius, Philosophumena, XXXV, 1 a-c.

> . . . τίνας ποτὲ Ὅμηρος ὀνομάζων χαίρει θεοῖς εἰκέλους . . .
> τίνας γὰρ ἀλλ᾿ ἢ τοὺς ἀρίστους . . . τί δέ, εἰ μὴ τῷ Διὶ εἴκαζεν
> αὐτούς, ἀλλ᾿ ἢ Μαχάονι τῷ ἰατρῷ, ἢ . . .

. . . whom is Homer pleased to greet as the equals of the
gods . . . whom but the best? . . . What then if he does
not liken them to Zeus but rather to Machaon, the physician,
or to . . .

196. Theophylactus, Historiae, II, 6, 5.

> Εἶτά οἱ ἐφέλκονται τὰ μὲν ἄλλα βέλη τὰ ἐμπεπηγότα τῷ σώματι καὶ παρεξάγουσιν, τὸ δ' ἀνὰ τὴν πλευρὰν ἐξελέσθαι οὐκ εἶχον· διεψιθύριζον γὰρ ὡς ἑαυτοὺς οἱ ταῦτα δὴ τὰ Χείρωνος καὶ Μαχάονος τεχνώμενοι μελετήματα, ὡς συνεκδημήσει αὐτῷ καὶ ἡ ψυχὴ τῇ τοῦ βέλους ἐξόδῳ.

They [*sc.*, the followers of Heraclius] then pull and draw out the other arrows implanted in his [*sc.*, the soldier's] body, but they dare not remove the arrow that lodged in his chest. For the whisper circulates among those who are skilled in these practices of Chiron and Machaon that a man's life would withdraw from him with the withdrawal of the arrow.

PODALIRIUS

197. Eustathius, Commentarii ad Homeri Iliadem, XIII, 830.

> Δόρυ δὲ μακρὸν πρὸς διαστολὴν τῶν μὴ τοιούτων, ὁποῖόν τι καὶ ἡ αἰγανέα. λειριόεις δὲ χροῦς ἐνταῦθα σκωπτικῶς εἴρηται ὁ ἐσκιατραφημένος καὶ ἀνθηρὸς διὰ τὸ ἐκ βάρους ἀγύμναστον τοῦ ἥρωος. λείριον γὰρ ἄνθος, ὅθεν καὶ φωνὴ λειριόεσσα, καὶ Ποδαλείριος, οὐ μόνον, ὡς καὶ ἀλλαχοῦ κεῖται, ὁ ἀνθηρόπους διὰ τὸ εὐτυχές, ἀλλὰ καὶ τὰ ἀνθοῦντα περιοδεύων· ἰατρὸς γὰρ ἦν ῥιζοτόμος τὰ πολλά, ὡς εἰκός.

Long spear: as distinguished from those which are not of this type, as for example the javelin. Lily-like color is here said mockingly of one who has been tenderly reared and is flowering because he is inexperienced in the violence of the hero. For the lily is a flower, whence also comes the expression "lily-like." Moreover Podalirius is the flower-footed (lily-footed) [1] not only, as is also stated elsewhere, because of his good fortune, but also because he diligently studied flowers. For as a physician he was for the most part an herb-gatherer, as was fitting.

[1] **197a.** Eustathius, *ibid.*, III, 150: . . . ἄνθους γὰρ εἶδος τὸ λείριον, ὅθεν καὶ Ποδαλείριος κύριον, οἱονεὶ ἀνθηρόπους; cf. T. 139.

198. Baebius Italicus, Ilias Latina, 350-51.

> . . . quem doctus ab arte paterna
> Paeoniis curat iuvenis Podalirius herbis.

. . . whom [sc., Menelaus] Podalirius, the youth, taught by his father's art, cured with Paeonian simples.

199. Quintus Smyrnaeus, Posthomerica, IV, 396-404.

Τάχα δέ σφι τετυμμένα τύμματα πάντα
ἠκέσατ' ἐνδυκέως Ποδαλείριος, οὕνεκ' ἄρ' αὐτὸς
πρῶτα μὲν ἐκμύζησεν, ἔπειτα δὲ χερσὶν ἑῇσι
ῥάψεν ἐπισταμένως, καθύπερθε δὲ φάρμακ' ἔθηκε
κεῖνα, τά οἱ τὸ πάροιθε πατὴρ ἐὸς ἐγγυάλιξε·
τοῖσι δ' ἄρ' ἐσσυμένως καὶ ἀναλθέα τύμματα φωτῶν
αὐτῆμαρ μορόεντος ὑπὲκ κακοῦ ἰαίνονται.
τῶν δ' ἄφαρ ἀμφὶ πρόσωπα καὶ εὐκομόωντα κάρηνα
τύμματ' ἀπαλθαίνοντο, κατηπιόωντο δ' ἀνῖαι.

Quickly Podalirius in friendly spirit healed all the wounds inflicted upon them [sc., Epeius and Acamas], for he first squeezed them out, then deftly stitched them together with his hands; then over them he spread those salves which his father once placed in his hands and by which even the unhealing wounds of men are quickly healed of their deadly evil on the very day. Straightway the wounds on their faces and on their long-haired heads were healed, while their pain was soothed. (**)

200. Quintus Smyrnaeus, Posthomerica, IV, 538-40.

Ἀντίθεον δὲ Θόαντα καὶ Εὐρύπυλον μενεχάρμην
ἠκέσατ' ἐσσυμένως Ποδαλείριος ἕλκεα πάντα,
ὅσσα περιδρύφθησαν ἀπὲκ δίφροιο πεσόντες.

Godlike Thoas and valiant Eurypylus Podalirius hastened to heal of all their wounds which they suffered when they fell from their chariot. (**)

201. Quintus Smyrnaeus, Posthomerica, IX, 461-66.

. . . τὸν δὲ στερεὸν καὶ ἄνουσον
ὠκύτερον ποίησε νοήματος αἰψηροῖο
ἶσος ἐπουρανίοις Ποδαλείριος, εὖ μὲν ὕπερθε
πάσσων φάρμακα πολλὰ καθ' ἕλκεος, εὖ δὲ κικλήσκων
οὔνομα πατρὸς ἑοῖο· θοῶς δ' ἰάχησαν Ἀχαιοὶ
πάντες κυδαίνοντες ὁμῶς Ἀσκληπιοῦ υἷα.

. . . swifter than swift thought Podalirius, the equal of the gods, made him [sc., Philoctetes] strong and sound,[1] to good effect spreading many salves over the wound, to good effect invoking his father's name. Presently the Achaeans raised a shout, all together praising the son of Asclepius. (**)

[1] Cf. T. 202; cf., however, T. 86; 92; 152; 175.

202. Apollodorus, Epitoma, V, 8.

Ὁ δὲ παραγενόμενος καὶ θεραπευθεὶς ὑπὸ Ποδαλειρίου Ἀλέξανδρον τοξεύει.

After he [sc., Philoctetes] had come [sc., to Troy] and had been cured by Podalirius,[1] he shot Alexander. (**)

[1] Cf. T. 201 and 201, n. 1.

203. Quintus Smyrnaeus, Posthomerica, VI, 458-93.

Περὶ δ' ἔντεα δύσατο πάντα
θυμὸν ἀδελφειοῖο χολούμενος· ἐν δέ οἱ ἀλκὴ
460　σμερδαλέον στέρνοισιν ἀέξετο μαιμώωντι
ἐς πόλεμον στονόεντα· μέλαν δέ οἱ ἔζεεν αἷμα
λάβρον ὑπὸ κραδίῃ· τάχα δ' ἔνθορε δυσμενέεσσι
χερσὶ θοῇσιν ἄκοντα ταννγλώχινα τινάσσων·
εἷλε δ' ἄρ' ἐσσυμένως Ἀγαμήστορος υἱέα δῖον
465　Κλεῖτον, ὃν ἠΰκομος Νύμφη τέκεν ἀμφὶ ῥεέθροις
Παρθενίου, ὅς τ' εἶσι διὰ χθονὸς ἠΰτ' ἔλαιον
πόντον ἐπ' Εὔξεινον προχέων καλλίρροον ὕδωρ.
ἄλλον δ' ἀμφὶ κασιγνήτῳ κτάνε δήιον ἄνδρα
Λᾶσσον, ὃν ἀντίθεος Προνόη τέκεν ἀμφὶ ῥεέθροις
470　Νυμφαίου ποταμοῖο μάλα σχεδὸν εὐρέος ἄντρου,
ἄντρου θηητοῖο, τὸ δὴ φάτις ἔμμεναι αὐτῶν
ἱρὸν Νυμφάων, ὁπόσαι περὶ μακρὰ νέμονται
οὔρεα Παφλαγόνων καὶ ὅσαι περὶ βοτρυόεσσαν
ναίουσ' Ἡράκλειαν· ἔοικε δὲ κεῖνο θεοῖσιν
475　ἄντρον, ἐπεί ῥα τέτυκται ἀπειρέσιον μὲν ἰδέσθαι
λάινεον, ψυχρὸν δὲ διὰ σπέος ἔρχεται ὕδωρ
κρυστάλλῳ ἀτάλαντον, ἐνὶ μυχάτοισι δὲ πάντῃ
λάινεοι κρητῆρες ἐπὶ στυφελῇσι πέτρῃσιν
αἰζηῶν ὡς χερσὶ τετυγμένοι ἰνδάλλονται·
480　ἀμφ' αὐτοῖσι δὲ Πᾶνες ὁμῶς Νύμφαι τ' ἐρατειναί,

ἱστοί τ' ἠλακάται τε, καὶ ἄλλ' ὅσα τεχνήεντα
ἔργα πέλει θνητοῖσι, τὰ καὶ περὶ θαῦμα βροτοῖσιν
εἴδεται ἐρχομένοισιν ἔσω ἱεροῖο μυχοῖο·
τῷ ἔνι δοιαὶ ἔνεισι καταιβασίαι τ' ἄνοδοί τε,
485   ἡ μὲν πρὸς βορέαο τετραμμένη ἠχήεντος
πνοιάς, ἡ δὲ νότοιο καταντίον ὑγρὸν ἀέντος,
τῇ θνητοὶ νίσσονται ὑπὸ σπέος εὐρὺ θεάων·
ἡ δ' ἑτέρη μακάρων πέλεται ὁδός, οὐδέ μιν ἄνδρες
ῥηιδίως πατέουσιν, ἐπεὶ χάος εὐρὺ τέτυκται
490   μέχρις ἐπ' Ἀϊδονῆος ὑπερθύμοιο βέρεθρον·
ἀλλὰ τὰ μὲν μακάρεσσι πέλει θέμις εἰσοράασθαι.
τῶνδ' αὖτ' ἀμφὶ Μαχάον' ἰδ' Ἀγλαΐης κλυτὸν υἷα
μαρναμένων ἑκάτερθεν ἀπέφθιτο πουλὺς ὅμιλος·

Embittered in heart by his brother's fate, he donned all his armor; in his breast his strength increased mightily, panting for the mournful fray; black blood surged furiously in his heart. Swiftly he leaped in, brandishing his javelin in ready hands against the enemies. Then quickly he slew Cleitus, the godlike son of Agamestor, whom the fair-haired Nymph bore by the streams of Parthenius who goes through the earth like an oily sea, emptying his fair-flowing waters into the Euxine. Then for his brother's sake he struck down another hostile warrior: Lassus, whom godlike Pronoë bore beside the streams of the river Nymphaeus, near a broad cave, a wondrous cave which, it is said, is the shrine of the Nymphs, all those who dwell in the tall mountains of the Paphlagonians and in the region of clustering Heracleia. That cave was like the work of the gods, for it was fashioned of stone, endless to the sight, and through the cave ran water cold and crystal clear. In inner recesses everywhere stone craters appeared on solid rocks as if devised by the hands of men. Around them were Pans as well as lovely Nymphs, looms, distaffs, and all the other artful contrivances men have, which are viewed by men with wonder when they enter the holy recess. Within the cave are two paths, one descending and the other ascending, one facing the blasts of roaring Boreas, the other toward the cold of hard-blowing Notos; by the latter mortals enter down into the broad cave of the goddesses. The first is the road of the immortals; not easily do mortals travel it, for a broad chasm has been constructed midway leading to the pit of overweening Hades, but it is lawful for the immortal gods to look upon these things. Thus, with such men doing battle around Machaon and the renowned son of Aglaia, on either side a great host perished. (**)

204. Quintus Smyrnaeus, Posthomerica, XII, 314-21.

Πρῶτος μὲν κατέβαινεν ἐς ἵππον κητώεντα
υἱὸς Ἀχιλλῆος, σὺν δὲ κρατερὸς Μενέλαος
. . . .
σὺν δ᾽ ἄρ᾽ ἐϋμμελίης Ποδαλείριος Εὐρύμαχός τε
. . . .

First Achilles's son climbed into the hollow horse, with him mighty Menelaus . . . and also well-armed Podalirius and Eurymachus . . .(**)

205. Strabo, Geographica, VI, 3, 9.

Δείκνυται δὲ τῆς Δαυνίας περὶ λόφον, ᾧ ὄνομα Δρίον, ἡρῷα, τὸ μὲν Κάλχαντος ἐπ᾽ ἄκρᾳ τῇ κορυφῇ (ἐναγίζουσι δ᾽ αὐτῷ μέλανα κριὸν οἱ μαντευόμενοι, ἐγκοιμώμενοι ἐν τῷ δέρματι), τὸ δὲ Ποδαλειρίου κάτω πρὸς τῇ ῥίζῃ, διέχον τῆς θαλάττης ὅσον σταδίους ἑκατόν. ῥεῖ δ᾽ ἐξ αὐτοῦ ποτάμιον πάνακες πρὸς τὰς τῶν θρεμμάτων νόσους.

In Daunia [i. e. Apulia], on a hill by the name of Drium, are to be seen two hero-temples: one, to Calchas, on the very summit (where those who consult the oracle sacrifice to him a black ram and sleep in the hide); and the other, to Podalirius,[1] down near the base of the hill, about one hundred stadia distant from the sea; and from it flows a stream which is a cure-all for diseases of animals. (*)

[1] Cf. T. **159**; **160**.

206. Scholia in Lycophronem, Ad Alexandram, 1050.

Εἰώθασιν οἱ Δαύνιοι . . . ἐν μηλωταῖς καθεύδειν ἐν τῷ τάφῳ τοῦ Ποδαλειρίου καὶ καθ᾽ ὕπνους λαμβάνειν χρησμοὺς ἐξ αὐτοῦ, εἰώθασι δὲ καὶ ἐν τῷ πλησίον ποταμῷ Ἀλθαίνῳ ἀπολούεσθαι καὶ αὐτοὶ καὶ τὰ θρέμματα αὐτῶν καὶ ἐπικαλεῖσθαι τὸν Ποδαλείριον καὶ ὑγιάζεσθαι, ὅθεν καὶ ὁ ποταμὸς ἔσχε τὴν ὀνομασίαν.

The Daunians were wont . . . to sleep in sheepskins on the tomb of Podalirius and in their dreams to receive prophecies from him. Both they and their flocks also were wont to bathe in the nearby river Althaenus and to invoke Podalirius and be made well. From this the river derived its name.

207. Scholia in Lycophronem, Ad Alexandram, 1047.

Οὕτω γοῦν ψεύδεται καὶ περὶ Ποδαλειρίου λέγων αὐτὸν ἐν Ἰταλίᾳ τελευτῆσαι. ὡς δὲ ὄπισθεν αὐτὸς ἔγραψα, ἀπὸ Τροίας εἰς Κολοφῶνα πεζοὶ ἦλθον Ἀμφίλοχος, Κάλχας, Λεοντεύς, Πολυποίτης καὶ οὗτος ὁ Ποδαλείριος. θάψαντες δὲ ἐκεῖ τὸν Κάλχαντα τὸν Θέστορος ὑπὸ τοῦ Μόψου ἀνῃρημένον ᾧ προείπομεν τρόπῳ Πολυποίτης καὶ Λεοντεὺς μετ᾽ ὀλίγον εἰς Ἑλλάδα διεσώθησαν, Ἀμφίλοχος δὲ κατά τινας συναιρεῖται τῷ Μόψῳ τούτῳ περὶ τῆς ἀρχῆς τῆς μαντικῆς μονομαχήσας, κατά τινας δὲ εἰς τὴν Ἑλλάδα ὑποστραφεὶς τὸ Ἀμφιλοχικὸν Ἄργος ἔκτισεν, ὁ δὲ Ποδαλείριος ὁμοίως εἰς Ἄργος ὑποστραφεὶς ἐμαντεύετο ἐν Δελφοῖς. ποῦ κατοικήσει· χρήσαντος δὲ τοῦ θεοῦ οἰκῆσαι πόλιν, οὗ περιέχοντος οὐρανοῦ οὐδὲν δεινὸν πείσεται, τῆς Καρικῆς χερονήσου τὸν πέριξ οὐρανὸν ὄρεσι περικυκλούμενον ᾤκησε.

Thus at any rate he [sc., Lycophron] gives a false report even concerning Podalirius when he says that he died in Italy. For, as I myself wrote in a previous section, Amphilochus, Calchas, Leonteus, Polypoetus and this same Podalirius went on foot from Troy to Colophon. Having buried there Calchas, son of Thestor, who was slain by Mopsus in the manner which I have described above, Polypoetus and Leonteus shortly reached Greece in safety; but Amphilochus, according to some authorities, was also killed after fighting in single combat with this same Mopsus over the leadership in the art of prophecy, while according to other accounts, having returned to Greece he founded the Amphilochian Argos. Podalirius, too, on returning to Argos, consulted the Oracle at Delphi concerning where he should dwell. And when the god advised him to live in a city where he would suffer no harm from the enveloping sky, he took up his abode in the Carian peninsula in the region which is encircled by mountains all round the horizon.[1]

---

[1] Cf. **207a.** ibid., 1048: Ὁ δὲ Ποδαλείριος κατὰ μὲν τὸν Λυκόφρονα ἐν Ἰταλίᾳ τελευτᾷ, κατ᾽ ἐμὲ δὲ τὴν Καρικὴν χερόνησον ᾤκησε.

208. Apollodorus, Epitoma, VI, 18.

Ποδαλείριος δὲ ἀφικόμενος εἰς Δελφοὺς ἐχρᾶτο ποῦ κατοικήσει·
χρησμοῦ δὲ δοθέντος, εἰς ἣν πόλιν τοῦ περιέχοντος οὐρανοῦ
πεσόντος οὐδὲν πείσεται, τῆς Καρικῆς Χερρονήσου τὸν πέριξ
οὐρανοῦ κυκλούμενον ὄρεσι τόπον κατῴκησεν.

Podalirius went to Delphi and inquired of the Oracle where he
should settle; and when an oracle was given that he should
settle in the city where, if the encompassing heaven were to
fall, he would suffer no harm, he settled in that place of the
Carian Chersonese [1] which is encircled by mountains all round
the horizon. (*)

[1] Cf. T. 156.

209. Stephanus Byzantius, Ethnica, *s. v.* Σύρνα.

Σύρνα, πόλις Καρίας. ἔκτισται δὲ ὑπὸ Ποδαλειρίου. ἐκπεσόντα
γὰρ αὐτὸν εἰς Καρίαν σωθῆναι ὑπό τινος αἰγοβοσκοῦ καὶ
ἀχθῆναι πρὸς Δαμαιθὸν Καρίας βασιλέα, οὗ τὴν θυγατέρα
Σύρναν πεσοῦσαν ἀπὸ τοῦ τέγους ὑπ’ αὐτοῦ θεραπευθῆναι. φασὶ
δὲ οὕτως ἀθυμοῦντος τοῦ Δαμαιθοῦ τὸν Ποδαλείριον ἀφ’ ἑκατέρου
τῶν βραχιόνων αἷμα ἀφελόντα σῶσαι τὴν παῖδα, τὸν δὲ θαυ-
μάσαντα συνοικῆσαι αὐτῷ τὴν παῖδα καὶ δοῦναι τὴν Χερρόνησον·
ἐν ᾗ δύο πόλεις κτίσαντα τὴν μὲν ἀπὸ γυναικὸς Σύρναν, τὴν δὲ
ἑτέραν ἀπὸ τοῦ σώσαντος αὐτὸν νομέως.

Syrna, a city in Caria. It was founded by Podalirius, for when
he was driven out of his course to Caria, he was rescued by a
certain goatherd and brought to Damaethus, the king of Caria,
whose daughter Syrna had fallen from the roof and was healed
by Podalirius. They relate that when the king was so dis-
heartened Podalirius saved his daughter by drawing blood
from each arm; and the king, in admiration, married his
daughter to Podalirius and gave him the peninsula, the Chersonese; [1] there he founded two cities, naming the one after his
wife, Syrna, the other after the herdsman who had saved him. [2]

[1] Cf. T. 156; 208.
[2] Cf. **209a.** Stephanus Byzantius, *Ethnica, s. v.* Βύβασσος: Βύβασσος, πόλις Καρίας,
ἀπὸ Βυβάσσου νομέως, περισώσαντος ἀπὸ θαλάσσης καὶ χειμῶνος εἰς Καρίαν ἐμπεσόντα
Ποδαλείριον.

210. Lucianus, Alexander, 39.

Τρίτῃ δὲ ἡμέρᾳ Ποδαλειρίου ἦν καὶ τῆς μητρὸς Ἀλεξάνδρου γάμος.

On the third day [sc., of the mysteries] there was the union of Podalirius and the mother of Alexander [sc., of Abonoteichus].

211. Martialis, Epigrammata, X, 56, 7.

Enterocelarum fertur Podalirius Hermes.

Hermes is said to be the Podalirius of hernias.

212. Photius, Bibliotheca, 176, p. 509 A.[1]

Περί τε τῶν ἐν Κῷ καὶ Κνίδῳ ἰατρῶν, ὡς Ἀσκληπιάδαι, καὶ ὡς ἐκ Σύρνου οἱ πρῶτοι ἀφίκοντο ἀπόγονοι Ποδαλειρίου.

[Theopompus tells also] about the physicians of Cos and Cnidus, that they were Asclepiads, and that from Syrnus came the first descendants of Podalirius.

[1] = Theopompus of Chius, Fr. 103 (14) [Jacoby].

213. Ioannes Tzetzes, Chiliades, VII, 944-58.

  Οὗτος ὁ Κῷος ἰατρός, ὁ μέγας Ἱπποκράτης,
945 πατρὸς μὲν ἦν Ἡρακλειδᾶ, μητρὸς δὲ Φαιναρέτης,
  τελῶν ἑπτακαιδέκατος Ἀσκληπιοῦ σπερμάτων.
  μετὰ γὰρ Τροίας ἅλωσιν ἐν τῇ περαίᾳ Ῥόδου
  ὁ Ποδαλείριος υἱὸς Ἀσκληπιοῦ ὑπάρχων
  Ἱππόλοχον ἐγέννησεν, οὗ Σώστρατος ἐξέφυ,
950 οὗ Δάρδανος, οὗ Κρίσαμις, οὗπερ Κλεομυττάδης,
  οὗπερ υἱὸς Θεόδωρος, τοῦ δὲ Σώστρατος ἄλλος,
  οὗπερ Σωστράτου Κρίσαμις ὁ δεύτερος ἐξέφυ.
  Κρισάμιδος Θεόδωρος δεύτερος πάλιν ἔφυ.
  ἐκ Θεοδώρου τούτου δὲ ὁ Σώστρατος ὁ τρίτος,
955 οὗ Νέβρος, οὗ Γνωσίδικος, ἐξ οὗπερ Ἱπποκράτης.
  τοῦ πρώτου Ἱπποκράτους δὲ υἱοῦ τοῦ Γνωσιδίκου
  παῖς ἦν Ἡρακλειδᾶς, οὗπερ καὶ Φαιναρέτης
  ὁ μέγας, ὁ καὶ δεύτερος, γέγονεν Ἱπποκράτης.

This Coan physician, the great Hippocrates, had as father Heracleides and as mother Phainarete, being seventeenth in

line of the descendants of Asclepius. For after the capture
of Troy, Podalirius, the son of Asclepius, while living on the
coast of Rhodes,[1] became the father of Hippolochus, whose son
was Sostratus; and his son was Dardanus, and his Crisamis,
and his Cleomyttades, and his son Theodorus, and his a second
Sostratus, and the son of this Sostratus was a second Crisamis.
The son of Crisamis again was a second Theodorus, and of
this Theodorus a third Sostratus, and his son was Nebrus,
and his Gnosidicus, and his Hippocrates. Then the child of the
first Hippocrates, son of Gnosidorus, was Heracleides, and of
him and Phainarete was born the second Hippocrates, the
great.

[1] Cf. T. 157.

214. Stephanus Byzantius, Ethnica, *s. v.* Κῶς.

Ἦν δὲ Ἱπποκράτης τῶν καλουμένων Νεβριδῶν· Νεβρὸς γὰρ
ἐγένετο ὁ διασημότατος τῶν Ἀσκληπιαδῶν, ᾧ καὶ ἡ Πυθία
ἐμαρτύρησεν. οὗ Γνωσίδικος, Γνωσιδίκου δὲ Ἱπποκράτης καὶ
Αἴνειος καὶ Ποδαλείριος, Ἱπποκράτους Ἡρακλείδης, οὗ Ἱπποκράτης ὁ ἐπιφανέστατος . . .

Hippocrates was one of the so-called Nebridae, for Nebrus was
the most eminent of the Asclepiads, to which even Pythia bore
witness. His son was Gnosidicus, and the sons of Gnosidicus
were Hippocrates, Aeneius, and Podalirius; Hippocrates' son
was Heracleides, whose son was the most distinguished
Hippocrates . . .

215. Hippocrates, Epistulae, 2 [IX, p. 314, 5 L.].

Γίνεται μὲν οὖν ὁ θεῖος Ἱπποκράτης, ἔνατος μὲν ἀπὸ Κρισάμιδος
τοῦ βασιλέως, ὀκτωκαιδέκατος δὲ ἀπὸ Ἀσκληπιοῦ . . .

The divine Hippocrates was born, then, ninth in descent from
Crisamis, the king, eighteenth in descent from Asclepius . . .

216. Soranus, Vita Hippocratis, 1.

Ἱπποκράτης γένει μὲν ἦν Κῷος . . . εἰς Ἡρακλέα καὶ Ἀσκληπιὸν
τὸ γένος ἀναφέρων, ἀφ' οὗ μὲν εἰκοστός, ἀφ' οὗ δὲ ἐννεακαιδέκατος.

Hippocrates was a Coan by birth, . . . who traced his ancestry back to Heracles and Asclepius, the twentieth in descent from the former, the nineteenth from the latter.[1]

[1] Cf. 216a. Hippocrates, *Epistulae*, 3 [IX, p. 316, 3 L.]: Ἱπποκράτους ἰητροῦ Κῴου ἀπὸ Ἀσκληπιοῦ γεγονότος; cf. 216b. Hippocrates, *Decretum Atheniensium* [IX, p. 400, 17 L.]: Ἱπποκράτης . . . γεγονὼς ἀπὸ Ἀσκληπιοῦ; cf. 216c. Hippocrates, *Epistulae*, 10 [IX, p. 324, 6 L.]: Σὺ γὰρ Ἀσκληπιῷ προσπέπλεξαι γένος καὶ τέχνην . . . ; cf. also T. 357.

ASCLEPIADS

217. Plato, Phaedrus, 270 C.

Εἰ μὲν Ἱπποκράτει γε τῷ τῶν Ἀσκληπιαδῶν δεῖ τι πείθεσθαι . . .

At least if Hippocrates, the Asclepiad,[1] is to be trusted . . .

[1] Cf. 217a. Plato, *Protagoras*, 311 B: . . . Ἱπποκράτη τὸν Κῷον, τὸν τῶν Ἀσκληπιαδῶν . . .

218. Asclepius, In Aristotelis Metaphysica, E 2, 1027 a 19.

. . . ὥς φησιν ὁ τῶν Ἀσκληπιαδῶν ἡγεμὼν . . .

. . . as the leader[1] of the Asclepiads says [*sc.*, Hippocrates] . . .

[1] Cf. 218a. Julianus, *Epistulae*, 50, 444 D: . . . καὶ τῶν Ἀσκληπιαδῶν ὁ ἄριστος Ἱπποκράτης ἔφη . . .

219. Theognis, Elegiae, 432-34.

Εἰ δ᾽ Ἀσκληπιάδαις τοῦτό γ᾽ ἔδωκε θεός,
ἰᾶσθαι κακότητα καὶ ἀτηρὰς φρένας ἀνδρῶν,
πολλοὺς ἂν μισθοὺς καὶ μεγάλους ἔφερον.

If god had granted this to the Asclepiads, to heal the evils and the hearts of men blinded by reckless impulses,[1] they would reap a great store of wealth.

[1] Cf. 219a. Athenaeus, *Deipnosophistae*, VI, 68, 256c: . . . καθάπερ τινὲς εἰρήκασιν, ὧν ἰατρεῦσαι τὴν ἄγνοιαν οὐδ᾽ Ἀσκληπιάδαις τοῦτό γε νομίζω δεδόσθαι.

220. Euripides, Alcestis, 965-71.

. . . .

κρεῖσσον οὐδὲν Ἀνάγκας
ηὗρον, οὐδέ τι φάρμακον
Θρῄσσαις ἐν σανίσιν, τὰς
Ὀρφεία κατέγραψεν
γῆρυς, οὐδ᾽ ὅσα Φοῖβος Ἀσκληπιάδαις ἔδωκε
φάρμακα πολυπόνοις ἀντιτεμὼν βροτοῖσιν.

... but naught have I found mightier than Necessity—no spell written down on Thracian tablets by the singer Orpheus, nor any among all the simples which Phoebus shredded to cure the ills of toil-fraught mortals and gave to the sons of Asclepius.

221. Ps. Galenus, Introductio, Cp. 2 [XIV, p. 676 K.].

Τοῦ δὲ εἰς σύστημα τέχνης ἀγαγεῖν, ὡς τὴν τῶν Ἀσκληπιαδῶν ἰατρικήν, ταύτης δὲ ἀρχὴ λόγος καὶ πεῖρα.

To create a systematic art such as the medicine of the Asclepiads,[1] the principles of reason and experience are needed.

[1] Cf. **221a.** Plinius, *Naturalis Historia*, VII, 49 (50), 160: Aesculapi ... secta; cf. **221b.** Simplicius, *In Aristotelis Categorias*, 10a 11: ... τῶν Ἀσκληπιαδῶν ἡ τέχνη.

222. Eustratius, Commentarium in Aristotelis Analytica Posteriora, B 11, 95 a 3.

Πολλὰ μέντοι τῶν κατὰ προαίρεσιν γένοιντ᾽ ἂν καὶ ἀπὸ τύχης, οἷον ὑγεία· αὕτη γὰρ γίνεται μὲν νοσοῦντί τινι καὶ ἀπὸ τέχνης ἰατρικῆς, διαίτης προσφερομένης καὶ φαρμάκων τῶν προσηκόντων, καὶ ὅσα ἄλλα τοῖς Ἀσκληπιάδαις ἐπιτετήδευται πρὸς τὴν τῆς σφετέρας τέχνης κατόρθωσιν.

Many of the things, however, that occur according to deliberate choice may also occur by chance, as for example, health. For this comes to an ailing person also from medical skill, applied diet, and proper remedies and whatever else is practiced by the Asclepiads for the successful accomplishment of their art.

223. Olympiodorus, Commentaria in Aristotelem, Prolegomena, p. 8, 21-24.

... καθάπερ οἱ τῶν Ἀσκληπιαδῶν παῖδες οὐ μόνον τὴν τῶν ὑγιεινῶν ἀλλὰ καὶ τὴν τῶν νοσερῶν παιδεύονται γνῶσιν ὑπὲρ τοῦ τὰ μὲν ἑλεῖν τὰ δὲ φυγεῖν, οὕτω καὶ ὁ φιλόσοφος ...

... just as the sons of the Asclepiads[1] not only are trained in the knowledge of health but also of sickness, with a view to acquiring the one and avoiding the other, so too the philosopher ...

[1] Cf. **223a.** Aelianus, *De Natura Animalium*, VII, 14: ... παῖδες Ἀσκληπιαδῶν ...

224. Julianus, In Matrem Deorum, 178 C.

Τὸ δὲ ὅτι μάλιστα μὲν πάσας τὰς νόσους, εἰ δὲ μή, ὅτι τὰς πλείστας καὶ μεγίστας ἐκ τῆς τοῦ πνεύματος εἶναι τροπῆς καὶ

παραφορᾶς συμβέβηκεν, οὐδεὶς ὅστις οἶμαι τῶν Ἀσκληπιαδῶν
οὐ φήσει.

For I think not one of the sons of Asclepius would deny that
all diseases, or at any rate very many and those most serious,
are caused by the disturbance and disorder of the breathing. (*)

225. Sextus Empiricus, Hypotyposeis, III, 225.

Κυνείων τε γεύσασθαι δοκοῦμεν ἡμεῖς ἀνίερον εἶναι, Θρᾳκῶν δὲ
ἔνιοι κυνοφαγεῖν ἱστοροῦνται. ἴσως δὲ καὶ παρ' Ἕλλησι τοῦτο
ἦν σύνηθες· διόπερ καὶ Διοκλῆς ἀπὸ τῶν κατὰ τοὺς Ἀσκληπιάδας ὁρμώμενος τισὶ τῶν πασχόντων σκυλάκεια δίδοσθαι
κελεύει κρέα.

We think it is unholy to eat dog's flesh, but some of the
Thracians are reported to eat dog's flesh. Moreover, this was
probably customary even among the Greeks; hence Diocles
[physician, 4th c. B. C.],[1] starting from the presuppositions of
the Asclepiads, recommended that the flesh of puppies be given
to some of the patients.

[1] = Fr. 93 [Wellmann].

226. Choricius Gazaeus, III, 16.

Καθάπερ οὖν τῶν Ἀσκληπιαδῶν οἱ τεχνῖται φοιτῶσι πολλάκις
παρὰ τὸν κάμνοντα . . .

Just as the skilled workmen of the sect of the Asclepiads[1]
visit the sick many times . . .

[1] Cf. **226a.** Idem, VII, 30: . . . καὶ τῶν Ἀσκληπιαδῶν δεηθῆναι πυνθανομένων ὑμῶν . . .

227. Ps. Empedocles, Fr. 156 [Diels].

Παυσανίην ἰητρὸν ἐπώνυμον, Ἀγχίτεω υἱόν,
      φῶτ' Ἀσκληπιάδην πατρὶς ἔθρεψε Γέλα,
ὃς πολλοὺς μογεροῖσι μαραινομένους καμάτοισιν
      φῶτας ἀπέστρεψεν Φερσεφόνης ἀδύτων.

The physician Pausanias, rightly so named [Paus-anias =
he who ends pain], son of Anchitus, descendant of Asclepius,
was born and bred at Gela. Many a wight pining in fell
torments did he bring back from Persephone's inmost shrine.

228. Ioannes Tzetzes, Chiliades, XII, 637-39.

Ἀσκληπιάδαι λέγονται κατὰ κυρίαν λέξιν,
ὅσοι γονὴν ἐσχήκασιν Ἀσκληπιοῦ σπερμάτων.
νῦν δὲ κατὰ κατάχρησιν τοὺς ἰατροὺς εἰρήκειν.

In proper terminology Asclepiads are called whoever have had their descent from the offspring of Asclepius. But now by misuse the term has come to mean the physicians.

229. Galenus, De Anatomicis Administrationibus, II, 1 [II, pp. 281-2 K.].

Ἐκπεσοῦσα τοίνυν ἔξω τοῦ γένους τῶν Ἀσκληπιαδῶν ἡ τέχνη, κἄπειτα διαδοχαῖς πολλαῖς ἀεὶ χείρων γιγνομένη, τῶν διαφυλαξόντων αὐτῆς τὴν θεωρίαν ὑπομνημάτων ἐδεήθησαν. ἔμπροσθεν δ᾽ οὐ μόνον ἐγχειρήσεων ἀνατομικῶν, ἀλλ᾽ οὐδὲ συγγραμμάτων ἐδεῖτο τοιούτων . . .

Therefore, when the art of medicine passed outside the family of the Asclepiads and then became ever inferior in its successive stages, they stood in need of textbooks that would safeguard its theory. Of old, the art of medicine had no need for anatomical studies, nor even for books of such a kind . . .

230. Galenus, De Compositione Medicamentorum Secundum Locos, IX, 4 [XIII, p. 273 K.].

. . . δηλοῖ καὶ τοὺς Ἀσκληπιάδας οὕτως ὀνομάζειν. ἐκ Τρίκκης γὰρ τὸ γένος αὐτῶν ἐστιν, ὡς καὶ ὁ ποιητής φησιν. *

* φησιν Bussemaker ἐστιν Kühn.

. . . he [sc., Philo of Tarsus, time unknown] makes clear that even the Asclepiads named [honey] in this way. For their family is from Tricca, as also the poet [sc., Homer] says.[1]

[1] Cf. T. 10-13.

231. Suidas, Lexicon, s. v. Δημοκήδης.

Δημοκήδης, Καλλιφῶντος, ἱερέως ἐν Κνίδῳ γενομένου Ἀσκληπιοῦ . . .

Democedes [physician, 6th c. B. C.], the son of Calliphon, who was a priest of Asclepius in Cnidus . . .[1]

[1] Cf. however, Herodotus, Historiae, III, 129 ff.

III. DEIFICATION AND DIVINE NATURE

DIVINE MYTH

232. Theodoretus, Graecarum Affectionum Curatio, III, 24-28.

Χρόνῳ δὲ ὕστερον τοὺς εὖ τι δεδρακότας, ἢ ἐν πολέμοις ἀνδραγα-
θισαμένους, ἢ γεωργίας τινὸς ἄρξαντας, ἢ σώμασί τισι θεραπείαν
προσενηνοχότας, ἐθεοποίησάν τε καὶ νεὼς τούτοις ἐδείμαντο. καὶ
γὰρ δὴ καὶ τὸν Κρόνον ἄνθρωπον εἶναι Σαγχωνιάθων ἔφησε . . .
καὶ Ἡρακλέα δέ, ὡς γενναῖόν τε καὶ ἀνδρεῖον, ἐθεοποίησαν
Ἕλληνες· καὶ τὸν Ἀσκληπιόν, ὡς τῆς ἰατρικῆς ἐπιστήμης
εὑρετὴν γεγενημένον, θεὸν τελευτήσαντα προσηγόρευσαν. διὰ
δὲ τὴν αὐτὴν αἰτίαν καὶ τὸν Ἀπιν Αἰγύπτιοι θείας προσηγορίας
ἠξίωσαν. καίτοι φασὶν Ἕλληνες τὸν μὲν Ἡρακλέα διὰ τὴν
ξυμβᾶσαν ἐκ τῆς Δηιανείρας ἐπιβουλὴν ἑαυτὸν καταπρῆσαι,
καὶ τοῦτον τὸν τρόπον καταλῦσαι τὸν βίον· τὸν δὲ Ἀσκληπιόν,
ἄνθρωπον ὄντα, καὶ πολλοὺς ἀνθρώπους διὰ τῆς ἰατρικῆς ἐπιστήμης παμπόλλων ἀπαλλάξαντα παθημάτων, κεραυνῷ βληθῆναι
ὑπὸ τοῦ Διὸς καὶ διαφθαρῆναι. καὶ ταῦτα πρὸς ἑτέροις πολλοῖς ὁ Σικελιώτης Διόδωρος ἐν τῷ τετάρτῳ τῶν Βιβλιοθηκῶν
ξυνέγραψεν. ἀλλ᾽ ὅμως οὐδὲν ἧττον Ἕλληνες καὶ ταῦτα μεμα-
θηκότες καὶ τὸν Ἡρακλέα καὶ τὸν Ἀσκληπιὸν θεοὺς ὀνομάζουσι . . .

Later, they not only deified those who had accomplished something well, or acted bravely in battle, or begun any cultivation
of the land, or administered a cure to any bodies, but also they
built temples to them.[1] Indeed Sanchoniathon [1st century
A. D.] said that even Cronus was a mortal . . . and the
Greeks deified Heracles, too, because he was noble and valiant;
also Asclepius, because he had been the inventor of the medical
art,[2] they addressed as a god when he had died. For the same
reason the Egyptians deemed Apis deserving of divine appellation. And yet the Greeks say that Heracles burned himself to
death in consequence of the treachery of Dejaneira and in this
fashion lost his life, while Asclepius, a mortal, who relieved
many men of countless sufferings through his medical art, was
struck with a thunderbolt by Zeus, and perished.[3] And these
facts along with many others Diodorus, the Sicilian, relates in

108

the fourth book of his *Libraries*.[4] And yet, even though the
Greeks have learned these stories, none the less they call
Heracles and Asclepius gods . . .

[1] Cf. T. 245; 350.   [3] Cf. T. 105 ff.
[2] Cf. T. 349.   [4] Cf. T. 4.

233. Tertullianus, Ad Nationes, II, 14.

Hunc vos de pyra in caelum sublevastis, illa facilitate qua et
alium ignis ex divini vi confectum, qui pauca experientiae
ingenia commentus dicebatur mortuos ad vitam recurasse. is
Apollinis filius, tam homo quam Jovis nepos, Saturni pronepos
(vel potius spurius ut incerto patre, ut Argivus Socrates de-
tulit, quippe expositum repertum, turpius Jove educatum,
canino scilicet ubere), merito, id quod nemo negare potest,
fulmine haustus est. malus Juppiter Optimus hic rursus est,
impius in nepotem, invidus in artificem.

This one [*sc.*, Heracles] you raised from the pyre to the sky,
with the same facility with which [you raised] also another
one who was destroyed by the divine violence of fire. He,
having devised some clever experiments, was said to have
restored the dead to life by his cures.[1] He was the son of
Apollo, both human and the grandson of Jupiter, and the
greatgrandson of Saturn (or rather he was a bastard, inasmuch
as it was uncertain who his father was, and according to
Socrates of Argos,[2] he was exposed, then found and nurtured
even more meanly than Jupiter, namely suckled at the dugs of a
dog) ;[3] deservedly, as nobody can deny, he perished by a stroke
of lightning. Bad, again, is this most excellent Jupiter, without
feeling for his grandson, jealous of the skilled practitioner.[4] (**)

[1] Cf. T. 66 ff.   [3] Cf. T. 9.
[2] Cf. T. 9a ; 35; 52.   [4] Cf. T. 102.

234. Arnobius, Adversus Nationes, VII, 44.[1]

Si esset nobis animus scrupulosius ista tractare, vobis ipsis
obtineremus auctoribus, minime illum fuisse divum, qui conceptus et natus muliebri alvo esset, qui annorum gradibus ad
eum finem ascendisset aetatis in quo illum vis fulminis, vestris
quemadmodum litteris continetur, et vita expulisset et lumine.
sed quaestione ab ista discedimus, Coronidis filius sit, ut vultis,

ex immortalium numero et perpetua praeditus sublimitate caelesti.

If we were disposed to be very scrupulous in dealing with your assertions, we might prove by your own authority that he was by no means divine who had been conceived and born from a woman's womb, who had by yearly stages reached that term of life at which, as is related in your books, a thunderbolt drove him at once from life and light. But we leave this question: let the son of Coronis be, as you wish, one of the immortals and possessed of the everlasting blessedness of heaven.

[1] Cf. T. 852.

235. Tatianus, Adversus Graecos, 21, 1.

Οὐ γὰρ μωραίνομεν, ἄνδρες Ἕλληνες, οὐδὲ λήρους ἀπαγγέλλομεν, Θεὸν ἐν ἀνθρώπου μορφῇ γεγονέναι καταγγέλλοντες . . . τέθνηκεν ὑμῶν ὁ Ἀσκληπιός . . .

We do not act as fools, O Greeks, nor utter idle tales, when we announce that God was born in the form of a man . . . your Asclepius died . . .

DEIFICATION

236. Minucius Felix, Octavius, XXIII, 7.

. . . Castores alternis moriuntur ut vivant, Aesculapius ut in deum surgat fulminatur, Hercules . . .

. . . Castor and Pollux die alternately in order to live, Asclepius is struck by lightning in order to be raised into a god,[1] Heracles . . . (*)

[1] Cf. **236a.** Cyprianus, *Quod Idola Dii non sint*, 2: . . . Aesculapius ut in deum surgat fulminatur . . .

237. Hyginus, Fabulae, CCLI, 2.

Qui licentia Parcarum ab inferis redierunt
. . . Asclepius Apollinis et Coronidis filius . . .

Those who returned from the underworld
with the permission of the Fates
. . . Asclepius, son of Apollo and Coronis . . .

238. Hyginus, Fabulae, CCXXIV, 5.

> Qui facti sunt ex mortalibus immortales
>
> . . . Asclepius Apollinis et Coronidis filius.

> Those who from mortals changed into immortals
>
> . . . Asclepius, son of Apollo and Coronis.

239. Cicero, De Natura Deorum, II, 24, 62.[1]

> Suscepit autem vita hominum consuetudoque communis ut beneficiis excellentis viros in caelum fama ac voluntate tollerent. hinc Hercules hinc Castor et Pollux hinc Aesculapius hinc Liber etiam . . .

> Human ways, moreover, and general custom have made it a practice to raise into heaven through renown and gratitude men, who are distinguished by their benefits. This is the origin of Heracles, of Castor and Pollux, of Asclepius and also of Liber . . . (**)

[1] Quoted by Lactantius, *Divinae Institutiones*, I, 15, 5.

240. Cicero, De Legibus, II, 8, 19.

> Divos et eos, qui caelestes semper habiti, colunto et ollos, quos endo caelo merita locaverint, Herculem, Liberum, Aesculapium, Castorem, Pollucem, Quirinum . . .

> They shall worship as gods both those who have always been regarded as dwellers in heaven and also those whose merits have placed them in heaven,[1] Heracles, Liber, Asclepius, Castor, Pollux, Quirinus . . . (*)

[1] Cf. also T. 367.

241. Porphyrius, Epistula ad Marcellam, 7.

> Ἀκούεις δὲ καὶ τὸν Ἡρακλέα τούς τε Διοσκούρους καὶ τὸν Ἀσκληπιὸν τούς τε ἄλλους ὅσοι θεῶν παῖδες ἐγένοντο ὡς διὰ τῶν πόνων καὶ τῆς καρτερίας τὴν μακαρίαν εἰς θεοὺς ὁδὸν ἐξετέλεσαν. οὐ γὰρ ἐκ τῶν δι᾽ ἡδονῆς βεβιωκότων ἀνθρώπων αἱ εἰς θεὸν ἀναδρομαί, ἀλλ᾽ ἐκ τῶν τὰ μέγιστα τῶν συμβαινόντων γενναίως διενεγκεῖν μεμαθηκότων.

You hear of Heracles and the Dioscuri, of Asclepius and the others, who were children of gods, how they completed the blessed road to the gods through toil and strength. For not by men who have lived for pleasure are the upward paths to god attained, but by those who have learned nobly to endure the hardest circumstances.

242. Origenes, In Jeremiam Homilia, V, 3.

Ὁμολογοῦσι γὰρ καὐτοὶ περί τινων, ὅτι πρότερον θνητοὶ ἦσαν καὶ ἀπεθεώθησαν. Ἡρακλέα προσκυνοῦσιν οὐχ ὡς γεγενημένον θεόν, ἀλλὰ ὡς ἐξ ἀνθρώπου εἰς θεὸν μεταβληθέντα· Ἀσκληπιὸν προσκυνοῦσιν ὡς ἐξ ἀνθρώπων δι᾽ ἀρετὴν εἰς θεὸν μεταβεβηκότα.

For they themselves agree about some that they formerly were mortals and were deified. They make obeisance to Heracles not as to one who was born a god, but as to one who changed from a mortal to a god. To Asclepius they make obeisance as to one who changed from mortal to God through his virtue.

243. Xenophon, Cynegeticus, I, 6.[1]

. . . Ἀσκληπιὸς δὲ μειζόνων ἔτυχεν, ἀνιστάναι μὲν τεθνεῶτας, νοσοῦντας δὲ ἰᾶσθαι· διὰ δὲ ταῦτα θεὸς ὢν παρ᾽ ἀνθρώποις ἀείμνηστον κλέος ἔχει.

. . . Asclepius won yet greater preferment, to raise the dead and to heal the sick; and for these things being a god he has everlasting fame among men. (**)

[1] Cf. T. **56.**

244. Celsus, De Medicina, Prooemium, 2.[1]

Ut pote cum vetustissimus auctor Aesculapius celebretur, qui quoniam adhuc rudem et vulgarem hanc scientiam paulo subtilius excoluit, in deorum numerum receptus est.

Hence Asclepius, since he is celebrated as its most ancient founder and because he cultivated this science [sc., medicine], as yet rude and vulgar, with a little more exactness, was numbered among the gods. (**)

[1] Cf. T. **354.**

245. Galenus, Protrepticus, 9, 22.

Ἀσκληπιός γέ τοι καὶ Διόνυσος, εἴτ᾽ ἄνθρωποι πρότερον ἤστην
εἴτ᾽ ἀρχῆθεν θεοί, τιμῶν ἀξιοῦνται μεγίστων, ὁ μὲν διὰ τὴν
ἰατρικήν, ὁ δ᾽ ὅτι τὴν περὶ τὰς ἀμπέλους ἡμᾶς τέχνην ἐδίδαξεν.

Asclepius at least, and also Dionysus, whether they were men
formerly or whether they were gods from the outset,[1] are
deemed deserving of the highest honors,[2] the one by reason of
his medical art, the other because he taught us the art of the
vine.

[1] Cf. T. 255.   [2] Cf. T. 232; 350.

246. Athenagoras, Pro Christianis, 30, 1-2.

. . . τί θαυμαστὸν τοὺς μὲν ἐπὶ ἀρχῇ καὶ τυραννίδι . . . κληθῆναι
θεούς . . . τοὺς δ᾽ ἐπ᾽ ἰσχύϊ ὡς Ἡρακλέα καὶ Περσέα, τοὺς δὲ
ἐπὶ τέχνῃ, ὡς Ἀσκληπιόν;

. . . what wonder if some should be called gods . . . on the
ground of their rule and sovereignty . . . and others for their
strength, as Heracles and Perseus; and others for their art,
as Asclepius?

247. Georgius Hamartolus, Chronicon, I, 54.

Οὐκοῦν πρῶτοι θεοὺς ὠνόμασαν Αἰγύπτιοι καὶ Φοίνικες, ἀφ᾽
ὧν Ἕλληνες παρειληφότες ἥλιον καὶ σελήνην, ἀστέρας τε καὶ
οὐρανὸν καὶ γῆν καὶ τἆλλα στοιχεῖα προσκυνήσαντες, ὡς
θεοὺς ἐλάτρευσαν. χρόνῳ δ᾽ ὕστερον καὶ τοὺς εὖ τι δεδρακότας
ἢ ἐν πολέμοις ἀνδραγαθήσαντας ὡς τὸν Ἡρακλέα γενναῖον καὶ
ἀνδρεῖον ἐθεοποίησαν Ἕλληνες, καὶ τὸν Ἀσκληπιὸν διὰ τὴν
ἰατρικὴν ἐπιστήμην θεὸν τελευτήσαντα προσηγόρευσαν.

The Egyptians and Phoenicians, then, were the first to name
gods the sun and moon, the stars and heaven and earth and the
other elements, all of which the Greeks, following them, respected and served as gods. Later, however, those who had
accomplished something well or acted nobly and bravely in
battle, such as the noble and valiant Heracles, the Greeks
numbered among the gods too; and also Asclepius because of
his medical knowledge they addressed as a god after his death.

248. Augustinus, De Civitate Dei, IV, 27.

> Quae sunt autem illa, quae prolata in multitudinem nocent?
> "Haec," inquit, "non esse deos Herculem, Aesculapium,
> Castorem, Pollucem; proditur enim a doctis, quod homines
> fuerint et humana condicione defecerint."

> But what are those things which do harm when brought before
> the multitude? "These," he [*sc.*, Mucius Scaevola, 2nd-1st c.
> B. C.] says, "that Heracles, Asclepius, Castor, and Pollux,
> are not gods; for it is shown by learned men [1] that these were
> men, and perished according to their mortal state." [2](*)

[1] Cf. T. 262.   [2] Cf. *e. g.* T. 251.

249. Origenes, Contra Celsum, III, 22.

> Οὐδὲν δὲ εἶδος τοῦ περὶ ἡμῶν διασυρμοῦ καὶ καταγέλωτος κατα-
> λιπὼν ὁ βωμολόχος Κέλσος, ἐν τῷ καθ᾽ ἡμῶν λόγῳ Διοσκούρους
> καὶ Ἡρακλέα καὶ Ἀσκληπιὸν καὶ Διόνυσον ὀνομάζει, τοὺς ἐξ
> ἀνθρώπων πεπιστευμένους παρ᾽ Ἕλλησι γεγονέναι θεούς, καί
> φησιν ʽ οὐκ ἀνέχεσθαι μὲν ἡμᾶς τούτους νομίζειν θεούς, ὅτι
> ἄνθρωποι ἦσαν καὶ πρῶτοι, καίτοι πολλὰ ἐπιδειξαμένους καὶ
> γενναῖα ὑπὲρ ἀνθρώπων . . .ʼ

> But this low jester Celsus, omitting no species of mockery
> and ridicule which can be employed against us, in his treatise
> attacking us mentions the Dioscuri, and Heracles, and Ascle-
> pius, and Dionysus, who are believed by the Greeks to have
> become gods after having been men, and says that "we cannot
> bear to call such beings gods, because they were at first men,
> even though they manifested many noble qualities, which
> were displayed for the benefit of mankind. . . ." (*)

250. Lactantius, Divinae Institutiones, I, 19, 3-4.

> Clamat summus poeta eos omnes qui inventas vitam excoluere
> per artes aput inferos esse ipsumque illum repertorem medi-
> cinae talis et artis ad Stygias undas fulmine detrusum, ut
> intellegamus quantum valeat pater omnipotens, qui etiam deos
> fulminibus extinguat. sed homines ingeniosi hanc secum habe-
> bant fortasse rationem: quia deus fulminari non potest, apparet
> non esse factum. immo vero quia factum est, apparet hominem
> fuisse, non deum.

The greatest of poets [*sc.*, Vergil] exclaims,[1] that all those who refined life by the invention of arts are in the lower regions, and that even the discoverer himself of such a medicine and art[2] was thrust down by lightning to the Stygian waves,[3] that we may understand how great is the power of the Almighty Father, who can extinguish even gods by His lightnings. But ingenious men perchance thus reasoned with themselves: because a god cannot be struck with lightning, it is manifest that the occurrence never took place.[4] Nay, rather, because it did take place, it is manifest that the person in question was a man, not a god. (*)

[1] *Aeneid,* VI, 663.
[3] Cf. T. 349.
[2] T. 77.
[4] Cf. T. 99; 104.

251. Lactantius, Divinae Institutiones, I, 15, 26.

Testatus est videlicet Attici conscientiam, ex ipsis mysteriis intellegi posse quod omnes illi homines fuerint qui coluntur, et cum de Hercule Libero Aesculapio Castore Polluce incunctanter fateretur, de Apolline ac Jove patribus eorum, item de Neptuno Vulcano Marte Mercurio, quos maiorum gentium deos appellavit, timuit aperte confiteri.

He [*sc.*, Cicero] invoked, as it is plain, Atticus' own awareness,[1] that it could be understood from the very mysteries that all those who are worshipped were mortals; and while he acknowledged this without hesitation in the case of Heracles, Liber, Asclepius, Castor, and Pollux, he was afraid openly to make the same admission respecting Apollo and Jupiter their fathers, and likewise respecting Neptune, Vulcan, Mars, and Mercury, whom he termed the greater gods. (*)

[1] *Tusculanae Disputationes,* I, 13, 29; cf. T. 253.

252. Cicero, De Natura Deorum, III, 18, 45.

Quid Apollinem Volcanum Mercurium ceteros deos esse dices, de Hercule Aesculapio Libero Castore Polluce dubitabis? at hi quidem coluntur aeque atque illi, apud quosdam etiam multo magis. ergo hi dei sunt habendi mortalibus nati matribus.

Again, will you call Apollo, Vulcan, Mercury and the rest gods, and will you yet have doubts about Heracles, Asclepius,

Liber, Castor and Pollux? But these are worshipped just as much as those, and indeed in some places very much more than they. Therefore these must be considered gods born of mortal mothers. (**)

253. Augustinus, De Civitate Dei, VIII, 5.

> . . . quae Alexander Macedo scribit ad matrem sibi a magno antistite sacrorum Aegyptiorum quodam Leone patefacta, ubi non Picus et Faunus et Aeneas et Romulus vel etiam Hercules et Aesculapius et Liber Semela natus et Tyndaridae fratres et si quos alios ex mortalibus pro diis habent, sed ipsi etiam maiorum gentium dii, quos Cicero in Tusculanis tacitis nominibus videtur attingere, Juppiter, Juno, Saturnus, Vulcanus, Vesta et alii plurimi, quos Varro conatur ad mundi partes sive elementa transferre, homines fuisse produntur.

> [In this line is also] . . . what Alexander the Macedonian[1] writes to his mother as revealed to him by Leo, a high priest of the Egyptian rites, where not only Picus and Faunus are shown to have been mortals, and Aeneas and Romulus, or even Heracles and Asclepius and Liber, born of Semele, and the twin sons of Tyndareus, or any other mortals who have been deified, but even the greater gods themselves, to whom Cicero in his *Tusculan Disputations*[2] alludes without mentioning their names, Jupiter, Juno, Saturn, Vulcan, Vesta, and many others whom Varro attempts to identify with the parts or the elements of the world. (*)

[1] Cf. Plutarch, *Alexander*, 27, 8.   [2] I, 13, 29, cf. T. **251.**

254. Apuleius, De Deo Socratis, XV, 153.

> Quippe tantum eos deos appellant, qui ex eodem numero iuste ac prudenter curriculo vitae gubernato pro numine postea ab hominibus praediti fanis et caerimoniis vulgo advertuntur, ut in Boeotia Amphiaraus, in Africa Mopsus, in Aegypto Osiris, alius alibi gentium, Aesculapius ubique.

> Naturally of this same group [*sc.*, of daemons] they deem gods only those who, having guided the chariot of their lives wisely and justly, and having been endowed afterward by men as divinities with shrines and religious ceremonies, are com-

monly worshipped as Amphiaraus in Boeotia, Mopsus in
Africa, Osiris in Egypt, one in one part of the world and
another in another part, Asclepius everywhere.[1]

[1] Cf. T. 282, 21.

255. Pausanias, Descriptio Graeciae, II, 26, 10.

Θεὸν δὲ Ἀσκληπιὸν νομισθέντα ἐξ ἀρχῆς καὶ οὐκ ἀνὰ χρόνον
λαβόντα τὴν φήμην τεκμηρίοις καὶ ἄλλοις εὑρίσκω καὶ Ὁμήρου
μαρτυρεῖ μοι τὰ περὶ Μαχάονος ὑπὸ Ἀγαμέμνονος εἰρημένα

Ταλθύβι᾿, ὅττι τάχιστα Μαχάονα δεῦρο κάλεσσον
φῶτ᾿ Ἀσκληπιοῦ υἱόν,

ὡς ἂν εἰ λέγοι θεοῦ παῖδα ἄνθρωπον.

That Asclepius was considered a god from the first, and did
not receive the title only in course of time,[1] I infer from several
other testimonies, and also what Agamemnon says about
Machaon in Homer substantiates this for me: " Talthybius,
make haste to call hither the man Machaon, son of Asclepius," [2]
as though he said the " human son of a god." (**)

[1] Cf. T. 245.
[2] Cf. T. 164, v. 194. Cf. however T. 257.

256. Eustathius, Commentarii ad Homeri Iliadem, XI, 517-18.[1]

Καινὸν δ᾿ οὐδὲν θεογενῆ, ὡς ἂν Σοφοκλῆς εἴποι, νομισθῆναι τὸν
Μαχάονα, καὶ πρὸ αὐτοῦ τὸν πατέρα τὸν Ἀσκληπιόν· τὸ γὰρ
τῆς ἰατρικῆς χρήσιμον τοιούτους νοεῖσθαι ὑπέβαλε τοὺς ἰητῆρας
. . . ὡς δὲ ὁ μῦθος θεοῖς ἐντάττει τὸν Ἀσκληπιὸν . . . ἀλλαχοῦ
δεδήλωται.

It is nothing new to believe that Machaon and before him
his father Asclepius, as Sophocles would say, were of divine
race. For the usefulness of medicine was responsible for the
fact that the physicians were considered to be such [i. e.
divine]. . . . That the legend numbers Asclepius among the
gods . . . has been expounded elsewhere.[2]

[1] Cf. T. 171.   [2] Cf. T. 30; 272a; 139.

DIVINE RANK

257. Theodoretus, Graecarum Affectionum Curatio, VIII, 23.[1]

Ὅτι δὲ ἐπὶ τῶν Ὁμήρου χρόνων οὐδέπω οὗτος τῆς θεοποιίας
ἐτετυχήκει, μάρτυς ὁ ποιητής, οὐ τὸν Ἀσκληπιὸν ἀλλὰ τὸν
Παιήονα δείξας τὰ τοῦ Ἄρεως θεραπεύσαντα τραύματα. δεῖ γὰρ
τουτουσὶ τοὺς θεοὺς καὶ τραύματα ἔχειν καὶ ἰατρούς. καὶ τὸν
Μαχάονα δὲ οὐ θεοῦ λέγει υἱὸν οὐδέ γε ἡμιθέου, ἀλλ᾿ ἰατροῦ·
ἔφη γάρ· " φῶτ᾿ Ἀσκληπιοῦ υἱὸν ἀμύμονος ἰητῆρος."

But that at the time of Homer, Asclepius had not yet been
deified, the poet testifies in showing that Paeon, not Asclepius,
healed the wounds of Ares.[2] For gods of this kind must have
wounds and physicians. And he [sc., Homer] calls Machaon
the son neither of a god, nor of a demigod, but of a physician,
for he says: " [Call] the man, [the] son of Asclepius, the
blameless physician." [3]

[1] T. 5.   [2] *Iliad*, V, 899.
[3] Cf. T. 164; cf. 257a. Eustathius, *Commentarii ad Homeri Iliadem*, XI, 518: . . . περὶ
δὲ τοῦ ἀμύμονός φασιν οἱ παλαιοὶ ὡς τὸ τοιοῦτον ἐπίθετον θνητοῖς ὁ ποιητὴς ἐπιλέγει.

258. Dionysius Halicarnasensis, Antiquitates Romanae, VII, 72, 13.

Τελευταῖα δὲ πάντων αἱ τῶν θεῶν εἰκόνες ἐπόμπευον . . . οὐ μόνον
Διὸς καὶ Ἥρας καὶ Ἀθηνᾶς καὶ Ποσειδῶνος καὶ τῶν ἄλλων,
οὓς Ἕλληνες ἐν τοῖς δώδεκα θεοῖς καταριθμοῦσιν, ἀλλὰ καὶ τῶν
προγενεστέρων, ἐξ ὧν οἱ δώδεκα θεοὶ μυθολογοῦνται γενέσθαι,
Κρόνου καὶ Ῥέας καὶ Θέμιδος καὶ Λητοῦς καὶ Μοιρῶν καὶ
Μνημοσύνης . . . καὶ τῶν ὕστερον, ἀφ᾿ οὗ τὴν ἀρχὴν Ζεὺς
παρέλαβε, μυθολογουμένων γενέσθαι, Περσεφόνης Εἰλειθυίας
Νυμφῶν Μουσῶν Ὡρῶν Χαρίτων Διονύσου, καὶ ὅσων ἡμιθέων
γενομένων αἱ ψυχαὶ τὰ θνητὰ ἀπολιποῦσαι σώματα εἰς οὐρανὸν
ἀνελθεῖν λέγονται, καὶ τιμὰς λαχεῖν ὁμοίας θεοῖς, Ἡρακλέους
Ἀσκληπιοῦ Διοσκούρων Σελήνης Πανὸς ἄλλων μυρίων.

Last of all the statues of the gods came in procession . . . not
only those of Zeus and Hera and Athena and Poseidon and
the others whom the Greeks number among the twelve gods,
but also those of the more ancient deities, from whom the
twelve gods were sprung, according to legend, Cronos and
Rhea and Themis and Leto and the Fates and Mnemosyne . . .

and the statues of those fabled to have been born later, after Zeus seized the power, Persephone, Eleithyia, the Nymphs, Muses, Horae, Graces, Dionysus, and statues of all the demigods whose souls are said to have left the mortal bodies, to have risen to heaven, and to have acquired honors similar to the gods, namely Heracles, Asclepius, the Dioscuri, Selene, Pan, and countless others.

259. Artemidorus, Onirocritica, II, 34.

> Τῶν θεῶν οἱ μέν εἰσι νοητοὶ οἱ δὲ αἰσθητοί· νοητοὶ μὲν οἱ πλείους, αἰσθητοὶ δὲ ὀλίγοι . . . φαμὲν δὲ τῶν θεῶν τοὺς μὲν Ὀλυμπίους, . . . τοὺς δὲ οὐρανίους, τοὺς δὲ ἐπιγείους . . . τῶν δὲ ἐπιγείων αἰσθητοὶ μὲν Ἑκάτη καὶ Πὰν καὶ Ἐφιάλτης καὶ Ἀσκληπιός (οὗτος δὲ καὶ νοητὸς ἅμα λέγεται) . . . τούτων δὲ τῶν εἰρημένων θεῶν τοῖς μὲν μέγα δυναμένοις . . . οἱ Ὀλύμπιοι συμφέρουσιν ὁρώμενοι, τοῖς δὲ μετρίοις οἱ οὐράνιοι, τοῖς δὲ πένησιν οἱ ἐπίγειοι.

Some of the gods are intelligible, others perceptible: intelligible are most of them; perceptible only a few . . . Of the gods we call some Olympians, . . . some celestial, some terrestrial. . . . Of the terrestrial gods, Hecate, Pan, Ephialtes, and Asclepius can be perceived by the senses (but the latter is also considered intelligible at the same time). . . . Of the aforementioned gods the Olympians are helpful to the great . . . when appearing to them, the celestials to the middle classes, the terrestrials to the poor.

260. Maximus Tyrius, Philosophumena, IX, 7, a–i.

> Ἀλλ' οὐχὶ δαιμόνων πᾶς πάντα δρᾷ, ἀλλ' αὐτοῖς διακέκριται κἀκεῖ τὰ ἔργα, ἄλλο ἄλλῳ. καὶ τοῦτό ἐστιν ἀμέλει τὸ ἐμπαθές, ᾧ ἐλαττοῦται δαίμων θεοῦ. ὡς γὰρ εἶχον φύσεως, ὅτε περὶ γῆν ἦσαν, οὐκ ἐθέλουσιν ταύτης παντάπασιν ἀπαλλάττεσθαι· ἀλλὰ καὶ Ἀσκληπιὸς ἰᾶται νῦν, καὶ ὁ Ἡρακλῆς ἰσχυρίζεται, καὶ Διόνυσος βακχεύει, καὶ Ἀμφίλοχος μαντεύεται, καὶ οἱ Διόσκουροι ναυτίλλονται, καὶ Μίνως δικάζει, καὶ Ἀχιλλεὺς ὁπλίζεται . . . ἐγὼ δὲ τὸν μὲν Ἀχιλλέα οὐκ εἶδον, οὐδὲ τὸν Ἕκτορα εἶδον. εἶδον δὲ καὶ Διοσκούρους . . . εἶδον καὶ τὸν Ἀσκληπιόν, ἀλλ' οὐχὶ ὄναρ . . .

But not every one of the demigods does everything, for their duties are allotted to them even there, one to one demigod, another to another.[1] And it is doubtless through the possession of emotion that the demigod is inferior to the god. For what their natural characteristics were when they dwelt on earth, these they do not desire wholly to relinquish. But Asclepius practices his art of medicine now, and Heracles displays his strength, and Dionysus revels, and Amphilochus prophesies, and the Dioscuri sail, and Minos judges, and Achilles is a warrior . . . I, however, did not see Achilles, nor Hector,—I did see the Dioscuri . . . and I did see also Asclepius, and not in a dream . . .

[1] Cf. T. 290-290 f; cf. also T. 351.

261. Origenes, Contra Celsum, V, 2.

Πρόκειται οὖν νῦν τὴν οὕτως ἔχουσαν αὐτοῦ ἀνατρέψαι λέξιν· " Θεὸς μέν, ὦ Ἰουδαῖοι καὶ Χριστιανοί, καὶ θεοῦ παῖς οὐδεὶς οὔτε κατῆλθεν οὔτε κατέλθοι. εἰ δέ τινας ἀγγέλους φατέ, τίνας τούτους λέγετε, θεοὺς ἢ ἄλλο τι γένος; ἄλλο τι, ὡς εἰκός, τοὺς δαίμονας." . . . καθολικῶς ἀποφηνάμενος θεὸν οὐδένα πρὸς ἀνθρώπους κατεληλυθέναι, ἢ θεοῦ παῖδα, ἀναιρεῖ καὶ τὰ ὑπὸ τῶν πολλῶν περὶ Θεοῦ ἐπιφανείας δοξαζόμενα, καὶ ὑπ' αὐτοῦ ἐν τοῖς ἀνωτέρω προειρημένα. εἰ γὰρ καθόλου λελεγμένον τὸ θεὸς καὶ θεοῦ παῖς οὐδεὶς κατῆλθεν οὐδὲ κατέλθοι, ἀληθῶς τῷ Κέλσῳ εἴρηται· δηλονότι ἀνήρηται τὸ εἶναι θεοὺς ἐπὶ γῆς, κατελθόντας ἐξ οὐρανοῦ, ἵνα ἤτοι μαντεύσωνται τοῖς ἀνθρώποις, ἢ διὰ χρησμῶν θεραπεύσωσι· καὶ οὔτε ὁ Πύθιος, οὔτ' Ἀσκληπιός, οὔτ' ἄλλος τις τῶν νενομισμένων τὰ τοιαῦτα ποιεῖν, θεὸς ἂν εἴη καταβὰς ἐξ οὐρανοῦ· ἢ θεὸς μὲν εἴη ἄν, ἀεὶ δὲ λαχὼν οἰκεῖν τὴν γῆν, καὶ ὡσπερεὶ φυγὰς τοῦ τόπου τῶν θεῶν· ἤ τις τῶν μὴ ἐξουσίαν ἐχόντων κοινωνεῖν τοῖς ἐκεῖ θείοις εἴη ἄν· ἢ οὐ θεοὶ εἶεν ὁ Ἀπόλλων καὶ ὁ Ἀσκληπιός, καὶ ὅσοι ἐπὶ γῆς τι ποιεῖν πεπιστευμένοι εἰσίν, ἀλλά τινες δαίμονες τῶν ἐν ἀνθρώποις σοφῶν καὶ ἐπὶ τὴν ἀψῖδα τοῦ οὐρανοῦ διὰ τὴν ἀρετὴν ἀναβαινόντων πολλῷ χείρους.

It is our task now to refute his [sc., Celsus'] saying which runs as follows: " Neither a god, O Jews and Christians, nor the son of a god ever descended or would descend. But if you say that some of the angels did, who do you think they are,

gods or rather some other kind of beings? Another kind, of
course, namely daemons." . . . [Celsus] by denying in general
that any god or son of a god ever descended to live among
men, confutes both that which is held true by the many in
regard to the epiphany of God, and that which has been said
previously by himself. For if the general statement made by
Celsus is correct that no god or son of a god ever descended
or would descend, this obviously confutes the contention that
there are gods on earth who have descended from heaven in
order to forewarn men or to heal them through oracles. And
neither Pythius [Apollo], nor Asclepius, nor any other one of
those who are believed to do such things would be a god,
descended from heaven: either he would be a god, yet one
who has drawn the lot for ever to live on earth and as a fugitive, as it were, from the realm of the gods; or he would be
one of those who are not permitted to participate in the divine
that exists there; or they would be no gods, Apollo and Asclepius and whoever are believed to accomplish something on
earth; rather are they some daemons, by far inferior to the
wise among men who through their virtue have ascended to
the vault of heaven.

262.   Cicero, De Natura Deorum, III, 15, 39.

> . . . iam vero in Graecia multos habent ex hominibus deos
> . . . Herculem Aesculapium Tyndaridas Romulum nostrum
> aliosque compluris, quos quasi novos et adscripticios cives in
> caelum receptos putant. haec igitur indocti . . .

> . . . nay, even in Greece they worship a number of deified
> human beings . . . Heracles, Asclepius, the sons of Tyn-
> dareus, our Romulus, and many others whom they regard as
> newly enrolled citizens in heaven,[1] as it were. That is what
> even the uneducated[2] say . . . (**)

[1] Cf. **262a.** Justinus, *Apologia*, 25, 1: . . . οἱ παλαιοὶ σεβόμενοι Διόνυσον . . . ἢ Ἀσκληπιὸν
ἤ τινα τῶν ἄλλων ὀνομαζομένων θεῶν . . .
[2] Cf. T. 248.

263.   Arnobius, Adversus Nationes, III, 39.

> Sunt praeterea nonnulli, qui ex hominibus divos factos hac
> praedicant appellatione signari, ut est Hercules Romulus
> Aesculapius Liber Aeneas.

There are some, besides, who assert that those who having been men became gods, are denoted by this name [*sc., novensiles, i. e.* newcomers]—as Heracles, Romulus, Asclepius, Liber, Aeneas.

264. Lucianus, Juppiter Tragoedus, 21.

Καί μοι ἐνταῦθα, ὦ Ζεῦ—μόνοι γάρ ἐσμεν καὶ οὐδεὶς ἄνθρωπος πάρεστι τῷ συλλόγῳ, ἔξω Ἡρακλέους καὶ Διονύσου καὶ Γανυμήδους καὶ Ἀσκληπιοῦ, τῶν παρεγγράπτων τούτων—ἀπόκριναι μετ᾽ ἀληθείας . . .

And now, Zeus, as we are alone and there is no mortal in our gathering except Heracles and Dionysus and Ganymede and Asclepius, these naturalized aliens, answer me truly . . . (*)

265. Lucianus, Dialogi Deorum, 13, 1-2.

ΔΙΟΣ, ΑΣΚΛΗΠΙΟΤ ΚΑΙ ΗΡΑΚΛΕΟΤΣ

ΖΕΤΣ. Παύσασθε, ὦ Ἀσκληπιὲ καὶ Ἡράκλεις, ἐρίζοντες πρὸς ἀλλήλους ὥσπερ ἄνθρωποι· ἀπρεπῆ γὰρ ταῦτα καὶ ἀλλότρια τοῦ συμποσίου τῶν θεῶν.

ΗΡΑ. Ἀλλὰ ἐθέλεις, ὦ Ζεῦ, τουτονὶ τὸν φαρμακέα προκατακλίνεσθαί μου;

ΑΣΚ. Νὴ Δία· καὶ ἀμείνων γάρ εἰμι.

ΗΡΑ. Κατὰ τί, ὦ ἐμβρόντητε; ἢ διότι σε ὁ Ζεὺς ἐκεραύνωσεν ἃ μὴ θέμις ποιοῦντα, νῦν δὲ κατ᾽ ἔλεον αὖθις ἀθανασίας μετείληφας;

ΑΣΚ. Ἐπιλέλησαι γὰρ καὶ σύ, ὦ Ἡράκλεις, ἐν τῇ Οἴτῃ καταφλεγείς, ὅτι μοι ὀνειδίζεις τὸ πῦρ;

ΗΡΑ. Οὔκουν, ἴσα καὶ ὅμοια βεβίωται ἡμῖν, ὃς Διὸς μὲν υἱός εἰμι, τοσαῦτα δὲ πεπόνηκα ἐκκαθαίρων τὸν βίον, θηρία καταγωνιζόμενος καὶ ἀνθρώπους ὑβριστὰς τιμωρούμενος· σὺ δὲ ῥιζοτόμος εἶ καὶ ἀγύρτης, νοσοῦσι μὲν ἴσως ἀνθρώποις χρήσιμος ἐπιθήσειν τῶν φαρμάκων, ἀνδρῶδες δὲ οὐδὲν ἐπιδεδειγμένος.

ΑΣΚ. Εὖ λέγεις, ὅτι σου τὰ ἐγκαύματα ἰασάμην, ὅτε πρῴην ἀνῆλθες ἡμίφλεκτος ὑπ᾽ ἀμφοῖν διεφθαρμένος τὸ σῶμα, καὶ τοῦ χιτῶνος καὶ μετὰ τοῦτο τοῦ πυρός· ἐγὼ δὲ εἰ καὶ μηδὲν ἄλλο, οὔτε ἐδούλευσα ὥσπερ σὺ οὔτε ἔξαινον ἔρια ἐν Λυδίᾳ πορφυρίδα ἐνδεδυκὼς καὶ παιόμενος ὑπὸ τῆς Ὀμφάλης χρυσῷ σανδάλῳ, ἀλλὰ οὐδὲ μελαγχολήσας ἀπέκτεινα τὰ τέκνα καὶ τὴν γυναῖκα.

ΗΡΑ. Εἰ μὴ παύσῃ λοιδορούμενός μοι, αὐτίκα μάλα εἴσῃ ὡς

οὐ πολύ σε ὀνήσει ἡ ἀθανασία, ἐπεὶ ἀράμενός σε ῥίψω ἐπὶ
κεφαλὴν ἐκ τοῦ οὐρανοῦ, ὥστε μηδὲ τὸν Παιῶνα ἰάσασθαί σε
τὸ κρανίον συντριβέντα.

ΖΕΥΣ. Παύσασθε, φημί, καὶ μὴ ἐπιταράττετε ἡμῖν τὴν ξυνου-
σίαν, ἢ ἀμφοτέρους ἀποπέμψομαι ὑμᾶς τοῦ ξυμποσίου. καίτοι
εὔγνωμον, ὦ Ἡράκλεις, προκατακλίνεσθαί σου τὸν Ἀσκληπιὸν
ἅτε καὶ πρότερον ἀποθανόντα.

JUPITER, ASCLEPIUS, and HERACLES

JUP.   Heracles and Asclepius, for shame, stop that quarrelling
with one another, like mortals; for that is most improper and
ill becomes the banquet of the gods.

HER.   But would you, Jupiter, permit this quack to lie down
before me?

ASC.   By Jupiter, since I am indeed your superior.

HER.   In what, you thunderstruck? because Jupiter struck
you with his thunderbolt, for doing what you ought not to
have done, while now again you are participating in immortality out of sheer pity?

ASC.   So you have forgotten that you, too, were burnt to
death, on Oeta, or you would not reproach me with the fire.

HER.   By no means were our lives on a level, I who am the
son of Jove but toiled so much to purify life, subduing monsters
and taking vengeance on insolent men; while you are nothing
but a simpler and a mountebank, useful, may be, for sick men
to provide them with drugs, but never having shown any
bravery.

ASC.   That comes well from you whose burns I healed when
lately you came up, your body half consumed by the flames of
both, the garment and the pyre in addition. Anyhow, if nothing else, I was neither a slave as you were, nor did I comb
wool in Lydia, clad in a purple garment and being beaten by
Omphale with a golden slipper, nor did I kill my children and
my wife in a fit of melancholy.

HER.   If you don't stop insulting me you shall soon find out
that immortality is of little use for you, for I will lift you up
and throw you headlong out of heaven, so that not even Paeon
shall heal your smashed skull.

JUP.   Stop it, I say, and do not disturb our gathering, or I
shall banish you both from the banquet. Anyway, it is reason-

able, Heracles, that Asclepius lie down before you, since he also died before you.

THE GOD'S NAME

266. Plutarchus, Decem Oratorum Vitae, VIII, 845 B.

Ὤμνυε δὲ καὶ τὸν Ἀσκληπιόν, προπαροξύνων Ἀσκλήπιον· καὶ παρεδείκνυεν αὐτὸν ὀρθῶς λέγοντα· εἶναι γὰρ τὸν θεὸν ἤπιον· καὶ ἐπὶ τούτῳ πολλάκις ἐθορυβήθη.

He [sc., Demosthenes] used also to swear by Asklépios, putting the acute on the antepenult; and he offered a proof that he was right, for he said that the god was "mild" [épios].[1] For this also he often provoked a clamor from the audience. (*)

[1] Cf. **266a.** Eustathius, *Commentarii ad Homeri Odysseam*, II, 319: Τὸν δὲ παραλόγως ὀξύνεσθαι δοκοῦντα Ἀσκληπιὸν ὁ Δημοσθένης ἐθεράπευσεν, Ἀσκλήπιον προπαροξυτόνως τολμήσας προφέρειν αὐτὸν ὡς καὶ Πλούταρχος ἱστορεῖ; cf. **266b.** Herodianus, *De Prosodia Catholica*, V, p. 123, 1 [Lentz], same text as Eustathius; cf. also T. **272.**

267. Eustathius, Commentarii ad Homeri Odysseam, II, 319.

. . . καὶ τὸ ἤπιος ὀξυνθὲν ἐν τῷ Ἀσκληπιός· ἐκ τοῦ ἠπίου γὰρ ἔγνωσται συγκεῖσθαι ὁ Ἀσκληπιός.

. . . and the word *epios* is accented on the last syllable in the word Asklepiós, for Asklepios is known to be made up from *epios*.[1]

[1] Concerning the Latin equivalent of the name Asklepios cf. **267a.** Priscianus, *Institutiones Grammaticae*, I, 37: Saepe interponitur [sc., littera u] inter cl vel cm in Graecis nominibus, ut Ἡρακλῆς Hercules, Ἀσκληπιός Aesculapius; **267b.** *ibid.*, I, 51: Ponitur [sc., littera ae] pro e longa, ut scaena pro σκηνή, et pro a, ut Aesculapius pro Ἀσκληπιός.

268. Cornutus, Theologiae Graecae Compendium, Cp. 33.[1]

Ὠνομάσθη δὲ ὁ Ἀσκληπιὸς ἀπὸ τοῦ ἠπίως ἰᾶσθαι καὶ ἀναβάλλεσθαι τὴν κατὰ τὸν θάνατον γινομένην ἀπόσκλησιν.

Asclepius derived his name from healing soothingly and from deferring the withering that comes with death.

[1] Cf. T. **6.**

269. Porphyrius, Quaestiones Homericae, α, 68.

Τὸ ἀσκελές σημαίνει τὸ ἄγαν σκληρόν. σκέλλειν γάρ ἐστι τὸ σκληροποιεῖν, καὶ ὁ σκελετὸς ὁ κατεσκληκὼς διὰ τὴν ἀσαρκίαν, καὶ Ἀσκληπιὸς κατὰ στέρησιν μετὰ ἠπιότητος, ὁ διὰ τῆς ἰατρικῆς μὴ ἐῶν σκέλλεσθαι.

Dried up means what is too harsh. For *skellein* means to make harsh. Also the skeleton is that which is dried up through lack of flesh, and the name Asklepios comes from this word with an alpha privative, together with the word for gentleness, that is, he who by the agency of the medical art does not permit dryness.

270. Scholia in Homerum, Ad Iliadem, IV, 195.

Ἀσκληπιὸς δὲ εἴρηται παρὰ τὸ ἀσκεῖν καὶ ἤπια τὰ μέλη ποιεῖν.

Asklepios is named from applying [*askein*] and from making the limbs gentle [*epia*].[1]

[1] Cf. **270a.** *Scholia Genevensia in Homerum, Ad Iliadem*, IV, 194: Ἀσκληπιὸς παρὰ τὸ ἀσκεῖν τὰ ἤπια.

271. Scholia in Lycophronem, Ad Alexandram, 1054.

Τὸ πρὶν δὲ ὁ Ἀσκληπιὸς Ἤπιος ἐκαλεῖτο, διὰ τὸ πρᾶον καὶ ἥσυχον· θεραπεύσας δὲ Ἄσκλην τὸν Ἐπιδαύρου τύραννον ὀφθαλμιῶντα σφοδρῶς ἐκλήθη Ἀσκληπιός, διὰ τὸ θεραπεῦσαι τὸν Ἄσκλην. Ἀσκληπιὸς εἴρηται, ὅτι Ἄσκλην τὸν Ἐπιδαύρου τύραννον ὀφθαλμιῶντα θεραπεύσας ἐκλήθη Ἀσκληπιός· πρότερον γὰρ Ἤπιος ἐκαλεῖτο.

Ἤπιος· οὕτως πρότερον ὁ Ἀσκληπιὸς ἐκαλεῖτο ἢ ἀπὸ τῶν τρόπων ἢ ἀπὸ τῆς τέχνης καὶ τῆς τῶν χειρῶν ἠπιότητος. θεραπεύσας δὲ Ἄσκλην τὸν Ἐπιδαυρίων βασιλέα ὕστερον Ἀσκληπιὸς ὠνομάσθη.

ἢ ὅτι τὰ ἀσκελῆ τῶν νοσημάτων ἤπια ποιεῖ.	ἢ ἀσκελοποιός τις ὢν ὁ μὴ ἐῶν σκέλλεσθαι καὶ ξηραίνεσθαι καὶ νεκροῦσθαι τοὺς ἀνθρώπους ὑπὸ νόσων.

Formerly Asklepios was named Epios,[1] on account of his gentleness and calmness, but after he had cured Askles, the tyrant of Epidaurus, who suffered seriously from ophthalmia, he was called Asklepios, because he had healed Askles.[2] Asklepios he is called because he healed Askles, the tyrant of Epidaurus, who was suffering from ophthalmia, and therefore was called Asklepios. For formerly he had been called Epios.

Epios: thus Asklepios was formerly called, either from his manners, or from his art, or from the gentleness of his hands.[3] But after he had cured Askles, the king of the Epidaurians, he was later on called Asklepios.

| Or because he makes the harsh in illnesses mild; | or as the one who sees to the prevention of pining, i. e. the one who does not allow men to be parched and dried up and mortified by diseases. |

[1] Cf. T. 158.
[2] Cf. T. 272; 273; 275; cf. also T. 93.
[3] Cf. 271a. *Etymologicum Magnum, s. v.* Ἤπιος: Ἤπιος—ἠπιότητος.

272. Eustathius, Commentarii ad Homeri Iliadem, IV, 202.[1]

Ἐκλήθη δέ, φησίν, Ἀσκληπιὸς ἢ διότι Ἄσκλητον Ἐπιδαύριον ὀφθαλμιῶντα ἰάσατο ἤπιος ἐκείνῳ φανείς, ἢ πλεονασμῷ τοῦ λ παρὰ τὸ ἀσκεῖν ἠπίως τοὺς νοσοῦντας, ὅ ἐστιν ἐπιμελείας ἀξιοῦν, ἢ παρὰ τὸ μὴ σκελετεύεσθαι αὐτοὺς ἐὰν ἠπίως προσφερόμενος. ἡ δὲ ὀξυτόνησις τοῦ ὀνόματος ἀπορίαν ἔχει, εἴ περ καὶ ἄλλως αἱ συνθέσεις τοὺς τόνους ἀναβιβάζουσι. καλῶς οὖν ἐποίει Δημοσθένης, ὡς ἱστορεῖται, προπαροξύνων τὴν λέξιν καὶ ἀναγιγνώσκων, Ἀσκλήπιος.

He was named, he [sc., the poet] says, Asklepiós, either because he cured Askletos of Epidaurus, who suffered from ophthalmia, gently appearing to him; [2] or, with the superfluous addition of the lambda, [his name is derived] from *askein epios etc., i. e.,* from gently taking pains with the sick, that is, considering them deserving of watchful care, or from the fact that, administering to them gently, he did not allow them to be dried up.[3] The acute accent on the last syllable of the name raises a difficulty, since combined words usually throw back their accents. Demosthenes, therefore, did well, as the story goes, in changing the accentuation to an acute on the antepenult and pronouncing it Asklépios.[4]

[1] Cf. T. 139.   [2] Cf. e. g. T. 425.   [3] Cf. T. 273 ff.
[4] Cf. T. 266; 267; cf. also 272a. Eustathius, *Commentarii ad Homeri Iliadem*, XI, 518: Ὡs δὲ ὁ μῦθος θεοῖς ἐντάττει τὸν Ἀσκληπιόν, καὶ ὡς Ἤπιος τὰ πρῶτα καλούμενος Ἀσκληπιὸς μετεκλήθη, καὶ διὰ τί τοῦτο, καὶ ὡς καινότερον ὀξύνεται ὁ Ἀσκληπιὸς ἐν τῇ συνθέσει ὁ πρὸ ταύτης ἐν τῷ Ἤπιος προπαροξυτονούμενος, ἀλλαχοῦ δεδήλωται; cf. T. 256.

273.   Eudocia Augusta, Violarium, XI.

Ὠνομάσθη δὲ Ἀσκληπιός, παρὰ τὸ τὰ ἀσκελῆ τῶν νοσημάτων ἤπια ποιεῖν. τὸ δὲ ἀσκελὲς παρὰ τὸ σκέλλω, τὸ ἄγαν σκληραίνω, ὅθεν καὶ σκέλος. ἢ διότι Ἄσκλην τὸν Ἐπιδαύριον ἰάσατο ὀφθαλμιῶντα, ἤπιος ἐκείνῳ φανείς, ἢ πλεονασμῷ τοῦ λ * παρὰ τὸ ἀσκεῖν ἠπίως τοὺς νοσοῦντας, ὅ ἐστιν ἐπιμελείας ἀξιοῦν. ἢ παρὰ τὸ μὴ σκελετεύεσθαι αὐτοὺς ἐᾶν, ἠπίως προσφερόμενος. ἢ ἀπὸ τοῦ ἐσκλῆσθαι καὶ ἀναβάλλεσθαι τὴν κατὰ τὸν θάνατον γινομένην ἀπόσκλησιν.

* πλεονασμῷ τοῦ λ ἢ MSS.

He was named Asklepios from the fact that he assuaged harshness in illnesses. The word *askeles* comes from *skello*, which means to be excessively dry, whence also the word *skelos*, leg. Or [he was named Asklepios] because he healed Askles, the Epidaurian, who suffered from ophthalmia, gently appearing to him; or, with the superfluous addition of the lambda, from the fact that he gently took pains with the sick, that is, deemed them worthy of watchful care; or from the fact that he did not permit them to be withered, administering to them gently; or from being withered and deferring the cessation that comes with death.

274.   Etymologicum Gudianum, *s. v.* Ἀσκληπιός.

Ἀσκληπιός· παρὰ τὸ τὰ σκέλη ἤπια ποιεῖν· ἀπὸ μέρους δὲ ὅλον τὸ σῶμα. ἢ ὁ τὰ ἄγαν σκληρὰ ἤπια ποιῶν.

Asklepios: from making the limbs soft; then from a part [the word came to include] the whole body; or he who makes soft the excessively harsh.[1]

[1] Cf. **274a**. *ibid.*, p. 213, 16-18: Ἀσκληπιός· ὁ τὰ ἄγαν σκληρὰ ἤπια ποιῶν, τοῦ ᾱ ἐπίτασιν σημαίνοντος. Ἀσκληπιός, ὁ τὰ σκέλη καὶ πᾶν τὸ σῶμα ὑγιὲς ποιῶν καὶ ἀνώδυνον. οὕτως Ἡρωδιανός ⟨I, p. xxix, 36 [Lentz]⟩.

275. Etymologicum Magnum, s. v. Ἀσκελές.

> . . . ἢ παρὰ τὸ ἄγαν ἐσκληκέναι. ὅθεν καὶ Ἀσκληπιός, ὁ μὴ
> ἐῶν τὰ σκέλη ἐσκληκέναι καὶ ξηραίνεσθαι. ἀπὸ μέρους δὲ τὸ
> ὅλον σῶμα δηλοῖ. καὶ γὰρ τὸν Ἐπιδαύρου τύραννον ὀφθαλμιῶντα
> θεραπεύσας, ἐκλήθη Ἀσκληπιός. πρότερον γὰρ Ἤπιος ἐκαλεῖτο.
> ἢ ὅτι τὰ ἀσκελῆ (ὅ ἐστι σκληρὰ) τῶν νοσημάτων ἤπια ποιεῖ.

> . . . or from being dried up too much. Whence also the name
> of Asklepios, who does not allow the limbs to be parched and
> dried up. From a part it refers to the whole body. And also
> because he healed the tyrant of Epidaurus who was suffering
> from ophthalmia, he was named Asklepios.[1] For formerly he
> was called Epios. Or because he makes the hard (that is the
> harsh) in illnesses mild.[2]

[1] Cf. 271 ff.   [2] Cf. T. 273-74.

276. Suidas, Lexicon, s. v. Ἀσκληπιάδης.

> . . . καὶ Ἀσκληπιάδαι, οἱ ἰατροί, ἀπὸ Ἀσκληπιοῦ. οὗτος δὲ
> παρὰ τὸ ἀσκελῆ καὶ ἤπια φυλάττειν τὰ σώματα.

> . . . and Asclepiads, the doctors, [get their name] from Askle-
> pios. And he derived his name from the process of keeping
> bodies soft and not dried up.

DIVINE RETINUE

277. Aristophanes, Plutus, 639-40.[1]

> ΧΟ. Ἀναβοάσομαι τὸν εὔπαιδα καὶ
> μέγα βροτοῖσι φέγγος Ἀσκληπιόν.

> CH. (singing) Sing we with all our might Asclepius first
> and best,
> To men a glorious light, Sire in his offspring blest.

[1] Cf. T. 421.

278. Scholia in Aristophanem, Ad Plutum, 639.

> Πολλοὶ γὰρ παῖδες τοῦ Ἀσκληπιοῦ · Ποδαλείριος, Μαχάων,
> Ἰασώ, Πανάκεια, Ὑγεία.

For the children of Asclepius were many: Podalirius, Machaon,[1] Iaso, Panacea, Hygeia.[2]

[1] Cf. T. 162; 162a.
[2] 278a. Eudocia Augusta, *Violarium*, XIb, same text, cf. T. 421, v. 701-02; T. 641.

279. Scholia in Aristophanem, Ad Plutum, 701.

> . . . ἐπεὶ καὶ Ἕρμιππος ἐν τῷ πρώτῳ ἰάμβῳ τῶν τριμέτρων Ἀσκληπιοῦ καὶ Λαμπετίας τῆς Ἡλίου λέγει Μαχάονα καὶ Ποδαλείριον καὶ Ἰασὼ καὶ Πανάκειαν καὶ Αἴγλην νεωτάτην. ἔνιοι δὲ προστιθέασιν Ἰανίσκον καὶ Ἀλεξήνορα.

> . . . for Hermippus[1] too in the first iamb of his trimeters says that Machaon, Podalirius, Iaso, Panacea, and Aegle, the youngest, were the children of Asclepius and Lampetia, the daughter of Helius.[2] Some add also Ianiscus and Alexenor.[3]

[1] = Fr. 73 [Kock].
[2] Cf. however, T. 280-81.
[3] Cf. however, T. 187-89.

280. Scholia in Lycophronem, Ad Alexandram, 1054.

> Παραδιδόασι δὲ τούτῳ γυναῖκα Ἠπιόνην, ἐξ ἧς αὐτῷ γενέσθαι Ἀκεσὼ καὶ Πανάκειαν· Σεξτίων ἐν ὑπομνήματι Λυκόφρονος.

> They attribute to him [*sc.*, Asclepius] Epione as his wife, from whom sprang Aceso and Panacea, his children,[1] (says) Sextion [time unknown] in his *Commentary on Lycophron.*

[1] For other lists of children of Asclepius and Epione, cf. 280a. Anonymus, *Paean Erythraeus*, 6 ff. (= T. 592): . . . Μαχάων καὶ Ποδαλείριος . . . Ἰασὼ Ἀκεσώ . . . Αἴγλα . . . Πανάκειά τε Ἠπιόνας παῖδες σὺν . . . Ὑγιείᾳ; cf. also T. 592a, and T. 593; cf. 280b. Aristides, *Oratio* VII, 46: Ἰασώ . . . Πανάκεια . . . Αἴγλη . . . Ὑγίεια . . . Ἠπιόνης δὲ παῖδες ἐπώνυμοι; cf. 280c. Suidas, *Lexicon, s. v.* Ἠπιόνη: γυνὴ Ἀσκληπιοῦ. καὶ θυγάτηρ αὐτῆς Ὑγεία, Αἴγλη, Ἰασώ, Ἀκεσώ, Πανάκεια. Cf. 280d. *Etymologicum Magnum, s. v.* Ἤπιος· ᾧ καὶ γυναῖκα παραδίδωσιν Ἠπιόνην, ἐξ ἧς αὐτῷ γενέσθαι Ἰάσονα, Πανάκειαν.

281. Cornutus, Theologiae Graecae Compendium, Cp. 33.[1]

> Παραδέδοται δὲ καὶ γυνὴ τοῦ Ἀσκληπιοῦ, Ἠπιόνη, τοῦ ὀνόματος οὐκ ἀργῶς εἰς τὸν μῦθον παρειλημμένου, δηλοῦντος δὲ τὸ πραϋντικὸν τῶν ὀχλήσεων διὰ τῆς ἠπίου φαρμακείας.

> There is mentioned in legend also the wife of Asclepius, Epione,[2] whose name is not idly introduced into the legend,

for it signifies the alleviation of troubles through the agency of soothing simples.[3]

[1] Cf. T. 6.
[2] Cf. **281a.** Hippocrates, *Epistulae*, 10 [IX, 324, 16 L.] : Ἡρακλέους θυγατρὸς Ἠπιόνης ; **281b.** *Scholia in Homerum, Ad Iliadem*, 1V, 195 : Ἠπιόνης τῆς Μέροπος. Other wives of Asclepius : **281c.** *Scholia in Aristophanem, Ad Plutum*, 701 : Λαμπετίας τῆς Ἡλίου ; **281d.** *Scholia in Homerum, Ad Iliadem*, IV, 195 : Ξάνθης ; **281e.** *Orphei Hymni*, LXVII, 7 : Ὑγίειαν ; **281f.** Ioannes Tzetzes, *Homerica*, 615 : Ἱππονόη ; **281g.** Pausanias, *Descriptio Graeciae*, II, 10, 3 : Ἀριστοδάμαν.
[3] Cf. Eudocia Augusta, *Violarium*, XI, same text.

282. Aristides, Oratio XXXVIII, 1-24.

ΑΣΚΛΗΠΙΑΔΑΙ.

' Κλῦτε φίλοι, θεῖός μοι ἐνύπνιον ἦλθεν ὄνειρος,' ἔφη αὐτὸ τὸ ὄναρ · ταύτην γὰρ δὴ ἐδόκουν ἀρχὴν ποιεῖσθαι τοῦ λόγου, ὡς ὕπαρ τὸ ὄναρ σκοπῶν ἐπ' ἐμαυτοῦ · ἐχέτω δὴ καὶ τὸ ἐνύπνιον ὡς ὕπαρ καὶ τὸ δρώμενον ὡς ἡ πρόρρησις εἶχεν. ἦν δὲ ὁ νοῦς ἄρα ἐγκώμιον ποιῆσαι Ποδαλειρίῳ · τὸ μὲν πρῶτον Ποδαλειρίῳ, μετὰ

2 δὲ τοῦτο ὑπήχθην εἰς Μαχάονα. ἀποροῦντι δέ μοι ὁπότερον χρὴ ἐπαινεῖν, τέλος ἔδοξεν ἀμφοτέρους · οὔτε γὰρ ἔτι θεμιτὸν εἶναι παραλείπειν τὸν ἕτερον, ἐπειδή γε ἀμφοῖν ἔσχον ἔννοιαν, ὁπότερός τε εἴη ὁ ἐκ τοῦ ὀνείρατος, ἐνέσεσθαι πάντως ἀμφοῖν ἐπαινουμένων · καὶ ἐμοὶ πρός τε ἐκεῖνον καὶ πρὸς ἀμφοτέρους

3 ἕξειν καλῶς. εἶεν. θεοῦ δὴ προβαλόντος, πότερον δεδιέναι χρὴ μειζόνως ἢ θαρρεῖν; δεδιέναι μὲν ἀνάγκη, μὴ χεῖρον ἐπιδειξώμεθα ἐν τοσούτῳ κριτῇ, αἱ δὲ ἐλπίδες καλλίους · αὐτῷ γὰρ καὶ μελήσειν εἰκός. οὐ γὰρ ἂν προὔβαλεν, εἰ μὴ κατὰ νοῦν

4 ἔμελλεν αὐτῷ γενήσεσθαι [τὸ σύμπαν ἐπὶ τοῦ λόγου]. πρὸς δὲ καὶ καλέσαι βοηθὸν τῷ λόγῳ θαυμαστῶς ὡς εἰς ἐπιτήδειον καθέστηκεν. εἰ γὰρ οἱ ποιηταὶ τὰ δοκοῦνθ' ἑαυτοῖς ὥστ' εἰπεῖν τὸν Ἀπόλλω καὶ τὰς Μούσας καλοῦσιν, ἦ που καλλίων ἡ κλῆσις ἡμῖν, ὅταν αὐτοῦ τοῦ προβαλόντος δεώμεθα μουσηγέτην ἅμα τῷ πατρὶ γενέσθαι. ἀλλ' ὦ πολλὰ δὴ πολλάκις κληθείς, ἔχω δ' εἰπεῖν ὅτι καὶ παραδείξας καὶ ἄλλα καὶ ⟨τὰ⟩ περὶ αὐτοὺς τοὺς λόγους, ἔξαγε καὶ νῦν ὅπη σοι φίλον τὸν λόγον.

5 Σχεδὸν δὲ οὐ πόρρω ἡ ἀρχὴ τῷ ἐγκωμίῳ, ἀλλ' ὁ αὐτὸς πατὴρ τοῦ τε λόγου καὶ τῶν νεανίσκων. τετραγονίαν γὰρ τοιάνδε οὐδείς πω Ἑλλήνων ἤκουσεν οὐδὲ διηγήσατο · ὥσπερ γὰρ οἱ θεσμοθέται, διὰ τεττάρων εἰσὶν εὐπατρίδαι, μᾶλλον δὲ ὡς οὐδένες θεσμοθέται οὐδ' ἄλλο γε οὐδὲν γένος ἀνθρώπων. τέταρτοι γάρ εἰσιν ἐκ Διὸς

6 διὰ πάντων ἄκρων. Ἀπόλλων μὲν γὰρ ἐξ αὐτοῦ Διός, Ἀσ-

κληπιὸς δὲ ἐξ Ἀπόλλωνος, οἱ δὲ ἐξ Ἀσκληπιοῦ, διὰ πάσης τῆς
κρατίστης φύσεως τὴν εὐγένειαν ἀνῃρημένοι. Ἀχιλλεὺς μὲν γὰρ
τέταρτος ἀπὸ Διὸς διὰ Πηλέως καὶ Αἰακοῦ, Μίνως δὲ καὶ
Ῥαδάμανθυς Διὸς παῖδες καὶ Θησεὺς Ποσειδῶνος, οὐδέτεροι
μόνοι ὁποτέρου, ἀλλὰ σὺν πολλοῖς καὶ θεοῖς καὶ ἥρωσιν, ὧν τῶν
7 μὲν ἡττῶνται, τοῖς δ' εἰς ἴσον καθίστανται. μόνοι δὲ οὗτοι
πλήθει καὶ ἀρετῇ προγόνων εἰσὶν ἀνανταγώνιστοι, τοὺς ἐκ Διός
τε καὶ Ἀπόλλωνος Ἀσκληπιοῦ προσθήκῃ νικῶντες, ὅσοι γε ἥρωες
αὐτῶν· γενομένους δ' αὐτοὺς τρέφει ὁ πατὴρ ἐν Ὑγιείας κήποις,
καὶ ἐπειδὴ ἐδέχετο ἡ ἡλικία, οὐκ ἐδιδάξατο τὴν τέχνην τὴν
ἰατρικήν, ἀλλ' ἐδίδαξεν αὐτός· οὐ γὰρ ἔτι εἰς Χείρωνος ἔδει
βαδίζειν ⟨τοὺς⟩ ἔχοντας οἴκοι τὸν ἐπιστάτην, οὗ πολὺ δὴ κατ'
8 ἐπωνυμίαν ὁ Χείρων ἤδη ἦν δεύτερος. Οὕτω δὴ φύντες καὶ
τραφέντες καὶ παιδευθέντες μέχρι μὲν ἡσύχαζε τὰ τῶν Ἀχαιῶν,
οἵδε Θετταλίᾳ τε ἦσαν κόσμος πολὺ λιμνῶν τε καὶ πεδίων καὶ
ποταμῶν ἐπιφανέστερος καὶ τοὺς ταύτῃ Ἕλληνας ὤρθουν, εἴς τε
τὸ κοινὸν πολιτευόμενοι τὰ ἐκείνοις πρέποντα καὶ τὰς ἰδίας
ἑκάστοις συμφορὰς ἐπανορθοῦντες, οὐδ' ἦν νοσεῖν ὅπου φανείη
9 Μαχάων ἢ Ποδαλείριος. κινηθέντων δὲ Ἑλλήνων διὰ τὴν
Τρώων ἀδικίαν οὔτε ἐξωνεῖσθαι τὴν οἴκοι μονὴν ἠξίουν οὔτε
ἀπέκρυπτον ἑαυτοὺς ὥσπερ ἄλλοι τινές, ἀλλὰ γνόντες ὡς αὐτῶν
ὁ καιρὸς εἴη καὶ προϊδόμενοι τὰς τοῦ πολέμου τύχας προὔστησαν
10 ἑκόντες τῆς ἁπάντων σωτηρίας. ἀφικόμενοι δ' εἰς Τροίαν
διπλῆν χρείαν παρείχοντο τοῖς Ἀχαιοῖς καὶ οὐχ ὅσον ἰατροὶ
συνῆσαν, ἀλλὰ καὶ τοῖς ὅπλοις ὠφέλουν, καὶ πολλάκις μὲν δὴ
τῶν πολεμίων τροπὴν ποιήσασθαι λέγονται, τὸ δ' ἁλῶναι τὴν
Τροίαν καὶ παντάπασιν ἦλθεν εἰς αὐτοὺς τῇ τε ἄλλῃ καὶ ⟨διὰ⟩
τὴν Φιλοκτήτου νόσον, ἣν Ὀδυσσεὺς καὶ Ἀτρεῖδαι προ-
καταγνόντες ἀνίατον εἶναι Φιλοκτήτην οὐχὶ δικαίως ἐν Λήμνῳ
κατέλιπον, οὗτοι δὲ δέκα ἔτεσιν αὐξηθεῖσαν ἰάσαντο. καὶ γίγνεται
Φιλοκτήτης τε ἐνεργὸς τοῖς Ἀχαιοῖς καὶ Φιλοκτήτῃ τὰ βέλη
11 ⟨τὰ⟩ τοῦ Ἡρακλέους χρήσιμα τῇ ἐκείνων τέχνῃ. ἐπεὶ δὲ
ἑάλω Τροία, προειδότες τὰς ἐσομένας ὕστερον Ἑλλήνων εἰς τὴν
Ἀσίαν ἀποικίας καὶ ἅμα βουλόμενοι τῆς αὐτῶν εὐεργεσίας ὅτι
πλείστους ἀπολαύειν, τοῦτο μὲν Τευθρανίαν ἡμεροῦσιν εἰς ὑπο-
δοχὴν τοῦ πατρός, τοῦτο δέ, ὡς ὁ Κῴων λόγος, πλεύσαντες εἰς
Κῶ τὴν Μεροπίδα, ἣν οἰκουμένην ὑπὸ Μερόπων Ἡρακλῆς ἐκ
Τροίας ἀνιὼν ἔτι πρότερον, ἀδικίαν ἐγκαλέσας, ἐπόρθησεν, οἰ-
κίζουσί τε καὶ εἰς ἤθη πρέποντα τῇ φύσει τῆς χώρας κατέστησαν,
τὸ λεγόμενον περὶ τοῦ προγόνου τοῦτο, ὡς ἔοικε, μιμησάμενοι.

12 Ἀπόλλω τε γάρ φασιν οἱ ποιηταὶ τὴν Δῆλον φερομένην
πρότερον στῆσαι κατὰ τοῦ πελάγους ἐρείσαντα, ἐπειδὴ πρῶτον
ἐν αὐτῇ ἐγένετο, καὶ οὗτοι τῆς Μεροπίδος τότε ἐπιβάντες. προκρίναντες ἁπασῶν εἶναι καλλίστην, ὅσαι παραπλήσιαι μέγεθος,
ἰάσαντό τε καὶ ἀπέφηναν ἐμβατὸν πᾶσιν Ἕλλησι καὶ βαρβάροις,
πρότερον σφαλερὰν καὶ ὕποπτον οὖσαν, καὶ τὴν εὐδαιμονίαν
13 κυρίαν τῇ νήσῳ κατέστησαν. οἶμαι δὲ καὶ Ῥοδίοις πολλῶν ἐκ
πολλοῦ σεμνῶν ὑπαρχόντων ἐν πρώτοις εἶναι τὴν ἐκείνων ἀρχήν,
ἣν αὐτοὶ προκρίναντες εἵλοντο σφῶν ἄρχειν Ἀσκληπιάδας.
Ἡρακλειδῶν ποιησάμενοι διαδόχους. ἔσχον δὲ καὶ τὸν Καρικὸν
τόπον καὶ Κνίδον τὴν τῆς Ἀφροδίτης ἱεράν· ὅ τι γὰρ ἢ ὁ ἕτερος
ἢ ἀμφότεροι, κεκοινώσθω τὰ νῦν· ἀπέλαυσε δέ τι καὶ Κύρνος
14 αὐτῶν. Ἀλλὰ ταῦτα μέν ἐστι λόγων πολλῶν. ποιησάμενοι δὲ
παῖδας συνεργούς τε καὶ διαδόχους τῆς σφετέρας ἐπιστήμης
ὡσπερεὶ κεφάλαιον τοῦτο ἐπέθηκαν τῶν εἰς τοὺς Ἕλληνας εὐεργεσιῶν, προσθήσω δὲ καὶ εἰς ἅπαντας ἤδη, ἵνα μή ποτε ἐπιλείπῃ
τὸ τῶν ἀνθρώπων γένος ἡ παρ' αὐτῶν ἐπικουρία καὶ χάρις, ἀλλ'
ὦσιν ἐκ προγόνων τε καὶ ἐκγόνων σωτῆρες ἀθάνατοι τῆς φύσεως,
τὸν ἅπαντα χρόνον κατ' ἀνθρώπους πολιτευόμενοι καὶ συνόντες
15 ἅπασιν ὥσπερ τοῖς ἐφ' αὑτῶν. καὶ γάρ τοι κατέλυσαν μὲν τοὺς
ἐν Αἰγύπτῳ βεβοημένους, τὰς δ' εὐεργεσίας σύμβολον τοῦ γένους
ἐποιήσαντο. ἵδρυνται δὲ οὔτε ἐν Θετταλίᾳ οὔτε ἐν τοῖς τῶν Κῴων
προαστίοις, ἀλλὰ πάντα ἰατρικῆς ἐνέπλησαν, ὥσπερ ὁ Τριπτόλεμος σίτου διὰ τῶν σπερμάτων· πάντα γὰρ ἐκ τούτων καὶ
παρὰ τούτων ἐξεφοίτησεν. καὶ καθάπερ τὴν Μήδειάν φασι διὰ
τοῦ Θετταλῶν πεδίου φεύγουσαν, ἐκχυθέντων τῶν φαρμάκων,
ποιῆσαι Θετταλίαν ἅπασαν πολυφάρμακον, οὕτως ἡ τούτων
ἐπιστήμη τε καὶ φιλανθρωπία χυθεῖσα ἐπὶ πλεῖστον πάσας μὲν
πόλεις Ἑλλήνων, πολλοὺς δὲ καὶ τῶν βαρβάρων τόπους ἐκόσ-
μησέν τε καὶ κοσμεῖ, καὶ προσέτι γε ἔσωσε καὶ σῴζει, καὶ ὁ
16 τρίτος ἐπ' ἀμφοῖν προσέστω χρόνος. εἰ δ' ἀνὴρ εἷς ἀπὸ τούτων
Ἱπποκράτης ἔφυ κληρονόμος τῆς τέχνης διὰ πάντων ἰδιωτῶν τῶν
μέσων, ἱκανὴ τῇ τε γῇ φορὰ καὶ χάρις τούτοις παρ' ἀνθρώπων
ἦν ἂν τῆς σπορᾶς· νῦν δ' ὥσπερ ἔθνος τὸ τῶν Ἀσκληπιαδῶν
κατεσκευάσθη δι' αἵματος τὴν τέχνην σῷζον, οὕτω θεία μοῖρα
17 ἡγήσατο Μαχάονι καὶ Ποδαλειρίῳ τῆς γενέσεως. μάθοι δ' ἄν
τις τοῖς Ἡρακλείδαις αὐτοὺς ἀντεξετάζων τῆς χρείας ἕνεκα τῆς
κοινῆς καὶ τῆς ἰδίας τύχης. Ἡρακλεῖδαι γὰρ ἓν μὲν καὶ πρῶτον
ἐσκεδάσθησαν καὶ οὐχ ἓν σύνταγμα αὐτῶν ἐγένετο, ὡς δ' εἰπεῖν
οὐδὲ ἓν φῦλον· οὐ γὰρ ἦσαν ὁμότιμοι, ὥστε ἀλλήλοις ἀντὶ ξένων

καθειστήκεσαν· ἔπειθ' ὅσον αὐτῶν ἦν κράτιστον, οὔτε συμφορῶν
ἄμοιρον λέγεται γενέσθαι τήν τε πατρῴαν οὐ παντελῶς δια-
σώσασθαι τέχνην. οὐ γὰρ ταῖς εἰς τὸ κοινὸν εὐεργεσίαις, ἀλλὰ
18 τῇ καθ' αὑτοὺς δυνάμει τὴν λαμπρότητα ἐκτήσαντο. Ἀσκληπιάδαι δὲ ἐκ Μαχάονος ἀρξάμενοι καὶ Ποδαλειρίου κοινῇ πᾶσιν
ἀσφάλεια καὶ σωτηρία γεγόνασιν, τὴν τοῦ προγόνου δια-
σωσάμενοι τέχνην, ὥσπερ ἄλλο τι σύμβολον τοῦ γένους. πρὸς
δὲ καὶ ἡ τύχη τῆς προαιρέσεως ἀξία· οὔτε γὰρ ἠλάθησαν οὔθ'
ἱκέτευσαν εἰς πόλιν οὐδεμίαν, ἀλλὰ διεξῆλθον καθαροὶ συμφορῶν,
μιᾷ φατρίᾳ καὶ μιᾷ γνώμῃ καὶ τύχῃ χρησάμενοι διὰ τέλους.
19 Ἐπάνειμι δὲ ὅθεν ἐξέβην (§ 15), ἐπὶ τοὺς ἀρχηγέτας τε καὶ
τοὔνομα πρώτους λαβόντας τὸ τῶν Ἀσκληπιαδῶν. οἱ δὲ ἕως
μὲν ἦσαν ἐν ἀνθρώποις, στρατείαις καὶ ὁμιλίαις καὶ γενέσει
παίδων πρεπόντων αὐτοῖς καὶ συλλήβδην ἁπάσῃ τῇ πολιτικῇ
δυνάμει τὰς πόλεις ὠφέλουν, οὐ μόνον τὰς τοῦ σώματος νόσους
ἐξαιροῦντες, ἀλλὰ καὶ τὰ τῶν πόλεων νοσήματα ἰώμενοι, μᾶλλον
δὲ οὐδ' ἐγγίγνεσθαι τὴν ἀρχὴν ἐῶντες, ἀπ' ἀμφοῖν σῴζοντες τοὺς
ὑπηκόους, [καὶ] τῇ τέχνῃ τὴν ἀρχὴν ἀκόλουθον κατασκευ-
20 ασάμενοι. ἐπεὶ δὲ κρείττους ἦσαν ἢ παρ' ἡμῖν μένειν, οἱ δὲ ὑπὸ
τοῦ πατρός τε καὶ τῶν προγόνων ἀποδύντες τὰ σώματα εἰς ἕτερον
θεσμὸν ἔρχονται, οὐχ ὥσπερ Μενέλεώς τε καὶ ὁ 'ξανθὸς Ῥαδάμανθυς,' εἰς τὸ Ἠλύσιον πεδίον καὶ τὰς ἔξω νήσους, ἀλλ'
ἀθάνατοι γενόμενοι τὴν γῆν διέρχονται, τοσοῦτον τῆς ἀρχαίας
21 φύσεως ἀποστάντες ὅσον τὴν ἡλικίαν φυλάττουσι. καὶ αὐτοὺς
πολλοὶ μὲν ἤδη ἐν Ἐπιδαύρῳ εἶδόν τε καὶ ἔγνωσαν ἐμφανῆ
κινουμένους, πολλοὶ δὲ ἄλλοθι πολλαχοῦ· ὃ καὶ μέγιστον ἔστω
κατ' αὐτῶν. Ἀμφιάραος μὲν γὰρ καὶ Τροφώνιος ἐν Βοιωτίᾳ καὶ
Ἀμφίλοχος ἐν Αἰτωλίᾳ χρησμῳδοῦσί τε καὶ φαίνονται, οὗτοι δὲ
πανταχοῦ τῆς γῆς διᾴττουσιν ὥσπερ ἀστέρες, περίπολοι κοινοὶ
καὶ πρόδρομοι τοῦ πατρός. ὁσαχοῖ δὲ Ἀσκληπιῷ εἴσοδοι, καὶ
τούτοις κλισιάδες τε [αὐτοῖς] ἀνεῖνται πανταχοῦ γῆς, καὶ διὰ
πάντων ἡ κοινωνία τῷ πατρὶ σῴζεται νεῶν θυσιῶν παιάνων
22 προσόδων ἔργων ἃ πράττουσιν. ὦ μακαριστοὶ μὲν ὑμεῖς τῶν
ἄνω προγόνων ἐπ' ἀμφότερα, εὐδαίμονες δὲ τῶν ἀφ' ὑμῶν φύντων,
ἔτι δὲ ὑμῶν τε αὐτῶν καὶ ἀδελφῶν, οἷς Ἰασώ τε καὶ Πανάκεια
καὶ Αἴγλη σύνεστιν καὶ Ὑγίεια, ἡ πάντων ἀντίρροπος, Ἠπιόνης
δὴ παῖδες ἐπώνυμοι· οὐδ' ὑμῖν θᾶκοι χωρὶς ἀλλήλων οὐδὲ διεσ-
23 κηνήσατε. Ὦ κάλλιστος μὲν αὐτοὶ χορὸς τῷ πατρί, πολλοὺς
δὲ ἀνάγοντες παρὰ ἀνθρώπων, χοροποιοὶ μακρῷ πάντων ἄριστοι,
καὶ προσέτι ἱεροποιοί τε καὶ ἐπιστάται κρατήρων καὶ χαρίτων

ἀπασῶν. αἱ μὲν ἄλλαι θυσίαι τε καὶ ἑορταὶ νόμῳ καθεστᾶσιν,
ὡς εἰπεῖν, αἱ δ' ἀφ' ὑμῶν καὶ τοῦ περὶ ὑμᾶς ἐργαστηρίου πλεῖσται
μὲν ἐφ' ἡμέραν καὶ πρὸς ἁπάσας τὰς ἄλλας, καθαρῶς δὲ ἀπὸ
καρδίας ἔρχονται τὴν εὐθυμίαν ἀφ' ὧν σύνισμεν φέρουσαι. ὑμῶν
ἴχνη πλεῖστα καὶ φανερώτατα, αἰεὶ δ', ὥσπερ ἀνθρώπῳ σκιά,
24   φῶς ὅποι κινοῖσθε ἕπεται. ὦ Διοσκούροις ἰσόμοιροι καὶ ἡλικιῶται
ἐν ἑτέρῳ χρόνῳ τῆς γενέσεως, οἳ πολλὰς μὲν ἤδη τρικυμίας κατε-
παύσατε, πολλοὺς δὲ καὶ στιλπνοὺς λαμπτῆρας ἔν τε νήσοις καὶ
κατ' ἤπειρον ἀνήψατε. οὗτος ὑμῖν ὁ παρ' ἡμῶν λόγος ἐξ ὕπνου τε
καὶ ἐνυπνίου συντεθεὶς εὐθύς. ὑμεῖς δὲ τῇ ὑμετέρᾳ πραότητι καὶ
φιλανθρωπίᾳ . . .* θέντες αὐτὸν εἰς καλλίους, τῆς τε νόσου
παύετε καὶ διδοίητε ὑγιείας τε ὅσον οἷς ἡ ψυχὴ βούλεται τὸ σῶμα
ὑπακούειν, καὶ τὸ σύμπαν εἰπεῖν βίου ῥᾳστώνην.

* Lacunam indicavit Keil.

The Asclepiads

"Listen friends, a dream came to me, a vision sent from the gods,"[1] said the dream itself. For I seemed in my dream to make this beginning of the discourse, seeing the dream before me as though it were waking reality. May the dream then become waking reality, and may the actual performance be like the prediction.

It was my intention, therefore, to make a eulogy of Podalirius; first of Podalirius, afterwards I was led on to Machaon. Wondering which of the two I should praise, it finally seemed to me, both. For it did not seem right to neglect one of them since I had a notion of both, and whichever was the one in my dream, he would be included anyhow if both were praised; and it would turn out well for me in relation to that one and to both. Well then, the god [sc., Asclepius] having set the task for me, should I be afraid rather than courageous? It is necessary to fear that we may appear inferior before such a judge, but our hopes are all the greater, since it is likely that he will take interest himself in the matter. For he would hardly have set the task for me unless he willed that it should be accomplished in accordance with his own intentions [the whole affair about the discourse]. And, moreover, to invoke his aid for the speech is wonderfully fitting. For if the poets, in order to say what they have in mind, invoke Apollo and the Muses, how much more appropriate is our invocation when we ask of

him who himself has set this task for us that he together with
his father may act as the leader of the Muses. But you who
are invoked for many things and on many occasions [2] — for I
can say that you have pointed out to me many things, and even
the subject of my own discourses—guide my speech now also,
as it is agreeable to you.

I need hardly seek far for a beginning of the eulogy: namely
the father alike of the speech and the youths. Of four such
generations no Greek ever heard or spoke. For just as the
legislators, so they,. too, have been patricians through four
generations, only in a higher degree than any legislators and
in fact any class of men. They are fourth from Zeus, and on
the same high level throughout. For Apollo comes from Zeus
himself, Asclepius in turn from Apollo, the two again from
Asclepius, picking up nobility of birth throughout due to the
supreme nature [of their ancestors]. For Achilles is fourth
from Zeus through Peleus and Aeacus; Minos and Rhadamanthys are children of Zeus; and Theseus is the son of
Poseidon; none of them, however, are the only children of
either, but are together with many other gods and heroes, to
some of whom they are inferior, while with others they are
on the same footing. These alone, however, in the number
and virtue of their ancestors are without rivals, since they are
superior to the children of Zeus and Apollo, at least to those
who are heroes, through the addition of Asclepius. After they
were born their father reared them in the gardens of Health,
and when they had grown old enough, he did not have them
instructed in the medical art; he educated them himself. For
it was no longer necessary for them to go to Chiron,[3] since
they had their instructor at home to whom Chiron was now
much inferior, as is indicated even by his name [" Inferior "].

Thus they were born and reared and educated, and while
the commonwealth of the Achaeans was still at peace, they
were for Thessaly an adornment much more conspicuous than
its lakes and plains and streams, and they guided wisely the
Greeks in that region, administering their cause for the common
good and straightening out their individual sufferings, and there
was no sickness wherever Machaon or Podalirius appeared.
But when the Greeks were roused by the injustice of the
Trojans, they [*sc.*, the Asclepiads] neither deemed it proper

simply to buy themselves the privilege of staying at home, nor did they hide themselves, as some others did, but recognizing that this was their opportunity and foreseeing the vicissitudes of the war, they voluntarily served as leaders in the common salvation. Having come to Troy, they were of twofold use to the Achaeans and were with them not only as physicians, but helped them also with their arms, and frequently they are said to have put the enemy to flight; the fall of Troy, moreover, was entirely due to them, for other reasons and particularly on account of Philoctetes' ailment. Odysseus and the Atrides had wrongly assumed beforehand that it was incurable, and had unjustly left Philoctetes behind in Lemnos, but they [the Asclepiads] healed it even though it had been augmented by ten years. And Philoctetes becomes fit for service with the Achaeans, and the missiles of Heracles are useful to Philoctetes through their art [sc., that of the Asclepiads].[4]

When Troy had fallen, foreseeing the subsequent Greek colonization of Asia, and wishing at the same time that as many as possible should enjoy the benefit of their good deeds, on the one hand they softened up Teuthrania for the reception of their father; on the other hand, as the Coan tradition has it, they sailed to the Meropian Cos,—which was inhabited by the Meropes and had been charged with injustice and destroyed by Heracles when he was returning from Troy at an earlier time,—and in Cos they settled and established customs proper to the nature of that country, imitating, as it seems, that which was told about their ancestor. For the poets say that Apollo firmly fixed Delos — which formerly was floating around — anchoring it on the sea when he was born there; and they, when they first arrived in the land of the Meropes, judging it to be the most beautiful of all that were of similar size, cured it and made accessible to all, Greeks and barbarians alike, the land that formerly was dangerous and suspicious, and established the happiness proper to that island. I believe that even among the Rhodians, where for a long time many noble institutions have prevailed, their rule was among the foremost which the Rhodians themselves held preferable: they chose the Asclepiads to rule over them, making them the successors of the Heraclids. They also held the Carian province and Cnidus, the holy place of Aphrodite. For whatever [country] one alone held and whatever both of them possessed together, shall be one and the same now. Also Corsica enjoyed something from them. But that is a long story.

Making their children co-workers and successors of their knowledge they added as the climax, so to speak, of the benefits which they bestowed upon the Greeks—and I shall add, upon all men—that the human race should never lack their care and grace, but that they be, both through their ancestors and through their offspring, the immortal saviors of nature, for all time dwelling among men and living with all generations to come just as with their own contemporaries. For they overthrew even those who were acclaimed in Egypt, making good deeds the warrant of their family. They settled down neither in Thessaly, nor in the suburbs of Cos, but every place they provided with the medical art, just as Triptolemus through his seeds provides every place with grain.[5] For everything was spread abroad through them and by them. And as it is said of Medea, that, while she fled through the Thessalian plain spilling her drugs, she made all Thessaly abound in herbs, thus their knowledge and their love of men, being poured out widely, adorned and is adorning, and moreover saved and is saving, all cities of the Greeks and many places of the barbarians, and let the third tense, the future, in addition to the past and present, be added.

But if Hippocrates were the only man sprung from them as the heir of their art, with all those in between laymen, this alone would have been a sufficiently great benefit for the earth and a blessing worth the gratitude of the human race toward them for their seed.[6] In fact, however, the Asclepiads were established as a family which by blood-relationship preserved the art; thus divine providence guided the children of Machaon and Podalirius. This one may learn by comparing them with the Heraclids in regard to their common usefulness and their personal fates. For the Heraclids were dispersed first of all, nor did they form one contingent, or one tribe, as it were. For they were not all of equal rank, so that they had become strangers to one another. Furthermore, their most outstanding representatives are said not to have been free of misfortune, nor to have succeeded in preserving their paternal skill. For they had founded their brilliance not on deeds for the common good, but rather on their own strength. The Asclepiads, on the contrary, starting with Machaon and Podalirius were a safeguard and a security common to all, having preserved the skill of their ancestor as another token of their family. Besides, even their fate was worthy of their character. For they were neither expelled, nor did they ever come as suppliants into any city, but they carried on free of misfortune, being of one brotherhood, one mind, and one destiny throughout.

Now I shall return to my starting-point, to the founders and those who first assumed the name Asclepiads. As long as they were dwelling among men, by participating in warfare and in community life, by begetting children worthy of themselves and in general by their whole political ability they helped the communities, alleviating not only the ailments of the body, but healing also the ills of the cities; or rather not even permitting evils of either kind to put in an appearance they saved their subjects from both and made their rule conform to their art. But since they were too mighty to stay with us, with the help of their father and their ancestors they left behind their bodies and entered another jurisdiction; yet they did not like Menelaus and the fair Rhadamanthys [7] go to the Elysian plain and the outward islands, but having become immortal they wander the earth, being removed from their former nature only in so far as they have preserved their youthfulness. And many have seen them in Epidaurus and have recognized them as they moved about in clear sight, and many have seen them in many other places, which should be remembered as the most remarkable thing about them. For Amphiaraus and Trophonius give oracles and appear in Boeotia, Amphilochus in Aetolia; these, however, are darting all over the earth, like stars, being both the servants and the heralds of their father. Wherever Asclepius has admittance, the doors are open to them, too, everywhere on earth, and in every respect is the unity with their father preserved, in temples, sacrifices, paeans, processions, and other things which are performed.

O, happy are you for your ancestors on both sides, blessed for those who have sprung from you, for yourselves and for your sisters, among whom are Iaso and Panacea and Aegle and Hygieia, who is worth as much as all the others, the famous children of Epione. You have no seats one without the other, nor have you taken separate quarters. O, you are yourselves a most beautiful chorus for your father, causing many others to be formed by men; you are by far the best chorus-masters of all; you are, too, temple-overseers and masters of the flowing bowl and of all thank-offerings. All the other sacrifices and festivals have been established by law, so to speak, but the ones that come from you and are concerned with your worship occur daily, most frequently even in comparison with all the others, but they spring purely from the heart, bringing, according to our own experience, contentment. Many and conspicuous are your traces; just as man is always followed by his shadow, so light follows you wherever you go. O you who are like unto the Dioscuri and their equals in age in another period of generation, you who have stopped mighty swells [of evil] and who have kindled many glittering torches on islands and on the mainland, this is our discourse which sprang from a dream in our sleep and is dedicated to you without delay. You in your kindness and love of men . . . counting it [the speech?] among the finer ones, relieve me of my disease and grant me as much health as is necessary in order that the body may obey that which the soul wishes, and, to say it in one word, a life lived with ease.

[1] Cf. Homer, *Iliad*, II, 56.
[2] Cf. T. 317, 1.
[3] Cf. T. 50 ff.
[4] Cf. T. 152.
[5] Cf. T. 467.
[6] Cf. T. 213 ff.
[7] Cf. Homer, *Odyssey*, IV, 564.

283. Plinius, Naturalis Historia, XXXV, 11 (40), 137.[1]

Tales sunt eius cum Aesculapio filiae, Hygia, Aegle, Panacea, Iaso . . .

Such works are his [*sc.*, of Socrates, the painter]: Asclepius with his daughters, Hygieia, Aegle, Panacea, Iaso . . .

[1] Cf. T. 665.

284. Scholia in Aristophanem, Ad Plutum, 701.

Πανάκεια δὲ παρὰ τὸ ἄκος, τὴν θεραπείαν. Ἄλλως· παρὰ τὸ ἰᾶσθαι τὴν Ἰασὼ πεποίηκε θυγατέρα Ἀσκληπιοῦ, καθάπερ καὶ τὴν Πανάκειαν καὶ τὴν Ὑγίειαν.

Panakeia was named from the word *akos* [remedy], the medical treatment. Variant: from the verb *iasthai* [to heal] he made Iaso the daughter of Asclepius just as also Panakeia and Hygieia.[1]

[1] Cf. **284a**. *Scholia in Aristophanem, Ad Plutum*, 639: Ἀναπέπλασται δὲ τὰ ὀνόματα παρὰ τὸ ἰᾶσθαι καὶ πάντας ἀκεῖσθαι καὶ παρὰ τὸ ὑγίειαν παρέχειν; cf. also Eudocia Augusta, *Violarium*, XI b, same text; cf. **284b**. *Scholia in Aristophanem, Ad Plutum* 701 [Rutherford]: Ἰασὼ μέν γε κ. τ. λ.: ἀναπέπλασται τὰ ὀνόματα παρὰ τὸ ἰᾶσθαι καὶ ⟨τὸ⟩ πάντας ἀκεῖσθαι. Ἰασώ: παρὰ τὸ ἰᾶσθαι· Ἰασώ: τὴν Ἰασὼ πεποίηκεν θυγατέρα Ἀσκληπιοῦ καθάπερ καὶ τὴν Πανάκειαν καὶ τὴν Ὑγείαν.

285. Scholia in Aristophanem, Ad Plutum, 701.

Ἀλλ᾽ Ἰασὼ μέν τις· οὐκ ἐῴκει. διότι προσῆκε τῷ Ἀσκληπιῷ ἡ Ἰασὼ παρὰ τὴν ἴασιν ὠνομασμένη. ἀλλὰ καὶ θυγατέρα τοῦ Ἀμφιαράου αὐτὴν εἶπεν ἐν ἐκείνοις· ᾽ἀλλ᾽ ὦ θύγατερ, ἔλεξ᾽, Ἰασοῖ, πρευμενής.᾽ εἰ δὲ καὶ τὴν Ἰασὼ Ἀσκληπιοῦ θυγατέρα, ὥσπερ καὶ τοῦ Ἀμφιαράου, ἄξιον ἀπορεῖν. . . . ἔστι δὲ καὶ Ἀμφιαράου θυγάτηρ Ἰασώ.

But someone understands: it was Iaso who blushed.[1] That is not likely, because Iaso is related to Asclepius, being named from [the word for] healing. But he also called her the daughter of Amphiaraus in these lines: "But gracious daughter, Iaso, he said." Even if he thinks that Iaso is the daughter of Asclepius, as she is also that of Amphiaraus, this remains difficult. . . . But Iaso is also the daughter of Amphiaraus.

[1] Cf. T. 421.

286. Pausanias, Descriptio Graeciae, I, 23, 4.

. . . τοῦ δὲ Διτρέφους πλησίον . . . θεῶν ἀγάλματά ἐστιν Ὑγείας τε, ἣν Ἀσκληπιοῦ παῖδα εἶναι λέγουσι . . .

. . . near the statue of Diitrephes . . . are figures of gods; of Health, whom they call daughter of Asclepius[1] . . .

[1] Cf. T. 641.

287. Inscriptiones Graecae, III, 1, no. 1159 [*ca.* 190-200 A. D.].

Τελεσφόρος Ἀσκληπιοῦ . . .

Telesphorus,[1] the son of Asclepius . . .

[1] Cf. T. 313.

288. Pausanias, Descriptio Graeciae, IV, 14, 8.

Μεσσήνιοι γὰρ οὐκ ἐσποιοῦσιν Ἀριστομένην Ἡρακλεῖ παῖδα ἢ
Διί, ὥσπερ Ἀλέξανδρον Ἄμμωνι οἱ Μακεδόνες καὶ Ἄρατον
Ἀσκληπιῷ Σικυώνιοι.

The Messenians do not make Aristomenes the son of Heracles
or of Zeus, as the Macedonians make Alexander Ammon's son
and the Sicyonians make Aratus the son of Asclepius.[1] (*)

[1] Cf. **288a.** Pausanias, *Descriptio Graeciae*, II, 10, 3: . . . τὴν δὲ . . . Ἀριστοδάμαν
Ἀράτου μητέρα εἶναι λέγουσι καὶ Ἄρατον Ἀσκληπιοῦ παῖδα εἶναι νομίζουσιν.

289. Sidonius Apollinaris, Carmina, II, 121-26.

. . . .

magnus Alexander nec non Augustus habentur
concepti serpente deo Phoebumque Jovemque
divisere sibi: namque horum quaesiit unus
Cinyphia sub Syrte patrem; maculis genetricis
alter Phoebigenam sese gaudebat haberi,
Paeonii iactans Epidauria signa draconis.

. . . Alexander the Great and Augustus are deemed to have been
conceived of a serpent god, and they claimed between them
Phoebus and Jupiter as their progenitors; for one of them
sought his sire near the Cinyphian Syrtes, the other rejoiced
that from his mother's marks he was deemed the offspring of
Phoebus, and he vaunted the imprints of the healing serpent
of Epidaurus.

RANGE OF POWER

290. Lucianus, Deorum Concilium, 6.

Οὗτοι γὰρ ὁ μὲν αὐτῶν ἰᾶται καὶ ἀνίστησιν ἐκ τῶν νόσων καὶ
ἔστι " πολλῶν ἀντάξιος ἄλλων," ὁ δὲ Ἡρακλῆς . . .

As far as they [*sc.*, Asclepius and Heracles] are concerned, one of them is a doctor who cures people of their illnesses and is " of the worth of many other men " [1] whilst Heracles . . .[2]

[1] Cf. T. 165, v. 514.
[2] Cf. T. 260; T. 351. Cf. also 290a. Lucianus, *Dialogi Deorum*, 26, 2: . . . ὁ δὲ Ἀσκληπιὸς ἰᾶται, σὺ δὲ [sc., Ἑρμῆς] παλαίειν διδάσκεις . . . ; 290b. Marcus Antoninus, *In Semet Ipsum*, VI, 43: Μήτι ὁ ἥλιος τὰ τοῦ ὑετοῦ ἀξιοῖ ποιεῖν; μήτι ὁ Ἀσκληπιὸς τὰ τῆς καρποφόρου; τί δὲ τῶν ἄστρων ἕκαστον; οὐχὶ διάφορα μέν, συνεργὰ δὲ πρὸς ταὐτόν; cf. 290c. Lactantius, *Institutionum Epitoma*, 2, 7: Non Vulcanus sibi aquam vindicabit aut Neptunus ignem, non Ceres artium peritiam nec Minerva frugum, non arma Mercurius nec Mars lyram, non Juppiter medicinam nec Asclepius fulmen; cf. 290d. Arnobius, *Adversus Nationes*, II, 65: Si enim patrem creditis Liberum dare posse vindemiam, medicinam non posse, si Cererem fruges, si Aesculapium sanitatem . . . ; 290e. Arnobius, *Adversus Nationes*, VII, 22: . . . ergo et musicis Apollo, quod musicus, et quod medicus Aesculapius, medicis, et quod faber Vulcanus est, fabris, et quod Mercurius eloquens, eloquentibus . . . ; 290f. *ibid.* I, 30: Apollo vobis pluit, Mercurius vobis pluit, Aesculapius, Hercules aut Diana rationem imbrium tempestatumque finxerunt?

291. Lucianus, Deorum Concilium, 16.

> Ἐργάζεσθαι δὲ τὰ αὑτοῦ ἕκαστον, καὶ μήτε τὴν Ἀθηνᾶν ἰᾶσθαι μήτε τὸν Ἀσκληπιὸν χρησμῳδεῖν μήτε τὸν Ἀπόλλω τοσαῦτα μόνον ποιεῖν, ἀλλὰ ἕν τι ἐπιλεξάμενον ἢ μάντιν ἢ κιθαρῳδὸν ἢ ἰατρὸν εἶναι.

> And be it further resolved that each ply his own trade; that Athena shall not heal the sick or Asclepius give oracles or Apollo combine in himself so many activities; but, selecting one, he shall be either seer or singer or physician.(*)

292. Origenes, Contra Celsum, III, 3.[1]

> . . . ὡς καὶ αὐτὸς ἐν τοῖς ἑξῆς παρατίθεται Ἀσκληπιὸν εὐεργετοῦντα, καὶ τὰ μέλλοντα προλέγοντα ὅλαις πόλεσιν ἀνακειμέναις αὐτῷ, οἷον τῇ Τρίκκῃ, καὶ τῇ Ἐπιδαύρῳ, καὶ τῇ Κῷ, καὶ τῇ Περγάμῳ . . .

> . . . as he [sc., Celsus] himself admits later that Asclepius conferred many benefits and foretold future events to entire cities, which were dedicated to him, such as Tricca, and Epidaurus, and Cos, and Pergamum . . .[2] (*)

[1] Cf. T. 711.
[2] Cf. T. 307; 328.

293. Origenes, Contra Celsum, III, 24.

> Καὶ πάλιν ἐπὰν μὲν περὶ τοῦ Ἀσκληπιοῦ λέγηται, ὅτι πολὺ ἀνθρώπων πλῆθος Ἑλλήνων τε καὶ βαρβάρων ὁμολογεῖ πολλάκις

ἰδεῖν, καὶ ἔτι ὁρᾶν, οὐ φάσμα αὐτὸ τοῦτο ἀλλὰ θεραπεύοντα καὶ
εὐεργετοῦντα, καὶ τὰ μέλλοντα προλέγοντα, πιστεύειν ἡμᾶς ὁ
Κέλσος ἀξιοῖ . . . οὐκ ἔχοντι παραστῆσαι ἀμύθητον (ὥς
φησι) πλῆθος ἀνθρώπων Ἑλλήνων καὶ βαρβάρων ὁμολογούντων
Ἀσκληπιῷ.

And again, when it is said of Asclepius that a great multitude
both of Greeks and barbarians acknowledge that they have
frequently seen, and still see, no mere phantom, but Asclepius
himself, healing and doing good, and foretelling the future;
Celsus requires us to believe this, . . . although he cannot
demonstrate that an incalculable number, as he asserts, of
Greeks and barbarians acknowledge the existence of Asclepius.

294. Eusebius Hieronymus, In Isaiam Commentaria, XVIII, 65.

Nihil fuit sacrilegii quod Israel populus praetermitteret, non
solum in hortis immolans, et super lateres thura succendens,
sed sedens quoque vel habitans in sepulcris, et in delubris
idolorum dormiens, ubi stratis pellibus hostiarum incubare
soliti erant, ut somniis futura cognoscerent. quod in fano
Aesculapii usque hodie error celebrat ethnicorum multorumque
aliorum, quae non sunt aliud, nisi tumuli mortuorum.

Nothing of a sacrilegious nature did the people of Israel refrain
from, not only performing sacrifices in gardens and lighting
incense on stones but also sitting or dwelling in sepulchres and
sleeping in shrines of idols, where they were wont to lie on the
outspread skins of sacrificial victims,[1] for the purpose of learning the future through dreams. The heathens, in their delusion,
celebrate this in the shrine of Asclepius up to the present day,
and [in the shrines] of many others which are nothing but
tombs of dead men.

[1] Cf. T. **158.**

295. Inscriptiones Graecae, IV², 1, no. 128, v, 57-79 [3rd c. B. C.].[1]

Καὶ τόδε σῆς ἀρετῆς, Ἀσκληπιέ, τοὖργον ἔδειξας
ἐγ κείνοισι χρόνοις, ὅκα δὴ στρατὸν ἆγε Φίλ[ι]ππος
εἰς Σπάρτην ἐθέλων ἀνελεῖν βασιληίδα τιμήν.
60 τοῖς δ' Ἀσκληπιὸς ἦλθε βοαθόος ἐξ Ἐπιδαύρου,
τιμῶν Ἡρακλέος γενεάν, ἃς φείδετο ἄρα Ζεύς.

τουτάκι δ' ἦλθε, ὄχ' ὁ παῖς ἐκ Βουσπόρου † ἦλθεν κάμνων.
τῶι τύγα ποστείχοντι συνάντησας σὺν ὅπλοισιν
λαμπόμενος χρυσέοισ', 'Ασκλαπιέ. παῖς δ' ἐσιδών σε
65 λίσσετο χεῖρ' ὀρέγων ἱκέτηι μύθωι σε προσαντῶν·
'ἄμμορός εἰμι τεῶν δώρων, 'Ασκληπιὲ Παιάν,
ἀλλά μ' ἐποίκτειρον.' τὺ δέ μοι τάδε ἔλεξας ἐναργῆ·
'θάρσει· καιρῶι γάρ σοι ἀφίξομαι—ἀλλὰ μέν' αὐτεῖ—
τοῖς Λακεδαιμονίοις χαλεπὰς ἀπὸ κῆρας ἐρύξας,
70 οὕνεκα τοὺς Φοίβου χρησμοὺς σώιζοντι δικαίως,
οὓς μαντευσάμενος παρέταξε πόληι Λυκοῦργος.'
ὣς ὁ μὲν ὤιχετο ἐ[[ε]]πὶ Σπάρτην· ἐμὲ δ' ὦρσε νόημα
ἀγγεῖλαι Λακεδαιμονίοις ἐλθόντα τὸ θεῖον
πάντα μάλ' ἐξείας. οἳ δ' αὐδήσαντος ἄκουσαν
75 σώτειραν φήμαν, 'Ασκλαπιέ, καί σφε σάωσας.
οἳ δὲ ἐκάρυξαν πάντας ξενίαις σε δέκεσθαι,
σωτῆρα εὐρυχόρου Λακεδαίμονος ἀγκαλέοντες.
ταῦτά τοι, ὦ μέγ' ἄριστε θεῶν, ἀνέθηκεν Ἴσυλλος
τιμῶν σὴν ἀρετήν, ὦναξ, ὥσπερ τὸ δίκαιον.

And of your power, Asclepius, you gave this example in those days when Philip, wishing to destroy the royal authority, led his army against Sparta. To them from Epidaurus Asclepius came as a helper, honoring the race of Heracles, which consequently Zeus spared. He came at the time when the sick boy came from Bosporos. Shining in your golden armor, you met him as he approached, Asclepius; and when the boy beheld you, he drew near to you, stretching forth his hand, and entreated you in suppliant words: "I have no share in your gifts, Asclepius Paean; have pity on me." Then you addressed these words to me clearly: "Take heart, for I shall come to you in due time — just wait here — after I have rescued the Lacedaemonians from grievous doom because they justly guard the precepts of Apollo which Lycurgus ordained for the city, after he had consulted the oracle." And so he went to Sparta. But my thoughts stirred me to announce the divinity's advent to the Lacedaemonians, everything in exact order. They listened to me as I spoke the message of safety, Asclepius, and you saved them. And they called upon all to welcome you with honors due a guest, proclaiming you the Savior of spacious Lacedaemon. These words, O far the best of all the gods,

Isyllus set up for you, honoring your power, O Lord, as is seemly.

¹ Hymn of Isyllus.

296. Inscriptiones Graecae, IV², 1, no. 128, i, 1–ii, 26 [3rd c. B. C.].

Ἴσυλλος Σωκράτευς Ἐπιδαύριος ἀνέθηκε
Ἀπόλλωνι Μαλεάται καὶ Ἀσκλαπιῶι.

Δᾶμος εἰς ἀριστοκρατίαν ἄνδρας αἰ προάγοι καλῶς,
αὐτὸς ἰσχυρότερος· ὀρθοῦται γὰρ ἐξ ἀνδραγαθίας.
5 αἰ δέ τις καλῶς προαχθεὶς θιγγάνοι πονηρίας
πάλιν ἐπαγκρούων, κολάζων δᾶμος ἀσφαλέστερος.
τάνδε τὰν γνώμαν τόκ᾿ ἦχον καὶ ἔλεγον καὶ νῦν λέγω.
εὐξάμαν ἀνγράψεν, αἴ κ᾿ εἰς τάνδε τὰν γνώμαν πέτη
ὁ νόμος ἁμίν, ὃν ἐπέδειξα. ἔγεντο δ᾿ οὐκ ἄνευ θεόν.

10 Τόνδ᾿ ἱαρὸν θείαι μοίραι νόμον ηὗρεν Ἴσυλλος
ἄφθιτον ἀέναον γέρας ἀθανάτοισι θεοῖσιν,
καί νιν ἅπας δᾶμος θεθμὸν θέτο πατρίδος ἁμᾶς,
χεῖρας ἀνασχόντες μακάρεσσιν ἐς οὐρανὸν εὐρύ[ν]·
οἵ κεν ἀριστεύωσι πόληος τᾶσδ᾿ Ἐπιδαύρου,
15 λέξασθαί τε ἄνδρας καὶ ἐπαγγεῖλαι κατὰ φυλάς,
οἷς πολιοῦχος ὑπὸ στέρνοις ἀρετά τε καὶ αἰδώς,
τοῖσιν ἐπαγγέλλεν καὶ πομπεύεν σφε κομῶντας
Φοίβωι ἄνακτι υἱῶι τε Ἀσκλαπιῶι ἰατῆρι
εἵμασιν ἐν λευκοῖσι, δάφνας στεφάνοις ποτ᾿ Ἀπόλλω,
20 ποὶ δ᾿ Ἀσκλαπιὸν ἔρνεσι ἐλαίας ἡμεροφύλλου
ἁγνῶς πομπεύειν καὶ ἐπεύχεσθαι πολιάταις
πᾶσιν ἀεὶ διδόμεν τέκνοις τ᾿ ἐρατὰν ὑγίειαν,
τὰν καλοκαγαθίαν τ᾿ Ἐπιδαυροῖ ἀεὶ ῥέπεν ἀνδρῶν
εὐνομίαν τε καὶ εἰράναν καὶ πλοῦτον ἀμεμφῆ,
25 ὥραις ἐξ ὡρᾶν νόμον ἀεὶ τόνδε σέβοντας·
οὕτω τοί κ᾿ ἁμῶν περιφείδοιτ᾿ εὐρύοπα Ζεύς.

Isyllus, son of Socrates, the Epidaurian, dedicated this to Apollo Maleatas and to Asclepius.

If the city properly educates men for aristocracy, it becomes itself mightier, for it is raised up by manly virtue. But if one who is properly educated sets his course back, falling again into baseness, then the city will be safer in chastising him. This opinion I held before and I pronounced it and I pronounce it

now. I vowed to inscribe it in stone if our law which I introduced would confirm that opinion. This came about, not
without the help of the gods.

This law, sacred by divine Fate, Isyllus composed, an imperishable, everlasting gift to the immortal gods; and all the
people, lifting their hands to the wide heaven, to the blessed
gods, set it up as a binding rule of our fatherland: to select
and to summon by tribes whichever men may be best in this
city of Epidaurus, those who have in their hearts virtue and
reverence that safeguard the city; to summon them and to
have them lead a procession to lord Phoebus and to his son
Asclepius, the physician, dressed in white raiment and with
flowing hair; to lead a solemn procession to the temple of
Apollo bearing garlands of laurel and then to the temple of
Asclepius bearing branches of tender olive shoots; to pray
them to grant forever to all citizens and to their children fair
health and to grant that the noble character of the men of
Epidaurus always prevail, together with good order and peace
and blameless wealth, from season to season so long as they
revere this law. So may Zeus the far-seeing spare us.

PHILOSOPHICAL INTERPRETATION

297. Pausanias, Descriptio Graeciae, VII, 23, 7-8.

Τῆς δὲ Εἰλειθυίας οὐ μακρὰν Ἀσκληπιοῦ τέ ἐστι τέμενος καὶ
ἀγάλματα Ὑγείας καὶ Ἀσκληπιοῦ. . . . ἐν τούτῳ τοῦ Ἀσκληπιοῦ
τῷ ἱερῷ ἐς ἀντιλογίαν ἀφίκετο ἀνήρ μοι Σιδόνιος, ὃς ἐγνωκέναι
τὰ ἐς τὸ θεῖον ἔφασκε Φοίνικας τά τε ἄλλα Ἑλλήνων βέλτιον
καὶ δὴ καὶ Ἀσκληπιῷ πατέρα μὲν σφᾶς Ἀπόλλωνα ἐπιφημίζειν,
θνητὴν δὲ γυναῖκα οὐδεμίαν μητέρα. Ἀσκληπιὸν μὲν γὰρ ἀέρα
γένει τε ἀνθρώπων εἶναι καὶ πᾶσιν ὁμοίως ζῴοις ἐπιτήδειον πρὸς
ὑγίειαν, Ἀπόλλωνα δὲ ἥλιον, καὶ αὐτὸν ὀρθότατα Ἀσκληπιῷ
πατέρα ἐπονομάζεσθαι, ὅτι ἐς τὸ ἁρμόζον ταῖς ὥραις ποιούμενος
ὁ ἥλιος τὸν δρόμον μεταδίδωσι καὶ τῷ ἀέρι ὑγιείας. ἐγὼ δ᾽ ἀπο-
δέχεσθαι μὲν τὰ εἰρημένα, οὐδὲν δέ τι Φοινίκων μᾶλλον ἢ καὶ
Ἑλλήνων ἔφην τὸν λόγον, ἐπεὶ καὶ ἐν Τιτάνῃ τῆς Σικυωνίων τὸ
αὐτὸ ἄγαλμα Ὑγείαν τε ὀνομάζεσθαι καὶ † παιδὶ εἶναι * δῆλα ὡς
τὸν ἡλιακὸν δρόμον ἐπὶ γῆς ὑγίειαν ποιοῦντα ἀνθρώποις.

* εἶναι L² ἦν cett.

Not far from Eileithyia is a precinct of Asclepius [*sc.*, in Aegium], with images of him and of Health. . . . In this sanctuary of Asclepius a man of Sidon entered upon an argument with me. He declared that the Phoenicians had better notions about the gods than the Greeks, giving as an instance, among other reasons, that to Asclepius they assign Apollo as father, but no mortal woman as mother. For Asclepius, he went on, is air, bringing health to mankind and to all animals likewise; Apollo is the sun, and most rightly is he named the father of Asclepius, because the sun, by adapting his course to the seasons, imparts to the air its healthfulness. I replied that I accepted his statements, but that the argument was as much Greek as Phoenician: for at Titane in Sicyonia the same image is called both Health and †, thus clearly showing even to a child that it is the course of the sun that brings health to mankind.(**)

298. Eusebius, Praeparatio Evangelica, III, 13, 15-16.

Μεταβὰς δὲ ἐκ τούτων καὶ τὰ λοιπὰ κατὰ τὸν ὅμοιον τρόπον τῆς γενναίας ἀπελέγξεις φυσιολογίας, εὐλόγως τῆς ἀναισχυντίας ἐπιμεμψάμενος τοῖς, φέρε, τὸν ἥλιον αὐτὸν εἶναι τὸν Ἀπόλλω, καὶ πάλιν τὸν Ἡρακλέα, καὶ αὖθις τὸν Διόνυσον, καὶ τὸν Ἀσκληπιὸν ὁμοίως ἀποφαινομένοις. πῶς γὰρ ὁ αὐτὸς πατὴρ ἂν γένοιτο καὶ υἱός, Ἀσκληπιὸς ὁμοῦ καὶ Ἀπόλλων;

Turning then from these things [the discussion of Dionysus, Attis, and Adonis], you will also refute the rest of this noble science of nature in a similar way, with good reason chiding for their shamelessness those, alas, who declare that the sun is Apollo himself and again Heracles and Dionysus as well and also Asclepius. For how could the same be both father and son, Asclepius and at the same time Apollo?

299. Eusebius, Praeparatio Evangelica, III, 13, 19-20.

Εἰ δὲ καὶ Ἀσκληπιὸς πάλιν αὐτοῖς εἴη ὁ ἥλιος, πῶς οὗτος κεραυνοῦται ὑπὸ τοῦ Διὸς ῥυπαρᾶς ἕνεκεν αἰσχροκερδείας, κατὰ τὸν Βοιώτιον μελοποιὸν Πίνδαρον ὧδέ πως λέγοντα . . . τίνες δὲ καὶ οἱ ἐξ ἡλίου Ἀσκληπιάδαι, εἰς μακρὸν τοῦ βίου διαφυλαχθέντες καὶ θνητῶν ἀνδρῶν γένεσιν πᾶσιν ἀνθρώποις παραπλησίαν ὑπο-

στησάμενοι; πλὴν ἀλλὰ πάλιν αὐτοῖς, ὥσπερ διὰ μηχανῆς, τὰς
αἰσχρὰς καὶ μυθικὰς περὶ θεῶν διηγήσεις ἀποφεύγουσιν, πρὸς
ἥλιον καὶ σελήνην καὶ τὰ λοιπὰ τοῦ κόσμου μέρη ἀνατρέχοι ἂν
ὁ λόγος.

Again if also Asclepius were for them [sc., the gentiles] the
sun,[1] how was he struck with a thunderbolt by Zeus because
of his sordid love of gain, as is related by the Boeotian lyric
poet Pindar, who says this[2] . . . And who are the Asclepiads,
descendants of the sun, who have survived for many ages and
who have taken it upon themselves to bring forth mortal men
similar to all other human beings?[3] However, since through
some machination, so to speak, they [sc., the gentiles] evade
the shameful mythical legends about the gods, it seems natural
that their argument should fall back upon the sun and the
moon and the other parts of the cosmos.

[1] Cf. **299a.** Eusebius, *Praeparatio Evangelica*, III, 11, 26: Τῆς δὲ σωστικῆς αὐτοῦ
[sc., τοῦ ἡλίου] δυνάμεως Ἀσκληπιὸς τὸ σύμβολον, cf. T. 301; 706.
[2] T. 1, v. 54.
[3] Cf. T. 282, 14.

300. Joannes Lydus, De Mensibus, IV, 45.

Φασὶ δὲ ταύτην καὶ ὑγείας εἶναι δότειραν, καθάπερ ἡμεῖς τὸν
Ἀσκληπιόν. ταὐτὸν δ' ἂν εἴη· ὥσπερ γὰρ ἡμεῖς τὸν ἥλιον εἰς
τὸν Ἀσκληπιὸν λαμβάνοντες μετὰ τὰς ἐκείνου δυσμὰς τῆς νυκτὸς
ἐπιγινομένης αἴτιον αὐτὸν τοῦ ὕπνου καὶ νυκτὸς καὶ ἀναπαύσεως,
αἴτιον καὶ δοτῆρα τῆς ὑγείας . . .* μυθεύουσιν.

* Lacunam indicavit Schowius.

They [sc., the Egyptians] say that she [sc., Isis] is also the
dispenser of health, just as we say Asclepius is. It would
amount to the same thing, for just as we, identifying the sun
with Asclepius, when night comes on after the setting of the
sun, call him the cause of sleep and night and rest, the cause
and dispenser of health, so they have a tradition [that Isis,
too, ?]. . . .

301. Macrobius, Saturnalia, I, 20, 1-4.

Hinc est, quod simulacris et Aesculapii et Salutis draco sub-
iungitur, quod hi ad solis naturam lunaeque referuntur. et est
Aesculapius vis salubris, de substantia solis subveniens animis

corporibusque mortalium. Salus autem natura lunaris effectus est quo corpora animantium iuvantur salutifero firmata temperamento. ideo ergo simulacris eorum iunguntur figurae draconum, quia praestant, ut humana corpora velut infirmitatis pelle deposita ad pristinum revirescant vigorem, ut revirescunt dracones per annos singulos pelle senectutis exuta. propterea et ad ipsum solem species draconis refertur, quia sol semper velut a quadam imae depressionis senecta in altitudinem suam ut in robur revertitur iuventutis. esse autem draconem inter praecipua solis argumenta etiam nominis fictione monstratur, quod sit nuncupatus ἀπὸ τοῦ δέρκειν id est videre. nam ferunt hunc serpentem acie acutissima et pervigili naturam sideris huius imitari atque ideo aedium adytorum oraculorum thesaurorum custodiam draconibus adsignari. Aesculapium vero eundem esse atque Apollinem non solum hinc probatur, quod ex illo natus creditur, sed quod ei et vis divinationis adiungitur. nam Apollodorus in libris, quibus titulus est Περὶ θεῶν, scribit quod Aesculapius divinationibus et auguriis praesit.

The following is the reason that the serpent is placed at the foot of the statues of Asclepius and Health, because these gods are traced back to the nature of the sun[1] and moon. And Asclepius is the healthful force which comes to the aid of mortal spirits and bodies from the sun's essence. Health, moreover, is by nature a lunar influence, by which the bodies of living creatures are aided, being strengthened by its salubrious quality. For this reason, therefore, images of serpents are attached to the statues of these gods, because they symbolize that human bodies, shedding the skin of infirmity, as it were, return to their original vigor, just as serpents grow young again every year by shedding the skin of old age. Therefore the image of the serpent is also referred to the sun itself[1] because the sun always returns, as it were, from an old age of the profoundest depth [at sunset] to its highest point [at midday], the strength of its youth, as it were. Even by the fashioning of its name the serpent is shown to be one of the special attributes of the sun, because it derives its name from *derkein*, meaning " to see," " for they say that this serpent has the keenest vision and counterfeits the nature of the star that is ever watchful; for this reason, furthermore, the protection of shrines, holy places, oracles, and treasures is entrusted to serpents. Not only on this account is it proved that Asclepius is the same as Apollo, because he is supposed to have originated from him, but because the power of divination is also attributed to him. For Apollodorus writes in the books which are entitled *On the Gods* that Asclepius superintended divination and augury.[2]

[1] Cf. also T. 701; 706.   [2] = Apollodorus Atheniensis, Fr. 116 [Jacoby].

302. Aristides, Oratio L, 56.

Καὶ προελθὼν μικρὸν ἀνασχὼν τὴν χεῖρα δείκνυσί μοι τόπον τινὰ τοῦ οὐρανοῦ, καὶ ἅμα δεικνὺς ἔφη· ‘οὗτος δή σοί ἐστιν ὃν καλεῖ Πλάτων τοῦ παντὸς ψυχήν.’ ἀναβλέπω τε δὴ καὶ ὁρῶ Ἀσκληπιὸν τὸν ἐν Περγάμῳ ἐνιδρυμένον ἐν τῷ οὐρανῷ, καὶ ἅμα τε ἀφυπνιζόμην ἐπὶ τούτοις . . .

And coming forward a little he [*sc.*, Pyrallianus] stretches out his hand and shows me a certain spot in the sky, and while pointing at it he said: "This is the one that Plato calls the soul of the universe."[1] I look and I see enthroned in the sky Asclepius of Pergamum; and immediately I woke up after that . . .

[1] = *Timaeus*, 34B.

303. Aristides, Oratio XLII, 4.[1]

. . . οὗτός ἐσθ᾽ ὁ τὸ πᾶν ἄγων καὶ νέμων σωτὴρ τῶν ὅλων καὶ φύλαξ τῶν ἀθανάτων, εἰ δὲ θέλοις τραγικώτερον εἰπεῖν, ‘ἔφορος οἰάκων,’ σῴζων τά τε ὄντα ἀεὶ καὶ τὰ γιγνόμενα. εἰ δ᾽ Ἀπόλλωνος παῖδα καὶ τρίτον ἀπὸ Διὸς νομίζομεν αὐτόν, αὖθις ⟨δ᾽⟩ αὖ καὶ συνάπτομεν τοῖς ὀνόμασιν, . . .* ἐπεί τοι καὶ αὐτὸν τὸν Δία γενέσθαι λέγουσίν ποτε, πάλιν δὲ αὐτὸν ἀποφαίνουσιν ὄντα τῶν ὄντων πατέρα καὶ ποιητήν.

* Lacunam indicavit Keil.

. . . he [*sc.*, Asclepius] is the one who guides and rules the universe, the savior of the whole and the guardian of the immortals, or if you wish to put it in the words of a tragic poet,[2] "the steerer of government," he who saves that which always exists and that which is in the state of becoming. But if we believe him to be the son of Apollo, and the third from Zeus, and if again we link him with these names. . . . ; since sometimes they maintain that even Zeus is born, and then again they show that he is the father and maker of everything.

[1] Cf. T. 317.   [2] *Tragicorum Graecorum Fragmenta, Adespota*, 39 [Nauck²].

304. Proclus, In Platonis Timaeum, I, 49 C.

Πορφύριος δὲ εἰκότως φησὶ καὶ ἰατρικὴν ἀπὸ τῆς Ἀθηνᾶς ἥκειν, διότι καὶ ὁ Ἀσκληπιὸς νοῦς ἐστι Σεληνιακός, ὥσπερ ὁ Ἀπόλλων Ἡλιακὸς νοῦς. οἷς καὶ ὁ θεῖος ἐπέπληξεν Ἰάμβλιχος ὡς οὐ καλῶς συγχέουσι τὰς τῶν θεῶν οὐσίας οὐδὲ ὀρθῶς ἀεὶ κατὰ τὸ παρὸν τοὺς νοῦς καὶ τὰς ψυχὰς τῶν ἐγκοσμίων διανέμουσιν· ἐπεὶ καὶ Ἀσκληπιὸν ἐν Ἡλίῳ θετέον καὶ ἀπ' ἐκείνου προϊέναι περὶ τὸν γενητὸν τόπον ἵν' ὥσπερ ὁ οὐρανός, οὕτω δὴ καὶ ἡ γένεσις κατὰ δευτέραν μετοχὴν ὑπὸ τῆς θεότητος ταύτης συνέχηται, συμμετρίας καὶ εὐκρασίας ἀπ' αὐτῆς πληρουμένη.

Porphyry declares it reasonable that even the art of healing comes from Athena, because also Asclepius is lunar intellect, just as Apollo is solar intellect. The divine Iamblichus, too, criticizes these statements since they unfairly confound the essences of the gods and do not always, according to the circumstances, rightly distribute among them the intellect and the soul of the mundane powers. In fact, one should place also Asclepius in the Sun letting him proceed from there to the world of becoming; in this way just as the heavens, so becoming itself may be held together in secondary partnership by this divinity, and be filled by it with symmetry and balanced union.

305. Julianus, In Helium Regem, 144 B.

Ἐπεὶ δὲ καὶ ὅλην ἡμῖν τὴν τῆς εὐταξίας ζωὴν συμπληροῖ, γεννᾷ μὲν ἐν κόσμῳ τὸν Ἀσκληπιόν, ἔχει δὲ αὐτὸν καὶ πρὸ τοῦ κόσμου παρ' ἑαυτῷ.

And since he [sc., the Sun] fills the whole of our life with fair order, he begets Asclepius in the world, though he has him by his side even before the beginning of the world. (*)

306. Julianus, In Helium Regem, 153 B.

Τί ἔτι σοι λέγω, πῶς τῆς ὑγιείας καὶ σωτηρίας πάντων προὐνόησε τὸν σωτῆρα τῶν ὅλων ἀπογεννήσας Ἀσκληπιόν . . .

Shall I now go on to tell you how Helius took thought for the health and safety of all by begetting Asclepius to be the savior of the whole world . . . (*)

307. Julianus, Contra Galilaeos, 200 A-B.[1]

Ἔλαθέ με μικροῦ τὸ μέγιστον τῶν Ἡλίου καὶ Διὸς δώρων . . . ὁ γάρ τοι Ζεὺς ἐν μὲν τοῖς νοητοῖς ἐξ ἑαυτοῦ τὸν Ἀσκληπιὸν ἐγέννησεν, εἰς δὲ τὴν γῆν διὰ τῆς Ἡλίου γονίμου ζωῆς ἐξέφηνεν. οὗτος ἐπὶ γῆς ἐξ οὐρανοῦ ποιησάμενος τὴν πρόοδον, ἐνοειδῶς μὲν ἐν ἀνθρώπου μορφῇ περὶ τὴν Ἐπίδαυρον ἀνεφάνη, πληθυνόμενος δὲ ἐντεῦθεν ταῖς προόδοις ἐπὶ πᾶσαν ὤρεξε τὴν γῆν τὴν σωτήριον ἑαυτοῦ δεξιάν. ἦλθεν εἰς Πέργαμον, εἰς Ἰωνίαν, εἰς Τάραντα μετὰ ταῦθ᾽, ὕστερον ἦλθεν εἰς τὴν Ῥώμην. ᾤχετο δὲ εἰς Κῶ, ἐνθένδε εἰς Αἰγάς. εἶτα πανταχοῦ γῆς ἐστι καὶ θαλάσσης. οὐ καθ᾽ ἕκαστον ἡμῶν ἐπιφοιτᾷ, καὶ ὅμως ἐπανορθοῦται ψυχὰς πλημμελῶς διακειμένας καὶ τὰ σώματα ἀσθενῶς ἔχοντα.

I had almost forgotten the greatest of the gifts of Helius and Zeus . . . I mean to say that Zeus engendered Asclepius from himself among the intelligible gods, and through the life of generative Helius he revealed him to the earth. Asclepius, having made his visitation to earth from the sky, appeared at Epidaurus singly, in the shape of a man; but afterwards he multiplied himself, and by his visitations stretched out over the whole earth his saving right hand. He came to Pergamum,[2] to Ionia, to Tarentum afterwards; and later he came to Rome.[3] And he travelled to Cos[4] and thence to Aegae.[5] Next he is present everywhere on land and sea. He visits no one of us separately, and yet he raises up souls that are sinful and bodies that are sick.

[1] Cf. T. 843.
[2] Cf. T. 292; 328; 801 ff.
[3] Cf. T. 845 ff.
[4] Cf. T. 794 ff.; cf. also T. 595.
[5] Cf. T. 816 ff.; cf. also T. 615.

308. Sallustius, De Dis, VI.

Τούτων δὲ πρώτως ἐχόντων τὸν κόσμον, καὶ τοὺς ἄλλους ἐν τούτοις ἡγητέον εἶναι θεούς, οἷον Διόνυσον μὲν ἐν Διὶ Ἀσκληπιὸν δὲ ἐν Ἀπόλλωνι Χάριτας δὲ ἐν Ἀφροδίτῃ.

While these gods possess the universe in a primary way, the other gods must be supposed to be contained in them, as for instance Dionysus in Zeus, Asclepius in Apollo, and the Graces in Aphrodite.

309. Proclus, In Platonis Rem Publicam, I, p. 69, 7 [Kroll].

. . . ὁ δὲ πάντα κατὰ φύσιν ἔχοντα δεικνὺς Ἀσκληπιός, δι' ὃν οὐ νοσεῖ ⟨τὸ⟩ πᾶν, οὐ γηράσκει, τὰ στοιχεῖα οὐκ ἀνίησι τῶν ἀλύτων δεσμῶν.

. . . he who causes everything to act according to nature is Asclepius who keeps the universe from falling sick or growing old, and the elements from relaxing their indestructible bonds.

310. Proclus, In Platonis Cratylum, LXXXI.

. . . (οὕτω γὰρ ἄν, οἶμαι, καὶ Διόνυσοι καὶ Ἀσκληπιοὶ καὶ Ἑρμαῖ καὶ Ἡρακλέες ὁμώνυμοι τοῖς ἐφόροις αὐτῶν θεοῖς ἐπ' εὐεργεσίᾳ τῶν τῇδε τόπων προεληλύθασιν σὺν πομπῇ θεῶν) . . .

. . . (for in this way, I think, the Dionysi also, and the Asclepii and the Hermes and the Heracles, having the same name as their guardian deities, have come forth for the benefit of the regions here below through divine ordinance) . . .

311. Proclus, In Platonis Timaeum, V, 290 C.

Ἢ πόθεν ἡμῖν Ἀσκληπιοὶ καὶ Διόνυσοι καὶ Διόσκουροι προσηγορεύθησαν; ὥσπερ οὖν ἐπὶ τῶν οὐρανίων θεῶν, οὕτω χρὴ καὶ ἐπὶ τῶν γενεσιουργῶν θεωρεῖν περὶ ἕκαστον αὐτῶν πλῆθος ὁμόστοιχον ἀγγελικόν, δαιμόνιον, ἡρωικόν . . .

Or whence have the Asclepii and the Dionysi and the Dioscuri received their names? Just as in the case of the heavenly deities, then, so we must proceed in the case of those who are concerned with generation, that is, we must investigate in regard to each of them the number of messengers, demigods, heroes attached to them . . .

312. Proclus, In Platonis Timaeum, I, 49 A.

Καὶ δὴ καὶ ἰατρικὴν ὡσαύτως ἐν μὲν τοῖς θεοῖς αὐτὴν τὴν παιώνιον θετέον, ἐν δὲ τοῖς δαίμοσι τὴν διακονικὴν καὶ ὑπουργικὴν . . .

ὡς γάρ εἰσι περὶ τὸν Ἔρωτα πολλοὶ δαίμονες, οὕτω καὶ περὶ τὸν
Ἀσκληπιόν, οἱ μὲν ὀπαδῶν, οἱ δὲ προπομπῶν τοῦ θεοῦ τάξιν
λαχόντες. ἐν δὲ ταῖς ἀνθρωπίναις ζωαῖς τὴν ἐκ θεωρημάτων καὶ
πείρας ποριζομένην, καθ᾿ ἣν οἱ μὲν μᾶλλον, οἱ δὲ ἧττον οἰκειοῦνται
πρὸς τὴν θείαν ἰατρικήν.

So, too, in the medical art [sc., as in mantics] the Paeonian
power itself must be assigned to the gods, while the function
of serving and helping belongs to the demigods . . . for just
as there are many divinities associated with Eros, so, too,
many are associated with Asclepius, some taking their place
behind the god, others in front of him. But to mortals must
be assigned the medical art resulting from theory and experience by means of which some master the divine art of healing
to a greater, others to a lesser, degree.

313. Damascius, Dubitationes et Solutiones, 245.

Καὶ ἡ μὲν ὁλότης οὐσία μᾶλλον, ἡ δὲ τελειότης τὸ τέλος ἐπάγει
καὶ τὸ εὖ τῆς οὐσίας. ὅθεν καὶ ὁ Τελεσφόρος τοῦ Ἀσκληπιοῦ
καταδεέστερος, ὅτι τὸ ἐλλεῖπον ἀναπληροῖ καὶ τὸ μὴ φθάνον τὴν
Ἀσκληπιοῦ παιάνιον ὁλότητα, ἐπανακαλεῖται πρὸς αὐτήν, καὶ
ἐπιτελειοῖ τὴν ὑγίειαν τῷ δεχομένῳ συμμέτρως.

And wholeness is virtually substance, but perfection, in addition, provides the aim and the virtue of substance. Hence,
though inferior to Asclepius, Telesphorus,[1] because he supplies
the missing element which is not previously present in the
Paeonian wholeness of Asclepius, is invoked in addition to
Asclepius, and Telesphorus perfects the health of one who
admits him properly.[2]

[1] Cf. T. 287; 749; 808.
[2] Cf. 313a. Themistius, De Anima, A, p. 5, 16-19 [Heinze]: . . . τὸ μέντοι γε
κινεῖσθαι ἀεί, καὶ τὸ προλέγειν μὲν τὸν Ἀπόλλω, θεραπεύειν δὲ τὸν Ἀσκληπιόν, ἱκανὸν ἤδη
προοίμιον τοῦ τί ἦν εἶναι, οἷον ὅτι θεός ἐστι ζῷον ἀίδιον εὐποιητικὸν ἀνθρώπων.

314. Proclus, In Platonis Timaeum, III, 158 E.[1]

. . . διὸ καὶ οἱ θεολόγοι τὴν μὲν εἰς Ἀσκληπιὸν ἀναφέρουσιν
ὑγείαν, τὴν ἰατρικὴν πᾶσαν τοῦ παρὰ φύσιν, εἴτε ἀεὶ τὸ παρὰ
φύσιν ἀναστέλλουσαν εἴτε ποτέ, τὴν δὲ πρὸ Ἀσκληπιοῦ γεννῶσι
τῇ δημιουργίᾳ συνυφεστῶσαν τῶν πραγμάτων, ἣν παράγουσιν
ἀπὸ Πειθοῦς καὶ Ἔρωτος, διότι τὸ πᾶν ἐκ νοῦ καὶ ἀνάγκης ἐστίν

. . . ὁ δ' οὖν δημιουργός, ὡς ἐκ τούτων δῆλον, τὴν πηγὴν περιέχει τῆς ὑγείας καὶ τῆς Ἀσκληπιακῆς καὶ τῆς δημιουργικῆς.

. . . wherefore the theologians ascribe to Asclepius the one kind of health, namely that which results from the whole process of healing whatever is contrary to nature, checking whatever is contrary to nature either always or at times; the other kind of health they [sc., the theologians] assume to have been created before Asclepius and to be coexistent with the creation of things, which they derive from Peitho and Eros because everything comes from reason and necessity . . . The demiurge, as is clear from this, is the source of health, of the Asclepiadic as well as of the demiurgic.

¹ = *Orphicorum Fragmenta*, 202 [Kern].

315. Orphicorum Fragmenta, 297a [Kern].

Ἑρμῆς . . . / Νύμφαι . . . Ἥφαιστος . . . Δημήτηρ / . . .
Ποσειδάων . . . / . . . Ἄρης . . . Ἀφροδίτη / . . . Διόνυσος . . . /
. . . Θέμις . . .
Ἥλιος ὃν καλέουσιν Ἀπόλλωνα κλυτότοξον,
Φοῖβον ἐκηβελέτην, μάντιν πάντων, ἑκάεργον,
ἰητῆρα νόσων Ἀσκληπιόν· ἐν τάδε πάντα.

Hermes . . . the Nymphs . . . Hephaestus . . . Demeter . . . Poseidon . . . Ares . . . Aphrodite . . . Dionysus . . . Themis . . . the sun whom they term Apollo, the renowned archer, Phoebus, the far-shooting, prophet of all things, working from afar; healer of diseases, Asclepius: all these are one.

316. Aristides, Oratio XLVIII, 4.

Καλοῦμεν δ' αὐτὸν καὶ πρὸς αὐτὰ ταῦτα ὥσπερ πρὸς ἅπαντα· πάντως δ' ἐστὶν πρὸς ἅπαντα κλητός, εἰ δή τις θεῶν.

We invoke him even for this, just as for everything; for always he, if any of the gods, is to be invoked.

317. Aristides, Oratio XLII, 1-15.

ΛΑΛΙΑ ΕΙΣ ΑΣΚΛΗΠΙΟΝ

Ὦ πολλὰ δὴ πολλάκις ἐν νυξί τε καὶ ἡμέραις ἰδίᾳ τε καὶ δημοσίᾳ κληθεὶς ὑφ' ἡμῶν, Ἀσκληπιὲ δέσποτα, ὡς ἀσμένοις καὶ

ὑπερποθοῦσιν ἔδωκας ἡμῖν οἷον ἐκ πελάγους πολλοῦ καὶ κατη-
φείας λιμένος τε λαβέσθαι γαληνοῦ καὶ προσειπεῖν τὴν κοινὴν
τῶν ἀνθρώπων ἑστίαν, ἧς ἀτέλεστος μὲν οὐδεὶς δήπου τῶν ὑφ᾽
ἡλίῳ, διισχυρίσασθαι δὲ ἔστιν ὡς Ἑλλήνων γε οὐδείς πω πλείω
μέχρι τοῦδε ἀπέλαυσεν. καὶ γὰρ εἰ σφόδρα εἰωθότι ταῦτα ἐμοὶ
2 λέγειν, ὀκνητέον γε οὐδὲν μᾶλλον. οὔκουν τάς γε προσρήσεις
τὰς ἐφ᾽ ἡμέραν ταύτας ἐλλείπομεν φεύγοντες τὴν συνήθειαν, ἀλλὰ
καὶ κατ᾽ αὐτὸ τοῦτο φυλάττομεν, ὅτι εἰθίσθημεν ἐξ ἀρχῆς. ἐμοὶ
δὲ ἐπιμελὴς μὲν δήπου καὶ ἡ διὰ τῶν θυμάτων τε καὶ θυμιαμάτων
χάρις τε καὶ τιμή, εἴτε κατὰ τὴν Ἡσιόδου παραίνεσιν γιγνομένη
εἴτε καὶ προθυμότερον τῆς δυνάμεως. ἡ δ᾽ ἐπὶ τοῦ λόγου μοι
3 πολὺ δὴ μάλιστα προσήκειν φαίνεται. εἰ γὰρ οὖν ὅλως μὲν
κέρδος ἀνθρώπῳ τοῦ βίου καὶ ὡσπερεὶ κεφάλαιον ἡ περὶ τοὺς
λόγους διατριβή, τῶν δὲ λόγων οἱ περὶ τοὺς θεοὺς ἀναγκαιότατοι
καὶ δικαιότατοι, φαίνεται δὲ ἡμῖν γε καὶ τὸ κατ᾽ αὐτούς τοὺς λόγους
παρ᾽ αὐτοῦ τοῦ θεοῦ γενόμενον, οὔτε τῷ θεῷ καλλίων χάρις, οἶμαι,
τῆς ἐπὶ τῶν λόγων οὔτε τοῖς λόγοις ἔχοιμεν ἂν εἰς ὅ τι κρεῖττον
χρησαίμεθα. καὶ δὴ λέγωμεν ἀπ᾽ ἀρχῆς ἀρξάμενοι, κοινὰ μὲν οἶδ᾽
ὅτι καὶ βοώμενα—πῶς γὰρ οὔ;—τοσούτῳ δ᾽ ἡμῖν δικαιότερα,
ὅσῳ προστιθέντες καὶ πλεονάζοντες ταῖς θεραπείαις ἀμείνους ἂν
εἴημεν ἢ παραλείποντες ἃ μηδεὶς τῶν πάντων ἀξιοῖ σιγᾶν.

4 Ἀσκληπιοῦ δυνάμεις μεγάλαι τε καὶ πολλαί, μᾶλλον δ᾽ ἅπασαι,
οὐχ ὅσον ὁ τῶν ἀνθρώπων βίος χωρεῖ. καὶ Διὸς Ἀσκληπιοῦ νεὼν
οὐκ ἄλλως οἱ τῇδε ἱδρύσαντο· ἀλλ᾽ εἴπερ ἐμοὶ σαφὴς ὁ διδάσ-
καλος, εἰκὸς δὲ παντὸς μᾶλλον—ἐν ὅτῳ δὲ ταῦτ᾽ ἐδίδαξεν τρόπῳ
καὶ ὅπως, ἐν τοῖς ἱεροῖς λόγοις εἴρηται—, οὗτός ἐσθ᾽ ὁ τὸ πᾶν
ἄγων καὶ νέμων σωτὴρ τῶν ὅλων καὶ φύλαξ τῶν ἀθανάτων, εἰ
δὲ θέλοις τραγικώτερον εἰπεῖν, 'ἔφορος οἰάκων,' σῴζων τά τε
ὄντα ἀεὶ καὶ τὰ γιγνόμενα. εἰ δ᾽ Ἀπόλλωνος παῖδα καὶ τρίτον
ἀπὸ Διὸς νομίζομεν αὐτόν, αὖθις ⟨δ᾽⟩ αὖ καὶ συνάπτομεν τοῖς
ὀνόμασιν, ...* ἐπεί τοι καὶ αὐτὸν τὸν Δία γενέσθαι λέγουσίν
ποτε, πάλιν δὲ αὐτὸν ἀποφαίνουσιν ὄντα τῶν ὄντων πατέρα καὶ
ποιητήν. ἀλλὰ ταῦτα μέν, ὥς φησι Πλάτων, ὅπως αὐτοῖς τοῖς
θεοῖς φίλον, οὕτως ἐχέτω καὶ λεγέσθω. ἐπανέλθωμεν δὲ ὅθεν
5 ἐξέβημεν. πάσας δὲ ἔχων ὁ θεὸς τὰς δυνάμεις διὰ πάντων ἄρα
εὐεργετεῖν προείλετο τοὺς ἀνθρώπους ἑκάστῳ τὰ προσήκοντα
ἀποδιδούς. μεγίστην δὲ καὶ κοινοτάτην εὐεργεσίαν εἰς ἅπαντας
κατέθετο ἀθάνατον ποιήσας τὸ γένος τῇ διαδοχῇ, γάμους τε καὶ
παίδων γενέσεις καὶ τροφῶν ἀφορμὰς καὶ πόρους διὰ τῆς ὑγιείας
ἐργασάμενος. τὰ δ᾽ ἐν μέρει 'πρὸς ἄνδρα ὁρῶν' ἤδη διεδίδου.

οἷον δὴ τέχνας καὶ ἐπιτηδεύματα καὶ βίους πάντας, κοινῷ τινι
φαρμάκῳ πρὸς ἅπαντας πόνους καὶ πράξεις πάσας τῇ ὑγιείᾳ
χρώμενος. ἰατρεῖα δ᾽ εἰς τὸ μέσον κατεστήσατο, καὶ φιλοτεχνεῖν
ἀνέθηκεν ἑαυτῷ νύκτα καὶ ἡμέραν ὑπὲρ εὐθυμίας τῶν αἰεὶ δεο-
6 μένων τε καὶ δεησομένων. ἄλλοι μὲν οὖν ἄλλα ᾄδουσίν τε καὶ
ᾄσονται τὸν αἰεὶ χρόνον, ἐγὼ δὲ τῶν εἰς ἐμαυτὸν οὑτωσὶ μνησθῆναι
βούλομαι. εἰσὶν οἵ φασιν ἀναστῆναι κείμενοι, ὁμολογούμενα
δήπου λέγοντες καὶ πάλαι τῷ θεῷ μελετώμενα· ἡμεῖς τοίνυν οὐχ
ἅπαξ, ἀλλ᾽ οὐδὲ ῥᾴδιον εἰπεῖν ὁσάκις τῆς εὐεργεσίας ταύτης
ἐτύχομεν. ἔτη καὶ χρόνους ἔστιν οἷς ἐπέδωκεν ἐκ προρρήσεως·
7 τούτων ἡμεῖς ἐσμέν· τοῦτο γὰρ εἰπεῖν ἀλυπότατον. ἀλλὰ καὶ
μέλη τοῦ σώματος αἰτιῶνταί τινες, καὶ ἄνδρες λέγω καὶ γυναῖκες,
προνοίᾳ τοῦ θεοῦ γενέσθαι σφίσι, τῶν παρὰ τῆς φύσεως διαφ-
θαρέντων, καὶ καταλέγουσιν ἄλλος ἄλλο τι, οἱ μὲν ἀπὸ στόματος
οὑτωσὶ φράζοντες, οἱ δὲ ἐν τοῖς ἀναθήμασιν ἐξηγούμενοι· ἡμῖν
τοίνυν οὐχὶ μέρος τοῦ σώματος, ἀλλ᾽ ἅπαν τὸ σῶμα συνθείς τε
καὶ συμπήξας αὐτὸς ἔδωκε δωρεάν, ὥσπερ Προμηθεὺς τἀρχαῖα
λέγεται συμπλάσαι τὸν ἄνθρωπον. πολλὰς ὀδύνας τε καὶ ἀλγη-
δόνας καὶ ἀπορίας μεθημερινάς τε καὶ νυκτερινὰς ἀφεῖλεν πολλοῖς,
οὐ μὲν οὖν ἔχοι τις ἂν εἰπεῖν ὅσοις· τὰς δέ γε ἡμετέρας περὶ
ταῦτα τρικυμίας αὐτὸς μὲν ἄριστα σύνοιδεν, αὐτὸς δὲ καὶ παύσας
8 φαίνεται. καὶ μὴν τό γε παράδοξον πλεῖστον ἐν τοῖς ἰάμασι τοῦ
θεοῦ, οἷον τὸν μὲν γύψου πίνειν, τὸν δὲ κωνείου, τὸν δὲ γυμ-
νοῦσθαι καὶ λούειν ψυχρῷ, θέρμης † οὐδόλως, ὡς ἄν τις δόξαι,
δεόμενον. ἡμᾶς τοίνυν καὶ τοῦτον τὸν τρόπον τετίμηκεν, κατάρ-
ρους καὶ ψύξεις ποταμοῖς καὶ θαλάττῃ παύων, κατακλίσεις
ἀπόρους ὁδῶν μήκεσιν ἰώμενος, τροφῆς δ᾽ ἐνδείᾳ συνεχεῖ τὰς
ἀμυθήτους καθάρσεις προστιθείς, ἀναπνεῖν δὲ ἀποροῦντι λέγειν
καὶ γράφειν προστάττων, ὥστ᾽ εἴ τι καὶ τοῖς οὕτω θεραπευθεῖσιν
9 ἔπεστιν αὔχημα, μηδ᾽ ἡμᾶς ἀμοίρους εἶναι τούτου. καὶ μὴν οἱ
μὲν καρτερήσεις ἑαυτῶν διηγοῦνται καὶ ὅσα καὶ οἷα ὑπέμειναν
τοῦ θεοῦ καθηγουμένου, οἱ δὲ ὡς ῥᾳστώνην εὕροντο ὧν ἐδέοντο·
ἡμῖν δὲ πλεῖστα μὲν δήπου κεκαρτέρηται κατὰ πολλοὺς καὶ παντοδαποὺς τρόπους, τὰ δὲ πάνυ κούφως ἐν ἡδονῇ γεγένηται, ὡς
μηδαμοῦ τοὺς τρυφῶντας ἂν εἶναι, εἰ βούλοιο ἀντεξετάζειν. καὶ
τὰς μὲν ἄλλας ἂν ἔχων εἰπεῖν πόλεις Ἀσίας καὶ Εὐρώπης . . .**
καὶ τὰς περὶ ταῦτα ὁμιλίας καὶ συνευφραινομένων ὡς ἐπ᾽ οἰκείοις
ἀγαθοῖς, πῶς οὐκ ἐπέκεινα τρυφῆς θήσομαι; τί δ᾽ ἂν εἴποις
θορύβους ἐν βουλευτηρίοις καὶ σπουδὰς ἔξω παραδείγματος;
τὸ δὲ καὶ πρὶν εἰπεῖν ὁτιοῦν πεπιστεῦσθαι προέχειν, ἆρ᾽ οὐ θεία

τις χάρις καὶ τὰ πρῶτα τῆς ῥᾳστώνης ἔχουσα; φαίην ἂν ἔγωγε,
10 εἴπερ εἴη μεμνῆσθαι *** τῶν κρειττόνων. ἤδη τοίνυν τινῶν ἤκουσα
λεγόντων ὡς αὐτοῖς πλέουσι καὶ θορυβουμένοις φανεὶς ὁ θεὸς
χεῖρα ὤρεξεν, ἕτεροι δέ γε φήσουσιν ὡς πράγματα ἄττα κατώρ-
θωσαν ὑποθήκαις ἀκολουθήσαντες τοῦ θεοῦ· οὐδὲ ταῦτα ἀκούειν
μᾶλλον ἢ λέγειν ἔχομεν πεπειραμένοι. ὅσα δ' αὐτῶν οἷόν τε
11 ἀπομνημονεῦσαι, ἐν τοῖς ἱεροῖς καὶ ταῦτα ἔνεστι λόγοις. ἀλλὰ
καὶ σοφίσματα πυκτικὰ πύκτῃ τινὶ τῶν ἐφ' ἡμῶν ἐγκαθεύδοντι
προειπεῖν λέγεται τὸν θεόν, οἷς ἔδει χρησάμενον καταβαλεῖν τινα
τῶν πάνυ λαμπρῶν ἀνταγωνιστῶν· μαθήματα δὲ ἡμῖν γε καὶ
μέλη καὶ λόγων ὑποθέσεις καὶ πρὸς τούτοις ἐννοήματα αὐτὰ καὶ
τὴν λέξιν, ὥσπερ οἱ τοῖς παισὶ τὰ γράμματα. ἐπιθεὶς τοίνυν
ὥσπερ κεφάλαιον τῶν εὐεργεσιῶν τοῦ θεοῦ καὶ δὴ κατακλείσω
τὸν λόγον ἐνταῦθά που.
12   Ἐμοὶ γάρ, ὦ δέσποτα Ἀσκληπιέ, πολλὰ καὶ παντοῖα, ὥσπερ
ὑπεῖπον, παρὰ σοῦ καὶ τῆς σῆς φιλανθρωπίας γεγένηται, μέγιστον δὲ καὶ πλείστης χάριτος ἄξιον καὶ σχεδὸν ὡς εἰπεῖν οἰ-
κειότατον οἱ λόγοι. τὸ γὰρ τοῦ Πινδάρου μετέβαλες· ἐκείνου
μὲν γὰρ ὁ Πὰν τὸν παιᾶνα ὠρχήσατο, ὡς λόγος, ἐγὼ δέ, εἰ θέμις
εἰπεῖν, ὧν . . . ** ὑποκριτὴς εἶναι· προύτρεψάς τε γὰρ αὐτὸς
13 ἐπ' αὐτοὺς καὶ τῆς ἀσκήσεως κατέστης ἡγεμών. καὶ οὐκ ἀπέχρη
ταῦτα, ἀλλ' ἃ καὶ τούτοις εἰκὸς ἦν ἀκολουθῆσαι, καὶ τούτων
ἐπεμελήθης, ὅπως ἔσται σοι τὸ ἔργον ἐν δόξῃ. καὶ οὐκ ἔστιν οὐ
πόλις, οὐκ ἰδιώτης, οὐ τῶν εἰς ἄρχοντας τελούντων, ὃς οὐ καὶ
κατὰ μικρὸν ἡμῖν ὁμιλήσας οὐκ ἠσπάσατο εἰς ὅσον οἷός τε ἦν
τὸν ἔπαινον ἐκτείνων, οὐ τῶν ἐμῶν, οἶμαι, λόγων ταῦτα ἐργαζο-
14 μένων, ἀλλὰ σοῦ τοῦ κυρίου. τὸ δὲ δὴ μέγιστον τῶν περὶ ταῦτα
τὸ καὶ τοῖς θείοις βασιλεῦσιν εἰς τοσοῦτον οἰκειοῦσθαι καὶ χωρὶς
τῆς διὰ τῶν γραμμάτων συνουσίας ἐπιδείξασθαι λέγοντα ἐν αὐτοῖς
καὶ σπουδαζόμενον ἃ μηδεὶς πώποτε, καὶ ταῦτα ὁμοίως μὲν παρὰ
τῶν βασιλέων, ὁμοίως δὲ τῶν βασιλίδων γενέσθαι, καὶ παντὸς
δὴ τοῦ βασιλείου χοροῦ. Ὀδυσσεῖ δὲ ὑπῆρξεν παρ' Ἀθηνᾶς ἐν
Ἀλκινόου καὶ Φαίαξιν ἐπιδείξασθαι,—μέγα δήπου καὶ τοῦτο καὶ
μάλα ἐν καιρῷ—καὶ ταῦτά τε οὕτως ἐπέπρακτο καὶ τὸ σύνθημα
παρῆν ἀνακαλοῦν, ἔργῳ σοῦ δείξαντος ὅτι πολλῶν εἴνεκα προ-
ήγαγες εἰς μέσον, ὡς φανείημεν ἐν τοῖς λόγοις καὶ γένοιτο
15 αὐτήκοοι τῶν κρειττόνων οἱ τελεώτατοι. τούτων καὶ πολλῶν
ἑτέρων οὔτε . . . * παρ' αὐτοῖς ἡμῖν ἰδίᾳ κἂν ταῖς ὁμιλίαις ταῖς
πρὸς τοὺς ἐντυγχάνοντας τὴν δυνατὴν ἔχοντες χάριν οὐ παυ-
σόμεθα, ἕως ἄν τι μνήμης καὶ τοῦ φρονεῖν μετὸν ἡμῖν τυγχάνῃ.

φαίην δ᾽ ἂν ἔγωγε καὶ ταύτην παρὰ σοῦ κεκομίσθαι τὴν χάριν,
τὸ σὲ τὸν πάντα ἄριστον παρεῖναί τε ἡμῖν καὶ ἐπιψηφίζειν τοὺς
λόγους.

* Lacunam indicavit Keil.   ** Lacunam indicavit Reiske.
*** εἴπερ εἴη μεμνῆσθαι Ο εἰ περιείη μεμνῆσθαι S¹C   εἴπερ εἴην μεμνημένος coni. Keil.

Speech in Honor of Asclepius

O Lord Asclepius, whom we have invoked for many things
and on many occasions [1] both at night and during the day, in
private and in public, it was you who to our satisfaction and
in fulfillment of our excessive desire granted us the opportunity
of reaching a calm haven, as it were, from the vast sea and utter
dejection, and allowed us to offer our greetings to the common
hearth of mankind, [2] in which there is no one under the sun who
has not been initiated; but I venture to assert that no Greek to
this day has had more benefit from it than I have. Although
I am quite accustomed to saying all this—still, I must not
hesitate to acknowledge it again. Therefore we do not shrink
from our habit and omit these daily addresses of ours, but we
retain this habit precisely for the reason that we have been
used to it from the beginning.

I am concerned, of course, with the gratitude and honor
expressed through sacrifices and incense, whether this devotion
be offered according to the counsel given by Hesiod, [3] or whether
it be more generous than one can afford; but the offering that
consists in speeches seems by far the most appropriate for me.
For if the gain of a man's life and its quintessence, so to speak,
consist entirely in his occupation with speeches, and if of all
the speeches those concerning the gods are the most imperative
and righteous ones, and if furthermore we believe that our
concern with speeches is inspired by the god himself, then, I
think, there is hardly a better way of thanking the god than
through speeches, nor could we possibly use our words to a
better purpose. Beginning at the beginning let us say the
things that are, I know, common and celebrated—for how could
it be otherwise?—but so much the more appropriate for us
inasmuch as it is better to be copious and to abound in our
service to the god than to omit what no one at all would think
it right to be silent about.

Asclepius has great and many powers, or rather he has every

power, and not alone that which concerns human life. And it is not by chance that the people here [*sc.*, at Pergamum] have built a temple of Zeus Asclepius. But if my teacher is right, which he is most likely to be—in what way he taught me this and how, has been stated in the *Sacred Speeches*—he is the one who guides and rules the universe, the savior of the whole and the guardian of the immortals, or if you wish to put it in the words of a tragic poet, "the steerer of government," [4] he who saves that which always exists and that which is in the state of becoming. But if we believe him to be the son of Apollo, and the third from Zeus, and if again we link him with these names . . . ; since sometimes they maintain that even Zeus is born, and then again they show that he is the father and maker of everything. Yet this should be viewed and expressed in a manner which is pleasing to the gods themselves, as Plato says. [5] But let us come back to the point from which we started. The god having all powers has chosen to be men's benefactor in every respect, granting every one that which is his due. The greatest and most common benefit, however, he bestowed upon all by making their race immortal through succession, by means of health bringing about marriage and the begetting of children and the procurement of the resources of nourishment. "Looking after man," [6] he granted him other things individually, such as the arts and professions and all the various modes of life, using one common remedy for all pains and troubles, namely health. Furthermore he established places of treatment in their midst, and he took it upon himself to pursue his art night and day in order to bring cheer to whoever always are and will be in need of it.

Some sing and will forever sing about one thing, others about another, but I wish to record in this way that which concerns myself. There are some who claim that they have risen after lying dead, stating something, to be sure, on which all agree and which has been one of the old-established practices of the god. [7] Now, not once alone, but it would not be easy to say how many times we have experienced this benefit. Years and decades has he granted for some through his oracles. Of these we are one ourselves; this much it is safe to say. But also limbs of the body, some declare—I mean men and women alike—have been restored to them through the god's providence after they had been destroyed by nature, and they enumerate, one this, the other that, some of them expressing it by word of mouth, others by their votive offerings. Now for us, he has put together and fastened not part of the body, but the whole frame, and has given it to us as a present, just as of old Prometheus is said to have fashioned man. From many pains and sufferings and distresses, by day and by night, he has delivered many people; no one could tell how many. Our own threefold suffering in this respect he himself knows best, and manifestly he himself has stopped it. And indeed it is the paradoxical which is paramount in the cures of the god, for example, one drinks chalk, another hemlock, another one is stripped of his clothes and takes cold ablutions when one would think him in need of warmth[?].[8] Now ourselves he has likewise distinguished in this way, stopping catarrhs and colds by baths in rivers and in the sea, healing us through long walks when we were helplessly bedridden, administering terrible cleansings on top of continuous abstinence from food, prescribing that I should speak and write when I could hardly breathe, so that if there is any cause for boasting for those who have been healed in such a way, we certainly have our share in this boast.

And furthermore some elaborate on their patient endurance and on what and how they have suffered under the god's guidance, others tell how they found with ease that for which they were asking. With us, most things have been endured in many and various ways, others have come about with perfect ease and happiness, so that nowhere braggarts would have a chance, in case you wish to make a comparison. And I may mention the other cities of Asia and Europe . . . and the social contacts regarding these things [?] and their rejoicing as though their own good were concerned—how shall I not consider this beyond all boasting? What would you think about the applause and the unprecedented zeal on the part of the officials? But to have won their confidence in one's own preeminence even before one has said anything, is this not almost divine favor reaching the peak of easiness? I would say so if there were no greater things to mention. Now I have heard some people saying that, when they were at sea and in the midst of a storm, the god appeared to them and stretched forth his hand; others again will tell how they settled their affairs following the advice of the god. Even these things we know not from hear-say, but we can talk about them from our own experience; and whichever of these incidents are worth remembering, these, too, are contained in the *Sacred Speeches*. But it is said that the god revealed even boxing tricks to one of our contemporary boxers while he was asleep, by the use of which it is no wonder that he knocked out one of his outstanding competitors.[9] To us, however, he has revealed knowledge and melodies and the subjects of speeches and in addition the ideas themselves and even the wording, just like those who teach children to read and write. Having reached, then, the climax, as it were, of the god's benefits, I shall conclude the discourse here.

To me, O Lord Asclepius, many and various things indeed, as I have said before, have come from you and your love of men, the greatest of all, however, and that which deserves the highest gratitude, and which I think is the most suitable one, are my speeches. For you have reversed that which was Pindar's lot:[10] his paean was danced by Pan, so the story goes, but I—if it is permissible to say so—I am the performer [of that which you yourself have taught me]. For you yourself have made me turn towards speeches, and of my profession you have been the leader. And even this did not suffice you, but whatever was likely to come of this [practice], even of that you have taken care so that your work should result in glory. And consequently there is no community, no private person, not even among those who have attained leadership, who, after having had some slight contact with us, has not rejoiced in extending his praise as best he could, I mean not that my own words brought this about, but rather the fact that you were my guardian. The greatest achievement in this regard is the fact that I have become so well acquainted even with their Sacred Majesties and apart from my intercourse with them by letter have spoken before them and have been treated with respect as no one ever was before, and this in an equal measure by emperors and empresses and by the whole imperial court. To Odysseus it was granted by Athena to discourse in the house of Alcinous and among the Phaeacians— and this was, to be sure, a great event and most timely—and in our case this [*sc.*, our appearance in court] was brought about in a similar way: the sign was given calling us forth, and you proved through deeds that for many reasons you wished to drag us into the spotlight in order that we might distinguish ourselves in words and that the élite might hear those best versed in the art. Of these and many others there was neither [?] . . . privately by ourselves and in our association with those whom we happen to meet we shall not cease to express our most sincere thanks, as long as we shall have any memory and thought left. May I say that even this benefit we have received from you, that you, the best of all, are protecting us and approve of our speeches.

[1] Cf. T. **282**, 4.   [2] Cf. T. **402**.   [3] *Opera*, 336.
[4] *Tragicorum Graecorum Fragmenta, Adespota*, 39 [Nauck²] ; cf. T. 303.
[5] *Phaedrus*, 246 D.   [8] Cf. T. **423**, 37 ; cf. also T. **408**.
[6] Aristophanes, *Aves*, 1334.   [9] Cf. T. **423**, 29.
[7] Cf. T. **69** ; **419**.   [10] Fr. 85 [Bowra].

ETHICAL AND MYSTICAL IMPLICATIONS

318. Porphyrius, De Abstinentia, II, 19.

> Δεῖ τοίνυν καθηραμένους τὸ ἦθος ἰέναι θύσοντας, τοῖς θεοῖς θεοφιλεῖς τὰς θυσίας προσάγοντας, ἀλλὰ μὴ πολυτελεῖς. νῦν δὲ ἐσθῆτα μὲν λαμπρὰν περὶ σῶμα μὴ καθαρὸν ἀμφιεσαμένοις οὐκ ἀρκεῖν νομίζουσιν πρὸς τὸ τῶν θυσιῶν ἁγνόν· ὅταν δὲ τὸ σῶμα μετὰ τῆς ἐσθῆτός τινες λαμπρυνάμενοι μὴ καθαρὰν κακῶν τὴν ψυχὴν ἔχοντες ἴωσιν πρὸς τὰς θυσίας, οὐδὲν διαφέρειν νομίζουσιν, ὥσπερ οὐ τῷ θειοτάτῳ γε τῶν ἐν ἡμῖν χαίροντα μάλιστα τὸν θεὸν διακειμένῳ καθαρῶς, ἅτε συγγενεῖ πεφυκότι. ἐν γοῦν Ἐπιδαύρῳ προεγέγραπτο·
> > ἁγνὸν χρὴ ναοῖο θυώδεος ἐντὸς ἰόντα
> > ἔμμεναι· ἁγνεία δ᾽ ἐστὶ φρονεῖν ὅσια.

It is seemly then that those who want to sacrifice go purified in their moral character, bringing sacrifices that are dear to the gods, not such that are lavish. Nowadays they believe that it does not suffice for the holiness of the sacrifice to throw a glittering garment around an unclean body. But if some have a shiny body along with a shiny garment, yet proceed to sacrifice with their souls not cleansed of evil, they think it makes no difference, as though the god did not delight most of all if what is most godlike within ourselves is clean, since it is homogeneous with him. In Epidaurus, at any rate, there was the inscription: "Pure must be he who enters the fragrant temple; purity means to think nothing but holy thoughts." [1]

[1] Cf. T. 336.

319. Corpus Inscriptionum Latinarum, VIII, 1, no. 2584 [209-11 A. D.].[1]

Bonus intra, melior exi.

Enter a good man, leave a better one.

[1] Inscription on the Asclepieion at Lambaesis, Africa.

320. Julianus, Epistulae, 78, 419 B.

Οὐδὲ γὰρ ὁ Ἀσκληπιὸς ἐπ᾽ ἀμοιβῆς ἐλπίδι τοὺς ἀνθρώπους ἰᾶται, ἀλλὰ τὸ οἰκεῖον αὐτῷ φιλανθρώπευμα πανταχοῦ πληροῖ.

Asclepius, again, does not heal mankind in the hope of repayment, but everywhere fulfills his own function of beneficence to mankind.

321. Aelianus, Fragmenta, 100.[1]

Ἀσκληπιὸς Παύσωνα καὶ Ἶρον κἂν ἄλλον τινὰ τῶν ἀπόρων ἰάσαιτο.

Asclepius may heal Pauson and Iros [typical names of paupers in Greek comedy] and any other of the poor people.

[1] Cf. T. 405.

322. Diogenes Laertius, Vitae Philosophorum, III, 45.

Φοῖβος ἔφυσε βροτοῖς Ἀσκληπιὸν ἠδὲ Πλάτωνα,
τὸν μὲν ἵνα ψυχήν, τὸν δ᾽ ἵνα σῶμα σάοι.

Phoebus gave to mortals Asclepius and Plato, the one to save their souls, the other to save their bodies.[1]

[1] Cf. **322a.** Olympiodorus, Vita Platonis, p. 4, 39 ff.: Ἀποθανόντος δ᾽ αὐτοῦ [sc. τοῦ Πλάτωνος] πολυτελῶς αὐτὸν ἔθαψαν οἱ Ἀθηναῖοι καὶ ἐπέγραψαν ἐν τῷ τάφῳ αὐτοῦ 'τοὺς δύ᾽ Ἀπόλλων φῦσ᾽, Ἀσκληπιὸν ἠδὲ Πλάτωνα/ τὸν μὲν ἵνα ψυχήν, τὸν δ᾽ ἵνα σῶμα σάοι.' Cf. **322b.** Anonymus, Vita Platonis, p. 9, 36 ff. [Westermann]: Ἄλλος δὲ χρησμὸς ἐδόθη, ὡς δύο παῖδες τεχθήσονται, Ἀπόλλωνος μὲν Ἀσκληπιός, Ἀρίστωνος δὲ Πλάτων, ὧν ὁ μὲν ἰατρὸς ἔσται σωμάτων, ὁ δὲ ψυχῶν.

323. Aristides, Oratio XXXVIII, 24.[1]

'Υμεῖς δὲ τῇ ὑμετέρᾳ πραότητι καὶ φιλανθρωπίᾳ . . .* θέντες αὐτὸν εἰς καλλίους, τῆς τε νόσου παύετε καὶ διδοίητε ὑγιείας τε ὅσον οἷς ἡ ψυχὴ βούλεται τὸ σῶμα ὑπακούειν, καὶ τὸ σύμπαν εἰπεῖν βίου ῥᾳστώνην.

*Lacunam indicavit Keil.

You in your kindness and love of men . . . counting it [the speech?] among the finer ones, relieve me of my disease and grant me as much health as is necessary in order that the body may obey that which the soul wishes, and, to say it in one word, a life lived with ease.

[1] Cf. T. 282.

324. Julianus, Contra Galilaeos, 235 B.

'Ιᾶται 'Ασκληπιὸς ἡμῶν τὰ σώματα, παιδεύουσιν ἡμῶν αἱ Μοῦσαι σὺν 'Ασκληπιῷ καὶ 'Απόλλωνι καὶ 'Ερμῇ λογίῳ τὰς ψυχάς . . .

Asclepius heals our bodies, the Muses train our souls with the aid of Asclepius and Apollo and Hermes, the god of eloquence . . .(*)

325. Aristides, Oratio XLIX, 46-48.

46   'Εγένετο δὲ καὶ φῶς παρὰ τῆς "Ισιδος καὶ ἕτερα ἀμύθητα φέροντα εἰς σωτηρίαν. ἐφάνη δὲ καὶ ὁ Σάραπις τῆς αὐτῆς νυκτός, ἅμα αὐτός τε καὶ ὁ 'Ασκληπιός, θαυμαστοὶ τὸ κάλλος καὶ τὸ μέγεθος
47   καί τινα τρόπον ἀλλήλοις ἐμφερεῖς. γενομένης δὲ τῆς περὶ Ζώτιμον συμφορᾶς—ἃ γὰρ μελλούσης προεῖπεν καὶ παρεμυθήσατο ὁ θεὸς παρίημι—, ἀλλ' ἐπειδὴ ἐγένετο καὶ χαλεπῶς εἶχον ἀπὸ τῆς λύπης, ἐδόκει μοι σμίλην τιν' ἔχων ὁ Σάραπις, ὥσπερ κάθηται τῷ σχήματι, περιτέμνειν μου τὰ κύκλῳ τοῦ προσώπου ὑπ' αὐτό πως τὸ † ὁρίζηλον, οἷον λύματ' ἀφαιρῶν καὶ καθαίρων καὶ μεταβάλλων εἰς τὸ προσῆκον. ὥστε καὶ ὕστερον ὄψις μοι γίγνεται παρὰ τῶν χθονίων θεῶν, εἰ τὸ σφόδρα οὕτω λυπεῖσθαι
48   ἐπὶ τοῖς τελευτῶσιν ἀνείην, συνοίσειν μοι ἐπὶ τὸ βέλτιον. πολὺ δέ τι τούτων φρικωδέστερον εἶχεν τὰ χρόνῳ ὕστερον φανθέντα, ἐν οἷς αἵ τε δὴ κλίμακες ἦσαν αἱ τὸ ὑπὸ γῆς τε καὶ ὑπὲρ γῆς ἀφορίζουσαι καὶ τὸ ἑκατέρωθι κράτος τοῦ θεοῦ, καὶ ἕτερα ἔκπληξιν θαυμαστὴν φέροντα, καὶ οὐδὲ ῥητὰ ἴσως εἰς ἅπαντας, ὥστε

ἀσμένῳ μοι φανῆναι ⟨τὰ⟩ σύμβολα [τοῦ Ἀσκληπιοῦ]. κεφάλαιον
δ᾽ ἦν περὶ τῆς τοῦ θεοῦ δυνάμεως, ὅτι καὶ χωρὶς ὀχημάτων καὶ χωρὶς
σωμάτων ὁ Σάραπις οἷός τ᾽ εἴη κομίζειν ἀνθρώπους ὅπη βούλοιτο.
τοιαῦτα ἦν τὰ τῆς τελετῆς, καὶ ἀνέστην οὐ ῥᾴδιος γνωρίσαι.

Light came forth from Isis, and other secret manifestations
conducive to salvation. Sarapis, too, appeared that same night,
he and Asclepius together, both of wondrous beauty and size,
and in a manner resembling each other. When misfortune
overtook me in the matter of Zosimus—for I pass over the
warning and advice which the god gave when the misfortune
was impending—, at any rate, when it occurred and I was
grievously suffering with pain, Sarapis in seated form, bearing
in his hand a kind of knife, seemed to me to make an incision
around my face near the hairline[?], as if removing and destroying the defilement and restoring me to my proper state; then later
a vision came to me from the gods of the underworld saying
that it would work to my greater advantage if I should cease
to be so sorely grieved over the dead. But much more awe-
inspiring than these events were those that occurred later,
among which were the ladders which mark the boundary between the upper world and the nether world and the power of
the god in either region, as well as other events which caused
extraordinary consternation and probably are not told to everyone, so that I was gratified to have the tokens disclosed to me.
The main one concerned the power of the god, in that Sarapis
was able to convey men without vehicles and without bodies
wherever he wished. Such were the mystic rites, and I rose up,
not easily to be recognized.

THE MAGICIAN

326. Catalogus Codicum Astrologorum, VIII, 4, 256-57.

"Vis loqui cum deo vel alicuius mortui animae?" cui dixi:
"Animae Asclepii loqui, si vos praecipitis, a quo expecto par-
ticipare perfectionem artis et scientiae."

"Do you desire to talk with a god or to the soul of someone
who is dead?" To him I [sc., Thessalus, the physician, 1st
c. A. D.] replied: "I desire to speak to the soul of Asclepius,

if you instruct me how, through whom I hope to share in the excellence of art and science."

327. Achilles Tatius, De Clitophontis et Leucippes Amoribus [Erotici Graeci, I, p. 126, 14-18].

Ἐπεὶ οὖν καιρὸς ἦν αὐτῇ πιεῖν τὸ φάρμακον, ἐγχέας προσηυχόμην αὐτῷ " ὦ γῆς τέκνον, φάρμακον, ὦ δῶρον Ἀσκληπιοῦ, ἀλήθευσόν σου τὰ ἐπαγγέλματα . . . καὶ σῶζέ μοι τὴν φιλτάτην . . ."

Since, then, it was time for her to drink the potion, filling the cup, I addressed a prayer to it [sc., the potion]: "O child of earth, remedy, O gift of Asclepius, let your promises come true . . . and save my dear one for me . . ."

328. Origenes, Contra Haereses, IV, 32.

. . . σκότος δὲ ἐν οἴκῳ ποιήσας ἐπεισάγειν φάσκει θεοὺς ἢ δαίμονας, καὶ εἴ πῃ ἀπαιτεῖ τις Ἀσκληπιὸν δεικνύναι, ἐπικαλεῖται οὕτως λέγων·

" Υἷα πάλαι φθίμενον, πάλιν ἄμβροτον Ἀπόλλωνος
κικλήσκω λοιβαῖσι μολεῖν ἐπίκουρον ἐμαῖσιν.
ὅς ποτε καὶ νεκύων ἀμενηνῶν μυρία φῦλα
Ταρτάρου εὐρώεντος ἀεικλαύτοισι μελάθροις
δύσνοστον πλώοντα ῥόον, κελανόν * τε δίαυλον
πᾶσιν ἴσον τελέσαντ' ἄνδρεσσι καταθνητοῖσιν
λίμνῃ πὰρ γοόωντα καὶ ἄλλιτα κωκύοντα
αὐτὸς ἀμειδήτοιο ἐρύσαο Φερσεφονείης.
εἶτ' ἐφέπεις Τρίκης ἱερῆς ἔδος, εἶτ' ἐρατεινὴν
Πέργαμον, εἶτ' ἐπὶ τοῖσιν Ἰαονίαν Ἐπίδαυρον,
δεῦρο, μάκαρ, καλέει σε μάγων πρόμος ὧδε παρεῖναι."

ἐπὰν δὲ χλευάζων λήξῃ, φαίνεται κατὰ τοῦ ἐδάφους πυρώδης Ἀσκληπιός.

* κελανόν = κελαινόν Meineke   κελαδόν MSS.

. . . when he [the leader of the magicians] has darkened the house, he says he brings in the gods or spirits, and if anyone by chance demands that he show Asclepius, he summons the god with these words:

" Son of Apollo, once subject to death, then immortal, I call you to come and preside over my libations. You who once

ransomed from gloomy Persephone even countless throngs of
fleeting corpses in the oft bewailed halls of dark Tartarus, when
they were lamenting beside the marshy lake and moaning in-
exorably, crossing the stream from which there is no return
and completing the murky course which awaits all mortal men.
Whether you dwell in holy Tricca, or lovely Pergamum, or
besides, Ionian Epidaurus,[1] hither, blessed one, the leader of
the magicians bids you approach."

When he ceased talking thus impertinently, Asclepius appeared like a firebrand down on the floor.

[1] Cf. T. 292; 307; T. 711.

329. Augustinus, De Civitate Dei, VIII, 26.

... et dicit: "Avus enim tuus, o Asclepi, medicinae primus
inventor, cui templum consecratum est in monte Libyae circa
litus crocodilorum, in quo eius iacet mundanus homo, id est
corpus; reliquus enim, vel potius totus, si est homo totus in
sensu vitae, melior remeavit in caelum, omnia etiam nunc
hominibus adjumenta praestans infirmis numine nunc suo,
quae solebat medicinae arte praebere." ecce dixit mortuum
coli pro deo in eo loco, ubi habebat sepulcrum, falsus ac fallens,
quod remeavit in caelum. . . . ecce duos deos dicit homines
fuisse, Aesculapium et Mercurium. sed de Aesculapio et
Graeci et Latini hoc idem sentiunt . . . verum et iste, sicut
Aesculapius, ex homine deus secundum testimonium tanti apud
suos viri, huius Trismegisti, nepotis sui.

... and he [sc., Hermes] says: "Thy grandsire, O Asclepius,
the first discoverer of medicine, to whom a temple was consecrated in a mountain of Libya, near the shore of the crocodiles,
in which temple lies his earthly man, that is, his body,—for
the better part of him, or rather the whole of him, if the whole
man is in the intelligent life, went back to heaven,—he who
even now by his divinity offers all those helps to infirm men
which formerly he was wont to afford to them by the art of
medicine." He said, therefore, that a dead man was worshipped as a god in that place where he had his sepulchre. He
is false and a deceiver, for the man " went back to heaven." ...
So here are two gods whom he [sc., Hermes] affirms to have
been men, Asclepius and Mercury. Now concerning Asclepius,

both the Greeks and the Latins think the same thing; . . .
It is sufficient to know that this Mercury [of whom Hermes
speaks] is, as well as Asclepius, a god who once was a man,
according to the testimony of this same Trismegistus, esteemed
so great by his countrymen, and the grandson of Mercury
himself.

330. Eusebius Hieronymus, De Vita Hilarionis, 2.

De eodem Gazensis emporii oppido, virginem Dei vicinus
juvenis deperibat. qui cum frequenter tactu, jocis, nutibus,
sibilis, et caeteris huiusmodi, quae solent moriturae virginitatis
esse principia, nihil profecisset, perrexit Memphim, ut confesso
vulnere suo, magicis artibus rediret armatus ad virginem.
igitur post annum doctus ab Aesculapii vatibus, non reme-
diantis animas, sed perdentis, venit praesumptum animo stu-
prum gestiens. . . .

A nun from the same trading center, the city of Gaza, was
desperately loved by a youth from the vicinity. When he had
accomplished nothing by frequent touching, joking, nodding,
whistling and other things of a similar kind which are wont
to be the first steps in the loss of virginity, he went to Memphis
so that after confessing his unhappy plight he might return to
the nun armed with magic devices. When, therefore, he had
been instructed for a year by the priests of Asclepius, the god
who does not cure souls but destroys them, he came back,
flaunting the dishonor which he had taken upon himself. . . .

331. Oxyrhynchus Papyrus XI, 1381 [2nd century A. D.].

Col. i.

[. .]ν τα[ῦτ]α ἀκούσας ὁ Νε[κτε-
[νε]ῖβις καὶ παροξυνθεὶς [σ]φό-
δρα μὲν ἐπὶ τοῖς ἀποστατ[ο]ῦ-
σιν τοῦ ἱεροῦ, βουλόμεν[ο]ς
5   δὲ ἐξ ἀναγραφῆς τὸ πλῆ[θ]ος αὐτῶν ἐπικρεῖναι θᾶτ[τ]ον, πα-
ρεκελεύετο Νεχαύτι [τ]ῷ διέ-
ποντι τότε τὴν ἀρχιδ[ι]κ[ασ]τε[ί-
αν ἔραυναν τῆς βίβλου μη⟨νὶ⟩

Col. ii.

σαν ἑκάστῳ π[ροφ]ητείαν. οὐ
μὴν ἀλλὰ καὶ [. . π]οιήσας τὴν
βίβλον ἀναν[εώ]σεως αὐτὸν   25
Ἀ[σ]κλήπιον [ἐπλο]ύτισεν ἄλ-
λ[αι]ς πυροφόροις ἀρούραις τρια-
κοσίαις τριάκοντα, καὶ μάλ[ι-]
στα ἀκούσας διὰ τῆς βίβλου
τὸν θεὸν ὑπὸ Μεγχορέους   30
[εἰ]ς μέγεθος ἠσκημένον σε-

10 ἐνὶ μάλιστα ποιήσασθαι. ὁ δὲ
ἐκτενέστερον αὐτὴν ἀναζη-
τήσας ἐκόμισε τῷ βασιλεῖ,
δ[ύ]ο [ἀν]τὶ τριάκοντα ἡμερῶν
μόνον ἀναλώσας εἰς τὴν
15 [ζ]ήτησιν. ἀναγνοὺς δὲ ὁ βασι-
[λε]ὺς πάνυ μὲν ἠγάσθη ἐπὶ
τῷ τῆς ἱστορίας θείῳ ἐξ δὲ
καὶ εἴκοσι εὑρὼν ἱερεῖς [τ]οὺς
ἀπὸ Ἡλίου πόλεως προπ[ο]μ-
20 πεύσαντας τὸν θεὸ[ν] εἰς τὴ[ν
Μέμφιν ἀπένειμεν αὐτῶν
τοῖς ἐγγόνοις τὴν προ[σ]ήκου-

[β]ασμῶν. ἐγὼ δὲ πολλάκις τῆς
[α]ὐτῆς βίβλου τὴν ἑρμηνείαν
[ἀρ]ξάμενος Ἑλληνίδι γλ[ώ]σσῃ
[ἔμ]αθον ὃν * αἰῶνι κηρῦξαι, καὶ　　　35
ἐν μέσῃ ῥέων τῇ γραφῇ
ἐπεσχέθην τὴν προθυμίαν
τῷ τῆς ἱστορίας [[τω]] μεγέθει,
δ[ι]ότι ἔξω ἑλεῖν ἔμελλο[ν] αὐ-
τήν· θε[οῖ]ς γὰρ μόνοι[ς] ἀλλ' οὐ　　　40
[θν]ητοῖς ἐ[[φ]]φικ[[..]]τ[ὸ]ν τὰς θεῶν διηγεῖσθα[ι] δυνάμεις. οὐ
γὰρ ἀποτυχό[ν]τι μοι μόνον
αἰδὼς ἦν πρὸς ἀνδρῶν ἀλλὰ

*1. ἐν for ὄν.

Col. iii.

45 καὶ ἐκώλυσέ [μετ]ὰ κατιό[ντα ...]
δια ἀγανακτήσαντος [καὶ ἀθα-
νάτου ἀρετῆς αὐτοῦ τ[ὸ τῆς γρ]α-
φῆ[ς] σ[υ]νπληρουμέν[ης] τ[απεί-]
νωμα, ὀφελήσαντι δ[ ]ὲ ὁ βί[ο]ς
50 μὲν εὐδαίμων, ἡ δὲ [ ] φήμη
[ἀ]θάνα[τ]ος. ἑτοιμότε[ ]ρος γὰρ ὁ
θεὸς πρὸ[ς] ε[ὐε]ργεσία[ ]ν εἴ γε καὶ
τοὺς αὐτ⟨ίκ⟩α μόνον εὐ[ ]σεβεῖς
τῇ προθυμίᾳ πολλά[ ]κις ἀπηυ-
55 δηκυίης τῆς ἰατρικ[ ]ῆς πρὸς
τὰς κατεχούσας αὐτοῦ[ ]ς νόσους
ἔσωσεν. ὅθεν φυγὼν [τὸ ῥ]ειψοκίν-
δυνον [εἰ]ς καιρὸν ἐτ[ ]ήρουν
τὸν τοῦ ⟨γ⟩ήρ[ο]υς,* ἀνε[β]α[ ]λ-
λό[μ]ην ⟨δὲ⟩
60 τὴν ὑπόσχεσιν. τότ[ ]ε γὰ[ρ] μάλιστα περισσόν τι τὴ[ ]ν ἡλικίαν
φρονεῖν πέφυκε, τ[α]χὺ γὰρ ἡ
νε[ότ]ης καὶ ἐφ[ορ]μὴ φ[ ]θάνει
ὀρέγουσα τὴν προθυ[μί]αν. ἐπεὶ
65 δὲ τ[ρ]ιετὴς πα[ρ]ῴχητ[ο] χρόνος
μ[ηδ]ὲν ἔτι μ[ο]υ κάμν[ ]οντος,
τρ[ιε]τὴς δ[ὲ..] τῇ μητ[ ]ρὶ ἐπι-
*1. ⟨γ⟩ήρως.

Col. iv.

σκ[ήψασα ἄ]θεος τεταρταία ἡ
φρείκη αὐτὴν ἐστρόβει, ὀψὲ
μόλις νοήσαντες ἱκέτ[α]ι πα-　　　70
ρῆμεν ἐπὶ τὸν θεὸν τῇ ⟨μ⟩ητρὶ [ ]
ωμενοι ἄκεσιν ἐπινεῦσαι
τῆς νόσου. ὁ δ' οἷα καὶ πρὸς πάντας χρηστὸς δι' ὀνειράτων
φανεὶς εὐτελέσιν αὐτὴν　　　75
ἀπήλλαξεν βοηθήμασιν,
ἡμεῖς δὲ [[μη]] τὰς ἐοικυίας
δ[ι]ὰ θυσιῶν τῷ σώσαντι
ἀπεδίδομεν χάριτας. ἐπεὶ
δὲ κἀμοὶ μετὰ ταῦτα αἰφνί-　　　80
δι[ο]ν ἄλγημα κατὰ δεξιοῦ
ἐρύη πλευροῦ, ταχὺς ἐπὶ
τὸν βοηθὸν τῆς ἀνθρω-
πίνης ὤ[ρ]μησα φύσεως,
[καὶ] πάλιν ἑτοιμότερος　　　85
ὑπακούσας εἰς ἔλεον
[ἐ]νεργέστε[ρ]ον τὴν ἰδίαν
ἀπεδείξατο εὐεργεσίαν,
ἣν ἐπαληθεύω μέλλων
τὰς αὐτοῦ φρικτὰς δυ-　　　90

Col. v.

ν[ά]με[ι]ς ἀπαγγέλλειν. νὺξ
ἦν ὅτε πᾶν [ἐ]κεκοίμητο
ζῷον πλὴν τῶν ἀλγ[ο]ύν-
των, τὸ δὲ θεῖον ἐνεργέ-
95 στερον ἐφαίνετο, καί με
σφοδρὸς ἔφλεγε πυρ⟨ετ⟩ός, ἄσθμα-
τί τε καὶ βηκὶ τῆς ἀπὸ τοῦ
πλευρ[οῦ] ἀναγομέν[η]ς ὀδύ-
νης ἐσφαδάϊζον· καρηβα-
100 ρηθεὶς [δ]ὲ τοῖς πόνοις {ἀ}λή-
θαργος [ε]ἰς ὕπνον ἐφερό-
μην· [ἡ] δὲ μήτηρ ὡς ἐπὶ
παιδί, κα[ὶ] φύ[σ]ει φιλόστοργος
γάρ ἐστιν, ταῖς ἐμαῖς ὑπερ-
105 αλγ[ο]ῦσα βασάνοις ἐκαθέ-
ζετο μηδὲ καθ᾽ ὀλίγον ὕπνου
μετ[α]λαμβάνουσα. εἶτ᾽ ἐξαπ[ί]-
νης ἑώρα—οὔτ᾽ ὄναρ οὔθ᾽ ὕ-
πνος, ὀφθαλμοὶ γὰρ ἦσαν
110 ἀκείνητοι διηννυγμένοι,*
βλέποντες μὲν οὐκ ἀκρει-
βῶς, θ[[.]]εία γὰρ αὐτὴν μετὰ

* 1. διηνοιγμένοι.

Col. vi.

δέ[ο]υς εἰσῄει φαντασία[[ν]],
καὶ ἀκό[π]ως κατ[ο]πτεύειν
κωλύουσα εἴτε αὐτὸν τὸν   115
θεὸν εἴτε αὐτοῦ θεράπον-
τας. πλὴν ἦν τις ὑπερμή-
κης μὲν ἢ κατ᾽ ἄνθρω-
πον λαμπ[ρ]αῖς ἠμφιεσμέ-
νος ὀθόναις τῇ εὐωνύ-   120
μῳ χειρὶ φέρων βίβλον,
ὃς μόνον ἀπὸ κεφαλ[ῆ]ς
ἕως ποδῶν δὶς καὶ τρ[ὶ]ς
ἐπισκοπήσας με ἀφανὴς
ἐγ[έ]νετο. ἡ δὲ ἀνανήψασα   125
ἔτι τρομώδης ἐγείρειν με
ἐπειρᾶτο. εὑροῦ[σ]α δέ με
τοῦ μὲν [π]υρετοῦ ἀπηλ[λ]α-
γμένον [ἱ]δρῶτα δέ μοι πολ-
λοῦ ἐπαπ[ο]λισθάνοντος **   130
τὴν μὲ[ν] τοῦ θε[ο]ῦ προσε-
κύνησε[ν] ἐπιφάνειαν, ἐ-
μὲ δὲ ἀπ[ο]μάσσουσα ν[η]φα-
λιώτε[ρο]ν ἐποίησεν. καὶ
διαλα[λή]σαντί μοι τὴν τοῦ   135

** 1. μου πολὺν ἐπαπ[ο]λισθάνοντα.

Col. vii.

θεοῦ πρ[οε]λομένη μηνύειν ἀρετὴν προλαβὼν ἐγὼ πάντα ἀ-
πήγγελον αὐτῇ· ὅσα[[γ]ὰρ δι[ὰ] τῆς
ὄψεως εἶδεν ταῦτα ἐγ[ὼ] δι᾽ ὀ-
140 νειράτων ἐφαντασιώθην.
καὶ τῶνδε τῆς πλευρᾶς λωφη-
σάντων μοι ἀλγηδόνων, ἔτι
μοι μί[α]ν δόντος τοῦ θεοῦ ἀκε-
σώδυνον ἰατρείαν, ἐκήρυσσον
145 αὐτοῦ [τ]ὰς εὐεργεσίας. πάλιν δ᾽ ἡ-
μῶν ταῖς κατὰ δύναμιν αὐτὸν

Col. viii.

ρ[ο]υ[ν, κα]ὶ μόλις ταπεινοῦν-
τί μοι τοῦτο τὸ θεῖο[ν] τῆ[ς] γραφῆς ὑπῄει με χρέος. ἐπεὶ   160
δ᾽ ἅπαξ ἐπεγνώκει[ς] με [ἀ]με[[λ]]-
λεῖν, δέσποτα, τῆς θεί[α]ς βίβλου, τὴν σὴν ἐπικαλεσάμε-
νος πρόνοιαν καὶ [ ]πλη[[ρ]]-
ρωθεὶς τῆς σῆς θε[ι]ότητος   165
ἐπὶ τὸν τῆς ἱστορία[ς] ὥρμη-
σα θεήλατον ἆθλον. καὶ
οἶμαι κατα[πλ]ώσειν * [τ]ὴν

ἐξευμενισαμένων θυ[σ]ίαις
αὐτὸς ἀπῄτει διὰ [τ]οῦ ἐν ἀγνείαις
αὐτῷ προσπολοῦντ[ο]ς ἱερέως
150 τὴν πάλαι κατηγγελμένην αὐτῷ
ὑπόσχεσιν. ἡμεῖς δὲ μηδὲ θυσιῶν μήτε ἀναθήματ[ο]ς χρε-
ώστας αὐτοὺς εἰδότες ὅμως
τού[το]ις αὐτὸν πάλιν ἱκετεύ-
155 ομ[ε]ν. ὡς [δ'] οὐ τούτοις πο[λ]-
λάκις
εἶπε{ι}ν ἥδεσθαι ἀλλὰ τῷ προ-
καθωμολ[ο]γημένῳ διηπό-

σὴν προφη[τε]ύων ἐπίνοι-
αν· καὶ γὰρ [τὸ]ν τῆς κοσμο-　170
ποιίας πιθ[α]νολ[ο]γηθέν-
τα μῦθον ἐν ἑτέρᾳ β[ί]βλῳ
φυσικῷ πρὸ[ς] ἀλήθειαν ἀνή-
πλωσα λόγῳ. καὶ ἐν τῇ ὅλῃ
γραφῇ τ[ὸ] μὲν ὕστερον προσ-　175
επλήρωσα, τὸ δὲ περ[ι]σσεῦ-
ον ἀφεῖλον, διήγημα δέ

που μακρολογούμ[ε]νο[ς]
συντόμως ἐλάλησα
*l. καθα [πλ]ώσειν

Col. ix.

180 καὶ ἀλλαττόλογο[ν μῦθ]ον
ἅπαξ ἔφρασα, ὅθεν, [δέσ]ποτα,
κατὰ τὴν σὴν εὐμ[ένει]αν
ἀλλ' οὐ κατὰ τὴν ἐμ[ὴν φρ]ό-
νησιν τετελεσιουρ[γ]ῆ[σ]θαι
185 τεκμαίρομαι τὴν β[ίβλ]ον.
τῇ γ[ὰ]ρ σῇ θειότητι [το]ιαύ-
τη ἁ[ρ]μόζει γ[ρ]αφή. τ[αύτ]ης
δ' εὑρετής, μέγιστε [θε]ῶν
Ἀσκλήπιε καὶ διδάσ[κ]αλε,
190 κα[ὶ] ταῖς ἀπ[άν]των δί[κ]νυ-
σαι χάρισι. [πᾶ]σα γὰρ [ἀ]να-
θήματος ἢ [θ]υσίας δ[ω]ρεὰ
τὸν παραυτ[ί]κα μ[ό]ν[ο]ν
ἀκμάζει κα[ιρ]όν, ἔφθαρ-
195 ται δὲ τὸν μέλλοντα, γραφῇ δὲ ἀθάνατος χάρ[ι]ς κατὰ καιρὸν ἀνηβάσκ[ο]υσα
τὴ[ν] μνήμην. Ἕλλην[ὶ]ς δὲ
π[ᾶ]σα γλῶσσα τὴν σὴν λα-
200 λ[ή][[..]]σε[ι] ἱστορίαν κ[αὶ] πᾶς
Ἕλ[λ]ην ἀνὴρ τὸν τ[ο]ῦ Φθᾶ
σεβήσεται Ἰμού[θ]ην.

Col. x.

σύνι[τε δε]ῦρο, [ὦ ἄν]δρες
εὐμ[ενεῖς] κα[ὶ ἀγα]θοί, ἄπι-
τε, βάσκα[νοι] κ[αὶ] ἀσεβεῖς·　205
σύν[ι]τε, ὦ [...]ο[..].[.], ὅσοι θη-
τεύ[σ]αντε[ς] τὸν [θ]εὸν νόσῳ[ν] ἀπηλλάγητε, [ὅ]σοι
τὴν ἰατρικὴν με[ταχ]ειρί-
ζεσθε ἐπι[σ]τήμη[ν, ὅσ]οι　210
πονήσετε ζηλ[ωτα]ὶ ἀρε-
τῆς, ὅσο[ι] πολλῷ πλήθει
ἐπηύξή[θ]ητε ἀγαθῶν,
ὅσοι κινδύνους θαλάσσης
πε[ρ]ιεσώθητε. εἰς πάν-　215
τα γὰρ τόπον διαπεφοίτη-
κεν ἡ τοῦ θεοῦ δύναμις
σωτήριος. μέλλω γὰρ αὐτοῦ
τερατώδεις ἀπαγγέλλειν
ἐπ[ι]φανείας δυνάμεως　220
τε μεγέθη εὐε[ρ]γετημά-
των ⟨τε⟩ δωρήματα. ἔχει δὲ οὕ-
τως· [ὁ] βασιλεὺς Μενε-
χέρης τριῶν θεῶν κη-
δε[ί]αν [εὐ]σεβήσας αἰωνίαν　225

Col. xi.

εἴληφε δόξαν, [καὶ διὰ τῆς?
βίβλου τὴν φ[ήμην εὐτυ-?
χήσας. τὴν τ[οῦ Ἀσκλη-
πίου παιδὸς Ἡφ[αίστου τα-
230 φὴν καὶ τὴν τ[οῦ Ὥ]ρου Ἐρ[-
μ[ο]ῦ ἔτι δὲ Καλεοίβιος
Ἀπόλλωνος παιδὸς ἀφθό-
νο[ι]ς χρήμασιν δωρησά-
μενος ἀντάποιναν ἔσ-
235 χεν εὐδαιμονίας πλή-
θος. ἀπολέμητος γὰρ τό-

τε Αἴγυπτος διὰ τοῦτο κ[α]ὶ
καρποῖς ἀφ⟨θ⟩όνοις εὐθη-
νεῖτο. τῇ γὰρ τοῦ προεσ-   240
τῶτος εὐσεβείᾳ ὑποτε-
ταγμέναι εὐπ[ορ]οῦσι χώ-
[ρ]αι, καὶ τοὐνα[ντί]ον ἐφ' οἷς
ἐκεῖνος δυσσ[εβε]ῖ ἐπὶ
τούτοις κακοῖς [ἀ]ναλίσκον-
ται. ὃν δὲ τρόπον ἔχρη-   245
σεν αὐτῷ ὁ θεὸ[ς Ἀ]σκλήπιος
σπουδάζειν αὐτ[ο]ῦ περὶ

[PRAISE OF IMOUTHES–ASCLEPIUS]

'Nectenibis on hearing this, being extremely vexed with the deserters from the temple and wishing to ascertain their number speedily by a list, ordered Nechautis, who then performed the duties of archidicastes, to investigate the book within a month, if possible. Nechautis conducted his researches with much strenuousness, and brought the list to the king after spending only two days instead of thirty upon the inquiry. On reading the book the king was quite amazed at the divine power in the story, and finding that there were twenty-six priests who conducted the god from Heliopolis to Memphis, he assigned to each of their descendants the due post of prophet. Not content with this, after completing the renewal of the book (?), he enriched Asclepius himself with three hundred and thirty arurae more of corn-land, especially because he had heard through the book that the god had been worshipped with marks of great reverence by Mencheres.

Having often begun the translation of the said book in the Greek tongue, I learnt at length how to proclaim it, but while I was in the full tide of composition my ardor was restrained by the greatness of the story, because I was about to make it public; for to gods alone, not to mortals, is it permitted to describe the mighty deeds of the gods. For if I failed, not only was I ashamed before men, but also hindered by the reproaches (?) that I should incur if the god were vexed, and by the poverty of my description, in course of completion, of his undying power. But if I did the god a service, both my life would be happy and my fame undying; for the god is disposed to confer benefits, since even those whose pious ardor is only for the moment are repeatedly pre-

served by him after the healing art has failed against diseases which have
overtaken them. Therefore avoiding rashness I was waiting for the
favorable occasion afforded by old age, and putting off the fulfillment
of my promise; for then especially is youth wont to aim too high, since
immaturity and enterprise too quickly extend our zeal. But when a
period of three years had elapsed in which I was no longer working, and
for three years my mother was distracted by an ungodly quartan ague
which had seized her, at length having with difficulty comprehended we
came as suppliants before the god, entreating him to grant my mother
recovery from the disease. He, having shown himself favorable, as he
is to all, in dreams, cured her by simple remedies; and we rendered due
thanks to our preserver by sacrifices. When I too afterwards was suddenly seized with a pain in my right side, I quickly hastened to the helper
of the human race, and he, being again disposed to pity, listened to me,
and displayed still more effectively his peculiar clemency, which, as I am
intending to recount his terrible powers, I will substantiate.

It was night, when every living creature was asleep except those in
pain, but divinity showed itself the more effectively; a violent fever
burned me, and I was convulsed with loss of breath and coughing, owing
to the pain proceeding from my side. Heavy in the head with my troubles
I was lapsing half-conscious into sleep, and my mother, as a mother
would for her child (and she is by nature affectionate), being extremely
grieved at my agonies was sitting without enjoying even a short period
of slumber, when suddenly she perceived—it was no dream or sleep, for
her eyes were open immovably, though not seeing clearly, for a divine
and terrifying vision came to her, easily preventing her from observing
the god himself or his servants, whichever it was. In any case there was
some one whose height was more than human, clothed in shining raiment
and carrying in his left hand a book, who after merely regarding me two
or three times from head to foot disappeared. When she had recovered
herself, she tried, still trembling, to wake me, and finding that the fever
had left me and that much sweat was pouring off me, did reverence to
the manifestation of the god, and then wiped me and made me more
collected. When I spoke with her, she wished to declare the power of
the god, but I anticipating her told her all myself; for everything that
she saw in the vision appeared to me in dreams. After these pains in my
side had ceased, and the god had given me yet another assuaging cure,
I proclaimed his benefits. But when we had again besought his favors
by sacrifices to the best of our ability, he demanded through the priest
who serves him in the ceremonies the fulfillment of the promise long ago

announced to him, and we, although knowing ourselves to be debtors in neither sacrifices nor votive offering, nevertheless supplicated him again with them. But when he said repeatedly that he cared not for these but for what had been previously promised, I was at a loss, and with difficulty, since I disparaged it, felt the divine obligation of the composition. But since thou hadst once noticed, master, that I was neglecting the divine book, invoking thy providence and filled with thy divinity I hastened to the inspired task of the history. And I hope to extend by my proclamation the fame of thy inventiveness; for I unfolded truly by a physical treatise in another book the convincing account of the creation of the world. Throughout the composition I have filled up defects and struck out superfluities, and in telling a rather long tale I have spoken briefly and narrated once for all a complicated story. Hence, master, I conjecture that the book has been completed in accordance with thy favor, not with my aim; for such a record in writing suits thy divinity. And as the discoverer of this art, Asclepius, greatest of gods and my teacher, thou art distinguished by the thanks of all men. For every gift of a votive offering or sacrifice lasts only for the immediate moment, and presently perishes, while a written record is an undying meed of gratitude, from time to time renewing its youth in the memory. Every Greek tongue will tell thy story, and every Greek man will worship the son of Ptah, Imouthes. Assemble hither, ye kindly and good men; avaunt ye malignant and impious! Assemble, all ye . . . , who by serving the god have been cured of diseases, ye who practise the healing art, ye who will labor as zealous followers of virtue, ye who have been blessed by great abundance of benefits, ye who have been saved from the dangers of the sea! For every place has been penetrated by the saving power of the god.

I now purpose to recount his miraculous manifestations, the greatness of his power, the gifts of his benefits. The history is this. King Mencheres by displaying his piety in the obsequies of three gods, and being successful in winning fame through the book, has won eternal glory. He presented to the tombs of Asclepius son of Hephaestus, Horus son of Hermes, and also Caleoibis son of Apollo money in abundance, and received as recompense his fill of prosperity. For Egypt was then free from war for this reason, and flourished with abundant crops, since subject countries prosper by the piety of their ruler, and on the other hand owing to his impiety they are consumed by evils. The manner in which the god Asclepius bade Mencheres busy himself with his tomb . . .'[1] (*)

[1] The rest is missing.

CHRIST AND ASCLEPIUS

332. Justinus, Apologia, 54, 10.

Ὅτε δὲ πάλιν ἔμαθον προφηθευτέντα, θεραπεύσειν αὐτὸν πᾶσαν νόσον καὶ νεκροὺς ἀνεγερεῖν, τὸν Ἀσκληπιὸν παρήνεγκαν.

And again when they [sc., the gentiles] learned about the prophecies to the effect that He [sc., Christ] would heal every disease and would raise the dead, they brought forward Asclepius.[1] (**)

[1] Cf. T. **94**; **95**; cf. also 332a. Origenes, *Contra Celsum*, III, 25: Ἵνα δὲ καὶ δῶ, ἰατρόν τινα δαίμονα θεραπεύειν σώματα τὸν καλούμενον Ἀσκληπιόν, εἴποιμ' ἂν πρὸς τοὺς θαυμάζοντας τὸ τοιοῦτο . . . εἰ δὲ μηδὲν θεῖον αὐτόθεν ἐμφαίνεται ἀπὸ τῆς Ἀσκληπιοῦ ἰατρικῆς καὶ Ἀπόλλωνος μαντικῆς, πῶς εὐλόγως ἄν τις . . . ὡς θεοὺς αὐτοὺς σέβοι ἂν καθαρούς τινας;

333. Lactantius, Divinae Institutiones, IV, 27, 12.

Ecce aliquis instinctu daemonis percitus dementit effertur insanit: ducamus hunc in Iovis Optimi Maximi templum vel, quia sanare homines Iuppiter nescit, in Aesculapi vel Apollinis fanum. iubeat utriuslibet sacerdos dei sui nomine ut nocens ille spiritus excedat ex homine: nullo id pacto fieri potest.

Behold, some one excited by the impulse of the demon is out of his senses, raves, is mad: let us lead him into the temple of Jupiter Optimus Maximus; or since Jupiter knows not how to cure men, into the fane of Asclepius or Apollo. Let the priest of either, in the name of his god, command the wicked spirit to come out of the man: that can in no way come to pass.

334. Acta Pilati, A, I, p. 216.

Λέγουσιν αὐτῷ· γόης ἐστίν, καὶ ἐν Βεελζεβοὺλ ἄρχοντι τῶν δαιμονίων ἐκβάλλει τὰ δαιμόνια, καὶ πάντα αὐτῷ ὑποτάσσεται. λέγει αὐτοῖς ὁ Πιλᾶτος· τοῦτο οὐκ ἔστιν ἐν πνεύματι ἀκαθάρτῳ ἐκβάλλειν τὰ δαιμόνια, ἀλλ' ἐν θεῷ τῷ Ἀσκληπιῷ.

They say to him [sc., Pilate]: he is a sorcerer and he casts out the devils in the name of the Devil who rules the devils, and everything is obedient to him. Pilate says to them: it is not possible to cast out devils in the name of an impure spirit but rather in the name of the god Asclepius.

335. Justinus, Apologia, 21, 1-2.

Τῷ δὲ καὶ τὸν λόγον, ὅ ἐστι πρῶτον γέννημα τοῦ θεοῦ, ἄνευ ἐπι-
μιξίας φάσκειν ἡμᾶς γεγεννῆσθαι Ἰησοῦν Χριστὸν τὸν διδάσκαλον ἡμῶν, καὶ τοῦτον σταυρωθέντα καὶ ἀποθανόντα καὶ
ἀναστάντα ἀνεληλυθέναι εἰς τὸν οὐρανόν, οὐ παρὰ τοὺς παρ᾽
ὑμῖν λεγομένους υἱοὺς τῷ Διὶ καινόν τι φέρομεν . . . Ἀσκληπιὸν
δέ, καὶ θεραπευτὴν γενόμενον, κεραυνωθέντα ἀνεληλυθέναι εἰς
οὐρανόν . . .

And when we say also that the Word, who is the first-birth
of God, Jesus Christ, our teacher, was produced without
sexual union, and that He was crucified and died, and rose
again, and ascended into heaven, we propound nothing new
and different from what you believe regarding those whom you
esteem sons of Jupiter . . . Asclepius, who, though he was a
great healer, was struck by a thunderbolt, and ascended to
heaven . . . (*)

336. Clemens Alexandrinus, Stromateis, V, 1, 13.

Οὔκουν εἰκῇ τοῖς παιδίοις παρακελευόμεθα τῶν ὤτων λαμβανο-
μένοις φιλεῖν τοὺς προσήκοντας, τοῦτο δήπου αἰνιττόμενοι δι᾽
ἀκοῆς ἐγγίγνεσθαι τῆς ἀγάπης τὴν συναίσθησιν, " ἀγάπη δὲ
ὁ θεὸς " ὁ τοῖς ἀγαπῶσι γνωστός, ὡς " πιστὸς ὁ θεὸς " ὁ τοῖς
πιστοῖς παραδιδόμενος διὰ τῆς μαθήσεως. καὶ χρὴ ἐξοικειοῦσθαι
ἡμᾶς αὐτῷ δι᾽ ἀγάπης τῆς θείας, ἵνα δὴ τὸ ὅμοιον τῷ ὁμοίῳ
θεωρῶμεν κατακούοντες τοῦ λόγου τῆς ἀληθείας ἀδόλως καὶ
καθαρῶς δίκην τῶν πειθομένων ἡμῖν παίδων. καὶ τοῦτο ἦν
ὁ ἠνίξατο ὅστις ἄρα ἦν ἐκεῖνος ὁ ἐπιγράψας τῇ εἰσόδῳ τοῦ ἐν
Ἐπιδαύρῳ νεώ·

ἁγνὸν χρὴ νηοῖο θυώδεος ἐντὸς ἰόντα
ἔμμεναι, ἁγνείη δ᾽ ἐστὶ φρονεῖν ὅσια.

" κἂν μὴ γένησθε ὡς τὰ παιδία ταῦτα, οὐκ εἰσελεύσεσθε " φησὶν
" εἰς τὴν βασιλείαν τῶν οὐρανῶν· " ἐνταῦθα γὰρ ὁ νεὼς τοῦ θεοῦ,
τρισὶν ἡδρασμένος θεμελίοις, πίστει, ἐλπίδι, ἀγάπῃ, φαίνεται.

Do we not rightly exhort our children taking them by their
ears that they should love their neighbors, intimating thereby
that understanding of charity comes through hearing; that
"God is Charity" means He who is known to those who

have charity, just as " God is Faith " means He who through learning imparts Himself to the faithful. And we must adapt ourselves to Him through the divine charity, so that we may see the like through the like, listening to the truth of the word without fraud and purely after the manner of the children who obey us. And that is what he referred to as in a riddle, whoever he was, who inscribed over the entrance of the temple of Epidaurus: " Pure must be he who enters the fragrant temple; purity means to think nothing but holy thoughts."[1] And " except ye become as little children ye shall not enter in the kingdom of heaven." For there is the sanctuary of God, set up over the three foundation-stones, Faith, Hope, Charity.[2]

[1] Cf. T. 318.
[2] Cf. however T. 103; cf. also 336a. Alcuinus, *Epistulae*, 245, p. 397, 24 f.: Aut ipse beatus Martinus, verus Dei cultor, in Christiano imperio minus venerari fas est, quam Scolapius falsator in paganorum potestate habuit?

IV. MEDICINE

PATRON AND LEADER OF MEDICINE

337. Hippocrates, Ius Iurandum, 1.

> Ὀμνύω Ἀπόλλωνα ἰητρὸν καὶ Ἀσκληπιὸν καὶ Ὑγείαν καὶ
> Πανάκειαν καὶ θεοὺς πάντας τε καὶ πάσας . . .

> I swear by Apollo Physician, and Asclepius, and Health, and
> Panacea and all the gods and goddesses . . .

338. Galenus, De Sanitate Tuenda, I, 8, 20.[1]

> Οὐ σμικρὸς δὲ τοῦ λόγου μάρτυς καὶ ὁ πάτριος ἡμῶν θεὸς
> Ἀσκληπιός . . .

> No slight witness of the statement is also our ancestral god
> Asclepius . . .[2]

[1] Cf. T. 413.
[2] Cf. T. 458: τὸν πάτριον θεὸν Ἀσκληπιόν; cf. also T. 348: ὁ ἡμέτερος πρόγονος Ἀσκληπιός; cf. T. 463.

339. Eunapius, Vitae Philosophorum, 498.

> . . . αὐτὸς . . . πρὸς τὸ ἄκρον ἐκδραμὼν τῆς ἰατρικῆς, τὸν
> πάτριον ἐμιμεῖτο θεόν, ὅσον ἀνθρώπῳ δυνατὸν ἐς τὴν μίμησιν
> ὑπελθεῖν τοῦ θείου.

> . . . by attaining the first rank in medicine he ⌊sc., Oribasius⌋
> imitated his ancestral god so far as it is possible for a mortal
> to progress towards the imitation of the divine.(*)

340. Plutarchus, Quaestiones Convivales, IX, 14, 4.

> . . . καὶ τοὺς ἰατροὺς Ἀσκληπιὸν ἔχοντας ἴσμεν ἡγεμόνα . . .

> . . . and we know that the physicians have Asclepius for their
> leader . . .[1]

[1] Cf. T. 704: Ἀσκληπιός . . . ὁ τῆς ἰατρικῆς ἔφορος.

179

341. Sextus Empiricus, Adversus Mathematicos, I, 260.[1]

. . . τὸν ἀρχηγὸν ἡμῶν τῆς ἐπιστήμης Ἀσκληπιόν . . .

. . . Asclepius, the founder of our science . . .[2]

[1] Cf. T. 69.   [2] Cf. T. 355.

342. Cornutus, Theologiae Graecae Compendium, Cp. 33.[1]

. . . τὸν Ἀσκληπιὸν . . . τὸν δοκοῦντα τοῖς ἀνθρώποις ὑποδεδειχέναι τὴν ἰατρικήν. ἐχρῆν γὰρ καὶ τούτῳ τῷ τόπῳ θεῖόν τι
ἐπιστῆσαι.

. . . Asclepius . . . reputed to have taught the medical art to
mankind. For even this branch of learning had to be under
the tutelage of something divine.

[1] Cf. T. 6.

343. Arnobius, Adversus Nationes, III, 23.

Aesculapius officiis et medendi artibus praeest: et cur plura
morborum et valetudinum genera ad sanitatem nequeunt in-
columitatemque perduci, immo sub ipsis fiunt curantium manibus atrociora?

Asclepius presides over the duties and arts of medicine: why
then cannot more kinds of disease and sickness be restored to
health and soundness, why in fact do they become worse under
the very hands of the physicians? (*)

344. Servius, Commentarii in Aeneidem, XII, 405.

. . . Aesculapius praeest medicinae, quam Apollo invenit, qui
in Ovidio de se ait: inventum medicina meum est.

. . . Asclepius presides over the art of medicine which Apollo
invented, who says in Ovid about himself, " Medicine is my
invention." [1]

[1] *Metamorphoses*, I, 521.

345. Anonymus, Expositio Totius Mundi et Gentium, 37.

Et est in omnibus et civitas et regio incomprehensibilis: et
totius orbis terrae paene de veritate philosophiae ipsa sola

habundat, in qua inveniuntur plurima genera philosophorum. itaque et Aesculapius dare ei voluit medicinae peritiam; ut habeat in toto mundo medicos optimos praestare dignatus est, et quamplurime initium salutis omnibus hominibus illa civitas constat.

Indeed in all things the city [*sc.*, of Alexandria] as well as the district is extraordinary: and of almost all the earth that city alone is rich in the truth of philosophy, for in that city are found most types of philosophers. Consequently Asclepius also desired to assign to it the skill of medicine: he thought it worthy to ensure that it have the best physicians in the whole world, and this city is well known among all men as being very often the beginning of recovery.

346. Choricius Gazaeus, Spartiates, XXIX (Decl. 8), 53.

> Εἰ τοίνυν οὔτε μέλος αἰσχρὸν ἐπὶ Μούσας ἀνάγομεν οὔτε τὸν Πᾶνά φαμεν ἀκολάστοις ᾄσμασι θέλγεσθαι καὶ φαῦλον ἰατρὸν ὑπηρέτην οὐ καλοῦμεν Ἀσκληπιοῦ καὶ τῆς Ἄρεως διακονίας ἐκβάλλομεν τὸν ἀδίκοις ἐπιχειροῦντα σφαγαῖς καὶ προπέτειαν ἀθλητοῦ τὸν Δία μισεῖν ὑπειλήφαμεν καὶ ταῦτα Μουσῶν μὲν καὶ Πανὸς ᾠδαῖς προσκειμένων, Ἀσκληπιοῦ δὲ σωμάτων θεραπείαις . . . οὐ γὰρ δίκαιον ὅσοι χρῶνται κακῶς τοῖς ἐκ θεῶν δεδομένοις, τούτους ἐπὶ προστασίαν θεοῦ καταφεύγειν.

If then we do not refer shameful song to the Muses nor say that Pan is enchanted with intemperate songs, if we do not call the worthless physician a servant of Asclepius, if we eject from the service of Mars him who employs unjust slaughter, if we assume that Zeus hates the reckless haste of the athlete, and these things we say, notwithstanding that the Muses and Pan are devoted to songs, that Asclepius is concerned with healing bodies, . . . for it is not right that those who employ unjustly the gifts of the gods should seek refuge in the patronage of the god.

347. Choricius Gazaeus, Spartiates, XXIX (Decl. 8), 49.

> Οὐδὲ γὰρ ἰατρικήν τις σχηματιζόμενος εἰ προσάγοι τοῖς κάμνουσιν ὅσα λυμαίνεται φάρμακά τε καὶ ποτὰ καὶ σιτία, φήσομεν ἥδεσθαι τούτοις τὸν Ἀπόλλωνος παῖδα, οὐδ᾽ ἂν καλοῖτο γνησίως θεράπων Ἀσκληπιοῦ τῆς ἐκείνου τέχνης παραβαίνων τοὺς νόμους.

Neither if anyone, pretending medical skill, should supply the sick with whatever drugs, potions, and provisions have harmful effects, shall we say that the son of Apollo is pleased with him nor would such a man rightly be called a servant of Asclepius, when transgressing the laws of the god's art.

CONTRIBUTIONS TO MEDICINE

348. Plato, Symposium, 186 D.

Δεῖ γὰρ δὴ τὰ ἔχθιστα ὄντα ἐν τῷ σώματι φίλα οἷόν τ᾽ εἶναι ποιεῖν καὶ ἐρᾶν ἀλλήλων . . . τούτοις ἐπιστηθεὶς ἔρωτα ἐμποιῆσαι καὶ ὁμόνοιαν ὁ ἡμέτερος πρόγονος Ἀσκληπιός, ὥς φασιν οἵδε οἱ ποιηταὶ καὶ ἐγὼ πείθομαι, συνέστησεν τὴν ἡμετέραν τέχνην.

Indeed he [sc., the physician] must be able to make the most hostile elements in the body friendly and loving towards each other . . . It was by knowing how to create love and unanimity in these that, as these poets here say and I [sc., Eryximachus, the physician] believe it, our forefather [1] Asclepius established this science of ours.

[1] Cf. T. 338; 458.

349. Hippocrates, Epistulae, 20 [IX, p. 386, 14-19 L.].

Ἐγὼ μὲν γὰρ ἰητρικῆς ἐς τέλος οὐκ ἀφῖγμαι, καίπερ ἤδη γηραλέος καθεστώς. οὐδὲ γὰρ ὁ τῆσδε εὑρετὴς Ἀσκληπιός, ἀλλὰ καὶ αὐτὸς ἐν πολλοῖς διεφώνησε, καθάπερ ἡμῖν αἱ τῶν ξυγγραφέων βίβλοι παραδεδώκασιν.

For I have not reached the perfection of the medical art, in spite of the fact that I am already an old man. Indeed, not even the discoverer of this art,[1] Asclepius [reached perfection], but he, too, failed in many instances as the books of the prose writers have imparted to us.

[1] Cf. **349a.** Lactantius, *Divinae Institutiones*, I, 19, 3: . . . ipsumque illum repertorem medicinae; cf. T. **250.**

350. Isidorus, Etymologiae, VIII, 11, 3.

Nam quorundam et inventiones artium cultum * peperisse dicuntur, ut Aesculapio medicina, Vulcano fabrica.

* cultum C   cultu Lindsay.

For even the inventions of the arts are said to have produced the cult of some of them [sc., those gods who formerly were men], such as medicine for Asclepius,[1] the smith's trade for Vulcan.

[1] Cf. T. 239 ff.

351. Georgius Hamartolus, Chronicon, I, 44.

Διὰ τοῦτο θεοὺς αὐτοὺς ἡγούμεθα, ὅτι τοῖς ἀνθρώποις χρήσιμοι γεγόνασι. Ζεὺς μὲν γὰρ λέγεται πλαστικὴν τέχνην ἐξευρηκέναι, Ποσειδῶν δὲ τὴν κυβερνητικήν, καὶ Ἥφαιστος μὲν χαλκευτικήν, Ἀσκληπιὸς δὲ ἰατρικήν, καὶ ὁ Ἀπόλλων τὴν μουσικήν, καὶ Ἀθηνᾶ μὲν τὴν ὑφαντικήν, Ἄρτεμις δὲ τὴν κυνηγετικήν, καὶ Ἥρα μὲν στολισμόν, Δημήτηρ δὲ γεωργίαν καὶ ἄλλοι ἄλλας.

For this reason we believe them gods because they have been useful to mankind.[1] For Zeus is said to have invented the plastic art, Poseidon that of sailing, Hephaestus the art of bronze work, Asclepius the medical, and Apollo the musical, and Athena the art of weaving; Artemis the art of the chase, Hera the art of dressing, Demeter the art of farming, and other gods other arts.[2]

[1] Cf. T. 239 ff.　　　　　　　　　[2] Cf. T. 290-290f; cf. also T. 260.

352. Lactantius Placidus, Commentarii in Statium, Ad Thebaidem, III, 398.

Epidaurius: civitas in Graecia, unde Aesculapius fuisse dicitur, medicinae artis inventor, et bene perito medico illius civitatis tribuit gloriam, a qua inventor artis ipsius dicitur sumpsisse principium.

Epidaurius: a city in Greece, whence Asclepius is said to have come,[1] the inventor of the medical art.[2] And to the well-skilled physician [sc., Idmon] he [sc., Statius] attributes the glory of that city from which the inventor of the art itself is said to have taken his origin.

[1] Cf. T. 20.
[2] Cf. 352a. Placidus, *Liber Glossarum*, s.v. Hescolapius: Asclepius medicine inventor.

353. Tertullianus, De Corona, 8.

Primus medelas Aesculapius exploraverit . . .

Supposing that Asclepius first investigated the healing art . . .

354. Celsus, De Medicina, Prooemium, 2.[1]

> . . . cum vetustissimus auctor Aesculapius celebretur, . . .
> quoniam adhuc rudem et vulgarem hanc scientiam paulo sub-
> tilius excoluit . . .

> . . . Asclepius celebrated as its most ancient founder . . .
> because he cultivated this science [*sc.*, of medicine] as yet rude
> and vulgar, with a little more exactness . . . (**)

[1] Cf. T. 244.

355. Diodorus, Bibliotheca Historica, V, 74, 6.

> Ἀπόλλωνος δὲ καὶ Κορωνίδος Ἀσκληπιὸν γενηθέντα, καὶ πολλ ι
> παρὰ τοῦ πατρὸς τῶν εἰς ἰατρικὴν μαθόντα, προσεξευρεῖν τήν τε
> χειρουργίαν καὶ τὰς τῶν φαρμάκων σκευασίας καὶ ῥιζῶν δυνάμεις,
> καὶ καθόλου προβιβάσαι τὴν τέχνην ἐπὶ τοσοῦτον, ὥστε ὡς
> ἀρχηγὸν αὐτῆς καὶ κτίστην τιμᾶσθαι.

> Asclepius, son of Apollo and Coronis,[1] who had learned many
> things pertaining to medicine from his father,[2] discovered in
> addition the art of surgery and the preparations of drugs [3] and
> the potencies of herbs and in general he advanced the art to
> such an extent that he was honored as its leader and founder.[4]

[1] Cf. T. 29; cf. also T. 4.   [2] Cf. T. 356.
[3] Cf. 355a. Theophylactus, *Historiae*, II, 6, 12: . . . ὁ στρατηγὸς . . . εἴς τε τὰς πόλεις
καὶ τὰ παρακείμενα φρούρια τὸ τετρωμένον ἐξέπεμπεν, ὅπως ἀκέσοιτο καὶ ἠπίοις τισὶ μαγ-
γανεύμασιν, ἔργοις Ἀσκληπιοῦ, τὰς τῶν τραυμάτων ἀκμὰς κατευνάσοιτο.
[4] Cf. T. 341.

356. Ps. Galenus, Introductio, Cp. 1 [XIV, p. 674 K.].

> Ἕλληνες τῶν τεχνῶν τὰς εὑρέσεις ἢ θεῶν παισὶν ἀνατιθέασιν,
> ἤ τισιν ἐγγὺς αὐτῶν οἷς πρῶτοι οἱ θεοὶ πάσης τέχνης ἐκοινώνησαν.
> οὕτως οὖν καὶ τὴν ἰατρικὴν πρῶτον μὲν Ἀσκληπιὸν παρ' Ἀπόλ-
> λωνος τοῦ πατρός φασιν ἐκμαθεῖν καὶ ἀνθρώποις μεταδοῦναι, διὸ
> καὶ δοκεῖ εὑρετὴς γεγονέναι αὐτῆς. πρὸ δὲ Ἀσκληπιοῦ τέχνη
> μὲν ἰατρικὴ οὔπω ἦν ἐν ἀνθρώποις, ἐμπειρίαν δέ τινα οἱ παλαιοὶ
> εἶχον φαρμάκων καὶ βοτανῶν, οἷα παρ' Ἕλλησι Χείρων ὁ
> κένταυρος ἠπίστατο καὶ οἱ ὑπὸ τούτου παιδευθέντες ἥρωες, ὅσα
> τε εἰς Ἀρισταῖον * καὶ Μελάμποδα καὶ Πολύειδον ἀναφέρεται.

* Ἀρισταῖον Edelstein Ἀριστέα Kühn.

The Greeks ascribe the inventions of the arts either to the children of the gods or to certain men akin to them to whom the gods first imparted every art. Thus also the art of medicine, they say, Asclepius first learned from Apollo, his father,[1] and transmitted it to mankind; for this reason he is considered to be the discoverer of it. Before the time of Asclepius, the medical art did not yet exist among men, but the men of old had some experience with herbs and drugs, such simples as among the Greeks Chiron, the Centaur, knew well, and the heroes who were instructed by him, and all those [simples] that are attributed to Aristaeus and Melampus and Polyidus.

[1] Cf. T. 355; cf. 356a. *Anthologia Latina*, I, 2, 719e, 2-3: Quod natum Phoebus docuit; cf. T. 147a; 614.

357. Isidorus, Etymologiae, IV, 3, 1-2.

De Inventoribus Medicinae

Medicinae autem artis auctor ac repertor apud Graecos perhibetur Apollo. hanc filius eius Aesculapius laude vel opere ampliavit. sed postquam fulminis ictu Aesculapius interiit, interdicta fertur medendi cura; et ars simul cum auctore defecit, latuitque per annos pene quingentos usque ad tempus Artaxerxis regis Persarum. tunc eam revocavit in lucem Hippocrates Asclepio patre genitus in insula Coo.

Concerning the Inventors of Medicine

The originator and discoverer of the art of medicine is among the Greeks reputed to be Apollo. This art his son Asclepius furthered in glory and accomplishment.[1] But after Asclepius died from the stroke of the thunderbolt,[2] healing is said to have been forbidden. The art ceased together with its founder and remained obscured for almost 500 years until the time of Artaxerxes, the King of the Persians.[3] Then Hippocrates, begotten of Asclepius[4] on the island of Cos, restored it to light.[5]

[1] Cf. T. 4.   [2] Cf. T. 105-15.   [3] Cf. T. 129a.   [4] Cf. T. 213 ff.   [5] Cf. T. 795.

358. Clemens Alexandrinus, Stromateis, I, 16, 75.

Οἱ δὲ Φοίνικας καὶ Σύρους γράμματα ἐπινοῆσαι πρώτους λέγουσιν, ἰατρικὴν δὲ Ἆπιν Αἰγύπτιον αὐτόχθονα πρὶν εἰς Αἴγυπτον ἀφικέσθαι τὴν Ἰώ. μετὰ δὲ ταῦτα Ἀσκληπιὸν τὴν τέχνην αὐξῆσαι λέγουσιν.

Some say that the Phoenicians and Syrians first invented the letters while Apis, a native Egyptian, invented medicine before Io came to Egypt. Subsequently, they say, Asclepius amplified the art.[1]

[1] Cf. **358a**. Theodoretus, *Graecarum Affectionum Curatio*, I, 20, p. 796A: Ἰατρικῆς δὲ ἄρξαι τὸν Ἆπιν φασὶ τὸν Αἰγύπτιον, εἶτα τὸν Ἀσκληπιὸν αὐξῆσαι τὴν τέχνην; cf. **358b**. Suidas, *Lexicon, s. v.* Γράμματα: . . . ἰατρικὴν δὲ ὁ Ἆπις ὁ Αἰγύπτιος [*sc., ἐς τὴν Ἑλλάδα ἐκόμισεν*]. ὁ δὲ Ἀσκληπιὸς ηὔξησε τὴν τέχνην; cf. **358c**. Georgius Hamartolus, *Chronicon*, I, 53: . . . ἰατρικῆς . . . καὶ τὸν Ἀσκληπιὸν αὐξῆσαι τὴν τέχνην; cf. also T. **8**.

359. Isidorus, Etymologiae, IV, 4, 1.

Prima Methodica inventa est ab Apolline, quae remedia sectatur et carmina. secunda Empirica, id est experientissima, inventa est ab Aesculapio, quae non indiciorum signis, sed solis constat experimentis. tertia Logica, id est rationalis, inventa ab Hippocrate.

First, methodical medicine was invented by Apollo, which pursues remedies and incantations. Second, empiric medicine was invented by Asclepius, that is, the most tested medicine, which is founded not on indications and signs but on experience alone. Third, logical, that is, rational medicine, was invented by Hippocrates.

360. Hyginus, Fabulae, CCLXXIV, 9.

Chiron centaurus Saturni filius artem medicinam chirurgicam ex herbis primus instituit; Apollo artem oculariam medicinam primus fecit; tertio autem loco Asclepius Apollinis filius clinicen repperit.

Chiron the Centaur, son of Saturn, first instituted the medical art of surgery after that of herbs; Apollo first created the medical art of ophthalmology; in the third place, moreover, Asclepius, son of Apollo, discovered " bedside " medicine.[1]

[1] Cf. however T. **795**.

361. Iamblichus, De Vita Pythagorica, XXXI, 208.

Εἶναι δὲ ταύτην τὴν ἐπιστήμην τὸ μὲν ἐξ ἀρχῆς Ἀπόλλωνός τε καὶ Παιῶνος, ὕστερον δὲ τῶν περὶ Ἀσκληπιόν.

They say that this knowledge [*sc.*, of dietetics] was at the beginning Apollo's and Paeon's, but later that of the associates of Asclepius.

362. Plautus, Menaechmi, V, 3, 885-86.

Ait se obligasse crus fractum Aesculapio,
Apollini autem brachium.

He [*sc.*, the doctor] says he set a broken leg for Asclepius, and for Apollo an arm. (*)

363. Augustinus, De Civitate Dei, III, 17.[1]

Atque in tanta strage bellorum etiam pestilentia gravis exorta est mulierum. nam priusquam maturos partus ederent, gravidae moriebantur. ubi se, credo, Aesculapius excusabat, quod archiatrum, non obstetricem profitebatur.

And while such disastrous wars were being waged, even a terrible disease broke out among the women. For those who were pregnant died before delivery. And Asclepius, I fancy, excused himself in this matter on the ground that he professed to be a general practitioner, not a midwife. (*)

[1] Cf. T. 853; 854.

364. Caelius Aurelianus, De Morbis Chronicis, Praefatio, 2.

Chronicae autem vel tardae passionis morbi, qui jam praejudicio quodam corpora possederint, solius medici peritiam poscunt, cum neque natura, neque fortuna solvantur. Podagram denique, vel Phthisin, aut Elephantiasin, vel similes tarditate passiones resolvi, nulli sectarum principes meminerunt, sicut saepe febres acutae solvuntur. hinc denique Graeci Asclepium nomen sumpsisse dixerunt, quod dura curando primus superaverit vitia.

Diseases of chronic or tardy nature which have taken possession of the body and somehow damaged it demand the experience of the physician alone since they are relieved neither by nature nor by chance. That gout,[1] then, or consumption, or elephantiasis or similar lingering illnesses are relieved, as acute fevers are often relieved, none of the leaders of the sects mention. Hence, then, the Greeks said Asclepius derived his name,[2] because he was the first to overcome diseases that are hard to heal.

[1] Cf. **364a.** Quintus Serenus Sammonicus, *Liber Medicinalis*, XLI, 767-69 : Quaedam †
etiam rabidae medicamina digna podagrae,/ Cui ter tricenas species Epidaurius ipse/
Dixit inesse deus?
[2] Cf. however T. **266-75.**

365. Plinius, Naturalis Historia, XXV, 2 (5), 13.

. . . Pythagoras clarus sapientia primus volumen de effectu earum composuit, Apollini, Aesculapio et in totum dis immortalibus inventione et origine adsignata.

. . . Pythagoras, renowned for his wisdom, was the first to write a volume on the effect of these [*sc.*, plants], with their discovery and origin attributed to Apollo, Asclepius, and the immortal gods in general.

366. Philostratus, Vita Apollonii, III, 44.

. . . ἐπανῆγεν ὁ Ἰάρχας ἐς τὸν περὶ τῆς μαντικῆς λόγον, καὶ
πολλὰ μὲν αὐτὴν ἀγαθὰ ἔλεγε τοὺς ἀνθρώπους εἰργάσθαι, μέγιστον δὲ τὸ τῆς ἰατρικῆς δῶρον· οὐ γὰρ ἄν ποτε τοὺς σοφοὺς
Ἀσκληπιάδας ἐς ἐπιστήμην τούτου παρελθεῖν, εἰ μὴ παῖς
Ἀπόλλωνος Ἀσκληπιὸς γενόμενος, καὶ κατὰ τὰς ἐκείνου φήμας
τε καὶ μαντείας ξυνθεὶς τὰ πρόσφορα ταῖς νόσοις φάρμακα, παισί
τε ἑαυτοῦ παρέδωκε, καὶ τοὺς ξυνόντας ἐδιδάξατο, τίνας μὲν δεῖ
προσάγειν πόας ὑγροῖς ἕλκεσι, τίνας δὲ αὐχμηροῖς καὶ ξηροῖς,
ξυμμετρίας τε ποτίμων φαρμάκων, ὑφ' ὧν ὕδεροι ἀποχετεύονται,
καὶ αἷμα ἴσχεται, φθόαι τε παύονται καὶ τὰ οὕτω κοῖλα. καὶ τὰ
τῶν ἰοβόλων δὲ ἄκη καὶ τὸ τοῖς ἰοβόλοις αὐτοῖς ἐς πολλὰ τῶν
νοσημάτων χρῆσθαι τίς ἀφαιρήσεται τὴν μαντικήν; οὐ γάρ μοι
δοκοῦσιν ἄνευ τῆς προγιγνωσκούσης σοφίας θαρσῆσαί ποτε
ἄνθρωποι τὰ πάντων ὀλεθριώτατα φαρμάκων ἐγκαταμῖξαι τοῖς
σώζουσιν.

. . . Iarchas led back the argument to the subject of divination, and said this art had conferred many blessings upon mankind, most important the gift of healing. " For," said he, " the wise sons of Asclepius would never have attained this wisdom if Asclepius had not been the son of Apollo, and had not in

accordance with the latter's responses and oracles concocted
drugs, appropriate for diseases; these he not only handed on
to his own sons, but he taught his companions what herbs must
be applied to running wounds, and what to inflamed and dry
wounds, and in what doses to administer liquid drugs, by
means of which dropsical patients are drained, and bleeding
is checked,[1] and diseases of decay and the cavities due to their
ravages are put an end to. And who (he said) can deprive the
art of divination of the credit of discovering simples which heal
the bites of venomous creatures and in particular of using the
virus itself as a cure of many diseases? For I do not think
that men without the forecasts of a prophetic wisdom would
ever have ventured to mingle with medicines that save life these
most deadly of poisons." (*)

[1] Cf. **366a.** Photius, *Bibliotheca*, 241, p. 1241 C: Ἰάρχας—ἴσχεται.

367. Arnobius, Adversus Nationes, I, 38.

Si enim vos Liberum, quod usum reppererit vini, si quod panis,
Cererem, si Aesculapium, quod herbarum, si Minervam, quod
. . . si Triptolemum, quod . . . Herculem, quod . . . divorum
retulistis in censum . . .

For if you have listed in the register of the gods[1] Liber, because
he discovered the use of wine; Ceres, because she discovered
the use of bread; Asclepius, because he discovered the use of
herbs; Minerva, because . . . ; Triptolemus, because . . . ;
Heracles, because . . .

[1] Cf. T. **236** ff., esp. T. **240.**

368. Arnobius, Adversus Nationes, I, 41.

Nonne Aesculapium medicaminum repertorem post poenas et
supplicia fulminis custodem nuncupavistis et praesidem sanitatis valetudinis et salutis?

Have you not, after his punishment and his death by lightning,
named Asclepius, the discoverer of medicines,[1] as the guardian
and protector of health,[2] of strength, and of well-being? (*)

[1] Cf. **368a.** Tertullianus, *Apologeticus*, XXIII, 6: . . . Aesculapius, medicinarum de-
monstrator. . . .
[2] Cf. T. **337-42.**

369. Plinius, Naturalis Historia, XXX, 8 (22), 69.

> Sedis vitiis efficacissima sunt oesypum . . . canini capitis cinis, senecta serpentis ex aceto, si rhagades sint, cinis fimi canini candidi cum rosaceo—aiunt inventum Aesculapii esse eodemque et verruccas efficacissime tolli.

> The most efficient remedies for diseases of the rectum are wool-grease . . . the ashes of a dog's head; a serpent's slough, with vinegar. In cases where there are chaps, the ashes of the white portion of dogs' dung, mixed with oil of roses; they say that this is an invention of Asclepius and that by the same treatment also warts are most efficiently removed.

370. Plinius, Naturalis Historia, XXV, 4 (11), 30.

> Panaces ipso nomine omnium morborum remedia promittit, numerosum et dis inventoribus adscriptum. unum quippe Asclepion cognominatur, a quo is filiam Panaciam appellavit.

> The panaces,[1] by its very name, promises remedies for all diseases: it is manifold and its discovery has been attributed to the gods. One kind, in fact, has the additional name "Asclepion," wherefore he called his daughter Panacia.[2]

[1] Cf. **370a.** Oribasius, *Collectiones Medicae*, XIV, 14, 6: πάνακες Ἀσκληπίειον καὶ πάνακες Χειρώνειον ἔλαττον πάνακος Ἡρακλείας.

[2] Cf. T. 280 ff.

371. Anonymus, Carmen De Herbis, 114-23.

> Νῦν δ' ἤτοι διερῶ μάλ' ἀριφραδές, οὐδέ σε κεύσω
> κενταύρου Κρονίδαο φερώνυμον εὕρεμα ῥίζαν
> Χείρωνος σθεναροῦ, τὴν πὰρ νάπῃ εἰνοσιφύλλῳ
> Πηλίου ἢ νιφόεντι κιχὼν ἐφράσσατο δειρῇ.
> ἣν βαθύρους Ἀχελῷος ἐγείνατο, καὶ φάτο πᾶσιν
> ἐσθλὴν Παιήων Ἀσκληπιὸς ἠπιόδωρος
> κικλήσκειν πανάκειαν, ἐπεὶ πάνθ' ὅσσα βροτοῖσι
> φλεγμαίνοντα πάθη παύει καὶ κρούσματα πληγῶν,
> ὑστερικαῖς τε γυναιξὶ λίην σωτήριόν ἐστι,
> στραγγουροῦντά τε παύει ἐν ἤματι καὶ λιθιῶντας.

> Now truly I shall explain at length very clearly and shall not conceal from you the well-named root discovered by mighty

Chiron, the Centaur, son of Cronus, the root which he found and observed along the leafy glen of Pelion or the snow-capped ridge. The deep-flowing stream of Achelöus produced this root and Paeon Asclepius, giver of gentle gifts, gave orders that, being good for everything, it should be called " panacea," all-healer, since it stops all inflamed conditions that befall men, as well as the effects of blows; it is exceedingly helpful to women suffering in their wombs and it releases within a day not only anyone suffering from strangury but also those afflicted with stones.

372. Galenus, De Compositione Medicamentorum per Genera, VII, Cp. 7 [XIII, p. 986 K.].

Καλεῖται δὲ Ἀσκληπιὸς ἡ δύναμις.

The drug [of which Galen gives the formula] is called Asclepius.[1]

[1] Cf. **372a.** Suidas, *Lexicon*, *s. v.* Ἀσκληπιάδης: καὶ Ἀσκληπίειον φάρμακον καὶ Ἀσκλήπειον φάρμακον, Ἀσκληπήϊον δὲ ἱερόν; cf. **372b.** *Etymologicum Magnum*, *s. v.* ἀσκελές: καὶ ἀσκληπιάς, βοτάνη τις.

373. Oribasius, Synopsis ad Eustathium, III, 162.

ρ̄ξ̄β̄. Σμήγματα καὶ δρώπακες.—Σμῆγμα ὁ Ἀσκληπιός.

Ἀλῶν κοινῶν, ἁλῶν ἀμμωνιακῶν, ἁλῶν Καππαδοκικῶν πεφωγμένων, κισήρεως, ἀφρονίτρου λευκοῦ, δαφνίδων ξηρῶν ἀνὰ ҩ ᾱ, ἐλλεβόρου λευκοῦ, στρουθίου, σταφίδος ἀγρίας, νάπυος Ἀλεξανδρίνου, θείου ἀπύρου, πυρέθρου, φέκλης, σχίνου ἄνθους, κυπέρου ἀνὰ ⌐ο ϛ̄, σαμψύχου, στυπτηρίας σχιστῆς, κηκῖδος, κόμμεως, λιβανωτοῦ, ἁλκυωνίου, πεπέρεως, κάχρυος, σικύου ἀγρίου ῥίζης, χαμαιλέοντος, ἴρεως, πρασίου ξηροῦ ἀνὰ ⌐ο γ̄. κόψας καὶ σήσας χρῶ.

162. Unguents and pitch-plasters. Unguent Asclepius.[1]

Common salts, rock salts, Cappadocian salts roasted, pumicestone, white coarse soda (potass), laurel berries, at the rate of 1 pound; white hellebore, soap-wort, wild raisin, Alexandrian mustard, unburnt brimstone, feverfew, salt of tartar, mastichtree blossoms, marsh-plant, at the rate of 6 ounces; marjoram, powdered astringent earth, gall-nut, gum, frankincense, bastardsponge, pepper, parched barley, raw cucumber root, chameleon thistle, iris, dry horehound, at the rate of 3 ounces; chopping and sifting these, use.

¹ Cf. **373a.** Paulus Aegineta, *Opera*, IV, 1, 7: καὶ τὸ τοῦ Ἀσκληπιοῦ δὲ ξηρὸν σμῆγμα; **373b.** *ibid.*, VII, 13, 17: σμῆγμα ξηρὸν ὁ Ἀσκληπιός; cf. also **373c.** *ibid.*, III, 24, 7: ἢ καὶ τῷ καλουμένῳ Ἀσκληπιῷ χρηστέον.

374. Antonius Musa, De Herba Vettonica, 47, v. 186.

Herba vettonica, quae prima inventa es ab Aesculapio vel Cirone Centauro, his precibus adesto.

Herb Betony, first discovered by Asclepius or by Chiron the Centaur, heed these prayers.

375. Appendix ad Ps. Apuleium, CXIX [CML IV, p. 296].

Herba apium, te deprecor per inventorem tuum Aescolapium, ut venias ad me . . .

Herb Parsley, I pray you, in the name of your discoverer Asclepius, to come to me . . .

376. Appendix ad Ps. Apuleium, CXX [CML IV, p. 296].

Sancta herba crisocantus, te deprecor per Aesculapium, herbarum inventorem, ut venias huc . . .

Holy plant Crisocantus, I pray you by Asclepius, discoverer of herbs, to come hither. . . .

377. Appendix ad Ps. Apuleium, CXXVI [CML IV, p. 297].

Herba erifion, ut adsis mihi roganti, et cum gaudio virtus tua praesto sit, et ea omnia persanes, quae Aesculapius aut Ciro Centaurus, magister medicinae, de te adinvenit.

Herb Erifion, may you attend my request, may your power be present bringing happiness with it, may you effect all those cures which Asclepius or Chiron the Centaur, teacher of medicine, devised from you.

378. Corpus Hippiatricorum Graecorum, I, 9.

Κεκλήσθων δὲ ἡμῖν συμφορεῖς τοῦ λόγου τοῦδε Ποσειδῶν τε ἵππειος καὶ ὁ τοῦ τῶν ἀνθρώπων γένους σωτὴρ Ἀσκληπιός, ᾧ

,πάντως που καὶ ἵππων μέλει, εἰ δεῖ τι Χείρωνος τοῦ Κενταύρου
καὶ τῶν ἐν Πηλίῳ διατριβῶν ἀξίως μεμνῆσθαι.

Let there be summoned as advocates in this argument Poseidon
the horseman and the savior of the human race, Asclepius, who
certainly has regard even for horses, if he duly remembers—as
he ought to—anything about Chiron the Centaur and his
education on Pelion.[1]

[1] Cf. T. 50 ff.

379. Cicero, De Natura Deorum, III, 22, 57.

Aesculapiorum primus Apollinis, quem Arcades colunt, qui
specillum invenisse. primusque volnus dicitur obligavisse, secundus secundi Mercurii frater : is fulmine percussus dicitur
humatus esse Cynosuris; tertius Arsippi et Arsinoae, qui
primus purgationem alvi dentisque evolsionem ut ferunt invenit, cuius in Arcadia non longe a Lusio flumine sepulcrum
et lucus ostenditur.

Of the various Asclepii the first is the son of Apollo, whom
the Arcadians worship; he is reputed to have invented the
probe and to have been the first surgeon to employ bandages.
The second is the brother of the second Mercury; he is said
to have been struck by lightning and buried at Cynosura.[1] The
third is the son of Arsippus and Arsinoë,[2] and is said to have
first invented the use of purges and the extraction of teeth;
his tomb and grove are shown in Arcadia, not far from the
river Lusius.[3] (*)

[1] Cf. T. 116; cf. also T. 9; 101.	[2] Cf. T. 40.	[3] Cf. T. 118.

380. Ioannes Lydus, De Mensibus, IV, 142.

Ἀσκληπιοὶ τρεῖς λέγονται γενέσθαι. πρῶτος Ἀπόλλωνος τοῦ
Ἡφαίστου, ὃς ἐξεῦρε μήλην. δεύτερος Ἴσχυος τοῦ Ἐλάτου καὶ
Κορωνίδος, ⟨ὃς ἐν τοῖς Κυνοσουρίδος⟩ ὁρίοις ἐτάφη. τρίτος
Ἀρσίππου καὶ Ἀρσινόης τῆς Λευκίππου. οὗτος εὗρε τομὴν καὶ
ὀδοντάγραν, καὶ τάφος αὐτῷ ἐν Ἀρκαδίᾳ. οἱ δὲ ἀστρονόμοι αὐτόν
φασιν εἶναι τὸν Ὀφιοῦχον τὸν ἐπὶ τοῦ Σκορπίου ἑστῶτα.

There are said to have been three Asclepii.[1] The first was the
son of Apollo, the son of Hephaestus, who invented the probe.

The second was the son of Ischys, the son of Elatus, and of Coronis; he was buried in the Cynosurean territories. The third was the son of Arsippus and Arsinoë, the daughter of Leucippus. He invented surgery and an instrument for the extraction of teeth,[2] and his tomb is in Arcadia. The astronomers claim that he is Ophiouchus [the serpent constellation] which stands above Scorpion.[3]

[1] Cf. **380a.** Lucius Ampelius, *Liber Memorialis*, 9, 8: Aesculapii tres: primus Apollo dictus Vulcani filius; secundus Elati filius; tertius † Aristeti et Alcippes filius; cf. **380b.** Arnobius, *Adversus Nationes*, IV, 15: . . . aiunt idem theologi, quattuor esse Volcanos et tris Dianas, Aesculapios totidem et Dionysos quinque. . . .

[2] Cf. T. 379.   [3] Cf. T. 121-22.

381. Ps. Galenus, Introductio, Cp. 1 [XIV, p. 676 K.].

Τελείαν δὲ ἰατρικὴν καὶ τοῖς ἑαυτῆς μέρεσι συμπεπληρωμένην, τὴν μὲν ὡς ἀληθῶς θείαν Ἀσκληπιὸν μόνον εὑρεῖν, τὴν δὲ ἐν ἀνθρώποις τοὺς Ἀσκληπιάδας παρὰ τούτου διαδεξαμένους τοῖς ἔπειτα παραδοῦναι, . . .

The perfect art of medicine, complete in all its parts, as far as it is really divine, Asclepius alone discovered, but as far as it is medicine among mortals, the Asclepiads, having received the art from him, transmitted it to their successors . . .

TEMPLE MEDICINE

EPIPHANY

382. Strabo, Geographica, VIII, 6, 15.

Καὶ αὕτη δ' οὐκ ἄσημος ἡ πόλις καὶ μάλιστα διὰ τὴν ἐπιφάνειαν τοῦ Ἀσκληπιοῦ θεραπεύειν νόσους παντοδαπὰς πεπιστευμένου καὶ τὸ ἱερὸν πλῆρες ἔχοντος ἀεὶ τῶν τε καμνόντων καὶ τῶν ἀνακειμένων πινάκων, ἐν οἷς ἀναγεγραμμέναι τυγχάνουσιν αἱ θεραπεῖαι, καθάπερ ἐν Κῷ τε καὶ Τρίκκῃ.

And this city [sc., Epidaurus] is not without distinction, and particularly because of the epiphany of Asclepius, who is believed to cure diseases of every kind and always has his temple full of the sick, and also of the votive tablets on which the treatments are recorded,[1] just as at Cos[2] and Tricca.(*)

[1] Cf. T. 384; 423; 735; 739.   [2] Cf. T. 794-96.

383. Livius, Ab Urbe Condita, XLV, 28, 3.

> . . . inde haud parem opibus Epidaurum, sed inclutam Aesculapi nobili templo, quod quinque milibus passuum ab urbe distans nunc vestigiis revolsorum donorum, tum donis dives erat, quae remediorum salutarium aegri mercedem sacraverant deo.

> . . . then [Aemilius Paulus, in 167 B. C., visited] Epidaurus, which, though unequal to them [*sc.*, Sicyon and Argos] in opulence, was yet renowned for a splendid temple of Asclepius, which, standing at five miles' distance from the city,[1] is now rich in traces of broken votives but then was rich in offerings, which the sick had dedicated to the god as an acknowledgment for the remedies which restored them to health. (*)

[1] Cf. T. 708.

384. Pausanias, Descriptio Graeciae, II, 27, 3.

> Στῆλαι δὲ εἱστήκεσαν ἐντὸς τοῦ περιβόλου τὸ μὲν ἀρχαῖον καὶ πλέονες, ἐπ' ἐμοῦ δὲ ἓξ λοιπαί. ταύταις ἐγγεγραμμένα καὶ ἀνδρῶν καὶ γυναικῶν ἐστιν ὀνόματα ἀκεσθέντων ὑπὸ τοῦ Ἀσκληπιοῦ, προσέτι δὲ καὶ νόσημα ὅ τι ἕκαστος ἐνόσησε, καὶ ὅπως ἰάθη. γέγραπται δὲ φωνῇ τῇ Δωρίδι.

> Tablets stood within the enclosure.[1] Of old, there were more of them: in my time six were left. On these tablets are engraved the names of men and women who were healed by Asclepius, together with the disease from which each suffered, and how he was cured. The inscriptions are in the Doric dialect. (*)

[1] Cf. T. 382; 388; 739. Cf. also **384a.** Pausanias, *Descriptio Graeciae*, II, 36, 1:
. . . ἡ δὲ Ἁλίκη τὰ μὲν ἐφ' ἡμῶν ἐστιν ἔρημος, ᾠκεῖτο δὲ καὶ αὕτη ποτέ, καὶ Ἁλικῶν λόγος ἐν στήλαις ἐστὶ ταῖς Ἐπιδαυρίων αἳ τοῦ Ἀσκληπιοῦ τὰ ἰάματα ἐγγεγραμμένα ἔχουσιν.

385. Themistius, Oratio XXVII [p. 402, 12-18 Dind.].

> Εἰ τὰ σώματα ἐνοσοῦμεν καὶ ἐδεόμεθα τῆς παρὰ τοῦ θεοῦ βοηθείας, ὁ δὲ ἐνταῦθα παρῆν ἐν τῷ νεῷ καὶ τῇ ἀκροπόλει καὶ παρεῖχεν ἑαυτὸν τοῖς κάμνουσιν, ὥσπερ δήποτε καὶ λέγεται, πότερον ἦν ἀναγκαῖον εἰς Τρίκκην βαδίζειν καὶ διαπλεῖν εἰς Ἐπίδαυρον κατὰ τὸ παλαιὸν κλέος, ἢ δύο βήματα κινηθέντας ἀπηλλάχθαι τοῦ νοσήματος;

If we were ill in body and required the help of the god, and he were present here in the temple and the acropolis and were offering himself to the sick, just as even of old he is said to have done, would it be necessary to go to Tricca and sail to Epidaurus on account of their ancient fame, or to move two steps and get rid of our illness?

386. Statius, Silvae, III, 4, 21-25.

Dicitur . . . Venus . . .
Pergameas intrasse domos, ubi maximus aegris
auxiliator adest et festinantia sistens
fata salutifero mitis deus incubat angui.

It is said that Venus . . . entered the shrine at Pergamum,[1] where the great helper of the sick is present to aid, and, a gentle god who stays the hurrying fates, rests on the health-bringing snake.[2] (*)

[1] Cf. T. **801** ff.　　　　　[2] Cf. T. **688** ff.

387. Philostratus, Vita Apollonii, I, 8.

Ἐκπεπληγμένων δὲ αὐτὸν τῶν περὶ τὸ ἱερὸν καὶ τοῦ Ἀσκληπιοῦ ποτε πρὸς τὸν ἱερέα φήσαντος, ὡς χαίροι θεραπεύων τοὺς νοσοῦντας ὑπὸ Ἀπολλωνίῳ μάρτυρι . . .

And the people round about the temple [sc., at Aegae] were struck with admiration for him [sc., Apollonius], and the god Asclepius one day said to the priest that he was delighted to have Apollonius as witness for his cures of the sick . . .

388. Libanius, Epistulae, 695, 2.

. . . νῦν μὲν τὴν τοῦ θεοῦ δύναμιν δεικνὺς ἐκ τῶν ἐπιγραμμάτων ἃ ἦν τῶν ὑγιανάντων, . . .

. . . now showing the power of the god by means of the inscriptions [1] of those who had become healthy [sc., at Aegae], . . .

[1] Cf. T. **382**; **384**; **423** ff.; **735**; **739**.

389. Sozomenus, Historia Ecclesiastica, II, 5.[1]

Καθότι Αἰγεᾶται μὲν ηὔχουν τοὺς κάμνοντας τὰ σώματα νόσων ἀπαλλάττεσθαι παρ' αὐτοῖς, ἐπιφαινομένου νύκτωρ καὶ ἰωμένου τοῦ δαίμονος.

As to the Aegeatae, they boasted that with them those sick in body were freed of their diseases since at night the demon [sc., Asclepius] appeared and healed.

[1] Cf. T. 819.

390. Eusebius Caesariensis, De Vita Constantini, III, 56.[1]

> . . . μυρίων ἐπτοημένων ἐπ' αὐτῷ ὡς ἂν ἐπὶ σωτῆρι καὶ ἰατρῷ, ποτὲ μὲν ἐπιφαινομένῳ τοῖς ἐγκαθεύδουσι, ποτὲ δὲ τῶν τὰ σώματα καμνόντων ἰωμένῳ τὰς νόσους.

. . . with thousands excited over him [sc., Asclepius] as if over a savior [2] and physician, who now revealed himself to those sleeping [in the temple at Aegae], and again healed the diseases of those ailing in body.

[1] Cf. T. 818.　　　　　　[2] Cf. T. 322.

391. Pausanias, Descriptio Graeciae, VII, 27, 11.[1]

> . . . ἱερόν ἐστιν Ἀσκληπιοῦ . . . καὶ ἰάματα ἀνθρώποις παρὰ τοῦ θεοῦ γίνεται.

. . . a sanctuary of Asclepius [sc., near Pellene] . . . where cures of patients are effected by the god.

[1] Cf. T. 782.

HEALING

HEALER AND HEALED

392. Lucianus, Bis Accusatus, 1.

> Ὁ μὲν γὰρ Ἀσκληπιὸς ὑπὸ τῶν νοσούντων ἐνοχλούμενος " ὁρῇ τε δεινὰ θιγγάνει τε ἀηδέων ἐπ' ἀλλοτρίῃσί τε ξυμφορῇσιν ἰδίας καρποῦται λύπας."

For Asclepius, pestered by the sick, " sees dire sights, and touches unpleasant things, and in the woes of others reaps sorrow for himself." [1]

[1] Quotation from Hippocrates, De Flatibus, 1.

393. Maximus Tyrius, Philosophumena, VI, 4 d.

> . . . ὁ μάντις σοφός, καὶ ὁ τέκτων σοφός, καὶ ὁ Ἀπόλλων δήπου, καὶ ὁ ἰατρός, καὶ ὁ Ἀσκληπιὸς δήπου ὁμοίως τίμιος . . .

. . . the soothsayer is wise, and the builder is wise, and Apollo obviously, and the physician, and Asclepius clearly is equally regarded with honor. . . .[1]

[1] Paraphrase of Homer, *Odyssey*, XVII, 384.

394. Philostratus, Vita Apollonii, I, 11.

Καὶ ἅμα ἐς τὸν Ἀσκληπιὸν βλέψας " φιλοσοφεῖς," ἔφη, " ὦ Ἀσκληπιέ, τὴν ἄρρητόν τε καὶ συγγενῆ σαυτῷ φιλοσοφίαν μὴ συγχωρῶν τοῖς φαύλοις δεῦρο ἥκειν, μηδ' ἂν πάντα σοι τὰ ἀπὸ Ἰνδῶν καὶ Σαρδῴων ξυμφέρωσιν. οὐ γὰρ τιμῶντες τὸ θεῖον θύουσι ταῦτα καὶ ἀνάπτουσιν, ἀλλ' ὠνούμενοι τὴν δίκην, ἣν οὐ συγχωρεῖτε αὐτοῖς δικαιότατοι ὄντες."

And at the same moment he [*sc.*, Apollonius] looked towards Asclepius, and said: " O Asclepius, you teach a philosophy that is secret and congenial to yourself, in that you suffer not the wicked to come hither, not even if they bring to you all the wealth of India and Sardis.[1] For it is not out of reverence for the divinity that they sacrifice these victims and kindle these fires, but in order to purchase a verdict, which you will not concede to them in your perfect justice."

[1] Cf. T. 320: οὐδὲ . . . ἐπ' ἀμοιβῆς ἐλπίδι τοὺς ἀνθρώπους ἰᾶται . . . ; cf. also T. 455; cf. however T. 99-100.

395. Cassius Dio, Historia Romana, LXXVIII, 15, 6-7.

Οὔτε γὰρ ὁ Ἀπόλλων ὁ Γράννος οὔθ' ὁ Ἀσκληπιὸς οὔθ' ὁ Σάραπις καίπερ πολλὰ ἱκετεύσαντι αὐτῷ πολλὰ δὲ καὶ προσκαρτερήσαντι ὠφέλησεν. ἔπεμψε γὰρ αὐτοῖς καὶ ἀποδημῶν καὶ εὐχὰς καὶ θυσίας καὶ ἀναθήματα, καὶ πολλοὶ καθ' ἑκάστην οἱ τοιοῦτό τι φέροντες διέθεον. ἦλθε δὲ καὶ αὐτὸς ὡς καὶ τῇ παρουσίᾳ τι ἰσχύσων, καὶ ἔπραξεν πάνθ' ὅσα οἱ θρησκεύοντές τι ποιοῦσιν, ἔτυχε δ' οὐδενὸς τῶν ἐς ὑγίειαν τεινόντων.

He [*sc.*, Alexander Severus] received no help from Apollo Grannus, nor yet from Asclepius or Sarapis, in spite of his many supplications and his unwearying persistence. For even while abroad he sent to them prayers, sacrifices and votive offerings, and many couriers ran hither and thither every day carrying something of this kind; and he also went to them himself, hoping to prevail by appearing in person, and did all that devotees are wont to do; but he obtained nothing that contributed to health.

396. Cicero, De Natura Deorum, III, 35, 84.

Hunc igitur nec Olympius Iuppiter fulmine percussit **nec** Aesculapius misero diuturnoque morbo tabescentem interemit, atque in suo lectulo mortuus in tyrannidis rogum inlatus est, eamque potestatem quam ipse per scelus erat nanctus quasi iustam et legitimam hereditatis loco filio tradidit.

Well, [Dionysius] was not struck dead with a thunderbolt by Olympian Jupiter, nor did Asclepius cause him to waste away and perish of some painful and lingering disease. He died in his bed and was laid upon a royal pyre, and the power which he had himself secured by crime he handed on as an inheritance to his son as a just and lawful sovereignty.

397. Philostratus, Vita Apollonii, I, 9.

Ἄξιον δὲ μηδὲ τὰ ἐν τῷ ἱερῷ παρελθεῖν βίον γε ἀφηγούμενον ἀνδρός, ὃς καὶ τοῖς θεοῖς ἦν ἐν λόγῳ· μειράκιον γὰρ δὴ Ἀσσύριον παρὰ τὸν Ἀσκληπιὸν ἧκον ἐτρύφα νοσοῦν καὶ ἐν πότοις ἔζη, μᾶλλον δὲ ἀπέθνησκεν. ὑδέρῳ δὲ ἄρα εἴχετο καὶ μέθῃ χαῖρον αὐχμοῦ ἠμέλει. ἠμελεῖτο δὴ ὑπὸ τοῦ Ἀσκληπιοῦ διὰ ταῦτα, καὶ οὐδὲ ὄναρ αὐτῷ ἐφοίτα. ἐπιμεμφομένῳ δὲ ταῦτα ἐπιστὰς ὁ θεὸς " εἰ Ἀπολλωνίῳ," ἔφη, " διαλέγοιο, ῥάων ἔσῃ." προσελθὼν οὖν τῷ Ἀπολλωνίῳ " τί ἄν," ἔφη, " τῆς σῆς σοφίας ἐγὼ ἀπολαύσαιμι; κελεύει γάρ με ὁ Ἀσκληπιὸς συνεῖναί σοι." " ὅ," ἦ δ᾽ ὅς, " ἔσται σοι πρὸς τὰ παρόντα πολλοῦ ἄξιον. ὑγιείας γάρ που δέῃ;" " νὴ Δί᾽," εἶπεν, " ἥν γε ὁ Ἀσκληπιὸς ἐπαγγέλλεται μέν, οὐ δίδωσι δέ." " εὐφήμει," ἔφη, " τοῖς γὰρ βουλομένοις δίδωσι, σὺ δὲ ἐναντία τῇ νόσῳ πράττεις, τρυφῇ γὰρ διδοὺς ὀψοφαγίαν ἐπεσάγεις ὑγροῖς καὶ διεφθορόσι τοῖς σπλάγχνοις καὶ ὕδατι ἐπαντλεῖς πηλόν."

Now it is well that I should not pass over what happened in the temple while relating the life of a man who was held in esteem even by the gods. For an Assyrian stripling came to Asclepius, and though he was sick, yet he lived the life of luxury, and being continually drunk, he did not live but rather he was ever dying. He suffered then from dropsy, and finding

his pleasure in drunkenness he did not care for dryness. On
this account then Asclepius took no care of him, and did not
visit him even in a dream.[1] When the youth grumbled at this,
the god approached him and said, " If you were to consult
Apollonius, you would be easier." He therefore went to
Apollonius, and said: " What is there in your wisdom that
I can profit by? For Asclepius bids me consult you." And he
replied: " I can advise you of what, under the circumstances,
will be most valuable to you; for I suppose you want to get
well." " Yes, by Zeus," answered the other, " I want the health
which Asclepius promises, but does not give." " Hush," said
the other, " for he gives to those who desire it, but you do
things that irritate and aggravate your disease, for you give
yourself up to luxury, and you accumulate luxurious meals
upon your water-logged and worn-out stomach and, as it were,
choke water with a flood of mud." (**)

[1] Cf. T. 415.

398. Philostratus, Epistulae, 18.

Μαλακώτερον διετέθης ὑπὸ τῆς σανδάλου θλιβείς, ὡς πέπεισμαι,
δειναὶ γὰρ δακεῖν σάρκας ἁπαλὰς αἱ τῶν δερμάτων καινότητες.
διὰ τοῦτο ὁ Ἀσκληπιὸς τὰ μὲν ἐκ πολέμου καὶ θήρας τραύματα
καὶ πάσης τοιαύτης τύχης ἰᾶται ῥᾳδίως, ταῦτα δὲ ἐᾷ διὰ τὸ
ἑκούσιον, ὡς ἀνοίᾳ μᾶλλον ἢ ἐπηρείᾳ δαιμόνων γενόμενα.

The rubbing of your sandal has softened you up, I believe.
For new leathers are prone to bite tender flesh. That is why
Asclepius easily heals wounds from war and the chase and
from all such occurrences of chance, but such as this he disregards as self-inflicted, since they arise from ignorance rather
than from the capricious dealing of the demons.

399. Aelianus, Fragmenta, 89.

Ἀνὴρ Εὐφρόνιος, κακοδαίμων ἀνήρ, καὶ ἔχαιρεν ἐπὶ ταῖς Ἐπικούρου φλυαρίαις καὶ ἐξ ἐκείνων κακὰ εἰρύσατο δύο, ἄθεός τε
καὶ ἀκόλαστος εἶναι.

ὁ δὲ ἐν τοσούτῳ κακῷ ὢν οὐκ ἐπελάθετο τῆς βδελυρᾶς ἐκείνης
καὶ ἀθέου συγγραφῆς, ἣν ὁ Γαργήττιος, ὥσπερ οὖν τὰ ἐκ Τιτανι-
κῶν σπερμάτων φύντα, τῷ βίῳ τῶν ἀνθρώπων κηλῖδα προσε-
τρίψατο.

ὁ δὲ ἀθλίως νόσῳ (περιπνευμονίαν καλοῦσιν Ἀσκληπιαδῶν παῖδες αὐτήν) πιεζόμενος τὰ μὲν πρῶτα ἐδεῖτο τῆς ἀνθρώπων ἰατρικῆς καὶ ἐκείνων ἤρτητο.

τῆς τῶν ἰατρῶν ἐπιστήμης βιαιότερον ἦν τὸ νόσημα.

ἐπεὶ τοίνυν ὑπὲρ τῶν ἐσχάτων ἐσάλευεν ἤδη, κομίζουσιν αὐτὸν οἱ προσήκοντες ἐς Ἀσκληπιοῦ.

καὶ καταδαρθόντι οἱ τῶν τις ἱερέων ἐδόκει λέγειν μίαν εἶναι σωτηρίας ὁδὸν τῷ ἀνδρὶ καὶ ἓν τῶν ἐφεστώτων κακῶν φάρμακον, εἴπερ οὖν τὰ Ἐπικούρου βιβλία καταφλέξας, καὶ τῶν ἀθέων τε καὶ ἀσεβῶν καὶ ἐκτεθηλυμμένων στιγμάτων τὴν σποδὸν ἀναδεύσει κηρῷ ὑγρῷ, καὶ ἐπιπλασάμενος τὴν νηδὺν καὶ τὸν θώρακα πάντα καταδήσει ταινίαις.

ὁ δὲ ὅσα ἤκουσε τοῖς οἰκείοις ὁμολογεῖ, καὶ ἐκεῖνοι περιχαρείας αὐτίκα ὑπεπλήσθησαν τῷ μὴ ἐκφρησθῆναι ἐκφαυλισθέντα καὶ ἀτιμασθέντα ὑπὸ τοῦ θεοῦ αὐτόν.

καί τινα ἐξ αὐτοῦ διδασκαλίαν ἐναυσάμενοι, κᾆτα ἐμιμήσαντο ἐς τὸ εὖ καὶ καλῶς.

The man Euphronius, a wretched creature, took pleasure in the silly talk of Epicurus and acquired two evils from this: being impious and intemperate.

He did not forget, when in such a wicked state, that shameless and impious treatise which the Gargettian [*sc.*, Epicurus], like an offspring of the Titan brood, inflicted as a blot upon the life of men.

Being grievously afflicted with a disease (the sons of the Asclepiads call it pneumonia), he first besought the healing aid of mortals and clung to them.

The illness was stronger than the knowledge of the physicians.

When he was already tottering close to the brink of death, his friends brought him to the temple of Asclepius. And as he fell asleep one of the priests seemed to say to him that there was one road of safety for the man, and only one remedy for the evils upon him, namely, if he burned the books of Epicurus, moistened the ashes of the impious, unholy, and effeminate books with melted wax and, spreading the plaster all over his stomach and chest, bound bandages around them.

What he had heard he communicated to his friends and they were straightway filled with excessive joy because he did not come out, disdained and dishonored by the god.

And having learned a lesson from him, they followed him forthwith in a good and honorable life.

400. Aristides, Oratio LII, 1.

. . . δωδεκάτῳ δὲ ἀφ' οὗ πρῶτον ἔκαμον πολλὰ καὶ θαυμαστὰ ἐφοίτα φαντάσματα ἄγοντα εἰς Ἐπίδαυρον τὴν ἱερὰν τοῦ θεοῦ.

. . . in the twelfth [sc., year] from the time when I first became ill many wondrous visions came repeatedly, leading [me] to Epidaurus, the holy city of the god.

401. Galenus, Commentarius in Hippocratis Epidemias, VI, iv, Sectio IV, 8 [XVII b, p. 137 K.].

Οὕτω γέ τοι καὶ παρ' ἡμῖν ἐν Περγάμῳ τοὺς θεραπευομένους ὑπὸ τοῦ θεοῦ πειθομένους ὁρῶμεν αὐτῷ πεντεκαίδεκα πολλάκις ἡμέρας προστάξαντι μηδ' ὅλως πιεῖν, οἳ τῶν ἰατρῶν μηδενὶ προστάττοντι πείθονται. μεγάλην γὰρ ἔχει ῥοπὴν εἰς τὸ πάντα ποιῆσαι τὰ προστατόμενα τὸ πεπεῖσθαι τὸν κάμνοντα βεβαίως ἀκολουθήσειν ὠφέλειαν ἀξιόλογον αὐτῷ.

Thus at any rate even among ourselves in Pergamum we see that those who are being treated by the god obey him when on many occasions he bids them not to drink at all for fifteen days, while they obey none of the physicians who give this prescription. For it has great influence on the patient's doing all which is prescribed if he has been firmly persuaded that a remarkable benefit to himself will ensue.

402. Aristides, Oratio XXIII, 15-18.

Ἑστία γὰρ Ἀσκληπιοῦ τῆς Ἀσίας ἐνταῦθα ἱδρύθη, κἀνταῦθα δὴ φρυκτοὶ φίλιοι πᾶσιν ἀνθρώποις αἴρονται παρὰ τοῦ θεοῦ καλοῦν-
16 τός τε ὡς αὐτὸν καὶ μάλα ἀληθινὸν φῶς ἀνίσχοντος. καὶ οὔτε χοροῦ σύλλογος πρᾶγμα τοσοῦτον οὔτε πλοῦ κοινωνία οὔτε διδασκάλων τῶν αὐτῶν τυχεῖν, ὅσον χρῆμα καὶ κέρδος εἰς Ἀσκληπιοῦ τε συμφοιτῆσαι καὶ τελεσθῆναι τὰ πρῶτα τῶν ἱερῶν ὑπὸ τῷ καλλίστῳ καὶ τελεωτάτῳ δᾳδούχῳ καὶ μυσταγωγῷ καὶ ᾧ πᾶς ἀνάγκης εἴκει θεσμός. ἐγὼ μὲν οὖν καὶ αὐτός εἰμι τῶν οὐ δὶς [βεβιωκότων] ὑπὸ τῷ θεῷ, ἀλλὰ πολλούς τε καὶ παντοδαποὺς βίους βεβιωκότων καὶ τὴν νόσον κατὰ τοῦτο εἶναι λυσιτελῆ

νομιζόντων, πρὸς δὲ καὶ ψήφους εἰληφότων, ἀνθ᾽ ὧν ἔγωγ᾽ ἂν
οὐδὲ σύμπασαν δεξαίμην τὴν ἐν ἀνθρώποις λεγομένην εὐδαιμονίαν.
17 οὐκοῦν οὐδὲ ἀλίμενα φήσαι τις ἂν εἶναι τὰ τῇδε, ἀλλ᾽ ἐκεῖνο ὀρθό-
τατον καὶ δικαιότατον λέγειν, ὡς ἄρα οὗτος λιμένων ἁπάντων
ὀχυρώτατος καὶ βεβαιότατος καὶ πλείστους δεχόμενος καὶ γαλήνῃ
πλεῖστον προέχων, ἐν ᾧ πᾶσιν ἐξ Ἀσκληπιοῦ τὰ ἐπίγυια τῆς
18 σωτηρίας ἤρτηται. ὥστε τίς οὐκ ἂν τοσαύτῃ καὶ τοιαύτῃ πόλει
συναιρόμενός τε καὶ συνευχόμενος καὶ πᾶν ὅσον εὐφημίας οἷός
τ᾽ εἴη νέμων σωφρονοίη; ἐγὼ μὲν γὰρ ἡγοῦμαι ὥσπερ αὕτη κοινὸν
ἅπασιν ἀνθρώποις ἀγαθόν ἐστι τῇ δυνάμει τοῦ θεοῦ, οὕτω καὶ
αὐτὴν δικαίως ἂν παρὰ πάντων κοινῆς εὐνοίας ἅμα καὶ τιμῆς
τυγχάνειν.

For the hearth of Asclepius was established in this part of Asia
Minor [sc., in Pergamum].[1] And here fire-brands friendly to
all men are raised on high by the god, who summons to himself
and raises aloft the truest light. And neither belonging to a
chorus nor sailing together nor having the same teacher is as
great a thing as the boon and profit of being a fellow pilgrim
to the temple of Asclepius and being initiated into the first of
the holy rites by the fairest and most perfect torchbearer and
leader of the mysteries, to whom every rule of necessity yields.
I myself am one of those who have lived not twice but many
varied lives through the power of the god,[2] and consequently
one of those who think that sickness for this reason is advantageous and who moreover have acquired precious gems in
return for which I would not accept all that which is considered happiness among men. Therefore no one would say
that these regions have no harbor, but most correctly and justly
is it said that this is the most secure and steadfast of all ports,
receiving the greatest number of people and affording the most
in tranquility. Here, the stern-cable of salvation for all is
anchored in Asclepius; so, how would he not be right-minded
who has part in so great a city and joins in prayer with it and
contributes as much praise as he is able. For I think that just
as this community is a good common to all men through the
power of the god, so it should justly enjoy the common goodwill and esteem of all.

[1] Cf. T. 801 ff.　　　　　　　　　　[2] Cf. T. 317, 6; 419.

403. Aristides, Oratio L, 23.

. . . Εὐάρεστος Κρής, τῶν ἐν φιλοσοφίᾳ διατριβόντων ἐλθὼν
ἀπὸ Αἰγύπτου καθ' ἱστορίαν τῶν περὶ τὸν θεόν, . . .

. . . Euarestos the Cretan, one of the philosophers who came
from Egypt, in order to learn about the god and his deeds, . . .

TREATMENT

404. Anthologia Palatina, VI, 330.

Αἰσχίνου ῥήτορος εἰς Ἀσκληπιὸν χαριστήριον

Θνητῶν μὲν τέχναις ἀπορούμενος, εἰς δὲ τὸ θεῖον
ἐλπίδα πᾶσαν ἔχων, προλιπὼν εὔπαιδας Ἀθήνας
ἰάθην ἐλθών, Ἀσκληπιέ, πρὸς τὸ σὸν ἄλσος,
ἕλκος ἔχων κεφαλῆς ἐνιαύσιον, ἐν τρισὶ μησίν.

Thank-offering to Asclepius by Aeschines the Rhetor

Having despaired of the skill of mortals, but with every hope
in the divine, forsaking Athens, blessed with children, coming
to your sacred grove, Asclepius, I was healed in three months
of a festering wound which I had had on my head for a whole
year.

405. Aelianus, Fragmenta, 100.

Ὁ Ἀσκληπιὸς Παύσωνα καὶ Ἶρον κἂν ἄλλον τινὰ τῶν ἀπόρων
ἰάσαιτο. ὀφθαλμὼ γάρ τις ἐνόσει. εἶτα ἐπιστὰς ὅδε λέγει ὄξει
λύσαντα κάπρου πιμελὴν κᾆτα ὑπαλείψασθαι. ὁ δὲ κοινοῦται
τῷ συνήθει ἰατρῷ. ὁ δὲ ἐπειρᾶτο τὰς αἰτίας λέγειν· τὸ μὲν γὰρ
συστέλλειν τὸ οἴδημα τῇ δριμύτητι, τὸ δὲ ἐπιλιπαίνειν καὶ ἡσυχῇ
ὑποτρέφειν ὁ εἴρων ἔλεγε.

Asclepius may heal Pauson and Iros [1] and any other of the
poor people. For instance, someone was afflicted with a
disease of the eyes; so, standing beside him, he instructs him
to free the fat of a boar with vinegar and anoint his eyes with
it. The patient imparted this information to the regular physician and the physician tried to explain the reasons [for this
remedy]. One part of the treatment contracted the tumor

through its acidity and the other anointed the eyes and nourished them gently, said the doctor sarcastically.

¹ Cf. **405a.** Suidas, *Lexicon, s. v.* Παύσων καὶ Ἴρος: ὀνόματα κύρια πενήτων, same text.
² Cf. T. 321.

406. Epictetus, Dissertationes, IV, 8, 28-29.

Οὕτως οὐκ ἀρκεῖ σοι τὸ μηδὲν ἀλγεῖν, ἂν μὴ κηρύσσῃς " συνέλθετε πάντες οἱ ποδαγρῶντες, οἱ κεφαλαλγοῦντες, οἱ πυρέσσοντες, οἱ χωλοί, οἱ τυφλοί, καὶ ἴδετέ με ἀπὸ παντὸς πάθους ὑγιᾶ "; τοῦτο κενὸν καὶ φορτικόν, εἰ μή τι ὡς ὁ Ἀσκληπιὸς εὐθὺς ὑποδεῖξαι δύνασαι, πῶς θεραπεύοντες αὖθις ἔσονται ἄνοσοι κἀκεῖνοι, καὶ εἰς τοῦτο φέρεις παράδειγμα τὴν ὑγίειαν τὴν σεαυτοῦ.

So is it not enough for you yourself to feel no pain unless you proclaim, " Come together, all you who are suffering from gout, headaches, fever, who are lame and blind, and look at me who is free from every suffering "? That is a vain and vulgar thing to say, unless, like Asclepius, you are able at once to show by what treatment those others will also become well again, and for this end set your own good health as an example. (*)

407. Marcus Antoninus, In Semet Ipsum, V, 8.

Ὁποῖόν τί ἐστι τὸ λεγόμενον, ὅτι " συνέταξεν ὁ Ἀσκληπιὸς τούτῳ ἱππασίαν, ἢ ψυχρολουσίαν, ἢ ἀνυποδησίαν," τοιοῦτόν ἐστι καὶ τὸ " συνέταξε τούτῳ ἡ τῶν ὅλων φύσις νόσον, ἢ πήρωσιν, ἢ ἀποβολὴν ἢ ἄλλο τι τῶν τοιούτων." καὶ γὰρ ἐκεῖ τὸ " συνέταξε " τοιοῦτόν τι σημαίνει· ἔταξε τούτῳ τοῦτο ὡς κατάλληλον πρὸς ὑγίειαν· καὶ ἐνταῦθα τὸ συμβαῖνον ἑκάστῳ τέτακταί πως αὐτῷ κατάλληλον πρὸς τὴν εἱμαρμένην . . . δεχώμεθα οὖν αὐτά, ὡς ἐκεῖνα, ἃ ὁ Ἀσκληπιὸς συντάττει. πολλὰ γοῦν καὶ ἐν ἐκείνοις ἐστὶ τραχέα· ἀλλὰ ἀσπαζόμεθα τῇ ἐλπίδι τῆς ὑγιείας.

We have all heard, " Asclepius has prescribed for so and so riding, or cold baths, or walking barefoot." Precisely so it may be said that " the Universal Nature has prescribed for so and so sickness or maim or loss or what not of the same kind." For, in the former case, " prescribed " has some such meaning as this: He ordained this for so and so as conducive to his health; while in the latter what befalls each man has been ordained in some way as conducive to his destiny. . . . Let us then accept our fate, as we accept the prescriptions of Asclepius. And in fact in these, too, there are many " bitter pills," but we welcome them in hope of health.

408. Aristides, Oratio XLVII, 65.

Πολλὰ μὲν οὖν καὶ παράδοξα ἐπετάχθημεν· ὧν δὲ ἀπομνημονεύω, δρόμος τέ ἐστιν ὃν ἔδει δραμεῖν ἀνυπόδητον χειμῶνος ὥρᾳ, καὶ πάλιν ἱππασία, πραγμάτων ἀπορώτατον, καί τι καὶ τοιοῦτον μέμνημαι. τοῦ γὰρ λιμένος κυμαίνοντος ἐξ ἀνέμου λιβὸς καὶ τῶν πλοίων ταραττομένων ἔδει διαπλεύσαντα εἰς τὸ ἀντιπέρας μέλιτος καὶ δρυὸς βαλάνων φαγόντα ἐμέσαι, καὶ γίγνεται δὴ κάθαρσις ἐντελής. πάντα δὲ ταῦτα ἐν ἀκμῇ τῆς φλεγμονῆς οὔσης ἐπράττετο καὶ δὴ πρὸς αὐτὸν ἀναχωρούσης τὸν ὀμφαλόν.

We were ordered to do many paradoxical things;[1] among those which I recall there is a race which I had to run barefoot in winter-time, and again horse riding,[2] the hardest of undertakings, and I also recall an exercise of the following kind: when the harbor waves were swollen by the south wind and ships were in distress, I had to sail across to the opposite side, eating honey and acorns from an oak tree, and vomit; then complete purification is achieved. All these things were done when the inflammation was at its peak and had even spread right to the navel.

[1] Cf. **408a.** Aristides, *Oratio* XLVIII, 80: Ἐφόρεσα δὲ καὶ ἱμάτιον χιτωνίσκου χωρὶς οὐκ οἶδ' ὁπόσας τινὰς ἡμέρας. καὶ μὴν ὅσα αὖ ποταμοῖς ⟨ἢ⟩ πηγαῖς ἢ καὶ θαλάττῃ προσέταξεν χρήσασθαι, . . . τὰ μὲν πρὸς Ἐλαίᾳ, τὰ δὲ Σμύρνῃ καὶ ἐν οἷστισι τοῖς καιροῖς ἕκαστα αὐτῶν, ἀμύθητα ἂν εἴη λέγειν; cf. also T. 317, 8.
[2] Cf. T. 413.

409. Aristides, Oratio XXXIX, 14-15.[1]

. . . οὕτω τοῦ μεγάλου θαυματοποιοῦ καὶ πάντα ἐπὶ σωτηρίᾳ πράττοντος ἀνθρώπων εὕρημα τοῦτο καὶ κτῆμά ἐστι· συμπράττει δὴ πρὸς ἅπαντα αὐτῷ καὶ γίγνεται πολλοῖς ἀντὶ φαρμάκου. πολλοὶ μὲν γὰρ τούτῳ λουσάμενοι ὀφθαλμοὺς ἐκομίσαντο, πολλοὶ δὲ πιόντες στέρνον ἰάθησαν καὶ τὸ ἀναγκαῖον πνεῦμα ἀπέλαβον, τῶν δὲ πόδας ἐξώρθωσεν, τῶν δὲ ἄλλο τι· ἤδη δέ τις πιὼν ἐξ ἀφώνου φωνὴν ἀφῆκεν, ὥσπερ οἱ τῶν ἀπορρήτων ὑδάτων πιόντες μαντικοὶ γιγνόμενοι· τοῖς δὲ καὶ αὐτὸ τὸ ἀρύτεσθαι ἀντ' ἄλλης σωτηρίας καθέστηκεν. καὶ τοῖς τε δὴ νοσοῦσιν οὕτως ἀλεξι-

φάρμακον καὶ σωτήριόν ἐστιν καὶ τοῖς ὑγιαίνουσιν ἐνδιαιτωμένοις παντὸς ἄλλου χρῆσιν ὕδατος οὐκ ἄμεμπτον ποιεῖ.

. . . of the great miracle-worker who does everything for
the salvation of men this [sc., well] is the discovery and
possession: it works with him in all matters and for many it
comes to take the place of a drug. For when bathed with it
many recovered their eyesight, while many were cured of ailments of the chest and regained their necessary breath by
drinking from it. In some cases it cured the feet, in others
something else. One man upon drinking from it straightway
recovered his voice after having been a mute, just as those who
drink sacred waters become prophetic. For some the drawing
of the water itself took the place of every other remedy. Furthermore, not only is it remedial and beneficial to the sick but
even for those who enjoy health it makes the use of any other
water improper.

¹ Cf. T. 804.

410. Aristides, Oratio XLIX, 29-30.

Φίλωνος δέ τίς ἐστιν, οἶμαι, κρᾶσις [φάρμακον ἕτερον]. ταύτης
ἐγὼ μὲν οὐδ᾽ ὀσφρᾶσθαι οἷός τ᾽ ἦν πρὸ τοῦ, τοῦ δὲ θεοῦ ση-
μήναντος χρῆσθαί τε καὶ ἅμα τὴν ὥραν ἣν ἔδει, οὐ μόνον εὐχερῶς
ἔπιον, ἀλλὰ καὶ πιὼν ἡδίων καὶ ῥᾴων εὐθὺς ἦν. ἄλλα τοίνυν
μυρία ἂν εἴη λέγειν φαρμάκων ἐχόμενα, ὧν τὰ μὲν αὐτὸς συντιθείς,
τὰ δὲ τῶν ἐν μέσῳ καὶ κοινῶν ἐδίδου θεραπείαν τοῦ σώματος, ὡς
ἑκάστοτε συμβαίνοι.

There is, I think, a compound [another drug] of Philo's. This
I was not even able to smell previously, but when the god gave
me a sign to use it and also signified the time at which it was
necessary to do this, not only did I drink it easily, but, even as
I drank it, I was at once happier and better. Moreover, it
would be possible to relate ten thousand other things concerning drugs—some of which he himself compounded, others
belonging to the common ordinary varieties which he prescribed as cure of the body, as it was appropriate in each
particular case.

411. Aristides, Oratio XLIX, 28.

Αὖθις δὲ ἐκέλευσεν πρὸς ἄρτῳ φαγεῖν τοῦ αὐτοῦ τούτου φαρμάκου, καὶ ἔφαγον πρὸς τῷ τρίποδι τῷ ἱερῷ, ἀφορμήν τινα ταύτην ἀσφαλείας ποιούμενος.

Again, he gave me instructions to eat of this same drug together with wheat-bread, and I ate it near the holy tripod, thus making a beginning of my well-being.[1]

[1] Cf. **411a.** Aristides, *Oratio* XLVIII, 80 : Ἀδελφὰ δὲ τούτων καὶ αἱ τῶν χειμώνων ἀνυποδησίαι αἱ συνεχεῖς καὶ διὰ παντὸς τοῦ ἱεροῦ κατακλίσεις ἐν ὑπαίθρῳ τε καὶ ὅπου τύχοι, καὶ οὐχ ἥκιστα δὴ ἡ ἐν τῇ ὁδῷ τοῦ νεὼ ὑπ᾽ αὐτὴν τὴν ἱερὰν λαμπάδα τῆς θεοῦ γενομένη.

412. Aristides, Oratio L, 64.

. . . ταύτῃ μοι ἐδόκει ὁ ἱερεὺς ὁ τοῦ Ἀσκληπιοῦ οὗτος, ὁ ἔτι νῦν ὤν, καὶ ὁ τούτου πάππος, ἐφ᾽ οὗ τὰ πολλὰ καὶ μεγάλα, ὡς ἐπυνθανόμεθα, ἐχειρούργησεν ὁ θεός, . . .

. . . on that day appeared to me the priest of Asclepius, the one who is still in office, and his grandfather in whose time, as we learned, the god performed so many great operations . . .

413. Galenus, De Sanitate Tuenda, I, 8, 19-21.

Καὶ οὐκ ὀλίγους ἡμεῖς ἀνθρώπους νοσοῦντας ὅσα ἔτη διὰ τὸ τῆς ψυχῆς ἦθος ὑγιεινοὺς ἀπεδείξαμεν, ἐπανορθωσάμενοι τὴν ἀμετρίαν τῶν κινήσεων. οὐ σμικρὸς δὲ τοῦ λόγου μάρτυς καὶ ὁ πάτριος ἡμῶν θεὸς Ἀσκληπιός, οὐκ ὀλίγοις * μὲν ᾠδάς τε γράφεσθαι καὶ μίμους γελοίων καὶ μέλη τινὰ ποιεῖν ἐπιτάξας, οἷς αἱ τοῦ θυμοειδοῦς κινήσεις σφοδρότεραι γενόμεναι θερμοτέραν τοῦ δέοντος ἀπειργάζοντο τὴν κρᾶσιν τοῦ σώματος, ἑτέροις δέ τισιν, οὐκ ὀλίγοις οὐδὲ τούτοις, κυνηγετεῖν καὶ ἱππάζεσθαι καὶ ὁπλομαχεῖν. εὐθὺς δὲ τούτοις διώρισε τό τε τῶν κυνηγεσίων εἶδος, οἷς τοῦτο προσέταξε, τό τε τῆς ὁπλίσεως, οἷς δι᾽ ὅπλων ἐκέλευσε τὰ γυμνάσια ποιεῖσθαι. οὐ γὰρ μόνον ἐπεγείρειν αὐτῶν τὸ θυμοειδὲς ἐβουλήθη, ἄρρωστον ὑπάρχον, ἀλλὰ καὶ μέτρον ὡρίσατο τῇ τῶν γυμνασίων ἰδέᾳ.

* ὀλίγοις Clift ὀλίγας MSS.

And not a few men, however many years they were ill through the disposition of their souls, we have made healthy by correcting the disproportion of their emotions. No slight witness of the statement is also our ancestral god Asclepius [1] who ordered not a few to have odes written as well as to compose comical mimes and certain songs (for the motions of their passions, having become more vehement, have made the temperature of the body warmer than it should be) ; and for others, these not a few either, he ordered hunting and horse riding [2] and exercising in arms; and at the same time he appointed the kind of hunting for those for whom he prescribed this; and the type of armor for those whom he enjoined to take exercise with armor. For he not only desired to awake the passion of these men because it was weak but also defined the measure by the form of exercises.

[1] Cf. T. 338; 458; 463.　　　　　　　　　[2] Cf. T. 408.

INCUBATIONS

GENERAL REMARKS

414. Iamblichus, De Mysteriis, 3, 3.

Οὕτως ἐν Ἀσκληπιοῦ μὲν τὰ νοσήματα τοῖς θείοις ὀνείροις παύεται · διὰ δὲ τὴν τάξιν τῶν νύκτωρ ἐπιφανειῶν ἡ ἰατρικὴ τέχνη συνέστη ἀπὸ τῶν ἱερῶν ὀνειράτων.

Thus in the Asclepieion illnesses are healed by divine dreams.[1] Through the ordinances of visions that occur at night the medical art was composed from divinely inspired dreams.

[1] Cf. **414a.** Pausanias, *Descriptio Graeciae*, II, 27, 2: Τοῦ ναοῦ [*sc.*, τοῦ ἐν Ἐπιδαύρῳ] δέ ἐστι πέραν ἔνθα οἱ ἱκέται τοῦ θεοῦ καθεύδουσιν.

415. Philostratus, Vita Apollonii, IV, 11.

Βαδίσας οὖν ἐς τὸ Πέργαμον . . . τοῖς τε ἱκετεύουσι τὸν θεὸν ὑποθέμενος, ὁπόσα δρῶντες εὐξυμβόλων ὀνειράτων τεύξονται . . .

Having made his way then to Pergamum [1] . . . he [*sc.*, Apollonius] gave hints to the suppliants of the god, what to do in order to obtain favorable dreams [2] . . .

[1] Cf. T. 802a.　　　　　　　　　[2] Cf. T. 397.

416. Cicero, De Divinatione, II, 59, 123.

Qui igitur convenit aegros a coniectore somniorum potius quam a medico petere medicinam? an Aesculapius an Serapis

potest nobis praescribere per somnium curationem valetudinis,
Neptunus gubernantibus non potest? et si sine medico medicinam dabit Minerva, Musae scribendi legendi ceterarum
artium scientiam somniantibus non dabunt? at si curatio dare-
tur valetudinis, haec quoque quae dixi darentur; quae quoniam
non dantur, medicina non datur.

What would be the sense in the sick seeking treatment from an
interpreter of dreams [1] rather than from a physician? [2] Or do
you think that Asclepius and Sarapis can prescribe a cure for
our illness through a dream, but that Neptune cannot aid pilots
through the same means? Or if Minerva will give treatment
in a dream without the aid of a physician, will not the Muses
impart a knowledge of writing, reading, and the other arts to
dreamers? But if remedies of illness were so given, the arts,
too, which I have just mentioned would thus be given. However, since they are not so conveyed, medicine is not either. (**)

[1] Cf. T. **430**, vv. 246 ff.
[2] Cf. **416a.** Cicero, *De Natura Deorum*, III, 38, 91: . . . nec ego multorum aegrorum
salutem non ab Hippocrate potius quam ab Aesculapio datam iudico . . .

417. Aristides, Oratio XLVIII, 31-35.

Ἐδηλώθη δὲ ὡς ἐναργέστατα, ὥσπερ οὖν καὶ μυρία ἔτερα ἐναργῆ
τὴν παρουσίαν εἶχε τοῦ θεοῦ. καὶ γὰρ οἷον ἅπτεσθαι δοκεῖν ἦν
καὶ διαισθάνεσθαι ὅτι αὐτὸς ἥκοι, καὶ μέσως ἔχειν ὕπνου καὶ ἐγρηγόρσεως, καὶ βούλεσθαι ἐκβλέπειν καὶ ἀγωνιᾶν μὴ προαπαλλαγείη,
καὶ ὦτα παραβεβληκέναι καὶ ἀκούειν, τὰ μὲν ὡς ὄναρ, τὰ δὲ ὡς
ὕπαρ, καὶ τρίχες ὀρθαὶ καὶ δάκρυα σὺν χαρᾷ καὶ γνώμης ὄγκος
ἀνεπαχθής, καὶ τίς ἀνθρώπων ταῦτά γ' ἐνδείξασθαι λόγῳ δυνατός;
εἰ δέ τις τῶν τετελεσμένων ἐστίν, σύνοιδέν τε καὶ γνωρίζει.
ὀφθέντων δὲ τούτων, ἐπειδὴ ἕως ἐγένετο, καλῶ τὸν ἰατρὸν Θεόδοτον·
καὶ ὡς ἧκε, διηγοῦμαι τὰ ὀνείρατα. ὁ δ' ἐθαύμαζε μὲν ὡς εἶχε
δαιμονίως, ἠπόρει δὲ ὅ τι χρήσοιτο, χειμῶνός τε ὥρᾳ καὶ ἅμα τὴν
ἄγαν ἀσθένειαν τοῦ σώματος δεδιώς· μηνῶν γὰρ ἤδη συχνῶν
ἐκείμην ἔνδον. ἔδοξεν οὖν ἡμῖν οὐ χεῖρον εἶναι μεταπέμψασθαι
καὶ τὸν νεωκόρον τὸν Ἀσκληπιακόν.

It [*sc.*, the remedy] was revealed in the clearest way possible,
just as countless other things also made the presence of the
god manifest. For I seemed almost to touch him and to perceive that he himself was coming, and to be halfway between

sleep and waking and to want to get the power of vision and to be anxious lest he depart beforehand, and to have turned my ears to listen, sometimes as in a dream, sometimes as in a waking vision, and my hair was standing on end and tears of joy (came forth), and the weight of knowledge was no burden— what man could even set these things forth in words? But if he is one of the initiates, then he knows and understands. Having seen these things, when morning dawned, I call the physician Theodotus; and as he comes, I describe to him my dreams. He was astonished at how strange they were, and he was at a loss as to what to make of them, since it was wintertime and, too, he was anxious over the great weakness of my body; for I had already been confined to my home for many months. For this reason it seemed to us to be a good idea to summon the sacristan Asclepiacus also.

418. Aristides, Oratio XLVII, 57.

Ταῦτα δὴ ἐφάνθη τὰ ὀνείρατα ἰατροῦ τε ἥκοντος καὶ παρεσκευασμένου βοηθεῖν ὅσα ἐπενόει. ὡς δὲ ἤκουσε τῶν ὀνειράτων, αὐτός τε ὑπεχώρει νοῦν ἔχων τῷ θεῷ καὶ ἡμεῖς ἐγνωρίζομεν τὸν ἀληθινὸν καὶ προσήκοντα ἡμῖν ἰατρὸν καὶ ἐποιοῦμεν ἃ ἐπέταξεν.

These dreams were revealed when the doctor came ready to aid according to his own lights. When he heard the dreams he was wise enough to yield place to the god, and we recognized the true physician, fitting to us, and we did what he prescribed.

419. Menander, Papyrus Didotiana, b, 1-15.

Πρόλογος
Ἐρημία μέν ἐστι, κοὐκ ἀκούσεται
οὐδεὶς παρών μου τῶν λόγων ὧν ἂν λέγω.
ἐγὼ τὸν ἄλλον, ἄνδρες, ἐτεθνήκειν βίον
ἅπανθ᾽ ὃν ἔζην, τοῦτό μοι πιστεύετε.
πάνυ ταὐτὸ τὸ καλόν, τἀγαθόν, τὸ σεμνὸν ⟨ἦν⟩
τὸ κακόν· τοιοῦτον ἦν τί μου πάλαι σκότος
περὶ τὴν διάνοιαν, ὡς ἔοικε, κ⟨ε⟩ίμενον,
ὃ πάντ᾽ ἔκρυπτε ταῦτα κἠφάνιζέ μοι.
νῦν δ᾽ ἐνθάδ᾽ ἐλθών, ὥσπερ εἰς Ἀσκληπιοῦ

ἐγκατακλιθεὶς σωθείς τε, τὸν λοιπὸν χρόνον
ἀναβεβίωκα· περιπατῶ, λαλῶ, φρονῶ.
τὸν τηλικοῦτον καὶ τοιοῦτον ἥλιον
νῦν πρῶτον εὗρον, ἄνδρες, ἐν τῇ τήμερον
ὑμᾶς ὁρῶ νῦν αἰθρίαι, τὸν ἀέρα,
τὴν ἀκρόπολιν, τὸ θέατρον.

Prologue

The place is deserted and no one near me will hear the words
which I speak. Believe me, men, I had been dead during all
the years of life that I was alive. The beautiful, the good, the
holy, the evil were all the same to me; such, it seems, was the
darkness that formerly enveloped my understanding and concealed and hid from me all these things. But now that I have
come here, I have become alive again for all the rest of my
life, as if I had lain down in the temple of Asclepius and had
been saved.[1] I walk, I talk, I think. This sun, so great, so
beautiful I have now discovered, men, for the first time; now
today I see under the clear sky you, the air, the acropolis, the
theater.

[1] Cf. T. 402.

CURES

Athens

420. Aristophanes, Plutus, 400-14.

ΧΡΕΜΤΛΟΣ.	μὰ Δία. δεῖ γὰρ πρῶτα—
400 ΒΛΕΨΙΔΗΜΟΣ.	τί;
ΧΡ.	βλέψαι ποιῆσαι νὼ—
ΒΛ.	τίνα βλέψαι; φράσον.
ΧΡ.	τὸν Πλοῦτον, ὥσπερ πρότερον ἑνί γέ τῳ τρόπῳ.
ΒΛ.	τυφλὸς γὰρ ὄντως ἐστί;
ΧΡ.	νὴ τὸν οὐρανόν.
ΒΛ.	οὐκ ἐτὸς ἄρ' ὡς ἔμ' ἦλθεν οὐδεπώποτε.
405 ΧΡ.	ἀλλ' ἢν θεοὶ θέλωσι, νῦν ἀφίξεται.
ΒΛ.	οὔκουν ἰατρὸν εἰσαγαγεῖν ἐχρῆν τινά;
ΧΡ.	τίς δῆτ' ἰατρός ἐστι νῦν ἐν τῇ πόλει;
	οὔτε γὰρ ὁ μισθὸς οὐδέν ἐστ' οὔθ' ἡ τέχνη.
ΒΛ.	σκοπῶμεν.

XP.　　　ἀλλ' οὐκ ἔστιν.

ΒΛ.　　　　　　　οὐδ' ἐμοὶ δοκεῖ.

410　　XP.　μὰ Δί', ἀλλ' ὅπερ πάλαι παρεσκευαζόμην
ἐγώ, κατακλίνειν αὐτὸν εἰς Ἀσκληπιοῦ
κράτιστόν ἐστι.

ΒΛ.　　　　πολὺ μὲν οὖν νὴ τοὺς θεούς.
μή νυν διάτριβ', ἀλλ' ἄνυε πράττων ἕν γέ τι.

CHREMYLUS.　　　　Aye, we've first to—

BLEPSIDEMUS.　　　　　　　　What?

CHR.　Restore the sight—

BL.　　　　　　Restore the sight of whom?

CHR.　The sight of Wealth by any means we can.

BL.　What, is he really blind?

CHR.　　　　　　　He really is.

BL.　O that is why he never came to me.

CHR.　But now he'll come, if such the will of Heaven.

BL.　Had we not better call a doctor in?

CHR.　Is there a doctor now in all the town?
There are no fees, and therefore there's no skill.

BL.　Let's think awhile.

CHR.　　　　　There is none.

BL.　　　　　　　No more there is.

CHR.　Why then, 'tis best to do what I intended,
To let him lie inside Asclepius' temple
A whole night long.

BL.　　　　　　That's far the best, I swear it,
So don't be dawdling: quick, get something
done.

421.　Aristophanes, Plutus, 633–747.

ΚΑΡΙΩΝ.　ὁ δεσπότης πέπραγεν εὐτυχέστατα,
μᾶλλον δ' ὁ Πλοῦτος αὐτός· ἀντὶ γὰρ τυφλοῦ

635　　ἐξωμμάτωται καὶ λελάμπρυνται κόρας,
Ἀσκληπιοῦ παιῶνος εὐμενοῦς τυχών.

ΧΟΡΟΣ.　λέγεις μοι χαράν, λέγεις μοι βοάν.

ΚΑ.　πάρεστι χαίρειν, ἤν τε βούλησθ' ἤν τε μή.

ΧΟ.　ἀναβοάσομαι τὸν εὔπαιδα καὶ

640　　μέγα βροτοῖσι φέγγος Ἀσκληπιόν.

ΓΥΝΗ.　τίς ἡ βοή ποτ' ἐστίν; ἆρ' ἀγγέλλεται

χρηστόν τι; τοῦτο γὰρ ποθοῦσ' ἐγὼ πάλαι
ἔνδον κάθημαι περιμένουσα τουτονί.

ΚΑ.   ταχέως ταχέως φέρ' οἶνον, ὦ δέσποιν', ἵνα
645   καὐτὴ πίῃς· φιλεῖς δὲ δρῶσ' αὐτὸ σφόδρα·
ὡς ἀγαθὰ συλλήβδην ἅπαντά σοι φέρω.

ΓΥ.   καὶ ποῦ 'στιν;

ΚΑ.   ἐν τοῖς λεγομένοις εἴσει τάχα.

ΓΥ.   πέραινε τοίνυν ὅ τι λέγεις ἀνύσας ποτέ.

ΚΑ.   ἄκουε τοίνυν, ὡς ἐγὼ τὰ πράγματα
650   ἐκ τῶν ποδῶν ἐς τὴν κεφαλήν σοι πάντ' ἐρῶ.

ΓΥ.   μὴ δῆτ' ἔμοιγ' ἐς τὴν κεφαλήν.

ΚΑ.   μὴ τἀγαθὰ
ἃ νῦν γεγένηται;

ΓΥ.   μὴ μὲν οὖν τὰ πράγματα.

ΚΑ.   ὡς γὰρ τάχιστ' ἀφικόμεθα πρὸς τὸν θεὸν
ἄγοντες ἄνδρα τότε μὲν ἀθλιώτατον,
655   νῦν δ' εἴ τιν' ἄλλον μακάριον κεὐδαίμονα,
πρῶτον μὲν αὐτὸν ἐπὶ θάλατταν ἤγομεν,
ἔπειτ' ἐλοῦμεν.

ΓΥ.   νὴ Δί' εὐδαίμων ἄρ' ἦν
ἀνὴρ γέρων ψυχρᾷ θαλάττῃ λούμενος.

ΚΑ.   ἔπειτα πρὸς τὸ τέμενος ᾖμεν τοῦ θεοῦ.
660   ἐπεὶ δὲ βωμῷ πόπανα καὶ προθύματα
καθωσιώθη, πέλανος Ἡφαίστου φλογί,
κατεκλίναμεν τὸν Πλοῦτον, ὥσπερ εἰκὸς ἦν·
ἡμῶν δ' ἕκαστος στιβάδα παρεκαττύετο.

ΓΥ.   ἦσαν δέ τινες κἄλλοι δεόμενοι τοῦ θεοῦ;

665 ΚΑ.   εἷς μέν γε Νεοκλείδης, ὅς ἐστι μὲν τυφλός,
κλέπτων δὲ τοὺς βλέποντας ὑπερηκόντισεν·
ἕτεροί τε πολλοὶ παντοδαπὰ νοσήματα
ἔχοντες· ὡς δὲ τοὺς λύχνους ἀποσβέσας
ἡμῖν παρήγγειλεν καθεύδειν τοῦ θεοῦ
670   ὁ πρόπολος, εἰπών, ἤν τις αἴσθηται ψόφου,
σιγᾶν, ἅπαντες κοσμίως κατεκείμεθα.
κἀγὼ καθεύδειν οὐκ ἐδυνάμην, ἀλλά με
ἀθάρης χύτρα τις ἐξέπληττε κειμένη
ὀλίγον ἄπωθεν τῆς κεφαλῆς του γραδίου,
675   ἐφ' ἣν ἐπεθύμουν δαιμονίως ἐφερπύσαι.
ἔπειτ' ἀναβλέψας ὁρῶ τὸν ἱερέα
τοὺς φθοῖς ἀφαρπάζοντα καὶ τὰς ἰσχάδας

ἀπὸ τῆς τραπέζης τῆς ἱερᾶς. μετὰ τοῦτο δὲ
περιῆλθε τοὺς βωμοὺς ἅπαντας ἐν κύκλῳ,
εἴ που πόπανον εἴη τι καταλελειμμένον·
ἔπειτα ταῦθ᾽ ἥγιζεν εἰς σάκταν τινά.
κἀγὼ νομίσας πολλὴν ὁσίαν τοῦ πράγματος
ἐπὶ τὴν χύτραν τὴν τῆς ἀθάρης ἀνίσταμαι.

ΓΥ.　ταλάντατ᾽ ἀνδρῶν, οὐκ ἐδεδοίκεις τὸν θεόν;
ΚΑ.　νὴ τοὺς θεοὺς ἔγωγε μὴ φθάσειέ με
　　　ἐπὶ τὴν χύτραν ἐλθὼν ἔχων τὰ στέμματα.
　　　ὁ γὰρ ἱερεὺς αὐτοῦ με προὐδιδάξατο.
　　　τὸ γρᾴδιον δ᾽ ὡς ᾔσθετο δή μου τὸν ψόφον,
　　　τὴν χεῖρ᾽ ὑφῆκε κᾆτα συρίξας ἐγὼ
　　　ὀδὰξ ἐλαβόμην, ὡς παρείας ὢν ὄφις.
　　　ἡ δ᾽ εὐθέως τὴν χεῖρα πάλιν ἀνέσπασε,
　　　κατέκειτο δ᾽ αὑτὴν ἐντυλίξασ᾽ ἡσυχῇ,
　　　ὑπὸ τοῦ δέους βδέουσα δριμύτερον γαλῆς.
　　　κἀγὼ τότ᾽ ἤδη τῆς ἀθάρης πολλὴν ἔφλων·
　　　ἔπειτ᾽ ἐπειδὴ μεστὸς ἦν, ἀνεπαυόμην.
ΓΥ.　ὁ δὲ θεὸς ὑμῖν οὐ προσῄειν;
ΚΑ.　　　　　　　　　　　οὐδέπω,
　　　μετὰ τοῦτο δ᾽ ἤδη· καὶ γελοῖον δῆτά τι
　　　ἐποίησα· προσιόντος γὰρ αὐτοῦ μέγα πάνυ
　　　ἀπέπαρδον· ἡ γαστὴρ γὰρ ἐπεφύσητό μου.
ΓΥ.　ἦ πού σε διὰ τοῦτ᾽ εὐθὺς ἐβδελύττετο.
ΚΑ.　οὔκ, ἀλλ᾽ Ἰασὼ μέν τις ἀκολουθοῦσ᾽ ἅμα
　　　ὑπηρυθρίασε χἠ Πανάκει᾽ ἀπεστράφη
　　　τὴν ῥῖν᾽ ἐπιλαβοῦσ᾽· οὐ λιβανωτὸν γὰρ βδέω.
ΓΥ.　αὐτὸς δ᾽ ἐκεῖνος;
ΚΑ.　　　　　οὐ μὰ Δί᾽, οὐδ᾽ ἐφρόντισεν.
ΓΥ.　λέγεις ἄγροικον ἄρα σύ γ᾽ εἶναι τὸν θεόν.
ΚΑ.　μὰ Δί᾽ οὐκ ἔγωγ᾽, ἀλλὰ σκατοφάγον.
ΓΥ.　　　　　　　　　　αἶ τάλαν.
ΚΑ.　μετὰ ταῦτ᾽ ἐγὼ μὲν εὐθὺς ἐνεκαλυψάμην
　　　δείσας, ἐκεῖνος δ᾽ ἐν κύκλῳ τὰ νοσήματα
　　　σκοπῶν περιῄει πάντα κοσμίως πάνυ.
　　　ἔπειτα παῖς αὐτῷ λίθινον θυείδιον
　　　παρέθηκε καὶ δοίδυκα καὶ κιβώτιον.
ΓΥ.　λίθινον;
ΚΑ.　　　μὰ Δί᾽ οὐ δῆτ᾽, οὐχὶ τό γε κιβώτιον.
ΓΥ.　σὺ δὲ πῶς ἑώρας, ὦ κάκιστ᾽ ἀπολούμενε,

680

685

690

695

700

705

710

ὃς ἐγκεκαλύφθαι φῄς;

KA.   διὰ τοῦ τριβωνίου.

715   ὀπὰς γὰρ εἶχεν οὐκ ὀλίγας μὰ τὸν Δία.
πρῶτον δὲ πάντων τῷ Νεοκλείδῃ φάρμακον
καταπλαστὸν ἐνεχείρησε τρίβειν, ἐμβαλὼν
σκορόδων κεφαλὰς τρεῖς Τηνίων. ἔπειτ’ ἔφλα
ἐν τῇ θυείᾳ συμπαραμιγνύων ὀπὸν

720   καὶ σχῖνον· εἶτ’ ὄξει διέμενος Σφηττίῳ,
κατέπλασεν αὐτοῦ τὰ βλέφαρ’ ἐκστρέψας, ἵνα
ὀδυνῷτο μᾶλλον. ὁ δὲ κεκραγὼς καὶ βοῶν
ἔφευγ’ ἀνᾴξας· ὁ δὲ θεὸς γελάσας ἔφη·
ἐνταῦθα νῦν κάθησο καταπεπλασμένος,

725   ἵν’ ὑπομνύμενον παύσω σε τῆς ἐκκλησίας.

ΓΥ.   ὡς φιλόπολίς τίς ἐσθ’ ὁ δαίμων καὶ σοφός.

KA.   μετὰ τοῦτο τῷ Πλούτωνι παρεκαθέζετο,
καὶ πρῶτα μὲν δὴ τῆς κεφαλῆς ἐφήψατο,
ἔπειτα καθαρὸν ἡμιτύβιον λαβὼν

730   τὰ βλέφαρα περιέψησεν· ἡ Πανάκεια δὲ
κατεπέτασ’ αὐτοῦ τὴν κεφαλὴν φοινικίδι
καὶ πᾶν τὸ πρόσωπον· εἶθ’ ὁ θεὸς ἐπόππυσεν.
ἐξῃξάτην οὖν δύο δράκοντ’ ἐκ τοῦ νεὼ
ὑπερφυεῖς τὸ μέγεθος.

ΓΥ.   ὦ φίλοι θεοί.

735 KA.   τούτω δ’ ὑπὸ τὴν φοινικίδ’ ὑποδύνθ’ ἡσυχῇ
τὰ βλέφαρα περιέλειχον, ὥς γ’ ἐμοὐδόκει·
καὶ πρίν σε κοτύλας ἐκπιεῖν οἴνου δέκα,
ὁ Πλοῦτος, ὦ δέσποιν’, ἀνειστήκει βλέπων·
ἐγὼ δὲ τὼ χεῖρ’ ἀνεκρότησ’ ὑφ’ ἡδονῆς,

740   τὸν δεσπότην τ’ ἤγειρον. ὁ θεὸς δ’ εὐθέως
ἠφάνισεν αὑτὸν οἵ τ’ ὄφεις εἰς τὸν νεών.
οἱ δ’ ἐγκατακείμενοι παρ’ αὐτῷ πῶς δοκεῖς
τὸν Πλοῦτον ἠσπάζοντο καὶ τὴν νύχθ’ ὅλην
ἐγρηγόρεσαν, ἕως διέλαμψεν ἡμέρα.

745   ἐγὼ δ’ ἐπῄνουν τὸν θεὸν πάνυ σφόδρα,
ὅτι βλέπειν ἐποίησε τὸν Πλοῦτον ταχύ,
τὸν δὲ Νεοκλείδην μᾶλλον ἐποίησεν τυφλόν.

CARION.   With happiest fortune has my master sped,
Or rather Wealth himself; no longer blind,
He hath relumed the brightness of his eyes,
So kind a Healer hath Asclepius proved.[1]

CHORUS.   (*singing*) Joy for the news you bring.

       Joy! Joy! with shouts I sing.

CA.   Aye, will you, nill you, it is joy indeed.

CH.   (*singing*) Sing we with all our might Asclepius
  first and best,

  To men a glorious light, Sire in his offspring blest.

WIFE.   What means this shouting? Has good news arrived?
  For I've been sitting till I'm tired within
  Waiting for *him*, and longing for good news.

CA.   Bring wine, bring wine, my mistress; quaff yourself
  The flowing bowl; (you like it passing well).
  I bring you here all blessings in a lump.

WIFE.   Where?

CA.   That you'll learn from what I am going to say.

WIFE.   Be pleased to tell me with what speed you can.

CA.   Listen. I'll tell you all this striking business
  Up from the foot on to the very head.

WIFE.   Not on *my* head, I pray you.

CA.   Not the blessings
  We have all got?

WIFE.   Not all that striking business.

CA.   Soon as we reached the Temple of the God
  Bringing the man, most miserable then,
  But who so happy, who so prosperous now?
  Without delay we took him to the sea
  And bathed him there.[2]

WIFE.   O what a happy man,
  The poor old fellow bathed in the cold sea!

CA.   Then to the precincts of the God we went.
  There on the altar honey-cakes and bakemeats[3]
  Were offered, food for the Hephaestian flame.
  There laid we Wealth as custom bids; and we
  Each for himself stitched up a pallet near.

WIFE.   Were there no others waiting to be healed?

CA.   Neocleides was, for one; the purblind man,
  Who in his thefts out-shoots the keenest-eyed.
  And many others, sick with every form
  Of ailment. Soon the Temple servitor
  Put out the lights, and bade us fall asleep,
  Nor stir, nor speak, whatever noise we heard.

So down we lay in orderly repose.
And I could catch no slumber, not one wink,
Struck by a nice tureen of broth which stood
A little distance from an old wife's head,
Whereto I marvellously longed to creep.
Then, glancing upwards, I behold the priest [4]
Whipping the cheese-cakes and the figs from off
The holy table; thence he coasted round
To every altar, spying what was left.
And everything he found he consecrated
Into a sort of sack; so I, concluding
This was the right and proper thing to do,
Arose at once to tackle that tureen.

WIFE. Unhappy man! Did you not fear the God?

CA. Indeed I did, lest he should cut in first,
Garlands and all, and capture my tureen.
For so the priest forewarned me he might do.
Then the old lady when my steps she heard
Reached out a stealthy hand; I gave a hiss,
And mouthed it gently like a sacred [5] snake.
Back flies her hand; she draws her coverlets
More tightly round her, and, beneath them, lies
In deadly terror like a frightened cat.
Then of the broth I gobbled down a lot
Till I could eat no more, and then I stopped.

WIFE. Did not the God approach you?

CA.   Not till later.
And then I did a thing will make you laugh.
For as he neared me, by some dire mishap
My wind exploded like a thunder-clap.

WIFE. I guess the God was awfully disgusted.

CA. No, but Iaso [6] blushed a rosy red
And Panacea [7] turned away her head
Holding her nose: my wind's not frankincense.

WIFE. But he himself?

CA.   Observed it not, nor cared.

WIFE. O why, you're making out the God a clown!

CA. No, no; an ordure-taster.

WIFE.   Oh! you wretch.

CA. So then, alarmed, I muffled up my head,

Whilst *he* went round, with calm and quiet tread,
To every patient, scanning each disease.
Then by his side a servant placed a stone
Pestle and mortar; and a medicine chest.[8]

WIFE.   A stone one?

CA.   Hang it, not the medicine chest.

WIFE.   How saw you this, you villain, when your head,
You said just now, was muffled?

CA.   Through my cloak.
Full many a peep-hole has that cloak, I trow.
Well, first he set himself to mix a plaster
For Neocleides, throwing in three cloves
Of Tenian garlic; and with these he mingled
Verjuice and squills; and brayed them up together.
Then drenched the mass with Sphettian vinegar,
And turning up the eyelids of the man
Plastered their inner sides, to make the smart
More painful. Up he springs with yells and roars
In act to flee; then laughed the God, and said,
Nay, sit thou there, beplastered; I'll restrain thee,
Thou reckless swearer, from the Assembly now.

WIFE.   O what a clever, patriotic God![9]

CA.   Then, after this, he sat him down by Wealth,
And first he felt the patient's head, and next
Taking a linen napkin, clean and white,
Wiped both his lips, and all around them, dry.
Then Panacea with a scarlet cloth
Covered his face and head; then the God clucked,
And out there issued from the holy shrine
Two great enormous serpents.[10]

WIFE.   O good heavens!

CA.   And underneath the scarlet cloth they crept
And licked his eyelids, as it seemed to me;
And, mistress dear, before you could have drunk
Of wine ten goblets, Wealth arose and saw.
O then for joy I clapped my hands together
And woke my master, and, hey presto! both
The God and serpents vanished in the shrine.
And those who lay by Wealth, imagine how
They blessed and greeted him, nor closed their eyes

The whole night long till daylight did appear.
And I could never praise the God enough
For both his deeds, enabling Wealth to see,
And making Neocleides still more blind.

¹ Cf. also **421a**. Sophocles, *Phineus*, Fr. 644 [Nauck²] = Fr. 710 [Pearson] : Ἐξωμμάτωται καὶ λελάμπρυνται κόρας / Ἀσκληπιοῦ παιῶνος εὐμενοῦς τυχών; cf. also T. 271.

² Cf. T. **512**.   ⁵ Cf. T. **700**.   ⁸ Cf. T. **448**.
³ Cf. T. **514**.   ⁶ Cf. T. **285**.   ⁹ Cf. T. **124**: πολιτικόν.
⁴ Cf. T. **490**.   ⁷ Cf. T. **284**.   ¹⁰ Cf. T. **448**.

Epidaurus

422. Aelianus, De Natura Animalium, IX, 33.

Ζῴῳ δ᾽ οὖν πονηρῷ πολέμιόν ἐστι, καὶ ἀναιρεῖ τὴν ἕλμινθα, ἥπερ οὖν ἐπὶ πλέον ἰοῦσα θηρίον γίνεται σπλάγχνοις μὲν ἐντικτόμενον, ἀνθρωπείαις δὲ νόσοις ἐναριθμούμενον, καὶ ταῦτα ταῖς ἄγαν ἀνιάτοις τε καὶ ὑπὸ χειρὸς θνητῆς ἐς ἄκεσιν ἥκειν ἀδυνάτοις. τεκμηριῶσαι τοῦτο καὶ Ἵππυς ἱκανός. ὃ δὲ λέγει ὁ συγγραφεὺς ὁ Ῥηγῖνος, τοιοῦτόν ἐστι. γυνὴ εἶχεν ἕλμινθα, καὶ ἰάσασθαι αὐτὴν ἀπεῖπον οἱ τῶν ἰατρῶν δεινοί. οὐκοῦν ἐς Ἐπίδαυρον ἦλθε, καὶ ἐδεῖτο τοῦ θεοῦ ἐξάντης γενέσθαι τοῦ συνοίκου πάθους. οὐ παρῆν ὁ θεός· οἱ μέντοι ζάκοροι κατακλίνουσι τὴν ἄνθρωπον ἔνθα ἰᾶσθαι ὁ θεὸς εἰώθει τοὺς δεομένους. καὶ ἡ μὲν ἄνθρωπος ἡσύχαζε προσταχθεῖσα, οἵ γε μὴν ὑποδρῶντες τῷ θεῷ τὰ ἐς τὴν ἴασιν αὐτῆς ἐποίουν, καὶ τὴν κεφαλὴν μὲν ἀπὸ τῆς δέρης ἀφαιροῦσι, καθίησι δὲ τὴν χεῖρα ὁ ἕτερος, καὶ ἐξαιρεῖ τὴν ἕλμινθα, θηρίου μέγα τι χρῆμα. συναρμόσαι δὲ καὶ ἀποδοῦναι τὴν κεφαλὴν ἐς τὴν ἀρχαίαν ἁρμονίαν οὐκ ἐδύναντο οὐκέτι. ὁ τοίνυν θεὸς ἀφικνεῖται, καὶ τοῖς μὲν ἐχαλέπηνεν ὅτι ἄρα ἐπέθεντο ἔργῳ δυνατωτέρῳ τῆς ἑαυτῶν σοφίας· αὐτὸς δὲ ἀμάχῳ τινὶ καὶ θείᾳ δυνάμει ἀπέδωκε τῷ σκήνει τὴν κεφαλήν, καὶ τὴν ξένην ἀνέστησε. καὶ οὔ τι που, ὦ βασιλεῦ καὶ θεῶν φιλανθρωπότατε Ἀσκληπιέ, ἀβρότονον ἔγωγε ἀντικρίνω τῇ σοφίᾳ τῇ σῇ. μὴ μανείην ἐς τοσοῦτον. ἀλλὰ ἐπελθὸν ἐμνήσθην εὐεργεσίας τε σῆς καὶ ἰάσεως ἐκπληκτικῆς. ὡς δὲ καὶ ἥδε ἡ πόα σὸν δῶρόν ἐστιν οὐδὲ ἀμφιβάλλειν χρή.

[Southernwood] is destructive to harmful organisms and removes the tapeworm, which increases in size and becomes an animal, produced in the intestines, but counted among the diseases of men, and moreover, among the absolutely incurable diseases and such as are incapable of reaching a cure under mortal hands. Hippys [5th or 3rd c. B. C.] gives sufficient proof of this. What the historian of Rhegium says is something to this effect: [1] A woman had a tapeworm and the cleverest of the physicians failed to cure her.[2] Then she came to Epidaurus and begged the god that she might become free of the ailment that lived within her. The god was not present. The attendants at the temple, however, made the woman lie down where the god was accustomed to heal the suppliants. And the woman rested quietly, as prescribed, while the servants of the god made the preparations for her cure. They removed her head from her neck. One stretched down his hand and pulled forth the worm, an animal of great size. But fit together and attach her head to its original joint, they could not do. The god then approached and was provoked at them because they set themselves to a task beyond their wisdom. But with a certain effortless divine power he himself attached her head to her body and raised up the stranger-woman.[3] Not in any way whatsoever, O Asclepius, King and most man-loving of the gods, do I match the southernwood against your wisdom. May I not be so rash as that. But when it came up, I was reminded of your benefits and your astounding cure. So also the plant is your gift,[4] and one must not doubt it.

[1] = Fr. 8 [Müller].
[2] Cf. T. 438; 582; 585.
[3] Cf. T. 423, 23.
[4] Cf. T. 374.

423. Inscriptiones Graecae, IV², 1, nos. 121-22 [2nd half of 4th c. B. C.].[1]

A

Θεός　　　Τύχα [ἀγ]αθά

['Ἰά]ματα τοῦ Ἀπόλλωνος καὶ τοῦ Ἀσκλαπιοῦ

I. [Κλ]εὼ πένθ᾽ ἔτη ἐκύησε. αὔτα πέντ᾽ ἐνιαυτοὺς ἤδη κυοῦσα ποὶ τὸν | [θε]ὸν ἱκέτις ἀφίκετο καὶ ἐνεκάθευδε ἐν τῶι ἀβάτωι· ὡς δὲ τάχισ||[τα] ἐξῆλθε ἐξ αὐτοῦ καὶ ἐκ τοῦ ἱαροῦ ἐγένετο, κόρον ἔτεκε, ὃς εὐ|[θ]ὺς γενόμενος αὐτὸς ἀπὸ τᾶς κράνας ἐλοῦτο καὶ ἅμα τᾶι ματρὶ | [π]εριῆρπε. τυχοῦσα δὲ τούτων ἐπὶ τὸ ἄνθεμα ἐπεγράψατο· Ὀὐ μέγε|[θο]ς πίνακος θαυμαστέον, ἀλλὰ τὸ θεῖον, πένθ᾽ ἔτη ὡς ἐκύησε ἐγ γασ|τρὶ Κλεὼ βάρος,

II. ἔστε ἐγκατεκοιμάθη καί μιν ἔθηκε ὑγιῆ.᾽ Τριέτης || [φο]ρά. Ἰθμονίκα Πελλανὶς ἀφίκετο εἰς τὸ ἱαρὸν ὑπὲρ γενεᾶς. ἐγ-

[κατα|κοι]μαθεῖσα δὲ ὄψιν εἶδε· ἐδόκει αἰτεῖσθαι τὸν θεὸν
κυῆσαι κό|[ραν], τὸν δ' Ἀσκλαπιὸν φάμεν ἔγκυον ἐσσεῖσθαί
νιν, καὶ εἴ τι ἄλλο | α[ἰτ]οῖτο, καὶ τοῦτό οἱ ἐπιτελεῖν, αὐτὰ δ'
οὐθενὸς φάμεν ἔτι ποι|δε[ῖ]σθαι. ἔγκυος δὲ γενομένα ἐγ
γαστρὶ ἐφόρει τρία ἔτη, ἔστε πα||ρέβαλε ποὶ τὸν θεὸν ἱκέτις
ὑπὲρ τοῦ τόκου· ἐγκατακοιμαθεῖσα | δὲ ὄψ[ι]ν εἶδε· ἐδόκει
ἐπερωτῆν νιν τὸν θεόν, εἰ οὐ γένοιτο αὐτᾶι | πάντα ὅσσα
αἰτήσαιτο καὶ ἔγκυος εἴη· ὑπὲρ δὲ τόκου ποιθέμεν | νιν οὐθέν,
καὶ ταῦτα πυνθανομένου αὐτοῦ, εἴ τινος καὶ ἄλλου δέ|οιτο,
λέγειν, ὡς ποιησοῦντος καὶ τοῦτο. ἐπεὶ δὲ νῦν ὑπὲρ τούτου ||
παρείη ποτ' αὐτὸν ἱκέτις, καὶ τοῦτό οἱ φάμεν ἐπιτελεῖν. μετὰ
δὲ | τοῦτο σπουδᾶι ἐκ τοῦ ἀβάτου ἐξελθοῦσα, ὡς ἔξω τοῦ ἱαροῦ
III. ἦς, ἔτε|κε κόραν. Ἀνὴρ τοὺς τᾶς χηρὸς δακτύλους ἀκρατεῖς
ἔχων πλὰν | ἑνὸς ἀφίκετο ποὶ τὸν θεὸν ἱκέτας· θεωρῶν δὲ τοὺς
ἐν τῶι ἱαρῶι | πίνακας ἀπίστει τοῖς ἰάμασιν καὶ ὑποδιέσυρε
τὰ ἐπιγράμμα||[τ]α. ἐγκαθεύδων δὲ ὄψιν εἶδε· ἐδόκει ὑπὸ
τῶι ναῶι ἀστραγαλίζον|[τ]ος αὐτοῦ καὶ μέλλοντος βάλλειν
τῶι ἀστραγάλωι, ἐπιφανέντα | [τ]ὸν θεὸν ἐφαλέσθαι ἐπὶ τὰν
χῆρα καὶ ἐκτεῖναί οὐ τοὺς δακτύ⟨λ⟩|λους· ὡς δ' ἀποβαίη,
δοκεῖν συγκάμψας τὰν χῆρα καθ' ἕνα ἐκτείνειν | τῶν δακτύλων·
ἐπεὶ δὲ πάντας ἐξευθύναι, ἐπερωτῆν νιν τὸν θεόν, || εἰ ἔτι ἀπι-
στησοῖ τοῖς ἐπιγράμμασι τοῖς ἐπὶ τῶμ πινάκων τῶν | κατὰ
τὸ ἱερόν, αὐτὸς δ' οὐ φάμεν. ὅτι τοίνυν ἔμπροσθεν ἀπίστεις |
αὐτο[ῖ]ς οὐκ ἐοῦσιν ἀπίστοις, τὸ λοιπὸν ἔστω τοι,' φάμεν,
' Ἄπιστος | ὄν[ομα].' ἀμέρας δὲ γενομένας ὑγιὴς ἐξῆλθε.
IV. Ἀμβροσία ἐξ Ἀθανᾶν | [ἀτερό]ππτ[ι]λλος. αὐτα ἱκέτις
ἦλθε ποὶ τὸν θεόν· περιέρπουσα δὲ || [κατὰ τ]ὸ ἱαρὸν τῶν
ἰαμάτων τινὰ διεγέλα ὡς ἀπίθανα καὶ ἀδύνα|[τα ἐόν]τα,
χωλοὺς καὶ τυφλοὺ[ς] ὑγιεῖς γίνεσθαι ἐνύπνιον ἰδόν|[τας
μό]νον. ἐγκαθεύδουσα δὲ ὄψιν εἶδε· ἐδόκει οἱ ὁ θεὸς ἐπιστὰς
| [εἰπεῖν], ὅτι ὑγιῆ μέν νιν ποιησοῖ, μισθὸμ μάντοι νιν δεησοῖ
ἀν|[θέμεν ε]ἰς τὸ ἱαρὸν ὗν ἀργύρεον ὑπόμναμα τᾶς ἀμαθίας.
εἴπαν||[τα δὲ ταῦτ]α ἀνσχίσσαι οὐ τὸν ὄπτιλλον τὸν νοσοῦντα
καὶ φάρμ[α|κόν τι ἐγχέ]αι· ἀμέρας δὲ γενομένας ὑγιὴς
V. ἐξῆλθε. Παῖς ἄφωνος. | [οὗτος ἀφί]κετο εἰς τὸ ἱαρὸν ὑπὲρ
φωνᾶς· ὡς δὲ προεθύσατο καὶ | [ἐπόησε τὰ] νομιζόμενα, μετὰ
τοῦτο ὁ παῖς ὁ τῶι θεῶι πυρφορῶν | [ἐκέλετο, π]οὶ τὸμ πατέρα
τὸν τοῦ παιδὸς ποτιβλέψας, ὑποδέκεσ||[θαι ἐντὸς² ἐ]νιαυτοῦ,
τυχόντα ἐφ' ἃ πάρεστι, ἀποθυσεῖν τὰ ἴατρα. | [ὁ δὲ παῖς ἐξ]α-
πίνας ' ὑποδέκομαι,' ἔφα· ὁ δὲ πατὴρ ἐκπλαγεὶς πάλιν | [ἐκέ-

λετο αὐ]τὸν εἰπεῖν· ὁ δὲ ἔλεγε πάλιν· καὶ ἐκ τούτου ὑγιὴς ἐγέ|-

VII. [νετο. Πάνδαρ]ος Θεσσαλὸς στίγματα ἔχων ἐν τῶι μετώπωι.
οὗτος | [ἐγκαθεύδων ὄ]ψιν εἶδε· ἐδόκει αὐτοῦ τα[ι]νίαι κατα-
δῆσαι τὰ στί| |[γματα ὁ θεὸς κ]αὶ κέλεσθαί νιν, ἐπεί [κα ἔξω]
γένηται τοῦ ἀβάτου, | [ἀφελόμενον τὰ]ν ταινίαν ἀνθέμε[ν
εἰ]ς τὸν ναόν· ἁμέρας δὲ γενο|[μένας ἐξανέστα] καὶ ἀφήλετο
τ[ὰν ται]νίαν, καὶ τὸ μὲν πρόσωπον | [κενεὸν εἶδε τῶ]ν
στιγμάτω[ν, τ]ὰν δ[ὲ τ]αινίαν ἀνέθηκε εἰς τὸν να||[όν,

VII. ἔχουσαν τὰ γρ]άμματ[α] τὰ ἐκ τοῦ μετώπου. Ἐχέδωρος τὰ
Πανδά||[ρου στίγματα ἔλ]αβε ποὶ τοῖς ὑπάρχουσιν. οὗτος
λαβὼν πὰρ [Παν|δάρου χρήματα], ὥστ᾽ ἀνθέμεν τῶι θεῶι εἰς
Ἐπίδαυρον ὑπὲρ αὐ[τοῦ, | οὐκ] ἀπεδίδου ταῦτα· ἐγκαθεύδων
δὲ ὄψιν εἶδε· ἐδόκει οἱ ὁ θε[ὸς] | ἐπιστὰς ἐπερωτῆν νιν, εἰ
ἔχοι τινὰ χρήματα πὰρ Πανδάρου ε[ἰς Ἀ] [3] |θηνᾶν ἄνθεμα εἰς
τὸ ἱαρόν· αὐτὸς δ᾽ οὐ φάμεν λελαβήκειν οὐθὲ[ν] ||τοιοῦτον
παρ᾽ αὐτοῦ· ἀλλ᾽ αἴ κα ὑγιῆ νιν ποήσαι, ἀνθησεῖν οἱ εἰκό | να
γραψάμενος· μετὰ δὲ τοῦτο τὸν θεὸν τὰν τοῦ Πανδάρου
ταινί|αν περιδῆσαι περὶ τὰ στίγματά οὐ καὶ κέλεσθαί νιν, ἐπεί
κα ἐξ|έλθηι ἐκ τοῦ ἀβάτου, ἀφελόμενον τὰν ταινίαν ἀπονίψα-
σθαι τὸ | πρόσωπον ἀπὸ τᾶς κράνας καὶ ἐγκατοπτρίξασθαι εἰς
τὸ ὕδωρ· ἁ||μέρας δὲ γενομένας ἐξελθὼν ἐκ τοῦ ἀβάτου τὰν
ταινίαν ἀφήλετο, | τὰ γράμματα οὐκ ἔχουσαν· ἐγκαθιδὼν δὲ
εἰς τὸ ὕδωρ ἑώρη τὸ αὐτοῦ | πρόσωπον ποὶ τοῖς ἰδίοις στίγμα-
σιν καὶ τὰ τοῦ Πανδ(ά)ρου γρά[μ]|ματα λελαβηκός.

VIII. Εὐφάνης Ἐπιδαύριος παῖς. οὗτος λιθιῶν ἐνε[κά]|θευδε· ἔδοξε
δή αὐτῶι ὁ θεὸς ἐπιστὰς εἰπεῖν· ᾽τί μοι δωσεῖς, αἴ τύ||κα
ὑγιῆ ποιήσω᾽; αὐτὸς δὲ φάμεν ᾽δέκ᾽ ἀστραγάλους.᾽ τὸν δὲ
θεὸν γελά|σαντα φάμεν νιν παυσεῖν· ἁμέρας δὲ γενομένας

IX. ὑγιὴς ἐξῆλθε. | Ἀνὴρ ἀφίκετο ποὶ τὸν θεὸν ἱκέτας ἀτερό-
πτιλλος οὕτως, ὥστε τὰ | βλέφαρα μόνον ἔχειν, ἐνεῖμεν δ᾽ ἐν
αὐτοῖς μηθέν, ἀλλὰ κενεὰ ε[ἴ]|μεν ὅλως. ἐ(γ)έ(λω)ν δή
τινες τῶν ἐν τῶι ἱαρῶι τὰν εὐηθίαν αὐτοῦ, τὸ ||νομίζειν βλε-
ψεῖσθαι ὅλως μηδεμίαν ὑπαρχὰν ἔχοντος ὀπτίλ|λου ἀλλ᾽ ἢ
χώραμ μόνον. ἐγκαθ[εύδο]ντι οὖν αὐτῶι ὄψις ἐφάνη· ἐδό|κει
τὸν θεὸν ἑψῆσαί τι φά[ρμακον, ἔπε]ιτα διαγαγόντα τὰ
βλέφα|ρα ἐγχέαι εἰς αὐτά· ἁμέρ[ας δὲ γενομέν]ας βλέπων

X. ἀμφοῖν ἐξῆλθε. | Κώθων. σκευοφόρος εἰ[ς τὸ] ἱαρ[ὸν]
ἕρπων, ἐπεὶ ἐγένετο περὶ τὸ δε||καστάδιον, κατέπετε· [ὡς δὲ]
ἀνέστα, ἀνῶιξε τὸγ γυλιὸν κα[ὶ ἐ]πεσκό|πει τὰ συντετριμμένα
σκ[ε]ύη· ὡς δ᾽ εἶδε τὸγ κώθωνα κατε[αγ]ότα, | ἐξ οὗ ὁ

δεσπότας εἴθιστο [π]ίνειν, ἐλυπεῖτο καὶ συνετίθει [τὰ] ὄ|στ-
ρακα καθιζόμενος. ὁδο[ι]πόρος οὖν τις ἰδὼν αὐτόν· 'τί, ὦ
ἄθλιε,' [ἔ]|φα, 'συντίθησι τὸγ κώθωνα [μά]ταν; τοῦτον γὰρ
οὐδέ κα ὁ ἐν Ἐπιδαύ||ρωι Ἀσκλαπιὸς ὑγιῆ ποῆσαι δύναιτο.'
ἀκούσας ταῦτα ὁ παῖς συν|θεὶς τὰ ὄστρακα εἰς τὸγ γυλιὸν
ἧρπε εἰς τὸ ἱερόν. ἐπεὶ δ' ἀφίκε|το, ἀνῶιξε τὸγ γυλιὸν καὶ
ἐξαιρεῖ ὑγιῆ τὸγ κώθωνα γεγενημέ|νον καὶ τῶι δεσπόται
ἡρμάνευσε τὰ πραχθέντα καὶ λεχθέ[[ε]]ντα· ὡ|ς δὲ ἄκουσ',

XI.   ἀνέθηκε τῶι θεῶι τὸγ κώθωνα. vac. || Αἰσχίνας ἐγκεκοιμι-
σμένων ἤδη τῶν ἱκετᾶν ἐπὶ δένδρεόν τι ἀμ|βὰς ὑπερέκυπτε εἰς
τὸ ἄβατον. καταπετὼν οὖν ἀπὸ τοῦ δένδρεος |περὶ σκόλοπάς
τινας τοὺς ὀπτίλλους ἀμφέπαισε. κακῶς δὲ δια|κείμενος
καὶ τυφλὸς γεγενημένος καθικετεύσας τὸν θεὸν ἐνε|κάθευδε καὶ

XII.   ὑγιὴς ἐγένετο. vac. || Εὔιππος λόγχαν ἔτη ἐφόρησε ἓξ ἐν
τᾶι γνάθωι. ἐγκοιτασθέντος |δ' αὐτοῦ ἐξελὼν τὰν λόγχαν ὁ
θεὸς εἰς τὰς χῆράς οἱ ἔδωκε· ἀμέρας |δὲ γενομένας ὑγιὴς

XIII.   ἐξῆρπε τὰν λόγχαν ἐν ταῖς χερσὶν ἔχων. |Ἀνὴρ Τορωναῖος
δεμελέας. οὗτος ἐγκαθεύδων ἐνύπνιον εἶδε· |ἔδοξέ οἱ τὸν θεὸν
τὰ στέρνα μαχαίραι ἀνσχίσσαντα τὰς δεμε||λέας ἐξελεῖν καὶ
δόμεν οἱ ἐς τὰς χεῖρας καὶ συνράψαι τὰ στή|θη· ἀμέρας δὲ
γενομένας ἐξῆλθε τὰ θηρία ἐν ταῖς χερσὶν ἔχων |καὶ ὑγιὴς ἐγέ-
νετο· κατέπιε δ' αὐτὰ δολωθεὶς ὑπὸ ματρυιᾶς ἐγ κυ|κᾶνι ἐμβε-

XIV.   βλημένας ἐκπιών. vac. |Ἀνὴρ ἐν αἰδοίωι λίθον. οὗτος ἐνύπνιον
εἶδε· ἐδόκει παιδὶ καλῶι||συγγίνεσθαι, ἐξονειρώσσων δὲ τὸλ
λίθον ἐγβάλλει καὶ ἀνελόμε|νος ἐξῆλθε ἐν ταῖς χερσὶν ἔχων.

XV.   vac. |Ἑρμόδικος Λαμψακηνὸς ἀκρατὴς τοῦ σώματος. τοῦτον
ἐγκαθεύ|δοντα ἰάσατο καὶ ἐκελήσατο ἐξελθόντα λίθον ἐνεγκεῖν
εἰς τὸ |ἱαρὸν ὁπόσσον δύναιτο μέγιστον· ὁ δὲ τὸμ πρὸ τοῦ

XVI.   ἀβάτου κείμε||νον ἤνικε. vac. |Νικάνωρ χωλός. τούτου καθη-
μένου παῖς τις ὕπαρ τὸν σκίπωνα ἁρ|πάξας ἔφευγε. ὁ δὲ

XVII.   ἀστὰς ἐδίωκε καὶ ἐκ τούτου ὑγιὴς ἐγένετο. |Ἀνὴρ δάκτυλον
ἰάθη ὑπὸ ὄφιος. οὗτος τὸν τοῦ ποδὸς δάκτυλον ὑ|πό του ἀγρίου
ἕλκεος δεινῶς διακείμενος μεθάμερα ὑπὸ τῶν θε||ραπόντων
ἐξενειχθεὶς ἐπὶ ἐδράματός τινος καθῖζε· ὕπνου δέ νιν |λαβόντος
ἐν τούτωι δράκων ἐκ τοῦ ἀβάτου ἐξελθὼν τὸν δάκτυλον |
ἰάσατο τᾶι γλώσσαι καὶ τοῦτο ποιήσας εἰς τὸ ἄβατον
ἀνεχώρησε |πάλιν. ἐξεγερθεὶς δὲ ὡς ἦς ὑγιής, ἔφα ὄψιν ἰδεῖν,
δοκεῖν νεανίσ|κον εὐπρεπῆ τὰμ μορφὰν ἐπὶ τὸν δάκτυλον

XVIII.   ἐπιπῆν φάρμακον. ||Ἀλκέτας Ἁλικός. οὗτος τυφλὸς ἐὼν
ἐνύπνιον εἶδε· ἐδόκει οἱ ὁ θεὸς ποτελθὼν τοῖς δα|κτύλοις

διάγειν τὰ ὄμματα καὶ ἰδεῖν τὰ δένδρη πρᾶτον τὰ ἐν τῶι ἰαρῶι.

XIX. ἀμέρας δὲ γε|νομένας ὑγιὴς ἐξῆλθε. Ἡραιεὺς Μυτιληναῖος.
οὗτος οὐκ εἶχε ἐν τᾶι κεφαλᾶι | τρίχας, ἐν δὲ τῶι γενείωι
παμπόλλας. αἰσχυνόμενος δὲ [ὡς] καταγελάμενος ὑπ[ὸ] |
τῶν ἄλλων ἐνεκάθευδε. τὸν δὲ ὁ θεὸς χρίσας φαρμάκωι τὰν

XX. κεφαλὰν ἐπόησε || τρίχας ἔχειν. vac. Λύσων Ἑρμιονεὺς παῖς
ἀϊδής. οὗ[τος] ὕπαρ ὑπὸ κυνὸς τῶν | κατὰ τὸ ἰαρὸν θε[ραπ]-
ευόμενος τοὺς ὀπτίλλους ὑγ[ιὴ]ς ἀπῆλθε.

B

XXI. Ἀράτα [Λά]καινα ὕδρωπ[α. ὑπ]ὲρ ταύτας ἁ μάτηρ ἐνεκάθευ-
δεν ἐλ Λακεδαίμο|νι ἔσσα[ς] καὶ ἐνύπνιον [ὁ]ρῆι· ἐδόκει τᾶς
θυγατρός οὐ τὸν θεὸν ἀποταμόν|τα τὰν κ[ε]φαλὰν τὸ σῶμα
κραμάσαι κάτω τὸν τράχαλον ἔχον· ὡς δ᾽ ἐξερρύα συ|χνὸν
ὑγρ[ό]ν, καταλύσαντα τὸ σῶμα τὰν κεφαλὰν πάλιν ἐπιθέμεν
ἐπὶ τὸν αὐ||χένα· ἰδο[ῦ]σα δὲ τὸ ἐνύπνιον τοῦτο ἀγχωρήσασα
εἰς Λακεδαίμονα κατα|λαμβάνε[ι τ]ὰν θυγατέρα ὑγιαίνουσαν

XXII. καὶ τὸ αὐτὸ ἐνύπνιον ὡρακυῖαν. | Ἕρμων Θ[άσιος. τοῦτο]ν
τυφλὸν ἐόντα ἰάσατο· μετὰ δὲ τοῦτο τὰ ἴατρα οὐκ ἀ|πάγοντ[α
ὁ θεός νιν] ἐπόησε τυφλὸν αὖθις· ἀφικόμενον δ᾽ αὐτὸν καὶ

XXIII. πάλιν | ἐγκαθε[ύδοντα ὑγι]ῆ κατέστασε. || Ἀριστα[γόρα
Τροζ]ανία. αὕτα ἕλμιθα ἔχουσα ἐν τᾶι κοιλίαι ἐνεκάθευδε |
ἐν Τροζ[ᾶνι ἐν τῶι] τοῦ Ἀσκλαπιοῦ τεμένει καὶ ἐνύπνιον εἶδε·
ἐδόκει οὐ | τοὺς υἱ[οὺς τοῦ θ]εοῦ, οὐκ ἐπιδαμοῦντος αὐτοῦ, ἀλλ᾽
ἐν Ἐπιδαύρωι ἐόντος, | τὰγ κεφα[λὰν ἀπο]ταμεῖν, οὐ δυνα-
μένους δ᾽ ἐπιθέμεν πάλιν πέμψαι τινὰ πο[τὶ] | τὸν Ἀσκλ[απιόν,
ὅ]πως μόληι· μεταξὺ δὲ ἀμέρα ἐπικαταλαμβάνει καὶ ὁ ἰαρ||
ρεὺς ὁρῆι [σάφα⁴ τ]ὰν κεφαλὰν ἀφαιρημέναν ἀπὸ τοῦ
σώματος· τᾶς ἐφερπού|σας δὲ νυκτ[ὸς Ἀρ]ισταγόρα ὄψιν εἶδε·
ἐδόκει οἱ ὁ θεὸς ἵκων ἐξ Ἐπιδαύρου | ἐπιθεὶς τ[ὰν κε]φαλὰν
ἐπὶ [τὸ]ν τράχαλον, μετὰ ταῦτα ἀνσχίσσας τὰγ κοιλ[ί|α]ν
τὰν αὐτ[ᾶς ἐξ]ελεῖν τὰν ἕ[λμ]ιθα καὶ συρράψαι πάλιν, καὶ

XXIV. ἐκ τούτου ὑγ[ι|ὴ]ς ἐγένετ[ο. vac. Ὑ]π[ὸ π]έτραι παῖς
Ἀριστόκριτος Ἁλικός. οὗτος || ἀποκολυμ[βάσ]ας εἰς τὰν
θά[λασ]σαν ἔπειτα δενδρύων εἰς τόπον ἀφίκετο | ξηρόν,
κύκ[λωι] πέτραις περ[ιεχό]μενον, καὶ οὐκ ἐδύνατο ἔξοδον
οὐδεμί|αν εὑρεῖν. [με]τὰ δὲ τοῦτο ὁ πατ[ὴρ α]ὐτοῦ, ὡς
οὐθαμεῖ περιετύγχανε μασ|τεύων, παρ᾽ [Ἀ]σκλαπιῶι ἐν τῶι
ἀ[βάτ]ωι ἐνεκάθευδε περὶ τοῦ παιδὸς καὶ ἐ|νύπνιον ε[ἶ]δε·
ἐδόκει αὐτὸν ὁ θ[εὸς] ἄγειν εἴς τινα χώραν καὶ δεῖξαί οἱ,

δ[ι]||ότι τουτ[ε]ῖ ἐστι ὁ υἱὸς αὐτοῦ. ἐξε[λθὼ]ν δ᾽ ἐκ τοῦ
ἀβάτου καὶ λατομήσας τὰ[ν] | πέτραν ἀ[ν]ηῦρε τὸμ παῖδα
ἑβδεμα[ῖο]ν. vac. Σωστράτα Φεραί[α θηρί᾽]⁵ | ἐκύησε.
α[ὔ]τα ἐμ παντὶ ἐοῦσα φοράδαν εἰς τὸ ἱαρὸν ἀφικομένα
ἐνε[κά]|θευδε. ὡς δὲ οὐθὲν ἐνύπνιον ἐναργ[ὲ]ς ἑώρη, πάλιν
οἴκαδε ἀπεκομίζ[ε]|το. μετὰ δὲ τοῦτο συμβολῆσαί τις περὶ
Κόρνους αὐτᾶι καὶ τοῖς ἑ[πομέ]||νοις ἔδοξε τὰν ὄψιν εὐπρεπὴς
ἀνήρ, ὃς πυθόμενος παρ᾽ αὐτῶν τ[ὰς δυσπρα]|ξίας τὰς αὐτῶν
ἐκελήσατο θέμεν τὰν κλίναν, ἐφ᾽ ᾆς τὰν Σωστρ[άταν ἔφε]|ρον.
ἔπειτα τὰγ κοιλίαν αὐτᾶς ἀνσχίσας ἐξαιρεῖ πλῆθος ζ[ῴων
πάμ]|πολυ,⁶ [δύ]ε ποδανιπτῆρας· συνράψας δὲ τὰ[ν γ]α-
στέρα καὶ ποήσας ὑ[γιῆ] | τὰν γυναῖκα τάν τε παρουσίαν τὰν
αὐτο[ῦ π]αρενεφάνιξε ὁ Ἀσκλαπιὸς || καὶ ἴατρα ἐκέλετο

XXVI.　ἀπ[ο]πέμπειν εἰς Ἐπί[δα]υρ[ον. vac.] Κύων ἰά|σατο παῖδα
Αἰ[γιν]άταν. οὗτος φῦμα ἐν τῶ[ι τρα]χάλωι εἶχε· ἀφικό-
μενο[ν] | δ᾽ αὐτὸν ποὶ τ[ὸν] θε[ὸ]ν κύων τῶν ἱαρῶν ὕ[παρ
τ]ᾶι γλώσσαι ἐθεράπευσε | καὶ ὑγιῆ ἐπόη[σ]ε. vac. Ἀνὴρ

XXVII.　ἐ[ντὸ]ς τᾶς κοιλίας ἕλκος ἔχων. οὖ|τος ἐγκαθεύδων ἐν-
[ύπνιο]ν εἶδε· ἐδόκ[ε]ι αὐτῶι ὁ θεὸς ποιτάξαι τοῖς || ἑπομένοις
ὑπηρέτα[ις συλ]λαβόντας αὐτὸν ἴσχειν, ὅπως τάμηι οὐ τὰν |
κοιλίαν, αὐτὸς δὲ φεύ[γει]ν, τοὺς δὲ συλλαβόντας νιν ποι-
δῆσαι ποὶ ῥό|πτον· μετὰ δὲ τοῦτο τὸν [Ἀσ]κλαπιὸν ἀνσχίσ-
σαντα τὰγ κοιλίαν ἐκτα|μεῖν τὸ ἕλκος καὶ συρρά[ψαι] πάλιν,
καὶ λυθῆμεν ἐκ τῶν δεσμῶν· καὶ ἐ|κ τούτου ὑγιὴς ἐξῆ[λθ]ε,
τὸ δὲ δάπεδον ἐν τῶι ἀβάτωι αἵματος κατά||πλεον ἦς. vac.

XXVIII.　Κλεινάτας Θηβαῖος ὁ τοὺς φθεῖρας· οὗτος π[λῆ]|θός τι
πάμπολυ φθε[ιρ]ῶν ἐν τῶι σώματι [ἔ]χων ἀφικόμενος ἐνεκά-
[θευ]|δε καὶ ὁρῆι ὄψιν· ἐδόκει αὐτόν νιν ὁ θεὸς ἐγδύσας καὶ
γυμνὸν κα|ταστάσας ὀρθὸν σάρ[ω]ι τινὶ τοὺς φθεῖρας ἀπὸ
τοῦ σώματος ἀποκα|θαίρειν· ἀμέρας δὲ γ[ε]νομένας ἐκ τοῦ

XXIX.　ἀβάτου ὑγιὴς ἐξῆλθε. vac. || Ἁγέστρατος κεφαλᾶς [ἄ]λγος·
οὗτος ἀγρυπνίαις συνεχόμενος διὰ | τὸμ πόνον τᾶς κεφαλᾶ[ς],
ὡς ἐν τῶι ἀβάτωι ἐγένετο, καθύπνωσε καὶ ἐν[ύ]|πνιον εἶδε·
ἐδόκει αὐτὸν ὁ θεὸς ἰασάμενος τὸ τᾶς κεφαλᾶς ἄλγος ὀρ|θὸν
ἀστάσας γυμνὸν παγκρατίου προβολὰν διδάξαι· ἀμέρας δὲ
γενη|θείσας ὑγιὴς ἐξῆλθε καὶ οὐ μετὰ πολὺγ χρόνον τὰ Νέμεα

XXX.　ἐνίκασε || παγκράτιον. vac. Γοργίας Ἡρακλειώτας πύος.
οὗτος ἐμ μάχαι | τινὶ τρωθεὶς εἰς τὸμ πλεύμονα τοξε[ύ]μ[α]τι
ἐνιαυτὸγ καὶ ἑξάμηνον | ἔμπυος ἦς οὕτω σφοδρῶς, ὥστε ἑπτὰ

καὶ ἑξήκοντα λεκάνας ἐνέπλησε | πύους· ὡς δ᾽ ἐνεκάθευδε, ὄψιν
εἶδε· ἐδόκει οἱ ὁ θεὸς τὰν ἀκίδα ἐξε|λεῖν ἐκ τοῦ πλεύμονος·
ἀμέρας δὲ γενομένας ὑγιὴς ἐξῆλθε τὰν ἀκί||δα ἐν ταῖς χερσὶ

XXXI.   φέρων. vac. Ἀνδρομάχα ἐξ Ἀπείρο[υ] περὶ παί|δων. αὕτα
ἐγκαθεύδουσα ἐνύπνιον εἶδε· ἐδόκει αὐτᾶι π[α]ῖς τις ὡραῖ|ος
ἀγκαλύψαι, μετὰ δὲ τοῦτο τὸν θεὸν ἅψασθαί οὐ τᾶι [χη]ρί·
ἐκ δὲ τού|του τᾶι Ἀνδρομάχαι υ[ἱ]ὸς ἐξ Ἀρύββα ἐγένετο.

XXXII.   Ἀ[ντικ]ράτης Κνίδι|ος ὀφθαλμούς. οὗτος ἔν τινι μάχαι ὑπὸ
δό[ρα]τος πλα[γεὶ]ς δι᾽ ἀμφοτέ||ρων τῶν ὀφθαλμῶν τυφλὸς
ἐγένετο καὶ τὰν λόγχαν [ἐνε]⁸οῦσαν ἐν τῶι | προσώπωι περιέ-
φερε· ἐγκαθεύδων [δ]ὲ ὄψιν εἶδε· ἐδ[όκε]ι οὐ τὸν θεὸν |
ἐξελκύσαντα τὸ βέλος εἰς τὰ β[λέ]φαρα τὰς καλουμ[έν]ας
κόρας πά|λιν ἐναρμόξαι· ἀμέρας δὲ γενομένας ὑγιὴς ἐξῆλθ[ε].

XXXIII. | [Θ]έρσανδρος Ἁλικὸς φθίσιν. οὗτος, ὡς ἐγκαθεύδων [οὐ]-
δεμίαν ὄψιν || [ἑ]ώρη, ἐφ᾽ ἁμάξας [ἄμπαλ]ιν ἀπεκομίζετο εἰς
Ἁλιεῖς, δράκων δέ τις | [τ]ῶν ἱαρῶν ἐπὶ τ[ᾶς ἁμ]άξας καθι-
δρυμένος ἦς, τὸ πο[λ]ὺ τᾶς ὁδοῦ περιη||[λι]γμένος περ[ὶ
τ]ὸν ἄξονα διετέλεσε. μολόντων δ᾽ [α]ὐ[τ]ῶν εἰς Ἁλιεῖς |
[κα]ὶ τοῦ Θερσ[ά]νδρου κατακλιθέντος οἴ[κο]ι, ὁ δράκων
ἀπὸ τᾶς ἁμά|[ξα]ς καταβὰ[ς τ]ὸν Θέρσανδρον ἰάσατο. [τᾶς
δ]ὲ πόλιος τῶν Ἁλικῶν || [ἐξετ]α[ζ]ούσας⁹ τὸ γεγενημένον
καὶ διαπορ[ουμένας] περὶ τοῦ ὄφι|[ος, πό]τερον εἰς Ἐπί-
δαυρον ἀποκομίζωντι [ἢ αὐτὸν κα]τὰ χώραν ἐῶν|[τι, ἔ]δοξε
τᾶι πόλι εἰς Δελφοὺς ἀποστεῖλα[ι χρησομέ]νους, πότερα |
[π]οιῶντι· ὁ δὲ θεὸς ἔχρησε τὸν ὄφιν ἐῆν αὐ[τεῖ καὶ ἱ]δρυσα-
μένου[ς | Ἀ]σκλαπιοῦ τέμενος καὶ εἰκόνα αὐτοῦ πο[ιησαμέ]-
νους ἀνθέμεν [εἰς] || τὸ ἱαρόν. ἀγγελθέντος δὲ τοῦ χρησ[μοῦ
ἁ πόλι]ς ἁ τῶν Ἁλικῶν [ἀφ]|ιδρύσατο τέμενος Ἀσκλαπιοῦ
[καὶ τἄλλα τὰ ὑπὸ το]ῦ θεοῦ μαντ[ευ]|σθέντα ἐπετέλεσε.

XXXIV.   [Ἁ δεῖνα Τροζα]νία ὑπὲρ τ[έ]|κνων. αὕτα ἐγκαθεύδου[σα]
ἐν[ύπνιον εἶδε· ἐδόκει ο]ἱ φᾶσαι ὁ [θεὸς] | ἐσσεῖσθαι γενεὰγ
καὶ ἐ[π]ερ[ωτῆν πότερ᾽ ἐπιθυμοῖ ἔ]ρσεν[α ἢ θηλυ]||τέραν,
αὐτὰ δὲ φάμεν¹⁰ ἐπι[θυμεῖν ἄρσενα· μετὰ δὲ τού]το ἐ[ντὸς

XXXV.   ἐνι] | αυτοῦ ἐγένετο αὐτᾶι υἱ[ός]. vac. - - - Ἐπιδαύριος
χωλός· οὗτος [ἱκέτας ἀφίκετο εἰς τὸ ἱαρὸν ἐπὶ κλίνας]·¹¹
| ἐγκαθεύδων δὲ ὄψιν εἶδε· [ἐδόκει οἱ ὁ θεὸς τὸν σκίπωνα
κατᾶξαι καὶ]¹¹ | ποιτάσσειν αὐτῶι κλίμα[κα ποτενεγκεῖν καὶ
ὡς ὑψοτάτω ἀμβᾶμεν ἐ]¹¹||πὶ τὸν ναόν· αὐτὸς δὲ τὸ μὲ[ν
πρᾶτον πειρῆσθαι, ἔπειτα τὸ θάρσος ἀ|ποβ]¹¹αλὲν καὶ ἄνω
ἐπὶ τοῦ θ[ριγκοῦ ἀμπαύεσθαι, τέλος δὲ ἀπειπεῖν | κ]¹¹αὶ τὰγ

κλίμακα μικρὸν κα[τὰ μικρὸν καταβᾶμεν· ὁ δὲ Ἀσκλαπιὸς
τὸ]¹¹ | πρᾶτον ἀγανακτῶν τ[ᾶ]ι πρά[ξι ἔπειτα ἐπιγελάσαι
αὐτῶι οὕτω δει]¹¹|λῶι ἐόντι· ἀποτολμ[ῶν]¹¹ δὲ ἀμ[έρας
XXXVI. γενομένας ἐπιτελέσαι ἀσκηθὴς ἐ]¹¹||ξῆλθε. vac. Καφισί[ας
― ― ― τὸμ πόδα. οὗτος τοῖς τοῦ Ἀσ]¹¹|κλαπιοῦ θεραπεύμασιν
ἐπ[ιγελῶν ‘χωλούς,’ ἔφα, ‘ἰάσασθαι ὁ θεὸς ψεύ]¹¹|δεται
λέγων· ὡς, εἰ δύναμιν ε[ἶχε, τί οὐ τὸν Ἅφαιστον ἰάσατο’;
ὁ δὲ θεὸς]¹¹ | τᾶς ὕβριος ποινὰς λαμβάνω[ν οὐκ ἔλαθε· ἱππεύων
γὰρ ὁ Καφισίας ὑπὸ]¹¹ | τοῦ βουκεφάλα ἐν τᾶι ἕδραι [γαργα-
λισθέντος ἐπλάγη ὥστε πηρωθῆ]¹¹||μεν τὸμ πόδα παραχρῆμα
καὶ [φοράδαν εἰς τὸ ἱαρὸν ἀγκομισθῆμεν.]¹¹ | ὕ[σ]τερον δὲ
πολλὰ καθικετεύ[σαντα αὐτὸν ὁ θεὸς ὑγιῆ ἐπόησε.]¹¹ vac.
XXXVII. | Κλειμένης Ἀργεῖος ἀκρατὴς [τοῦ σώματος· οὗτος ἐλθὼν εἰς
τὸ ἄβα]¹¹|τον ἐνεκάθευδε καὶ ὄψιν εἶδ[ε· ἐδόκει οἱ ὁ θεὸς
φοινικίδα ἐρεᾶν πε]¹¹|ριελίξαι περὶ τὸ σῶμα καὶ μικ[ρὸν ἔξω
τοῦ ἱαροῦ εἰς λουτρὸν ἄγειν]¹¹|| νιν ἐπί τινα λίμναν, ᾆς τὸ
ὕδωρ [εἶμεν καθ’ ὑπερβολὰν ψυχρόν· δειλῶς]¹¹ | δ’ αὐτοῦ ἔθεν
διακειμένου τὸν Ἀ[σκλαπιὸν οὐκ ἰασεῖσθαι τοὺς δει]¹¹|λοὺς
τῶν ἀνθρώπων εἰς ταῦτα φά[μεν, ἀλλ’ ἢ οἵτινές κα ποτ’ αὐτόν
νιν ἀ]¹¹|φικνῶνται εἰς τὸ τέμενος ἐόντ[ες εὐέλπιδες, ὡς οὐθὲν
κακὸν τὸν]¹¹ | τοιοῦτον ποιησοῖ, ἀλλ’ ὑγιῆ ἀποπ[εμψοῖ·
XXXVIII. ἐξεγερθεὶς δ’ ἐλοῦτο καὶ ἀσ]¹¹||κηθὴς ἐξῆλθε. vac. Δίαιτος
Κυρρ[αῖος· οὗτος ἀκρατὴς ἐὼν ἐτύγχα]¹¹|νε τῶγ γονάτων·
ἐγκαθεύδων δὲ ἐνύ[πνιον εἶδε· ἐδόκει οἱ ὁ θεὸς] | τοὺς ὑπη-
ρέτας κέλεσθαι ἀειραμέν[ους νιν ἐξενεγκεῖν ἐκ τοῦ ἀ]|δύτου
καὶ καταθέμεν πρὸ τοῦ ναοῦ· ἐ[πεὶ δὲ αὐτὸν οὗτοι ἔξω ἤνεγ-]
|καν, ἅρμα ζεύξαντα τὸν θεὸν ἵππων πε[ριελαύνειν τρὶς περί
νιν κύ]¹¹||κλωι καὶ καταπατεῖν (ν)ιν τοῖς ἵπποις, [καὶ ἐγκρατῆ
τῶγ γονάτων γε]¹¹|νέσθαι εὐθύς. ἀμέρας δὲ γενομένας ὑγ[ιὴς
XXXIX. ἐξῆλθε. vac.] [Ἀγαμή]¹¹|δα ἐκ Κέου. αὗτα περὶ παίδων
ἐγκαθεύδ[ουσα ἐνύπνιον εἶδε· ἐδό]|κει οἱ ἐν τῶι ὕπνωι δράκων
ἐπὶ τᾶς γαστ[ρὸς κεῖσθαι· καὶ ἐκ τούτου] | παῖδές οἱ ἐγένοντο
XL. πέντε. vac. Τίμω[ν ― ― ― ― ― ― ― λόγχαι τρω]||θεὶς ὑπὸ τὸν
ὀφθαλμόν· οὗτος ἐγκαθεύδ[ων ἐνύπνιον εἶδε· ἐδό]|κει οἱ ὁ
θεὸς ποίαν τρίψας ἐγχεῖν εἰς τ[ὸν ὀφθαλμόν οὐ·¹² καὶ ὑγι]|ὴς
XLI. ἐγένετο. vac. Ἐρασίππα ἐκ Καφνιᾶν ἰ[ούλους. αὗτα ὠγκωμέ-
ναν]¹¹ | εἶχε τὰγ γαστέρα καὶ ἐπέπρητο ὅλα καὶ ο[ὐ κατεῖχε
οὐθέν· ἐγκα]¹¹|θεύδουσα δὲ ἐνύπνιον εἶδε· ἐδόκει οἱ ὁ θεὸ[ς
τὰγ γαστέρα ἀντρί]||βων φιλῆσαί νιν, μετὰ δὲ τοῦτο φιάλαν
οἱ δό[μεν, ἐν ἆι φάρμακον], | καὶ κέλεσθαι ἐκπιεῖν, ἔπειτα

ἐμεῖν κέλεσ[θαί νιν, ἐξεμέσασα¹³] | δὲ ἐμπλῆσαι τὸ λώπιον
τὸ αὐτᾶς· ἀμέρας δὲ γ[ενομένας ἐώρη πᾶν] | τὸ λώπιον μεστὸν
ὧν ἐξήμεσε κακῶν, καὶ ἐκ το[ύτου ὑγιὴς ἐγένε]|το. vac.

XLII.   Νικασιβούλα Μεσσανία περὶ παίδω[ν ἐγκαθεύδουσα] || ἐνύ-
πνιον εἶδε· ἐδόκει οἱ ὁ θεὸς δράκοντα μεθ[έρποντα ἵκειν]¹¹ |
φέρων παρ᾽ αὐτάν, τούτωι δὲ συγγενέσθαι αὐτά· [καὶ ἐκ
τούτου] | παῖδές· οἱ ἐγένοντο εἰς ἐνιαντὸν ἔρσενες δύ[ε. vac.

XLIII.   – – –] | Κιανὸς ποδάγραν. τούτου ὕπαρ χὰν ποτιπορευο[μέ-
νου δάκνων] | αὐτοῦ τοὺς πόδας καὶ ἐξαιμάσσων ὑγιῆ
ἐπόη[σε].

¹ Nos. 123-4 have been omitted because they are too badly damaged to yield reliable
information. For the text, cf. also R. Herzog, *Die Wunderheilungen von Epidaurus,
Philologus, Supplement*, III, 3, 1931.
² ἐντὸς Herzog αὐτὸν Hiller.
³ εἰς Ἀθηνᾶν Herzog ἐξ Εὐθηνᾶν Hiller.
⁴ σάφα Hiller ὕπαρ Kavvadias-Herzog.
⁵ θηρί· Herzog παρ Hiller.
⁶ ζῴων Herzog ζωϊυφίων Hiller.
⁷ Supplevit Herzog.
⁸ ἐνεοῦσαν Herzog παροῦσαν Hiller.
⁹ ἐξεταζούσας Herzog ἀγγελλούσας Hiller.
¹⁰ ἃ τῶν Ἁλικῶν—φάμεν cf. Herzog ad no. XXXIV.
¹¹ Supplevit Herzog.
¹² οὐ Herzog τι Hiller.
¹³ ἐξεμέσασα Herzog ἐξεμεσάσας Hiller.

Stele I.

God and Good Fortune.
Cures of Apollo and Asclepius.

1.   Cleo was with child for five years. After she had been
pregnant for five years she came as a suppliant to the god and
slept in the Abaton. As soon as she left it and got outside the
temple precincts ¹ she bore a son who, immediately after birth,
washed himself at the fountain and walked about with his
mother. In return for this favor she inscribed on her offering:
" Admirable is not the greatness of the tablet, but the Divinity,
in that Cleo carried the burden in her womb for five years,
until she slept in the Temple and He made her sound."

2.   A three-years' pregnancy. Ithmonice of Pellene came to the
Temple for offspring. When she had fallen asleep she saw a
vision. It seemed to her that she asked the god that she might
get pregnant with a daughter and that Asclepius said that she
would be pregnant and that if she asked for something else he

would grant her that too, but that she answered she did not need anything else. When she had become pregnant she carried in her womb for three years, until she approached the god as a suppliant concerning the birth. When she had fallen asleep she saw a vision. It seemed to her that the god asked her if she had not obtained all she had asked for and was pregnant; about the birth she had added nothing, and that, although he had asked if she needed anything else, she should say so and he would grant her this too. But since now she had come for this as a suppliant to him, he said he would accord even it to her. After that, she hastened to leave the Abaton, and when she was outside the sacred precincts [2] she gave birth to a girl.

3.   A man whose fingers, with the exception of one, were paralyzed, came as a suppliant to the god. While looking at the tablets in the temple he expressed incredulity regarding the cures and scoffed at the inscriptions. But in his sleep he saw a vision. It seemed to him that, as he was playing at dice below the Temple and was about to cast the dice, the god appeared, sprang upon his hand, and stretched out his [the patient's] fingers. When the god had stepped aside it seemed to him [the patient] that he [the patient] bent his hand and stretched out all his fingers one by one. When he had straightened them all, the god asked him if he would still be incredulous of the inscriptions on the tablets in the Temple. He answered that he would not. " Since, then, formerly you were incredulous of the cures, though they were not incredible, for the future," he said, " your name shall be ' Incredulous.' " When day dawned he walked out sound.

4.   Ambrosia of Athens, blind of one eye. She came as a suppliant to the god. As she walked about in the Temple she laughed at some of the cures as incredible and impossible, that the lame and the blind should be healed by merely seeing a dream. In her sleep she had a vision. It seemed to her that the god stood by her and said that he would cure her, but that in payment he would ask her to dedicate to the Temple a silver pig [3] as a memorial of her ignorance. After saying this, he cut the diseased eyeball and poured in some drug. When day came she walked out sound.

5.   A voiceless boy. He came as a suppliant to the Temple for his voice. When he had performed the preliminary sacrifices

and fulfilled the usual rites,[4] thereupon the temple servant who brings in the fire[5] for the god, looking at the boy's father, demanded he should promise to bring within a year the thank-offering for the cure if he obtained that for which he had come. But the boy suddenly said, " I promise." His father was startled at this and asked him to repeat it. The boy repeated the words and after that became well.

6. Pandarus, a Thessalian, who had marks on his forehead. He saw a vision as he slept. It seemed to him that the god bound the marks round with a headband and enjoined him to remove the band when he left the Abaton and dedicate it as an offering to the Temple. When day came he got up and took off the band and saw his face free of the marks; and he dedicated to the Temple the band with the signs which had been on his forehead.

7. Echedorus received the marks of Pandarus in addition to those which he already had. He had received money from Pandarus to offer to the god at Epidaurus in his name, but he failed to deliver it. In his sleep he saw a vision. It seemed to him that the god stood by him and asked if he had received any money from Pandarus to set up as an offering an Athena in the Temple. He answered that he had received no such thing from him, but if he [the god] would make him well he would have an image painted and offer it to him [the god]. Thereupon the god seemed to fasten the headband of Pandarus round his marks, and ordered him upon leaving the Abaton to take off the band and to wash his face at the fountain and to look at himself in the water. When day came he left the Abaton, took off the headband, on which the signs were no longer visible. But when he looked into the water he saw his face with his own marks and the signs of Pandarus in addition.

8. Euphanes, a boy of Epidaurus. Suffering from stone he slept in the temple. It seemed to him that the god stood by him and asked: " What will you give me if I cure you?" " Ten dice," he answered. The god laughed and said to him that he would cure him. When day came he walked out sound.

9. A man came as a suppliant to the god. He was so blind that of one of his eyes he had only the eyelids left—within them was nothing, but they were entirely empty. Some of those in the Temple laughed at his silliness to think that he could recover his sight when one of his eyes had not even a trace of the ball,

but only the socket. As he slept a vision appeared to him. It seemed to him that the god prepared some drug, then, opening his eyelids, poured it into them. When day came he departed with the sight of both eyes restored.

10.   The goblet. A porter, upon going up to the Temple, fell when he was near the ten-stadia stone. When he had gotten up he opened his bag and looked at the broken vessels. When he saw that the goblet from which his master was accustomed to drink was also broken, he was in great distress and sat down to try to fit the pieces together again. But a passer-by saw him and said: " Foolish fellow, why do you put the goblet together in vain? For this one not even Asclepius of Epidaurus could put to rights again." The boy, hearing this, put the pieces back in the bag and went on to the Temple. When he got there he opened the bag and brought the goblet out of it, and it was entirely whole; and he related to his master what had happened and had been said; when he [the master] heard that, he dedicated the goblet to the god.

11.   Aeschines, when the suppliants were already asleep, climbed up a tree and tried to see over into the Abaton. But he fell from the tree on to some fencing and his eyes were injured. In a pitiable state of blindness, he came as a suppliant to the god and slept in the Temple and was healed.

12.   Euhippus had had for six years the point of a spear in his jaw. As he was sleeping in the Temple the god extracted the spearhead and gave it to him into his hands. When day came Euhippus departed cured, and he held the spearhead in his hands.

13.   A man of Torone with leeches. In his sleep he saw a dream. It seemed to him that the god cut open his chest with a knife and took out the leeches, which he gave him into his hands, and then he stitched up his chest again. At daybreak he departed with the leeches in his hands and he had become well. He had swallowed them, having been tricked by his stepmother who had thrown them into a potion which he drank.

14.   A man with a stone in his membrum. He saw a dream. It seemed to him that he was lying with a fair boy and when he had a seminal discharge he ejected the stone and picked it up and walked out holding it in his hands.

15.   Hermodicus of Lampsacus was paralyzed in body. This one,

when he slept in the Temple, the god healed and he ordered him upon coming out to bring to the Temple as large a stone as he could.[6] The man brought the stone which now lies before the Abaton.

16.   Nicanor, a lame man. While he was sitting wide-awake, a boy snatched his crutch from him and ran away. But Nicanor got up, pursued him, and so became well.

17.   A man had his toe healed by a serpent. He, suffering dreadfully from a malignant sore in his toe, during the daytime was taken outside by the servants of the Temple and set upon a seat. When sleep came upon him, then a snake issued from the Abaton and healed the toe with its tongue, and thereafter went back again to the Abaton. When the patient woke up and was healed he said that he had seen a vision : it seemed to him that a youth with a beautiful appearance had put a drug upon his toe.

18.   Alcetas of Halieis. This blind man saw a dream. It seemed to him that the god came up to him and with his fingers opened his eyes, and that he first saw the trees in the sanctuary. At daybreak he walked out sound.

19.   Heraieus of Mytilene. He had no hair on his head, but an abundant growth on his chin. He was ashamed because he was laughed at by others. He slept in the Temple. The god, by anointing his head with some drug, made his hair grow thereon.

20.   Lyson of Hermione, a blind boy. While wide-awake he had his eyes cured by one of the dogs in the Temple and went away healed.

[1] T. 488; 739.	[3] Cf. T. 545.	[5] Cf. T. 498, n. 1.
[2] T. 488; 739.	[4] Cf. T. 511.	[6] Cf. T. 431.

Stele II.

21.   Arata, a woman of Lacedaemon, dropsical. For her, while she remained in Lacedaemon, her mother slept in the temple and sees a dream. It seemed to her that the god cut off her daughter's head and hung up her body in such a way that her throat was turned downwards. Out of it came a huge quantity of fluid matter. Then he took down the body and fitted the head back on to the neck.[1] After she had seen this dream she went back to Lacedaemon, where she found her daughter in good health; she had seen the same dream.

22.   Hermon of Thasus. His blindness was cured by Asclepius.

But, since afterwards he did not bring the thank-offerings, the god made him blind again. When he came back and slept again in the Temple, he [*sc.*, the god] made him well.

23. Aristagora of Troezen. She had a tapeworm in her belly, and she slept in the Temple of Asclepius at Troezen[2] and saw a dream. It seemed to her that the sons of the god, while he was not present but away in Epidaurus, cut off her head, but, being unable to put it back again, they sent a messenger to Asclepius asking him to come. Meanwhile day breaks and the priest clearly sees her head cut off from the body. When night approached, Aristagora saw a vision. It seemed to her the god had come from Epidaurus and fastened her head on to her neck. Then he cut open her belly, took the tapeworm out, and stitched her up again. And after that she became well.[3]

24. Aristocritus, a boy of Halieis, under a rock. He, after having dived and swum away into the sea, came then to a dry hiding place which was surrounded by rocks, and he could not find any way out. Thereafter his father, since he did not find him anywhere on his search, came to Asclepius and slept in the Abaton on behalf of his son and saw a dream. It seemed to him that the god led him to a certain place and showed him that there was his son. When he came out of the Abaton and quarried through the cliffs he found his son after seven days.

25. Sostrata, a woman of Pherae, was pregnant with worms. Being in a very bad way, she was carried into the Temple and slept there. But when she saw no distinct dream she let herself be carried back home. Then, however, near a place called Kornoi, a man of fine appearance seemed to come upon her and her companions. When he had learned from them about their bad luck, he asked them to set down on the ground the litter in which they were carrying Sostrata. Then he cut open her abdomen and took out a great quantity of worms—two wash-basins full. After having stitched her belly up again and made the woman well, Asclepius revealed to her his presence and enjoined her to send thank-offerings for her treatment to Epidaurus.

26. A dog cured a boy from Aegina. He had a growth on the neck. When he had come to the god, one of the sacred dogs healed him—while he was awake—with its tongue and made him well.

27. A man with an abscess within his abdomen. When asleep in the Temple he saw a dream. It seemed to him that the god ordered the servants who accompanied him to grip him and hold him tightly so that he could cut open his abdomen. The man tried to get away, but they gripped him and bound him to a door knocker. Thereupon Asclepius cut his belly open, removed the abscess, and, after having stitched him up again, released him from his bonds. Whereupon he walked out sound, but the floor of the Abaton was covered with blood.

28. Cleinatas of Thebes with the lice. He came with a great number of lice on his body, slept in the Temple, and sees a vision. It seems to him that the god stripped him and made him stand upright, naked, and with a broom brushed the lice from off his body. When day came he left the Temple well.

29. Hagestratus with headaches. He suffered from insomnia on account of headaches. When he came to the Abaton he fell asleep and saw a dream. It seemed to him that the god cured him of his headaches and, making him stand up naked, taught him the lunge used in the pancratium. When day came he departed well, and not long afterwards he won in the pancratium at the Nemean games.[4]

30. Gorgias of Heracleia with pus. In a battle he had been wounded by an arrow in the lung and for a year and a half had suppurated so badly that he filled sixty-seven basins with pus. While sleeping in the Temple he saw a vision. It seemed to him the god extracted the arrow point from his lung. When day came he walked out well, holding the point of the arrow in his hands.

31. Andromache of Epeirus, for the sake of offspring. She slept in the Temple and saw a dream. It seemed to her that a handsome boy uncovered her, after that the god touched her with his hand, whereupon a son was born to Andromache from Arybbas.

32. Anticrates of Cnidos, eyes. In a battle he had been hit by a spear in both eyes and had become blind; and the spear point he carried with him, sticking in his face. While sleeping he saw a vision. It seemed to him that the god pulled out the missile and then fitted into his eyelids again the so-called pupils. When day came he walked out sound.

33. Thersandrus of Halieis with consumption. He, when in his temple sleep he saw no vision, was carried back to Halieis on a

wagon; one of the sacred serpents, however, was sitting on the
wagon and remained for the greater part of the journey coiled
around the axle. When they came to Halieis and Thersandrus
was resting on his bed at home, the serpent descended from the
wagon and cured Thersandrus. When the city of Halieis made
an inquiry as to what had happened and was at a loss regarding
the serpent, whether to return it to Epidaurus or to leave it in
their territory, the city resolved to send to Delphi for an oracle
as to what they should do. The god decided they should leave
the serpent there and put up a sanctuary of Asclepius, make an
image of him, and set it up in the temple. When the oracle
was announced the city of Halieis erected the sanctuary of
Asclepius [5] and followed the rest of the god's commands.

34.   . . . of Troezen for offspring. She slept in the Temple and
saw a dream. The god seemed to say to her she would have
offspring and to ask whether she wanted a male or a female,
and that she answered she wanted a male. Whereupon within
a year a son was born to her.

35.   . . . of Epidaurus, lame. He came as a suppliant to the
sanctuary on a stretcher. In his sleep he saw a vision. It
seemed to him that the god broke his crutch and ordered him to
go and get a ladder and to climb as high as possible up to the
top of the sanctuary. The man tried it at first, then, however,
lost his courage and rested up on the cornice; finally he gave up
and climbed down the ladder little by little. Asclepius at first
was angry about the deed, then he laughed at him because he
was such a coward. He dared to carry it out after it had
become daytime and walked out unhurt.

36.   Cephisias . . . with the foot. He laughed at the cures of
Asclepius and said: " If the god says he has healed lame people
he is lying; for, if he had the power to do so, why has he not
healed Hephaestus? " [6] But the god did not conceal that he was
inflicting penalty for the insolence. For Cephisias, when riding,
was stricken by his bullheaded horse which had been tickled in
the seat, so that instantly his foot was crippled and on a
stretcher he was carried into the Temple. Later on, after he
had entreated him earnestly, the god made him well.

37.   Cleimenes of Argus, paralyzed in body. He came to the
Abaton and slept there and saw a vision. It seemed to him
that the god wound a red woolen fillet around his body and led
him for a bath a short distance away from the Temple to a lake

of which the water was exceedingly cold.[7] When he behaved in a cowardly way Asclepius said he would not heal those people who were too cowardly for that, but those who came to him into his Temple, full of hope that he would do no harm to such a man, but would send him away well. When he woke up he took a bath and walked out unhurt.

38.   Diaetus of Cirrha. He happened to be paralyzed in his knees. While sleeping in the Temple he saw a dream. It seemed to him that the god ordered his servants to lift him up and to carry him outside the Adyton and to lay him down in front of the Temple. After they had carried him outside, the god yoked his horses to a chariot and drove three times around him in a circle and trampled on him with his horses and he got control over his knees instantly. When day came he walked out sound.

39.   Agameda of Ceos. She slept in the Temple for offspring and saw a dream. It seemed to her in her sleep that a serpent lay on her belly. And thereupon five children were born to her.

40.   Timon . . . wounded by a spear under his eye. While sleeping in the Temple he saw a dream. It seemed to him that the god rubbed down an herb and poured it into his eye. And he became well.

41.   Erasippe of Caphyiae with worms. She had her stomach swollen and was burning all over [?] and could not keep anything inside. While sleeping in the Temple she saw a dream. It seemed to her that the god massaged her stomach and kissed her[8] and then gave her a vessel which contained a drug and ordered her to drink it and then ordered her to vomit; when she had vomited, her dress was full with it. When day came she saw that her whole dress was full with the evil matter which she had vomited, and thereupon she became well.

42.   Nicasibula of Messene for offspring slept in the Temple and saw a dream. It seemed to her that the god approached her with a snake which was creeping behind him; and with that snake she had intercourse. Within a year she had two sons.

43.   . . . of Cios with gout. While awake he was walking towards a goose who bit his feet and by making him bleed made him well.

[1] Cf. below, No. 23, and T. 422.
[2] Cf. 753, n. 1.
[3] Cf. T. 422.
[4] Cf. T. 317, 11.
[5] Cf. T. 753, n. 1.
[6] Cf. T. 427.
[7] Cf. T. 317, 8.
[8] Cf. T. 446.

424. Inscriptiones Graecae, IV², 1, no. 127 [224 A. D.].

Τιβ. Κλ. Σενῆρος
Σινωπεὺς Ἀπόλ–
λωνι Μαλεάτᾳ καὶ
Σωτῆρι Ἀσκληπιῷ
κατ᾽ ὄναρ, ὃν ὁ θεὸς
εἰάσατο ἐν τῷ ἐν–
κοιμητηρίῳ, χοι–
ράδας ἔχοντα ἐπ[ὶ]
τοῦ τραχή[λου] καὶ
καρκίνον [τ]ο̣[ῦ ὠ̣]τός,
ἐπιστὰς ἐ̣[ν]αργῶς,
οἷος ἐστ – – – –.
ἐπὶ ἱερέω[ς] Μάρ.
Αὐρ. Πυ[θοδώρ]ου
ἔτους ἑ[κ]α̣[το]σ–
στοῦ πρώτου.

Tib. Cl. Severus of Sinope to Apollo Maleatas and the Savior Asclepius according to a dream; him, the god healed in the sleeping hall of the temple, when he had scrofulous swellings in the glands of the neck and an ulcer of the ear, appearing to him clearly, as he is . . . under the priesthood of Marcus Aurelius Pythodorus. In the 101st year.

Pergamum

425. Oribasius, Collectiones Medicae, XLV, 30, 10-14.

Σπασμὸς δ᾽ ἄρα καὶ ἡ ἐπιληψία· ταύτης οὖν τεταρταῖος πυρετὸς ἴαμά ἐστιν, ὥστε ἤν τε ὕστερον ἐπιγένηται, λύεται ἡ ἐπιληψία, ἤν τε πρόσθεν, οὐκ ἂν ἔτι τούτῳ τῷ ἀνθρώπῳ γένοιτο. ὅπως δὲ καὶ Τεύκρῳ τῷ Κυζικηνῷ ἔσχεν, εἰπεῖν ἄξιον· ἐπεὶ γὰρ ἥλω τῇ ἐπιληψίᾳ, ἧκε μὲν εἰς Πέργαμον παρὰ τὸν Ἀσκληπιόν, αἰτῶν λύσιν τῆς νόσου· ὁ δ᾽ αὐτῷ φανεὶς εἰς λόγους ἀφικνεῖται, καὶ ἐρωτᾷ εἰ ἐθέλει τῶν παρόντων ἕτερα ἀλλάξασθαι. καὶ ὃς μάλιστα μὲν οὐκ ἐθέλειν ἔφη, ἀλλά τινα εὐθεῖαν ἀπαλλαγὴν σχεῖν τοῦ κακοῦ· εἰ δ᾽ ἄρα, μὴ χείρω τὰ γενησόμενα εἶναι τῶν παρόντων. φήσαντος δὲ τοῦ θεοῦ ῥᾴω τε ἔσεσθαι καὶ παντὸς ἄλλου σαφέστε–ρον θεραπεύσειν, οὕτω δὲ ὑφίσταται τὴν νόσον, καὶ αὐτῷ ἥκει τεταρταῖος πυρετός, καὶ τὸ ἀπὸ τοῦδε τῆς ἐπιληψίας ἐξάντης γίνεται.

Also epilepsy is a cramp; of this quartan fever, therefore, is a
cure, so that if it supervenes afterwards epilepsy is broken up,
while if it comes previously epilepsy does not befall that man
any more. How it happened to Teucer, the Cyzicenean [*ca.*
100 A. D.], is worth telling : when he was afflicted with epilepsy
he came to Pergamum to Asclepius, asking for liberation from
the disease. The god appearing to him holds converse with him
and asks if he wants to exchange his present disease against
another one.[1] And he said he surely did not want that but
would rather get some immediate relief from the evil. But if
at all, he wished that the future might not be worse than the
present. When the god had said it would be easier and this
would cure him more plainly than anything else, he [*sc.*,
Teucer] consents to the disease, and a quartan fever attacks
him, and thereafter he is free from epilepsy.[2]

[1] Cf. T. 436.
[2] The passage is taken from Rufus, a physician of the 1st cent. A. D.

Lebena

426. Inscriptiones Creticae, I, xvii, no. 9 [2nd c. B. C.].

Δήμανδρον Καλάβιος Γορτύνιον ἰσ[χι-
α]λγικὸν γενόμενον προσέταξε ἀπο[μο-
λὲν ἐς Λεβήναν ὅτι θεραπεύσειν· αἶ[ψα
δ' εὐθόντα ἔταμε καθ' ὕπνον χὺγιὴς ἐ[γέ-
νετο. vac. Φαλάρει Εὐθυχίωνος Λεβη[να-
ίωι οὐ γινομένω τέκνω ἰόντος ἐν π[εντή-
κ]οντα ἤδη ϝέτεθι προσέταξε τὰν γυ[ναῖ-
κα ἐφευδησίονσαν ἀποσστῆλαι καὶ [ἐπ-
ευθ⟨όνσ⟩αν ἐς τὸ ἄδυτον ἐπέθηκε τὰν σικ[ύαν
ἐ]πὶ τὰν γαυτέρα κἠκέλετο ἀπέρπεν [ἐν
τά]χει κἠκύσατο. vac. Κύννιον Σοάρχω Γορ[τύ-
νι]ον τ[ῶ] κοίλω ϝοι ἐσχέζοντος κατὰ τὰν ε[– –
– –

When Demandrus, son of Calabis, of Gortyn, had become
subject to sciatica, he [*sc.*, the god] ordered him to come to
Lebena so as to cure him. As soon as he had arrived there the
god operated on him in his sleep and he became well. – – –
The god ordered Phalaris, the son of Euthychion, of

Lebena, who had no children and was already in his fiftieth year, to send his wife to sleep in the Temple, and when she entered the Adyton he put the cupping instrument on her belly and ordered her to leave in a hurry and she became pregnant. – – –

Cynnius, the son of Soarchus, of Gortyn, who in his abdomen had . . .

ORACLES

Athens

427. Suidas, Lexicon, *s. v. Δομνῖνος.*

. . . ἦν δὲ οὐδὲ τὴν ζωὴν ἄκρος οἷον ἀληθῶς φιλόσοφον εἰπεῖν· ὁ γὰρ Ἀθήνησιν Ἀσκληπιὸς τὴν αὐτὴν ἴασιν ἐχρησμῴδει Πλουτάρχῳ τε τῷ Ἀθηναίῳ καὶ τῷ Σύρῳ Δομνίνῳ, τούτῳ μὲν αἷμ' ἀποπτύοντι πολλάκις καὶ τοῦτο φέροντι τῆς νόσου τὸ ὄνομα, ἐκείνῳ δὲ οὐκ οἶδα ὅ τι νενοσηκότι. ἡ δὲ ἴασις ἦν ἐμπίπλασθαι χοιρείων κρεῶν. ὁ μὲν δὴ Πλούταρχος οὐκ ἠνέσχετο τῆς τοιαύτης ὑγιείας καίτοι οὐκ οὔσης αὐτῷ παρανόμου κατὰ τὰ πάτρια, ἀλλὰ διαναστὰς ἀπὸ τοῦ ὕπνου καὶ διαγκωνισάμενος ἐπὶ τοῦ σκίμποδος ἀποβλέπων εἰς τὸ ἄγαλμα τοῦ Ἀσκληπιοῦ (καὶ γὰρ ἐτύγχανεν ἐγκαθεύδων τῷ προδόμῳ τοῦ ἱεροῦ), 'ὦ δέσποτα,' ἔφη, 'τί δὲ ἂν προσέταξας Ἰουδαίῳ νοσοῦντι ταύτην τὴν νόσον; οὐ γὰρ ἂν καὶ ἐκείνῳ ἐμφορεῖσθαι χοιρείων κρεῶν ἐκέλευσας.' ταῦτα εἶπεν, ὁ δὲ Ἀσκληπιὸς αὐτίκα ἀπὸ τοῦ ἀγάλματος ἐμμελέστατον δή τινα φθόγγον, ἑτέραν ὑπεγράψατο θεραπείαν τῷ πάθει. Δομνῖνος δὲ οὐδὲ κατὰ θέμιν πεισθεὶς τῷ ὀνείρῳ, θέμιν τὴν Σύροις πάτριον, οὐδὲ παραδείγματι τῷ Πλουτάρχῳ χρησάμενος ἔφαγέ τε τότε καὶ ἤσθιεν ἀεὶ τῶν κρεῶν. λέγεταί που μίαν εἰ διέλειπεν ἡμέραν ἄγευστος ἐπιτίθεσθαι τὸ πάθημα πάντως, ἕως ἐνεπλήσθη.

. . . [Domninus, 5th c. A. D.] was not perfect in his manner of life, so as to call him a true philosopher. For the Asclepius at Athens revealed the same cure for Plutarch the Athenian [*ca.* 400 A. D.] and for Domninus the Syrian, of whom the latter continually coughed up blood and had the sickness of this name, the former was ill with some disease, I know not what. The treatment prescribed was to keep sated with pork. Plutarch could not abide the health thus acquired although it was not contrary to his ancestral laws, but rising up from the dream and leaning on his elbow on the couch, he looked at the statue of Asclepius (for he happened to be sleeping in the vestibule of the shrine) and said, "My lord, what would you have prescribed to a Jew suffering this same illness, for certainly you would not bid him to take his fill of pork." [1] Thus he spoke, and straightway Asclepius spoke from the statue in a very harmonious voice, prescribing another remedy for the illness. But Domninus, trusting the dream, even if it was not in accordance with the law—the ancestral law of the Syrians—and not availing himself of the example of Plutarch, ate of the meat at that time and ever after. It is said somewhere that if he omitted one day, fasting from the meat, the illness unfailingly returned, until he was sated with pork again.

[1] Cf. T. 423, 36; cf. also T. 433.

428. Inscriptiones Graecae, II², no. 4514 [2nd c. A. D.].

Διοφάντου Σφηττίου.
Τάδε σοὶ ζάκορος φίλιος λέγω
'Ασκληπιέ, Λητοΐδου πάϊ·
πῶς χρύσεον ἐς δόμον ἵξομαι
τὸν σόν, μάκαρ ὦ πεποθημένε,
θεία κεφαλά, πόδας οὐκ ἔχων,
οἷς τὸ πρὶν ἐς ἱερὸν ἤλυθον,
εἰ μὴ σὺ πρόφρων ἐθέλοις ἐμὲ
ἰασάμενος ⟨π⟩άλιν εἰσάγειν,
ὅππως σ' ἐσίδω, τὸν ἐμὸν θεόν,
τὸν φαιδρότερον χθονὸς εἰαρινᾶς;

Τάδε σοὶ Διόφαντος ἐπεύχομαι·
σῶσόν με, μάκαρ, σθεναρώτατε,
ἰασάμενος ποδάγραν κακήν,
πρὸς σοῦ πατρός, ὧι μεγάλ' εὔχομαι·
οὐ γάρ τις ἐπιχθονίων βροτῶν
τοιῶνδε πόροι λύσιν ἀλγέων.
μόνος εἶ σύ, μάκαρ θεῖε, σθένων·
σὲ γὰρ θεοὶ οἱ πανυπείροχοι
δῶρον μέγα, τὸν φιλελήμονα,
θνητοῖς ἔπορον, λύσιν ἀλγέων.

[Τρισμ]άκαρ, ὦ Παιὰν 'Ασκληπιέ, σῆς ὑπὸ τέχνης
[ἰα]θεὶς Διόφαντος ἀνίατον κακὸν ἕλκος

οὐκέτι καρκινόπους ἐσορώμενος οὐδ' ἐπ' ἀκάνθας
ὡς ἀγρίας βαίνων, ἀλλ' ἀρτίπος, ὥσπερ ὑπέστης.

[Dedicated by] Diophantus of Sphettus.

I, a beloved temple attendant, say these things to you, Asclepius, son of Leto's child. How shall I come to your golden abode, O blessed longed-for, divine head, since I do not have the feet with which I formerly came to the shrine, unless by healing me you graciously wish to lead me there again so that I may look upon you, my god, brighter than the earth in springtime.

So I, Diophantus, pray you, save me, most powerful and blessed one, by healing my painful gout: in the name of your father, to whom I offer earnest prayer. For no mortal man may give release from such sufferings. Only you, blessed divine one, have the power. For the gods who are eminent above all gave you to mortal men as a great gift, the compassionate one, the deliverance from sufferings.

Thrice-blessed Paeon Asclepius, by your skill Diophantus was healed of his painful incurable ailment; no longer does he appear crab-footed nor as if walking on cruel thorns, but sound of foot, just as you promised.

Epidaurus

429. Plautus, Curculio, I, 1, 61-62.

> . . . hic leno, hic qui aegrotus incubat
> in Aesculapi fano . . .

> . . . the pimp, who's ill and taking the cure in the shrine
> of Asclepius here . . .[1]

[1] Cf. also **429a.** *ibid.* V, 3, 699: Aesculapio huic habeto . . . gratiam; cf. also T. **738; 741.**

430. Plautus, Curculio, II, 1, 216–2, 273.

CAPPADOX. Migrare certumst iam nunc e fano foras,
quando Aesculapi ita sentio sententiam,
ut qui me nihili faciat nec salvom velit.
valetudo decrescit, adcrescit labor;
220 nam iam quasi zona liene cinctus ambulo,

geminos in ventre habere videor filios.
nil metuo nisi ne medius disrumpar miser.

PALINURUS. Si recte facias, Phaedrome, auscultes mihi
atque istam exturbes ex animo aegritudinem.
paves, parasitus quia non rediit Caria.
adferre argentum credo; nam si non ferat,
tormento non retineri potuit ferreo,
quin reciperet se huc esum ad praesepem suam.

CA. Quis hic est qui loquitur?

PAL.   Quoiam vocem ego audio?

CA. Estne hic Palinurus Phaedromi?

230 PAL.   Quis hic est homo
cum collativo ventre atque oculis herbeis?
de forma novi, de colore non queo
novisse. iam iam novi: leno est Cappadox.
congrediar.

CA.   Salve, Palinure.

PAL.   O scelerum caput,
salveto. quid agis?

CA.   Vivo.

PAL.   Nempe ut dignus es.
sed quid tibi est?

CA.   Lien enecat, renes dolent,
pulmones distrahuntur, cruciatur iecur,
radices cordis pereunt, hirae omnes dolent.

PAL. Tum te igitur morbus agitat hepatiarius.

CA. Facile est miserum inridere.

240 PAL.   Quin tu aliquot dies
perdura, dum intestina exputescunt tibi,
nunc dum salsura sat bonast. si id feceris,
venire poteris intestinis vilius.

CA. Lien dierectust.

PAL.   Ambula, id lieni optumumst.

CA. Aufer istaec, quaeso, atque hoc responde quod rogo.
potin coniecturam facere, si narrem tibi
hac nocte quod ego somniavi dormiens?

PAL. Vah, solus hic homost qui sciat divinitus.
quin coniectores a me consilium petunt;

250 quod eis respondi, ea omnes stant sententia.

COCUS. Palinure, quid stas? quin depromuntur mihi

quae opus sunt, parasito ut sit paratum prandium,
quom veniat?
PAL.   Mane sis, dum huic conicio somnium.
COC.   Tute ipse, si quid somniasti, ad me refers.
PAL.   Fateor.
COC.   Abi, deprome.
PAL.   Age tu interea huic somnium
narra, meliorem quam ego sum suppono tibi.
nam quod scio omne ex hoc scio.
CA.   Operam ut det.
PAL.   Dabit.
CA.   Facit hic quod pauci, ut sit magistro obsequens.
da mi igitur operam.
COC.   Tam etsi non novi, dabo.
260   CA.   Hac nocte in somnis visus sum viderier
procul sedere longe a me Aesculapium,
neque eum ad me adire neque me magni pendere
visumst.
COC.   Item alios deos facturos scilicet;
sane illi inter se congruont concorditer.
nihil est mirandum, melius si nil fit tibi,
namque incubare satius te fuerat Iovi,
qui tibi auxilio in iure iurando fuit.
CA.   Siquidem incubare velint qui periuraverint,
locus non praeberi potis est in Capitolio.
270   COC.   Hoc animum advorte: pacem ab Aesculapio
petas, ne forte tibi evenat magnum malum,
quod in quiete tibi portentumst.
CA.   Bene facis.
ibo atque orabo.

CA.   Yes, I am resolved to quit the temple this moment,
since I see for sure that Asclepius cares nothing for
me, has no wish to cure me. My strength is decreasing
and my pain is increasing. Why, already my spleen is
wound around me like a girdle as I walk along—
anyone would think I was carrying twins. Oh dear!
All I am afraid of is that I shall blow up in the middle.
PAL.   (to Phaedromus) You'd do well to listen to me, sir,
and shake off that doleful spirit of yours. You're
panic-struck just because the parasite hasn't got back

from Caria! He's bringing the money, I reckon. For otherwise he couldn't be kept by fetters of iron from hying himself back here to eat at his own manger.

CA. Who is that talking here?

PAL. Whose voice is that?

CA. Palinurus, is it, Phaedromus's man?

PAL. Who's that fellow with the comprehensive belly and the grass-green eyes? His figure looks familiar, but I don't recognize that color scheme. Now! Now I recognize him! It's the pimp, Cappadox. I'll up to him.

CA. Good day, Palinurus.

PAL. Ah there, you fount of iniquity! Good day to you. How are you?

CA. Living.

PAL. As you deserve, no doubt. What ails you, though?

CA. My spleen is killing me, my kidneys ache, my lungs are torn to tatters, my liver is in agony, my heart-strings are clean gone, and all my small intestines pain me.

PAL. Ah, then you must be suffering from some hepatic affection.

CA. It is easy to laugh at a poor wretch.

PAL. I say, hold out for a few days longer while your intestines go rotten, now while the pickling is good enough. You do this, and you can sell your intestines for more than your whole carcass.

CA. My spleen is racked.

PAL. Take walks—best thing in the world for the spleen.

CA. For mercy's sake, drop your joking and do answer me this. Supposing I told you a dream I had when I was asleep last night, could you interpret it?[1]

PAL. Hah! Why, here's your one and only expert at divination. Man alive, professional clairvoyants come to me for advice, and the answers I give 'em they all abide by.

COOK. Palinurus! What are you standing still for? Why don't you fetch the things I need, so that lunch will be prepared for the parasite when he appears?

PAL. You just kindly wait till I interpret a dream for this chap.

COOK. You! Why, you yourself refer all your dreams to me.

PAL. Admitted.

COOK. Be off; fetch the stuff.

PAL. (to Cappadox) Here, you! Meanwhile you tell your dream to this fellow. I leave you to my substitute—a better man than I am. Why, all I know I owe to him.

CA. If he would only help me.

PAL. He will.

CA. He [sc., Palinurus] does what few do, in letting his teacher have his way. Well, then, you help me.

COOK. I don't know you, but help you I will.

CA. Last night in my sleep I seemed to see Asclepius sitting a long way off from me, and he seemed not to come near me or to think much of me.

COOK. That means the other gods will do the same; they pull together perfectly, you know. No wonder you get no better; why, the thing for you to do was to lie in the temple of Jove, the god that's been your backer in those solemn oaths of yours.

CA. But if all the perjurers wanted to lie there, they could not find accommodations in the Capitol.

COOK. Mark my words now—go sue Asclepius for grace, or you may chance to meet with the dreadful disaster your dream portended.

CA. Thanks! Thanks! I'll go in and pray.

¹ Cf. T. 416.

431. Inscriptiones Graecae, IV², 1, no. 125 [3rd c. B. C.].

Ἑρμόδικ[ος Λαμψακ]ηνός
σῆς ἀρετῆς [παράδειγμ]', Ἀσκληπιέ, | τόνδε ἀνέ[θηκα
π]έτρον ἀειρά|μενος, πᾶσι[ν ὁρᾶν] φανερόν, ||
ὄψιν σῆς τέχνης· πρὶν γὰρ | σὰς εἰς χέρας ἐλθεῖν |
σῶν τε τέκνων κεῖμαι | νούσου ὑπὸ στυγερᾶς |
ἔμπυος ὢν στῆθος χει||ρῶν τε ἀκρατής· σὺ δέ, | Παιάν,
πείσας με ἄρασθαι | τόνδε, ἄνοσον διάγειν.

Hermodicus of Lampsacus

As an example of your power, Asclepius, I have put up this rock which I had lifted up, manifest for all to see, an evidence of your art. For before coming under your hands and those of your children I was stricken by a wretched illness, having an abscess in my chest and being paralyzed in my hands. But you, Paean, by ordering me to lift up this rock[1] made me live free from disease.

[1] Cf. T. 423. 15.

432. Inscriptiones Graecae, IV², 1, no. 126 [*ca.* 160 A. D.].

Ἐπὶ ἱερέως Π. Αἰλ. Ἀντιόχου

Μ̄. Ἰούλιος Ἀπελλᾶς Ἰδριεὺς Μυλασεὺς μετεπέμφθην
ὑπὸ τοῦ θεοῦ, πολλάκις εἰς νόσους ἐνπίπτων καὶ ἀπεψί-
αις χρώμενος. κατὰ δὴ τὸν πλοῦν ἐν Αἰγείνῃ ἐκέλευσέν
με μὴ πολλὰ ὀργίζεσθαι. ἐπεὶ δὲ ἐγενόμην ἐν τῷ ἱερῷ, ἐ-
κέλευσεν ἐπὶ δύο ἡμέρας συνκαλύψασθαι τὴν κεφαλήν,
ἐν αἷς ὄμβροι ἐγένοντο, τυρὸν καὶ ἄρτον προλαβεῖν, σέλει-
να μετὰ θρίδακος, αὐτὸν δι' αὐτοῦ λοῦσθαι, δρόμῳ γυμνάζε-
σθαι, κιτρίου προλαμβάνειν τὰ ἄκρα, εἰς ὕδωρ ἀποβρέξαι, πρὸς
ταῖς ἀκοαῖς ἐν βανείῳ προστρίβεσθαι τῷ τοίχωι, περιπάτῳ χρῆσθαι ὑπερῴῳ, αἰώραις, ἀφῇ πηλώσασθαι, ἀνυπόδητον περι-
πατεῖν, πρὶν ἐνβῆναι ἐν τῶι βαλανείῳ εἰς τὸ θερμὸν ὕδωρ
οἶνον περιχέασθαι, μόνον λούσασθαι καὶ Ἀττικὴν δοῦναι
τῶι βαλανεῖ, κοινῇ θῦσαι Ἀσκληπιῷ, Ἠπιόνῃ, Ἐλευσεινίαις,
γάλα μετὰ μέλιτος προλαβεῖν· μιᾷ δὲ ἡμέρᾳ πιόντός μου γάλα μόνον, εἶπεν· 'μέλι ἔμβαλλε εἰς τὸ γάλα, ἵνα δύνηται διακό-
πτειν.' ἐπεὶ δὲ ἐδεήθην τοῦ θεοῦ θᾶττόν με ἀπολῦσαι, ᾤμην (ν)ά-
πϋϊ καὶ ἁλσὶν κεχρειμένος ὅλος ἐξιέναι κατὰ τὰς ἀκοὰς ἐκ τοῦ
ἀβάτου, παιδάριον δὲ ἡγεῖσθαι θυμιατήριον ἔχον ἀτμίζον
καὶ τὸν ἱερέα λέγειν 'τεθεράπευσαι, χρὴ δὲ ἀποδιδόναι τὰ ἴατρα.'
καὶ ἐποίησα, ἃ εἶδον, καὶ χρείμενος μὲν τοῖς ἁλσὶ καὶ τῶι νάπυ-
ϊ ὑγρῶι ἤλγησα, λούμενος δὲ οὐκ ἤλγησα. ταῦτα ἐν ἐννέα ἡμέραις ἀφ' οὗ ἦλθον. ἥψατο δέ μου καὶ τῆς δεξιᾶς χιρὸς καὶ τυῦ
μαστοῦ, τῇ δὲ ἑξῆς ἡμέρᾳ ἐπιθύοντός μου φλὸξ ἀναδραμοῦ-
σα ἐπέφλευσε τὴν χεῖρα, ὡς καὶ φλυκταίνας ἐξανθῆσαι· μετ' ὀ-
λίγον δὲ ὑγιὴς ἡ χεὶρ ἐγένετο. ἐπιμείναντί μοι ἄνηθον με-
τ' ἐλαίου χρήσασθαι πρὸς τὴν κεφαλαλγίαν εἶπεν. οὐ μὴν ἤλ-
γουν τὴν κεφαλήν. συνέβη οὖν φιλολογήσαντί μοι συνπλη-
ρωθῆναι· χρησάμενος τῷ ἐλαίῳ ἀπηλάγην τῆς κεφαλαλγί-
ας. ἀναγαργαρίζεσθαι ψυχρῷ πρὸς τὴν σταφυλὴν—καὶ γὰρ περὶ
τούτου παρεκάλεσα τὸν θεὸν—τὸ αὐτὸ καὶ πρὸς παρίσθμια. ἐκέλευσεν δὲ καὶ ἀναγράψαι ταῦτα. χάριν εἰδὼς καὶ ὑγιὴς γενόμενος ἀπηλλάγην.

In the priesthood of Poplius Aelius Antiochus

I, Marcus Julius Apellas, an Idrian from Mylasa, was sent for by the god, for I was often falling into sickness and was suffering from dyspepsia. In the course of my journey, in Aegina,[1] the god told me not to be so irritable. When I arrived at the temple, he told me for two days to keep my head covered, and for these two days it rained; to eat cheese and bread, celery with lettuce, to wash myself without help, to practise running, to take lemonpeels, to soak them in water, near the (spot of the) *akoai* in the bath to press against the wall, to take a walk in the upper portico, to take some passive exercise, to sprinkle myself with sand, to walk around barefoot, in the bathroom, before plunging into the hot water, to pour wine over myself, to bathe without help and to give an Attic drachma to the bath attendant, in common to offer sacrifice to Asclepius, Epione and the Eleusinian goddesses,[2] to take milk with honey. When one day I had drunk milk alone he said, " Put honey in the milk so that it can get through." When I asked of the god to relieve me more quickly I thought I walked out of the abaton near the (spot of the) *akoai* being anointed all over with mustard and salt, while a small boy was leading me holding a smoking censer, and the priest said: " You are cured but you must pay up the thank-offerings." And I did what I had seen, and when I anointed myself with the salts and the moistened mustard I felt pains, but when I bathed I had no pain. That happened within nine days after I had come. He touched my right hand and also my breast. The following day as I was offering sacrifice the flame leapt up and scorched my hand, so that blisters appeared. Yet after a little the hand got well. As I stayed on he said I should use dill along with olive oil against my headaches. I usually did not suffer from headaches. But it happened that after I had studied, my head was congested. After I used the olive oil I got rid of the headache. To gargle with a cold gargle for the uvula—since about that too I had consulted the god—and the same also for the tonsils. He bade me also inscribe this. Full of gratitude I departed well.

[1] Cf. T. 733-34.
[2] T. 519.

Pergamum

433. Philostratus, Vitae Sophistarum, I, 25, 4.

Ἥκων δὲ ἐς τὸ Πέργαμον, ὅτε δὴ τὰ ἄρθρα ἐνόσει, κατέδαρθε μὲν ἐν τῷ ἱερῷ, ἐπιστάντος δὲ αὐτῷ τοῦ Ἀσκληπιοῦ καὶ προειπόντος ἀπέχεσθαι ψυχροῦ ποτοῦ ὁ Πολέμων " βέλτιστε," εἶπεν, " εἰ δὲ βοῦν ἐθεράπευες ";

Again, when he [sc. Polemo, ca. 100 A. D.] came to Pergamum suffering from a disease of the joints, he slept in the temple, and when Asclepius appeared to him and told him to abstain from drinking anything cold, " My good sir," said Polemo,[1] " but what if you were doctoring a cow? "[2]

[1] Cf. T. 807.　　　　　[2] Cf. T. 378; 466; cf. also 427.

434. Philostratus, Vitae Sophistarum, II, 25, 5.

Ἐπεὶ δὲ ἐστί μοι προστεταγμένον ὑπὸ τοῦ κατὰ τὸ Πέργαμον Ἀσκληπιοῦ πέρδικα σιτεῖσθαι λιβανωτῷ θυμιώμενον, τὸ δὲ ἄρωμα τοῦτο οὕτω τι σπανιστὸν καθ᾽ ἡμᾶς νῦν, ὡς ψαιστὸν καὶ δάφνης φύλλα τοῖς θεοῖς θυμιᾶσθαι, δέομαι λιβανωτοῦ ταλάντων πεντήκοντα, ἵνα θεραπεύοιμι μὲν τοὺς θεούς, θεραπευοίμην δὲ αὐτός.

However, I [Hermocrates, 2nd c. A. D.] have been ordered by Asclepius of Pergamum to eat partridge stuffed with frankincense, and this seasoning is now so scarce in our country that we have to use barley meal and laurel leaves for incense to the gods. I therefore ask for fifty talents' worth of frankincense, that I may treat the gods properly and get proper treatment myself.

435. Aristides, Oratio L, 17.

. . . καὶ ἅμα διηγεῖταί μοι ἔργον τοῦ θεοῦ θαυμαστόν, ὥς τινι κάμνοντι προστάξας οὕτω διαγωνίσασθαι συμβάντος ἱδρῶτος δι᾽ ἀγωνίαν λύσειεν τὸ νόσημα πᾶν.

. . . and at the same time he [sc., Sedatus] describes to me the marvellous work of the god—how he instructed a certain sick man thus to struggle hard and when perspiration resulted from the struggle, he broke up the entire sickness.

436. Galenus, Subfiguratio Empirica, Cp. X, p. 78 [Deichgräber].

Ἄλλος δέ τις ἀνὴρ πλούσιος οὐχ ἡμεδαπὸς οὗτός γε, ἀλλ' ἐκ
μέσης Θρᾴκης ἧκεν ὀνείρατος προτρέψαντος αὐτὸν εἰς τὴν
Πέργαμον· εἶτα τοῦ θεοῦ προστάξαντος ὄναρ αὐτῷ πίνειν τε τοῦ
διὰ τῶν ἐχιδνῶν φαρμάκου καθ' ἑκάστην ἡμέραν καὶ χρίειν ἔξωθεν
τὸ σῶμα μετέπεσε τὸ πάθος οὐ μετὰ πολλὰς ἡμέρας εἰς λέπραν
ἐθεραπεύθη τε πάλιν οἷς ὁ θεὸς ἐκέλευσε φαρμάκοις καὶ τοῦτο τὸ
νόσημα.

Another wealthy man, this one not a native but from the
interior of Thrace, came, because a dream had driven him, to
Pergamum. Then a dream appeared to him, the god prescribing that he should drink every day of the drug produced
from the vipers and should anoint the body from the outside.
The disease after a few days turned into leprosy; and this
disease, in turn, was cured by the drugs which the god
commanded.[1]

[1] Cf. T. 425.

437. Herodianus, Ab Excessu Divi Marci, IV, 8, 3.

Ταῦτα δὲ ποιήσας . . . ἐπείχθη ἐς Πέργαμον τῆς Ἀσίας, χρήσασθαι βουλόμενος θεραπείαις τοῦ Ἀσκληπιοῦ. ἀφικόμενος δὴ ἐκεῖ,
καὶ ἐς ὅσον ἤθελε τῶν ὀνειράτων ἐμφορηθείς, ἧκεν ἐς Ἴλιον.

When he [sc., Caracalla] had accomplished this . . . he hurried
to Pergamum in Asia Minor, desiring to avail himself of the
treatment of Asclepius. Arriving there and having taken his
fill of the dreams as extensively as he wished, he went to Ilium.

Rome

438. Inscriptiones Graecae, XIV, no. 966 [2nd c. A. D.].[1]

Αὐταῖς ταῖς ἡμέραις Γαΐῳ τινὶ τυφλῷ ἐχρημάτισεν ἐλθεῖν ἐπ[ὶ
τὸ] ἱε|ρὸν βῆμα καὶ προσκυνῆσαι, εἶ[τ]α ἀπὸ τοῦ δεξιοῦ ἐλθεῖν
ἐπὶ τὸ ἀριστερὸν | καὶ θεῖναι τοὺς πέντε δακτύλους ἐπάνω τοῦ
βήματος καὶ ἆραι τὴν χεῖ|ρα καὶ ἐπιθεῖναι ἐπὶ τοὺς ἰδίους
ὀφθαλμούς· καὶ ὀρθὸν ἀνέβλεψε τοῦ | δήμου παρεστῶτος καὶ
συνχαιρομένου, ὅτι ζῶσαι ἀρεταὶ ἐγένοντο ἐπὶ | τοῦ Σεβαστοῦ
ἡμῶν Ἀντωνείνου. |
Λουκίῳ πλευρειτικῷ καὶ ἀφηλπισμένῳ ὑπὸ παντὸς ἀνθρώπου
ἐχρησμάτι|σεν ὁ θεὸς ἐλθεῖν καὶ ἐκ τοῦ τριβώμου ἆραι τέφραν

καὶ μετ' οἴνου ἀνα|φυρᾶσαι καὶ ἐπιθεῖναι ἐπὶ τὸ πλευρόν· καὶ
ἐσώθη καὶ δημοσίᾳ ηὐχαρίστησεν | τῷ θεῷ καὶ ὁ δῆμος συνεχάρη
αὐτῷ. |

αἷμα ἀναφέροντι Ἰουλιανῷ, ἀφηλπισμένῳ ὑπὸ παντὸς ἀνθρώπου
ἐχρησμά|τισεν ὁ θεὸς ἐλθεῖν καὶ ἐκ τοῦ τριβώμου ἆραι κόκκους
στροβίλου καὶ | φαγεῖν μετὰ μέλιτος ἐπὶ τρεῖς ἡμέρας· καὶ
ἐσώθη καὶ ἐλθὼν δημοσίᾳ | ηὐχαρίστησεν ἔμπροσθεν τοῦ δήμου. |

Οὐαλερίῳ Ἄπρῳ στρατιώτῃ τυφλῷ ἐχρημάτισεν ὁ θεὸς ἐλθεῖν
καὶ λαβεῖν αἷμα | ἐξ ἀλεκτρυῶνος λευκοῦ μετὰ μέλιτος καὶ κολ-
λύριο[ν] συντρῖψαι καὶ ἐπὶ | τρεῖς ἡμέρας ἐπιχρεῖσαι ἐπὶ τοὺς
ὀφθαλμούς· καὶ ἀνέβλεψεν καὶ ἐλήλυθεν | καὶ ηὐχαρίστησεν
δημοσίᾳ τῷ θεῷ.

In those days he [*sc.*, the god] revealed to Gaius, a blind
man, that he should go to the holy base [*sc.*, of the statue] [2]
and there should prostrate himself; then go from the right to
the left and place his five fingers on the base and raise his
hand and lay it on his own eyes. And he could see again clearly
while the people stood by and rejoiced that glorious deeds
lived again under our Emperor Antoninus.

To Lucius who suffered from pleurisy and had been despaired of by all men [3] the god revealed that he should go and
from the threefold altar lift ashes [4] and mix them thoroughly
with wine and lay them on his side. And he was saved and
publicly offered thanks to the god, and the people rejoiced
with him.

To Julian who was spitting up blood and had been despaired
of by all men [3] the god revealed that he should go and from
the threefold altar take the seeds of a pine cone and eat them
with honey for three days. And he was saved and went and
publicly offered thanks before the people.

To Valerius Aper, a blind soldier, the god revealed that he
should go and take the blood of a white cock along with honey
and compound an eye salve and for three days should apply it
to his eyes. And he could see again and went and publicly
offered thanks to the god.

[1] Cf. also Dittenberger, Sylloge[3], III, 1173.
[2] Cf. T. 451: Ἀσκληπιὸς . . . ἐπὶ βάσεως ὁρώμενος καὶ προσκυνούμενος (βῆμα is usually
translated: altar).
[3] Cf. T. 422; 582; 585.
[4] Cf. T. 439.

Lebena

439. Inscriptiones Creticae, I, xvii, no. 17 [1st c. B. C.].

’Ασκληπιῶ[ι
Πόπλιος Γράνιος ['Ροῦφος
κατ᾽ ἐπιταγήν.

ἐκ διετίας βήσσοντά με ἀδ[ιαλεί-
πτως, ὥστε σάρκας ἐνπύου[ς καὶ
ἠμαγμένας δι᾽ ὅλης ἡμέρας ἀ[πο-
βάλλειν, vac. ὁ θεὸς ἐπεδέξατο θερ[α-
πεῦσαι.   vac.
ἔδωκεν εὔζωμον νήστη τρώγειν,
εἶτα πεπερᾶτον Ἰταλικὸν πείνειν,
πάλιν ἄμυλον διὰ θερμοῦ ὕδατος,
εἶτα κονίαν ἀπὸ τῆς ἱερᾶς σποδοῦ
καὶ τοῦ ἱεροῦ ὕδατος, εἶτα ῷὸν καὶ
ῥητείνην, πάλιν πίσσαν ὑγράν,
εἶτα εἴρην μετὰ μέλιτος, εἶτα μῆλον
Κυδώ[νιον κ]αὶ πεπ[λ]ίδα συνεψή-
σαντα τὸ μὲν χύμα πεί]νειν τὸ δὲ μῆλον
τρώγειν, εἶτα τρώγει]ν σῦκα μετὰ σποδοῦ ἱερᾶς τῆς ἐκ τοῦ] βωμοῦ ὅπου θύουσι τῷ θεῷ.]   vac.
– – – – – – – – – – ἀπὸ τῆς ἐν τῷ δε-
– – – – – – – – – – –]ς πολὺ αἷμα . [– – –
– – – – – – – – – – –] οὖντα [ἱ]κέ[την – – –
– –

To Asclepius
Poplius Granius Rufus
according to command.

When for two years I had coughed incessantly so that I dis-
charged purulent and bloody pieces of flesh all day long, the
god took in hand to cure me. . . . He gave me rocket to nibble
on an empty stomach, then Italian wine flavored with pepper to
drink, then again starch with hot water, then powder of the
holy ashes[1] and some holy water, then an egg and pine-resin,
then again moist pitch, then iris with honey, then a quince and
a wild purslane to be boiled together—the fluid to be drunk,

while the quince was to be eaten—then to eat a fig with holy
ashes taken from the altar where they sacrifice to the god. – – –

¹ Cf. T. 438.

440. Inscriptiones Creticae, I, xvii, no. 18 [1st c. B. C.].

’Ασκληπιῶι
Πόπλιος Γράνιος ῾Ροῦφο[ς
κα]τ’ ἐπιτα[γήν.

τ[οῦ δεξ]ιοῦ ὤμου χ[. . . . .]κους
κα[ὶ . . .]ντος καὶ σύμ[παντο]ς ἀπο
σ[. . . . .]ο . ου ἀφορήτου[ς δόντος] ἀ[λ-
γηδόνας ὁ [θε]ὸς ἐκέλευσέν με π[ροσ-
καρτερεῖν κ[αὶ ἔδ]ωκεν θεραπείαν·
ἄλευρον κρ[ίθινο]ν μετὰ παλαιοῦ οἴ[νου
καταπλάσα[ντα κα]ὶ στρόβειλον λε[οτρι-
βήσαντα μ[ετ’ ἐλαίο]υ ἐπιθεῖναι, ὁμ[οῦ δὲ
σῦκον καὶ σ[τέαρ τρά?]γειον, εἶτα θήν[ιον,
πέπερι, κηρό[πισσον?] καὶ ἔλαιον συ[νεψή-
σαντα ωσ[. . . . . . .]ν μαλακῷ ο . . . .
καταραψα[. . . . . . . τ]οῦ θώρακο[ς . . . .
τρεασῃ. [– – – – τ]ῆς σμύρνα[ς . . . .
κ . . ηναι [– – – – ἔλ]αιον ἀπο τ . . . . .
των λυχ[ν – – – –]ω ἡλίῳ τω . . . . .
τ]οῦ μύρ[– – – –]ς ἀπεθε – – – – –

To Asclepius
Poplius Granius Rufus
according to command.

My right shoulder – – – and – – – and the whole from – – –
giving me unendurable pains, the god ordered me to be confi-
dent and gave me relief. I should apply a plaster of barley-meal
mixed with old wine and of a pine cone ground down with
olive oil, and at the same time a fig and goat's fat, then milk
with pepper, wax-pitch and olive oil boiled together – – –

441. Inscriptiones Creticae, I, xvii, no. 19 [2nd-1st c. B. C.].

– – – – – – – – – – – – – – – – – – – –
– – – – – – – – ασα – – – – – – – – – – –
– – – – – κεφαλὴν κα[ὶ – – – – – – – – –

..εὐ[χ]αριστεῖ Ἀσκληπιὸ[ν Σωτῆρα λαβοῦσα
ἐπὶ τοῦ μεικροῦ δακτύλο[υ ἕλκωσίν τινα
ἀ]γρίαν καὶ θεραπευθεῖσ[α, τοῦ θεοῦ ἐπιτά-
ξαντος ἐπιθεῖναι ὀστ[ρέου τὸ ὄστρακον
κατακαύσασαν καὶ λεο[τριβήσασαν μετὰ
ῥοδίνου καὶ μολόχῃ μ[ετ᾽ ἐλαίου χρίσασ-
θαι· καὶ οὕτως ἐθεράπ[ευσεν. ἰδοῦσαν
δέ με πλείονας ἀρετὰ[ς τοῦ θεοῦ καθ᾽ ὕπνον [ἀν]αγράφειν ὁ θεὸ[ς ἐκέλευσε τὰς
ὄψ]εις ọ....σι ταῖς. – – – – – – – – –
..περ . . . . δακτυ[λο – – – – – καθ᾽ ὕ-
πν]ον ἐπιτάξαντο[ς τοῦ θεοῦ – – – – – –
..... ρον ἀπὸ δυο – – – – – – – – – – –

– – – (a certain woman) – – – at the head and – – – gives
thanks to Asclepius the Savior; having suffered from a malignant sore on her little finger she was healed by the god
who ordered her to apply the shell of an oyster, burnt and
ground down by her with rose-ointment, and to anoint [sc.,
her finger] with mallow, mixed with olive oil. And thus he
cured her. After I had seen many more glorious deeds of the god
in my sleep the god ordered me to inscribe my visions – – –
in my sleep the god ordered – – –

442. Inscriptiones Creticae, I, xvii, no. 24 [3rd c. A. D.].[1]

Δοιούς σοι Διόδω|ρος ἐθήκατο, Σῶτερ, | Ὀνείρους |
ἀντὶ διπλῶν ὄσσων | φωτὸς ἐπαυράμενος.

Diodorus dedicated two statues of the Dream-God, Savior,
for his two eyes, since he now enjoys the light of day.[2]

[1] = Epigrammata Graeca, 839 [Kaibel].
[2] Cf. T. 455; 520-22.

EPIPHANIES AND DREAMS
IN OTHER PLACES

443. Statius, Silvae, III, 4, 65-71.

Olim etiam, ne prima genas lanugo nitentes
carperet et pulchrae fuscaret gratia formae,
ipse deus patriae celsam trans aequora liquit
Pergamon. haud ulli puerum mollire potestas

credita, sed tacita iuvenis Phoebeius arte
leniter haud ullo concussum vulnere corpus
de sexu transire iubet.

Once, lest the first down should spoil thy [sc., Euarinus']
radiant cheeks and the charm of thy comeliness be darkened,
the god of thy land left his lofty Pergamum and crossed the
sea. To none else was trusted the power to unman the lad,
but the son of Phoebus with quiet skill gently bids his body
lose its sex, unmarred by any wound. (*)

444. Pausanias, Descriptio Graeciae, X, 38, 13.[1]

Τοῦ δὲ Ἀσκληπιοῦ τὸ ἱερὸν . . . ἐξ ἀρχῆς . . . ᾠκοδόμησεν
αὐτὸ ἀνὴρ ἰδιώτης Φαλύσιος. νοσήσαντι γάρ οἱ τοὺς ὀφθαλμοὺς
καὶ οὐ πολὺ ἀποδέον τυφλῷ ὁ ἐν Ἐπιδαύρῳ πέμπει θεὸς Ἀνύτην
τὴν ποιήσασαν τὰ ἔπη φέρουσαν σεσημασμένην δέλτον. τοῦτο
ἐφάνη τῇ γυναικὶ ὄψις ὀνείρατος, ὕπαρ μέντοι ἦν αὐτίκα. καὶ
εὗρέ τε ἐν ταῖς χερσὶ ταῖς αὑτῆς σεσημασμένην δέλτον, καὶ
πλεύσασα ἐς τὴν Ναύπακτον ἐκέλευσεν ἀφελόντα τὴν σφραγῖδα
Φαλύσιον ἐπιλέγεσθαι τὰ γεγραμμένα. τῷ δὲ ἄλλως μὲν οὐ δυνατὰ
ἐφαίνετο ἰδεῖν τὰ γράμματα ἔχοντι οὕτω τῶν ὀφθαλμῶν. ἐλπίζων
δέ τι ἐκ τοῦ Ἀσκληπιοῦ χρηστὸν ἀφαιρεῖ τὴν σφραγῖδα, καὶ
ἰδὼν ἐς τὸν κηρὸν ὑγιής τε ἦν, καὶ δίδωσι τῇ Ἀνύτῃ τὸ ἐν τῇ
δέλτῳ γεγραμμένον, στατῆρας δισχιλίους χρυσοῦ.

The sanctuary of Asclepius [sc., at Naupactus] was . . . originally built by a private person called Phalysius. For he had a
complaint of the eyes, and when he was almost blind the god
at Epidaurus sent to him the poetess Anyte [ca. 300 B.C.],
who brought with her a sealed tablet. This appeared to the
woman as the vision of a dream, but it proved at once to be a
waking reality. For she found in her own hands a sealed
tablet; so sailing to Naupactus she bade Phalysius take away
the seal and read what was written. He did not think it
possible under other circumstances to read the writing with his
eyes in such a condition, but hoping to get some benefit from
Asclepius he took away the seal, and when he looked at the
wax he was sound, and gave to Anyte what was written on the
tablet, two thousand staters of gold. (*)

[1] Cf. T. 717.

445. Marinus, Vita Procli, Cp. 30.

Τήν γε μὴν περὶ τὸν Ἀσκληπιὸν αὐτοῦ οἰκειότητα ἔδειξε μὲν καὶ
τὸ πρῴην ἔργον, ἔπεισε δὲ ἡμᾶς καὶ ἐν τῇ τελευταίᾳ νόσῳ
ἐπιφάνεια τοῦ θεοῦ. μεταξὺ γὰρ ὢν ὕπνου καὶ ἐγρηγόρσεως, εἶδε
δράκοντα περὶ τὴν κεφαλὴν αὐτοῦ ἕρποντα, ἀφ᾽ ἧς αὐτῷ τὴν
ἀρχὴν ἐπέθετο τὸ τῆς παρέσεως νόσημα, καὶ οὕτω ἐκ τῆς ἐπιφανείας ἀνακωχῆς τινὸς τοῦ νοσήματος ᾔσθετο, καὶ εἰ μὴ προθυμία
καὶ πολλὴ ἔφεσις τοῦ θανάτου ἐκώλυσεν, ἐπιμελείας δὲ τῆς
προσηκούσης ἠξίωσε τὸ σῶμα, ὑγιές, οἶμαι, τέλεον αὖθις
ἐγεγόνει.

The afore-mentioned happenings[1] demonstrated his [sc.,
Proclus'] intimate connection with Asclepius and the epiphany
of the god even in his extreme illness also convinced us. For
between sleeping and waking he saw a serpent creeping around
his head, at which point the affliction of his paralysis had its
origin. And thus from the moment of the epiphany he noticed
a subsidence of the illness and if his eagerness and strong
desire for death had not prevented it and if he had deemed
the body worthy of the proper cure, he would, I think, have
become perfectly healthy again.

[1] Cf. T. 582.

446. Marinus, Vita Procli, Cp. 31.

Ἐδεδίει γάρ, ἀκμαζούσης αὐτῷ τῆς ἡλικίας, μήποτε ἡ τοῦ πατρὸς
ἀρθρῖτις νόσος, ἅτε φιλοῦσα καὶ εἰωθυῖα δὲ τὰ πολλὰ εἰς παῖδας
ἐκ πατέρων χωρεῖν, οὕτω καὶ ἐπ᾽ αὐτὸν ἔλθοι. καὶ οὐκ ἀδεές,
οἶμαι, ἐδεδίει. ἤδη γὰρ ἦν πρὸ τοῦ, ὅπερ καὶ ἔδει πρότερον
ἱστορῆσαι, ἀλγηδόνος τοιαύτης αἰσθόμενος, ἡνίκα δὴ καὶ ἄλλο
παράδοξον ἐγεγόνει περὶ αὐτόν. συμβουλευθεὶς γὰρ παρά τινων
ἐπέθηκε τὸ λεγόμενον πτυγμάτιον τῷ ἀλγοῦντι ποδί, καὶ κειμένου
αὐτοῦ ἐπὶ τῆς κλίνης, στρουθὸς ἐξαίφνης καταπτὰς ὑφήρπασε τὸ
πτυγμάτιον. ἦν μὲν οὖν καὶ ὁ σύμβολος θεῖος καὶ ὄντως παιώνειος
ἱκανός τε θάρρος ἐμποιῆσαι περὶ τοῦ μέλλοντος. ὁ δέ, ὥσπερ
ἔφην, καὶ ἐς ὕστερον οὐδὲν ἧττον φόβῳ τῆς νόσου κατείχετο.
ἱκετεύσας δὴ τὸν θεὸν περὶ τούτου καὶ δεηθεὶς φῆναί τι αὐτῷ
σαφές, καθευδήσας εἶδε (τολμηρὸν μὲν καὶ ἐνθυμηθῆναι, τολμη-
τέον δ᾽ οὖν ὅμως καὶ οὐκ ἀποδειλιατέον τὸ ἀληθὲς εἰς φῶς ἀγαγεῖν)
εἶδε δέ, ὡς ἐδόκει, ἥκοντά τινα ἐξ Ἐπιδαύρου καὶ ἐπικύψαντα εἰς

τὰ σκέλη, καὶ οὐδὲ τὰ γόνατα διὰ φιλανθρωπίαν ἀπαρνησάμενον
φιλεῖν. διετέλεσεν οὖν τὸ ἐντεῦθεν πάντα τὸν βίον περὶ τούτου
θαρρῶν, καὶ εἰς γῆρας ἀφίκετο βαθύ, μηδενὸς ἔτι πάθους τοιούτου
ἐπαισθανόμενος.

For when he [sc., Proclus] was at the prime of life he feared
that arthritis, the disease of his father, which is prone and
accustomed in many cases to pass from father to children,
might thus come upon him also. And not without reason was
he afraid, I think. For before this, as should have been
related previously, he experienced such a pain, when still
another unexpected thing happened to him. For upon the
advice of certain men he placed the so-called pledget on his
ailing foot and while he was lying on the couch, all of a
sudden a sparrow [1] seized the bandage and snatched it off.
This was, then, a symbol divine and truly paeonian and strong
enough to encourage him about the future. But he, as I have
said, even later was none the less possessed by a fear of the
disease. And imploring the god about this and beseeching
him to speak clearly to him, when he fell asleep he saw (a
daring thing to reflect upon, and yet, it must be dared and
one must not flinch from bringing the truth to light), he
saw, so he thought, someone come from Epidaurus, bend over
his legs, and, for his love of mankind, not even decline to
kiss his legs. [2] And thenceforth he lived his whole life uncon-
cerned about this disease and he arrived at a ripe old age
without having experienced again such an ailment.

[1] Cf. T. 730.　　　　　　　　　　[2] Cf. T. 423, 41.

447.　Libanius, De Vita Sua, 143.

Καὶ ὁ κλύδων οὗτος ἔτη τέτταρα ἐπεκράτει, καὶ καταφεύγω δι᾽
οἰκέτου πρὸς τὸν ἕτοιμον ἀμύνειν, τὸν μέγαν Ἀσκληπιόν, καὶ
φράσαντος οὐ καλῶς ἀφεστάναι με τῶν εἰωθότων πίνω τε οὗ
πάλαι φαρμάκου, καὶ ἦν μέν τι κέρδος, οὐ μὴν παντελῶς γε
ἐξελήλατο τὸ κακόν. ἔφη δὲ ὁ θεὸς καὶ τοῦτο χαριεῖσθαι. ἐγὼ δὲ
ᾔδειν μέν, ὡς οὐκ εὐσεβὲς ἀπιστεῖν ἐγγυητῇ τοιούτῳ, θαυμάζειν δὲ
ὅμως παρῆν, εἰ καὶ ταύτης εἶναί ποτε δόξαιμι τῆς χάριτος ἄξιος.
καὶ ἦν μὲν ἔτος ἕβδομον ἐπὶ τοῖς πεντήκοντα λῆγον ἤδη, τρισὶ
δὲ ἐνυπνίοις ὁ θεός, ὧν τὼ δύο μεθημερινώ, μέρος οὐ μικρὸν

ἑκάστῳ τοῦ νοσήματος ἀφῄρει καὶ κατέστησεν εἰς τοῦτο, ὃ μή
ποτε ἀφέλοιτο.

And this wave of disaster prevailed for four years. Through
a friend of mine I had recourse to him who is always ready to
succor, the great Asclepius, and when he said that I was not
right in abstaining from my customary treatment, I drank of
the old medicine, and there was a certain improvement, but
the illness was not completely banished. Yet the god said that
this, too, he would graciously grant. I knew that it is not
reverent to distrust such a guarantor, but nevertheless I wondered whether I should seem at all worthy of this gratification.
And the seventh year in addition to fifty was already coming
to an end when in three dreams, of which two were in the
daytime, the god took away no small part of the sickness in
each one and brought it to this state, which may he never take
away.[1]

¹ Cf. 447a. Libanius, *Epistulae*, 1300, 1-4: Καὶ σοὶ τῷ πρεσβευτῇ χάρις καὶ τῷ τὸν
ὕπνον ὑμῖν ἐπιδόντι καὶ τῷ φήναντι τὴν γυναῖκα τὴν μεγάλην τε καὶ καλήν, καὶ σοὶ πάλιν
χάρις, ὅτι ταύτην οἴει τὴν Ὑγίειαν εἶναι. . . . ἐλπίζω δέ τι πλέον· οὐδὲ γὰρ τὸ νῦν μικρόν,
ὃ καὶ αὐτὸ τοῦ θεοῦ τίθεμαι, παρ' οὗ τὴν νύμφην λαμβάνεις. ἤδη γὰρ ὁ ποὺς δύο μοίρας
ἀπείληφε τῆς δυνάμεως ἥν ποτε εἶχεν. αἱ μὲν χεῖρες τοῖν Ἠπειρώταιν, τὸ δὲ δῶρον Ἀσκληπιοῦ.
πιστεύειν οὖν χρὴ καὶ περὶ τοῦ λειπομένου. τοῦτο δὲ εἰ γένοιτο, δραμούμεθα παρὰ τὸν φιλόδωρον
θεὸν βεβαιωσόμενοί τε τὸ δοθὲν καὶ σοὶ δᾷδα ἄψοντες ἐν τοῖς γάμοις· πρὶν δὲ κομίσασθαι τὸ
πᾶν, οὐκ ἀσφαλές, οἶμαι, μεῖζω τῆς δυνάμεως τολμᾶν.

448. Hippocrates, Epistulae, 15 [IX, p. 340, 1 ff. L.].

Ἐδόκεον γὰρ αὐτὸν τὸν Ἀσκληπιὸν ὁρῆν φαίνεσθαί τε αὐτὸν πλησίον . . . ὁ δὲ Ἀσκληπιός, οὐχ ὡς εἰώθεσαν αὐτέου αἱ εἰκόνες, μεί-
λιχός τε καὶ πρᾶος ἰδέσθαι κατεφαίνετο, ἀλλὰ διεγηγερμένος τῇ
σχέσει καὶ ἰδέσθαι φοβερώτερος. εἵποντο δὲ αὐτῷ δράκοντες,
χρῆμά τι ἑρπετῶν ὑπερφυές, ἐπειγόμενοι δὲ καὶ αὐτοὶ μακρῷ τῷ
ἐπισύρματι, καί τι φρικῶδες ὡς ἐν ἐρημίῃ καὶ νάπῃσι κοίλῃσιν
ὑποσυρίζοντες. οἱ δὲ κατόπιν ἑταῖροι κίστας φαρμάκων εὖ μάλα
περιεσφηκωμένας ἔχοντες ᾔεσαν. ἔπειτα ὤρεξέ μοι τὴν χεῖρα ὁ
θεός. κἀγὼ λαβόμενος ἀσμένως ἐλιπάρεον ξυνέρχεσθαι καὶ μὴ
καθυστερέειν μου τῆς θεραπείης. ὁ δέ, οὐδέν τι, ἔφη, ἐν τῷ
παρεόντι ἐμεῦ χρήζεις, ἀλλά σε αὕτη τὰ νῦν ξεναγήσει θεὸς κοινὴ
ἀθανάτων τε καὶ θνητῶν . . . καὶ ὁ μὲν δαίμων ἐχωρίσθη.

I thought I saw Asclepius himself and that he appeared near
me . . . Asclepius did not appear, as the statues of him are wont

to do,[1] gentle and calm, but in a lively posture and rather frightening to behold. Serpents followed him, enormous sort of reptiles,[2] they too hurrying on, with their tremendous train of coils, making a whistling noise as in the wilderness and woodland glens. His associates followed him carrying boxes of drugs, tightly bound.[3] Then the god stretched forth his hand[4] to me. And taking it gladly I begged him to join me and not to be too late to aid me in my treatment. He replied: "At the moment you have no need of me at all, but this goddess here [sc., Truth], who holds sway over mortals and immortals alike, for the present will herself guide you." . . . And the god left.

[1] Cf. T. 627.   [2] Cf. T. 421, vv. 733-34.   [3] Cf. T. 421, vv. 710-11.   [4] Cf. T. 456.

449. Artemidorus, Onirocritica, V, 89.

Ἔδοξέ τις νοσῶν τὸν στόμαχον καὶ συνταγῆς δεόμενος παρὰ τοῦ Ἀσκληπιοῦ εἰς τὸ ἱερὸν τοῦ θεοῦ εἰσιέναι, καὶ τὸν θεὸν ἐκτείναντα τῆς δεξιᾶς ἑαυτοῦ χειρὸς τοὺς δακτύλους παρέχειν αὐτῷ ἐσθίειν. φοίνικας πέντε ἐσθίων ἐθεραπεύθη. καὶ γὰρ αἱ τοῦ φοίνικος βάλανοι αἱ σπουδαῖαι δάκτυλοι καλοῦνται.

Someone suffering a stomach ailment and wanting a prescription from Asclepius dreamed he entered the temple of the god, and the god, stretching forth the fingers of his right hand, offered them to him to eat. Having eaten five dates he was cured. For the excellent dates of the palm tree also are called fingers.[1]

[1] Cf. **449a.** Aristides, Oratio XLVIII, 40-3: . . . στρέφει με ὁ Σωτὴρ Ἀσκληπιὸς . . . ἔπειτα οὐ πολὺ ὕστερον ἡ Ἀθηνᾶ φαίνεται τήν τε αἰγίδα ἔχουσα . . . οὕτως ἐφάνη τε ἡ θεὸς καὶ παρεμυθήσατο καὶ ἀνέσωσεν καὶ δὴ κείμενον καὶ τῶν εἰς τὴν τελευτὴν οὐδενὸς ἔτι ἐλλείποντα. καὶ δῆτα εὐθύς με εἰσῆλθεν κλύσματι χρήσασθαι μέλιτος Ἀττικοῦ, καὶ ἐγένετο κάθαρσις χολῆς.

450. Artemidorus, Onirocritica, V, 61.

Ἔδοξέ τις ὑπὸ τοῦ Ἀσκληπιοῦ ξίφει πληγεὶς εἰς τὴν γαστέρα ἀποθανεῖν. οὗτος ἀπόστημα γενόμενον κατὰ τῆς γαστρὸς ἰάσατο τομῇ χρησάμενος.

A certain man dreamed that, struck in the belly by Asclepius with a sword, he died; this man, by means of an incision, healed the abscess which had developed in his belly.

451. Artemidorus, Onirocritica, II, 37.

Ἀσκληπιὸς ἱδρυμένος μὲν ἐν νεῷ καὶ ἑστὼς ἐπὶ βάσεως ὁρώμενος καὶ προσκυνούμενος ἀγαθὸς πᾶσι. κινούμενος δὲ ἢ προσιὼν ἢ εἰς οἰκίαν εἰσιὼν νόσον καὶ λοιμὸν μαντεύεται· τότε γὰρ μάλιστα τοῖς ἀνθρώποις δεῖ τοῦ θεοῦ τούτου. τοῖς δὲ ἤδη νοσοῦσι σωτηρίαν προαγορεύει· Παιήων γὰρ ὁ θεὸς λέγεται. ἀεὶ δὲ ὁ Ἀσκληπιὸς τοὺς ἐν ταῖς χρείαις συλλαμβανομένους καὶ τοὺς οἰκονομοῦντας τὸν οἶκον τοῦ ἰδόντος δηλοῖ. ἐν δὲ ταῖς δίκαις συνηγόρους σημαίνει.

Asclepius, if seen and reverenced when he is placed in his temple and standing on a base,[1] bodes good for all. When in motion, however, either approaching or entering a home, he forebodes sickness and plague; for at that time especially men stand in need of this god.[2] To those already stricken with illness, he foretells deliverance; for the god is called Paieon.[3] Always Asclepius indicates those who help in time of need and those who assist the household of the one who dreams of him. In lawsuits he indicates the supporters [of the one who dreams of him].

[1] Cf. T. 438, n. 1.　　　[2] Cf. T. 460.　　　[3] Cf. T. 482, v. 1.

452. Artemidorus, Onirocritica, II, 33.

Ἄλλους δὲ θύοντας ἰδεῖν πονηρὸν τῷ νοσοῦντι, κἂν Ἀσκληπιῷ θύοντας ἴδῃ, διὰ τὴν τοῦ θυομένου ἱερείου ἀναίρεσιν· θάνατον γὰρ σημαίνει.

To see in a dream others sacrificing is disastrous for the sick person, even if he sees them sacrificing to Asclepius, because of the destruction of the sacrificial animal, for this means death.

453. Artemidorus, Onirocritica, V, 13.

Ἔδοξε παῖς παλαιστὴς περὶ τῆς ἐγκρίσεως πεφροντικὼς τὸν Ἀσκληπιὸν κριτὴν εἶναι καὶ παροδεύων ἅμα τοῖς ἄλλοις παισὶν ἐν παρεξαγωγῇ ὑπὸ τοῦ θεοῦ ἐκκεκρίσθαι. καὶ δὴ πρὸ τοῦ ἀγῶνος ἀπέθανεν· ὁ γὰρ θεὸς οὐ τοῦ ἀγῶνος ἀλλὰ τοῦ ζῆν, οὗπερ εἶναι κριτὴς νομίζεται, ἐξέβαλεν αὐτόν.

A boy wrestler, anxious about admission to the contest, dreamed that Asclepius was the judge and, when marching together with the other boys in the parade before the contest, he was rejected by the god. And he died even before the contest, for the god disqualified him—not from the contest but from life, of which he is thought to be the judge.

454. Artemidorus, Onirocritica, V, 66.

Ἔδοξέ τις λέγειν αὐτῷ τινὰ " θῦσον τῷ Ἀσκληπιῷ." τῇ ὑστεραίᾳ μεγάλη συμφορᾷ ἐχρήσατο· κατηνέχθη γὰρ ἀπὸ ὀχήματος περιστρεφθέντος καὶ συνετρίβη τὴν χεῖρα τὴν δεξιάν, καὶ τοῦτο ἦν ἄρα, ὅπερ ἐσήμαινεν αὐτῷ τὸ ὄναρ, δεῖν φυλάττεσθαι καὶ θύειν ἀποτρόπαια τῷ θεῷ.

A certain man dreamed that someone said to him, " Sacrifice to Asclepius." On the next day he experienced a great misfortune: he fell from a carriage which overturned and his right hand was crushed. And this it was, therefore, that the dream signified to him—that he must be on his guard and perform averting sacrifices [1] to the god.

[1] Cf. T. 502 ff.

UNDETERMINABLE CASES

455. Aelianus, Fragmenta, 101.[1]

Ἀρίσταρχος, Τεγεάτης, ὁ τῶν τραγῳδιῶν ποιητής, νοσεῖ τινα νόσον. εἶτα αὐτὸν ἰᾶται ὁ Ἀσκληπιὸς καὶ προστάσσει χαριστήρια τῆς ὑγείας. ὁ δὲ ποιητὴς τὸ δρᾶμα τὸ ὁμώνυμόν οἱ νέμει. θεοὶ δὲ ὑγείας μὲν οὐκ ἄν ποτε μισθὸν αἰτήσαιεν οὐδ' ἂν λάβοιεν. ἢ πῶς ἄν; εἴ γε τὰ μέγιστα ἡμῖν φρενὶ φιλανθρώπῳ καὶ ἀγαθῇ παρέχουσι προῖκα, ἥλιόν τε ὁρᾶν καὶ τοῦ θεοῦ τοῦ τοσούτου τῆς παναρκοῦς ἀμισθὶ μεταλαμβάνειν ἀκτῖνος, καὶ χρῆσιν ὕδατος καὶ πυρὸς συντέχνου μυρίας ἐπιγονάς, καὶ ποικίλας ἅμα καὶ συνεργοὺς ἐπικουρίας, καὶ ἀέρος σπᾶν καὶ ἔχειν τροφὴν ζωῆς τὸ ἐξ αὐτοῦ πνεῦμα. ἐθέλουσι δὲ ἄρα ἐν τοῖσδε τοῖς μικροῖς μήτε ἀχαρίστους εἶναι μήτε ἀμνήμονας ἡμᾶς, καὶ ἐν τούτοις ἀμείνονας ἀποφαίνοντες.

Aristarchus of Tegea, the tragic poet [5th century B. C.], is ill of some disease. Then Asclepius cures him and commands thank-offerings [2] for the restoration of his health. The poet offers the play with the same name as the god. Never would the

gods demand money for health, nor would they even accept it.[3]
How could it be otherwise? Since with men-loving and kindly
hearts they freely offer us the greatest gifts: to see the sun
and to have a share in the all-pervading rays of such a god
without paying; and the advantage of water, and the countless
products of the helpful fire, together with its varied helping
aids; and to breathe the air and draw from it the breath, the
nourishment of life. By exacting from us such minor things
[*sc.*, as thank-offerings], then, they wish us to be neither
ungrateful nor forgetful, even in such matters making us
better.

[1] = 455a. Suidas, *Lexicon, s. v.* Ἀρίσταρχος: Ἀρίσταρχος—ἀποφαίνοντες.
[2] Cf. T. 520 ff.
[3] Cf. T. 320; cf. however T. 99–100.

456. Suidas, Lexicon, *s. v.* Θεόπομπος.[1]

Θεόπομπος, Θεοδέκτου ἢ Θεοδώρου, Ἀθηναῖος, κωμικός. ἐδίδαξε
δράματα κδ΄. ἔστι δὲ τῆς ἀρχαίας κωμῳδίας κατὰ Ἀριστοφάνην.
δράματα δὲ αὐτοῦ εἰσι . . . * καὶ ἄλλα πολλά. ὅτι Ἀσκληπιὸς καὶ
τῶν ἐν παιδείᾳ ἦν προμηθής. φθόη γοῦν Θεόπομπον ῥινώμενόν τε
καὶ λειβόμενον ἰάσατο καὶ κωμῳδίαις αὖθις διδάσκειν ἐπῆρεν, ὁλό-
κληρόν τε καὶ σῶν καὶ ἀρτεμῆ ἐργασάμενος. καὶ δείκνυται καὶ
νῦν ὑπὸ λίθῳ Θεοπόμπου· πατρόθεν ὁμολογοῦντος αὐτὸν τοῦ
ἐπιγράμματος· Τισαμενοῦ γὰρ ἦν υἱός· εἴδωλον Παρίας λίθου.
καὶ ἔστι τὸ ἴνδαλμα τοῦ πάθους μάλα ἐναργές, κλίνη καὶ αὐτὴ
λίθου. ἐπ' αὐτῆς κεῖται νοσοῦν τὸ ἐκείνου φάσμα, χειρουργίᾳ
φιλοτέχνῳ· παρέστηκε δὲ ὁ θεὸς καὶ ὀρέγει οἱ τὴν παιώνιον χεῖρα,
καὶ παῖς νεαρὸς ὑπομειδιῶν καὶ οὗτος. τί δὲ ἄρα νοεῖ ὁ παῖς; ἐγὼ
συννίημι, [τοῦ] φιλοπαίστην ⟨τὸν⟩ ** ποιητὴν ὑποδηλοῦν. γελᾷ
γὰρ καὶ τῆς κωμῳδίας τὸ ἴδιον διὰ συμβόλων αἰνίττεται. εἰ δὲ ἄλλος
νοεῖ ἑτέρως, κρατείτω τῆς ἑαυτοῦ γνώμης, ἐμὲ δὲ μὴ ἐνοχλείτω.

* Lacunam indicavit E. F. Krause, *Byzantinische Zeitschrift*, XIII, 1904, p. 113.
** [τοῦ] φιλοπαίστην ⟨τὸν⟩ Friedländer.

Theopompus, the son of Theodectes or Theodorus, an Athenian, a comic poet, presented twenty-four plays. He was a poet
of the "old comedy," a contemporary of Aristophanes. Plays
of his are . . . and many others. [It is evident] that Asclepius
was also the protector of the cultured people: at least when
Theopompus was wasting and pining away with consumption,
he cured him and enabled him to present a comedy again,
making him whole and sound and well. And on the base of

an altar there is shown to the present day an image of Theopompus, made of Parian marble, with an epigram identifying him through the name of his father—he was the son of Tisamenus. And the representation of his sufferings is very lively —even the sickbed itself being made of stone. On this lies his wasted figure [carved] with artistic skill. Beside him stands the god, stretching forth his healing hand to him,[2] and a young boy, and he too smiles a little. What do you suppose the boy means? I suspect it indicates that the poet was fond of boys; for he laughs and symbolically intimates the characteristics of comedy. But if someone else thinks otherwise, let him keep his own opinion, but not annoy me.

[1] Cf. **456a.** Aelianus, Fr. 99: ὅτι—ἐνοχλείτω same text; cf. also **456b.** Suidas, *Lexicon, s.v.* Φθόη: . . . καὶ Αἰλιανός, φθόῃ γοῦν Θεόπομπον τὸν Ἀθηναῖον ῥινώμενόν τε καὶ λειβόμενον ἰάσατο ὁ Ἀσκληπιός.
[2] Cf. T. **448.**

457. Pausanias, Descriptio Graeciae, II, 26, 8.[1]

. . . τοῦτο δὲ Ἀρχίας ὁ Ἀρισταίχμου, τὸ συμβὰν σπάσμα θηρεύοντί οἱ περὶ τὸν Πίνδασον ἰαθεὶς ἐν τῇ Ἐπιδαυρίᾳ, τὸν θεὸν ἐπηγάγετο ἐς Πέργαμον.

. . . again, when Archias [1st half 4th c. B. C.], son of Aristaechmus, was healed in Epidauria after spraining himself while hunting about Pindasus, he brought the god to Pergamum. (*)

[1] Cf. T. 709.

458. Galenus, De Libris Propriis, Cp. 2 [II, p. 99 M.].

. . . τὸν πάτριον θεὸν Ἀσκληπιόν, οὗ καὶ θεραπευτὴν ἀπέφαινον ἐμαυτόν, ἐξ ὅτου με θανατικὴν διάθεσιν ἀποστήματος ἔχοντα διέσωσε . . .

. . . the ancestral god Asclepius[1] of whom I declared myself to be a servant since he saved me when I had the deadly condition of an abscess . . .

[1] Cf. T. 338; 348; 413.

459. Galenus, De Morborum Differentiis, Cp. 9 [VI, p. 869 K.].

Νικομάχῳ δὲ τῷ Σμυρναίῳ πᾶν ἀμέτρως ηὐξήθη τὸ σῶμα, καὶ οὐδὲ κινεῖν ἔτι δυνατὸς ἦν ἑαυτόν· ἀλλὰ τοῦτον μὲν ὁ Ἀσκληπιὸς ἰάσατο.

The whole body of Nicomachus of Smyrna swelled excessively
and it was impossible for him to move himself. But this man
Asclepius healed.

460. Aristides, Oratio L, 9.

Καὶ χρόνοις δὴ ὕστερον ἡ λοιμώδης ἐκείνη συνέβη νόσος, ἧς
ὅ τε Σωτὴρ καὶ ἡ δέσποινα Ἀθηνᾶ περιφανῶς ἐρρύσαντό με.

And some time later that pestilential disease[1] occurred, from
which both the Savior and the Lady Athena[2] manifestly
protected me.

[1] Cf. T. 451.   [2] Cf. T. 449a.

461. Libanius, Epistulae, 316, 3.

Καὶ χάριν ἔχω τῷ μὲν Ἀσκληπιῷ τοῦ στῆσαι τὴν ἀσθένειαν,
σοὶ δὲ τοῦ μηνῦσαι.

And I feel grateful to Asclepius for having stayed the disease,
to you [sc., Acacius], however, for having disclosed this to me.

462. Julianus, Contra Galilaeos, 235 C.

Ἐμὲ γοῦν ἰάσατο πολλάκις Ἀσκληπιὸς κάμνοντα ὑπαγορεύσας
φάρμακα.

At any rate, when I have been sick, Asclepius has often cured
me[1] by prescribing remedies.

[1] Cf. **462a.** Cicero, *Epistulae, Ad Fam.*, XIV, 7, 1: Χολὴν ἄκρατον noctu eieci: statim
ita sum levatus ut mihi deus aliquis medicinam fecisse videatur. cui quidem tu deo,
quem ad modum soles, pie et caste satis facies [id est Apollini et Aesculapio].

ASCLEPIUS THE SAVIOR

463. Hippocrates, Epistulae, 17 [IX, p. 372, 13 L.].

Ὁ σὸς πρόγονος Ἀσκληπιὸς νουθεσίη σοι γινέσθω, σώζων
ἀνθρώπους κεραυνοῖσιν ηὐχαρίστηται.

Let your ancestor[1] Asclepius be a warning to you [sc., Hip-
pocrates] in that he was requited with a thunderbolt[2] for
saving mankind.

[1] Cf. T. 338; 348; 413; 458.   [2] Cf. T. 105 ff.

464. Aelianus, De Natura Animalium, X, 49.

... καὶ αὐτὸν σώζειν εἰδότα καὶ μέντοι καὶ τὸν σωτῆρα καὶ νόσων ἀντίπαλον Ἀσκληπιὸν φύσαντα.

... not only did he [sc., Apollo] know himself how to save but also was the father of Asclepius, the Savior[1] and the adversary of diseases.

[1] Cf. **464a.** Ps. Julianus, *Epistulae*, 79, 406 D : ... σὺ δ' [sc., ὦ Ἰάμβλιχε] ὥσπερ ἐπὶ σωτηρίᾳ τοῦ κοινοῦ τῶν ἀνθρώπων γένους τεχθείς, τὴν Ἀσκληπιοῦ χεῖρα πανταχοῦ ζηλῶν, ἅπαντα ἐπέρχῃ λογίῳ τε καὶ σωτηρίῳ νεύματι. Cf. **464b.** Aristides, *Oratio* XXVIII, 132: Ἡμεῖς τοι καὶ εἰς τὸ σῶμα πληγέντες οὐκ ἐπ' ἀγεννεῖς ἱκετείας ἰατρῶν ἀφικόμεθα, ἀλλὰ καίτοι σὺν θεοῖς εἰπεῖν τοὺς ἀρίστους τῶν ἰατρῶν φίλους κεκτημένοι κατεφύγομεν εἰς Ἀσκληπιοῦ, νομίσαντες εἴτε δέοι σῴζεσθαι δι' ἐκείνου κάλλιον εἶναι, εἴτε μὴ ἐγχωροῖ, καιρὸν εἶναι τεθνάναι.

465. Suidas, Lexicon, *s. v.* Ἰάκωβος.

Πάντας δὲ ὡς εἰπεῖν ἀπήλλαττε τῶν ἐνοχλούντων παθῶν παραυτίκα ἢ μικρὸν ὕστερον. διόπερ οἱ μὲν ἄλλοι σωτῆρα τὸν Ἰάκωβον ἀπεκάλουν, οἷά ποτε καὶ τὸν Ἀσκληπιόν· οἱ δὲ ἰατροὶ διέβαλλον ἀεὶ καὶ ἐλοιδόρουν, ὡς οἷόν τινα οὐκ ἰατρόν, ἀλλὰ θεοφιλῆ τινα καὶ ἱερόν· καὶ οὐκ ἐψεύδοντο. ἦν γὰρ ἐπιεικὴς ὁ ἀνὴρ καὶ θεῷ τῷ ὄντι κεχαρισμένος. εἰ δὲ δεῖ τὸ τοῦ φιλοσόφου εἰπεῖν, Ἀσκληπιαδικὴν ᾤετο τὴν τοῦ Ἰακώβου ψυχὴν καὶ κατὰ φύσιν Παιώνειον.

Practically everybody he [sc., Jacob, 5th c. A. D.] liberated from the sufferings that beset them, either immediately or after a little while. Wherefore people called Jacob savior, just as once they called Asclepius that, but the doctors always slandered and reviled him for being not a physician but a favorite of the gods and holy. And they were not wrong, for the man was gentle and very pleasing indeed to God. If one must quote the saying of the philosopher [Isidorus?], he thought that the soul of Jacob was Asclepiadic[1] and by nature Paeonic.

[1] Cf. T. 314; cf. also T. 310.

466. Aelianus, Fragmenta, 98.[1]

Ἀλεκτρύονα ἀθλητὴν Ταναγραῖον. ᾄδονται δὲ εὐγενεῖς οὗτοι. ὁ δὲ ἐμοὶ δοκεῖν ὁρμῇ τῇ παρὰ τοῦ Ἀσκληπιοῦ ἐς τὸν δεσπότην ἀσκωλιάζων θάτερον τῶν ποδῶν ἔρχεται, καὶ ὄρθριον ᾀδομένου τοῦ παιᾶνος τῷ Ἀσκληπιῷ ἑαυτὸν ἀποφαίνει τῶν χορευτῶν ἕνα, καὶ ἐν τάξει στὰς ὥσπερ οὖν παρά τινος λαβὼν χορολέκτου τὴν στάσιν, ὡς οἷός τε ἦν συνᾴδειν ἐπειρᾶτο τῷ ὀρνιθείῳ μέλει, συνῳδόν τε καὶ σύμμελὲς ἀναμέλπων.

ὁ δὲ ἀλεκτρύων ἑστὼς ἐπὶ θατέρου ποδὸς προύτεινε τὸν λελωβημένον καὶ κυλλόν, ὥσπερ οὖν μαρτυρόμενος καὶ ἐμφαίνων οἷα ἐπεπόνθει.

ὁ δὲ ἀλέκτωρ ὑμνεῖ τὸν σωτῆρα ᾗπερ οὖν ἔσθενε φωνῇ, καὶ ἐδεῖτο ἀρτίπουν θεῖναι αὐτόν.

καὶ ὁ μὲν ἔδρασε τὸ προσταχθέν, ὁ δὲ ὄρνις πρὸ βουλυτοῦ ἐπ᾽ ἀμφοῖν βαδίζων καὶ τὼ πτέρυγε κρούων καὶ βαίνων μακρὰ καὶ αἴρων τὸν τράχηλον καὶ τὸν λόφον ἐπισείων, οἷον ὁπλίτης γαῦρος, τὴν ἐς τὰ ἄλογα προμήθειαν ἐπεδείκνυτο.

ἀφίησι τῷ Ἀσκληπιῷ ἀνάθημά τε καὶ ἄθυρμα εἶναι, οἱονεὶ θεράποντα καὶ οἰκέτην περιπολοῦντα τῷ νεῷ τὸν ὄρνιν, ὁ Ἀσπένδιος ἐκεῖνος.

The champion cock from Tanagra. These are celebrated as highbred animals. The cock, inspired by Asclepius, I think, goes to its Lord hopping on one foot, and when the paean[2] for Asclepius is sung at daybreak, it appears as one of the members of the god's chorus,[3] and standing in line, as if taking its position at the command of a chorus-leader, it tried, as far as possible, with its bird's voice, to accompany the singing in harmony and in tune.

The cock, standing on one foot, stretched forth the mutilated and crippled one as if bearing witness to and disclosing what sort of things it suffered.

The cock praised the savior with whatever voice it could muster and begged that it be made sound of foot.

And Asclepius accomplished what was demanded. Before evening, walking on both feet, flapping its wings and taking large strides, poking its neck forward and shaking its crest, like a majestic heavy-armed soldier, the bird demonstrated the consideration of the god towards animals.[4]

The man from Aspendus sends it to be a votive offering and plaything for Asclepius, a bird to be, as it were, a servant and attendant, roaming about the temple.

---

[1] Cf. 466a. Suidas, Lexicon, s.v. Ἀσκωλιάζοντες: ὁ δὲ ἐμοὶ—μέλει; cf. 466b. ibid., s.v. Ἀλεκτρυόνα: ἀλεκτρυόνα—οὗτοι. ἀφίησι τῷ Ἀσκληπιῷ ἀνάθημά τε καὶ ἄθυρμα εἶναι, οἱονεὶ θεράποντα τῷ θεῷ τὸν ὄρνιν ὁ Ἀσπένδιος ἐκεῖνος.
[2] Cf. T. 587 ff.   [3] Cf. T. 605.   [4] Cf. T. 378; 433.

ASCLEPIUS AS SYMBOL OF MEDICINE

467. Hippocrates, Epistulae, 2 [IX, p. 314, 16 L.].

Καθαίρει δὲ οὐ θηρίων μὲν γένους,* θηριωδῶν δὲ νοσημάτων καὶ
ἀγρίων πολλὴν γῆν καὶ θάλασσαν, διασπείρων πανταχοῦ, ὥσπερ
ὁ Τριπτόλεμος τὰ τῆς Δήμητρος σπέρματα, τὰ τοῦ Ἀσκληπιοῦ
βοηθήματα. τοιγαροῦν ἐνδικώτατα καὶ αὐτὸς ἀνιέρωται πολλαχοῦ
τῆς γῆς, ἠξίωταί τε τῶν αὐτῶν Ἡρακλεῖ τε καὶ Ἀσκληπιῷ ὑπὸ
Ἀθηναίων δωρεῶν.

* γένους Van der Linden γένοςMSS.

He [sc., Hippocrates] rids the wide earth and sea not of wild
animals but rather of diseases savage and wild, disseminating
everywhere the remedies of Asclepius just as Triptolemus dis-
seminates the seeds of Demeter.[1] Therefore he most justly
receives divine reverence in many places of the world and is
considered by the Athenians deserving of the same honorary
gifts as Heracles and Asclepius.

[1] Cf. T. 282, 15.

468. Crinagoras, Epigrammata, LI.

[Εἰς εἰκόνα Πραξαγόρου ἰατροῦ.]

Αὐτός σοι Φοίβοιο πάϊς λαθικηδέα τέχνης
　ἰδμοσύνην πανάκη χεῖρα λιπηνάμενος,
Πρηξαγόρη, στέρνοις ἐνεμάξατο. τοιγὰρ ἀνῖαι
　ὄρνυνται δολιχῶν ὁππόσαι ἐκ πυρετῶν,
καὶ ὁπόσα τμηθέντος ἐπὶ χροὸς ἄρκια θεῖναι
　φάρμακα, πρηείης οἶσθα παρ᾽ Ἠπιόνης.
θνητοῖσιν δ᾽ εἰ τοῖοι ἐπήρκεον ἰητῆρες,
　οὐκ ἂν ἐπορθμεύθη νεκροβαρὴς ἄκατος.

[To the statue of Praxagoras, the physician.]

The son of Phoebus himself, anointing his hand with the all-
healing herb, Praxagoras, rubbed into your breast the skill of
art that banishes care. For whatever the griefs that arise from
protracted fevers, and whatever the drugs that are fit to apply
to the wounded flesh, these you know from gentle Epione.
If physicians like you assisted mortals, the boat laden with the
dead would not be ferried across to Hades. (**)

469. Athenaeus, Deipnosophistae, X, 44, 434 d.

Καλλισθένης δὲ ὁ σοφιστής . . . ἐν τῷ συμποσίῳ τοῦ Ἀλεξάνδρου τῆς τοῦ ἀκράτου κύλικος εἰς αὐτὸν ἐλθούσης ὡς διωθεῖτο, εἰπόντος τέ τινος αὐτῷ " διὰ τί οὐ πίνεις; " " οὐδὲν δέομαι," ἔφη, " Ἀλεξάνδρου πιὼν τοῦ Ἀσκληπιοῦ δεῖσθαι."

But the sophist Callisthenes [4th c. B. C.] . . . having pushed aside the cup of unmixed wine when it came to him at Alexander's symposium, replied when somebody said to him, "Why don't you drink?" "I don't want to be in need of one of Asclepius' cups after drinking from one of Alexander's.[1] (*)

[1] Cf. **469a.** Plutarchus, *Quaestiones Convivales*, I, 6, 1 (624): Ἐπεὶ καὶ κύλικα λεγομένην Ἀλεξάνδρου μεγάλην, ἐλθοῦσαν ἐπ' αὐτόν [sc., τὸν Καλλισθένην], ἀπεώσατο φήσας οὐκ ἐθέλειν Ἀλεξάνδρου πιὼν Ἀσκληπιοῦ δεῖσθαι.

470. Maximus Tyrius, Philosophumena, XIV, 8 g.

Ἐκολάκευσεν ἀνθρώπους καὶ ἰατρικὴ νόθος, ὅτε τὴν Ἀσκληπιοῦ καὶ τὴν Ἀσκληπιαδῶν ἴασιν καταλιπόντες οὐδὲν διαφέρουσαν ἀπέφηναν τὴν τέχνην ὀψοποιϊκῆς . . .

Pseudo-medicine also flattered men when, neglecting the healing of Asclepius and the Asclepiads,[1] they made the art in no way different from the art of delicate cookery . . .

[1] Cf. **470a.** Maximus Tyrius, *Philosophumena*, XVII, 3 g: τὴν Ἀσκληπιοῦ καὶ τὴν Ἀσκληπιαδῶν τέχνην.

471. Maximus Tyrius, Philosophumena, V, 4 f.

Τί γὰρ εἰ καὶ τὰ μόρια τοῦ σώματος φωνὴν λαβόντα, ἐπειδὰν κάμνῃ τι αὐτῶν ὑπὸ τοῦ ἰατροῦ τεμνόμενον ἐπὶ σωτηρίᾳ τοῦ ὅλου, εὔξαιτο τῇ τέχνῃ μὴ φθαρῆναι; οὐκ ἀποκρινεῖται ὁ Ἀσκληπιὸς αὐτοῖς, ὡς · " Οὐχ ὑμῶν ἕνεκα, ὦ δείλαια, χρὴ οἴχεσθαι τὸ πᾶν σῶμα, ἀλλὰ ἐκεῖνο σωζέσθω, ὑμῶν ἀπολλυμένων."

For what if also the parts of the body should acquire the power of speech and, whenever any of these parts, being sick, is cut off by the physician in order to save the whole body, should make a plea to the art that it may not be destroyed? Will not Asclepius reply to the parts of the body in this fashion: "Not for your sake, fools, must the whole body be destroyed, but let that be saved, even if you perish."

472. Apollonius Tyanensis, Epistulae, VIII.

" Καὶ σωμάτων δὲ ὀδύνας ἀφαιρεῖ καὶ πάθη παύει." τοῦτό που καὶ πρὸς τὸν Ἀσκληπιὸν κοινὸν τὸ ἔγκλημα.

"And he [sc., Apollonius] eases the body of its agonies and stops suffering." This charge he has in common even with Asclepius.[1] (**)

[1] Cf. **472a.** *Scholia in Pindarum, Ad Pythias,* III, 11a: τὸν κατασκευαστὴν τῆς νωδυνίας· ἰατρὸς γάρ [sc., ὁ Ἀσκληπιός].

473. Galenus, De Sanitate Tuenda, I, 12, 15.

Ἔνια γὰρ οὕτως εὐθὺς ἐξ ἀρχῆς κατεσκεύασται κακῶς, ὡς μηδ' εἰς ἑξηκοστὸν ἔτος ἀφικέσθαι δύνασθαι, κἂν αὐτὸν ἐπιστήσῃς αὐτοῖς τὸν Ἀσκληπιόν.

For some [sc., bodies] were so badly formed right from the beginning that they cannot reach the age of sixty even if you should put Asclepius himself in charge of them.

474. Carmina Priapea, XXXVII, 3-7.

Cum penis mihi forte laesus esset
chirurgique manum miser timerem,
dis me legitimis nimisque magnis,
ut Phoebo puta filioque Phoebi,
curatum dare mentulam verebar.

When perchance my penis was hurt and in my unhappiness I feared the surgeon's hands, I was reluctant to entrust my membrum virile to the cure of the true and very great gods, such as for instance Phoebus and the son of Phoebus.

475. Dio Chrysostomus, Oratio VI, 23-24.

Τοὺς δὲ ἀνθρώπους οὕτως μὲν πάνυ φιλοζῴους ὄντας, τοσαῦτα δὲ μηχανωμένους πρὸς ἀναβολὴν τοῦ θανάτου, τοὺς μὲν πολλοὺς αὐτῶν μηδὲ εἰς γῆρας ἀφικνεῖσθαι, ζῆν δὲ νοσημάτων γέμοντας, ἃ μηδὲ ὀνομάσαι ῥάδιον, τὴν δὲ γῆν αὐτοῖς μὴ ἐξαρκεῖν παρέχουσαν φάρμακα, δεῖσθαι δὲ καὶ σιδήρου καὶ πυρός. καὶ μήτε Χείρωνος μήτε Ἀσκληπιοῦ μήτε τῶν Ἀσκληπιαδῶν ἰωμένων μηδὲν αὐτοῖς ὄφελος εἶναι διὰ τὴν αὐτῶν ἀκολασίαν καὶ πονηρίαν, μηδὲ μάντεων μαντευομένων μηδὲ ἱερέων καθαιρόντων.

As for men, however, who are so very fond of life and devise so many ways to postpone death, most of them do not even reach old age, but live infested by a host of maladies,—which it is no easy task even to name,—and the earth does not supply them with sufficient drugs, but they require the knife and cautery as well. Nor are Chiron and Asclepius and Asclepius' sons, with all their healing power, of any use to them because of their excesses and wickedness, nor are prophetic seers and purifying priests. (**)

476. Maximus Tyrius, Philosophumena, XXXVI, 5 f.

... μηδὲν φαρμάκων δεηθείς, μὴ σιδήρου, μὴ πυρός, μὴ Χείρωνος, μὴ Ἀσκληπιοῦ, μὴ Ἀσκληπιαδῶν, μὴ μάντεων μαντευομένων, μὴ ἱερέων καθαιρόντων, μὴ γοήτων ἐπᾳδόντων.

... [Diogenes] wanting nothing in the way of drugs, or iron, or fire, or Chiron, or Asclepius, or the Asclepiads, or prophesying seers, or purifying priests, or chanting sorcerers.

477. Maximus Tyrius, Philosophumena, XL, 3 d-e.

... ὁ μὲν διώκων ἀκρότητα ἐν ὑγιεινῷ διώκει πρᾶγμα φεῦγον, καὶ οὔτε Ἀσκληπιῷ οὔτε Χείρωνι ἐξ ἐπιδρομῆς ἁλώσιμον.

... he who pursues the peak of health pursues a fleeting phantom which cannot be overtaken in the running even by Asclepius or Chiron.

478. Athenaeus, Deipnosophistae, I, 51, 28 e.[1]

Θάσιον ἔγχει ...
ὁ γὰρ λαβών μου καταφάγῃ τὴν καρδίαν,
ὅταν πίω τοῦδ', εὐθὺς ὑγιὴς γίνεται·
Ἀσκληπιὸς κατέβρεξε ...

Fill a cup of Thasian: for the sorrow that has gripped me gnaws at my heart; once I get a drink of that, straightway my heart is sound again. Asclepius has drenched me ... (*)

---

[1] = Antidotus, II, 411 [Kock].

479. Nicephorus Gregoras, Byzantinae Historiae, XXVI, 20.

> Καὶ ἦν μὴ σύ μοι λύσαις, Ἀσκληπιός τις καὶ Ἱπποκράτης οὑτωσί
> πως καμνούσῃ φανεὶς διανοίᾳ, οὐκ ἂν ἐμοὶ τῶν ἐφεξῆς ἐνείη λέγειν
> οὐδαμῆ . . .

> And if you [sc., Agathangelus] should not release me just like
> some Asclepius or Hippocrates, appearing to my laboring mind,
> it would in no way be possible for me to speak of the
> following . . .

480. Phalaris, Epistulae, I.

> Πολύκλειτος ὁ Μεσσήνιος . . . ἰάσατό μου νόσον ἀνήκεστον
> . . . σὲ δ᾽ οὐκ ἂν οὐδ᾽ αὐτὸς ὁ τῆς τέχνης ἡγεμὼν Ἀσκληπιὸς
> μετὰ πάντων ἰάσαιτο τῶν θεῶν. σώματος μὲν γὰρ ἀρρωστίαν
> θεραπεύει τέχνη, ψυχῆς δὲ νόσον ἰατρὸς ἰᾶται θάνατος . . .

> Polyclitus the Messenian . . . cured my incurable sickness . . .
> but you [sc., Lycinius] not even the leader of the art himself,[1]
> Asclepius, with all the other gods, would heal; for the body's
> sickness skilful art can cure, but the soul's sickness only the
> physician death can cure . . .

[1] Cf. T. 340.

481. Anonymus, Duodecim Sapientes, 152, 23-25.

> Vivit pectore sub dolente vulnus,
> quod Chironia nec manus levarit
> nec Phoebus sobolesve clara Phoebi.

> Under his anguished heart [sc., of the envious] there persists
> a wound which neither Chiron's hand nor Phoebus nor the
> famous offspring of Phoebus could heal.

V. CULT

GENERAL DESCRIPTION

482. Herondas, Mimiambi, IV, 1-95.

ΑΣΚΛΗΠΙΩΙ ΑΝΑΤΙΘΕΙΣΑΙ ΚΑΙ ΘΥΣΙΑΖΟΥΣΑΙ ΚΥΝΝΩ

Χαίροις, ἄναξ Παίηον, ὃς μεδεῖς Τρίκκης
καὶ Κῶν γλυκεῖαν κἠπίδαυρον ᾤκηκας,
σὺν καὶ Κορωνὶς ἥ σ' ἔτικτε κὠπόλλων
χαίροιεν, ἧς τε χειρὶ δεξιῇ ψαύεις
5 Ὑγίεια κὠνπερ οἵδε τίμιοι βωμοί,
Πανάκη τε κἠπιώ τε κἰησὼ χαίροι,
κοἰ Λεωμέδοντος οἰκίην τε καὶ τείχη
πέρσαντες, ἰητῆρες ἀγρίων νούσων,
Ποδαλείριός τε καὶ Μαχάων χαιρόντων
10 κὤσοι θεοὶ σὴν ἑστίην κατοικεῦσιν
καὶ θεαί, πάτερ Παίηον· ἵλεῳ δεῦτε
τοῦ ἀλέκτορος τοῦδ', ὅντιν' οἰκίης τοίχων
κήρυκα θύω, τἀπίδορπα δέξαισθε.
οὐ γάρ τι πολλὴν οὐδ' ἑτοῖμον ἀντλεῦμεν,
15 ἐπεὶ τάχ' ἂν βοῦν ἢ νενημένην χοῖρον
πολλῆς φορίνης, κοὐκ ἀλέκτορ', ἴητρα
νούσων ἐποιεύμεσθα τὰς ἀπέψησας
ἐπ' ἠπίας σὺ χεῖρας, ὦ ἄναξ, τείνας.
ἐκ δεξιῆς τὸν πίνακα, Κοκκάλη, στῆσον
20 τῆς Ὑγιείης.[1]

ΚΟΚΚΑΛΗ

μᾶ καλῶν, φίλη Κυννοῖ,
ἀγαλμάτων· τίς ἦρα τὴν λίθον ταύτην
τέκτων ἐποίει καὶ τίς ἐστιν ὁ στήσας;

ΚΥΝΝΩ

οἱ Πρηξιτέλεω παῖδες· οὐχ ὁρῇς κεῖνα
ἐν τῇ βάσει τὰ γράμματ'; Εὐθίης δ' αὐτὴν
25 ἔστησεν ὁ Πρήξωνος.

ΚΟΚΚΑΛΗ

ἵλεως εἴη
καὶ τοῖσδ᾽ ὁ Παιὼν καὶ Εὐθίη καλῶν ἔργων.
ὅρη, φίλη, τὴν παῖδα τὴν ἄνω κείνην
βλέπουσαν ἐς τὸ μῆλον· οὐκ ἐρεῖς αὐτήν,
ἢν μὴ λάβῃ τὸ μῆλον ἐκ τάχα ψύξειν; —
30   κεῖνον δέ, Κυννοῖ, τὸν γέροντα· — πρὸς Μοιρέων
τὴν χηναλώπεκ᾽ ὡς τὸ παιδίον πνίγει.
πρὸ τῶν ποδῶν γοῦν εἴ τι μὴ λίθος, τοὖργον,
ἐρεῖς, λαλήσει. μᾶ, χρόνῳ κοτ᾽ ὥνθρωποι
κῆς τοὺς λίθους ἕξουσι τὴν ζοὴν θεῖναι —
35   τὸν Βατάλης γὰρ τοῦτον, οὐχ ὁρῇς, Κυννοῖ,
ὅκως βέβηκεν, ἀνδριάντα τῆς Μύττεω;
εἰ μή τις αὐτὴν εἶδε Βατάλην, βλέψας
ἐς τοῦτο τὸ εἰκόνισμα μὴ ἐτύμης δείσθω.

ΚΥΝΝΩ

ἕπευ, φίλη, μοι καὶ καλόν τί σοι δείξω
40   πρῆγμ᾽ οἷον οὐχ ὥρηκας ἐξ ὅτευ ζώεις.
Κύδιλλ᾽, ἰοῦσα τὸν νεωκόρον βῶσον.
οὐ σοὶ λέγω, αὕτη, τῇ ὧδε κὦδε χασκούσῃ;
μᾶ, μή τιν᾽ ὥρην ὧν λέγω πεποίηται·
ἕστηκε δ᾽ εἴς μ᾽ ὀρεῦσα καρκίνου μέζον.
45   ἰοῦσα, φημί, τὸν νεωκόρον βῶσον.
λαίμαστρον, οὔτ᾽ † ὀργή † σε κρηγύην οὔτε
βέβηλος αἰνεῖ, πανταχῆ δ᾽ ἴσου κεῖσαι.
μαρτύρομαι, Κύδιλλα, τὸν θεὸν τοῦτον
ὡς ἔκ με καίεις οὐ θέλουσαν οἰδῆσαι·
50   μαρτύρομαι, φήμ᾽· ἔσσετ᾽ ἡμέρη κείνη
ἐν ᾗ τὸ βρέγμα τοῦτο τὸ ἀσυρὲς κνήσῃ.

ΚΟΚΚΑΛΗ

μὴ πάνθ᾽ ἑτοίμως καρδιηβολεῦ, Κυννοῖ·
δούλη 'στί, δούλης δ᾽ ὦτα νωθρίη θλίβει.

ΚΥΝΝΩ

ἀλλ᾽ ἡμέρη τε κἠπὶ μέζον ὠθεῖται·
55   αὕτη σύ, μεῖνον· ἡ θύρη γὰρ ὤϊκται
κἀνεῖθ᾽ ὁ παστός·

ΚΟΚΚΑΛΗ

οὐχ ὁρῇς, φίλη Κυννοῖ;
οἶ᾿ ἔργα κεῖν᾿·—ἤν,[2] ταῦτ᾿ ἐρεῖς Ἀθηναίην
γλύψαι τὰ καλά—χαιρέτω δὲ δέσποινα.
τὸν παῖδα δὴ[3] τὸν γυμνὸν ἢν κνίσω τοῦτον
60 οὐχ ἕλκος ἕξει, Κύννα; πρὸς γάρ οἱ κεῖνται
αἱ σάρκες οἶα θερμὰ θερμὰ πηδεῦσαι
ἐν τῇ σανίσκῃ· τὠργυρεῦν δὲ πύραστρον
οὐκ ἦν ἴδῃ Μύελλος[4] ἢ Παταικίσκος
ὁ Λαμπρίωνος, ἐκβαλεῦσι τὰς κούρας
65 δοκεῦντες ὄντως ἀργυρεῦν πεποιῆσθαι;
ὁ βοῦς δὲ κὠ ἄγων αὐτὸν ἤ θ᾿ ὁμαρτεῦσα
κὠ γρυπὸς οὗτος κὠ ἀνάσιλλος ἄνθρωπος
οὐχὶ ζόην βλέπουσιν ἡμέρην πάντες;
εἰ μὴ ἐδόκευν τι[5] μέζον ἢ γυνὴ πρήσσειν,
70 ἀνηλάλαξ᾿ ἄν, μή μ᾿ ὁ βοῦς τι πημήνῃ·
οὕτω ἐπιλοξοῖ, Κυννί, τῇ ἑτέρῃ κούρῃ.

ΚΥΝΝΩ

ἀληθιναί, φίλη. γὰρ αἱ Ἐφεσίου χεῖρες
ἐς πάντ᾿ Ἀπελλέω γράμματ᾿, οὐδ᾿ ἐρεῖς "κεῖνος
ὤνθρωπος ἓν μὲν εἶδεν, ἓν δ᾿ ἀπηρνήθη,"
75 ἀλλ᾿ ὅ οἱ ἐπὶ νοῦν γένοιτο, καὶ θέων[6] ψαύειν
ἠπείγεθ᾿· ὃς δ᾿ ἐκεῖνον ἢ ἔργα τὰ ἐκείνου
μὴ παμφαλήσας ἐκ δίκης ὀρώρηκεν,
ποδὸς κρέμαιτ᾿ ἐκεῖνος ἐν γναφέως οἴκῳ.

ΝΕΩΚΟΡΟΣ

κάλ᾿ ὕμιν, ὦ γυναῖκες, ἐντελέως τὰ ἱρὰ
80 καὶ ἐς λῷον ἐμβλέποντα· μεζόνως οὔτις
ἠρέσατο τὸν Παιήον᾿, ἤπερ οὖν ὑμεῖς.
ἰὴ ἰὴ Παίηον, εὐμενὴς εἴης
καλοῖς ἐπ᾿ ἱροῖς ταῖσδε κεἴ τινες τῶνδε
ἔασ᾿ ὀπυιηταί τε καὶ γενῆς ἆσσον.
85 ἰὴ ἰὴ Παίηον· ὧδε ταῦτ᾿ εἴη.

ΚΥΝΝΩ

εἴη γάρ, ὦ μέγιστε, κὐγίη πολλῇ
ἔλθοιμεν αὖτις μέζον᾿ ἶρ᾿ ἀγινεῦσαι
σὺν ἀνδράσιν καὶ παισί—Κοκκάλη καλῶς

τεμοῦσα μέμνεο τὸ σκελύδριον δοῦναι
90 τῷ νεωκόρῳ τοὔρνιθος, ἔς τε τὴν τρώγλην
    τὸν πελανὸν ἔνθες τοῦ δράκοντος εὐφήμως
    καὶ ψαιστὰ δεῦσον· τἄλλα δ᾽ οἰκίης ἕδρῃ
    δαισόμεθα, καὶ ἐπὶ μὴ λάθῃ φέρειν.

ΝΕΩΚΟΡΟΣ

αὕτη
    τῆς ὑγιίης μοι⁷ πρόσδος· ἢ γὰρ ἱροῖσιν
95 μέζων † ἁμαρτίης ἡ ὑγίη ᾽στὶ † τῆς μοίρης.⁸

¹ Ὑγιείης Herzog, Knox Ὑγιίης μοι Headlam.
² οἱ᾽ ἔργα; καινήν Herzog οἱ᾽ ἔργα! ⟨ν⟩αὶ ⟨μ⟩ὴν Knox.
³ δὴ Herzog, Knox γοῦν Headlam.
⁴ ἴδῃ Μύελλος Herzog, Knox ἴδῃσι Μύλλος Headlam.
⁵ τι Herzog, Knox ἂν Headlam.
⁶ θεῶν Herzog, Knox.
⁷ τῆς ὑγιίης· δῶ Herzog τῆς ὑγιίης δ᾽, ὅ οἱ Knox.
⁸ v. 93 αὕτη—v. 95 μοίρης attribuunt Kynno Herzog, Knox.

Dedications and Sacrifices to Asclepius

CYNNO. Hail to thee, Lord Paieon, ruler of Tricca, who
hast got as thine habitation sweet Cos and Epidaurus, hail to
Coronis thy mother withal and Apollo; hail to her whom thou
touchest with thy right hand, Hygieia, and those to whom
belong these honored altars, Panace and Epio and Iaso; hail
ye twain which did sack the house and walls of Laomedon,
healers of savage sicknesses, Podalirius and Machaon, and
what gods and goddesses soever dwell by thine hearth, Father
Paieon; come hither with your blessings and accept the after-
course of this cock[1] whom I sacrifice, herald of the walls of
my house. For we draw no bounteous nor ready spring; else
might we, perchance, with an ox or stuffed pig[2] of much
fatness and no humble cock, be paying the price of cure from
diseases that thou didst wipe away, Lord, by laying on us thy
gentle hands.

Set the tablet, Coccale, on the right of Hygieia.

COCCALE. La! Cynno dear, what beautiful statues! What
craftsman was it who worked this stone, and who dedicated it?

CY. The sons of Praxiteles—only look at the letters on the
base, and Euthies, son of Prexon, dedicated it.

CO. May Paeon bless them and Euthies for their beautiful

works. See, dear, the girl yonder looking up at the apple; wouldn't you think she will swoon away suddenly, if she does not get it? Oh, and yon old man, Cynno. Ah, in the Fates' name, see how the boy is strangling the goose. Why, one would say the sculpture would talk, that is if it were not stone when one gets close. La! in time men will be able even to put life into stones. Yes, only look, Cynno, at the gait of this statue of Batale, daughter of Myttes. Anyone who has not seen Batale may look at this image and be satisfied without the woman herself.

CY. Come along, dear, and I will show you a beautiful thing such as you have never seen in all your life. Cydilla, go and call the sacristan.[3] It's you I am speaking to, you who are gaping up and down! La! not an atom of notice does she take of what I am saying, but stands and stares at me worse than a crab! Go, I tell you again, and call the sacristan. You glutton, there is not a patch of ground, holy[?] or profane, that would praise you as an honest girl—everywhere alike your value is the same. Cydilla, I call this god to witness, that you are setting my wrath aflame, little as I wish my passion to rise. I repeat, I call him to witness; the day will come when you shall have cause to scratch your filthy noodle.

CO. Don't take everything to heart so, Cynno: she is a slave, and a slave's ears are oppressed with dullness.

CY. But it is day-time and the crush is getting greater, so stop there! for the door is thrown open[4] and there is access to the sacristy.

CO. Only look, dear Cynno, what works are those there! See these, you would say, were chiselled by Athene herself— all hail, Lady! Look, this naked boy, he will bleed, will he not, if I scratch him, Cynno; for the flesh seems to pulse warmly as it lies on him in the picture; and the silver toasting-iron, if Myellos and Pataiciscos, son of Lamprion, see it, won't their eyes start from their sockets when they suppose it real silver!

And the ox and its leader, and the girl in attendance, and this hook-nosed and this snub-nosed fellow, have they not all of them the look of light and life? If I did not think it would be unbecoming for a woman, I should have screamed for fear the ox would do me a hurt: he is looking so sideways at me, dear Cynno, with one eye.

CY.   Yes, dear, the hands of Apelles of Ephesus are true in all his paintings, and you cannot say that "he looked with favor on one thing and fought shy of another": no, whatever came into his fancy, he was ready and eager to essay off-hand, and if any gaze on him or his works save from a just point of view, may he be hung up by the foot at the fuller's!

SACRISTAN.   Your sacrifice is entirely favorable, ladies, with still better things in store; no one has appeased Paieon in greater sort than you. Glory, glory to thee, Paieon, mayst thou look with favor for fair offerings on these, and all that be their husbands or near of kin. Glory, glory, Paieon. So be it.

CY.   So be it, Almighty, and may we come again in full health once more bringing larger offerings, and our husbands and children with us.

Coccale, remember to carve the leg of the fowl off carefully and give it to the sacristan, and put the mess into the mouth of the snake reverently, and souse the meat-offering. The rest we will eat at home;[5] and remember to take it away.

S.   Ho there! give me some of the holy bread; for the loss of this[?] is more serious to holy men than the loss of our portion. (*)

[1] Cf. T. 523 ff.   [3] T. 495 ff.   [5] Cf. however, T. 510; 739.
[2] Cf. T. 536; 534.   [4] T. 485.

TEMPLE

483. Aristoteles, *Fragmenta*, 491 [Rose].[1]

Ἐπίδαυρος: πόλις πρὸς τῷ Ἄργει θηλυκῶς λεγομένη . . . ἐκαλεῖτο καὶ Μειλισσία, καὶ Αἱμηρὰ διὰ τὸ συνεχῶς αἱμάσσεσθαι τὸν βωμὸν τοῦ Ἀσκληπιοῦ ὑπὸ τῶν θυσιῶν. εἶτα Ἐπίταυρος καὶ Ἐπίδαυρος.

Epidaurus: A city near Argos, in the feminine gender . . . It was also called Meilissia, and also Haimera because the altar of Asclepius was continually stained with blood (*haimassesthai*) from the sacrifices.[2] Then it was called Epitauros and Epidauros.

[1] Cf. also **483a**. Stephanus Byzantius, *Ethnica*, s. v. Ἐπίδαυρος, same text; cf. **483b**. Suidas, *Lexicon*, s. v. Ἐπίδαυρος: ὄνομα τοπικόν· ἐν ᾧ ὁ Ἀσκληπιὸς ἐτιμᾶτο.
[2] Cf. T. 517.

484. Athenaeus, Deipnosophistae, XV, 48, 693 e.

Ὅτι δὲ δοθείσης τῆς τοῦ Ἀγαθοῦ Δαίμονος κράσεως ἔθος ἦν βαστάζεσθαι τὰς τραπέζας ἔδειξεν διὰ τῆς αὐτοῦ ἀσεβείας ὁ Σικελιώτης Διονύσιος. τῷ γὰρ Ἀσκληπιῷ ἐν ταῖς Συρακούσαις ἀνακειμένης τραπέζης χρυσῆς προπιὼν αὐτῷ ἄκρατον Ἀγαθοῦ Δαίμονος ἐκέλευσεν βασταχθῆναι τὴν τράπεζαν.

After the mixture to the Good Daemon had been given it was customary to have the tables removed, as is shown in the case of Dionysius of Sicily [430-367 B. C.] by his own sacrilege. For in Syracuse there was a gold table[1] dedicated to Asclepius; when Dionysius had drunk in his honor unmixed wine of the Good Daemon he ordered the table to be removed.

[1] Cf. T. 490.

485. Aristides, Oratio XLVII, 11.

. . . καὶ γὰρ εἶναι περὶ λύχνους ἤδη τοὺς ἱεροὺς τάς τε δὴ κλεῖς ἀνακομίζειν τὸν νεωκόρον. καὶ τυχεῖν ἐν τούτῳ κλεισθὲν τὸ ἱερόν, . . .

. . . for it was now the time to light the sacred candles and the sacristan[1] was bringing up the keys. And the temple happened to be closed at this time . . .[2]

[1] Cf. T. 482, v. 58; T. 495 ff.
[2] 485a. Aristides, *Oratio* LI, 28: Καὶ ἦν μὲν τῆς ὥρας τὸ μετὰ λύχνους τοὺς ἱερούς; cf. T. 544.

486. Aristides, Oratio XLVIII, 31.

Ἐδόκουν ἐν τοῖς προπυλαίοις ἑστάναι τοῦ ἱεροῦ, συνειλέχθαι δὲ καὶ ἄλλους πολλούς τινας, ὥσπερ ἡνίκα ἂν τὸ [ἱερὸν] καθάρσιον γίγνηται, εἶναι δὲ λευχείμονάς τε καὶ τἆλλα πρέποντι τῷ σχήματι.

I thought that I stood within the entrance of the temple and that many others had assembled, just as when a purification takes place, and that they were clad in white[1] and otherwise too in suitable fashion.

[1] Cf. T. 296; 513; cf. however, T. 787.

487. Iamblichus, De Vita Pythagorica, 27, 126.

Ἕτερον δέ, ξένου τινὸς ἐκβεβληκότος ἐν Ἀσκληπιείῳ ζώνην χρυσίον ἔχουσαν, καὶ τῶν μὲν νόμων τὸ πεσὸν ἐπὶ τὴν γῆν

κωλυόντων ἀναιρεῖσθαι, τοῦ δὲ ξένου σχετλιάζοντος, κελεῦσαι τὸ
μὲν χρυσίον ἐξελεῖν, ὃ μὴ πέπτωκεν ἐπὶ τὴν γῆν, τὴν δὲ ζώνην
ἐᾶν· εἶναι γὰρ ταύτην ἐπὶ τῆς γῆς.

Another (story is) of a certain foreigner who had dropped in
the Asclepieion a wallet containing gold, and when the stranger
made indignant complaints because the laws forbade picking up
what fell upon the ground, he [sc., Pythagoras] commanded
him to take out the gold which did not fall upon the ground,
but to let the wallet go, for this was on the ground.[1]

[1] Cf. T. 842.

488. Pausanias, Descriptio Graeciae, II, 27, 6.[1]

Ἐπιδαυρίων δὲ οἱ περὶ τὸ ἱερὸν μάλιστα ἐταλαιπώρουν, ὅτι μήτε
αἱ γυναῖκες ἐν σκέπῃ σφίσιν ἔτικτον καὶ ἡ τελευτὴ τοῖς κάμνουσιν
ὑπαίθριος ἐγίνετο· ὁ δὲ καὶ ταῦτα ἐπανορθούμενος κατεσκευάσατο
οἴκησιν· ἐνταῦθα ἤδη καὶ ἀποθανεῖν ἀνθρώπῳ καὶ τεκεῖν γυναικὶ
ὅσιον.

The Epidaurians about the sanctuary were in great distress,
because their women were not allowed to bring forth under
shelter,[2] and their sick were obliged to die under the open sky.
To remedy the inconvenience he [sc., Antoninus] provided a
building where a man may die and a woman may give birth to
her child without sin. (**)

[1] Cf. T. 739.   [2] Cf. T. 423, 1; 423, 2.

489. Aelianus, De Natura Animalium, VII, 13.[1]

. . . τὰ λείποντα δὲ τῶν ἀναθημάτων ἐπόθουν αἱ γραφαί τε καὶ
αἱ χῶραι ἔνθα ἀνέκειντο.

. . . moreover, the catalogues indicated the missing votive
offerings and so did the empty spaces where the votive offerings
had been set up.

[1] Cf. T. 731.

PRIESTS AND ATTENDANTS

490. Aristophanes, Plutus, 676-81.[1]

Ἔπειτ᾿ ἀναβλέψας ὁρῶ τὸν ἱερέα
τοὺς φθοῖς ἀφαρπάζοντα καὶ τὰς ἰσχάδας

ἀπὸ τῆς τραπέζης τῆς ἱερᾶς. μετὰ τοῦτο δὲ
περιῆλθε τοὺς βωμοὺς ἅπαντας ἐν κύκλῳ,
εἴ που πόπανον εἴη τι καταλελειμμένον·
ἔπειτα ταῦθ᾿ ἥγιζεν εἰς σάκταν τινά.

Then, glancing upwards, I behold the priest
Whipping the cheese-cakes and the figs from off
The holy table;[2] thence he coasted round
To every altar, spying what was left.
And everything he found he consecrated
Into a sort of sack.

[1] Cf. T. 421.
[2] Cf. T. 484.

491. Inscriptio Pergamena [*Inschriften von Pergamon*, II, no. 251;
2nd c. B. C.?].

Ἐπὶ πρυτάνεως Καβείρου, μηνὸ(ς) Πανθείου
εἰκάδι· ἔγνω βουλὴ καὶ δῆμος γνώμηι
στρατηγῶν περὶ τῆς ἱερωσύνης τοῦ Ἀσκληπι[οῦ,
ὅπως ὑπάρχηι εἰς τὸν ἅπαντα χρόνον

5   Ἀσκληπιάδηι καὶ τοῖς ἀπογόνοις τοῖς
Ἀσκληπιάδου· τύχηι τῆι ἀγαθῆι· δεδόχθαι
τῆι βουλῆι καὶ τῶι δήμωι· τὴν μὲν ἱερωσύνην
τοῦ Ἀσκληπιοῦ καὶ τῶν ἄλλων θεῶν τῶν ἐν τῶ[ι
Ἀσκλη[πι]είωι ἱδρυμένων εἶναι Ἀσκληπιάδου

10   τοῦ [Ἀρχί]ου καὶ τῶν ἀπογόνων τῶν Ἀσκληπιάδου
εἰς ἄ[π]α[ντ]α [τ]ὸν χρόνον καὶ στεφανηφορεῖν αὐτῶν
ἀεὶ τὸν ἔχοντα τὴν ἱερωσύνην, λαμβάνειν δὲ
καὶ γέρα τῶν θυομένων ἱερείων ἐν τῶι ἱερῶι
πάντων σκέλος δεξιὸν καὶ τὰ δέρματα καὶ τἆλλ[α

15   τραπεζώματα πάντα τὰ παρατιθέμεν[α
    ....σ[.. καρ]πεύεσθαι [δ]ὲ α[ὐ]τὸν καὶ τὸ ἱερ[ὸν

------ ὑ[πάρ]χειν εἰς ἄπαν[τ]α [τ]ὸν χρό[νον.

20   εἶναι δ]ὲ καὶ ἀτέλειαν Ἀσκληπιάδηι πάντων
ὧν] ἡ πόλις κυρία, καὶ εἰς τὸ λοιπὸν
ἀ]εὶ τῶι τὸν στέφανον ἔχοντι, ἀναγορεύεσθαι δ[ὲ
εἰς προεδρίαν τὸν ἱερέα ἐν ἅπασι τοῖς ἀγῶσιν.
ἐπιμελεῖσθαι δὲ καὶ τῆς εὐκοσμίας τῆς κατὰ τὸ ἱερ[ὸν

25　πάσης τὸν ἱερέ[α] ὡς ἂν αὐτῶι δοκῆ[ι
　　καλῶς ἔχειν καὶ ὁσίως, κυριεύοντα τῶν ἱερῶν παίδων.
　　ὅπως δὲ ταῦτα εἰς τὸν ἅπαντα χρόνον διαμένηι
　　βέβαια ('Α)σκληπιάδηι καὶ τοῖς ἀπογόνοις τοῖς
　　Ἀσκληπιάδου, ἐπιτελεῖν ὁρκωμόσιον τὴν πόλιν
30　ἐν τῆι ἀγορᾶι ἐπὶ τοῦ Διὸς τοῦ σωτῆρος τῶι βωμῶι
　　καὶ ὁμόσαι τ(ὰ)ς τιμουχίας · ἦ μὴν ἐμμενεῖν ἐν οἷς
　　ἐψήφισται ἡ πόλις Ἀσκληπιάδηι καὶ τοῖς ἀπογόνοις
　　τοῖς Ἀσκληπιάδου. τοὺς δὲ στρατηγοὺς τοὺς ἐπ[ὶ
　　Καβείρου πρυτάνεως ἐπιμεληθῆναι, ὅπως
35　συντελεσθῇ ὁ ὅρκος καθάπερ γέγραπται.
　　ἀναγράψαι δὲ αὐτοὺς καὶ τὸ ψήφισμα τόδ[ε
　　εἰς στήλας λιθίνας τρεῖς καὶ στῆσαι αὐτῶν
　　μίαν μὲν ἐν τῶι ἱερῶι τοῦ Ἀσκληπιοῦ ἐμ Περγάμῳ,
　　ἄλλην δὲ ἐν τῶι ἱερῶι τῆς Ἀθηνᾶς ἐν ἀκροπόλ[ει,
40　τὴ]ν δὲ τρίτην ἐμ Μυτιλήνηι ἐν τῶι ἱερῶι τοῦ
　　Ἀσκλ]ηπιοῦ. ἐγγράψαι δὲ καὶ εἰς τοὺς νόμους
　　τοὺς τ]ῆς πόλεως τὸ ψήφισμα τόδε καὶ
　　χρήσθω]σαν αὐτῶι νόμωι κυρίωι εἰς ἅπαντα τὸν χρόνον.

In the magistracy of Cabeirus, on the 20th day of the month Pantheius: The Council and Demos on the motion of the generals decreed concerning the priesthood of Asclepius that it should belong to Asclepiades and to the descendants of Asclepiades for all time to come. With good fortune. Be it resolved by the Council and the Demos, that the priesthood of Asclepius and of the other gods who are residing in the Asclepieion shall belong to Asclepiades, the son of Archias,[1] and to the descendants of Asclepiades for all time to come, and that he who among them at any particular time is holding the priesthood shall wear a crown, and that he shall take as a perquisite the right leg and the skin of all the sacrificial animals offered in the temple, and all the other offerings which are dedicated on the holy table[2] along with them . . . and he shall enjoy the fruits and the holy . . . for all time to come. And Asclepiades shall be exempted from all obligations that the city is entitled to impose, and likewise for all time to come shall he be exempted whoever wears the crown, and the priest shall be entitled to a front seat at all the games.[3] And furthermore the priest shall have charge of the general good conduct within the sanctuary as it may seem right to him and fitting, and he shall have power over the temple servants. In order that this may be safeguarded for all time to come for Asclepiades and the descendants of Asclepiades, the city shall undertake a solemn oath in the Agora at the altar of Zeus Savior, and the officials shall swear to it: namely that the city shall abide by what it has decreed for Asclepiades and the descendants of Asclepiades.[2] It shall be the duty of the generals of the magistracy of Cabeirus to see to it that the oath be fulfilled as it is written. They must inscribe this decree on three stone tablets and set up one of them in the sanctuary of Asclepius at Pergamum, the other one in the sanctuary of Athena on the Acropolis, the third one at Mytilene in the sanctuary of Asclepius. And they must list this decree among the laws of the city and must treat it as a regular law for all time to come.

[1] Cf. T. 801.   [2] Cf. T. 490.
[3] Cf. 491a. *Inscriptiones Graecae*, II², no. 5045 [time of Hadrian]: Ἱερέως / Ἀσκληπιοῦ / Π[αί]ω⟨ν⟩ος? [inscribed on front seat in the theater of Dionysus at Athens].
[4] Cf. 491b. Aristides, *Oratio* L, 102: . . . ἐψηφίζετο [sc., ὁ δῆμος] δέ μοι τὴν ἱερωσύνην τοῦ Ἀσκληπιοῦ. Cf. also 491c. Aristides, *ibid.*, 64: . . . ὁ ἱερεὺς ὁ τοῦ Ἀσκληπιοῦ . . . καὶ ὁ τούτου πάππος . . . ; cf. also T. 791. Cf. however T. 494: . . . λαχὼν ἱερεὺς τοῦ Ἀσκληπιοῦ. Cf. also 491d. *Inscriptiones Graecae*, II², no. 4532 [beginning of 3rd c. A.D.]: . . . ἐπὶ ἱε[ρέ]ως διὰ βίου. Cf. also 491e. *Inscriptio Chalcedonia*, no. 3052 [Collitz-Bechtel; 2nd c. B.C.]: Ὠνείσθω δὲ ὅς [κα ἦι ὁλ]όκλαρος καὶ ὧι δαμοσιοργίας μ[έτε/στι]· ἐξέστω δὲ καὶ [π]αιδὶ ὠνείσθαι, ἄ[λ/λωι] δὲ μηθενὶ ἐξέστω τὰν ἱερωτεία[ν ἢ ἑαυτ]ῶι. Ὅς δέ κα εἴπηι ἢ προαισιμνάσηι [ἢ ἐν βου/λᾶι] ἢ ἐν δάμωι ἢ ἄλλει καὶ χ' ὁπειοῦν, [ὡς δεῖ ἀφε]λέσθαι τὸν πριάμενον τὰν ἱερω[τείαν, / χιλ]ίας δραχμὰς ἀποτεισάτω ἱερὰ[ς τοῦ / Ἀσ]κλαπιοῦ·

492. Tertullianus, De Pallio, I, 1-2.

Pallii extrinsecus habitus et ipse quadrangulus, ab utroque laterum regestus et cervicibus circumstrictus in fibulae morsu, humeris acquiescebat. instar eius hodie Aesculapio jam vestro sacerdotium est.

The outward arrangement of the mantle, itself too quadrangular, thrown back from either side, and meeting closely round the neck in the grip of the buckle, used to repose on the shoulders. Its counterpart is now the priestly dress sacred to your [sc., the Carthaginians'] Asclepius.[1] (*)

[1] Cf. **687**; cf. 492a. Ps. Aristides, *Oratio* XXX, 27: . . . τῆς ἁλουργίδος τῆς ἱερᾶς τοῦ σωτῆρος . . . .

493. Aristides, Oratio XLVII, 61.

Ὁ μὲν γὰρ θεὸς . . . ἔδωκεν ἀλεξιφάρμακα καὶ ὑποδήματα Αἰγύπτια, οἷσπερ οἱ ἱερεῖς χρῆσθαι νομίζουσιν.

For the god . . . prescribed antidotes and Egyptian sandals[1]
which the priests are accustomed to use.[2]

[1] Cf. however T. 687.　　　　　　　[2] For the fillet of the priest, cf. T. 850, v. 676.

494. Inscriptiones Graecae, II², no. 1163 [288-7 B. C.].[1]

Θεο[ί]
Πρόξενος Πυλαγόρου Ἀχερδούσι-
ος εἶπεν· ἐπειδὴ Φυλεὺς λαχὼν ἱερεὺς τοῦ Ἀσκληπιοῦ ἐπὶ Ἰσαίο[υ ἄ]-
ρχοντος τάς τε θυσίας ὅσας πρ[οσ]-
[ῆ]κεν αὐτὸν θῦσαι πάσας καλῶς κα-
[ὶ] φιλοτίμως τέθυκεν ὑπὲρ τοῦ δήμου τοῦ Ἀθηναίων, ἐπιμελεῖται δ-
ὲ καὶ τῆς κληρώσεως τῶν δικαστ[η]-
[ρ]ίων καὶ τῶν ἄλλων ὧν αὐτῶι οἵ τ[ε]
νόμοι καὶ τὰ ψηφίσματα προστάτ-
τουσιν δικαίως καὶ κατὰ το[ὺ]ς νόμους καὶ διὰ ταῦτα ἥ τε βουλὴ καὶ
ὁ δῆμος ἐπεινέκασιν αὐτόν· ἀγα[θ]-
[ε]ῖ τύχει δεδόχθαι Ἱπποθωντ[ί]δ[α]-
[ι]ς ἐπαινέσαι Φυλέα Χαιρίου Ἐλ[ε]-
[υσ]ίνιον εὐσεβείας ἕνεκα τῆς πρ-
ὸ[ς] τοὺς θεοὺς καὶ δικαιοσύνη[ς κ]-
α[ὶ] φιλοτιμίας τῆς εἰς τοὺς φυλέ-
τας καὶ τὸν δῆμον τὸν Ἀθηναί[ω]ν κ-
αὶ στεφανῶσαι αὐτὸν θαλλοῦ στε-
φάνωι. ἀναγράψαι δὲ τόδε τὸ ψήφισμα τοὺς ἐπιμελητὰς τῆς φυλῆς ἐ-
ν στήλαις λιθίναις καὶ στῆσαι τ-
ὴν μὲν ἐν τῶι Ἀσκληπιείωι, τὴν δὲ
ἐν τῶι Ἱπποθωντίωι· ὅ τι δ᾽ ἂν ἀνάλ-
ωμα γένηται λογίσασθαι τεῖ φ[υλ]-
εῖ.

GODS

Proxenus, son of Pylagoras, of Acherdous, proposed this
decree: Whereas, Phyleus, chosen by lot priest of Asclepius
in the archonship of Isaeus, fittingly and zealously performed
in behalf of the Demos of the Athenians all of the sacrifices so
many as it was proper that he perform; and whereas, he

supervises the allotment of the courts and discharges justly and according to the laws the other duties which the laws and the decrees enjoin upon him; and whereas, for these reasons the Council and the Demos have commended him; therefore, with good fortune, be it resolved, that Phyleus, son of Chaerias, of Eleusis be commended by the Tribesmen of Hippothontis because of his piety toward the gods and because of the qualities of justice and of zeal manifested to his fellow-tribesmen and to the Demos of the Athenians, and that he be crowned with an olive wreath;[2] further, that the *epimeletae* of the Tribe cause this decree to be inscribed on stone stelae and have one erected in the Asclepieion, the other in the Hippothontion, the cost of the action being charged to the Tribe.

[1] Dated 286-5 B. C. by Pritchett and Meritt.   [2] Cf. T. 503.

495. Aristides, Oratio XLVIII, 29.

. . . ἦν Φιλάδελφος τῶν νεωκόρων ἄτερος . . .

. . . Philadelphus was one of the two sacristans[1] . . .

[1] Cf. T. 482, v. 90; T. 485; 496-98; cf. also T. 731.

496. Herondas, Mimiambi, IV, 88-90.[1]

Κοκκάλη καλῶς
τεμοῦσα μέμνεο τὸ σκελύδριον δοῦναι
τῷ νεωκόρῳ τοὔρνιθος . . .

Coccale, remember to carve the leg of the fowl off carefully and give it to the sacristan . . .

[1] Cf. T. 482.

497. Aristides, Oratio XLVIII, 35.

Ἔδοξεν . . . μεταπέμψασθαι καὶ τὸν νεωκόρον . . . εἰώθειν κοινοῦσθαι τὰ πολλὰ τῶν ἐνυπνίων αὐτῷ.

It seemed good . . . to summon the sacristan also . . . I was wont to communicate to him most of the dreams.

498. Aristides, Oratio XLVIII, 47.

Οἵ τε γὰρ νεωκόροι . . . καὶ πάντες οἱ περὶ τὸν θεὸν θεραπευταὶ καὶ τάξεις ἔχοντες . . .

For the sacristans . . . and all the servants of the god and those who had posts in the temple [1] . . .

[1] Cf. T. 561; 562: ἀοιδοί; 553: ἀρρηφοροῦσαν; 432: βαλανεύς; 422: ζάκορος; 498a. *Inscriptiones Graecae*, II², no. 4073: ὑποζακορεύοντος; 544: θυρωρός; 562: ἱερομνήμονες; 787: ἱεροποιοί; 498b. *Inscriptiones Graecae*, IV², 1, no. 438: ἱεροφάντης; 553: κλειδοῦχος; 421, v. 670: πρόπολος; 423, 5: πυρφορῶν; 561: φρουροί.

499. Pausanias, Descriptio Graeciae, X, 32, 12.[1]

Ἐντὸς μὲν δὴ τοῦ περιβόλου τοῖς τε ἱκέταις καὶ ὅσοι τοῦ θεοῦ δοῦλοι, τούτοις μὲν ἐνταῦθά εἰσι καὶ οἰκήσεις.

Within the precincts are dwellings for both the suppliants of the god and his servants.

[1] Cf. T. 719.

500. Aelianus, De Natura Animalium, VII, 13.[1]

. . . κύων καὶ τῶν ζακόρων ἀμείνων ἐς ἀγρυπνίαν . . .

. . . a dog, better than the temple attendants in keeping sleepless vigil . . .

[1] Cf. T. 731.

RITUAL

PRIVATE WORSHIP

501. Theocritus, Epigrammata, VII.[1]

Εἰς ἄγαλμα Ἀσκληπιοῦ

Ἦλθε καὶ ἐς Μίλητον ὁ τοῦ Παιήονος υἱός,
    ἰητῆρι νόσων ἀνδρὶ συνοισόμενος
Νικίᾳ, ὅς μιν ἐπ᾽ ἦμαρ ἀεὶ θυέεσσιν ἱκνεῖται,
    καὶ τόδ᾽ ἀπ᾽ εὐώδους γλύψατ᾽ ἄγαλμα κέδρου,
Ἠετίωνι χάριν γλαφυρᾶς χερὸς ἄκρον ὑποστὰς
    μισθόν. ὁ δ᾽ εἰς ἔργον πᾶσαν ἀφῆκε τέχνην.

For a statue of Asclepius

Even to Miletus he hath come, the son of Paeon, to dwell with one that is a healer of all sickness, with Nicias, who with sacrifices approaches him day by day forever, and hath let carve this statue out of fragrant cedar-wood promising to Eëtion [of

Amphipolis, *ca.* 275 B. C.] a high guerdon for his skill of hand: on this work Eëtion has put forth all his craft. (**)

¹ = VIII [Wilamowitz].

AVERTING OBSERVANCES

502. Artemidorus, Onirocritica, V, 66.¹

. . . δεῖν φυλάττεσθαι καὶ θύειν ἀποτρόπαια τῷ θεῷ.

. . . that he must be on his guard and perform averting sacrifices to the god.

¹ Cf. T. 454.

503. Libanius, Epistulae, 1303, 1.

Τὸν μὲν ἐκ τοῦ ἱεροῦ θαλλὸν ἔχω, γέγονε δὲ ἀπ' αὐτοῦ μοι πλέον οὐδέν, πλὴν εἰ τὸ τῶν ἰατρῶν ἔργον δεῖ νομίζειν τοῦ θεοῦ. καὶ δοκείτω γε οὕτω καὶ ἔστω· καλὸν γὰρ ἅμα καὶ ἀσφαλὲς ἥδε ἡ δόξα.

I have the olive shoot¹ from the sanctuary, but nothing more has come to me from it unless the work of the physicians must be considered the god's. Thus let it be considered and so be it. For this belief is both fair and sure.

¹ Cf. T. 494.

504. Aristides, Oratio XLVIII, 27.

Ἀλλὰ δεῖν οὕτω ποιεῖν, πρῶτον μὲν ἀναβάντα ἐπὶ τὸ ζεῦγος ἐλθεῖν ἐπὶ τὸν ποταμὸν τὸν διὰ τῆς πόλεως ῥέοντα, γενόμενον δὲ οὗ ἐστιν ἤδη ἔξω τῆς πόλεως ἱερὰ δρᾶσαι ἐπιβόθρια, οὕτω γὰρ αὐτὰ προσεῖπεν· ἔδει δὲ ἄρα βόθρους ὀρύξαντα ἐπ' αὐτῶν δρᾶσαι τὰ ἱερὰ οἷστισιν δὴ καὶ ἔδει θεῶν· ἔπειτα ἀναστρέφοντα λαβόντα κέρματα διαβαίνειν τε τὸν ποταμὸν καὶ ἀπορρίπτειν, καὶ ἔτερ' ἄττα, οἶμαι, πρὸς τούτοις ἐκέλευσεν· μετὰ δὲ ταῦτα ἐλθόντα εἰς τὸ ἱερὸν θῦσαι τῷ Ἀσκληπιῷ τέλεια καὶ στῆσαι κρατῆρας ἱεροὺς καὶ νεῖμαι μοίρας ἱερὰς ἅπασιν τοῖς συμφοιτηταῖς· δεῖν δὲ καὶ τοῦ σώματος αὐτοῦ παρατέμνειν ὑπὲρ σωτηρίας τοῦ παντός. ἀλλὰ γὰρ εἶναι τοῦτο ἐργῶδες· τοῦτο μὲν καὶ δὴ παριέναι μοι, ἀντὶ δὲ τούτου τὸν δακτύλιον ὃν ἐφόρουν περιελόμενον ἀναθεῖναι τῷ Τελεσφόρῳ. τὸ γὰρ αὐτὸ ποιεῖν ὥσπερ ἂν εἰ τὸν δάκτυλον αὐτὸν

προείμην· ἐπιγράψαι δὲ εἰς τὴν σφενδόνην τοῦ δακτυλίου
‘ Κρόνου παῖ ’· ταῦτα ποιοῦντι σωτηρίαν εἶναι.

(I dreamed) that I should proceed in the following way:
first, mounting the chariot, I should go to the river which
flows through the city and then, when I reached the spot
where it leaves the city, I should perform the ἱερὰ ἐπιβόθρια
[*i. e.*, sacrifices in ritual pits]; for thus he [*sc.*, Asclepius]
named these rites. Having dug the pits, then, I should perform
the sacred rites over them to whomever of the gods it is fitting.
Next, turning back and taking up small coins, I should cross
the river and throw them away. And I believe he gave some
other instructions in addition to these. Afterwards, I should
go to the holy shrine and offer perfect sacrificial animals to
Asclepius and set up holy craters and distribute holy portions
to all the fellow-pilgrims. And (he indicated) that it was also
imperative to cut off part of the body itself in behalf of the
safety of the whole. This, however, would be too great a
demand, and from it he would exempt me. Instead, I should
take off the ring which I was wearing and offer it to Telesphorus.[1] For this would do the same as if I offered the finger
itself. Furthermore, I should inscribe on the band of the ring
“ Son of Cronus.” After this there would be salvation.

[1] Cf. T. 287.

505. Livius, Ab Urbe Condita, XL, 37, 2-3.

Postremo prodigii loco ea clades haberi coepta est. C. Servilius
pontifex maximus piacula irae deum conquirere iussus, decemviri libros inspicere, consul Apollini Aesculapio Saluti dona
vovere et dare signa inaurata; quae vovit deditque. decemviri
supplicationem in biduum valetudinis causa in urbe et per
omnia fora conciliabulaque edixerunt; maiores duodecim annis
omnes coronati et lauream in manu tenentes supplicaverunt.

Finally the disaster came to be regarded as a portent. Gaius
Servilius, chief pontiff [in 180 B. C.], was directed to inquire
into the manner of atoning the wrath of the gods and the
decemvirs to look into the Books; the consul was ordered to
vow gifts and to give gilded statues[1] to Apollo, Asclepius and
Salus; these he vowed and gave. The decemvirs proclaimed a

two-day period of prayer for health, not only in the City but
in all the rural settlements and communities; all people above
the age of twelve, wearing crowns [2] and carrying laurel branches
in their hands,[3] made the supplication.[4] (*)

[1] Cf. T. 545.   [2] Cf. however T. 296; 593.
[3] Cf. T. 572.   [4] Cf. T. 583.

506. Theodoretus, Graecarum Affectionum Curatio, XII, 96-97.

Τοῦτο γὰρ δὴ καὶ ὁ Πορφύριος ἐν οἷς καθ᾽ ἡμῶν ξυνέγραψεν
εἴρηκε· " νυνὶ δέ," φησί, " θαυμάζουσιν εἰ τοσούτων ἐτῶν
κατείληφε νόσος τὴν πόλιν, Ἀσκληπιοῦ μὲν ἐπιδημίας καὶ τῶν
ἄλλων θεῶν οὐκ ἔτι οὔσης. Ἰησοῦ γὰρ τιμωμένου, οὐδὲ μιᾶς
δημοσίας τις θεῶν ὠφελείας ᾔσθετο." ταῦτα ὁ πάντων ἡμῖν
ἔχθιστος Πορφύριος εἴρηκε, καὶ ἀναφανδὸν ὡμολόγησεν ὡς
πιστευόμενος ὁ Ἰησοῦς φρούδους ἀπέφηνε τοὺς θεούς, καὶ μετὰ
τὸν σταυρὸν καὶ τὸ σωτήριον πάθος οὐκ ἔτι φενακίζει τοὺς
ἀνθρώπους Ἀσκληπιός, οὐδὲ ἄλλος τις τῶν καλουμένων θεῶν.
ἅπαντα γὰρ αὐτῶν τὸν ὁρμαθὸν οἷόν τινας νυκτερίδας τῷ σκότῳ
παρέπεμψεν ἀνατεῖλαν τὸ φῶς.

For Porphyry, also, in the work which he wrote against us
said this: "Now, then," he says, "they are amazed if disease
has held the city for so many years, since Asclepius and the
other gods no longer sojourn there. For, since Christ is honored, none of the gods have enjoyed even one public service."
This is what Porphyry said, that most hateful to us of all men,
and openly he admitted that the belief in Christ has made the
gods vanish, and that since the Crucifix and the Passion Asclepius no longer deceives mankind, nor does any other of the
so-called gods. The whole chain of these the rising light has
chased into the darkness like bats.

507. Augustinus, De Civitate Dei, III, 17.[1]

Ubi ergo erant illi dii, qui propter exiguam fallacemque mundi
huius felicitatem colendi existimantur, cum Romani, quibus se
colendos mendacissima astutia venditabant, tantis calamitatibus
vexarentur? ubi erant ... quando post longas et graves Romae
seditiones, quibus ad ultimum plebs in Ianiculum hostili diremp-
tione secesserat, huius mali tam dira calamitas erat, ut eius rei
causa, quod in extremis periculis fieri solebat, dictator crearetur

Hortensius, qui plebe revocata in eodem magistratu exspiravit, quod nulli dictatori ante contigerat et quod illis diis iam praesente Aesculapio gravius crimen fuit?

Where, then, were those gods who supposedly are to be worshipped for the slight and delusive prosperity of this world, when the Romans, to whom they sold their cult by the most fallacious astuteness, were harassed by so many calamities? Where were they . . . when, after serious and long-continued rebellions in Rome, the plebs finally withdrew to the Janiculum in hostile secession? The danger of this evil was so grave that because of it Hortensius was created dictator, a measure which was customarily adopted only in extreme emergencies, and Hortensius brought back the plebs and died while still in office, an event without precedent in the case of any dictator and one which reflected more serious shame upon these gods because now even Asclepius was present. (**)

[1] Cf. T. 853.

LOCAL RITES

508. Plutarchus, Septem Sapientium Convivium, 16, 159 E-F.

Ἄλλον δὲ πρὸς τὴν ὑγίειαν οἶμαι χαλεπανεῖν· ‘δεινὸν γὰρ εἰ μηδενὸς νοσοῦντος οὐ στρωμνῆς ἔτι μαλακῆς ὄφελος οὐ κλίνης, οὐκ Ἀσκληπιῷ θύσομεν οὐκ ἀποτροπαίοις, ἰατρικὴ δὲ μετ’ ὀργάνων καὶ φαρμάκων ἀποκείσεται τοσούτων ἀκλεὴς καὶ ἀπόθεστος.’

Still another man, I imagine, may entertain a violent hatred against health: " For it will be a terrible thing if nobody is ill, and there is no longer any use for a soft bed or couch, and we shall not offer sacrifice to Asclepius or the averting deities, and the profession of medicine together with its numerous instruments and remedies shall be left inglorious and despised." [1] (*)

[1] Cf. T. 540.

509. Festus, De Verborum Significatu, 237 M [p. 268, Lindsay].

Peregrina sacra appellantur, quae aut evocatis dis in oppugnandis urbibus Romam sunt † conata † aut quae ob quasdam religiones per pacem sunt petita, ut ex Phrygia Matris Magnae, ex Graecia Cereris, Epidauro Aesculapi: quae coluntur eorum more, a quibus sunt accepta.

Foreign are called either those rites which, after the gods of besieged cities had been called forth, were brought[?] to Rome, or those which were sent for in peace-time on account of a certain religious awe, such as from Phrygia those of the Magna Mater, from Greece those of Ceres, from Epidaurus those of Asclepius:[1] these are performed according to the customs óf those from whom they were taken over.

[1] Cf. T. 737.

510. Pausanias, Descriptio Graeciae, II, 27, 1.[1]

Τὰ δὲ θυόμενα, ἤν τέ τις Ἐπιδαυρίων αὐτῶν ἤν τε ξένος ὁ θύων ᾖ, καταναλίσκουσιν ἐντὸς τῶν ὅρων· τὸ δὲ αὐτὸ γινόμενον οἶδα καὶ ἐν Τιτάνῃ.

The sacrifices, whether offered by one of the Epidaurians themselves or by a foreigner, are consumed within the bounds [sc., of the sanctuary].[2] I know that the same thing is done also at Titane. (*)

[1] Cf. T. 739.
[2] Cf. 510a. *Inscriptiones Graecae*, II², no. 1364 [1st c. A. D. or B. C.]: . . . τῶν δὲ κρεῶν μὴ / φέρεσθαι.   Cf. however, T. 482, vv. 92 f.

INITIAL CEREMONIES

511. Inscriptiones Graecae, IV², 1, no. 121, 5.[1]

. . . ὡς δὲ προεθύσατο καὶ [ἐπόησε τὰ] νομιζόμενα, μετὰ τοῦτο . . .

. . . when he [sc., the voiceless boy] had performed the preliminary sacrifices and fulfilled the usual rites, thereupon . . .

[1] Cf. T. 423. 5.

512. Pausanias, Descriptio Graeciae, V, 13, 3.

Τὸ δὲ αὐτὸ καὶ ἐν τῇ Περγάμῳ τῇ ὑπὲρ ποταμοῦ Καΐκου πεπόνθασιν οἱ τῷ Τηλέφῳ θύοντες. ἔστι γὰρ δὴ οὐδὲ τούτοις ἀναβῆναι πρὸ λουτροῦ παρὰ τὸν Ἀσκληπιόν.

The same rule applies to those who sacrifice to Telephus at Pergamum[1] on the river Caïcus; these too may not go up to the temple of Asclepius before they have bathed.[2]

[1] Cf. T. 607.
[2] T. 421, vv. 656-58.

513. Inscriptio Pergamena [*Inschriften von Pergamon*, II, no. 264; Imperial time].

– – εἰσπορενέσθ]ω εἰς – – – – – – – – – – – – – – – – – –
– – – – – – ἡμέ]ρας δέκ[α] ἀποδέξετ(α)[ι – – – – – – – –
– – – – – – – εἰσιὼν λουσάμενος, ἐὰ[ν – – – – – – – – – –
– – – – ἀπαλ]λάσσεσθαι περικαθαιρέ[τω – – – – – – – – –
– – – χιτῶ]νι λευκῶι καὶ [θ]είωι καὶ δ[άφνηι – – – – – –
– – – ται]νίας, ἅς περικ[α]θαιρέτω ω – – – – – – – – – –
– – – εἰσ]πορενέσθω πρὸς τὸν θεὸν τ – – – – – – – – – –
εἰς τὸ μ]έγα ἐνκοιμητήριον ὁ ἐγκο[ιμησόμενος – – – – – –
– – – ἱερε]ίοις λευκοῖς ἁγνοῖς ἐλάας ἔ[ρνεσιν ἐστεμμένοις
– – – μήτε δακτ]ύλιον μήτε ζώνην μ[ήτε – – – – – – – –
– – – – – – – – – – – – – ἀνυπ]ό[δ]ητο[ν – – – – – – – – –

. . . let him enter into . . .
. . . he will have ten days . . .
. . . entering, after bathing, if . . .
. . . to be set free, let him purify completely . . .
. . . in a white chiton[1] and with brimstone, and with laurel . . .
. . . with fillets which let him purify completely . . .
. . . let him go toward the god . . .
into the great incubation room, the incubant . . .
. . . with pure white sacrificial victims garlanded with olive shoots[2]
. . . neither seal-ring nor belt nor . . .
. . . barefoot . . .[3]

[1] Cf. T. 296 ; 486 ; cf. however T. 787.
[2] Cf. e. g. T. 296 ; 593.
[3] **513a.** *Inscriptio Ptolemaica* [*Revue Archéologique*, I, 2, 1883, p. 182; ca. 3rd c. B. C.]: τοὺς δὲ εἰσιόντας εἰς τ[ὸ] / ἀγνεύειν κατὰ ὑποκ / ἀπὸ πάθους ἰδίου καὶ / ἡμέρας ξ' ἀναπαλλ / χη ἐκτρωσμοῦ συν / τετοκυίας καὶ τρεφούσης / καὶ ἐὰν ἐχθῇ (?) ιδ' τοὺς δὲ ἄ[νδρας / ἀπὸ γυναικὸς β', τὰς δὲ γ[υναῖκας / ἀκολούθως τοῖς ἀνδράσ[ιν / αν ἐκτρωσμοῦ μ/τὴν δὲ τεκοῦσαν καὶ τρέ[φουσαν / ἐὰν δὲ ἐχθῇ τὸ βρέφος / ἀπόβατα μηνιῶν ξ' / ἀνδρὸς β' μυρσίνην δὲ [ends of lines broken off].

514. Aristophanes, Plutus, 660-61.[1]

> Ἐπεὶ δὲ βωμῷ πόπανα καὶ προθύματα
> καθωσιώθη, πέλανος Ἡφαίστου φλογί,
>
> . . . .

> There on the altar honey-cakes and bakemeats
> Were offered, food for the Hephaestian flame,
>
> . . . .

[1] Cf. T. 421; cf. also T. 490.

515. Inscriptiones Graecae, II², no. 4962 [beginning 4th c. B. C.].[1]

> Θεοί
>
> Κατὰ τάδε προθύεσθα-
> ι· Μαλεάτηι πόπανα τρ-
> ία· Ἀπόλλωνι πόπανα τ-
> ρία· Ἑρμῆι πόπανα τρί-
> α· Ἰασοῖ πόπανα τρία· Ἀ-
> κεσοῖ πόπανα τρία· Πα-
> νακείαι πόπανα τρία·
> κυσὶν πόπανα τρία· κυ-
> νηγέταις πόπανα τρί[α].

> Gods
>
> The initial offerings are to be made in the following way:
> to Maleas three sacrificial cakes, to Apollo three cakes, to
> Hermes three cakes, to Iaso three cakes, to Aceso three cakes,
> to Panacea three cakes, to the dogs three cakes, to the hunts-
> men three cakes.

[1] Dated 355-4 B. C. by Pritchett and Meritt.

516. Inscriptiones Graecae, IV², 1, no. 128, iii, 27-31 [Isyllus, *ca.*
300 B. C.].

> Πρῶτος Μᾶλος ἔτευξεν Ἀπόλλωνος Μαλεάτα
>   βωμὸν καὶ θυσίαις ἠγλάϊσεν τέμενος·
> οὐδέ κε Θεσσαλίας ἐν Τρίκκηι πειραθείης
> εἰς ἄδυτον καταβὰς Ἀσκληπιοῦ, εἰ μὴ ἀφ' ἁγνοῦ
> πρῶτον Ἀπόλλωνος βωμοῦ θύσαις Μαλεάτα.

> Malos first built the altar of Apollo Maleatas and made the
> precinct splendid with sacrifices. Not even in Thessalian

Tricca would you attempt to go down into the Adyton of Asclepius [1] unless you first sacrifice [2] on the holy altar of Apollo Maleatas.

[1] Cf. T. **423**, 38.   [2] Cf. T. **511**.

517. Philostratus, Vita Apollonii, I, 10.

Ἰδὼν δὲ ἀθρόον ποτὲ ἐν τῷ βωμῷ αἷμα, καὶ διακείμενα ἐπὶ τοῦ βωμοῦ τὰ ἱερά, τεθυμένους τε βοῦς Αἰγυπτίους καὶ σῦς μεγάλους, καὶ τὰ μὲν δέροντας αὐτούς, τὰ δὲ κόπτοντας, χρυσίδας δὲ ἀνα-κειμένας δύο καὶ λίθους ἐν αὐταῖς τῶν Ἰνδικωτάτων καὶ θαυμασίων, προσελθὼν τῷ ἱερεῖ "τί ταῦτα"; ἔφη, "λαμπρῶς γάρ τις χαρίζεται τῷ θεῷ." ὁ δὲ "θαυμάσῃ," ἔφη, "μᾶλλον, ὅτι μήτε ἱκετεύσας ποτὲ ἐνταῦθα μήτε διατρίψας, ὃν οἱ ἄλλοι χρόνον, μήτε ὑγιάνας πω παρὰ τοῦ θεοῦ, μηδ' ἅπερ αἰτήσων ἦλθεν ἔχων. χθὲς γὰρ δὴ ἀφιγμένῳ ἔοικεν, ὁ δὲ οὕτως ἀφθόνως θύει. φησὶ δὲ πλείω μὲν θύσειν, πλείω δὲ ἀναθήσειν, εἰ πρόσοιτο αὐτὸν ὁ Ἀσκληπιός. ἔστι δὲ τῶν πλουσιωτάτων· κέκτηται γοῦν ἐν Κιλικίᾳ βίον πλείω ἢ Κίλικες ὁμοῦ πάντες· ἱκετεύει δὲ τὸν θεὸν ἀποδοῦναί οἱ τὸν ἕτερον τῶν ὀφθαλμῶν ἐξερρυηκότα." ὁ δὲ Ἀπολλώνιος . . . "δοκεῖ μοι," ἔφη, "ὦ ἱερεῦ, τὸν ἄνθρωπον τοῦτον μὴ προσ-δέχεσθαι τῷ ἱερῷ, μιαρὸς γάρ τις ἥκει καὶ κεχρημένος οὐκ ἐπὶ χρηστοῖς τῷ πάθει, καὶ αὐτὸ δὲ τὸ πρὶν εὑρέσθαι τι παρὰ τοῦ θεοῦ πολυτελῶς θύειν οὐ θύοντός ἐστιν, ἀλλ' ἑαυτὸν παραιτου-μένου σχετλίων τε καὶ χαλεπῶν ἔργων." ταῦτα μὲν ὁ Ἀπολλώνιος. ὁ δ' Ἀσκληπιὸς ἐπιστὰς νύκτωρ τῷ ἱερεῖ "ἄπιτω," ἔφη, "ὁ δεῖνα τὰ ἑαυτοῦ ἔχων, ἄξιος γὰρ μηδὲ τὸν ἕτερον τῶν ὀφθαλμῶν ἔχειν."

One day he [*sc.*, Apollonius] saw a flood of blood upon the altar,[1] and victims laid out upon it, Egyptian bulls that had been sacrificed and great hogs, and some the attendants were flaying and others they were cutting up; and two gold vases had been dedicated set with jewels, the rarest and most beautiful that India can provide.[2] So he went up to the priest and said: "What is all this; surely some one is making a very handsome gift to the god?" And the priest replied: "Rather you may be surprised at a man's offering all this without ever having made his prayer here and without having stayed with us as long as other people do, and without ever obtaining health from the god nor securing all the things he came to ask for here. For he appears to have come only yesterday and yet he is sacrificing on this lavish scale. And he declares that he will sacrifice more victims, and dedicate more gifts, if Asclepius will hearken to him.[3] And he is one of the richest men; at any rate he owns in Cilicia an estate bigger than all the Cilicians together possess. And he is supplicating the god [4] to restore to him one of his eyes that has been destroyed." But Apollonius . . . said: "It seems to me, O Priest, that we ought not to welcome this fellow in the temple: for he is some ruffian who has come here; and that he is afflicted in this way is due to some sinister reason: nay, his very conduct in sacrificing on such a magnificent scale before he has gained anything from the god is not that of a votary, but rather of a man who is begging himself off from the penalty of some horrible and cruel deeds." This was what Apollonius said: and Asclepius appeared to the priest by night, and said: "This man shall go away at once with all his possessions, for he does not deserve even to have the other eye." (*)

[1] Cf. T. 483.   [2] Cf. T. 537 ff.   [3] Cf. T. 523.   [4] Cf. T. 576 ff.

INTERMEDIATE SACRIFICES

518. Aristides, Oratio L, 6.

Κἀνταῦθα δὴ καθαρμοί τε ἐγίγνοντο ἐπὶ τοῦ ποταμοῦ ὑπὸ σπονδῶν καὶ καθάρσεις οἴκοι δι᾽ ἐμέτου.

Thereupon purifications took place near the river by means of libations, and purifications at home by means of vomiting.

519. Inscriptiones Graecae, IV², 1, no. 126.[1]

. . . κοινῇ θῦσαι Ἀσκληπιῷ, Ἠπιόνῃ, Ἐλευσεινίαις, . . .

. . . in common to offer sacrifice to Asclepius, Epione and the Eleusinian goddesses, . . .

[1] Cf. T. 432.

THANK-OFFERINGS

520. Aelianus, Fragmenta, 101.[1]

. . . ὁ Ἀσκληπιὸς . . . προστάσσει χαριστήρια τῆς ὑγιείας.

... Asclepius ... commands thank-offerings[2] for the restoration of his [*sc.*, Aristarchus'] health.

[1] Cf. T. 455.　　　　　　　　　　　[2] Cf. *e. g.* T. 423, 4; 423, 25.

521. Inscriptio Erythraea [Wilamowitz, *Nordionische Steine*, pp. 40-41; 380-360 B. C.].[1]

<blockquote>
25　　　　ὅταν δὲ ἡ πόλις τὴν θυσίην
τῶι Ἀσκληπιῶι ποιῆι, τὰ τῆς πόλεως
προτεθύσθαι ὑπὲρ πάντων, ἰδιώτης δὲ
μηδεὶς προθυέτω ἐν τῆι ἑορτῆι, ἀνὰ δὲ
τὸν ἄλλογ χρόνον προθυέτω κατὰ τὰ
30　　　προγεγραμμένα· ὅσοι δὲ ἐγκατακοιμη-
θέντες θυσίην ἀποδιδῶσι τῶι Ἀσκλη-
πιῶι καὶ τῶι Ἀπόλλωνι ἢ εὐξάμενοι θυ-
σίην ἀποδιδῶσιν, ὅταν τὴν ἱρὴν μοῖ-
ραν ἐπιθῆι, παιωνίζειν πρῶτον περὶ
35　　　τὸμ βωμὸν τοῦ Ἀπόλλωνος τόνδε τὸμ
παιῶνα ἐστρίς· ἰὴ Παιών· ὤ, ἰὴ Παιών,
ἰὴ Παιών· ὤ, ἰὴ Παιών, ἰὴ Παιών, ὤ, ἰὴ Παιών.
[ὤ] (ἄ)ναξ Ἄπολλον φείδεο κούρων φείδ[εο]
</blockquote>

... Whenever the city makes sacrifice to Asclepius, let the offerings of the city be the first to be made in behalf of all and let no individual make a sacrifice before this at the festival. For the rest of the time, let him sacrifice in accordance with the afore-mentioned regulations.[2]

So many as, having slept in the temple, make a sacrifice to Asclepius in return and to Apollo, or so many as, having made a vow, make a sacrifice, whenever they set up the sacred portion, must first raise this paeon thrice around the altar of Apollo:

Hail Paeon, O, hail Paeon, hail Paeon, O, hail Paeon, hail Paeon, O, hail Paeon. O Lord Apollo, spare the youths spare ...

[1] *Abh. Berl. Akad.,* 1909, *philos.-hist. Klasse,* II, pp. 37 ff. Cf. also T. 592.
[2] This refers to lines 1-24 which are badly mutilated.

522. Callimachus, Epigrammata, 55.[1]

<blockquote>
Τὸ χρέος ὡς ἀπέχεις, Ἀσκληπιέ, τὸ πρὸ γυναικὸς
Δημοδίκης Ἀκέσων ὤφελεν εὐξάμενος,
</blockquote>

γινώσκειν. ἢν δ᾽ ἄρα λάθῃ ⟨πάλι⟩ καί μιν ἀπαιτῇς,
    φησὶ παρέξεσθαι μαρτυρίην ὁ πίναξ.

Know, Asclepius, that thou hast received the debt which Aceson
owed thee by his vow for his wife Demodice. But if thou dost
forget and demand payment again, the tablet says it will bear
witness.

¹ = Epigrammata, 54 [Wilamowitz]. Cf. also Anthologia Palatina, VI, 147.

Cock

523.   Artemidorus, Onirocritica, V, 9.

Ηὔξατό τις τῷ Ἀσκληπιῷ, εἰ διὰ τοῦ ἔτους ἄνοσος ἔλθοι, θύσειν
αὐτῷ ἀλεκτρυόνα. ἔπειτα διαλιπὼν ἡμέραν ηὔξατο πάλιν τῷ
Ἀσκληπιῷ, εἰ μὴ ὀφθαλμιάσειεν, ἕτερον ἀλεκτρυόνα θύσειν. καὶ
δὴ εἰς νύκτα ἔδοξε λέγειν αὐτῷ τὸν Ἀσκληπιὸν " εἷς μοι ἀλεκ-
τρυὼν ἀρκεῖ." ἄνοσος μὲν οὖν ἔμεινεν, ὠφθαλμίασε δὲ ἰσχυρῶς.
καὶ γὰρ μιᾷ εὐχῇ ὁ θεὸς ἀρκούμενος τὸ ἕτερον ἠρνεῖτο.

A certain man vowed to Asclepius that if he lived in health
throughout the year, he would sacrifice a cock to him. Then,
having waited a day, he vowed again to Asclepius that if he
did not suffer from ophthalmia, he would offer another cock
to him.¹ And at night he dreamed that Asclepius said to him,
" One cock is sufficient for me." Free from disease henceforth
he actually remained, but he suffered grievously from ophthalmia, for the god, being satisfied with one prayer, denied
the other.

¹ Cf. T. 517.

524.   Plato, Phaedo, 118 A.

. . . εἶπεν, ὃ δὴ τελευταῖον ἐφθέγξατο· Ὦ Κρίτων, ἔφη,
τῷ Ἀσκληπιῷ ὀφείλομεν ἀλεκτρυόνα· ἀλλὰ ἀπόδοτε καὶ μὴ
ἀμελήσητε.

. . . he [sc., Socrates] said—and these were his last words—
" Crito," he said, " we owe a cock to Asclepius. Pay it and
do not neglect it." (*)

525. Lucianus, Bis Accusatus, 5.

... ἔπιεν ἄθλιος τοῦ κωνείου, μηδὲ τὸν ἀλεκτρυόνα τῷ ᾿Ασκληπιῷ ἀποδεδωκώς.

... and drank hemlock, poor fellow [*sc.*, Socrates], before he had even paid the cock to Asclepius.

[1] Cf. **525a.** *Scholia in Lucianum, Ad Bis Accusatum,* 5, p. 138 [Rabe]: ῾Ο γὰρ Πλάτων ἐν τῷ Φαίδωνι τὸν θάνατον Σωκράτους διηγούμενος εἰσάγει αὐτὸν ἐν τῷ μέλλειν τελευτᾶν ἐπιτρέποντα τυθῆναι τῷ ᾿Ασκληπιῷ τὸν ἀλεκτρυόνα, ὃν συνετάξατο.

526. Olympiodorus, In Platonis Phaedonem Commentaria, p. 205, 24.

Διὰ τί τῷ ᾿Ασκληπιῷ τὸν ἀλεκτρυόνα ἀποδίδωσιν; ἢ ἵνα τὰ νενοσηκότα τῆς ψυχῆς ἐν τῇ γενέσει ταῦτα ἐξιάσηται. μήποτε δὲ κατὰ τὸ λόγιον καὶ αὐτὸς τὸν παιᾶνα ᾄδων βούλεται ἀναδραμεῖν εἰς τὰς οἰκείας ἀρχάς.

Why does he [*sc.*, Socrates] pay a cock to Asclepius? Is it in order that he may heal the parts of the soul that have become sick in the process of generation or is it not that in accordance with the oracle he too desires to return to his native beginnings singing the paean?

527. Olympiodorus, In Platonis Phaedonem Commentaria, p. 244, 17.

Διὰ τί ὀφείλειν ἔφη τῷ ᾿Ασκληπιῷ τὴν θυσίαν καὶ τοῦτο τελευταῖον ἐφθέγξατο; καίτοι, εἰ ὤφειλεν, ἐπιστρεφὴς ὢν ἀνὴρ οὐκ ἂν ἐπελάθετο· ἢ ὅτι Παιωνίου δεῖται προνοίας ἡ ψυχή, ἀπαλλαττομένη τῶν πολλῶν πόνων; διὸ καὶ τὸ λόγιόν φησι, τὰς ψυχὰς ἀναγομένας τὸν παιᾶνα ᾄδειν.

Why did he [*sc.*, Socrates] say he owed the sacrifice to Asclepius and why were these his last words? Naturally, if he owed it, being a watchful man, he would not have forgotten it; or is it because the soul, being liberated from its many cares, requires the foresight of Paeon? Wherefore the oracle also says the souls when returning sing the paean.

528. Tertullianus, Apologeticus, XLVI, 5.

Idem et cum aliquid de veritate sapiebat deos negans, Aesculapio tamen gallinaceum prosecari iam in fine iubebat, credo,

ob honorem patris eius, quia Socratem Apollo sapientissimum ·
omnium cecinit.

Socrates, again—though he did know something of the truth
and denied the gods—at the end of his life he ordered a cock
to be sacrificed to Asclepius—I suppose, out of compliment to
Asclepius' father; for Apollo declared Socrates to be the wisest
of men.[1]

[1] Cf. **528a**. Tertullianus, *Ad Nationes*, II, 2: Socrates ipse deos istos quasi certus
negabat. idem Aesculapio gallinaceum secari quasi certus iubebat; cf. **528b**. Tertullianus, *De Anima,* I: Haec sapientia de schola caeli deos quidem saeculi negare liberior,
quae nullum Aesculapio gallinaceum reddi iubens praevaricetur . . . .

529. Lactantius, Divinae Institutiones, III, 20, 16-17.

Illut vero nonne summae vanitatis, quod ante mortem famili-
ares suos rogavit, ut Aesculapio gallum quem voverat prose-
carent? timuit videlicet ne aput Rhadamanthum reciperatorem
reus voti fieret ab Asclepio. dementisse hominem putarem, si
morbo adfectus perisset. cum vero sanus hoc fecerit, est ipse
insanus qui eum putat fuisse sapientem.

But was it not a mark of consummate vanity, that before his
death he [*sc.*, Socrates] asked his friends to sacrifice for him
a cock which he had vowed to Asclepius? He evidently feared
lest he should be put upon his trial before Rhadamanthys, the
judge, by Asclepius on account of the vow. I should consider
him mad if he had died under the influence of disease. But
since he did this in his sound mind, one who thinks that he was
wise is himself of unsound mind.

530. Lactantius, Institutionum Epitoma, 32, 4-5.

Nam et deus, qui utique supra nos est, quaerendus est et
religio suscipienda, quae sola nos discernit a beluis: quam
quidem Socrates non modo repudiavit, verum etiam derisit per
anserem canemque iurando: quasi vero per Aesculapium non
posset, cui voverat gallum. en sapientis viri sacrificium! et
quia eum prosecare ipse non potuit, amicos moriturus oravit,
ut post se solverent votum, scilicet ne apud inferos velut debitor
teneretur. hic profecto et pronuntiavit quod nihil scierit et
probavit.

For God, who is certainly above us, is to be sought for; and religion is to be embraced, which alone separates us from the brutes, which indeed Socrates not only rejected, but even derided, in swearing by a goose and a dog,[1] as if in truth he could not have sworn by Asclepius, to whom he had vowed a cock. Behold the sacrifice of a wise man! And because he was unable to sacrifice this in his own person, when he was at the point of death, he entreated his friends to fulfil the vow after his death, lest forsooth he should be detained as a debtor in the lower regions. Thereby certainly he showed and proved that he knew nothing. (*)

[1] Cf. T. 622.

531.　Prudentius, Apotheosis, 203-6.

. . . .

hos omnes quamvis anceps labyrinthus et error
circumflexus agat, quamvis promittere et ipsi
gallinam soleant aut gallum, clinicus ut se
dignetur pracstarc deus morientibus aequum

. . . .

. . . although the perilous labyrinth and circuitous error guide all these [sc., philosophers], although even they are wont to promise a hen or a rooster that the god who attends patients may deign to be just to them when they are dying . . .

OTHER ANIMALS

532.　Pausanias, Descriptio Graeciae, II, 26, 9.

Διάφορον δὲ Κυρηναίοις τοσόνδε ἐς ᾿Επιδαυρίους ἐστίν, ὅτι αἶγας οἱ Κυρηναῖοι θύουσιν, ᾿Επιδαυρίοις οὐ καθεστηκότος.

There is this difference between the Cyreneans [1] and the Epidaurians, that whereas the Cyreneans sacrifice goats, it is against the custom of the Epidaurians to do so. (*)

[1] Cf. T. 831.

533.　Pausanias, Descriptio Graeciae, X, 32, 12.[1]

. . . θύειν δὲ αὐτῷ τὰ πάντα ὁμοίως νομίζουσι πλὴν αἰγῶν.

. . . they [*sc.*, the Tithoreans] [2] believe in sacrificing to the god any animal except goats. (*)

[1] Cf. T. 499.
[2] Cf. T. 719; 668b.

534. Sextus Empiricus, Hypotyposeis, III, 220-21.

Ὅθεν καὶ τὰ περὶ θυσιῶν καὶ τῆς περὶ τοὺς θεοὺς θρησκείας ὅλως πολλὴν ἀνωμαλίαν ἔχει. ἃ γὰρ ἔν τισιν ἱεροῖς ὅσια, ταῦτα ἐν ἑτέροις ἀνόσια . . . οἷον γοῦν Σαράπιδι χοῖρον οὐκ ἂν θύσειέ τις, Ἡρακλεῖ δὲ καὶ Ἀσκληπιῷ θύουσιν. . . . αἶγας Ἀρτέμιδι θύειν εὐσεβές, ἀλλ᾽ οὐκ Ἀσκληπιῷ.

Hence, too, sacrificial usages, and the ritual of worship in general, exhibit great diversity. For things which are in some cults accounted holy are in others accounted unholy. . . . Thus, for example, no one would sacrifice a pig to Sarapis, but they sacrifice it to Heracles and Asclepius.[1] . . . It is an act of piety to offer goats to Artemis, but not to Asclepius.

[1] Cf. T. 482, v. 15; cf. also T. 536; 555.

535. Servius, Commentarii in Georgica, II, 380.

Victimae numinibus aut per similitudinem aut per contrarietatem immolantur: per similitudinem, ut nigrum pecus Plutoni; per contrarietatem, ut porca . . . Cereri, ut caper . . . Libero, item capra Aesculapio, qui est deus salutis, cum capra numquam sine febre sit.

Victims are sacrificed to the gods either because of similarity or contrariety: for similarity a black ox to Pluto, for instance; for contrariety, a pig . . . to Ceres, a goat . . . to Bacchus, also to Asclepius, who is the god of health, a she-goat is offered because a goat is never free from fever.

536. Herondas, Mimiambi, IV, 14-16.[1]

Οὐ γάρ τι πολλὴν οὐδ᾽ ἑτοῖμον ἀντλεῦμεν,
ἐπεὶ τάχ᾽ ἂν βοῦν ἢ νενημένην χοῖρον
πολλῆς φορίνης, κοὐκ ἀλέκτορ᾽, ἰητρα
νούσων ἐποιεύμεσθα . . .

For we draw no bounteous nor ready spring; else might we,

perchance, with an ox or stuffed pig² of much fatness and no
humble cock, be paying the price of cure from diseases . . .

¹ Cf. T. 482.
² Cf. 536a. Pausanias, *Descriptio Graeciae*, II, 11, 7: . . . θυομένων δὲ τῷ θεῷ
ταύρου καὶ ἀρνὸς καὶ ὑὸς . . . ; cf. also T. 534; 555.

Various Objects

537. Libanius, Declamationes, XXXIV, Argumentum.

> Φιλαργύρου παῖς τοῦ πατρὸς αὐτῷ κάμνοντος ηὔξατο τῷ Ἀσ-
> κληπιῷ τάλαντον δώσειν, εἰ ὁ πατὴρ τὴν νόσον διαφύγοι. ὑγιάνας
> ὁ πατὴρ ἀποκηρύττει τὸν παῖδα.

When his father was ill, the son of an avaricious man vowed to
give a talent to Asclepius if his father escaped the sickness.
The father, having become well, disinherits his son.¹

¹ Foerster does not consider the speech genuine.

538. Libanius, Declamationes, XXXIV, 23-26.

> Καὶ τῶν μὲν Ἀσκληπιαδῶν ὑπερεώρα, ἐλάλει δὲ παρακαθήμενος,
> ' τί τὸν Ἀσκληπιόν, ὦ πάτερ, καταλελοίπαμεν; πολλοὺς πολλάκις
> ὁ θεὸς οὗτος ἀνέστησεν.' ἐδόκει τί μοι λέγειν, νὴ τοὺς θεούς,
> φιλάνθρωπον. ἐσκόπουν γὰρ κατ' ἐμαυτὸν ὅτι προῖκα τὰ παρὰ
> 24 τῶν θεῶν ἡμῖν ἀγαθὰ γίνεται. καὶ πρὸς αὐτὸν ἔλεγον· ' νῦν ἡμῖν
> τὸν ἰατρὸν ἀποδέδωκας, ὦ παῖ, νῦν τὸν ἰασόμενον εὕρηκας. αὐ-
> τόματοι πρὸς εὐποιΐαν οἱ θεοί, παρακαλούμενοι δὲ βελτίονες.
> οὐδὲν τῶν ἀνθρωπίνων ἐπιζητοῦσιν. οὐ πλοῦτος αὐτοῖς προσφιλής,
> οὐ κέρδος ἕτερον. μόνην τὴν εὐποιΐαν ἔχουσιν. ἐν ταύτῃ πλου-
> τοῦσιν οἱ θεοί. ἴθι γοῦν ἡμῖν αἰτήσων ὑγείαν, ὦ παῖ. εὐχὴν εὖξαι
> τὴν πρέπουσαν. πρόσθες τῇ τῆς ὑγείας εὐχῇ πλοῦτον πολὺν καὶ
> τὴν ἀπὸ τοῦ κέρδους εὐδαιμονίαν τῇ μετὰ τῆς ὑγείας καὶ χρήστας
> εὐγνώμονας, μᾶλλον δὲ τοῦτο μόνον, εἰ δοκεῖ, πρόσευξαι, τόκους
> πλέονας. ἀπὸ γὰρ τούτων ἡ σωτηρία, μᾶλλον δέ, τοῦτό μοι
> ζωή, τοῦτο φῶς. ἂν δ' ἄρω τὸ τάλαντον μετὰ τῆς ὑγείας, ἄμεινον.'
> 25 οὕτω μὲν καὶ φρονεῖν καὶ λαλεῖν ὑπελάμβανον, ὁ δὲ τὸν νεὼν
> καταλαβὼν ἁπλοῦν μὲν οὐδὲν οὐδὲ δίκαιον εὔχεται οὐδὲ ψιλὴν
> ὑγείαν ᾔτησε, ταλάντου δέ μοι ταύτην ὠνήσατο. ταῦτα γὰρ νυνὶ
> μανθάνω ὅτι πωλοῦσι τὴν ὑγείαν οἱ θεοὶ καὶ διὰ ταύτης ὁ Ἀσ-
> κληπιὸς πλουτεῖν ἐσπούδακε καινότερον εὐπορίας τρόπον τὴν

26 ἰατρείαν εὑρόμενος. ὡς δ' οὖν πάντα καθ' ἡμῶν εὐξάμενος
ἐπανήει, ἐγὼ δ' ἤδη τῆς ὑγείας, μᾶλλον δὲ τῆς πενίας ἠρχόμην,
ἐλάνθανον γὰρ κατ' ἐμαυτοῦ τούτῳ τὴν εὐχὴν ἐπιτρέπων, ὡς δ'
οὖν ἄμεινον ἐδόκουν ἔχειν, προσελθὼν οὗτος ἔφη· ʻτί μὴ τοὺς
ἱερέας εὐγνωμονοῦμεν, ὦ πάτερ; τί μὴ τὸ τάλαντον οἱ προφῆται
λαμβάνουσιν; ὁρᾷς ὡς τὴν ὑγείαν ἀπείληφας. γενώμεθα καὶ ἡμεῖς
περὶ τὸν θεὸν εὐγνώμονες. οὕτω γὰρ συνεθέμην.ʼ

And for the Asclepiads he [sc., my son] was full of disdain,
but sitting beside me he chatted in this fashion: " Why, father,
have we neglected Asclepius? Many men on many occasions
has this god raised again to life." It seemed to me, by the
gods, that there he mentioned something promising of benevolence. For I thought to myself that the goods which we receive from the gods are free of charge. And I said to him [sc.,
my son]: " Now, my son, you have revealed to us the physician, now you have found the one who will heal. The gods of
their own free will are prone to beneficence, but when invoked
they are still better. They do not strive after human goods.
Wealth is not gratifying to them, nor any other gain. They
are interested in beneficence alone. In this the gods find their
riches. Go, therefore, my son, to implore health for us. Offer
the prayer which is fitting. In addition to health pray for
great wealth, and for the happiness ensuing from gain along
with that which issues from health, and for benevolent credi-
tors; or pray only for more interest if you think it better. For
from those things comes salvation, more than that, they mean
life, they mean light. If, then, I carry off the talent along
with my health, so much the better." Thus I took it one should
argue and speak; yet he having reached the sanctuary prays
for nothing simple or sensible, nor did he even ask merely
for health, but instead he bought me health for one talent.
For this I now understand, that the gods offer health for sale,
and by this means Asclepius was eager to gather riches, having
invented medicine as a new way of becoming wealthy.[1] But as
soon as he [sc., my son] returned having vowed everything
for us, I began immediately to draw my share of health, or
rather of poverty—for of this I was as yet unaware myself
since I left the prayer to him—and when I seemed to get better
he came to me and said: " Why should we not reward the
priests, father? Why should not the attendants of the temple

get the talent? You see how you have acquired your health again. Let us, too, be fair to the god. For thus I agreed to act."

¹ Cf. T. 1, v. 54; 99.

539. Libanius, Declamationes, XXXIV, 35-36.

35 Τί οὖν; οὐκ ἦσαν εὐχῆς ἕτεροι τρόποι; οὐκ ἦν λιβανωτὸς ἄφθονος; χαίρουσι καὶ δάφναις οἱ θεοὶ καὶ θαλλοῖς ἐλαίας. πολλάκις ἤδη τινὲς τῶν παλαιῶν καὶ δρυῶν κόμαις ἀπήρχοντο. πολλοὶ καὶ
36 στεφάνοις τὸν θεὸν γεραίρουσιν, ὕμνοις ἕτεροι. τούτων οὐδὲν ἐνενόησας οὐδὲ θαλλοὺς ἐνεθυμήθης καὶ στέμματα. μέχρις ἀλεκτρυόνος τὴν εὐχὴν στῆσαι ἔδει. βαρὺ μὲν γὰρ καὶ φορτικὸν καὶ τοῦτο. πλὴν οἶσα ἂν ἀνεκτῶς. εἰ γὰρ μὴ τὸ τάλαντον ὅλον, μέρος δὲ τῆς οὐσίας ἀνέθηκας, οὐκ ἂν ἀπεκήρυξα. σὺ δ' οὐδὲ τὰς γραφὰς ἐνενόησας. μετὰ τὴν νόσον πολλάκις ἄνθρωποι τῆς εὐποιίας ἀμειβόμενοι τὸν θεὸν εἰκόνι τὸν Ἀσκληπιὸν ὡς εὐεργέτην γράφουσι. πόσην ἂν ἡ γραφὴ βλάβην ἤνεγκε; πόσος δ' ἂν ὁ τοῦ ζωγράφου μισθὸς ἐγένετο; ἑκὼν ἂν ἐπὶ τούτοις κατέβαλον.

Well, then: were there not other ways of vowing? Was not frankincense abundant?¹ The gods find delight also in laurel and in olive shoots.² On many occasions some of the men of old began the sacrifice even with oak leaves. And many honor the god with garlands,³ others with songs. None of these did you think of, nor did you consider branches and chaplets. As far as a cock you should have limited your vow. For even this is a heavy burden and troublesome, although I would have done my best to endure it. And if you had offered part of the money, not the whole talent, I would not have disinherited you. But you did not think of the pictures either.⁴ After an illness men often in exchange for the god's beneficence picture Asclepius as the well-doer on an image. How great an expenditure would the picture have entailed? How large would have been the pay of the painter? Willingly would I have spent the money for such things.

¹ Cf. T. 434; 317, 2.   ² Cf. T. 296.   ³ Cf. T. 543.   ⁴ Cf. T. 423, 7.

540. Libanius, Declamationes, XXXIV, 38-42.

38 Οὐκ ἀνέχομαι ζῆν ἐλεούμενος, μισῶ τὴν ὑγείαν πενόμενος. τί μοι τὸ μεῖζον ἀφεὶς καταλέγειν δυστύχημα μικρὸν εὐποιίας προφέρεις

ὄνομα, μᾶλλον δέ, κινεῖς πρὸς ἀγανάκτησιν; τίς ἰᾶται συλῶν τὸν
κάμνοντα; τίς εὖ ποιεῖσθαι βούλεται μείζονα τῆς δωρεᾶς ἀδι-
κούμενος; ἀπόδος ἡμῖν τὸ τάλαντον καὶ τὴν ὑγείαν οὐ μέμφομαι.

39   εἰ δὲ ἐπὶ τούτοις ὁ Ἀσκληπιὸς ἀγανακτεῖ, αὐτό μοι μόνον ἀπο-
διδοὺς τὸ τάλαντον φερέτω καὶ μετὰ τοῦ πλούτου τὴν νόσον.
ἕτοιμος ἀναλαμβάνειν μετὰ τοῦ ταλάντου τὸν κίνδυνον, καὶ πρόσ-
θοιτό τι πλέον, εἰ βούλοιτο, τῇ νόσῳ, μόνον ἡμῖν ἐπανελθέτω τὸ

40   τάλαντον. εἶτ᾽, ὦ κατάρατε καὶ ὑπὲρ τὸ δέον φιλότιμε, ἐγὼ μόνος
ἐδυστύχουν ἀφόρητα, ἵνα μοι δυστυχέστερον τῆς νόσου τὴν ὑγείαν
ἀντέδωκας; πόσους ἐκ χαλεπωτέρων οἶδα σωθέντας; πόσοι δὲ
ἤδη τελευτῆς ἐγγὺς γενόμενοι πάλιν προσωμίλησαν τῷ βίῳ;

41   ἆρα τούτους ἡμῖν πάντας ὁ Ἀσκληπιὸς ἀνέστησε; πόθεν οὖν τῷ
πλήθει τῶν ἰατρῶν στενοχωρούμεθα; τί δὲ μάτην οὗτοι κατὰ τὴν
πόλιν στρέφονται; πόθεν δὲ πλουτεῖν νομίζουσιν; ἢ πόθεν τὸ
κέρδος αὐτοῖς; οὐκ ἀφ᾽ ὧν παραμυθοῦνται τοὺς κάμνοντας;
οὐκοῦν τὰ κατ᾽ αὐτοὺς οἴχεται, εἰ μόνος ὁ Ἀσκληπιὸς χρήματα

42   λήψεται; ἀλλ᾽ ἐπὶ μεγάλοις, φησίν, ὁ Ἀσκληπιὸς ἰᾶται καὶ δεῖ
μεγάλας τὰς ἐπαγγελίας ἐπ᾽ αὐτῷ ποιεῖσθαι. τοῦτο νῦν πρῶτον
ἀκήκοα. ἐγὼ μὲν γὰρ αὐτὸν καὶ παρέχειν ἑτέροις ὑπελάμβανον.
οὐκοῦν οὐδὲν ἡμῶν οἱ θεοὶ βελτίους. εἶτα καὶ πόσον ἐχρῆν ἐπαγ-
γείλασθαι; μὴ γὰρ ἀθανασίαν ἡμῖν δεδώρηται; μὴ γὰρ συνῆψέ
με φυγόντα τὴν νόσον τοῖς κρείττοσι; πάλιν ἡμῖν ὁ θάνατος
ἐλπίζεται. προσδοκῶ καὶ δευτέραν νόσον ἄνθρωπος ὤν. οὐκοῦν
πάλιν καὶ ἐφ᾽ ἑτέρῳ δεῖ στένειν ταλάντῳ;

I cannot bear to live while I am pitied, I hate health while
I am poor. Why do you emphasize the name of a small benefit
while failing to mention the greater misfortune, or worse even,
why do you drive me to utter vexation? Who heals by strip-
ping his patient? Who wishes to receive benefits while suffering injustice far greater than the gift? Return to us the talent,
and I shall not find fault with my health. But if Asclepius is
irritated by this, let him only return the talent to me and along
with the wealth also the disease. I am willing to take upon
myself the danger along with the talent, and let him add even
more to the disease, if he pleases, only let the talent come back
to us! O, accursed and unduly greedy one, did I alone suffer
unbearably in order that you should give me in return health
which is even more disgraceful than disease? How many have
I known to have been saved from worse afflictions? How

many, at the very point of death, have clung to life? Was it
Asclepius who restored all these for us? Why are we cramped
with innumerable physicians? Is it in vain that they swarm
the city? How do they expect to get rich? Or where do they
find their gain? Not from the soothing words which they
address to the sick? Is not their share gone if only Asclepius
is to collect money?[1] But in important cases, he [sc., my
son] says, Asclepius heals and it is advisable to offer considerable promises to him. This I hear now for the first time.
For I used to assume that he makes presents to others. The
gods, then, are no better than we are! Besides, how much
were we to promise him? For has he granted us immortality?
Has he transported me among the gods after I escaped the
disease? Death seems to be in store for us again. Being only a
mortal, I am waiting for another disease. Or must we bemoan
yet another talent?

[1] Cf. T. 508.

541.	Libanius, Declamationes, XXXIV, 49.

> Ἀλλ’ ὦ Δήλιε καὶ Παιὰν Ἄπολλον, ἤδη γὰρ καὶ κατὰ σοῦ λέγειν
> ἄρχομαι, τί φιλοκερδῆ τῷ γένει τὸν Ἀσκληπιὸν ἔφηνας; οὐδὲν
> τοῦ πατρὸς μεμίμηται. σὺ μὲν γὰρ εὔωνον τὴν μαντείαν ἔχεις
> ὀβολοῦ πολλάκις τὸν χρησμὸν ἀντιδιδοὺς καὶ τὰ ἐκ τῆς μαντείας
> ἀγαθά, τοῦτον δὲ φιλάνθρωπόν τινες ὀνομάζουσιν, ἴσως οἱ νόσου
> μὴ πεπειραμένοι.

> O Delian and Paeonian Apollo, for now I begin to speak
> against you, why did you make Asclepius covetous by nature?[1]
> He in no way resembles his father. You sell a prophecy at low
> price, since you often give responses from the oracle and the
> advantages from the prophecy for an obol. But him [sc., Ascle-
> pius], some people call the friend of man,[2] in all likelihood
> those who have had no experience with illness.

[1] Cf. T. 1, v. 54; 99.
[2] Cf. e. g. T. 422.

542.	Theophrastus, Characteres, XXI, 10.

> Καὶ ἀναθεὶς δακτύλιον χαλκοῦν ἐν τῷ Ἀσκληπιείῳ τοῦτον ἐκ-
> τρίβειν στεφανοῦν ἀλείφειν ὁσημέραι.

If he [sc., the man of petty ambition] has dedicated a brass ring in the temple of Asclepius he will wear it to a wire with daily burnishings and oilings.

543. Aristides, Oratio XLVII, 44.

Μετὰ δὲ τοῦτο ὡς ἐν Περγάμῳ ἐδόκουν στέφανον πέμπειν τῷ θεῷ τῶν μακρῶν, οἷ⟨οί⟩ εἰσιν οὓς ἰδίᾳ τῷ Ἀσκληπιῷ κομίζουσιν, . . .

Afterwards it seemed to me as if I were in Pergamum and I sent to the god a wreath,[1] one of the big ones, such as those which they provide especially for Asclepius . . .

[1] Cf. 543a. Aristides, *Oratio* XLVII, 45: Ὁ δὲ στέφανος ἦν ἐκ τοῦ ἱεροῦ τοῦ Διὸς Ἀσκληπιοῦ.

544. Aristides, Oratio XLVII, 32.

Τρίτῃ λύχνων εἰσφορά τις ἐφαίνετο εἰς τὸ ἱερὸν ὑπὸ τοῦ θυρωροῦ κατ᾽ εὐχὴν ὑπὲρ ἐμοῦ, καὶ ἐμεῖν ἔδει· ἤμεσα.

On the third day I dreamed that the doorkeeper[1] brought the candles into the temple in accordance with the vow made on my behalf, and it was necessary for me to vomit; I vomited.

[1] Cf. T. 498, n. 1.

545. Polyaenus, Strategemata, V, 2, 19.[1]

. . . ὅσα ἦν ἀναθήματα ἐν Ἀσκληπιοῦ — πολλὰ δὲ ἦν ἀργύρου καὶ χρυσοῦ — ταῦτα ἔξω τοῦ ἱεροῦ κομίσαντας . . .

. . . what offerings there were in the temple of Asclepius— there were a great many in silver and gold[2]—to take those out of the sanctuary . . .

[1] Cf. T. 841.   [2] Cf. T. 423, 4; 505; 743.

OFFERINGS ON SPECIAL OCCASIONS

546. Polybius, Historiae, XXXII, 15, 1-5.[1]

Ὅτι Προυσίας μετὰ τὸ νικῆσαι τὸν Ἄτταλον καὶ * τὸ παρελθεῖν πρὸς τὸ Πέργαμον παρασκευασάμενος θυσίαν πολυτελῆ προσή- γαγε πρὸς τὸ τέμενος τ᾽ Ἀσκληπιοῦ, καὶ βουθυτήσας καὶ καλλι-

ερήσας τότε μὲν ἐπανῆλθεν εἰς τὴν παρεμβολήν, κατὰ δὲ τὴν
ἐπιοῦσαν καταστήσας τὴν δύναμιν ἐπὶ τὸ Νικηφόριον τούς τε
νεὼς ἅπαντας διέφθειρε καὶ τὰ τεμένη τῶν θεῶν, ἐσύλησε δὲ καὶ
τοὺς ἀνδριάντας καὶ τὰ λίθινα τῶν ἀγαλμάτων. τὸ δὲ τελευταῖον
καὶ τὸ τ᾿ Ἀσκληπιοῦ βαστάσας ἄγαλμα, περιττῶς ὑπὸ Φυρομάχου
κατεσκευασμένον, ἀπήνεγκεν ὡς αὐτόν, ᾧ τῇ πρότερον ἡμέρᾳ
κατασπένδων ἐβουθύτει καὶ κατηύχετο, δεόμενος, ὅπερ εἰκός,
ἵλεων αὐτῷ γενέσθαι καὶ εὐμενῆ κατὰ πάντα τρόπον.

* καὶ Schweighaeuser μετὰ MSS.

Prusias on reaching Pergamum after his victory over Attalus
[155-4 B. C.] prepared a magnificent sacrifice which he
brought to the temple of Asclepius,[2] and having offered the
oxen[3] and obtained favorable omens, returned on that day to
his camp; but on the next day directing his army to the
Nicephorium, he destroyed all the temples and sacred precincts
of the gods, and carried off the bronze and marble statues;
finally removing and carrying off for himself the statue of
Asclepius, an admirable work of art by Phyromachus,[4] that
very Asclepius to whom on the previous day he had offered
libations, sacrifices and prayers, supplicating him, as was
fitting, to be in every way merciful and gracious to him. (*)

[1] Cf. also 546a. Suidas, *Lexicon, s. v.* Προυσίας: μετὰ τὸ νικῆσαι—τρόπον.
[2] Cf. T. 802.   [3] Cf. T. 536.   [4] Cf. T. 658.

547. Arrianus, Alexandri Anabasis, II, 5, 8.

Ἀλέξανδρος δὲ ἐν Σόλοις θύσας τε τῷ Ἀσκληπιῷ καὶ πομπεύσας
αὐτός τε καὶ ἡ στρατιὰ πᾶσα, καὶ λαμπάδα ἐπιτελέσας καὶ ἀγῶνα
διαθεὶς γυμνικὸν καὶ μουσικὸν Σολεῦσι μὲν δημοκρατεῖσθαι
ἔδωκεν.

At Soli Alexander sacrified to Asclepius, and staged a procession, himself and his whole army, with a torch relay race[1]
and athletic and literary competitions.[2] Soli he allowed to
continue democratic. (*)

[1] Cf. **547a.** *Inscriptiones Graecae*, XII, 5, 1, no. 176 [Paros]: Λαμπάδαρχος [sc.,
τοῦ Ἀσκληπιοῦ].
[2] Cf. **547b.** Curtius Rufus, *Historiae Alexandri*, III, 7, 3: Vota deinde pro salute
suscepta per ludum atque otium reddens ostendit, quanta fiducia barbaros sperneret;
quippe Aesculapio et Minervae ludos celebravit; cf. also T. **556** ff.

548. Pausanias, Descriptio Graeciae, VIII, 28. 1.[1]

Λέγουσι δὲ οἱ ἐπιχώριοι καὶ τάδε, ὡς Ἀλέξανδρος ὁ Φιλίππου
τὸν θώρακα καὶ δόρυ ἀναθείη τῷ Ἀσκληπιῷ. καὶ ἐς ἐμέ γε ἔτι
ὁ θώραξ καὶ τοῦ δόρατος ἦν ἡ αἰχμή.

The natives [sc., of Gortys] also say that Alexander the son
of Philip dedicated to Asclepius his breastplate and spear.[2]
And up to my time the breastplate and the head of the spear
were still there. (*)

[1] Cf. T. 652.　　　　　　　　　　[2] Cf. T. 800.

549. Athenaeus, Deipnosophistae, VIII, 45, 351 e-f.

Νικήσας δ' ἐν Σικυῶνι τοὺς ἀνταγωνιστὰς ἀνέθηκεν εἰς τὸ Ἀσκληπιεῖον τρόπαιον ἐπιγράψας· ʼΣτρατόνικος ἀπὸ τῶν κακῶς
κιθαριζόντων.ʼ

Victorious over his competitors in Sicyon, he [sc., Stratonicus,
ca. 400 B. C.] dedicated in the temple of Asclepius a trophy
with the inscription: " Dedicated by Stratonicus from the
spoils of bad harp-players." [1]

[1] Cf. 549a. Inscriptiones Graecae, II[2], no. 3120b [between 190 and 200 A. D.] :
Ἄρχων Διονυσόδωρος Εὐκάρπου τέχνης / πάσης με κῦδος κωμικῆς τραγικῆς χορῶν / τὸν
δειθύραμβον τρίποδα θῆκ' Ἀσκληπιῷ.

550. Statius, Silvae, III, Praefatio.

Earinus praeterea . . . scis quamdiu desiderium eius moratus
sim, cum petisset ut capillos suos, quos cum gemmata pyxide
et speculo ad Pergamenum Asclepium mittebat, versibus
dedicarem.

Next Earinus . . . you know how long I have put off his
expressed desire that I should dedicate with a poem his tresses
which he was sending to Asclepius at Pergamum together with
a jewelled box and a mirror. (*)

551. Statius, Silvae, III, 4, 6-8.

Accipe laudatos, iuvenis Phoebeie, crines,
quos tibi Caesareus donat puer, accipe laetus
intonsoque ostende patri.

Accept, O son of Phoebus, these much-praised locks that
Caesar's favorite presents to thee, accept them joyfully and
show them to thy unshorn sire.

552. Inscriptiones Graecae, II², no. 772 [ca. 252-1 B. C.].[1]

```
. . . . . . . . . . . . . . . .
. . . . . . . . . . . ἔδοξεν τῶι δή-
μωι· Ἀκρότιμος Αἰσχίου Ἰκαριεὺς εἶ-
πεν· ἐπειδὴ πάτριόν ἐστιν τοῖς ἰατρο-
ῖς ὅσοι δημοσιεύουσιν θύειν τῶι Ἀσκ-
ληπιῶι καὶ τεῖ Ὑγιείαι δὶς τοῦ ἐνιαυ-
τοῦ ὑπέρ τε αὐτῶν καὶ τῶν σωμάτων ὧν ἕ-
καστοι ἰάσαντο, ἀγαθῆι τύχει δεδόχ-
θαι τεῖ βουλεῖ, τοὺς προέδρους οἳ ἂν
λάχωσιν εἰς τὴν ἐπιοῦσαν ἐκκλησίαν
χρηματίσαι περὶ τούτων ἐν ἱεροῖς, γν-
ώμην δὲ ξυμβάλλεσθαι τῆς βουλῆς εἰς
[τὸν δῆ]μο[ν ὅ]τι δοκεῖ τεῖ βουλεῖ τὸν ἱ-
[ερέα τὸν ἀεὶ λα]νχάνοντ[α . . . ]νγι – –
```

. . . it was decreed by the Demos; Acrotimus, son of Aes-
chius, of the deme of Icaria, put the motion: Whereas it is an
ancestral custom of the physicians who are in the service of
the state to sacrifice to Asclepius and to Hygieia twice each
year in behalf of themselves and of the people whom they healed:
with good fortune it was decreed by the Council that those
prohedroi who may be chosen by lot should bring up these
matters before the ensuing ecclesia when discussing the holy
affairs and should communicate the opinion of the Council to
the Demos: that the Council decrees that whoever is chosen
priest by lot . . .

[1] Dated 270-69 B. C. by Pritchett and Meritt.

PUBLIC SACRIFICES

553. Inscriptiones Graecae, II², no. 974 [138-7 B. C.].

```
. . . . . . . . . . . .
vac.   [ἔδοξεν] τεῖ [βουλεῖ καὶ τῶι δήμωι]·
Διογένης [Διοκ]λείδου Κυδαθ[ηναιεὺς εἶπεν· ἐπειδὴ Λεωνίδης]
Νικοκ[ράτου] Φλυεὺς ὁ γενόμεν[ος ἱερεὺς Ἀσκληπιοῦ καὶ Ὑγιεία]-
```

[ς ἐπ]ὶ τὸν Τιμάρχου ἄρχοντος ἐν[ιαυτὸν τά τε εἰσιτητήρια ἔθυσ}-
εν καλῶς καὶ εὐσεβῶς τῶι Ἀσκλη[πιῶι καὶ τῆι Ὑγιείαι καὶ τοῖς ἄλλ]-
οις θεοῖς οἷς πάτριον ἦν, καὶ ἐβουθ[ύτησεν Ἀσκληπιείοις]
καὶ Ἐπιδαυρίοις καὶ Ἡρώιοις παρὰ τ[ῶι Ἀσκληπιείωι τῶι ἐν ἄστει]
[κ]αὶ τὰς τούτων παννυχίδας συν[ετέλεσεν· θύσας δὲ καὶ ὑπὲρ]
[τ]ῆς βουλῆς καὶ τοῦ δήμου καὶ παίδ[ων καὶ γυναικῶν τὰς θυσ]-
[ί]ας ἐν ἅπασιν ἀπήγγειλεν τεῖ βο[υλῆι γεγονέναι τὰ ἱερὰ καλὰ καὶ]
σωτήρια· ἔστρωσεν δὲ καὶ τὰς κλ[ίνας — — — — — — — — — — —]
. . ει τῶν θυσιῶν ἐπιφανῶς καὶ ε — — — — — — — — — — — — — —
ἔδωκε δὲ καὶ τὴν ἑαυτοῦ θυγατ[έρα — — — — — — — — — εἰς τὰ]
Ἐπιδαύρια ἀρρηφοροῦσαν βουλό[μενος — — — — — — — — — τὰς]
πρὸς τοὺς θεοὺς τιμὰς καὶ τὴν τ — — — — — — — — — — — — — —
σεν καλῶς καὶ ἐνδόξως ταῦρον — — — — — — — — — — — — — — —
καὶ παννυχίδα(ς) συνετέλεσεν παρ' ὅ[λον τὸν ἐνιαυτόν(?)· κατέστησεν]
δὲ καὶ τὸν υἱὸν Δῖον κλειδοῦχον κα[ὶ — — — — — — — — — — — —]
[κα]θ' ἑκάστην ἡμέραν γενομένας θε[ραπείας — — — — — — — — — —]
[τῶ]ι θεῶι κεχορήγηκεν ἐκτενῶς τουτ[— — — — — τοῦ Ἀσκληπιοῦ]
[καὶ τῆ]ς Ὑγιείας καὶ τοῦ ναοῦ καὶ τῶν ἐν — — — — — — — — — —
[. . . . . τ]ὴν καθήκουσαν [ἐπιμέ]λειαν ἐποιή[σατο — — — — — — — —]

. . . . . .

. . . it was decreed by the Council and the Demos; Diogenes,
son of Diocleides, of the deme of Cydathenaion, put the motion:
whereas . . . [Leonides],[1] the son of Nicocrates, of Phlya,
who was priest of Asclepius and Hygieia in the year of the
archonship of Timarchus, performed in fair and pious fashion
the initial sacrifices of the year to Asclepius and Hygieia and
the other gods to whom it was an ancestral custom to make
offerings, and whereas at the Asclepieia, the Epidauria, and
the Heroa he sacrificed bulls in the Asclepius sanctuary in
town and performed the night-festivals of these celebrations;
moreover, having made sacrifices in behalf of the Council and
the Demos and the children and the women, he reported in all
cases to the Council that the sacrifices had been favorable and
that they assured safety; moreover, he spread the couches . . .
of the sacrifices clearly and . . . he had his daughter . . .
serve as "arrhephoros" at the Epidauria, wishing . . . the
honors to the gods, and the . . . with good and glorious omen
a bull . . . and he performed the night-festivals[2] throughout
the whole year [?], and he also appointed his son Dius the

custodian of the keys[3] . . . and the services which were held
each day . . . he zealously provided for the god a chorus[4] . . .
of Asclepius and Hygieia and the temple and the . . . he took
proper care. . . .

[1] Cf. Pritchett and Meritt, p. 79.   [3] Cf. T. 498, n. 1.
[2] Cf. T. 571.   [4] Cf. T. 561; 562; 605.

554. Cicero, Verrinae Orationes, IV, 57, 127-28.

Quid? signum Paeanis ex aede Aesculapi praeclare factum
sacrum ac religiosum non sustulisti? quod omnes propter
pulchritudinem visere, propter religionem colere solebant . . .
atque ille Paean sacrificiis anniversariis simul cum Aesculapio
apud illos colebatur.

Next, did you not take from the temple of Asclepius[1] a statue
of Paean, a fine piece of sculpture, holy and sacred? A statue
which everybody used to visit for its beauty and to worship
for its sanctity. . . . And that Paean was worshipped by them
[sc., the Syracusans] with annual sacrifices, together with
Asclepius. (**)

[1] Cf. T. 841.

555. Pausanias, Descriptio Graeciae, II, 11, 7.

Τῆς δὲ Κορωνίδος ἔστι μὲν καὶ ταύτης ξόανον, καθίδρυται δὲ
οὐδαμοῦ τοῦ ναοῦ· θυομένων δὲ τῷ θεῷ ταύρου καὶ ἀρνὸς καὶ
ὑὸς ἐς Ἀθηνᾶς ἱερὸν τὴν Κορωνίδα μετενεγκόντες ἐνταῦθα τιμῶσιν.
ὁπόσα δὲ τῶν θυομένων καθαγίζουσιν, οὐδὲ ἀποχρᾷ σφισιν ἐκ-
τέμνειν τοὺς μηρούς· χαμαὶ δὲ καίουσι πλὴν τοὺς ὄρνιθας,
τούτους δὲ ἐπὶ τοῦ βωμοῦ.

There [sc., in Titane][1] is also a wooden image of Coronis,
but it has no fixed position anywhere in the temple. While to
the god are being sacrificed a bull, a lamb, and a pig,[2] they
remove Coronis to the sanctuary of Athena and honor her
there. The parts of the victims which they offer as a burnt
sacrifice, and they are not content with cutting out the thighs,
they burn on the ground, except the birds, which they burn on
the altar.

[1] Cf. T. 749; 667.   [2] Cf. T. 536.

FESTIVALS AND GAMES

EPIDAURUS

556. Pindarus, Nemeae, V, 95-97.

> . . . .
> πύκταν τέ νιν καὶ παγκρατί-
>   ῳ φθέγξαι ἑλεῖν Ἐπιδαύρῳ διπλόαν
> νικῶντ' ἀρετάν...

. . . proclaim him [sc., Themistius] as a boxer, and tell that he hath won a double victory in the pancratium by his conquest in Epidaurus . . . [1]

[1] Cf. **556a**. Pindarus, *Nemeae*, III, 147: Νεμέας Ἐπιδαυρόθεν τ' ἄπο καὶ Μεγάρων δέδορκεν φάος; cf. **556b**. Pindarus, *Isthmiae*, VIII, 149: ἐπεί νιν Ἀλκαθόου τ' ἀγὼν σὺν τύχᾳ / ἐν Ἐπιδαύρῳ τε νεότας δέκετο πρίν.

557. Scholia in Pindarum, Ad Nemeas, V, 96.

> Ἐν δὲ Ἐπιδαύρῳ Ἀσκληπιοῦ ἱερὸς ἀγὼν ἄγεται.

In Epidaurus is held a sacred contest in honor of Asclepius.[1]

[1] Cf. **557a**. *Inscriptiones Graecae*, IV², 1, no. 98 [3rd c. B.C.?]: . . . ὁ ἀγωνοθέτας . . . καὶ οἱ Ἑλλανοδίκαι . . . .

558. Scholia in Pindarum, Ad Nemeas, V, 94b.

> Καὶ πύκτην οὖν αὐτὸν ὑμνεῖ καὶ παγκρατιαστήν, διπλόαν νικῶντα ἀρετὴν ἐν Ἐπιδαύρῳ, ὡς διττὰ Ἀσκληπιεῖα νενικηκότος αὐτοῦ ἢ ὡς δύο στεφάνους εἰληφότος ἑνὶ ἀγῶνι πυγμῆς καὶ παγκρατίου.

And therefore he [sc., Pindar] praises him [sc., Themistius] in song as a boxer and as a pancratiast who has gained a two-fold virtue in Epidaurus, because he has triumphed twice in the Asclepieia, or because he has gained two crowns at one festival, in boxing and in the pancratium.

559. Scholia in Pindarum, Ad Nemeas, III, 147.

> Ἐπιδαυρόθεν τ' ἀπό: φησὶ καὶ ἐν Ἐπιδαύρῳ αὐτὸν νενικηκέναι. τίθεται δὲ ἐν Ἐπιδαύρῳ ἀγὼν Ἀσκληπιῷ, τῶν Ἀσκληπιαδῶν πρώτων θέντων, μετὰ ταῦτα δὲ Ἀργείων Ἀσκληπιῷ, διὰ πεντα-ετηρίδος. τίθεται δὲ ἐν τῷ ἄλσει τοῦ Ἀσκληπιοῦ, ἄγεται δὲ μετὰ ἐννέα ἡμέρας τῶν Ἰσθμίων.

From Epidaurus: he [*sc.*, Pindar] says that he [*sc.*, Aristocleides of Aegina] also won a victory in Epidaurus. A contest in honor of Asclepius is held in Epidaurus, the Asclepiads having first inaugurated it; after this the Argives made a contest in honor of Asclepius at five year intervals. It is held in the grove [1] of Asclepius and is performed nine days after the Isthmian games.

[1] Cf. T. 568.

560. Plato, Ion, 530A.

ΣΩ. Τὸν Ἴωνα χαίρειν. πόθεν τὰ νῦν ἡμῖν ἐπιδεδήμηκας; ἢ
οἴκοθεν ἐξ Ἐφέσου;

ΙΩΝ. Οὐδαμῶς, ὦ Σώκρατες, ἀλλ᾽ ἐξ Ἐπιδαύρου ἐκ τῶν Ἀσκληπιείων.

ΣΩ. Μῶν καὶ ῥαψῳδῶν ἀγῶνα τιθέασι τῷ θεῷ οἱ Ἐπιδαύριοι;

ΙΩΝ. Πάνυ γε, καὶ τῆς ἄλλης γε μουσικῆς.

ΣΩ. Τί οὖν; ἠγωνίζου τι ἡμῖν; καὶ πῶς τι ἠγωνίσω;

ΙΩΝ. Τὰ πρῶτα τῶν ἄθλων ἠνεγκάμεθα, ὦ Σώκρατες.

ΣΩ. Εὖ λέγεις· ἄγε δὴ ὅπως καὶ τὰ Παναθήναια νικήσομεν.

ΙΩΝ. Ἀλλ᾽ ἔσται ταῦτα, ἐὰν θεὸς ἐθέλῃ.

soc. Welcome, Ion. Where have you come from now to pay us this visit? From your home in Ephesus?

ion. No, no, Socrates; from Epidaurus, from the festival there of Asclepius.

soc. Do you mean to say that the Epidaurians honor the god with a contest of rhapsodies also?

ion. Certainly, and of arts in general.

soc. Why then, you were competing in some contest for us, were you? And how went your competition?

ion. We carried off the first prize, Socrates.

soc. Well done; so now, mind that we win too at the Panathenaea.

ion. Why, so we shall, God willing. (*)

561. Inscriptiones Graecae, IV², 1, no. 40 [*ca.* 400 B. C.].

[τἀπόλλονι θύεν βôν ἔρσ]-
[ενα καὶ hομονάοις βôν ἔ]-
[ρσενα· ἐπὶ τ]ο[ῦ] β[ομοῦ τοῦ]
Ἀπόλλο[νο]ς τα[ῦτα] θ[ύεν κ]-

αἰ καλαΐδα τᾶι Λατοῖ κα-
ὶ τἀρτάμιτι ἄλλαν, φερν-
ὰν τῶι θιῶι κριθᾶν μέδι-
μμνον, σπυρὸν ἡμίδιμμ-
νον, οἴνου ἡμίτειαν κα-
ὶ τὸ σσκέλος τοῦ βοὸς το-
ῦ πράτου, τὸ δ' ἄτερον σκέλος τοὶ ἱαρομνάμονες
φερόσθο· τοῦ δευτέρου β-
οὸς τοῖς ἀοιδοῖς δόντο
τὸ σκέλος, τὸ δ' ἄτερον σκέλος τοῖς φρουροῖς δόν-
το καὶ τἐνδοσθίδια.

To Apollo sacrifice a bull, and a bull to the gods who share
his temple. On the altar of Apollo sacrifice these things, and
a hen to Leto and another to Artemis; the portion for the god
a medimnus of barley, a half medimnus of wheat, one-twelfth
medimnus of wine, and the leg of the first bull; let the *hieromnemones* take the other leg. Let them give one leg of the
second bull to the members of the choir [1] and the other to the
guards, as well as the entrails.

[1] Cf. T. 553; 562; 605.

562. Inscriptiones Graecae, IV², 1, no. 41 [*ca.* 400 B. C.].

τῶι Ἀσσκλαπιῶι θύεν βο-
ν ἔρσενα καὶ ὁμονάοις
βον ἔρσενα καὶ ὁμονάα-
ις βον θέλειαν. ἐπὶ τοῦ β-
ομοῦ τοῦ Ἀσκλαπιοῦ θύε-
ν ταῦτα καὶ καλαΐδα. ἀνθ-
έντο τῶι Ἀσκλαπιῶι φερ-
νὰν κριθᾶν μέδιμμνον σ-
πυρὸν ἡμίδιμμνον, οἴν-
ου ἡμίτειαν· σκέλος τὸ
πράτου βοὸς παρθέντο τ-
[ὸ]ι θιῶι, τὸ δ' ἄτερον τοὶ ἱ-
[αρο]μνάμονες φ[ε]ρόσθο· τ-
[οῦ δε]υτέρο τοῖς ἀοιδοῖ-
[ς δόντο, τ]ὸ δ' ἄτερον το[ὶ]ς

[φρουροῖς δόντο καὶ τὲν]-
[δοσθίδια].

To Asclepius sacrifice a bull, and a bull to the gods who share
his temple, and a cow to the goddesses who share his temple.
On the altar of Asclepius sacrifice these things and a cock.
Let them dedicate to Asclepius as his portion a medimnus of
barley, a half medimnus of wheat, one-twelfth medimnus of
wine. Let them set before the god one leg of the first bull.
Let the *hieromnemones* take the other leg. Let them give one
leg of the second to the members of the choir, and the other
to the guards, as well as the entrails.

563. Inscriptiones Graecae, IV², 1, no. 47 [4th c. B. C.].

θεός, τύχα ἀγαθά. ἔδοξε τοῖς
Ἐπιδαυρίοις Ἀστυπαλαιεῦ-
σιν ἀποίκοις Ἐπιδαρίων
ἐοῦσιν καὶ εὐεργέταις ἀτέλειαν εἶμεν πάντων καὶ ἀσυ-
λίαν καὶ ἐν ἰράναι καὶ ἐν πο-
λέμωι καὶ κατὰ γᾶν καὶ κατὰ
θάλασσαν καὶ τὰ ἰαρ[ε]ῖα τὰ
τῶν Ἀστυπαλ[αι]έων πέμπεσ-
[θ]αι σὺν τᾶ[ι τῶν] Ἐπιδαυρίων
[πο]μπᾶι καὶ θύεν τοῖς θεοῖ[ς]
[τοῖς] ἐν Ἐπι[δαύρωι — —

God, good fortune. It was decreed by the Epidaurians that
the inhabitants of Astypalaca, who are colonists of the Epidaurians and their well-wishers, have exemption from all state
burdens as well as asylum both in times of war and in times
of peace, both on land and on the sea, and that the sacrificial
animals of the people of Astypalaea be included in the procession of the Epidaurians and that they sacrifice to the
Epidaurian gods . . .

ATHENS

564. Pausanias, Descriptio Graeciae, II, 26, 8.[1]

Τοῦτο μὲν γὰρ Ἀθηναῖοι, τῆς τελετῆς λέγοντες Ἀσκληπιῷ μεταδοῦναι, τὴν ἡμέραν ταύτην Ἐπιδαύρια ὀνομάζουσι καὶ θεὸν ἀπ᾽
ἐκείνου φασὶν Ἀσκληπιόν σφισι νομισθῆναι.

In the first place, the Athenians, who say that they gave a share of their mystic rites to Asclepius, call this day of the festival Epidauria and they allege that Asclepius was reckoned a god by them from that date.[2] (*)

[1] Cf. T. 709.
[2] Cf. **564a**. Philostratus, *Epistulae*, 8: ... ξένοι καὶ ... ὁ Ἀσκληπιὸς Ἀθηναίων.

565. Philostratus, Vita Apollonii, IV, 18.

> Ἦν μὲν δὴ Ἐπιδαυρίων ἡμέρα. τὰ δὲ Ἐπιδαύρια μετὰ πρόρρησίν τε καὶ ἱερεῖα δεῦρο μυεῖν Ἀθηναίοις πάτριον ἐπὶ θυσίᾳ δευτέρᾳ, τουτὶ δὲ ἐνόμισαν Ἀσκληπιοῦ ἕνεκα, ὅτι δὴ ἐμύησαν αὐτὸν ἥκοντα Ἐπιδαυρόθεν ὀψὲ μυστηρίων.

It was then the day of the Epidaurian festival, at which it is customary for the Athenians to hold the initiation at a second sacrifice after both proclamation and victims have been offered; and this custom was instituted in honor of Asclepius,[1] because they still initiated him when he arrived from Epidaurus too late for the mysteries.[2] (*)

[1] Cf. T. 467.
[2] Cf. **565a**. *Inscriptiones Graecae*, II², no. 4781 [175 A. D.]: Μύστην Ἡρώδ[η]ς Ἀσκληπιὸ[ν ε]ἵσατο Δηοῖ / νοῦσον ἀλεξή[σ]αντ' ἀντιχα[ρι]ζόμενος.

566. Aeschines, Orationes, III, 66-67.

> ... Δημοσθένης ... γράφει ψήφισμα, τοὺς καιροὺς τῆς πόλεως ὑφαιρούμενος, ἐκκλησίαν ποιεῖν τοὺς πρυτάνεις τῇ ὀγδόῃ ἱσταμένου τοῦ ἐλαφηβολιῶνος μηνός, ὅτ' ἦν τῷ Ἀσκληπιῷ ἡ θυσία καὶ ὁ προαγών, ἐν τῇ ἱερᾷ ἡμέρᾳ, ὃ πρότερον οὐδεὶς μέμνηται γεγονός, ...

... Demosthenes ... stealing away the opportunities of the city [*sc.*, of Athens], brought in a decree that the prytanes call an assembly for the eighth day of Elaphebolion, when the sacrifice to Asclepius and the preceding dramatic contests were held, for the sacred day—a thing that no man remembers ever to have been done before, ... (**)

567. Aristoteles, Res Publica Atheniensium, 56, 4.

> Πομπῶν δ' ἐπιμελεῖται τῆς τε τῷ Ἀσκληπιῷ γιγνομένης, ὅταν οἰκουρῶσι μύσται, καὶ τῆς Διονυσίων τῶν μεγάλων ...

He [sc., the archon] supervises processions,[1] the one celebrated in honor of Asclepius when initiates keep a watch-night, and the one at the Great Dionysia . . .

[1] Cf. T. 568.

Cos

568. Hippocrates, Epistulae, 11 [IX, p. 324-6 L.].

. . . καὶ ἔτυχε τότ᾽ ἐοῦσα τῆς ῥάβδου ἡ ἀνάληψις ἐν ἐκείνῃ τῇ ἡμέρῃ καὶ ἐτήσιος ἑορτή, ὡς ἴστε, πανήγυρις ἡμῖν καὶ πομπὴ πολυτελὴς ἐς κυπάρισσον, ἣν ἔθος ἀνάγειν τοῖς τῷ θεῷ προσήκουσιν.

. . . and the assumption of the staff happened to be celebrated on that day and the annual feast, as you know, a solemn festival[1] for us [sc., the Coans], and a splendid procession[2] to the cypress grove,[3] which it was customary for those to lead who were related to the god.

[1] Cf. T. 569.   [2] Cf. T. 567.   [3] Cf. T. 559.

Pergamum

569. Lucianus, Icaromenippus, 24.

Ἐξ οὗ δὲ ἐν Δελφοῖς μὲν Ἀπόλλων τὸ μαντεῖον κατεστήσατο, ἐν Περγάμῳ δὲ τὸ ἰατρεῖον ὁ Ἀσκληπιός . . . ἐπὶ ταῦτα μὲν ἅπαντες θέουσι καὶ πανηγύρεις ἀνάγουσι καὶ ἑκατόμβας παριστᾶσιν . . .

But since Apollo founded his oracle at Delphi and Asclepius his hospital in Pergamum . . . these are the places where they all run[1] and celebrate solemn festivals[2] and bring hecatombs . . . (*)

[1] 569a. Aristides, Oratio XLVIII, 81: . . . ἦσαν γὰρ οἱ βαδίζοντες εἰς Πέργαμον τῶν θεωριῶν εἵνεκα . . . .
[2] Cf. T. 568.

570. Aristides, Oratio XLVIII, 74.

Ἦν ἰσημερία ἡ μετὰ χειμῶνα ὅτε πηλοῦνται τῷ θεῷ, . . .

It was the time of the equinox after winter had passed when they smear themselves with clay in honor of the god [sc., Asclepius] . . .

571. Aristides, Oratio XLVII, 6.

Διήγαγον δὲ περιπάτῳ χρώμενος κατ᾽ οἰκίαν καὶ ἅμα παιδιᾷ, οἷα ἑορτῶν οὐσῶν. καὶ γὰρ ἡ τοῦ θεοῦ παννυχὶς ἐγκατειλήφει τὴν προτέραν ἑορτήν, τὴν τοῦ Ποσειδῶνος.

I stayed at home, walking about and playing around, since these were holidays. For the night-festival of the god[1] immediately followed the preceding festival that of Poseidon.

[1] Cf. T. 553.

LAMPSACUS

572. Corpus Inscriptionum Graecarum, II, Add. no. 3641b [2nd c. B. C.].[1]

. . . .
. . . . ὅταν δὲ ὁ δῆμος βουλεύ(σ)ηται [περὶ τῆς μετὰ] ταῦτα συσσταθησομένης τραπεζιτείας ἀπ[οδε]ιξάτωσαν ἥ τε βο[υλὴ καὶ οἱ στρα]- τηγοὶ οὓς δεῖ χειρίζειν τὰ καθιερώμενα χρήματα τῷ Ἀσκληπιῷ, ὅ [ταν δὲ αἱ ἑορταὶ] συντελῶνται ἔν τε τῷ μηνὶ Ληναιῶνι καὶ ἐν τῷ Λευκαθιῶν[ι, ἀφεῖναι] μὲν τοὺς παῖδας ἐκ τῶν μαθημάτων, τοὺς δὲ οἰκέτας ἀπὸ τῶν [ἔργων, στε]- φανηφορεῖν δὲ Λαμψακηνοὺς πάντας· ὁ δὲ ἱερεὺς θυμιάτω (λι)βαν[ωτὸν ἐπὶ] τοῦ βωμοῦ τοῦ Ἀσκληπιοῦ, καὶ ἐπιμελείσθω ὅπως δάφνη καὶ νή[ριον ὦσιν ἐν] [τῷ] ἱερῷ εἰς τοὺς στεφάνους . . . .

. . . . when the people deliberate concerning the money to be deposited in the bank hereafter, the Council and the generals shall indicate those whose duty it is to handle the property consecrated to Asclepius; and when the festivals are performed in the months of Lenaion and Leucathion, the children shall be released from their schoolwork and the slaves from their labor and all the people of Lampsacus shall wear crowns.[2] Let the priest sacrifice frankincense on the altar of Asclepius and let him see that olive branches and oleander are in the temple for the crowns . . .

[1] Cf. L. Robert, *Bulletin de Correspondance Hellénique*, LII, 1928, p. 159.
[2] Cf. T. 505.

EPHESUS

573. Inscriptio Ephesia [*Österreichische Jahreshefte*, VIII, 1905, p. 128; 2nd-3rd c. A. D.).

. . . .]ω ἐ[π]ὶ ἱερέως τοῦ Ἀσκλ[η-

πιοῦ................α]νου Ἀσιάρχου ἄρχοντος τῶν
ἰατρῶν.............ἀγ]ωνοθετοῦντος τῶν μεγάλων
Ἀσκληπιείων?........ε]ινου ἀρχιατροῦ τὸ δ'. οἵδε ἐνεί-
κησαν τὸν ἀγῶνα τῶν ἰατρ]ῶν, συντάγματος· Πο. Οὐή(διος) Ῥουφεῖνος
...χειρουργίας· Πο. Αἴλιο]ς Μένανδρος ἀρχιατρός, προ-
βλήματος.................]ς ἀρχιατρός, ὀργάνων· Πο. Αἴλι-
ος..................ἀγαθῇ] τύχῃ
.... Ἐφ]εσίων τοῖς ἀπὸ τοῦ Μουσείου
ἰατροῖς.................]νη[...]ων ἡγεμὼν Ἀσκληπιός
........................]προ..... τὰς θυσίας ἡδέως προσ-
........................χομ[....] ἐγυμνασιάρχησεν
ταῖς δυσὶ τοῦ ἀγῶνος ἡμέραις?]. θαλλός ν'

. . . under the priest of Asclepius . . . priest of the Imperial
cult and president (of the association) of physicians . . . acting as judge in the games of the great Asclepieia, . . . being
city physician for the fourth time. The following triumphed
in the contest of the physicians: in composition: P. Vedius
Rufinus . . . in surgery: P. Aelius Menander, city physician;
in thesis: . . . city physician; in instruments: P. Aelius . . .
Good Fortune . . . of the Ephesians to the physicians of the
Museum . . . leader Asclepius [1] . . . the sacrifices gladly . . .
was gymnasiarch during the two days of the contest. The
fiftieth contest.

[1] Cf. T. 340.

ITALY

574. Arnobius, Adversus Nationes, VII, 32.

> Aesculapii geritur celebraturque vindemia. colunt enim dii
> vineas et ad suas usiones contractis exprimunt vindemiatoribus
> vinum.

> The vintage-season of Asclepius is in progress and is being
> celebrated. For the gods tend the vines and, having gathered
> together the vintagers, press the wine for their own use.

575. Pollux, Onomasticon, I, 37.

> Ἑορταὶ ἔντιμοι . . . Ἀσκληπιοῦ Ἀσκληπιεῖα . . .

Festivals in honor of (someone): . . . the Asclepieia in honor of Asclepius . . .

PRAYERS AND SUPPLICATIONS

576. Augustinus, De Civitate Dei, IV, 22.

Quoniam nihil prodest, inquit, hominis alicuius medici nomen formamque nosse, et quod sit medicus ignorare: ita dicit nihil prodesse scire deum esse Aesculapium, si nescias eum valetudini opitulari atque ita ignores cur ei debeas supplicare.

"Just as it is of no advantage," he [*sc.*, Varro] says, "to know the name and appearance of any man who is a physician, and not to know what a physician is, so," he says, "it is of no advantage to know that Asclepius is a god, if you are not aware that he bestows the gift of health,[1] and consequently do not know why you ought to supplicate him." (*)

[1] Cf. **576a.** Arnobius, *Adversus Nationes*, I, 49: Aesculapium . . . datorem . . . sanitatis; cf. T. **584**; cf. also T. **368**.

577. Fronto, Epistulae, III, 9, 1-2.

Deos igitur omnes, qui usquam gentium vim suam praesentem promptamque hominibus praebent, qui vel somniis vel mysteriis vel medicina vel oraculis usquam iuvant atque pollent, eorum deorum unumquemque mihi votis advoco, meque pro genere cuiusque voti in eo loco constituo, de quo deus ei rei praeditus facilius exaudiat.

Igitur iam primum Pergami arcem ascendo et Aesculapio supplico, uti valetudinem magistri mei bene temperet vehementerque tueatur.

I [*sc.*, Marcus Aurelius] call, therefore, with my vows to hear me each one of all the gods, who anywhere in the world provide present and prompt help for men; who anywhere give their aid and show their power in dreams or mysteries, or healing, or oracles; and I place myself according to the nature of each vow in that spot where the god who is invested with that power may the more readily hear.

Therefore I now first climb the citadel of Pergamum and

beseech Asclepius to bless my master's [*sc.*, Fronto's] health
and mightily protect it.[1] (*)

[1] Cf. **577a**. Lucianus, *Demonax*, 27: Εἰπόντος δέ τινος τῶν ἑταίρων, Ἀπίωμεν, Δημῶναξ, εἰς τὸ Ἀσκληπιεῖον καὶ προσευξώμεθα ὑπὲρ τοῦ υἱοῦ, Πάνυ, ἔφη, κωφὸν ἡγῇ τὸν Ἀσκληπιόν, εἰ μὴ δύναται κἀντεῦθεν ἡμῶν εὐχομένων ἀκούειν.

578. Terentius, Hecyra, III, 2, 337-38.

Male metuo ne Philumenae magis morbus adgravescat:
quod te, Aesculapi, et te, Salus, nequid sit huius oro.

I sadly fear that Philumena's sickness has become more serious:
to you, Asclepius, and you, Health, I pray that there may be
none of this.

579. Plautus, Curculio, III, 1, 389-90.

Quis hic est qui operto capite Aesculapium
salutat?

Who's this chap with his head covered doing homage to
Asclepius?

580. Diogenes Laertius, Vitae Philosophorum, VI, 37-38.

Θεασάμενός ποτε γυναῖκα ἀσχημονέστερον τοῖς θεοῖς προσ-
πίπτουσαν, βουλόμενος αὐτῆς περιελεῖν τὴν δεισιδαιμονίαν, καθά
φησι Ζωίλος ὁ Περγαῖος, προσελθὼν εἶπεν, " οὐκ εὐλαβῇ, ὦ
γύναι, μή ποτε θεοῦ ὄπισθεν ἑστῶτος—πάντα γάρ ἐστιν αὐτοῦ
πλήρη—ἀσχημονήσῃς; " τῷ Ἀσκληπιῷ ἀνέθηκε πλήκτην, ὃς
τοὺς ἐπὶ στόμα πίπτοντας ἐπιτρέχων συνέτριβεν.

One day he [*sc.*, Diogenes, 4th c. B. C.] saw a woman kneeling
before the gods [1] in an ungraceful attitude, and wishing to free
her of superstition, according to Zoïlus of Perga [otherwise
unknown], he came forward and said, "Are you not afraid,
my good woman, that a god may be standing behind you—
for all things are full of his presence—and you may be put to
shame?" He dedicated to Asclepius a bruiser who, whenever
people fell on their faces, used to run up to them and beat
them up. (*)

[1] Cf. T. 438; 584.

581. Philostratus, Vita Apollonii, I, 12.

... ἐξωρμήθη ἐς τὰς Αἰγὰς νοσεῖν τε ἑαυτὸν φήσας καὶ τοῦ
Ἀσκληπιοῦ δεῖσθαι, καὶ προσελθὼν τῷ Ἀπολλωνίῳ βαδίζοντι
ἰδίᾳ "σύστησόν με," ἔφη, "τῷ θεῷ." ὁ δὲ ὑπολαβὼν "καὶ τί
σοι δεῖ τοῦ συστήσοντος," εἶπεν, "εἰ χρηστὸς εἶ; τοὺς γὰρ
σπουδαίους οἱ θεοὶ καὶ ἄνευ τῶν προξενούντων ἀσπάζονται."
"ὅτι νὴ Δί'," ἔφη, "Ἀπολλώνιε, σὲ μὲν ὁ θεὸς πεποίηται ξένον,
ἐμὲ δὲ οὔπω." "ἀλλὰ κἀμοῦ," ἔφη, "καλοκἀγαθία προὐξένησεν,
ᾗ χρώμενος, ὡς δυνατὸν νέῳ, θεράπων τέ εἰμι τοῦ Ἀσκληπιοῦ καὶ
ἑταῖρος· εἰ δὲ καὶ σοὶ καλοκἀγαθίας μέλει, χώρει θαρρῶν παρὰ
τὸν θεὸν καὶ εὔχου, ὅ τι ἐθέλεις."

... and hurrying off to Aegae he [*sc.*, a man from Cilicia] pretended he was sick and must have the help of Asclepius. There
he came upon Apollonius walking alone and said to him,
"Recommend me to the god." But he replied: "Why do you
need anyone to recommend you if you are good? For the gods
love men of virtue and welcome them without any introduc-
tions." "Because, to be sure," said the other, "the god, O
Apollonius, has invited you to be his guest, but so far has not
invited me." "Nay," answered Apollonius, "'tis my good
qualities, so far as a young man can display good qualities,
which have introduced me to Asclepius, whose servant and
companion I am. If you too care for good qualities, go boldly
up to the god and tender what prayer you will."[1] (*)

[1] Cf. **581a.** Libanius, *Epistulae*, 839, 1-3: Καὶ Ἡρακλείδης τυχέτω τῆς παρὰ σοῦ συμ-
μαχίας, ἀνὴρ Μεμφίτης, Ἀσκληπιῷ φίλος, ἐπιεικὴς τὸν τρόπον, πατρῷα ταῦτα διαδεξάμενος.
καὶ γὰρ ἐκεῖνος ἦν ἑταῖρός τε τῷ θεῷ καὶ τἄλλα χρηστὸς καὶ δι' αὐτό γε τοῦτο τῷ θεῷ φίλος,
ὅτι χρηστός. ᾧ πολλὰ μὲν εἰς τὴν πόλιν χρήματα, πλείω δὲ εἰς ξένους ἀνάλωται ... τοῦτον
ἡμῖν, πρὸς Ἀσκληπιοῦ, χαίροντα ἀπόπεμψον, ὅπως σε τῇδε μὲν παρ' ἡμῖν, οἴκοι δὲ ἐπαινέσῃ
παρὰ τῷ θεῷ. ἔρρωσό μοι πανοικεσίᾳ, δικαιότατε.

582. Marinus, Vita Procli, Cp. 29.

Καὶ Ἀσκληπιγένειά ποτε ἡ Ἀρχιάδου μὲν καὶ Πλουτάρχης
θυγάτηρ, ... ἔτι κόρη οὖσα καὶ ὑπὸ τοῖς πατράσι τρεφομένη,
νόσῳ χαλεπῇ κατείχετο καὶ τοῖς ἰατροῖς ἰάσασθαι ἀδυνάτῳ. ὁ δὲ
Ἀρχιάδας ἐπ' αὐτῇ μόνῃ τὰς ἐλπίδας ἔχων τοῦ γένους, ἤσχαλλε
καὶ ὀδυνηρῶς διέκειτο, ὥσπερ ἦν εἰκός. ἀπογιγνωσκόντων δὲ τῶν
ἰατρῶν ἦλθεν, ὥσπερ εἰώθει ἐν τοῖς μεγίστοις, ἐπὶ τὴν ἐσχάτην
ἄγκυραν, μᾶλλον δὲ ὡς ἐπὶ σωτῆρα ἀγαθὸν τὸν φιλόσοφον, καὶ
λιπαρήσας αὐτὸν ἠξίου σπεύδοντα καὶ αὐτὸν εὔχεσθαι ὑπὲρ τῆς

θυγατρός. ὁ δὲ παραλαβὼν τὸν μέγαν Περικλέα τὸν ἐκ τῆς Λυδίας, ἄνδρα μάλα καὶ αὐτὸν φιλόσοφον, ἀνήει εἰς τὸ Ἀσκληπιεῖον προσευξόμενος τῷ θεῷ ὑπὲρ τῆς καμνούσης· καὶ γὰρ ηὐτύχει τούτου ἡ πόλις τότε, καὶ εἶχεν ἔτι ἀπόρθητον τὸ τοῦ Σωτῆρος ἱερόν. εὐχομένου δὲ αὐτοῦ τὸν ἀρχαιότερον τρόπον, ἀθρόα μεταβολὴ περὶ τὴν κόρην ἐφαίνετο καὶ ῥᾳστώνη ἐξαίφνης ἐγίγνετο· ῥεῖα γὰρ ὁ Σωτήρ, ὥστε θεός, ἰᾶτο. συμπληρωθέντων δὲ τῶν ἱερῶν, πρὸς τὴν Ἀσκληπιγένειαν ἐβάδιζε καὶ κατελάμβανεν αὐτὴν ἄρτι μὲν τῶν περιεστώτων τὸ σῶμα λελυμένην παθῶν, ἐν ὑγιεινῇ δὲ καταστάσει διάγουσαν. καὶ τοιοῦτον ἔργον διεπράξατο οὐκ ἄλλως ἢ κἀνταῦθα τοὺς πολλοὺς λανθάνων, καὶ οὐδεμίαν πρόφασιν τοῖς ἐπιβουλεύειν ἐθέλουσι παρασχών, συνεργησάσης αὐτῷ πρὸς τοῦτο καὶ τῆς οἰκίας ἐν ᾗ αὐτὸς ᾤκει· καὶ γὰρ πρὸς τοῖς ἄλλοις εὐτυχήμασιν, ἁρμοδιωτάτη αὐτῷ καὶ ἡ οἴκησις ὑπῆρξεν ἣν καὶ ὁ πατὴρ αὐτοῦ Συριανὸς καὶ ὁ προπάτωρ ὡς αὐτὸς ἐκάλει, Πλούταρχος, ᾤκησαν, γείτονα μὲν οὖσαν τοῦ ἀπὸ Σοφοκλέους ἐπιφανοῦς Ἀσκληπιείου, καὶ τοῦ πρὸς τῷ θεάτρῳ Διονυσίου, ὁρωμένην δὲ ᾗ καὶ ἄλλως αἰσθητὴν γιγνομένην τῇ ἀκροπόλει τῆς Ἀθηνᾶς.

When still a young maiden and under the care of her parents, Asclepigenia, the daughter of Archiades and Plutarche, . . . was stricken with a grievous illness, which the physicians were unable to cure. Since Archiades rested all his hopes for his family line upon her alone, he was grieved and greatly distressed, as was natural. When the physicians despaired, he went, as was his custom in critical situations, to the last resort, or rather as if to a noble savior, to the philosopher [sc., Proclus], and begged and entreated that he too make haste to pray for the daughter. And Proclus, taking along the great Pericles from Lydia, who was also a truly wise man, went up to the Asclepieion to pray to the god for the sick girl. For the city [sc., Athens] still enjoyed the god's presence at that time and still held the temple of the Savior[1] unravaged. While he prayed in the more ancient fashion, all of a sudden a change seemed to come over the maiden and all at once she was eased of pain. For easily did the Savior, inasmuch as he was a god, heal her. When the sacrifices were fulfilled, he went to Asclepigenia and found her now relieved of the suffering that had gripped her, and in a healthy state.

And such work he accomplished, even as in this case hiding from the many, and offering no excuse to those wishing to persecute him; the house in which he lived aided him in this, for in addition to other marks of good fortune the house, which his ' father ' Syrianus and his ' grandfather,' Plutarchus, as he called them, had dwelt in, was most suitable for him, being next door to the Asclepieion, famous through Sophocles,[2] and to the temple of Dionysus near the theater; and it was visible and otherwise noticeable from the Acropolis of Athena.[3]

[1] Cf. T. 463 ff.   [2] Cf. T. 587 ff.   [3] Cf. T. 728.

583. Livius, Ab Urbe Condita, X, 47, 7.

Inventum in libris Aesculapium ab Epidauro Romam arcessendum; neque eo anno, quia bello occupati consules erant, quicquam de ea re actum, praeterquam quod unum diem Aesculapio supplicatio habita est.

It was discovered in the Books that Asclepius must be summoned to Rome from Epidaurus; but nothing could be done about it that year [293 B. C.], because the consuls were occupied with the war, except that for one day a supplication to Asclepius was held.[1] (*)

[1] Cf. T. 505.

584. Arnobius, Adversus Nationes, I, 49.

Et quoniam beneficia salutis datae aliorum numinum comparatis et Christi, quot milia vultis a nobis debilium vobis ostendi, quot tabificis adfectos morbis nullam omnino rettulisse medicinam, cum per omnia supplices irent templa, cum deorum ante ora prostrati limina ipsa converrerent osculis, cum Aesculapium ipsum datorem ut praedicant sanitatis, quoad illis superfuit vita, et precibus fatigarent et invitarent miserrimis votis?

And since you compare the blessings of health bestowed by other deities and by Christ, how many thousands of infirm persons do you wish to be shown to you by us; how many persons affected with wasting diseases, whom no medicine whatever restored, although they went as suppliants through all the temples, although they prostrated[1] themselves before the gods, and swept the very thresholds with their lips[2]—

although with their prayers they wearied Asclepius himself, whom they call health-giver,[3] as long as life remained, and importuned him with most piteous vows? (*)

[1] Cf. T. 580.   [2] Cf. T. 438.   [3] Cf. T. 576a.

585. Ps. Lactantius, De Mortibus Persecutorum, 33, 5.

. . . nihil humanae manus promovent. confugitur ad idola: Apollo et Asclepius orantur, remedium flagitatur . . .

. . . but human hands bring nothing about. Refuge is taken to idols:[1] Apollo and Asclepius are besought importunately; a remedy is demanded . . .

[1] Cf. *e. g.* T. 582; 422; 438.

586. Augustinus, De Civitate Dei, IV, 21.

Cur enim esset invocanda propter fessos diva Fessona, propter hostes depellendos diva Pellonia, propter aegros medicus vel Apollo vel Aesculapius vel ambo simul, quando esset grande periculum?

For why should they have invoked for the weary the goddess Fessona; for driving away enemies, the goddess Pellonia; for the sick, as a physician, either Apollo or Asclepius, or both together if there was great danger? (*)

PAEANS AND OTHER SONGS

587. Inscriptiones Graecae, II², no. 4510 [Imperial time].[1]

Σοφοκλέους   [Παι]άν

[Φλεγύα] κούρα περιώνυμε, μᾶτερ ἀλεξιπό[ν]ο[ιο] θεοῦ,
[. . . . .]ᴸς ἀκειρεκόμασ ᵜ [·]εναρξομαι [ὕμ]νον ἐγερσιβόαν
[— — — —]νεσι[ν] εὐεπ/ [— — — —] ᵜ [. . .]ᴧ[.]αν[. . .]οβοα
[— — — — — — — — — — —] συρίγμασι μιγνύ[μεν]ον
[— — — — — — — — — — — —]σι Κεκροπιδῶν [ἐπ]ιτάρροθον
[— — — — — — — — — — — — — —]′ μόλοις τὸν [χρυσο]κόμα[ν] (?)
[— — — — — — — — — — — — — — — —]ν αὐτόν [— — — — — — —]
[— —]
[— —]

[– 'Ολύ]μπιον

vac. versus 6

[– –]τερα

Sophocles' Paean

O far-famed daughter of Phlegyas, mother of the god who
wards off pains . . . the unshorn [Phoebus]. I begin my loud-
voiced hymn . . . accompanied by flutes . . . the helper of the
sons of Cecrops . . . may you come . . . the golden-haired [?]
. . . him . . . the Olympian . . .

[1] Cf. J. H. Oliver, *Hesperia*, V, 1936, p. 112.

588. Lucianus, Demosthenis Encomium, 27.

Οὐδὲ γὰρ τἀσκληπιῷ μεῖόν τι γίνεται τῆς τιμῆς, εἰ, μὴ τῶν
προσιόντων αὐτῶν ποιησάντων παιᾶνα, τὰ Ἰσοδήμου τοῦ Τροιζη-
νίου καὶ Σοφοκλέους ᾄδεται, . . .

For no less honor comes to Asclepius if, in case his worshippers composed no paeans of their own, the songs of Isodemus
of Troezen [otherwise unknown] and of Sophocles are sung . . .

589. Philostratus, Vita Apollonii, III, 17.

Οἱ δὲ ᾖδον ᾠδήν, ὁποῖος ὁ παιὰν ὁ τοῦ Σοφοκλέους, ὃν Ἀθήνησι
τῷ Ἀσκληπιῷ ᾄδουσιν.

But they [*sc.*, the Indian sages] sang a song resembling the
paean of Sophocles which they sing at Athens in honor of
Asclepius.

590. Philostratus Iunior, Imagines, 13.

Ἀσκληπιὸς δέ, οἶμαι, οὗτος ἐγγὺς παιᾶνά που παρεγγυῶν γράφειν
καὶ κλυτομήτης οὐκ ἀπαξιῶν παρὰ σοῦ ἀκοῦσαι βλέμμα τε αὐτοῦ
πρὸς σὲ φαιδρότητι μεμιγμένον τὰς παρὰ ⟨σοὶ⟩ * μικρὸν ὕστερον
ἐπιξενώσεις αἰνίττεται.

*: σοὶ addidit P. Friedländer   παρὰ delevit Kayser.

This is Asclepius near by, I think, doubtless urging you [*sc.*,
Sophocles] to write a paean, and though " famed for his skill "
he does not disdain to hear it from you; and his gaze that is

fixed upon you, with joyousness intermingled, dimly fore-
shadows his hospitable reception at your house a little later.[1] (*)

[1] Cf. **590a.** Plutarchus, *Numa*, 4, 9: Σοφοκλεῖ δὲ καὶ ζῶντι τὸν Ἀσκληπιὸν ἐπιξενωθῆναι
λόγος ἐστί, πολλὰ μέχρι νῦν διασώζων τεκμήρια . . . ; cf. also T. **58.** Cf. **590b.** Plutarchus,
Non posse suaviter vivi secundum Epicurum, XXII, 1103 B : Ἡ Φορμίων τοὺς Διοσκόρους
ἢ τὸν Ἀσκληπιὸν Σοφοκλῆς ξενίζειν αὐτός τε πειθόμενος καὶ τῶν ἄλλων οὕτως ἐχόντων διὰ
τὴν γενομένην ἐπιφάνειαν;

591. Etymologicum Magnum, *s. v.* Δεξίων.

Οὕτως ὠνομάσθη Σοφοκλῆς ὑπὸ Ἀθηναίων μετὰ τὴν τελευτήν.
φασὶν ὅτι Ἀθηναῖοι τελευτήσαντι Σοφοκλεῖ, βουλόμενοι τιμὰς
αὐτῷ περιποιῆσαι, ἡρῷον αὐτῷ κατασκευάσαντες, ὠνόμασαν αὐτὸν
Δεξίωνα, ἀπὸ τῆς τοῦ Ἀσκληπιοῦ δεξιώσεως. καὶ γὰρ ὑπεδέξατο
τὸν θεὸν ἐν τῇ αὐτοῦ οἰκίᾳ, καὶ βωμὸν ἱδρύσατο. ἐκ τῆς αἰτίας
οὖν ταύτης Δεξίων ἐκλήθη.

Thus [*i. e. Dexion*] Sophocles was named by the Athenians
after his death. The Athenians, they say, wishing to do honor
to him, built a *heroon* to him now that he was dead and they
called him *Dexion* (he who receives) from the reception of
Asclepius; for he received the god into his house and set up
an altar[1] to him. For this reason, then, he was called *Dexion*.

[1] Cf. however T. **720** ff.

592. Anonymus, Paean Erythraeus in Asclepium [*ca.* 380-360 B. C.].

α′　[Παιᾶνα κλυτό]μητιν ἀείσατε
　　[κοῦροι Λατοΐδαν Ἕκ]ατον,
　　　　ἰὲ Παιάν,
　　ὃς μέγα χάρ[μα βροτοῖσ]ιν ἐγείνατο
　　μιχθεὶς ἐμ φι[λότητι Κορ]ωνίδι
　　ἐν γᾶι τᾶι Φλεγυείαι,
　　ἰὴ Παιάν, Ἀσκληπιὸν
　　δαίμονα κλεινό[τατ]ον,
　　　　ἰὲ Παιάν,

β′　[Το]ῦ δὲ καὶ ἐξεγένοντο Μαχάων
　　καὶ Πο[δα]λείριος ἠδ᾽ Ἰασώ,
　　　　ἰὲ Παιάν,
　　Αἴγλ[α τ᾽] ἐοῶπις Πανάκειά τε
　　Ἠπιόνας παῖδες σὺν ἀγακλυτῶι
　　ἐοαγεῖ Ὑγιείαι—

ἰὴ Παιάν, Ἀσκληπιὸν
δαίμονα κλεινότατον,
ἰὲ Παιάν,

γ′ Χαῖρέ μοι, ἵλαος δ᾽ ἐπινίσεο
τὰν ἐμὰν πόλιν εὐρύχορον,
ἰὲ Παιάν,
δὸς δ᾽ ἡμᾶς χαίροντας ὁρᾶν φάος
ἀελίου δοκίμους σὺν ἀγακλυτῶι
ἐοαγεῖ Ὑγιείαι—
ἰὴ Παιάν, Ἀσκληπιὸν
δαίμονα κλεινότατον,
ἰὲ Παιάν.

Sing, lads, the far-darting son of Leto, Paean, so famed for his skill—hail, Paean!—who begat great joy for man when he mingled in love with Coronis in the land of the Phlegyae. Hail, Paean, Asclepius, most renowned of demigods, hail, Paean!

Of him were begotten Machaon and Podalirius, and Iaso—O hail, Paean!—and Aegle, the fair-eyed, and Panacea, children of Epione,[1] along with bright Hygieia, the glorious—hail, Paean, Asclepius, most renowned of demigods, hail, Paean!

Welcome, and come thou propitious to my wide-spaced city—hail, Paean!—and grant we may see the sunlight in joy, acceptable with bright Hygieia, the glorious—hail, Paean, Asclepius, most renowned of demigods, hail, Paean![2]

---

[1] Cf. T. 278 ff.

[2] Cf. 592a. Anonymus, *Paean Ad Urbem Dium Repertus* [2nd c. A. D. (?)] : Παιᾶνα κλυτόμητιν ἀείσατε κοῦ[ροι / Λητοΐδην Ἕκατον, ἰὲ ὣ ἰὲ Παιάν, / ὃς μέγα χάρμα βροτοῖσιν ἐγείνατο / μειχθεὶς ἐν φιλότητι Κορωνίδι τῇ Φλεγύαο / Ἰὴ Παιᾶνα Ἀσκληπιὸν δαίμονα κλεινότατον, ἰὲ Παιάν. / Τοῦ δὲ καὶ ἐξεγένοντο Μαχάων καὶ Ποδαλείριος / ἠδ᾽ Ἰασὼ Ἀκεσώ τε πολύλλιτος, ὣ ἰὲ Παιάν, / Αἴγλη τε εὐῶπις Πανάκειά τε Ἠπιόνης παῖδες / σὺν ἀγακλύτῳ εὐαγεῖ Ὑγείᾳ, / Ἰὴ Παιὰν Ἀσκληπιέ, δαῖμον κλεινότατε, ἰὲ Παιάν. / Χαῖρέ μοι, εἵλαος δ᾽ ἐπινείσεο Δείων πόλιν / εὐρύχορον, ἰὲ ὣ ἰὲ ὣ ἰὲ Παιάν, / δὸς δ᾽ ἡμᾶς χαίροντας ὁρᾶν φάος ἠελίου / δοκίμους σὺν ἀγακλυτῷ εὐαγεῖ Ὑγείᾳ. / ἰὴ Παιὰν Ἀσκληπιέ, δαῖμον σεμνότατε, / ἰὲ Παιάν.

593. Inscriptiones Graecae, II², no. 4473 [1st c. A. D.].

Μακε[δόνιος]
ἐποίη[σεν]
Δήλιον εὐφαρέτρα[ν ἑκατηβόλον – – – – –]

εὔφρονι θυμῷ εὔφημ[εῖτε – – – ἰὴ Παιάν]
ἰκτῆρ[α] κλάδον ἐν παλά[μαισιν ἔχοντες – – – –]
ἀ[γλ]αὸν ἔρνος, κοῦροι Ἀθη[ναίων – –
.. υρ.. ἄμε[μπ] τος ὕμνος ἀεὶ – – – – –
κλυτὸν [.]| ...ον...... vac.
ἐπιτάρροθον ὅγ ποτ[ε γεί]νατο νούσω[ν ἰητῆρ᾽ ἠδὲ]
δύης, Ἀσκληπιὸν εὔφ[ρον]α κοῦρον. vac.
τὸν δ᾽ ἀνὰ Πηλιάδας κορυφὰς ἐδίδαξε [τέ]χνη[ν τε καὶ σο]-
φίαν Κένταυρος ἀλεξίπονον [μ]ερόπεσσιν, vac.
παῖδα Κορωνίδος, ἤπιον ἀν[δ]ράσι, δαίμονα σεμνότα[τον].
τοῦ δ᾽ ἐγένοντο κόροι Ποδαλείριος ἠδὲ Μαχάων, Ἕλλη[σιν δύ᾽ ἀκέστορε]
λόγχης vac.　　ἰὴ Παιάν
ἠδ᾽ Ἰασὼ Ἀκεσώ τε καὶ Αἴγλη καὶ Πανάκεια, Ἠπιόνης κ[οῦροι σὺν]
ἀριπρέπτῳ Ὑγιείᾳ, vac.　　ἰὴ Παιάν
χαῖρε, βροτοῖς μέγ᾽ ὄνειαρ, δαῖμον κλεινότατε, vac. [ἰὲ] ὦ [ἰὲ Παιάν]
Ἀσκληπιέ, σὴν δὲ δίδου σοφίαν ὑμνοῦντας ἐς αἰ[εὶ θ]άλλειν vac.
ἐν βιοτῇ σὺν τερπνοτάτῃ Ὑγιείᾳ, ἰὴ Παιά[ν]
σώζοις δ᾽ Ἀτθίδα Κεκροπίαν πόλιν αἰὲν ἐπερχόμ[εν]ος ἰὲ Παιάν
ἤπιος ἔσσο, μάκαρ, στυγερὰς δ᾽ ἀπέρυκε νούσους ἰὲ ὦ ἰὲ Παιάν.

Macedonius composed this song

To the Delian, the far-darter with the beautiful quiver . . .
with gladsome heart sing praises, O Athenian youths, . . . hail,
Paean, bearing in your hands a suppliant olive branch . . .
blameless song for ever . . . famous. . . . Whom once upon
a time his mother bore as a defender from illness and a healer
of pain, Asclepius, gladsome youth. Him the Centaur taught,
along the ridges of Pelion, the skill and wisdom that safeguard
men from pain, the son of Coronis, kindly to men, most august
of the demigods. By him were begotten the youths Podalirius
and Machaon, the two healers of Greek spearsmen . . . ,
hail, Paean, and Iaso, Aceso, Aegle, and Panacea, daughters of
Epione, together with noble Hygieia, . . . hail, Paean. Hail,
great boon to mankind, most renowned among the demigods,
hail, O hail, Paean. Asclepius, grant those who hymn your
wisdom[1] forever to abound in life with delightsome Hygieia,
hail, Paean. May you safeguard the Attic city of Cecrops, ever
visiting it, hail, Paean. Be gentle, blessed one, and ward off
loathsome diseases, hail, O hail, Paean.

[1] Cf. **593a**. *Inscriptiones Graecae*, IV², 1, no. 129 [P. Maas, *Epidaurische Hymnen*,
p. 128; Hellenistic period ?] : . . . ἠδ᾽ Ἀσκλαπιὸν ὑψιτέχναν. . . .

594. Inscriptiones Graecae, IV², 1, no. 128, iii, 32–iv, 56 [*ca.* 300 B. C.].

Ἴσυλλος Ἀστυλάιδαι ἐπέθηκε μαντεύσασθαί οἱ
περὶ τοῦ παιᾶνος ἐν Δελφοῖς, ὃν ἐπόησε εἰς τὸν Ἀπόλλωνα καὶ τὸν Ἀσκλαπιόν, ἢ λώιόν οἱ κα εἴη ἀγγρά-
φοντι τὸν παιᾶνα. ἐμάντευσε λώιόν οἱ κα εἶμεν ἀγ-
γράφοντι καὶ αὐτίκα καὶ εἰς τὸν ὕστερον χρόνον.

ἰεπαιᾶνα θεὸν ἀείσατε λαοί, | ζαθέας ἐνναέται
τᾶσδ' Ἐπιδαύρου. | ὧδε γὰρ φάτις ἐνέπουσ' ἤλυ-
θ' ἐς ἀκοὰς | προγόνων ἁμετέρων, ὦ Φοῖβε Ἀπόλ-
λων. | Ἐρατὼ Μοῦσαν πατὴρ Ζεὺς λέγεται Μά|λ[ωι]
δόμεν παράκοιτιν ὁσίοισι γάμοις. | Φλεγύας δ', [ὃς]
πατρίδ' Ἐπίδαυρον ἔναιεν, | θυγατέρα Μάλου γαμ[[ε]]-
εῖ, τὰν Ἐρατὼ γεί|νατο μάτηρ, Κλεοφήμα δ' ὀνομάσθη. | ἐγ
δὲ Φλεγύα γένετο, Αἴγλα δ' ὀνομάσθη· | τόδ' ἐπώνυμον·
τὸ κάλλος δὲ Κορωνὶς ἐπεκλήθη. | κατιδὼν δὲ ὁ χρυ-
σότοξος Φοῖβος ἐμ Μά|λου δόμοις παρθενίαν ὥραν
ἔλυσε· | λεχέων δ' ἱμεροέντων ἐπέβας, Λα|τώιε κόρε
χρυσοκόμα. | σέβομαί σε· ἐν δὲ θυώδει τεμένει τέ-
κε|τό νιν Αἴγλα, γονίμαν δ' ἔλυσεν ὠδῖ|να Διὸς
παῖς μετὰ Μοιρᾶν Λάχεσίς τε μαῖα ἀγανά. | ἐπίκλη-
σιν δέ νιν Αἴγλας ματρὸς Ἀσκλα|πιὸν ὠνόμαξε
Ἀπόλλων, τὸν νόσων παύ|στορα, δωτῆρα ὑγιείας,
μέγα δώρημα βροτοῖς. | ἰεπαιάν, ἰεπαιάν, χαῖρεν
Ἀσκλα|πιέ, τὰν σὰν Ἐπίδαυρον ματρόπολιν αὔ|-
ξων, ἐναργῆ δ' ὑγίειαν ἐπιπέμποις | φρεσὶ καὶ σώ-
μασιν ἀμοῖς· ἰεπαιάν, ἰεπαιάν.

Isyllus bade Astulaïdas consult the oracle in Delphi for him
concerning the paean which he composed in honor of Apollo
and Asclepius, whether it would be better for him to inscribe
the paean on stone. The oracle replied: "It would be better
for him to inscribe it on stone both for the present time and
for days to come."

O people, praise the god to whom "Hail, Paean" is sung,
you who dwell in this sacred Epidaurus. For thus the message
came to the ears of our forefathers, O Phoebus Apollo. Zeus
the Father is said to have given the Muse Erato to Malos as
his bride in holy wedlock. Then Phlegyas, who dwelt in Epi-

daurus, his fatherland, married the daughter of Malos whom
Erato, her mother, bore, and her name was Kleophema. By
Phlegyas then a child was begotten, and she was named Aigle;
this was her name, but because of her beauty she was also
called Coronis. Then Phoebus of the golden bow, beholding
her in the palace of Malos, ended her maidenhood. You went
into her lovely bed, O golden-haired son of Leto. I revere you.
Then in the perfumed temple Aigle bore a child, and the son
of Zeus, together with the Fates and Lachesis, the noble midwife, eased her birth pains. Apollo named him Asclepius from
his mother's name, Aigle, the reliever of illness, the granter
of health, great boon to mankind. Hail, Paean, hail, Paean,
Asclepius, increase your maternal city of Epidaurus, send
bright health to our hearts and bodies, hail, Paean, hail, Paean.

595. Galenus, De Antidotis, I, 6 [XIV, p. 42 K.].

Ἱλήκοις ὃς τήνδε μάκαρ τεκτήναο, Παιών,
　　εἴτε σε Τρικκαῖοι, δαῖμον, ἔχουσι λόφοι,
ἢ Ῥόδος ἢ Βουρίννα καὶ ἀγχιάλη Ἐπίδαυρος·
　　ἰλήκοις, ἱλαρὴν δ᾽ αἰὲν ἄνακτι δίδου
παῖδα τεὴν Πανάκειαν. ὁ δ᾽ εὐαγέεσσι θυηλαῖς
　　ἱλάσεται τὴν σὴν αἰὲν ἀνωδυνίην.

Be gracious, blessed Paeon, you who fashioned this remedy,
whether the Triccaean ridges hold you, O demigod, or Rhodes,
or Cos and Epidaurus on the sea;[1] be gracious, send your
always gracious daughter, Panacea, to the emperor, who will
propitiate you with pure sacrifices for the everlasting freedom
from pain which you can grant.

[1] Cf. T. 307; 615; 711.

596. Inscriptio Pergamena [*Berl. Sitzber.*, 1934, pp. 753 ff.].

. . . .
　[σωτὴρ . . . .]
[κληθεὶς ἐν νυξίν τε καὶ ἤμασι πολλ]άκι[ς ἦλθες]
[νούσοις ἀργαλέαις τρυομένω]ι κραδίην.
[εἰν ἁλὶ δὲ πρόφρων μου] κήδεαι οὐδὲ δ[αμῆναι]
　[οὔποτέ με προδίδως] πήμασι λευγαλ[έοις],
[ἀλλ᾽ ἀδεῶς μὲν πέμψας], ὅτε πλώοντά με Δ[ήλωι]

[ἴσχες ἄμ᾽ Αἰγαῖον τ᾽ οἶδ]μα καταστόρεσᾶς,
[ῥύσαο δ᾽ αὖ ναυηγό]ν, ὅτε στροφάλιγγι βαρεί[ηι]
   [κῦμα τρόπει μι]κρῆι στῆσας ὑφ᾽ ἡμετέρηι,
[πρήυνας δ᾽ ἀνέ]μους, ὅτ᾽ ἐπ᾽ ἀνδράσι μαίνετ᾽ ἀ[ήτης]
   [αἰνὴν ἀμ]φ᾽ αὐτοῖς αἶσαν ἄγων θανάτου.
[ἐκ δ᾽ ἀλέη]ς με σάωσᾶς ἀεικέος ἔκ τε ῥο[άων]
   [χειμε]ρίων ποταμῶν ἔκ τ᾽ ἀνέμοιο βίη[ς].
[αὐτὸς δ᾽] Αὐσονίων ἕταρον ποίησας ἀν[άκτων]
   [καὶ κλέ]ος ἐκ πολίων ἐσθλὸν ἔνευσ[ας ἄγειν]
[εὐλογίηι ζ]αθέης Βειθυνίδος ἔνδοθι [χώρης]
   [καὶ γῆν θεσπ]εσίην σὴν ἀνὰ Τευθρ[ανίην].
[ἀνθ᾽ ὧν σὸν τιμ]ῶ τε καὶ ἅζομαι οὔν[ομα, Σῶτερ],
   [κληίζω δέ σε] γῆς πε[ίρ]ατ[᾽ ἐς ἡμεδαπῆς].

. . . Savior . . . .
Summoned night and day, you came to me many times when
I was distressed in heart by painful illness. On the sea you
kindly protected me, and you never let me be overcome by baneful calamities, but you conducted me safely when you steered my
ship to Delos, smoothing the Aegean swell. Again, you rescued
the shipwrecked when you made the waves in a powerful eddy
rise under my little boat. You calmed the winds when the gale
raged against men and surrounded them with the grim destiny
of death. You saved me from the glaring heat and from the
currents of the wintery rivers and from the violence of the
wind. You made me a comrade of the Ausonian lords and
graciously promised on account of my rhetoric to spread my
good reputation from the cities within the sacred Bithynian
land even to your divine land of Teuthrania. In gratitude for
these kindnesses I revere and respect your name, Savior;
I sing your praises to the ends of our [i. e., the Roman]
Empire.

---

[1] This hymn is ascribed to Aristides by Herzog, *loc. cit.*

597. Aristides, Oratio XLIX, 4.

    . . . οἱ δ᾽ ἀνεγίγνωσκόν τε καὶ ᾖδον τὰ ἔπη τάδε ὑπηχοῦντες ὡς
ἥδιστον·

       πολλοὺς δ᾽ ἐκ θανάτοιο ἐρύσατο δερκομένοιο,
       ἀστραφέεσσι πύλῃσιν ἐπ᾽ αὐτῇσιν βεβαῶτας
       Ἀΐδεω.

ταῦτα δ᾽ ἐστὶ τῶν ἡμετέρων ἐπῶν, ἃ πρῶτα σχεδὸν ἐποιήσαμεν
τῷ θεῷ.

. . . they [*sc.*, the school children] read and sang the following words, responding sweetly: "From sharp-sighted death he
rescued many who had advanced right to the gates of Hades
whence none return." These are some lines of mine, almost
the first that I composed to the god.

598. Epigrammata Graeca, 1027 [Kaibel; 2nd-3rd c. A. D.].[1]

> Ἀγαθῇ Τύχῃ

Ἔγρεο, Παιήων Ἀσκληπιέ, κοίρανε λαῶν,
Λητοΐδου σεμνῆς τε Κορωνίδος ἠπιόφρων παῖ.
ὕπνον ἀπὸ βλεφάρων σκεδάσας εὐχῶν ἐπάκουε
σῶν μερόπων, οἳ πολλὰ γεγηθότες ἱλάσκονται
σὸν σθένος ἠπιόφρων Ἀσκληπιὲ πρῶτον Ὑγείαν.
ἔγρεο καὶ τεὸν ὕμνον ἰήϊε κέκλυθι χαίρων.

> Good Fortune

Wake, Paeon Asclepius, lord of men, gentle-minded child of the
son of Leto and noble Coronis, dispersing sleep from your eyes,
heed the prayers of your worshippers who, rejoicing greatly,
propitiate your prime power, Health, gentle-minded Asclepius;
wake and be pleased to give ear to your hymn, you whom we
invoke with the cry "Hail."

[1] Cf. P. Maas, *Epidaurische Hymnen*, p. 151.

599. Inscriptiones Graecae, XIV, no. 967a [2nd–3rd c. A. D.].

Τῷ [σωτ]ῆρι Ἀσκληπιῷ σῶστρα καὶ
χαριστήρια Νικομήδης ἰατρός.
Τὰν παιδὸς καλλίσταν εἰκὼ τάνδε θεοῖο,
Παιᾶνος κούρου ματρὸς ἀπ᾽ ἀρτιτόκου,
δαιδάλλων μερόπεσσιν ἐμήσαο, σεῖο, Βόηθε,
εὐπαλάμου σοφίης μνᾶμα καὶ ἐσσομένοις·
θῆκε δ᾽ ὁμοῦ νούσων τε κακῶν ζωάγρια Νικομήδης καὶ χειρῶν δεῖγμα παλαιγενέων.

To the Savior Asclepius Nicomedes the physician gives gifts
for deliverance and thank-offerings. This beautiful image of

the divine child, Paean, who has just been borne by his mother, you, Boëthus,[1] cunningly devised, a reminder of your inventive wisdom even to future generations. Nicomedes dedicated it as an offering for recovery from grievous illness and at the same time as a model of the craftsmanship of old.

[1] Cf. T. 664.

600. Inscriptiones Graecae, XIV, no. 967b [2nd–3rd c. A. D.].

> Τῷ βασιλεῖ Ἀσκληπιῷ σῶστρα καὶ [χα]ριστήρια
>   Νικομήδης Σμυρναῖος ἰατρός.
> Οἷον ἐμαιώσαντο νέον τόκον Εἰλείθυιαι
>   ἐκ Φλεγύου κούρης Φοίβῳ ἀκειρεκόμῃ,
> τοῖόν τοι, Παιὰν Ἀσκληπιέ, σεῖο Βόηθος
>   χειρὸς ἄγαλμ᾽ ἀγαθῆς τεῦξεν ἐαῖς πραπίσιν·
> νηῷ δ᾽ ἐν τῷδε ζωάγρια θῆκεν ὁρᾶσθαι,
>   πολλάκι σαῖς βουλαῖς νοῦσον ἀλευάμενος,
> σὸς θεράπων, εὐχῆς ὀλίγη[ν] δόσιν, οἷα θεοῖσιν
>   ἄνδρες ἐφημέριοι τῶν[δε] φέρουσι χάριν.

To the King Asclepius Nicomedes the physician from Smyrna gives gifts for deliverance and thank-offerings. Such a young child as the goddesses of childbirth brought forth from the maiden, daughter of Phlegyas, for unshorn Phoebus, just such a likeness of you, Paean Asclepius, did Boëthus[1] fashion by the skill of his good craftsmanship. In this temple your servant, according to his vow, placed it as a visible offering for recovery, since many times he avoided sickness by your counsel, a slight gift, such as short-lived men bring to the gods in return for their favors.

[1] Cf. T. 664.

601. Orphei Hymni, LXVII.

> Ἀσκληπιοῦ, θυμίαμα μάνναν
> Ἰητὴρ πάντων, Ἀσκληπιέ, δέσποτα Παιάν,
> θέλγων ἀνθρώπων πολυαλγέα πήματα νούσων,
> ἠπιόδωρε, κραταιέ, μόλοις κατάγων ὑγίειαν
> καὶ παύων νούσους χαλεπὰς θανάτοιό τε κῆρας,
> αὐξιθαλὲς κοῦρος καὶ ἀλεξίκακ᾽, ὀλβιόμοιρε,
> Φοίβου Ἀπόλλωνος κρατερὸν θάλος, ἀγλαότιμον,

ἐχθρὲ νόσων, Ὑγίειαν ἔχων σύλλεκτρον ἀμεμφῆ,
ἐλθέ, μάκαρ, σωτήρ, βιοτῆς τέλος ἐσθλὸν ὀπάζων.

To Asclepius, the burning of incense

Healer of all, Asclepius, Lord Paean, softening the painful
sufferings of men's diseases, giver of gentle gifts, mighty one,
may you come bringing health and checking illnesses and the
harsh fate of death. O youth who promote growth and ward
off evil, youth of happy fate, mighty scion of Phoebus Apollo,
splendidly honored, enemy of diseases, having as your wife
faultless Hygieia, come blessed one, savior, granting a good
end to life.

602. Aristides, Oratio XLVIII, 21.

. . . καὶ βοὴ πολλὴ τῶν τε παρόντων καὶ ἐπιόντων τὸ πολυύμνη-
τον δὴ τοῦτο βοώντων ‘ μέγας ὁ Ἀσκληπιός.’

. . . and the loud cry both of those who are present and of
those who are coming, shouting this widely renowned refrain,
" Great is Asclepius."

603. Athenaeus, Deipnosophistae, VI, 55, 250c.

Λήξαντος δὲ τῆς ὀργῆς τοῦ Διονυσίου πάλιν ὁ Δημοκλῆς ἔφη ·
" χαρίσαιο δ' ἄν μοί τι, Διονύσιε, κελεύσας τινὶ τῶν ἐπισταμένων
διδάξαι με τὸν πεποιημένον εἰς τὸν Ἀσκληπιὸν παιᾶνα · ἀκούω
γάρ σε πεπραγματεῦσθαι περὶ τοῦτον."

When the anger of Dionysius [sc., the Younger, of Sicily]
was thus allayed, Democles [the parasite of Dionysius] re-
sumed: " You would do me a favor, Dionysius, if you would
command someone who knows it to teach me the paean composed in honor of Asclepius; for I hear that you have been
occupied with that."

604. Aristides, Oratio L, 42.

Ἐδόκει γὰρ ᾄδειν ἐμὸν παιᾶνα, ἐν ᾧ ταύτην τὴν πρόσρησιν εἶναι ·
‘ Ἰὴ Παιὰν Ἡρακλες Ἀσκληπιέ.’ καὶ οὕτω δὴ τὸν παιᾶνα
ἀπέδωκα ἀμφοτέροις τοῖς θεοῖς κοινόν.

For he [sc., the Macedonian] seemed to sing my paean in

which there was this address: "Hail Paean Heracles Asclepius." And thus I rendered the paean to both these gods
together.

605. Aristides, Oratio L, 44.

Πάλιν δὲ ἐπὶ τῷ δεκάτῳ τελεσθέντι ἐγὼ μὲν ἔν τι τῶν ᾀσμάτων
ἔτυχον παρεικὼς διὰ τὸ παντελῶς αὐτοσχεδίως τε καὶ ἐκ τοῦ
ῥᾴστου πεποιῆσθαι καὶ ὅσον αὐτῷ φασιν, ἔπειθ' ἧκεν ὄναρ ἀπαι-
τοῦν καὶ τοῦτο, καὶ ἀπέδομεν.

Again, at the performance of the tenth chorus,[1] I happened to
pass over a certain one of the songs because it was composed
entirely on the spur of the moment and most casually and for
myself, as it were; then the dream came, demanding it too,
and I offered it.

[1] **605a.** Aristides, *Oratio* L, 43: Καὶ τοίνυν καὶ χοροὺς ἔστησα δημοσίᾳ, δέκα τοὺς σύμ-
παντας, παίδων, τοὺς δὲ ἀνδρῶν; cf. also T. 552.

606. Aristides, Oratio XLVII, 73.

. . . ποιῆσαι μέλη, γάμον τε Κορωνίδος καὶ γένεσιν τοῦ θεοῦ,
καὶ τὴν στροφὴν ὡς ἐπὶ μήκιστον ἀποτεῖναι.

. . . to compose songs, the marriage of Coronis and the birth
of the god, and to prolong the strophe to the greatest possible
length.

607. Pausanias, Descriptio Graeciae, III, 26, 10.

Διὸ καὶ τάδε αὐτὸς οἶδα περὶ τὸ Ἀσκληπιεῖον τὸ ἐν Περγάμῳ
γινόμενα. ἄρχονται μὲν ἀπὸ Τηλέφου τῶν ὕμνων, προσᾴδουσι
δὲ οὐδὲν ἐς τὸν Εὐρύπυλον, οὐδὲ ἀρχὴν ἐν τῷ ναῷ θέλουσιν ὀνομάζειν αὐτόν, οἷα ἐπιστάμενοι φονέα ὄντα Μαχάονος.

I myself know that to be the reason of the practice at the
temple of Asclepius in Pergamum; they begin their hymns
with Telephus[1] but they do not sing to Eurypylus, or care to
mention his name in the temple at all, as they know that he
was the slayer of Machaon.[2] (*)

[1] Cf. T. 512.
[2] Cf. T. 180; 186.

ORATIONS, PLAYS AND BOOKS

608. Apuleius, Florida, 18.[1]

> Eius dei hymnum Graeco et Latino carmine vobis ⟨etiam⟩
> canam [iam] illi a me dedicatum. sum enim non ignotus illi
> sacricola nec recens cultor nec ingratus antistes, ac iam et
> prorsa et vorsa facundia veneratus sum, ita ut etiam nunc
> hymnum eius utraque lingua canam, cui dialogum similiter
> Graecum et Latinum praetexui, . . . in principio libri facio
> quendam ex his, qui mihi Athenis condidicerunt, percontari
> . . . quae ego pridie in templo Aesculapi disseruerim.

> Of this god [*sc.*, Asclepius] I shall also sing you a hymn both
> in Greek and Latin, which I have dedicated to him. For I am
> well known to him as his priest and worshipper of long stand-
> ing and grateful minister, and I have worshipped him in prose
> and verse, so that even now I can sing a hymn for him in both
> languages; this I have prefaced with a dialogue likewise in
> Greek and Latin. . . . In the beginning of my book I make one
> of those who studied with me in Athens ask . . . what I said
> the other day in the temple of Asclepius. (*)

[1] T. 837.

609. Apuleius, Apologia, 55.

> . . . cum primis diebus quibus Oeam veneram p[l]ublice dis-
> serens de Aesculapii maiestate eadem ista prae me tuli et quot
> sacra nossem percensui. ea disputatio celebratissima est, vulgo
> legitur, in omnibus manibus versatur, non tam facundia mea
> quam mentione Aesculapii religiosis Oeensibus commendata.

> . . . in a public discourse on the greatness of Asclepius deliv-
> ered by me during the first days after I had come to Oea,
> I made the same boast [*sc.*, of having participated in several
> sacred mysteries] and recounted the number of the mysteries
> I knew. That discourse is very famous, is read far and wide,
> is in everybody's hands, and has been recommended to the
> pious inhabitants of Oca not so much through my eloquence
> as through its mention of Asclepius. (**)

610. Libanius, Epistulae, 695, 1-2.

Οὐχ ὅγ' ἄνευθε θεοῦ φησὶν Ὅμηρος οὐδὲ σὺ ταῦτα ἄνευ τῆς Ἀσκληπιοῦ ῥοπῆς, ἀλλ' ἀκριβῶς αὐτὸς συνεφήπτετο τῆς γραφῆς. εἰκὸς δὲ αὐτὸν ὄντα Ἀπόλλωνος ἔχειν τι τῆς τοῦ πατρὸς μουσικῆς καὶ νέμειν οἷς ἂν ἐθέλῃ. σοὶ δὲ πῶς οὐκ ἔμελλεν ἐν τοῖς ὑπὲρ αὐτοῦ συλλήψεσθαι λόγοις;

"And he not without the aid of the gods," says Homer,[1] nor do you [sc., Acacius] write these words without the influence of Asclepius, for manifestly he joined with you in the writing. It is, of course, fitting for him, as the son of Apollo, to have some of the cultural talent of his father and to apportion it to whomever he desires. How then would it be possible for him not to assist you in these discourses concerning himself?[2]

[1] *Iliad*, V, 185.   [2] Cf. e. g. T. 317, 12.

611. Athenaeus, Deipnosophistae, XI, 70, 485b.[1]

Ἀντιφάνης δὲ ἐν Ἀσκληπιῷ·
     τὴν δὲ γραῦν τὴν ἀσθενοῦσαν πάνυ πάλαι, τὴν βρυτικήν,
     ῥίζιον τρίψας τι μικρὸν δελεάσας τε γεννικῇ
     τὸ μέγεθος κοίλῃ λεπαστῇ, τοῦτ' ἐποίησεν ἐκπιεῖν.

So Antiphanes [1st half of 4th cent. B. C.] in 'Asclepius':[2] "(The doctor) pounded a little rootlet, and enticing her with the bait of a deep and generous-sized drinking cup, he made that old hag, the one who has been sick so very long, the one soaked with beer, drink it all up." (*)

[1] = Fr. 45 [Kock].
[2] A play in which a physician, possibly Asclepius himself, is the main character.

612. Athenaeus, Deipnosophistae, XI, 73, 487a.[1]

Φιλέταιρος Ἀσκληπιῷ·
     ἐνέσεισε μεστὴν ἴσον ἴσῳ μετανιπτρίδα
     μεγάλην, ἐπειπὼν τῆς Ὑγιείας τοὔνομα.

Philetaerus [1st half of 4th cent. B. C.] in 'Asclepius':[2] "He brandished a huge, brimming, after-dinner drink, mixed half and half, pronouncing over it the name of Hygieia."

[1] = Fr. 1 [Kock].
[2] **612a.** Athenaeus, *Deipnosophistae*, VIII, 27, 342a: Φιλέταιρος δ' ἐν Ἀσκληπιῷ [τὸν Ὑπερείδην] πρὸς τῷ ὀψοφαγεῖν καὶ κυβεύειν αὐτόν φησι. . . .

613. Athenaeus, Deipnosophistae, XIV, 7, 617b.[1]

Κομψῶς δὲ κἀν τῷ Ἀσκληπιῷ ὁ Τελέστης ἐδήλωσε τὴν τῶν αὐλῶν
χρείαν ἐν τούτοις·
ἢ Φρύγα καλλιπνόων αὐλῶν ἱερῶν βασιλῆα,
Λυδὸν ὃς ἥρμοσε πρῶτος
Δωρίδος ἀντίπαλον Μούσης νόμον αἰόλον, ὀμφᾷ
πνεύματος εὔπτερον αὔραν ἀμφιπλέκων καλάμοις.

Again, in 'Asclepius,' Telestes [ca. 400 B. C.] [2] elegantly set
forth the use of flutes in these lines:
"Or that Phrygian king of the sacred, fair-breathing flutes,
who first composed the quivering Lydian strain to match the
Dorian Muse, with tuneful breath trilling the winged airs on
his reeds."

[1] = Fr. 2 [Diehl].
[2] Cf. **613a.** Philodemus, *De Pietate,* 52 [Gomperz]: κἀν Ἀσκληπιῶ(ι Τελ)έστης; **613b.**
Suidas, *Lexicon, s. v.* Τελέστης: κωμικός· τούτου δράματά ἐστιν Ἀργὼ καὶ Ἀσκληπιός,
ὥς φησιν Ἀθήναιος ἐν τῷ ιδ' τῶν Δειπνοσοφιστῶν; cf. also T. **106.**

614. Anthologia Latina, I, 2, 719e, 1-7.

Quod natum Phoebus docuit, quod Chiron Achillem,
quod didicere olim Podalirius atque Machaon
a genitore suo, qui quondam versus in anguem
templa Palatinae subiit sublimia Romae

. . . .

hoc liber iste tenet diverso e dogmate sumptum.

What Phoebus taught his son, what Chiron taught Achilles,
what long ago Podalirius and Machaon learned from their
father,[1] who once in the guise of a serpent entered the lofty
temple of Rome [2] on the Palatine . . . such knowledge, derived
from various doctrines, this book here contains.

[1] Cf. T. **147a**; **356a.**
[2] Cf. T. **846** ff.

615. Q. Serenus Sammonicus, Liber Medicinalis, Prooemium, 1-10.

Phoebe, salutiferum, quod pangimus, adsere carmen
inventumque tuum prompto comitare favore.
tuque potens artis, reduces qui tradere vitas
nosti et in caelum manes revocare sepultos,
qui colis Aegeas, qui Pergama quique Epidaurum,

qui quondam placida tectus sub pelle draconis
Tarpeias arces atque inclita templa petisti
depellens taetros praesenti numine morbos:
huc ades et quicquid cupido mihi saepe locutus
firmasti, cunctum teneris expone papyris.

Phoebus, claim as your own the health-bringing song which
we are composing, and attend with ready favor your own
invention; and you, skilled in art, who know how to impart
new lives, and to recall buried shades to the light of heaven,[1]
you who cherish Aegae, Pergamum, and Epidaurus,[2] who one
time cloaked in the soft skin of a serpent sought the Tarpeian
rocks and the renowned shrines, dispersing ugly diseases by
your divine presence: be present here and what you have often
confirmed in speaking to me who was anxious to listen, expound
it all on these pages made of fine papyrus.

[1] Cf. e. g. T. 419.   [2] Cf. T. 307; 595.

616. Lucianus, Quomodo Historia Conscribenda, 16.

Καὶ νὴ Δία καὶ τὸ προοίμιον ὑπέρψυχρον ἐποίησεν οὕτως συν-
αγαγών· οἰκεῖον εἶναι ἰατρῷ ἱστορίαν συγγράφειν, εἴ γε ὁ
Ἀσκληπιὸς μὲν Ἀπόλλωνος υἱός, Ἀπόλλων δὲ Μουσηγέτης καὶ
πάσης παιδείας ἄρχων.

And, by Zeus, he [sc., Callimorphus] made also his preface
rather frigid, arguing thus: It is seemly for a doctor to write
history, at least if Asclepius is the son of Apollo, and Apollo
the leader of the Muses and the patron of all education.

617. Libanius, Epistulae, 1342, 1.

Αὐτὸς αὑτῷ μοι δοκεῖ διὰ τοῦ σοῦ στόματος ὁ θεὸς εἰργάσθαι
τὸν ἀπόλογον· οὐ γὰρ ἂν πάντα οὕτως εἶχε τὰ τοῦ λόγου φάρμακα
μὴ τοῦ θεοῦ σοι τὴν ψυχὴν κατασχόντος. πολλὰ δὲ τοιαῦτα ἐν
Ἀπόλλωνος οἰκίᾳ, καὶ θαυμαστὸν οὐδὲν Ἀσκληπιὸν τὸν αὑτοῦ
μιμεῖσθαι πατέρα.

It seems to me that the god himself produced the story
through your [sc., Acacius'] mouth; for it would not have all
the remedial effects of the word, had not the god possessed
your soul. There are many such remedies in Apollo's house;
and it is not strange that Asclepius imitates his father.

SWEARING BY ASCLEPIUS' NAME

618. Menander, Boeotia, Fr. 91 [Kock].

Οὐκ ἔστιν ἐκτεὺς τοῦτο, μὰ τὸν Ἀσκληπιόν.

By Asclepius,[1] this is no pint!

[1] Cf. **618a.** Menander, *Perikeiromene*, 146: Μὰ τὸν Ἀσκληπιόν.

619. Menander, Samia, 94-95.

Μὰ τὸν Διόνυσον, μὰ τὸν Ἀπ[ό]λλ[ω, 'γὼ μὲν οὔ,]
μὰ τὸν Δία τὸν σωτῆρα, μὰ τὸν Ἀ[σκληπιόν].

By Dionysus, by Apollo I would not,
By Zeus the savior, by Asclepius![1]

[1] Cf. **619a.** Athenaeus, *Deipnosophistae*, XIV, 49, 642d [= Alexis, Fr. 163 (Kock)]:
Οὐδὲ φιλόδειπνός εἰμι, μὰ τὸν Ἀσκληπιόν, / τραγήμασιν χαίρω δὲ μᾶλλον.

620. Galenus, De Compositione Medicamentorum Secundum Locos,
Cp. 3 [XIII, pp. 271-72 K.].

Προσέθηκε δὲ τῷ Διὶ τὴν Πίσσαν, ὡς εἰώθασι πολλοὶ καὶ χωρὶς
ποιητικῆς ἐν τῷ βίῳ λέγειν, μὰ τὸν ἐν Περγάμῳ Ἀσκληπιόν, μὰ
τὴν ἐν Ἐφέσῳ Ἄρτεμιν. . .

He [*sc.,* Philo][1] added Pissa [*i. e.,* Olympia] to the name of
Zeus as many are wont, even apart from poetry, to say in
everyday-life, "by Asclepius in Pergamum," "by Artemis in
Ephesus" . . .

[1] Cf. T. **230.**

621. Julianus, In Heraclium Cynicum, 234 D.

. . . νὴ τὸν Ἀσκληπιόν . . .

. . . by Asclepius . . .

622. Theophilus Antiochenus, Ad Autolycum, III, 2.

. . . ἢ Σωκράτην τὸ ὀμνύειν τὸν κύνα, καὶ τὸν χῆνα, καὶ τὴν
πλάτανον, καὶ τὸν κεραυνωθέντα Ἀσκληπιόν, καὶ τὰ δαιμόνια
ἃ ἐπεκαλεῖτο;

. . . or [what good did it do] for Socrates to swear by the dog and the goose[1] and the plane-tree and by thunderstruck Asclepius and the demons whom he used to invoke? (*)

[1] Cf. T. 530.

COGNOMINA OF ASCLEPIUS

623. Hesychius, Lexicon, *s. v.* Αἰγλάηρ.

Αἰγλάηρ· ὁ Ἀσκληπιός.

Aiglaër: Asclepius.

624. Hesychius, Lexicon, *s. v.* Ἀγλαόπης.

Ἀγλαόπης· ὁ Ἀσκληπιός. Λάκωνες

Aglaopes: Asclepius. (So say) the Laconians.

625. Suidas, Lexicon, *s. v.* Αἴγλη.

Αἴγλη· βόλος κυβευτικός, λαμπηδών, αὐγή . . . ἀλλὰ καὶ ἡ σελήνη οὕτω καλεῖται καὶ ὁ Ἀσκληπιός.

Aigle is a throw of dice, lustre, a beam of light. . . . But also the moon is called thus, as well as Asclepius.

626. Pausanias, Descriptio Graeciae, III, 14, 7.[1]

Ἀσκληπιοῦ δέ ἐστιν ἐπίκλησις ὁ Ἀγνίτας . . .

Agnitas is a surname of Asclepius . . .[2]

[1] Cf. T. 764.
[2] Cf. T. 719: Ἀρχαγέτας; T. 771: Ἀσκληπιοῦ . . . Αὐλωνίου; cf. T. 637; T. 668c: Ἀσκληπιὸς ἐπίκλησιν Γορτύνιος; T. 780: Ἀσκληπιοῦ . . . Δημαινέτου; T. 776: Καουσίου; T. 763: Ἀσκληπιὸν Κοτυλέα; T. 773: Ἀσκληπιοῦ Παιδός; T. 759: Φιλόλαον . . . ὀνομάζουσι.

VI. IMAGES

GENERAL REMARKS

627. Callistratus, Descriptiones, 10.

Εἰς τὸ ἄγαλμα τοῦ Παιᾶνος

Εἶτα τὸ μὲν Ἀργῷον σκάφος ἔμφωνον γενέσθαι πειθόμεθα τὸ
ὑπὸ τῶν Ἀθηνᾶς τεχνηθὲν χειρῶν, ὃ καὶ τὴν ἐν ἄστροις ἐκλη-
ρούχησε τύχην, ἄγαλμα δὲ οὐ πιστεύσομεν, εἰς ὃ τὰς δυνάμεις
Ἀσκληπιὸς ἀνίησι τὸν προνοητικὸν ἐπεισάγων νοῦν ἐπὶ τὴν
ἑαυτοῦ κοινωνίαν, τοῦ συνοικοῦντος τὴν δύναμιν πρέπειν, ἀλλ᾽ εἰς
μὲν ἀνθρώπινα ⟨σώματα⟩* κατάγεσθαι τὸ θεῖον δώσομεν, ἔνθα
καὶ μιανθῆναι παθήμασιν, οὐ πιστεύσομεν δέ, ᾗ μηδὲν ἔγγονον
κακίας παραπέφυκεν; ἐμοὶ μὲν οὖν οὐ τύπος εἶναι δοκεῖ τὸ ὁρώ-
μενον, ἀλλὰ τῆς ἀληθείας πλάσμα. ἰδοὺ γὰρ ὡς οὐκ ἀνηθοποίητος
ἡ τέχνη, ἀλλ᾽ ἐνεικονισαμένη τὸν θεὸν εἰς αὐτὸν ἐξίσταται. ὕλη
μὲν οὖσα θεοειδὲς ἀναπέμπει νόημα, δημιούργημα δὲ χειρὸς τυγ-
χάνουσα, ἃ μὴ δημιουργίαις ἔξεστι, πράττει τεκμήρια ψυχῆς
ἀρρήτως ἀποτίκτουσα. πρόσωπον δέ σοι θεασαμένῳ δουλοῦται
τὴν αἴσθησιν. οὐ γὰρ εἰς κάλλος ἐπίθετον ἐσχημάτισται, ἀλλὰ
πάναγνον καὶ ἵλεων ἀνακινῶν ὄμμα βάθος ἄφραστον ὑπαστράπτει
σεμνότητος αἰδοῖ μιγείσης. πλοκάμων δὲ ἕλικες ῥεόμενοι χάρισιν
οἱ μὲν εἰς νῶτα τεθηλότες ἄφετοι κέχυνται, οἱ δὲ ὑπὲρ μετώπου
πρὸς τὰς ὀφρῦς ἐπιβαίνοντες τοῖς ὄμμασιν εἰλοῦνται. οἷον δὲ
ἐκ ζωτικῆς αἰτίας καὶ αὐτοὶ καταρδόμενοι εἰς τὴν τῶν βοστρύχων
καμπὴν συνελίττονται, τῷ νόμῳ τῆς τέχνης μὴ πειθομένης τῆς
ὕλης, ἀλλὰ νοούσης ὅτι σχηματίζει θεὸν καὶ δεῖ δυναστεύειν. τῶν
δὲ γενομένων εἰωθότων φθείρεσθαι ἡ τοῦ ἀγάλματος ἰδέα, ἅτε
δὴ τῆς ὑγείας τὴν οὐσίαν ἐν ἑαυτῇ φέρουσα, ἀκμὴν ἀνώλεθρον
ἐπικτωμένη θάλλει. ἡμεῖς μὲν δή σοι καὶ λόγων, ὦ Παιάν, νεαρῶν
καὶ μνήμης ἐγγόνων ἀπηρξάμεθα· κελεύεις γὰρ οἶμαι· πρόθυμος
δέ σοι καὶ τὸν νόμον ᾄδειν, εἰ νέμοις ὑγείαν.

* σώματα addidit B.

On the statue of Paean

Are we then to believe that the vessel Argo, which was wrought
by the hands of Athena and later assumed its allotted place
among the stars, became capable of speech, and yet in the case

343

of a statue into which Asclepius infused his own powers, introducing purposeful intelligence therein and thus making it a partner with himself, not believe that the power of the indwelling god is clearly manifest therein? Nay, more, shall we admit that the divine spirit descends into human bodies, there to be even defiled by passions, and nevertheless not believe it in a case where there is no attendant engendering of evil? To me, at any rate, the object before our eyes seems to be, not an image, but a modelled presentment of truth; for see how Art not only is not without power to delineate character, but, after having portrayed the god in an image, it even passes over into the god himself. Matter though it is, it gives forth divine intelligence, and though it is the work of human hands, it succeeds in doing what handicrafts cannot accomplish, in that it begets in a marvellous way tokens of a soul. The face, as you look at it, enthralls the senses; for it has not been fashioned to an adventitious beauty, but, as it raises a chaste and propitious eye, it flashes forth an indescribable depth of dignity mixed with compassion. Curly locks abounding in grace, some fall luxuriant and unconfined on the back, while others come down over the forehead to the eyebrows and hang thick about the eyes. But, as if stirred by life and kept moist of themselves, they coil themselves into the bending curls, the material not rendering obedience to the law of art, but realizing that it represents a god and that he must work his own will. And although all things that are born are wont to die, yet the form of the statue, as though carrying within itself the essence of health, flourishes in the possession of indestructible youth. And so we, O Paean, have offered to you the first fruits of discourse, freshly made, and the offspring of memory; for you bid us do so, I think; and I am eager also to sing the strains to you if you allot me health. (*)

DESCRIPTION OF IMAGES

Middle Greece

628. Pausanias, Descriptio Graeciae, I, 21, 4.[1]

Τοῦ δὲ Ἀσκληπιοῦ τὸ ἱερὸν ἔς τε τὰ ἀγάλματά ἐστιν, ὁπόσα τοῦ θεοῦ πεποίηται καὶ τῶν παίδων, . . . θέας ἄξιον.

The sanctuary of Asclepius [sc., at Athens] is worth seeing
for the statues of the god and of his children . . .[2] (*)

[1] Cf. T. 725.
[2] As regards the image at Elateia (Phocis), cf. T. 662; near Tithorea (Phocis), cf.
T. 668b; at Lebadeia (Boeotia), cf. T. 650; at Megara (Megaris), cf. T. 656.

AEGINA

629. Pausanias, Descriptio Graeciae, II, 30, 1.

Τοῦ δὲ Ἀσκληπιοῦ . . . λίθου δὲ ἄγαλμα καθήμενον.

Of Asclepius . . . there is a stone image,[1] seated[2] [sc., at
Aegina].[3] (*)

[1] Cf. also T. 668.　　　　[2] Cf. T. 680.　　　　[3] Cf. T. 733-34.

ARGOLID

630. Pausanias, Descriptio Graeciae, II, 27, 2.[1]

Τοῦ δὲ Ἀσκληπιοῦ τὸ ἄγαλμα μεγέθει μὲν τοῦ Ἀθήνῃσιν Ὀλυμπίου Διὸς ἥμισυ ἀποδεῖ, πεποίηται δὲ ἐλέφαντος καὶ χρυσοῦ·
μηνύει δὲ ἐπίγραμμα τὸν εἰργασμένον εἶναι Θρασυμήδην Ἀριγνώτου Πάριον. κάθηται δὲ ἐπὶ θρόνου βακτηρίαν κρατῶν, τὴν δὲ
ἑτέραν τῶν χειρῶν ὑπὲρ κεφαλῆς ἔχει τοῦ δράκοντος, καί οἱ καὶ
κύων παρακατακείμενος πεποίηται. τῷ θρόνῳ δὲ ἡρώων ἐπειργασμένα Ἀργείων ἐστὶν ἔργα, Βελλεροφόντου τὸ ἐς τὴν Χίμαιραν καὶ
Περσεὺς ἀφελὼν τὴν Μεδούσης κεφαλήν.

The image of Asclepius [sc., at Epidaurus] is half the size[2]
of the image of Olympian Zeus at Athens: it is made of ivory
and gold.[3] An inscription sets forth that the sculptor was
Thrasymedes [ca. 375 B. C.], a Parian, son of Arignotus.[4]
The god is seated on a throne,[5] grasping a staff[6] while holding
the other hand over the head of the serpent;[7] and a dog,[8] lying
by his side, is also represented. On the throne are carved in
relief the deeds of Argive heroes: of Bellerophon against the
Chimaera, and Perseus who has cut off Medusa's head.[9] (*)

[1] Cf. T. 739.
[2] Cf. T. 849: *simulacrum ingens.*
[3] Cf. T. 649; 675; 683.
[4] Cf. however T. 645.
[5] Cf. T. 680.
[6] Cf. T. 850, v. 655.
[7] Cf. T. 688.
[8] Cf. T. 5; T. 691.
[9] Cf. T. 3.

631. Pausanias, Descriptio Graeciae, II, 29, 1.[1]

Τέμενος δή ἐστιν Ἀσκληπιοῦ καὶ ἀγάλματα ὁ θεὸς αὐτὸς καὶ Ἠπιόνη.

There is, of course [sc., at Epidaurus], a precinct of Asclepius and images, the god himself and Epione.

[1] Cf. T. 742; cf. also T. 673.

632. Pausanias, Descriptio Graeciae, V, 11, 11.

Ἐν Ἐπιδαύρῳ δὲ ἐρομένου μου καθ᾽ ἥντινα αἰτίαν οὔτε ὕδωρ τῷ Ἀσκληπιῷ σφίσιν οὔτε ἔλαιόν ἐστιν ἐγχεόμενον, ἐδίδασκόν με οἱ περὶ τὸ ἱερὸν ὡς καὶ τὸ ἄγαλμα τοῦ θεοῦ καὶ ὁ θρόνος ἐπὶ φρέατι εἴη πεποιημένα.

When I asked at Epidaurus why they pour neither water nor olive oil over (the statue of) Asclepius, the attendants at the sanctuary informed me that both the image of the god and the throne were built over a cistern.[1] (*)

[1] Cf. T. 643; cf. also T. 707; 739, 6.

633. Anonymus, De Incredibilibus, II, p. 89, 6-9.

Τὰ ἑπτὰ θεάματα

Τινὲς δὲ τάττουσι καὶ τὸν ἐν Ἐπιδαύρῳ Ἀσκληπιὸν καὶ τὸν ἐν Παρίῳ βωμὸν καὶ τοὺς κρεμαστοὺς κήπους καὶ τὴν ἱσταμένην Ἀθηνᾶν ἐν Ἀθήναις καὶ τὰ Κύρου βασίλεια.

The Severn Wonders

Some include also the (statue of) Asclepius in Epidaurus[1] and the altar in Parium and the hanging gardens and the standing (figure of) Athena in Athens and the palaces of Cyrus.

[1] Cf. also T. 845.

634. Clemens Alexandrinus, Protrepticus, IV, 52, 4.

Αἱ δὲ χελιδόνες καὶ τῶν ὀρνέων τὰ πλεῖστα κατεξερῶσιν αὐτῶν τῶν ἀγαλμάτων εἰσπετόμενα, οὐδὲν φροντίσαντα οὔτε Ὀλυμπίου Διὸς οὔτε Ἐπιδαυρίου Ἀσκληπιοῦ οὐδὲ μὴν Ἀθηνᾶς Πολιάδος ἢ Σαράπιδος Αἰγυπτίου.

Swallows and most birds, alighting on these statues, defile

them, having no reverence for either Olympian Zeus or Epidaurian Asclepius,[1] or the Athena Polias or the Egyptian Sarapis.

[1] As regards the image at Sicyon, cf. T. 649; at Troezen, cf. T. 654; at Argos, cf. T. 663; at Corinth, cf. T. 670; at Titane, cf. T. 678; at Phlius, cf. T. 681.

LACONIA

635. Pausanias, Descriptio Graeciae, III, 23, 10.

. . . θέας δὲ αὐτόθι ἄξια . . . τὸ δὲ Ἀσκληπιοῦ καὶ ἄγαλμα ὀρθὸν λίθου.

. . . worth seeing there [sc., at Epidaurus Limera] [1] are . . . also the sanctuary of Asclepius with a standing statue [2] made of stone.[3] (*)

[1] Cf. T. 756.　　　　　[2] Cf. T. 679.
[3] Cf. T. 668; as regards the image at Leuctra, cf. T. 668; at Cyphanta, cf. T. 668; 755; at Gythium, cf. T. 676; at Sparta, cf. T. 677.

MESSENIA

636. Pausanias, Descriptio Graeciae, IV, 31, 10.[1]

Πλεῖστα δέ σφισι καὶ θέας μάλιστα ἀγάλματα ἄξια τοῦ Ἀσκληπιοῦ παρέχεται τὸ ἱερόν. χωρὶς μὲν γὰρ τοῦ θεοῦ καὶ τῶν παίδων ἐστὶν ἀγάλματα, . . .

The most numerous statues and the most worth seeing are to be found in the sanctuary of Asclepius [sc., in Messene]. For besides the statues of the god and his sons,[2] . . .

[1] Cf. T. 770.　　　　　[2] Cf. also T. 657.

637. Pausanias, Descriptio Graeciae, IV, 36, 7.[1]

Ἐν δὲ Αὐλῶνι καλουμένῳ ναὸς Ἀσκληπιοῦ καὶ ἄγαλμά ἐστιν Αὐλωνίου.

In the so-called Aulon [glen] there is a temple and statue of Asclepius Aulonius.[2]

[1] Cf. T. 771.　　　　　[2] Cf. T. 626, n. 2; for the image at Corone, cf. T. 668a.

ARCADIA

638. Pausanias, Descriptio Graeciae, VIII, 32, 4.[1]

Ἐνταῦθα ἔστι μὲν ἱερὸν Ἀσκληπιοῦ καὶ ἀγάλματα αὐτός τε καὶ Ὑγεία.

Here [sc., at Megalopolis] there is a sanctuary of Asclepius and images, he himself and Health. (*)

[1] Cf. T. 773.

639. Pausanias, Descriptio Graeciae, VIII, 31, 1.

Ἐπειργασμένοι δὲ ἐπὶ τύπων πρὸ τῆς ἐσόδου τῇ μὲν Ἄρτεμις, τῇ δὲ Ἀσκληπιός ἐστι καὶ Ὑγεία.

Carved in relief before the entrance are, on one side Artemis, on the other Asclepius and Health [sc., at Megalopolis].[1]

[1] As regards the image at Mantinea, cf. T. 647; at Gortys, cf. T. 652; at Tegea, cf. T. 653.

640. Pausanias, Descriptio Graeciae, VIII, 54, 5.[1]

Ἔστι δὲ ἐπὶ τῆς ὁδοῦ πρῶτα μὲν ναὸς καὶ ἄγαλμα Ἀσκληπιοῦ.

On the road [sc., from Tegea to Argos] come first a temple and image of Asclepius.

[1] Cf. T. 772.

ELIS

641. Pausanias, Descriptio Graeciae, V, 20, 3.

Κατὰ δὲ ἑκατέραν πλευρὰν τῇ μὲν Ἀσκληπιὸς καὶ τῶν Ἀσκληπιοῦ θυγατέρων Ὑγεία ἐστίν, . . .

On one side [sc., of the table of Colotes at Olympia] are Asclepius and one of the daughters of Asclepius, Health,[1] . . .

[1] Cf. T. 278 ff.; 286.

642. Pausanias, Descriptio Graeciae, V, 26, 2.

Παρὰ δὲ τοῦ ναοῦ τοῦ μεγάλου τὴν ἐν ἀριστερᾷ πλευρὰν ἀνέθηκεν ἄλλα, Κόρην . . . καὶ Ἀφροδίτην . . . ποιητῶν δὲ Ὅμηρον καὶ Ἡσίοδον, καὶ θεοὺς αὖθις Ἀσκληπιὸν καὶ Ὑγείαν.

Along the left side of the great temple [sc., in Olympia] he [sc., Micythus] dedicated other offerings: the Maid . . . and Aphrodite . . . , of the poets Homer and Hesiod, and deities again, Asclepius and Health.[1] (*)

[1] As regards an image at Cyllene, cf. T. 648.

ACHAIA

643. Pausanias, Descriptio Graeciae, VII, 27, 11.

> . . . καὶ ἐπὶ τῇ μεγίστῃ τῶν πηγῶν τοῦ 'Ασκληπιοῦ τὸ ἄγαλμα
> ἴδρυται.

> . . . and at the largest of the springs[1] stands the image of
> Asclepius [sc., at Pellene].[2]

[1] Cf. T. 707.
[2] As regards the image at Aegium, cf. T. 660; at Patrae, cf. T. 669; at Aegira, cf.
T. 672.

ROME

644. Suetonius, Augustus, 59.

> Medico Antonio Musae, cuius opera ex ancipiti morbo con-
> valuerat, statuam aere conlato iuxta signum Aesculapi
> statuerunt.

> In honor of his physician, Antonius Musa, through whose
> care he [sc., Augustus] had recovered from a dangerous ill-
> ness, a sum of money was raised and Musa's statue set up
> beside that of Asclepius.[1]

[1] Cf. also T. 651; 661.

ARTISTS

645. Athenagoras, Pro Christianis, 17, 4.

> . . . ὁ ἐν 'Επιδαύρῳ 'Ασκληπιὸς ἔργον Φειδίου.

> . . . the Asclepius in Epidaurus—the work of Phidias [middle
> of 5th c. B. C.].[1]

[1] Cf. however T. 630, where Thrasymedes [ca. 375 B. C.] is attested as the sculptor
of the Epidaurian statue; cf. also T. 739.

646. Libanius, Oratio XXX, 22-23.

> . . . ἦν ἄγαλμα ἐν Βεροίᾳ τῇ πόλει χαλκοῦν, 'Ασκληπιὸς ἐν εἴδει
> τοῦ Κλεινίου παιδὸς τοῦ καλοῦ καὶ ἡ τέχνη τὴν φύσιν ἐμιμεῖτο,
> τοσοῦτον δὲ ἦν τὸ τῆς ὥρας, ὥστε καὶ οἷς ὑπῆρχεν αὐτὸν καθ'
> ἡμέραν ὁρᾶν, εἶναι τῆς θέας ὅμως ἐπιθυμίαν. τούτῳ θύεσθαι
> θυσίας οὐδεὶς οὕτως ἀναιδής, ὡς εἰπεῖν ἂν τολμῆσαι. τοῦτο τοίνυν,
> ὦ βασιλεῦ, τὸ τοιοῦτον πολλῷ μέν, ὡς εἰκός, πόνῳ, λαμπρᾷ δὲ

ἠκριβωμένον ψυχῇ κατακέκοπται καὶ οἴχεται, καὶ τὰς Φειδίου
χεῖρας πολλαὶ διενείμαντο. διὰ ποῖον αἷμα; διὰ ποίαν μάχαιραν;
διὰ ποίαν ἔξω τῶν νόμων θεραπείαν; ὥσπερ οὖν ἐνταῦθα καίτοι
θυσίαν οὐδεμίαν εἰπεῖν ἔχοντες ὅμως πολλὰ μέρη τὸν Ἀλκιβιάδην,
μᾶλλον δὲ τὸν Ἀσκληπιὸν ἔτεμνον ἀποκοσμοῦντες τὴν πόλιν τοῖς
περὶ τὸ ἄγαλμα, οὕτω χρὴ νομίζειν αὐτοῖς καὶ τὰ περὶ τοὺς ἀγροὺς
ἐσχηκέναι.

. . . there was a bronze statue in the city of Beroea—Asclepius
in the form of the beautiful son of Cleinias [i. e. Alcibiades];
and the work of art imitated the work of nature. Of such
youthful beauty was it that even those who were able to behold
him [sc., Alcibiades] daily still longed for the sight of it.
No one [sc., of the Christians] was so shameless that he would
dare to say that sacrifices were offered to this statue. Even
such an image as this, O King, perfected as was fitting with
great care as well as with bright genius, has been broken to
pieces and is gone, and many hands divided up what Phidias'
hands had put together. Because of what sacrificial blood?
Because of what sacrificial knife? Because of what illegal
service to the gods? Just as here, then, although they could
mention no sacrifice, still they cut up Alcibiades, or rather
Asclepius into many parts, disgracing the city in respect to the
statue, in the same way must we believe the matter stands with
them in regard to the happenings in the country.

647. Pausanias, Descriptio Graeciae, VIII, 9, 1.

Ἔστι δὲ Μαντινεῦσι ναὸς διπλοῦς μάλιστά που κατὰ μέσον τοίχῳ
διειργόμενος. τοῦ ναοῦ δὲ τῇ μὲν ἄγαλμά ἐστιν Ἀσκληπιοῦ, τέχνη
Ἀλκαμένους, . . .

The Mantineans possess a temple composed of two parts, being
divided almost exactly at the middle by a wall. In one part of
the temple is an image of Asclepius, the skilled art of Alcamenes
[a pupil of Phidias, 2nd half of 5th c. B. C.] . . . (*)

648. Strabo, Geographica, VIII, 3, 4.

Ἔστι δὲ κώμη μετρία, τὸν Ἀσκληπιὸν ἔχουσα τὸν Κολώτου,
θαυμαστὸν ἰδεῖν ξόανον ἐλεφάντινον.

[Cyllene in Elis] is a village of moderate size; it has the Asclepius made by Colotes [a pupil of Phidias, 2nd half of 5th c. B. C.]—an ivory image,[1] wonderful to behold. (*)

[1] Cf. T. **674**.

649. Pausanias, Descriptio Graeciae, II, 10, 3.

Ἐσελθοῦσι δὲ ὁ θεός ἐστιν οὐκ ἔχων γένεια χρυσοῦ καὶ ἐλέφαντος, Καλάμιδος δὲ ἔργον. ἔχει δὲ καὶ σκῆπτρον καὶ ἐπὶ τῆς ἑτέρας χειρὸς πίτυος καρπὸν τῆς ἡμέρου.

When you have entered [sc., the Asclepieion at Sicyon],[1] you see the god, a beardless[2] figure of gold and ivory[3] made by Calamis [middle of 5th c. B. C.]. He holds a scepter[4] in one hand, and a cone of the cultivated pine in the other. (*)

[1] Cf. T. **748**.   [2] Cf. T. **681**.   [3] Cf. T. **630, 675**.   [4] Cf. T. **690**.

650. Pausanias, Descriptio Graeciae, IX, 39, 4.

. . . Τροφωνίου . . . ἄγαλμά ἐστιν, Ἀσκληπιῷ καὶ τοῦτο εἰκασμένον. Πραξιτέλης δὲ ἐποίησε τὸ ἄγαλμα.

[At Lebadeia in Boeotia there is] . . . an image of Trophonius which also resembles Asclepius.[1] Praxiteles [1st half of 4th c. B. C.] made the image. (*)

[1] Cf. T. **654**; in regard to Trophonius, cf. also T. **690**.

651. Plinius, Naturalis Historia, XXXVI, 5 (4), 24.

Praxitelis filius Cephisodotus et artis heres fuit . . . Romae eius opera sunt Latona . . . Venus . . . et intra Octaviae porticus in Iunonis aede Aesculapius ac Diana.

Cephisodotus was the son of Praxiteles and the heir of his art . . . At Rome[1] there are of his works a Latona . . . a Venus . . . and within the colonnade of Octavia in the temple of Juno an Asclepius[2] and a Diana.

[1] Cf. T. **644**; **661**.
[2] Cf. **651a**. Tertullianus, *De Testimonio Animae*, Cp. 2: Sub Aesculapio stas, Iunonem in aere exornas . . . ; cf. also T. **712**.

652. Pausanias, Descriptio Graeciae, VIII, 28, 1.[1]

Ἔστι δὲ αὐτόθι ναὸς Ἀσκληπιοῦ λίθου Πεντελησίου, καὶ αὐτός τε οὐκ ἔχων πω γένεια καὶ Ὑγείας ἄγαλμα. Σκόπα δὲ ἦν ἔργα.

Here [*sc.*, at Gortys] there is a temple of Asclepius, made of Pentelic marble, with the god, beardless,[2] and an image of Health. They were the works of Scopas [*ca.* 400-350 B. C.]. (*)

[1] Cf. T. 774.   [2] Cf. T. 681.

653. Pausanias, Descriptio Graeciae, VIII, 47, 1.

Τῷ δὲ ἀγάλματι τῆς Ἀθηνᾶς τῇ μὲν Ἀσκληπιός. τῇ δὲ Ὑγεία παρεστῶσά ἐστι λίθου τοῦ Πεντελησίου. Σκόπα δὲ ἔργα Παρίου.

On one side of the image of Athena [*sc.*, at Tegea] stands Asclepius, on the other Health, in Pentelic marble,[1] works of Scopas of Paros [*ca.* 400-350 B. C.]. (*)

[1] Cf. T. 671.

654. Pausanias, Descriptio Graeciae, II, 32, 4.

Τοῦ δὲ Ἀσκληπιοῦ τὸ ἄγαλμα ἐποίησε μὲν Τιμόθεος, Τροιζήνιοι δὲ οὐκ Ἀσκληπιὸν ἀλλὰ εἰκόνα Ἱππολύτου φασὶν εἶναι.

Timotheus [1st half of 4th c. B. C.] made the image of Asclepius, but the Troezenians say that it is not Asclepius, but a likeness of Hippolytus.[1] (*)

[1] Cf. T. 671, where a statue of Asclepius is taken by some for that of Prometheus; T. 650, where a statue of Trophonius also resembles Asclepius.

655. Plinius, Naturalis Historia, XXXIV, 8 (19), 73.

Bryaxis Aesculapium et Seleucum fecit, . . .

Bryaxis [2nd half of 4th c. B. C.] made Asclepius and Seleucus, . . .

656. Pausanias, Descriptio Graeciae, I, 40, 6.

Τοῦ δὲ Ἀσκληπιοῦ τὸ ἄγαλμα Βρύαξις καὶ αὐτὸ καὶ τὴν Ὑγείαν ἐποίησεν.

Bryaxis [2nd half of 4th c. B. C.] made the image of Asclepius, itself and that of Health as well [*sc.*, in Megara]. (*)

657. Pausanias, Descriptio Graeciae, IV, 31, 12.[1]

Γέγραπται δὲ καὶ Ἀσκληπιός, . . . καὶ Μαχάων καὶ Ποδαλείριος . . . ταύτας τὰς γραφὰς ἔγραψεν Ὀμφαλίων, Νικίου τοῦ Νικομήδους μαθητής.

Asclepius too is painted [*sc.*, at Messene] . . . also Machaon
and Podalirius . . . These pictures were painted by Omphalion
[end of 4th c. B. C.], pupil of Nicias, the son of Nicomedes. (*)

[1] Cf. T. 39a.; 147b.

658. Polybius, Historiae, XXXII, 15, 4.

. . . τὸ τ' Ἀσκληπιοῦ . . . ἄγαλμα, περιττῶς ὑπὸ Φυρομάχου
κατεσκευασμένον, . . .

. . . the statue of Asclepius [*sc.*, at Pergamum],[1] an admirable
work of art by Phyromachus [1st half of 2nd c. B. C.] . . .

[1] Cf. T. 546.

659. Theocritus, Epigrammata, VII.[1]

Εἰς ἄγαλμα Ἀσκληπιοῦ

. . . .

καὶ τόδ' ἀπ' εὐώδους γλύψατ' ἄγαλμα κέδρου,
Ἠετίωνι χάριν γλαφυρᾶς χερὸς ἄκρον ὑποστὰς
μισθόν· ὁ δ' εἰς ἔργον πᾶσαν ἀφῆκε τέχνην.

For a statue of Asclepius
. . . and [Nicias, the physician] hath let carve this statue out
of fragrant cedar-wood[2] promising to Eëtion [of Amphipolis,
ca. 275 B. C.] a high guerdon for his skill of hand: on this
work Eëtion has put forth all his craft.[3] (**)

[1] Cf. T. 501.   [2] Cf. T. 677.   [3] Cf. T. 599–600.

660. Pausanias, Descriptio Graeciae, VII, 23, 7.

. . . Ἀσκληπιοῦ τέ ἐστι τέμενος καὶ ἀγάλματα Ὑγείας καὶ Ἀσκληπιοῦ. ἰαμβεῖον δὲ ἐπὶ τῷ βάθρῳ τὸν Μεσσήνιον Δαμοφῶντα
εἶναι τὸν εἰργασμένον φησίν.

. . . a precinct of Asclepius [*sc.*, in Aegium][1] with images
of him and Health. An iambic line on the pedestal says that
the artist was Damophon [1st half of 2nd c. B. C.], the
Messenian.

[1] Cf. T. 783.

661. Plinius, Naturalis Historia, XXXIV, 8 (19), 80.

. . . Niceratus Aesculapio et Hygia, qui sunt in Concordiae
templo Romae.

. . . Niceratus [1st half of 2nd c. B. C.] (is estimated) by his Asclepius and Hygieia which are in the temple of Concord at Rome.[1]

[1] Cf. T. 644; 651.

662. Pausanias, Descriptio Graeciae, X, 34, 6.

Τῷ δὲ Ἀσκληπιῷ ναὸς ᾠκοδόμηται καὶ ἄγαλμα γένεια ἔχον ἐστί· τοῖς ἐργασαμένοις τὸ ἄγαλμα ὀνόματα μὲν Τιμοκλῆς καὶ Τιμαρχίδης, γένους δέ εἰσι τοῦ Ἀττικοῦ.

A temple has been built to Asclepius [sc., at Elateia],[1] and there is a bearded image.[2] The names of the artists of the image are Timocles and Timarchides [2nd c. B. C.], who are of Attic birth. (*)

[1] Cf. T. 718.   [2] Cf. T. 682.

663. Pausanias, Descriptio Graeciae, II, 23, 4.

Τὸ δ' ἐπιφανέστατον Ἀργείοις τῶν Ἀσκληπιείων ἄγαλμα ἐφ' ἡμῶν ἔχει καθήμενον Ἀσκληπιὸν λίθου λευκοῦ, καὶ παρ' αὐτὸν ἔστηκεν Ὑγεία. κάθηνται δὲ καὶ οἱ ποιήσαντες τὰ ἀγάλματα Ξενόφιλος καὶ Στράτων.

The most famous of all the Asclepius sanctuaries at Argos[1] in our times contains an image, a seated Asclepius[2] of white marble,[3] and by his side stands Health. There are also seated figures of those who made the statues, Xenophilus [2nd half of 2nd c. B. C.] and Strato [a son of Xenophilus]. (**)

[1] Cf. T. 753.   [2] Cf. T. 680.   [3] Cf. T. 670.

664. Inscriptiones Graecae, XIV, no. 967a [2nd–3rd c. A. D.].[1]

. . . .
Τὰν παιδὸς καλλίσταν εἰκὼ τάνδε θεοῖο,
     Παιᾶνος κούρου ματρὸς ἀπ' ἀρτιτόκου,
δαιδάλλων μερόπεσσιν ἐμήσαο σεῖο, Βόηθε,
     εὐπαλάμου σοφίης μνᾶμα καὶ ἐσσομένοις.

. . . This beautiful image of the divine child,[2] Paean, who has just been borne by his mother, you, Boëthus [2nd cent. B. C.?],[3] cunningly devised, a reminder of your inventive wisdom even to future generations.

[1] Cf. T. 599.   [2] Cf. T. 679.   [3] Cf. T. 600.

665. Plinius, Naturalis Historia, XXXV, 11 (40), 137.[1]

> Nam Socrates iure omnibus placet; tales sunt eius cum Aesculapio filiae, Hygia, Aegle, Panacea, Iaso . . .

> Socrates [the painter, time unknown], to be sure, rightly pleases everybody; of such quality are his Asclepius with his daughters Hygieia, Aegle, Panacea, Iaso . . .

[1] Cf. T. 283.

666. Plinius, Naturalis Historia, XXXV, 11 (40), 147.

> Pinxere et mulieres . . . Aristarete, Nearchi filia et discipula, Aesculapium.

> Women also painted . . . Aristarete, daughter and pupil of Nearchus [time unknown], painted an Asclepius.

667. Pausanias, Descriptio Graeciae, II, 11, 6.

> Τὸ δὲ ἄγαλμα . . . οὔτε τὸν ποιήσαντα ἴσασι, πλὴν εἰ μή τις ἄρα ἐς αὐτὸν τὸν Ἀλεξάνορα ἀναφέροι.

> . . . nor do they know the name of the maker [sc., of the image at Titane],[1] unless anyone should attribute it to Alexanor himself.[2] (**)

[1] Cf. T. 749.　　　　　　　　　　[2] Cf. T. 187-189.

MATERIAL

668. Pausanias, Descriptio Graeciae, III, 26, 4.

> Λίθου δέ ἐστιν Ἀσκληπιοῦ τε ἄγαλμα . . .

> There is a stone statue[1] of Asclepius [sc., at Leuctra, Laconia] . . .

[1] Cf. T. 629: λίθου δὲ ἄγαλμα [sc., at Aegina]; T. 635: καὶ ἄγαλμα . . . λίθου [sc., at Epidaurus Limera, Laconia]; T. 755: λίθου δὲ τὸ ἄγαλμα [sc., at Cyphanta, Laconia]; **668a.** Pausanias, IV, 34, 6: τῷ μὲν δὴ Ἀσκληπιῷ . . . λίθου . . . ἄγαλμα ἐπὶ τῆς ἀγορᾶς πεποίηται [sc., at Corone, Messenia]; **668b.** Pausanias, X, 32, 12: καὶ ἄγαλμα λίθου πεποιημένον [sc., near Tithorea, Phocis], cf. T. 719; **668c.** Pausanias, II, 11, 8: λίθου δὲ Ἀσκληπιὸς ἐπίκλησιν Γορτύνιος [sc., at Titane].

669. Pausanias, Descriptio Graeciae, VII, 20, 9.

> Τὸ μὲν δὴ ἄγαλμα τοῦ Ἀσκληπιοῦ, πλὴν ἐσθῆτος, λίθου τὰ ἄλλα.

The image of Asclepius [*sc.*, at Patrae, Achaia], save for the drapery,[1] is all of stone. (*)

[1] Cf. T. 678.

670. Pausanias, Descriptio Graeciae, II, 4, 5.

Τὰ δὲ ἀγάλματα Ἀσκληπιὸς μὲν καὶ Ὑγεία λευκοῦ λίθου . . .

The images of Asclepius and of Health [*sc.*, at Corinth] are of white marble[1] . . .

[1] Cf. T. 663, concerning the image at Argos.

671. Pausanias, Descriptio Graeciae, X, 4, 4.

Πανοπεῦσι δέ ἐστιν ἐπὶ τῇ ὁδῷ πλίνθου τε ὠμῆς οἴκημα οὐ μέγα καὶ ἐν αὐτῷ λίθου τοῦ Πεντελῆσιν ἄγαλμα, ὃν Ἀσκληπιόν, οἱ δὲ Προμηθέα εἶναί φασι.

At Panopeus [Phocis] there is by the roadside a small building of unburnt brick, in which is an image of Pentelic marble,[1] said by some to be Asclepius, by others Prometheus.[2]

[1] Cf. T. 653, concerning the statue at Tegea.
[2] Cf. T. 654.

672. Pausanias, Descriptio Graeciae, VII, 26, 7.

Ἀσκληπιοῦ δὲ ἀγάλματα ὀρθά ἐστιν ἐν ναῷ καὶ Σαράπιδος ἑτέρωθι καὶ Ἴσιδος, λίθου καὶ ταῦτα Πεντελησίου.

There are in a temple [*sc.*, at Aegira, Achaia] standing images[1] of Asclepius and, on the other side, of Sarapis and of Isis, these latter too being of Pentelic marble. (*)

[1] Cf. T. 679.

673. Pausanias, Descriptio Graeciae, II, 29, 1.[1]

. . . ἀγάλματα ὁ θεὸς αὐτὸς καὶ Ἠπιόνη . . . ταῦτά ἐστιν ἐν ὑπαίθρῳ λίθου Παρίου.

. . . images, the god himself and Epione [*sc.*, at Epidaurus] . . . These are set up in the open and are made of Parian marble. (*)

[1] Cf. T. 742; cf. also T. 631.

674. Strabo, Geographica, VIII, 3, 4.[1]

   . . . τὸν Ἀσκληπιὸν . . . θαυμαστὸν ἰδεῖν ξόανον ἐλεφάντινον.

   . . . the Asclepius [sc., at Cyllene, Elis] . . . an ivory image wonderful to behold.

[1] Cf. T. 648.

675. Pausanias, Descriptio Graeciae, II, 10, 3.[1]

   . . . ὁ θεός ἐστιν . . . χρυσοῦ καὶ ἐλέφαντος . . .

   . . . the god . . . of gold and ivory[2] [sc., at Sicyon] . . .

[1] Cf. T. 649.
[2] Cf. T. 630, concerning the statue at Epidaurus.

676. Pausanias, Descriptio Graeciae, III, 21, 8.[1]

   . . . Ἀσκληπιοῦ χαλκοῦν ἄγαλμά ἐστιν, . . .

   . . . a bronze image of Asclepius [sc., at Gythium, Laconia], . . .

[1] Cf. T. 761.

677. Pausanias, Descriptio Graeciae, III, 14, 7.

   Ἀσκληπιοῦ δέ ἐστιν ἐπίκλησις ὁ Ἀγνίτας, ὅτι ἦν ἄγνου τῷ θεῷ ξόανον. ἡ δὲ ἄγνος λύγος καὶ αὐτὴ κατὰ ταὐτά ἐστι τῇ ῥάμνῳ.

   Agnitas is a surname[1] of Asclepius [sc., in Sparta], because the god had a wooden image[2] of *agnus castus*. The agnus is a willow-like tree and it is the same as the thorn. (*)

[1] Cf. T. 626.
[2] Cf. T. 501, concerning a statue at Miletus.

678. Pausanias, Descriptio Graeciae, II, 11, 6.

   Τὸ δὲ ἄγαλμα οὔτε ὁποίου ξύλου γέγονεν ἢ μετάλλου μαθεῖν ἔστιν . . . φαίνεται δὲ τοῦ ἀγάλματος πρόσωπον μόνον καὶ ἄκραι χεῖρες καὶ πόδες· χιτὼν γάρ οἱ λευκὸς ἐρεοῦς καὶ ἱμάτιον ἐπιβέβληται.

   One cannot learn of what wood or metal the image is [sc., at Titane] . . . Of the image can be seen only the face, the tips of the hands and feet. For it has wrapped about it a tunic of white wool and a cloak.[1] (*)

[1] Cf. T. 669.

POSTURE

679. Pausanias, Descriptio Graeciae, VIII, 32, 5.[1]

... τούτου μὲν δὴ τὸ ἄγαλμα ὀρθὸν πεποίηται πηχυαῖον
μάλιστα ...

... his image [sc., that of the Child Asclepius,[2] at Mega-
lopolis] is upright[3] and about a cubit in height ...

[1] Cf. T. 773.   [2] Cf. T. 664; 599; 600.
[3] Cf. T. 672; cf. also T. 635: ἄγαλμα ὀρθόν [sc., at Epidaurus Limera]; T. 690: Εἰσὶ
δὲ ἐν τῷ σπηλαίῳ τοῦ ποταμοῦ τε αἱ πηγαὶ καὶ ἀγάλματα ὀρθά, ... [sc., at Lebadeia,
Boeotia].

680. Pausanias, Descriptio Graeciae, II, 27, 2.[1]

Κάθηται δὲ ἐπὶ θρόνου ...

(The god) is seated[2] on a throne [sc., at Epidaurus] ...

[1] Cf. T. 630.
[2] Cf. T. 629: ἄγαλμα καθήμενον [sc., at Aegina]; T. 663: ἄγαλμα ... καθήμενον [sc.,
at Argos].

TYPES

681. Pausanias, Descriptio Graeciae, II, 13, 5.[1]

... καὶ ἄγαλμα οὐκ ἔχον πω γένεια.

... an image of the god not having a beard[2] [sc., at Phlius,
Argolid].

[1] Cf. T. 750.
[2] Cf. also T. 649: ὁ θεὸς ... οὐκ ἔχων γένεια [sc., at Sicyon]; T. 652: καὶ αὐτός τε
οὐκ ἔχων πω γένεια [sc., at Gortys, Arcadia].

682. Pausanias, Descriptio Graeciae, X, 34, 6.[1]

... ἄγαλμα γένεια ἔχον ...

... a bearded image[2] [sc., at Elateia, Phocis] ...

[1] Cf. T. 662.
[2] Cf. also 682a. Pausanias, X, 32, 12: ἄγαλμα ... γένεια ἔχον [sc., at Tithorea, Phocis].

683. Cicero, De Natura Deorum, III, 34, 83.

Idemque Aesculapi Epidauri barbam auream demi iussit;
neque enim convenire barbatum esse filium cum in omnibus
fanis pater imberbis esset.

He [*sc.*, Dionysius] also gave orders for the removal of the golden beard[1] of the Epidaurian Asclepius, saying it was not fitting for the son to wear a beard when his father appeared in all his temples beardless.[2] (*)

[1] Cf. T. 630.
[2] Cf. 683a. Valerius Maximus, *Facta et Dicta Memorabilia*, I, 1 ext. 3 : Idem [*sc.*, Dionysius] Epidauri Aesculapio barbam auream demi iussit, quod adfirmaret non convenire patrem Apollinem inberbem, ipsum barbatum conspici; 683b. Lactantius, *Divinae Institutiones*, II, 4, 18 = *De Origine Erroris*, 4 : Idem [*sc.*, Dionysius] auream barbam detrahens Aesculapio incongruens et inicum esse ait, cum Apollo pater eius inberbis adhuc esset ac levis, priorem filium quam patrem barbatum videri; 683c. Arnobius, *Adversus Nationes*, VI, 21 : Nam quid Aesculapii gravitatem ab eo [*sc.*, Dionysio] esse commemorem risam? quem cum barba spoliaret amplissima boni ponderis et philosophae densitatis, facinus esse dicebat indignum, ex Apolline procreatum patre levi et glabro simillimoque inpuberi ita barbatum filium fingi, ut in ancipiti relinquatur, uter eorum pater sit, uter filius, immo an sint generis et cognationis unius?

684. Lucianus, Iuppiter Tragoedus, 26.

Ὥστε μὴ μειρακιεύου πρὸς ἡμᾶς, ἀλλὰ λέγε θαρρῶν ἤδη τὰ δοκοῦντα μηδὲν αἰδεσθείς, εἰ ἀγένειος ὢν δημηγορήσεις, καὶ ταῦτα βαθυπώγωνα καὶ εὐγένειον οὕτως υἱὸν ἔχων τὸν Ἀσκληπιόν.

So don't play the boy with us (Apollo): Say what you think boldly and don't be sensitive about speaking without a beard when you have such a long-bearded,[1] hairy-faced son in Asclepius.

[1] Cf. T. 850, v. 656: longae . . . barbae, in reference to the Epidaurian Asclepius; cf. 684a. *Carmina Priapea*, XXXVI, 8: intonsa semper Aesculapio barba est; cf. also T. 685.

685. Minucius Felix, Octavius, XXIII, 5.

Quid? formae ipsae et habitus nonne arguunt ludibria et ⟨de-⟩ decora deorum vestrorum? Vulcanus claudus deus et debilis, Apollo tot aetatibus levis, Aesculapius bene barbatus, etsi semper adulescentis Apollinis filius . . .

Again, do not form and feature indicate the contemptibility and shamelessness of your gods? Vulcan is a lame and crippled god; Apollo after years and years still beardless; Asclepius full bearded, though the son of ever young Apollo. . . . (**)

GARMENT

686. Athenaeus, Deipnosophistae, VII, 33, 289 C.

Καὶ ἄλλος δ' αὐτῷ τῶν περισωθέντων Ἀσκληπιοῦ στολὴν ἀνα-
λαβὼν συμπεριεφέρετο.*

* συμπεριεφέρετο coni. Meyer   συμπεριεφθείρετο A.

Still another of his [*i. e.* Menecrates Zeus', 4th c. B. C.]
patients who had been restored to health moved about in his
company clad in the garb of Asclepius.[1]

[1] Cf. T. 687.

687. Tertullianus, De Pallio, IV, 10.

Cum ipsum hoc pallium morosius ordinatum et crepidae Grae-
catim Aesculapio adulantur, quanto tunc magis arguas illud et
urgeas oculis, et⟨si⟩ iam simplicis et inaffectatae, tamen super-
stitionis reum?

When this mantle itself,[1] arranged with more rigorous care,
and sandals after the Greek model,[2] serve to flatter Asclepius,
how much more should you then accuse and assail it with your
eyes as being guilty of superstition—albeit superstition simple
and unaffected?

[1] Cf. T. 492.   [2] Cf. however T. 493.

ATTRIBUTES

688. Pausanias, Descriptio Graeciae, II, 27, 2.[1]

Κάθηται δὲ ἐπὶ θρόνου βακτηρίαν κρατῶν, τὴν δὲ ἑτέραν τῶν
χειρῶν ὑπὲρ κεφαλῆς ἔχει τοῦ δράκοντος, καί οἱ καὶ κύων παρακατακείμενος πεποίηται.

He [*sc.*, Asclepius] is seated on a throne, grasping a staff, while
holding the other hand over the head of the serpent; and a
dog, lying by his side, is also represented. (*)

[1] Cf. T. 630.

689. Ovidius, Metamorphoses, XV, 654-59.[1]

. . . sed qualis in aede
esse solet, baculumque tenens agreste sinistra

caesariem longae dextra deducere barbae

. . . .

hunc modo serpentem, baculum qui nexibus ambit,
perspice . . .

. . . even as he is wont to appear in his own temple [*sc.*, at
Epidaurus], holding his rustic staff [2] in his left hand and with
his right stroking his flowing beard [3] . . . Only look upon this
serpent which twines about my staff [4] . . .

[1] Cf. T. 850.
[2] Cf. T. 630; 691; cf. also **689a.** Arnobius, *Adversus Nationes*, VI, 25: Aesculapius
baculo; **689b.** Sidonius Apollinaris, *Epistulae*, IV, 3, 5: cum Orpheo plectrum, cum
Aesculapio baculum; cf. however T. 649; 690.
[3] Cf. T. 682.
[4] Cf. T. 690; 706.

690. Pausanias, Descriptio Graeciae, IX, 39, 3.

> Εἰσὶ δὲ ἐν τῷ σπηλαίῳ τοῦ ποταμοῦ τε αἱ πηγαὶ καὶ ἀγάλματα
> ὀρθά, περιειλιγμένοι δέ εἰσιν αὐτῶν τοῖς σκήπτροις δράκοντες.
> ταῦτα εἰκάσαι μὲν ἄν τις Ἀσκληπιοῦ τε εἶναι καὶ Ὑγείας, εἶεν
> δ’ ἂν Τροφώνιος καὶ Ἕρκυνα, ἐπεὶ μηδὲ τοὺς δράκοντας Ἀσκλη-
> πιοῦ μᾶλλον ἢ καὶ Τροφωνίου νομίζουσιν ἱεροὺς εἶναι.

In the cave [*sc.*, at Lebadeia, Boeotia] are the sources of the
river and standing images,[1] and serpents are coiled around
their sceptres.[2] One might conjecture the images to be of
Asclepius and Health, but they might be Trophonius and
Hercyna, because they think that serpents are just as much
sacred to Trophonius [3] as to Asclepius.

[1] Cf. T. 679.　　　　[2] Cf. T. 649; cf. also T. 706.　　　　[3] Cf. T. 650.

691. Festus, De Verborum Significatu, 110 M [p. 98, Lindsay].

In insula Aesculapio facta aedes fuit, quod aegroti a medicis
aqua maxime sustententur. eiusdem esse tutelae draconem,
quod vigilantissimum sit animal; quae res ad tuendam vali-
tudinem aegroti maxime apta est. canes adhibentur eius templo,
quod is uberibus canis sit nutritus. bacillum habet nodosum,
quod difficultatem significat artis. laurea coronatur, quod ea
arbor plurimorum sit remediorum. huic gallinae immolabantur.

To Asclepius on the island [*sc.*, in the Tiber] a temple was
built [1] because sick people are aided by physicians particularly

through water. The serpent is the guard of this temple because it is a most vigilant beast and this faculty is especially appropriate in safeguarding the health of invalids. Dogs are also used in his temple[2] because he was nourished by the teats of a dog.[3] He has a gnarled staff which signifies the difficulty of his art.[4] He is crowned with laurel[5] because this tree is full of very many remedies. To him hens were sacrificed.[6]

[1] Cf. T. 846 ff.   [3] Cf. T. 5; 9.   [5] Cf. *e. g.* T. 539.
[2] Cf. T. 630.   [4] Cf. however T. 705; 706.   [6] Cf. however T. 523 ff.

692. Pausanias, Descriptio Graeciae, II, 28, 1.

Δράκοντες δὲ † οἱ λοιποὶ καὶ ἕτερον γένος ἐς τὸ ξανθότερον ῥέπον τῆς χρόας ἱεροὶ μὲν τοῦ Ἀσκληπιοῦ νομίζονται καὶ εἰσὶν ἀνθρώποις ἥμεροι, τρέφει δὲ μόνη σφᾶς ἡ τῶν Ἐπιδαυρίων γῆ.

The serpents, even including a peculiar kind of yellowish color [?], are considered sacred to Asclepius, and are tame with men. Only the land of the Epidaurians breeds them. (*)

693. Sidonius Apollinaris, Carmina, XXII, 79-80.

Pendet per teretes tripodas Epidaurius anguis
diffusus sanctum per colla salubria virus.

The serpent of Epidaurus[1] hangs about the shapely tripod, with a hallowed essence diffused throughout his health-giving neck. (*)

[1] Cf. 693a. Horatius, *Saturae*, I, 3, 27: serpens Epidaurius.

694. Plinius, Naturalis Historia, XXIX, 4 (22), 72.

Anguis Aesculapius Epidauro Romam advectus est vulgoque pascitur et in domibus, ac nisi incendiis semina exurerentur, non esset fecunditati eorum resistere in orbe terrarum.

The Asclepian snake was brought to Rome from Epidaurus, and is commonly reared even in private houses, and unless its seeds were destroyed by fire there would be no limit to their increase all over the earth. (**)

695. Pausanias, Descriptio Graeciae, II, 10, 3.[1]

Φασὶ δέ σφισιν ἐξ Ἐπιδαύρου κομισθῆναι τὸν θεὸν ἐπὶ ζεύγους ἡμιόνων δράκοντι εἰκασμένον . . .

(The Sicyonians) say that the god was carried to them from Epidaurus on a carriage drawn by two mules, that he was in the likeness of a serpent . . .

[1] Cf. T. 748.

696. Aelianus, De Natura Animalium, XVI, 39.

Λέγουσι δὲ Αἰγύπτιοι λόγοι καὶ ἐπὶ τοῦ Φιλαδέλφου ἐξ Αἰθιοπίας ἐς τὴν Ἀλεξάνδρου πόλιν κομισθῆναι δράκοντας δύο ζῶντας, καὶ τὸν μὲν αὐτῶν εἶναι πήχεων δεκατεττάρων τὸν δὲ δεκατριῶν· ἐπί γε μὴν τοῦ Εὐεργέτου τρεῖς κομισθῆναι, καὶ τὸν μὲν εἶναι πήχεων ἐννέα, τὸν δὲ ἑπτά, τὸν δὲ τρίτον ἑνὶ ἀπολείπεσθαι· καὶ τρέφεσθαί γε ἐν Ἀσκληπιοῦ σὺν πολλῇ τῇ κομιδῇ αὐτοὺς Αἰγύπτιοί φασι.

The Egyptian books say that under Philadelphus two living serpents were brought from Ethiopia to Alexandria, the one being fourteen, the other thirteen cubits long. Under Euergetes three (serpents) were brought (to Alexandria), one being nine, the other seven cubits long and the third one shorter by one cubit. And they are tended with great care in the Asclepieion, the Egyptians say.

697. Nicander, Theriaca, 438-57.

Φράζεο δὲ χλοάοντα δαεὶς κύανόν τε δράκοντα,
ὅν ποτε Παιήων λασίῃ ἐνεθρέψατο φηγῷ
Πηλίῳ ἐν νιφόεντι Πελεθρόνιον κατὰ βῆσσαν.
ἤτοι ὅγ' ἄγλαυρος μὲν ἐείδεται, ἐν δὲ γενείῳ
τρίστοιχοι ἑκάτερθε περιστιχόωσιν ὀδόντες·
πίονα δ' ἐν σκυνίοισιν ὑπ' ὄθματα· νέρθε δὲ πώγων
αἰὲν ὑπ' ἀνθερεῶνι χολοίβαφος. οὐ μὲν ὅγ' αὔτως
ἐγχρίμψας ἤλγυνε, καὶ ἢν ἔκπαγλα χαλεφθῇ·
βληχρὸν γὰρ μυὸς οἷα μυχηβόρου ἐν χροῒ νύγμα
εἴδεται αἱμαχθέντος ὑπὸ κραντῆρος ἀραιοῦ.
τῷ μέν τ' ἔκπαγλον κοτέων βασιλήϊος ὄρνις
αἰετὸς ἐκ παλαχῆς ἐπαέξεται, ἀντία δ' ἐχθρὴν
δῆριν ἄγει γενύεσσιν, ὅταν βλώσκοντα καθ' ὕλην
δέρκηται· πάσας γὰρ ὅγ' ἠρήμωσε καλιάς,
αὔτως ὀρνίθων τε τόκον κτίλα τ' ὠεὰ βρύκων.
αὐτὰρ ὁ τοῦ καὶ ῥῆνα καὶ ἠνεμόεντα λαγωὸν
ῥεῖα δράκων ἤμερσε νέον μάρψαντος ὄνυξι
θάμνου ὑπαΐξας· ὁ δ' ἀλεύεται· ἀμφὶ δὲ δαιτὸς

μάρνανθ᾽· ἱπτάμενον δὲ πέριξ ἀτέλεστα διώκει
σπειρηθεὶς καὶ λοξὸν ὑποδρὰξ ὄμμασι λεύσσων.

Tell what you know of the blue-green serpent which Paeon
once raised on bushy oaks along the Pelethronian glen on
snowy Pelion. Truly the serpent is bright to look upon, but
in its jaws are ranged three rows of teeth on each side. Its
eyes lie deep under shaggy brows. The beard beneath its chin
is yellow colored. When it attacks it does not cause pain even
if it is in wild rage. For it pricks the skin with its teeth,
drawing blood lightly like a mouse nibbling in its hole. From
of old the royal bird, the eagle, has been wrathful towards
the serpent and with its beak engages it in hostile battle, when
it sees the serpent crawling along the ground. For the serpent
strips bare all nests, greedily devouring the young of the birds
and their cherished eggs. And darting from under a shrub it
easily deprives the eagle even of a lamb or a swift hare which
it has snatched in its talons. The eagle retreats; they fight
over the prey. To no avail the serpent coils and pursues the
flying bird, looking up grimly with distorted eyes.

698. Scholia in Nicandrum, Ad Theriaca, 438.

Φράζεο· ὅλος ὁ νοῦς τοῦ στίχου οὗτος, ὁ δράκων ἐτράφη μὲν ὑπὸ
τοῦ Ἀσκληπιοῦ κατά τινα τόπον τοῦ Πηλίου καλούμενον Πελεθρόνιον, τὴν δὲ φύσιν ἐστὶ μέλας, ὑπόχλωρος τὴν κοιλίαν, καλὸς
δὲ τὴν μορφήν, τοὺς ὀδόντας ἔχων τριστοίχους, τὸ δὲ ἐπισκύνιον
παρεμφαίνων· καὶ κατὰ τὸν ἀνθερεῶνα λεπίδες ἐπεγηγερμέναι
πώγωνος παρέχουσι φαντασίαν χολοβαφινῷ σώματι προσεοι-
κότος. τὸ δὲ δῆγμα ἄπρακτον καὶ τῷ τοῦ μυὸς δήγματι προσε-
οικός, κἂν σφόδρα χαλεπήνῃ. πολεμεῖται δὲ καὶ πολεμεῖ συνεχῶς
τῷ ἀετῷ, πολλάκις μὲν καὶ ἐπὶ τὴν νοσσιὰν παραγενόμενος καὶ
τὰ ἐν αὐτῇ ὠὰ σιτούμενος, συνεχῶς δὲ κἀκείνῳ, ὅτε ἐπὶ ἁρπαγὴν
ἀρνὸς ἢ λαγωοῦ τράπηται. Πελεθρόνιον δὲ τόπος ἐστὶ τοῦ Πηλίου
ἀνθώδης· διὸ καὶ οὕτω καλεῖται ἀπὸ τοῦ πέλειν ἐκεῖσε τὰ ἄνθη·
θρόνα γὰρ τὰ ἄνθη λέγονται.

'Tell': the general meaning of the verse is this: the serpent
was nurtured by Asclepius in some part of Pelion which is
called Pelethronion. It is by nature black, with a pale yellow
belly, beautiful in shape, with triple rows of teeth, showing its
brow-ridges; and along its chin raised scales give the impres-

sion of a beard, yellow like the body. Its bite is harmless, just like the bite of a mouse, even when it is very angry. It is engaged in fights with the eagle and fights it continually, often, too, attacking the eagle's nest and eating the eggs inside it and continually fighting with the eagle whenever it swoops down on its prey of a lamb or a hare. Pelethronion is a flowery place on Pelion. And it is given this name because of the fact that flowers (*anthe*) are (*pelein*) there; and *anthe* is equivalent to *throna*.

699. Plinius, Naturalis Historia, XXIX, 4 (22), 71-72.

Neque anguis venenatus est nisi per mensem luna instigatus, et prodest vivus conprehensus et in aqua contusus, si foveantur ita morsus. quin et inesse ei remedia multa creduntur, ut digeremus, et ideo Aesculapio dicatur.

The snake is not venomous, except when according to (the time of) the month it is irritated by the moon: it is very useful when taken alive and pounded in water, if the bites are fomented with the preparation. Indeed, it is generally supposed that it [*sc.*, the snake] is full of many remedies, as we shall explain in detail, and therefore the snake is consecrated to Asclepius.[1] (**)

[1] Cf. **699a.** Prudentius, *Peristephanum*, X, 257: sacratus aspis Aesculapii.

700. Aelianus, De Natura Animalium, VIII, 12.

Ὁ παρείας ἢ παρούας (οὕτω γὰρ Ἀπολλόδωρος ἐθέλει) πυρρὸς τὴν χρόαν, εὐωπὸς τὸ ὄμμα, πλατὺς τὸ στόμα, δακεῖν οὐ σφαλερός, ἀλλὰ πρᾶος. ἔνθεν τοι καὶ τῷ θεῶν φιλανθρωποτάτῳ ἱερὸν ἀνῆκαν αὐτόν, καὶ ἐπεφήμισαν Ἀσκληπιοῦ θεράποντα εἶναι οἱ πρῶτοι ταῦτα ἀνιχνεύσαντες.

The Pareias[1] or Parouas (for thus Apollodorus [the author on poisons, *ca.* 300 B. C.] says) is yellowish-red in color, sharp of vision, has a flat mouth and is not likely to sting, but rather gentle. Wherefore those who first searched out these facts made it sacred to the most man-loving of the gods and said it was the servant of Asclepius.[2]

[1] Cf. T. **421**, v. 691.
[2] Cf. however **700a.** Pausanias, *Descriptio Graeciae*, II, 11, 8: Παρὰ δὲ τοὺς δράκοντας ἐσιέναι τοὺς ἱεροὺς οὐκ ἐθέλουσιν [*sc.*, ἐν Τιτάνῃ] ὑπὸ δείματος· καταθέντες δέ σφισι πρὸ τῆς ἐσόδου τροφὴν οὐκέτι πολυπραγμονοῦσι.

701. Scholia in Aristophanem, Ad Plutum, 733.

Ἐξηξάτην οὖν δύο δράκοντες· (κοινῶς μὲν καὶ τοῖς ἄλλοις ἥρωσι δράκοντες παρετίθεντο, ἐξαιρέτως δὲ τῷ Ἀσκληπιῷ). δράκοντες δὲ λέγονται ἀπὸ τοῦ δέρκω, ὅ ἐστι βλέπω· ὀξυδερκὲς γὰρ τὸ ζῷον. ἀφιέρωται δὲ τῷ Ἀσκληπιῷ, ἐπειδὴ τὸ γῆρας ἀποβάλλει καὶ ἡ ἰατρικὴ δὲ φυλάττει φύσει τὸ νέον, ἐξωθοῦσα τὰ νοσήματα . . . εἰκότως φησὶν ὑπηρέτας ἔχειν τὸν Ἀσκληπιὸν ὄφεις. ἐπειδὴ γὰρ οἱ ὄφεις τὸ παλαιὸν αὐτῶν ἀπεκδυόμενοι δέρμα ἀεὶ νεάζουσιν· οὕτως καὶ ὁ θεὸς τῶν ἀσθενούντων τὰς νόσους οἷά τινα λεβηρίδα ἀπορίπτων νεάζοντας δείκνυσιν.

Then two serpents rushed forth: serpents are commonly ascribed as attributes to other heroes[1] also, specifically, however, to Asclepius. They are called serpents from *derko*, which means to see, for the animal is sharp-sighted. To Asclepius it is sacred because it casts off old age just as the medical art safeguards youth in nature by banishing diseases. . . . Quite reasonably he [*sc.*, Aristophanes] says Asclepius has serpents for servants, inasmuch as serpents, shedding their old skin, constantly become young. So, too, by casting off the diseases of the infirm, just like a serpent's skin, the god shows them forth young again.[2]

[1] Cf. T. 703.   [2] Cf. also T. 301.

702. Festus, De Verborum Significatu, 67 M [p. 59, Lindsay].

Dracones dicti ἀπὸ τοῦ δέρκεσθαι, quod est videre. clarissimam enim dicuntur habere oculorum aciem: qua ex causa incubantes eos thesauris custodiae causa finxerunt antiqui. ideoque Aesculapio adtribuuntur, quod vigilantissimi generis putantur, quae res medicinae maxime necessaria est.

They are called serpents from *derkesthai* which means to see, for they are reputed to have very keen vision. Wherefore the ancients fancied them lying near treasures to guard them. And for this reason, too, they are given as attributes to Asclepius, because they are thought to be of a most watchful kind, a quality particularly requisite to the medical art.

703. Artemidorus, Onirocritica, II, 13.

Δράκων βασιλέα σημαίνει διὰ τὸ δυνατὸν [καὶ ἄρχοντα] καὶ
χρόνον διὰ τὸ μῆκος, καὶ διὰ τὸ ἀποδιδύσκεσθαι τὸ γῆρας καὶ
πάλιν νεάζειν. τὸ γὰρ αὐτὸ καὶ τῷ χρόνῳ συμβέβηκε κατὰ τὰς
τοῦ ἔτους ὥρας. καὶ πλοῦτον καὶ χρήματα διὰ τὸ ἐπὶ θησαυροὺς
ἱδρύεσθαι, καὶ θεοὺς πάντας, οἷς ἐστιν ἱερός. εἰσὶ δὲ οἵδε Ζεὺς
Σαβάζιος Ἥλιος Δημήτηρ καὶ Κόρη Ἑκάτη Ἀσκληπιὸς Ἥρωες.

The serpent indicates the king because of its power, and time
because of its length and because of its shedding old age and
becoming young again. For the same thing happens to time
in the course of the seasons of the year. Also it [sc., the
serpent] stands for wealth and possessions because it lies on
guard over treasures; and it stands for all the gods to whom
it is sacred. These are Zeus, Sabazius, Helius, Demeter, Kore,
Hecate, Asclepius, the Heroes.[1]

[1] Cf. T. 701.

704. Cosmas, Ad Carmen LII Gregorii Nazianzeni.[1]

Περὶ δὲ τοῦ Ὀφιούχου φασίν· οὗτος ὁ Ὀφιοῦχος Ἀσκληπιός
ἐστιν ὁ τῆς ἰατρικῆς ἔφορος. σύμβολον δὲ ὁ ὄφις τοῦ ἀγήρω.
λέγεται γὰρ ἀποδύεσθαι τὸ γῆρας ὁ ὄφις καὶ ἀνανεοῦσθαι. ἐπεὶ
οὖν Ἀσκληπιὸς διὰ τῆς ἰατρικῆς ἀνανεοῖ τὰ ἀνθρώπινα σώματα
ποιοῦσιν αὐτὸν μετὰ τοῦ δράκοντος. θέλοντες οὖν αὐτόν, φησίν,
οἱ θεοὶ ἀντὶ ἀγάλματος ἀνιερῶσαι δι' ἄστρων ἀνέθηκαν ἐν τῷ
οὐρανῷ.

About Ophiuchus they say: this Ophiuchus is Asclepius, who
is the guardian of medical science.[2] The serpent is a symbol
of perennial youth. For the serpent is said to shed old age
and grow young again. Therefore, since Asclepius renews the
human body by his medical skill, they associate him with the
serpent. Desiring, then, to endow him with holiness as an
honor, he [sc., Eratosthenes] says, the gods set him up in
heaven among the stars.[3]

[1] Cf. E. Maass, Analecta Eratosthenica, *Philologische Untersuchungen*, VI, 1883, p. 6.
[2] Cf. T. 340, n. 1.
[3] Cf. T. 121. Cf. also **704a.** Hyginus, *Astronomica*, II, 14: Complures etiam astrologi
hunc [sc., Ophiuchum] Aesculapium finxerunt, quem Iuppiter Apollinis causa inter astra
collocavit. Aesculapius enim cum esset inter homines, et tantum medicina ceteris
praestaret, ut non satis ei videretur hominum dolores levare, nisi etiam mortuos

revocaret ad vitam, novissime fertur Hippolytum, quod iniquitate novercae et inscientia parentis erat interfectus, sanasse, ita uti Eratosthenes dicit. nonnulli Glaucum Minoos filium eius opera revixisse dixerunt; pro quo, ut peccato, Iovem domum eius fulmine incendisse, ipsum autem propter artificium et Apollinem eius patrem inter sidera anguem tenentem constituisse. ut quidam dixerunt, hac de causa anguem dicitur tenere, quod cum Glaucum cogeretur sanare, conclusus quodam loco secreto, bacillum tenens manu, cum quid ageret cogitaret, dicitur anguis ad bacillum eius adrepsisse: quem Aesculapius mente commotus interfecit, bacillo fugientem feriens saepius. postea fertur alter anguis eodem venisse, ore ferens herbam, et in caput eius inposuisse; quo facto, utrosque loco fugisse. quare Aesculapium usum eadem herba et Glaucum revixisse. itaque anguis et in Aesculapii tutela et in astris dicitur collocatus. qua consuetudine ducti posteri eius tradiderunt reliquis, ut medici anguibus uterentur.

705. Cornutus, Theologiae Graecae Compendium, Cp. 33.[1]

Διὰ τοῦτο γὰρ δράκοντα αὐτῷ παριστᾶσιν, ἐμφαίνοντες ὅτι ὅμοιόν τι τούτῳ πάσχουσιν οἱ χρώμενοι τῇ ἰατρικῇ κατὰ τὸ οἱονεὶ ἀνανεάζειν ἐκ τῶν νόσων καὶ ἐκδύεσθαι τὸ γῆρας, ἅμα δ' ἐπεὶ προσοχῆς ὁ δράκων σημεῖον, ἧς πολλῆς δεῖ πρὸς τὰς θεραπείας. καὶ τὸ βάκτρον δὲ τοιούτου τινὸς ἔοικεν εἶναι σύμβολον· παρίσταται γὰρ δι' αὐτοῦ ὅτι, εἰ μὴ ταύταις ταῖς ἐπινοίαις ἐπεστηριζόμεθα ὅσον ἐπὶ τὸ συνεχῶς εἰς ἀρρωστίαν ἐμπίπτειν, κἂν θᾶττον τοῦ δέοντος σφαλλόμενοι κατεπίπτομεν.

For this reason, therefore, they give him a serpent as an attribute, indicating that those who avail themselves of medical science undergo a process similar to the serpent in that they, as it were, grow young again after illnesses and slough off old age; also because the serpent is a sign of attention, much of which is required in medical treatments. The staff also seems to be a symbol of some similar thing.[2] For by means of this it is set before our minds that unless we are supported by such inventions as these, in so far as falling continually into sickness is concerned, stumbling along we would fall sooner than necessary.

[1] Cf. T. 6; cf. also **705a**. Eudocia Augusta, *Violarium* XI, same text.
[2] Cf. T. 691; T. 706.

706. Eusebius, Praeparatio Evangelica, III, 11, 26.

Τῆς δὲ σωστικῆς αὐτοῦ δυνάμεως Ἀσκληπιὸς τὸ σύμβολον· ᾧ τὸ μὲν βάκτρον δεδώκασι, τῆς τῶν καμνόντων ὑπερείσεως καὶ ἀναπαύσεως, ὁ δὲ ὄφις περισπειρᾶται,* τῆς περὶ τὸ σῶμα καὶ τὴν ψυχὴν σωτηρίας φέρων σημεῖον. . . . πνευματικώτατον γὰρ τὸ ζῷόν ἐστι καὶ τὴν ἀσθένειαν τοῦ σώματος ἀποδύεται. δοκεῖ δὲ

καὶ ἰατρικώτατον εἶναι· τῆς γὰρ ὀξυδορκίας εὗρε τὸ φάρμακον
καὶ μυθεύεται τῆς ἀναβιώσεως εἰδέναι τινὰ βοτάνην.

* περισπειρᾶται Dindorf   περισπείρεται Gifford.

Of the safeguarding power (of the sun) a symbol is Asclepius [1]
to whom they attribute a staff, as a sign of support and relief
for invalids;[2] the serpent is twined about it,[3] being the sign of
the preservation of body and soul. . . . For this animal is most
animated and strips off the weakness of the body. It seems to
be also the most skilful in medicine. For it discovered the
remedy of sharp-sightedness and it is said to know some drug
for the return to life.

[1] Cf. T. 299; 301.
[2] Cf. however T. 691.
[3] Cf. T. 689; 690.

VII. SANCTUARIES

GENERAL REMARKS

707. Vitruvius, De Architectura, I, 2, 7.

Naturalis autem decor sic erit, si primum omnibus templis saluberrimae regiones aquarumque fontes in his locis idonei eligentur, in quibus fana constituantur, deinde maxime Aesculapio, Saluti, et eorum deorum, quorum plurimi medicinis aegri curari videntur. cum enim ex pestilenti in salubrem locum corpora aegra translata fuerint et e fontibus salubribus aquarum usus subministrabuntur, celerius convalescent. ita efficietur, uti ex natura loci maiores auctasque cum dignitate divinitas excipiat opiniones.

Moreover there will be a natural appropriateness if, for all temples in the first place, the most healthy sites be chosen and suitable springs of water in those places in which shrines are to be set up, and for Asclepius in particular and for Salus and for those gods by whose medical power a great many of the sick seem to be healed. For when sick persons are moved from a pestilent to a healthy place and the water supply is from wholesome fountains,[1] they will more quickly recover. So will it happen that the divinity from the nature of the site will gain a greater and higher reputation and authority. (**)

[1] Cf. T. 723 ff.; 739, 6; 755; 761; 782; 804; 805; cf. also T. 632; 643.

708. Plutarchus, Aetia Romana, 94, 286 D.

' Διὰ τί τοῦ Ἀσκληπιοῦ τὸ ἱερὸν ἔξω πόλεώς ἐστι; ' πότερον ὅτι τὰς ἔξω διατριβὰς ὑγιεινοτέρας ἐνόμιζον εἶναι τῶν ἐν ἄστει; καὶ γὰρ Ἕλληνες ἐν τόποις καθαροῖς καὶ ὑψηλοῖς ἐπιεικῶς ἱδρυμένα τὰ Ἀσκληπιεῖα ἔχουσιν· ἢ ὅτι τὸν θεὸν ἐξ Ἐπιδαύρου μετάπεμπτον ἥκειν νομίζουσιν; Ἐπιδαυρίοις δ' οὐ κατὰ πόλιν ἀλλὰ πόρρω τὸ Ἀσκληπιεῖόν ἐστιν. ἢ ὅτι τοῦ δράκοντος ἐκ τῆς τριήρους κατὰ τὴν νῆσον ἀποβάντος καὶ ἀφανισθέντος αὐτὸν ᾤοντο τὴν ἵδρυσιν ὑφηγεῖσθαι τὸν θεόν;

370

'Why is the shrine of Asclepius outside the city?'[1] Is it because they considered it more healthful to spend their time
outside the city than within its walls? In fact the Greeks, quite
reasonably, have their shrines of Asclepius situated in places
which are both clean and high.[2] Or is it because they believe
that the god came at their summons from Epidaurus, and the
Epidaurians have their shrine of Asclepius not in the city, but
at some distance?[3] Or is it because the serpent came out from
the trireme into the island [sc., in the Tiber], and there disappeared, and thus they [sc., the Romans] thought that the
god himself was indicating to them the site for building?[4] (*)

[1] Cf. T. 759; 776; 794; 839; 848; 856; cf. also 383.
[2] Cf. T. 762; 780; 793; 834.   [3] Cf. T. 848.   [4] Cf. T. 845 ff.

709. Pausanias, Descriptio Graeciae, II, 26, 8-9.

Μαρτυρεῖ δέ μοι καὶ τόδε ἐν Ἐπιδαύρῳ τὸν θεὸν γενέσθαι· τὰ
γὰρ Ἀσκληπιεῖα εὑρίσκω τὰ ἐπιφανέστατα γεγονότα ἐξ Ἐπιδαύρου. τοῦτο μὲν γὰρ Ἀθηναῖοι, τῆς τελετῆς λέγοντες Ἀσκληπιῷ
μεταδοῦναι, τὴν ἡμέραν ταύτην Ἐπιδαύρια ὀνομάζουσι καὶ θεὸν
ἀπ' ἐκείνου φασὶν Ἀσκληπιόν σφισι νομισθῆναι. τοῦτο δὲ
Ἀρχίας ὁ Ἀρισταίχμου τὸ συμβὰν σπάσμα θηρεύοντί οἱ περὶ
τὸν Πίνδασον ἰαθεὶς ἐν τῇ Ἐπιδαυρίᾳ, τὸν θεὸν ἐπηγάγετο ἐς
Πέργαμον. ἀπὸ δὲ τοῦ Περγαμηνῶν Σμυρναίοις γέγονεν ἐφ'
ἡμῶν Ἀσκληπιεῖον τὸ ἐπὶ θαλάσσῃ. τὸ δ' ἐν Βαλάγραις ταῖς
Κυρηναίων ἐστὶν Ἀσκληπιὸς καλούμενος Ἰατρὸς ἐξ Ἐπιδαύρου
καὶ οὗτος. ἐκ δὲ τοῦ παρὰ Κυρηναίοις τὸ ἐν Λεβήνῃ τῇ Κρητῶν
ἐστιν Ἀσκληπιεῖον.

There is other evidence that the god was born in Epidaurus;
for I find that the most famous sanctuaries of Asclepius had
their origin from Epidaurus. In the first place, the Athenians,
who say that they gave a share of their mystic rites to Asclepius, call this day (of the festival) Epidauria,[1] and they allege
that their worship of Asclepius dates from then. Again, when
Archias, son of Aristaechmus, was healed in Epidauria[2] after
spraining himself while hunting about Pindasus, he brought
the god to Pergamum.[3] From the one at Pergamum has been
built in our day the sanctuary of Asclepius by the sea at
Smyrna.[4] Further, at Balagrae of the Cyreneans there is an
Asclepius called Healer, he, too, from Epidaurus.[5] From the

one at Cyrene was founded the sanctuary of Asclepius at
Lebene, in Crete.⁶ (*)

¹ Cf. T. 564 ff.   ³ Cf. T. 801.   ⁵ Cf. T. 831.
² Cf. T. 457.   ⁴ Cf. T. 813-14.   ⁶ Cf. T. 792.

710. Julianus, Contra Galilaeos, 235 C.

. . . ἰατρικὴν τὴν ἐξ Ἀσκληπιοῦ, οὗ πανταχοῦ γῆς ἐστι χρησ-
τήρια . . .

. . . the art of healing, derived from Asclepius ¹ whose oracles
are found everywhere on earth . . .

¹ Cf. T. 307.

711. Origenes, Contra Celsum, III, 3.¹

. . . πόλεσιν ἀνακειμέναις αὐτῷ, οἷον τῇ Τρίκκῃ καὶ τῇ Ἐπιδαύρῳ
καὶ τῇ Κῷ καὶ τῇ Περγάμῳ.

. . . cities which were dedicated to him [sc., Asclepius], such
as Tricca, and Epidaurus, and Cos, and Pergamum.²

¹ Cf. T. 292.   ² Cf. T. 307; 328.

712. Tertullianus, De Idolatria, Cp. 20.

Deos nationum nominari lex prohibet non utique: nomina
eorum pronuntiemus, quae nobis ut dicamus conversatio extor-
quet. nam id plerumque dicendum est: in templo Aesculapii
illum habes, et, in vico Isidis habito, et sacerdos Iovis factus
est . . .

The law prohibits the gods of the gentiles from being named,
not of course that we are not to pronounce their names, the
speaking of which common intercourse extorts from us: for
this must very frequently be said, "You find him in the temple
of Asclepius"; and "I live in Isis Street"; and "He has been
made priest of Jupiter"¹ . . .

¹ Cf. T. 651a.

713. Epictetus, Dissertationes, II, 22, 17.

Ὅταν γοῦν εἰς τοῦτο ἐμποδίζειν ἡμῖν οἱ θεοὶ δοκῶσιν, κἀκείνους
λοιδοροῦμεν καὶ τὰ ἱδρύματα αὐτῶν καταστρέφομεν καὶ τοὺς ναοὺς

ἐμπιπρῶμεν, ὥσπερ Ἀλέξανδρος ἐκέλευσεν ἐμπρησθῆναι τὰ
Ἀσκλήπεια ἀποθανόντος τοῦ ἐρωμένου.

When, for instance, we think that the gods stand in the way of
our attainment of this [sc., our own interest], we revile even
them, cast their statues to the ground, and burn their temples,
as Alexander ordered the temples of Asclepius to be burned
when his loved one died.[1]

[1] Cf. T. 821.

NORTHERN AND MIDDLE GREECE

THESSALY

714. Strabo, Geographica, IX, 5, 17.[1]

Ἔστι δ' ἡ μὲν Τρίκκη, ὅπου τὸ ἱερὸν τοῦ Ἀσκληπιοῦ τὸ ἀρχαιό-
τατον καὶ ἐπιφανέστατον . . .

Now Tricca, where is the earliest and most famous temple of
Asclepius . . .

[1] Cf. **714a**. Eustathius, *Commentarii ad Homeri Iliadem*, II, 729, where this passage
is quoted. Cf. also **714b**. Herondas, *Mimiambi*, II, 97: κὡσκληπιὸς κῶς ἦλθεν ἐνθάδ'
[sc., εἰς τὴν Κῶν] ἐκ Τρίκκης . . .

715. Strabo, Geographica, VIII, 4, 4.

Δείκνυται δ' ἐν τῇ Γερηνίᾳ Τρικκαίου ἱερὸν Ἀσκληπιοῦ, ἀφίδρυμα
τοῦ ἐν τῇ Θετταλικῇ Τρίκκῃ.

In Gerenia [sc., in Messenia] [1] is to be seen a temple of Triccaean Asclepius, a reproduction of the one in the Thessalian
Tricca.

[1] Cf. T. 767.

EPIRUS

716. Polybius, Historiae, XXI, 27, 2.

. . . καὶ τρία μὲν ἔργα κατὰ τὸ Πύρρειον προσῆγεν διὰ τῶν
ἐπιπέδων [τόπων] . . . τέταρτον δὲ κατὰ τὸ Ἀσκληπιεῖον . . .

He [i. e. the consul Marcus Fulvius Nobilior, in 189 B. C.]
brought up three machines through the level country near
the Pyrrheium [sc., in Ambracia] . . . a fourth at the
Asclepieion [1] . . .

[1] Cf. **716a**. Livius, *Ab Urbe Condita*, XXXVIII, 5, 2: Tria [sc., opera] . . . adversus
Pyrrheum, quod vocant, admovit, unum e regione Aesculapii, . . .

LOCRIS OZOLIS

717. Pausanias, Descriptio Graeciae, X, 38, 13.[1]

Τοῦ δὲ Ἀσκληπιοῦ τὸ ἱερὸν ἐρείπια ἦν, ἐξ ἀρχῆς δὲ ᾠκοδόμησεν αὐτὸ ἀνὴρ ἰδιώτης Φαλύσιος.

The sanctuary of Asclepius [sc., at Naupactus] was in ruins; it was originally built by a private person called Phalysius. (*)

[1] Cf. T. 444.

PHOCIS

718. Pausanias, Descriptio Graeciae, X, 34, 6.[1]

Τῷ δὲ Ἀσκληπιῷ ναὸς ᾠκοδόμηται . . .

A temple has been built to Asclepius [sc., in Elateia] . . .

[1] Cf. T. 662; cf. also 718a. Inscriptiones Graecae, IX, 1, no. 39 [beginning of 2nd c. B.C.]: . . . Σωτηρίδας . . . ἀφίητι ἐλευθέρους Ξένωνα καὶ Παράμονον . . . καὶ ἀνατί-[θητι] αὐτοὺς τῷ θεῷ τῷ Ἀσσκλαπιῷ τῷ ἐν Στείρει . . . ; cf. T. 858.

719. Pausanias, Descriptio Graeciae, X, 32, 12.[1]

Σταδίοις δὲ ἀπωτέρω Τιθορέας ἑβδομήκοντα ναός ἐστιν Ἀσκλη-πιοῦ, καλεῖται δὲ Ἀρχαγέτας. τιμὰς δὲ παρὰ αὐτῶν ἔχει Τιθορέων καὶ ἐπ᾽ ἴσης παρὰ Φωκέων τῶν ἄλλων. ἐντὸς μὲν δὴ τοῦ περιβόλου τοῖς τε ἱκέταις καὶ ὅσοι τοῦ θεοῦ δοῦλοι, τούτοις μὲν ἐνταῦθά εἰσι καὶ οἰκήσεις. ἐν μέσῳ δὲ ὅ τε ναὸς . . . κλίνη δὲ ἐν δεξιᾷ κεῖται τοῦ ἀγάλματος . . .

Seventy stades distant from Tithorea is a temple of Asclepius called Archagetas. He receives honors from the Tithoreans themselves and no less from the other Phocians. Within the precincts are dwellings for both the suppliants of the god and his servants. In the middle is the temple of the god . . . A couch is set on the right of the image[2] . . . (*)

[1] C. T. 499.
[2] Cf. T. 668b; cf. also T. 533.

ATTICA

720. Inscriptiones Graecae, II², no. 4960a [beginning of 4th c. B.C.].

. . . .

[. . . . . . . ἀ]νελθὼν Ζεόθ[ε]-

[ν μυστηρί]οις τοῖς μεγά-
[λοις κατ]ήγετο ἐς τὸ Ἐλ-
[ευσίνιο]ν καὶ οἴκοθεν
[μεταπεμ]ψάμενος δ⟨ρ⟩ά[κ]-
[οντα ἤγ]αγεν δεῦρε ἐφ' [ἅ]-
[ρματος] Τηλ[ε]μάχο [..]α[.]-
[......ο]ς· ἅμα ἦλθεν Ὑγ-
[ίεια καὶ] οὕτως ἱδρύθη
[τὸ ἱερὸ]ν τόδε ἅπαν ἐπὶ
['Αστυφί]λο ἄρχοντος Κυ-
[δαντίδο]. Ἀρ[χ]έας· ἐπὶ το-
[ύτο οἱ Κ]ήρυκες ἠμφεσβ-
[ήτον τῷ χ]ωρίο καὶ ἔνια
[διεκώλυ]σαν ποῆσαι. Ἀν-
[τιφῶν...· ἐπὶ τ]ούτο ευ-
[...... Εὔφημος]· ἐπὶ τ-
[ούτο – – – – – – – – – –]

– – –

. . . when the god came from Zea at the time of the great
mysteries, he took up his residence in the Eleusinium, and
having summoned from his home the snake he brought it
hither in the chariot of Telemachus . . . At the same time
came Hygieia; and thus this temple was founded entirely in
the archonship of Astyphilus, of Kydantidae. Archeas: in his
time the *Kerykes* disputed over the land and prevented some
things from being done. Antiphon . . . in his time . . . ,
Euphemus: in his time . . .[1]

[1] Cf. **720a**. Apollophanes, Fr. 1 [Demianczuk]: Ἀσκληπιός, Κύννειος, Ἀφρόδιτος,
Τύχων. Cf. also **720b**. Hesychius, *Lexicon, s. v.* θεοὶ ξενικοί· παρὰ Ἀθηναίοις τιμῶνται,
οὓς καταλέγει Ἀπολλοφάνης ἐν Κρησί.

721. Aristophanes, Plutus, 620–21.

Ἐγὼ δὲ καὶ σύ γ' ὡς τάχιστα τὸν θεὸν
ἐγκατακλινοῦντ' ἄγωμεν εἰς Ἀσκληπιοῦ.

But you and I will take the God at once
To spend the night inside Asclepius' Temple.

722. Scholia in Aristophanem, Ad Plutum, 621.

Εἰς Ἀσκληπιοῦ· τὸν ἐν ἄστει λέγει Ἀσκληπιόν. δύο γάρ εἰσιν,
ὁ μὲν ἐν ἄστει, ὁ δὲ ἐν Πειραιεῖ.

Into the temple of Asclepius: he [sc., Aristophanes] means the Asclepius in town. For there are two, the one in town, the other in the Piraeus.

723. Xenophon, Memorabilia, III, 13, 3.

Πότερον δέ, ἔφη, τὸ παρὰ σοὶ ὕδωρ θερμότερον πιεῖν ἐστιν ἢ τὸ ἐν Ἀσκληπιοῦ; Τὸ ἐν Ἀσκληπιοῦ, ἔφη.

Which is the warmer to drink, he [sc., Socrates] said, the water in your house or that in the temple of Asclepius?[1] That in the temple of Asclepius [sc., at Athens], he [sc., the interlocutor] said. (*)

[1] Cf. T. 707, n. 1.

724. Plinius, Naturalis Historia, II, 103 (106), 225.

Subeunt terras rursusque redduntur Lycus in Asia, Erasinus in Argolica, Tigris in Mesopotamia et, quae in Aesculapi fonte Athenis mersa sunt, in Phalerico redduntur.

The Lycus in Asia, the Erasinus in Argolis, and the Tigris in Mesopotamia sink into the earth and burst out again, and what is thrown into the fountain of Asclepius at Athens is cast up at the fountain of Phalerum. (*)

725. Pausanias, Descriptio Graeciae, I, 21, 4-5.

Ἰόντων δὲ Ἀθήνησιν ἐς τὴν ἀκρόπολιν ἀπὸ τοῦ θεάτρου τέθαπται Κάλως. . . . τοῦ δὲ Ἀσκληπιοῦ τὸ ἱερὸν ἔς τε τὰ ἀγάλματά ἐστιν, ὁπόσα τοῦ θεοῦ πεποίηται καὶ τῶν παίδων, καὶ ἐς τὰς γραφὰς θέας ἄξιον. ἔστι δὲ ἐν αὐτῷ κρήνη, παρ' ᾗ λέγουσι Ποσειδῶνος παῖδα Ἁλιρρόθιον θυγατέρα Ἄρεως Ἀλκίππην αἰσχύναντα ἀποθανεῖν ὑπὸ Ἄρεως, καὶ δίκην ἐπὶ τούτῳ τῷ φόνῳ γενέσθαι πρῶτον. ἐνταῦθα ἄλλα τε καὶ Σαυροματικὸς ἀνάκειται θώραξ.

On the way to the Athenian Acropolis from the theater is the tomb of Calos. . . . The sanctuary of Asclepius is worth seeing both for the statues of the god and of his children[1] and for its paintings. In it there is a spring,[2] by which they say that Poseidon's son Halirrhothius, having deflowered Alcippe, the daughter of Ares, was killed by Ares, and this murder was the first to be prosecuted. Among the votive offerings there is a Sauromatic breastplate.[3] (*)

[1] Cf. T. 628.   [2] Cf. T. 707, n. 1.   [3] Cf. T. 548.

726. Pausanias, Descriptio Graeciae, I, 22, 1.

Μετὰ δὲ τὸ ἱερὸν τοῦ Ἀσκληπιοῦ ταύτῃ πρὸς τὴν ἀκρόπολιν ἰοῦσι Θέμιδος ναός ἐστι.

After the sanctuary of Asclepius, as you go by this way towards the Acropolis, there is a temple of Themis [*sc.*, at Athens].

727. Lucianus, Piscator, 42.

. . . παρὰ δὲ τὸ Πελασγικὸν ἄλλοι καὶ κατὰ τὸ Ἀσκληπιεῖον ἕτεροι καὶ παρὰ τὸν Ἄρειον πάγον ἔτι πλείους, . . .

. . . others are coming up beside the Pelasgicon, others by the precinct of Asclepius, even more of them along the Areopagus [*sc.*, at Athens] . . .

728. Marinus, Vita Procli, Cp. 29.[1]

. . . ἡ οἴκησις . . . ἦν καὶ ὁ πατὴρ αὐτοῦ Συριανὸς καὶ ὁ προπάτωρ . . . ᾤκησαν, γείτονα μὲν οὖσαν τοῦ ἀπὸ Σοφοκλέους ἐπιφανοῦς Ἀσκληπιείου, καὶ τοῦ πρὸς τῷ θεάτρῳ Διονυσίου, ὁρωμένην δὲ ἢ καὶ ἄλλως αἰσθητὴν γιγνομένην τῇ ἀκροπόλει τῆς Ἀθηνᾶς.

. . . the house . . .which his ' father ' Syrianus and his ' grandfather ' . . . had dwelt in, was next door to the Asclepieion, famous through Sophocles,[2] and to the temple of Dionysus near the theater; and it was visible and otherwise noticeable from the Acropolis of Athena.

[1] Cf. T. **582.**
[2] Cf. T. **591.**

729. Inscriptiones Graecae, II[2], no. 4969 [middle of 4th c. B. C.].

[Ὁ] θεὸς ἔχρησεν τῶι δήμωι τῶι Ἀθην[αίων ἀναθεῖναι]
[τ]ὴν οἰκίαν τὴν Δήμωνος καὶ τὸν κῆ[πον τὸν προσόντα]
τῶι Ἀσκληπιῶι καὶ αὐτὸν Δήμωνα [ἱερέα εἶναι αὐτοῦ].
          vac.
ἱερεὺς Δήμων Δημομέλους Παιαν[ιεὺς ἀνέθηκε]
καὶ τὴν οἰκίαν καὶ τὸν κῆπον προσ[τάξαντος τοῦ θεοῦ]
καὶ τοῦ δήμου τοῦ Ἀθηναίων δόν[τος ἱερέα εἶναι]
τοῦ Ἀσκληπιοῦ κατὰ τὴν μαν[τείαν].

The god declared in an oracle to the Athenian people that they should consecrate to Asclepius the house of Demo and the adjoining garden and that Demo himself should be his priest.

As priest, Demo, son of Demomeles, from the deme of Paeania, consecrated both the house and the garden, when the god commanded and the Athenian people granted that he be the priest of Asclepius in accordance with the oracle.

730. Aelianus, Varia Historia, V, 17.

Ὅτι τοσοῦτον ἦν Ἀθηναίοις δεισιδαιμονίας, εἴ τις πρινίδιον ἐξέκοψεν ἐξ ἡρῴου, ἀπέκτεινον αὐτόν. ἀλλὰ καὶ Ἀτάρβην, ὅτι τοῦ Ἀσκληπιοῦ τὸν ἱερὸν στρουθὸν ἀπέκτεινε πατάξας, οὐκ ἀργῶς τοῦτο Ἀθηναῖοι παρεῖδον, ἀλλ᾽ ἀπέκτειναν Ἀτάρβην, καὶ οὐκ ἔδοσαν οὔτε ἀγνοίας συγγνώμην οὔτε μανίας, πρεσβύτερα τούτων ἀμφοτέρων τὰ τοῦ θεοῦ ποιησάμενοι. ἐλέγετο γὰρ ἀκουσίως, οἱ δὲ μεμηνὼς τοῦτο δρᾶσαι.

Because the Athenians had such religious scruples, if anyone cut down an oak shoot from the shrine of a hero, they put him to death. Even Atarbes [otherwise unknown] because he struck and killed the sacred sparrow[1] of Asclepius,—the Athenians did not idly disregard this, but put Atarbes to death and did not grant him forgiveness on the grounds of ignorance or madness, considering the things pertaining to the god more important than either ignorance or madness. For he was said to have done this involuntarily—some said when he was in a state of fury.

[1] Cf. T. 446.

731. Aelianus, De Natura Animalium, VII, 13.

Ἐς Ἀσκληπιοῦ παρῆλθε θεοσύλης τό τε μεσαίτατον τῆς νυκτὸς παραφυλάξας καὶ τῶν καθευδόντων τὸν βαθύτατον ὕπνον ἐπιτηρήσας, εἶτα ὑφείλετο τῶν ἀναθημάτων πολλά, καὶ ὥς γε ᾤετο ἐλελήθει. ἦν δὲ ἄρα σκοπὸς ἀγαθὸς ἔνδον κύων καὶ τῶν ζακόρων ἀμείνων ἐς ἀγρυπνίαν, ὅσπερ οὖν εἵπετό οἱ διώκων, καὶ ὑλακτῶν οὐκ ἀνίει, ᾗπερ οὖν ἔσθενε δυνάμει τὸ πραχθὲν μαρτυρόμενος. τὰ μὲν οὖν πρῶτα ἔβαλλεν αὐτὸν λίθοις αὐτός τε καὶ οἱ τῆς κακῆς ἐκείνης πράξεως κοινωνοί, τὰ δὲ τελευταῖα προύσειεν

ἄρτους τε καὶ μάζας. ἐπήγετο δὲ ἄρα ταῦτα δέλεαρ κυνῶν προ-
μηθῶς, ὥς γε ὑπελάμβανεν. ἐπεὶ δὲ καὶ παρελθόντος ἐς τὴν οἰκίαν
οὗ κατήγετο ὑλάκτει καὶ πάλιν προϊόντος, ἐγνώσθη μὲν ὁ κύων
ἔνθεν ἦν, τὰ λείποντα δὲ τῶν ἀναθημάτων ἐπόθουν αἱ γραφαί τε
καὶ αἱ χῶραι ἔνθα ἀνέκειντο. συνέβαλον οὖν τοῦτον ἐκεῖνον
εἶναι οἱ Ἀθηναῖοι, καὶ στρεβλώσαντες τὸ πᾶν κατέγνωσαν. καὶ
ὁ μὲν ἐδικαιώθη τὰ ἐκ τοῦ νόμου, ὁ δὲ κύων ἐτιμήθη δημοσίᾳ
τροφῇ καὶ κηδεμονίᾳ, οἷα δήπου φύλαξ πιστὸς καὶ τῶν νεωκόρων
οὐδενὸς μείων τὴν ἐπιμέλειαν.

A thief entered the Asclepieion watching for the middle of the
night and looking out for the time of the deepest slumber of
those sleeping in the temple. Then he filched many of the votive
offerings, and, as he thought, escaped detection. Yet there was
at this time a good guard inside (the temple), a dog, better
than the temple attendants in keeping sleepless vigil; the dog,
therefore, followed in pursuit of the man and never stopped
barking; thus with all the power he possessed serving notice
of the crime. At first the thief and his confederates in the
impious deed pelted him with stones; finally they threw him
bread and cakes. With foresight, at least so he thought, he
had taken along such things, the bait of dogs; but since the
dog kept on barking even when the thief entered the dwelling
where he lodged and again when he came forth, it was known
where the dog was. Moreover, both the catalogues and the empty
spaces where the votive offerings had been set up indicated
those which were missing. The Athenians concluded that this
was the man and subjecting him to torture they learned everything, and he was punished in conformity with the law. But
the dog was honored with public meals and care as truly a
faithful guard and inferior to none of the sacristans[1] in
watchfulness.[2]

[1] Cf. T. 495.
[2] Cf. 731a. Plutarchus, *De Sollertia Animalium*, XIII, 11-13, 969E-970A: Παραρρυεὶς
γὰρ ἄνθρωπος εἰς τὸν νεὼν τοῦ Ἀσκληπιοῦ τὰ εὔογκα τῶν ἀργυρῶν καὶ χρυσῶν ἔλαβεν ἀναθημάτων καὶ λεληθέναι νομίζων ὑπεξῆλθεν· ὁ δὲ φρουρὸς κύων, ὄνομα Κάππαρος, ἐπεὶ μηδεὶς
ὑλακτοῦντι τῶν ζακόρων ὑπήκουσεν αὐτῷ, φεύγοντα τὸν ἱερόσυλον ἐπεδίωκε· καὶ πρῶτον μὲν
βαλλόμενος λίθοις οὐκ ἀπέστη· γενομένης δ' ἡμέρας, ἐγγὺς οὐ προσιὼν ἀλλ' ἀπ' ὀφθαλμοῦ
παραφυλάττων εἵπετο καὶ τροφὴν προβάλλοντος οὐκ ἐλάμβανεν· ἀναπαυομένῳ δὲ παρενυκτέρευε
καὶ βαδίζοντος πάλιν ἀναστὰς ἐπηκολούθει, τοὺς δ' ἀπαντῶντας ὁδοιπόρους ἔσαινεν, ἐκείνῳ δ'
ἐφυλάκτει καὶ προσέκειτο. ταῦτα δ' οἱ διώκοντες πυνθανόμενοι παρὰ τῶν ἀπαντώντων ἅμα καὶ
τὸ χρῶμα φραζόντων καὶ τὸ μέγεθος τοῦ κυνὸς προθυμότερον ἐχρήσαντο τῇ διώξει, καὶ κατα-
λαβόντες τὸν ἄνθρωπον ἀνήγαγον ἀπὸ Κρομμυῶνος. ὁ δὲ κύων ἀναστρέψας προηγεῖτο γαῦρος καὶ
περιχαρής, οἷον ἑαυτοῦ ποιούμενος ἄγραν καὶ θήραμα τὸν ἱερόσυλον. ἐψηφίσαντο δὴ σῖτον αὐτῷ
δημοσίᾳ μετρεῖσθαι καὶ παρεγγυᾶσθαι τοῖς ἱερεῦσιν εἰς ἀεὶ τὴν ἐπιμέλειαν, ἀπομιμησάμενοι τὸ
πρὸς τὸν ἡμίονον φιλανθρώπευμα τῶν παλαιῶν Ἀθηναίων.

732. Philostratus, Epistulae, 8.

… ξένοι καὶ ὄμβροι τῆς γῆς καὶ ποταμοὶ τῆς θαλάττης καὶ ὁ Ἀσκληπιὸς Ἀθηναίων καὶ ὁ Ζεὺς ἡμῶν καὶ ὁ Νεῖλος Αἰγυπτίων καὶ ὁ Ἥλιος πάντων.

… friends are: the rain to the earth and the rivers to the sea and Asclepius to the Athenians and Zeus to us and Neilos to the Egyptians and the Sun to everyone.

AEGINA

733. Aristophanes, Vespae, 122-23.

. . . .
διέπλευσεν εἰς Αἴγιναν· εἶτα ξυλλαβὼν
νύκτωρ κατέκλινεν αὐτὸν εἰς Ἀσκληπιοῦ.

He [sc., Bdelycleon] shipped him [sc., Philocleon] over to Aegina.[1] Then, he carried him off and laid him for the night inside the temple of Asclepius.

[1] Cf. T. 432.

734. Pausanias, Descriptio Graeciae, II, 30, 1.

Τοῦ δὲ Ἀσκληπιοῦ τὸ ἱερὸν ἔστι μὲν ἑτέρωθι καὶ οὐ ταύτῃ …

The sanctuary of Asclepius [sc., at Aegina] is in another place, not here …

PELOPONNESE

ARGOLID

735. Strabo, Geographica, VIII, 6, 15.[1]

Καὶ αὕτη δ' οὐκ ἄσημος ἡ πόλις καὶ μάλιστα διὰ τὴν ἐπιφάνειαν τοῦ Ἀσκληπιοῦ θεραπεύειν νόσους παντοδαπὰς πεπιστευμένου, καὶ τὸ ἱερὸν πλῆρες ἔχοντος ἀεὶ τῶν τε καμνόντων καὶ τῶν ἀνακειμένων πινάκων, ἐν οἷς ἀναγεγραμμέναι τυγχάνουσιν αἱ θεραπεῖαι, καθάπερ ἐν Κῷ τε καὶ Τρίκκῃ.

And this city [sc., of Epidaurus] is not without distinction, and particularly because of the epiphany of Asclepius, who is believed to cure diseases of every kind and always has his temple full of the sick, and also of the votive tablets[2] on which the treatments are recorded, just as at Cos[3] and Tricca.[4] (*)

[2] Cf. T. 384; 739, 3.　　　[1] Cf. T. 382.　　　[3] Cf. T. 794.　　　[4] Cf. T. 714.

736. Plinius, Naturalis Historia, IV, 5 (9), 18.

> . . . in eo Epidaurum oppidum, Aesculapi delubro celebre . . .

> . . . upon this [sc., the Saronian gulf] is the town of Epidaurus, famous for its temple of Asclepius . . .[1]

[1] Cf. **736a.** Ovidius, *Ibis*, 403-4: . . . quem [sc., Periphetem] . . . urbe Coronides vidit ab ipse sua; cf. **736b.** Pomponius Mela, *Chorographia*, II, 3 (49): Habitant ab Isthmo ad Scyllaeon Epidaurii, Aesculapi templo incluti . . . ; cf. **736c.** Lactantius Placidus, *Commentarii in Statium, Ad Thebaidem*, IV, 123: Epidauria: civitas Argivorum, ubi Aesculapius colitur. . . .

737. Minucius Felix, Octavius, VI, 1.

> Inde adeo per universa imperia, provincias, oppida videmus singulos sacrorum ritus gentiles habere et deos colere municipes, ut Eleusinios Cererem, Phrygas Matrem, Epidaurios Aesculapium . . .

> Hence it is that throughout wide empires, provinces and towns, we see each people having its own individual rites and worshipping its local gods, the Eleusinians Ceres, the Phrygians the Great Mother, the Epidaurians Asclepius[1] . . .

[1] Cf. T. **509.**

738. Solinus, Collectanea Rerum Memorabilium, Cp. 7, 10.

> Epidauro decus est Aesculapii sacellum, cui incubantes aegritudinum remedia capessunt de monitis somniorum.

> The adornment of Epidaurus is the precinct of Asclepius, where those who sleep in his temple get remedies for their diseases from the admonitions which they receive in their dreams.[1]

[1] Cf. T. **429** ff.

739. Pausanias, Descriptio Graecae, II, 27, 1-7.

> Τὸ δὲ ἱερὸν ἄλσος τοῦ Ἀσκληπιοῦ περιέχουσιν ὅροι πανταχόθεν· οὐδὲ ἀποθνήσκουσιν ἄνθρωποι οὐδὲ τίκτουσιν αἱ γυναῖκές σφισιν ἐντὸς τοῦ περιβόλου, καθὰ καὶ ἐπὶ Δήλῳ τῇ νήσῳ τὸν αὐτὸν νόμον. τὰ δὲ θυόμενα, ἤν τέ τις Ἐπιδαυρίων αὐτῶν ἤν τε ξένος ὁ θύων ᾖ, καταναλίσκουσιν ἐντὸς τῶν ὅρων· τὸ δὲ αὐτὸ γινόμενον
> 2 οἶδα καὶ ἐν Τιτάνῃ. τοῦ δὲ Ἀσκληπιοῦ τὸ ἄγαλμα μεγέθει μὲν τοῦ Ἀθήνῃσιν Ὀλυμπίου Διὸς ἥμισυ ἀποδεῖ, πεποίηται δὲ ἐλέ-

φαντος καὶ χρυσοῦ· μηνύει δὲ ἐπίγραμμα τὸν εἰργασμένον
εἶναι Θρασυμήδην Ἀριγνώτου Πάριον. κάθηται δὲ ἐπὶ θρόνου
βακτηρίαν κρατῶν, τὴν δὲ ἑτέραν τῶν χειρῶν ὑπὲρ κεφαλῆς ἔχει
τοῦ δράκοντος, καί οἱ καὶ κύων παρακατακείμενος πεποίηται. τῷ
θρόνῳ δὲ ἡρώων ἐπειργασμένα Ἀργείων ἐστὶν ἔργα, Βελλεροφόντου τὸ ἐς τὴν Χίμαιραν καὶ Περσεὺς ἀφελὼν τὴν Μεδούσης
κεφαλήν. τοῦ ναοῦ δέ ἐστι πέραν ἔνθα οἱ ἱκέται τοῦ θεοῦ καθεύδου-
3   σιν. οἴκημα δὲ περιφερὲς λίθου λευκοῦ καλούμενον Θόλος ᾠκοδόμηται πλησίον, θέας ἄξιον· ἐν δὲ αὐτῷ Παυσίου γράψαντος
βέλη μὲν καὶ τόξον ἐστὶν ἀφεικὼς Ἔρως, λύραν δὲ ἀντ' αὐτῶν
ἀράμενος φέρει. γέγραπται δὲ ἐνταῦθα καὶ Μέθη, Παυσίου καὶ
τοῦτο ἔργον, ἐξ ὑαλίνης φιάλης πίνουσα· ἴδοις δὲ κἂν ἐν τῇ
γραφῇ φιάλην τε ὑάλου καὶ δι' αὐτῆς γυναικὸς πρόσωπον. στῆλαι
δὲ εἰστήκεσαν ἐντὸς τοῦ περιβόλου τὸ μὲν ἀρχαῖον καὶ πλέονες,
ἐπ' ἐμοῦ δὲ ἓξ λοιπαί· ταύταις ἐγγεγραμμένα καὶ ἀνδρῶν καὶ
γυναικῶν ἐστιν ὀνόματα ἀκεσθέντων ὑπὸ τοῦ Ἀσκληπιοῦ, προσέτι
δὲ καὶ νόσημα ὅ τι ἕκαστος ἐνόσησε καὶ ὅπως ἰάθη· γέγραπται
4   δὲ φωνῇ τῇ Δωρίδι. χωρὶς δὲ ἀπὸ τῶν ἄλλων ἐστὶν ἀρχαία στήλη·
ἵππους δὲ Ἱππόλυτον ἀναθεῖναι τῷ θεῷ φησιν εἴκοσι. ταύτης
τῆς στήλης τῷ ἐπιγράμματι ὁμολογοῦντα λέγουσιν Ἀρικιεῖς, ὡς
τεθνεῶτα Ἱππόλυτον ἐκ τῶν Θησέως ἀρῶν ἀνέστησεν Ἀσκληπιός·
ὁ δὲ ὡς αὖθις ἐβίω, οὐκ ἠξίου νέμειν τῷ πατρὶ συγγνώμην, ἀλλὰ
ὑπεριδὼν τὰς δεήσεις ἐς Ἰταλίαν ἔρχεται παρὰ τοὺς Ἀρικιεῖς,
καὶ ἐβασίλευσέ τε αὐτόθι καὶ ἀνῆκε τῇ Ἀρτέμιδι τέμενος, ἔνθα
ἄχρι ἐμοῦ μονομαχίας ἆθλα ἦν καὶ ἱερᾶσθαι τῇ θεῷ τὸν νικῶντα·
ὁ δὲ ἀγὼν ἐλευθέρων μὲν προέκειτο οὐδενί, οἰκέταις δὲ ἀποδρᾶσι
5   τοὺς δεσπότας. Ἐπιδαυρίοις δέ ἐστι θέατρον ἐν τῷ ἱερῷ μάλιστα
ἐμοὶ δοκεῖν θέας ἄξιον· τὰ μὲν γὰρ Ῥωμαίων πολὺ δή τι ὑπερῆρκε
τῶν πανταχοῦ τῷ κόσμῳ, μεγέθει δὲ Ἀρκάδων τὸ ἐν Μεγάλῃ
πόλει· ἁρμονίας δὲ ἢ κάλλους ἕνεκα ἀρχιτέκτων ποῖος ἐς ἅμιλλαν
Πολυκλείτῳ γένοιτ' ἂν ἀξιόχρεως; Πολύκλειτος γὰρ καὶ θέατρον
τοῦτο καὶ οἴκημα τὸ περιφερὲς ὁ ποιήσας ἦν. ἐντὸς δὲ τοῦ ἄλσους
ναός τέ ἐστιν Ἀρτέμιδος καὶ ἄγαλμα Ἠπιόνης καὶ Ἀφροδίτης
ἱερὸν καὶ Θέμιδος καὶ στάδιον, οἷα Ἕλλησι τὰ πολλὰ γῆς χῶμα,
6   καὶ κρήνη τῷ τε ὀρόφῳ καὶ κόσμῳ τῷ λοιπῷ θέας ἀξία. ὁπόσα δὲ
Ἀντωνῖνος ἀνὴρ τῆς συγκλήτου βουλῆς ἐφ' ἡμῶν ἐποίησεν,
ἔστι μὲν Ἀσκληπιοῦ λουτρόν, ἔστι δὲ ἱερὸν θεῶν οὓς Ἐπιδώτας
ὀνομάζουσιν· ἐποίησε δὲ καὶ Ὑγείᾳ ναὸν καὶ Ἀσκληπιῷ καὶ
Ἀπόλλωνι ἐπίκλησιν Αἰγυπτίοις. καὶ ἦν γὰρ στοὰ καλουμένη
Κότυος, καταρρυέντος δέ οἱ τοῦ ὀρόφου διέφθαρτο ἤδη πᾶσα ἅτε

ὠμῆς τῆς πλίνθου ποιηθεῖσα· ἀνῳκοδόμησε καὶ ταύτην. Ἐπιδαυρίων δὲ οἱ περὶ τὸ ἱερὸν μάλιστα ἐταλαιπώρουν, ὅτι μήτε αἱ
γυναῖκες ἐν σκέπῃ σφίσιν ἔτικτον καὶ ἡ τελευτὴ τοῖς κάμνουσιν
ὑπαίθριος ἐγίνετο· ὁ δὲ καὶ ταῦτα ἐπανορθούμενος κατεσκευάσατο
οἴκησιν· ἐνταῦθα ἤδη καὶ ἀποθανεῖν ἀνθρώπῳ καὶ τεκεῖν γυναικὶ
ὅσιον.

7   Ὄρη δέ ἐστιν ὑπὲρ τὸ ἄλσος τό τε Τίτθιον καὶ ἕτερον ὀνομαζό-
μενον Κυνόρτιον, Μαλεάτου δὲ Ἀπόλλωνος ἱερὸν ἐν αὐτῷ. τοῦτο
μὲν δὴ τῶν ἀρχαίων· τὰ δὲ ἄλλα ὅσα περὶ τὸ ἱερὸν τοῦ Μαλεάτου
καὶ ἔλυτρον κρήνης, ἐς ὃ τὸ ὕδωρ συλλέγεταί σφισι τὸ ἐκ τοῦ
θεοῦ, Ἀντωνῖνος καὶ ταῦτα Ἐπιδαυρίοις ἐποίησεν.

The sacred grove of Asclepius is surrounded by bounds on
every side. No men die or women give birth within the
enclosure;[1] the same rule is observed in the island of Delos.
The sacrifices, whether offered by one of the Epidaurians
themselves or a foreigner, are consumed within the bounds.[2]
I know that the same thing is done also at Titane. The image
of Asclepius[3] is half the size of the image of Olympian Zeus
at Athens: it is made of ivory and gold. An inscription sets
forth that the sculptor[4] was Thrasymedes [ca. 375 B. C.],
a Parian, son of Arignotus. The god is seated on a throne,
grasping a staff, while holding the other hand over the head of
the serpent; and a dog, lying by his side, is also represented.
On the throne are carved in relief the deeds of Argive heroes:
of Bellerophon against the Chimaera, and Perseus who has cut
off Medusa's head.[5] Over against the temple is the place where
the suppliants of the god sleep. A round building of white
marble, called the Rotunda (Tholos) is near by, worth seeing.
In it is a picture by Pausias [4th c. B. C.]: Eros who has
thrown away his bow and arrows, and has picked up a lyre
instead. Here is also another painting, that, too, a work of
Pausias: Drunkenness drinking out of a crystal goblet; and in
the picture you can see the crystal goblet and through it the
woman's face.

Tablets stood within the enclosure.[6] Of old, there were more
of them: in my time six were left. On these tablets are engraved the names of men and women who were healed by
Asclepius, together with the disease from which each suffered,
and how he was cured. The inscriptions are in the Doric

dialect. Apart from the others stands an ancient tablet; it says
that Hippolytus dedicated twenty horses to the god. The people
of Aricia tell a tale that agrees with the inscription on this
tablet, that when Hippolytus was done to death by the curses
of Theseus, Asclepius raised him from the dead;[7] on coming
to life again, he [sc., Hippolytus] refused to forgive his father,
and disregarding his entreaties went away to Aricia in Italy.[8]
There he reigned, and consecrated to Artemis a precinct, where
down to my time there were contests in single combat and the
victor became a priest of the goddess. The competition was
not open to free men, but only to slaves who had run away
from their masters.

In the Epidaurian sanctuary there is a theater, in my opinion
most especially worth seeing. It is true, in splendor the Roman
theaters far transcend all the theaters in the world, and in size
the theater at Megalopolis in Arcadia is superior; but for symmetry and beauty, what architect could vie with Polyclitus?
For it was Polyclitus [middle of the 4th c. B. C.] who made
this theater and the round building also.

Within the grove is a temple of Artemis and an image of
Epione;[9] also a sanctuary of Aphrodite and Themis; and a
stadium formed, like most Greek stadiums, by banks of earth;
also a fountain [10] worth seeing for its roof and general splendor.

The buildings erected in our time by the Roman senator
Antoninus include a bath of Asclepius and a sanctuary of the
gods whom they name Bountiful.[11] Further, he built a temple
to Health, as well as to Asclepius, and Apollo, with the surname 'Egyptians'.[12] There was also a colonnade called the
Colonnade of Cotys: after the roof had collapsed the whole
edifice was in ruins, having been built of unburnt brick; this,
too, he restored. The Epidaurians about the sanctuary were
in great distress, because their women were not allowed to
bring forth under shelter, and their sick were obliged to die
under the open sky.[13] To remedy the inconvenience he provided a building where a man may die and a woman may give
birth to her child without sin.

Above the grove is Mount Titthium and another mountain
named Cynortium. On the latter is a sanctuary of Maleatian
Apollo. This is one of the ancient (buildings), but everything
about the sanctuary of Maleatas, including the cistern in which

the holy water is collected, is also a gift of Antoninus to the
Epidaurians.[14] (**)

[1] Cf. T. **488**; **423**, 1 and 2.
[2] Cf. T. **510**; cf. however T. **482**, vv. 92 f.
[3] Cf. T. **630**.
[4] Cf. however T. **645**.
[5] Concerning Asclepius' relation to Medusa, cf. T. **3**.
[6] Cf. T. **382**; **384**; **423**; **735**.
[7] Cf. T. **74** ff.
[8] Cf. T. **75**, vv. 755-56.
[9] Cf. T. **280** ff.; **742**.
[10] Cf. T. **707** and Footnote.
[11] Cf. T. **747**.
[12] Cf. T. **827** ff.
[13] Cf. n. 1.
[14] For a general description of the Epidaurian temple, cf. also T. **850**.

740. Plautus, Curculio, I, 1, 14.

Hoc Aesculapi fanumst.

Yonder is the shrine of Asclepius.

741. Plautus, Curculio, I, 1, 61-62.[1]

Id eo fit, quia hic leno, hic qui aegrotus incubat
in Aesculapi fano, is me excruciat.

That's all because the pimp, who's ill and is taking the cure in
the shrine of Asclepius here, is torturing me.

[1] Cf. T. **429**.

742. Pausanias, Descriptio Graeciae, II, 29, 1.[1]

Αὐτὴ δὲ τῶν Ἐπιδαυρίων ἡ πόλις παρείχετο ἐς μνήμην τάδε
ἀξιολογώτατα· τέμενος δή ἐστιν Ἀσκληπιοῦ καὶ ἀγάλματα ὁ θεὸς
αὐτὸς καὶ Ἠπιόνη, γυναῖκα δὲ εἶναι τὴν Ἠπιόνην Ἀσκληπιοῦ
φασι· ταῦτά ἐστιν ἐν ὑπαίθρῳ λίθου Παρίου.

The city of the Epidaurians offered this as most noteworthy
for memory: There is, of course, a precinct of Asclepius and
images, the god himself and Epione. Epione, they say, is the
wife of Asclepius.[2] These are set up in the open and are of
Parian marble. (*)

[1] Cf. T. **673**.　　　　　　　　　[2] Cf. T. **280** ff.; cf. also T. **739**.

743. Diodorus, Bibliotheca Historica, XXXVIII, 7.

Ὅτι ὁ Σύλλας χρημάτων ἀπορούμενος ἐπέβαλε τὰς χεῖρας τρισὶν
ἱεροῖς, ἐν οἷς ἀναθημάτων ἀργυρῶν τε καὶ χρυσῶν πλῆθος ἦν,

ἐν μὲν Δελφοῖς τῷ Ἀπόλλωνι καθιερωμένων, ἐν Ἐπιδαύρῳ δὲ τῷ Ἀσκληπιῷ, ἐν Ὀλυμπίᾳ δὲ τῷ Διί.

That Sulla, being in want of money [sc., in 86 B. C.], laid hands on three sanctuaries in which there were plenty of silver and golden offerings, in Delphi dedicated to Apollo, in Epidaurus to Asclepius, in Olympia to Zeus.[1]

[1] Cf. **743a**. Plutarchus, *Sulla*, 12, 3: Ἐπεὶ δὲ καὶ χρημάτων ἔδει [sc., ὁ Σύλλας] πολλῶν πρὸς τὸν πόλεμον, ἐκίνει τὰ τῆς Ἑλλάδος ἄσυλα, τοῦτο μὲν ἐξ Ἐπιδαύρου τοῦτο δ' ἐξ Ὀλυμπίας τὰ κάλλιστα καὶ πολυτελέστατα τῶν ἀναθημάτων μεταπεμπόμενος; cf. **743b**. Pausanias, *Descriptio Graeciae*, IX, 7, 5: Ἡνίκα ἤρχετο τοῦ πρὸς Μιθριδάτην πολέμου, χρημάτων ἐσπάνιζε· συνέλεξεν οὖν ἔκ τε Ὀλυμπίας ἀναθήματα καὶ τὰ ἐξ Ἐπιδαύρου καὶ τὰ ἐκ Δελφῶν, ὁπόσα ὑπελίποντο οἱ Φωκεῖς; cf. **743c**. Appianus, *Historia Romana*, XII, 54: . . . ἁψάμενος ἤδη τῶν ἐν Πυθοῖ καὶ Ὀλυμπίᾳ καὶ Ἐπιδαύρῳ χρημάτων . . . ; cf. also T. **545**; **841**.

744. Plutarchus, *Pompeius*, 24, 6.

Τῶν δ' ἀσύλων καὶ ἀβάτων πρότερον ἱερῶν ἐξέκοψαν ἐπιόντες τὸ Κλάριον, τὸ Διδυμαῖον, τὸ Σαμοθράκιον, τὸν ἐν Ἑρμιόνῃ τῆς Χθονίας νεών, καὶ τὸν ἐν Ἐπιδαύρῳ τοῦ Ἀσκληπιοῦ . . .

Of places of refuge and formerly inviolate sanctuaries they [sc., the pirates, before 67 B. C.] attacked and laid waste the Clarian, the Didymaean, the Samothracian, the temple of Chthonia in Hermione, and the one of Asclepius in Epidaurus . . .

745. Pausanias, *Descriptio Graeciae*, II, 4, 5.

Πρὸς τούτῳ τῷ γυμνασίῳ ναοὶ θεῶν εἰσιν ὁ μὲν Διός, ὁ δὲ Ἀσκληπιοῦ.

By this gymnasium [sc., in Corinth] are temples of Zeus and Asclepius.

746. Pausanias, *Descriptio Graeciae*, II, 2, 3.

Ἐν δὲ Κεγχρέαις Ἀφροδίτης τέ ἐστι ναὸς . . . , κατὰ δὲ τὸ ἕτερον πέρας τοῦ λιμένος Ἀσκληπιοῦ καὶ Ἴσιδος ἱερά.

In Cenchreae [the harbor of Corinth] are a temple of Aphrodite . . . , and at the other end of the harbor sanctuaries of Asclepius and of Isis.

747. Pausanias, *Descriptio Graeciae*, II, 10, 2.

Ἐντεῦθέν ἐστιν ὁδὸς ἐς ἱερὸν Ἀσκληπιοῦ. παρελθοῦσι δὲ ἐς τὸν περίβολον ἐν ἀριστερᾷ διπλοῦν ἐστιν οἴκημα. κεῖται δὲ Ὕπνος

ἐν τῷ προτέρῳ καί οἱ πλὴν τῆς κεφαλῆς ἄλλο οὐδὲν ἔτι λείπεται.
τὸ ἐνδοτέρω δὲ Ἀπόλλωνι ἀνεῖται Καρνείῳ, καὶ ἐς αὐτὸ οὐκ ἔστι
πλὴν τοῖς ἱερεῦσιν ἔσοδος. ˙ κεῖται δὲ ἐν τῇ στοᾷ κήτους ὀστοῦν
θαλασσίου μεγέθει μέγα καὶ μετ᾽ αὐτὸ ἄγαλμα Ὀνείρου καὶ
Ὕπνος κατακοιμίζων λέοντα, Ἐπιδώτης δὲ ἐπίκλησιν. ἐς δὲ τὸ
Ἀσκληπιεῖον ἐσιοῦσι καθ᾽ ἑκάτερον τῆς ἐσόδου τῇ μὲν Πανὸς
καθήμενον ἄγαλμά ἐστι, τῇ δὲ Ἄρτεμις ἕστηκεν.

From here [i. e., the gymnasium] is a way to a sanctuary of
Asclepius [sc., at Sicyon]. On passing into the enclosure you
see on the left a building with two rooms. In the outer room
lies a figure of Sleep, of which nothing remains now except the
head. The inner room is given over to the Carnean Apollo;
into it none may enter except the priests. In the portico lies a
huge bone of a sea-monster, and after it an image of the
Dream-god and Sleep, surnamed Epidotes (Bountiful),[1] lulling
to sleep a lion. Within the Asclepieion on either side of the
entrance is an image, on the one hand Pan seated, on the other
Artemis standing. (*)

[1] Cf. T. 739.

748. Pausanias, Descriptio Graeciae, II, 10, 3.

Φασὶ δέ σφισιν ἐξ Ἐπιδαύρου κομισθῆναι τὸν θεὸν ἐπὶ ζεύγους
ἡμιόνων δράκοντι εἰκασμένον, τὴν δὲ ἀγαγοῦσαν Νικαγόραν εἶναι
Σικυωνίαν Ἀγασικλέους μητέρα, γυναῖκα δὲ Ἐχετίμου. . . .
τὴν δὲ ἐπὶ τῷ δράκοντι Ἀριστοδάμαν Ἀράτου μητέρα εἶναι
λέγουσι, καὶ Ἄρατον Ἀσκληπιοῦ παῖδα εἶναι νομίζουσιν.

The Sicyonians say that the god was carried to them from
Epidaurus[1] on a carriage drawn by two mules, that he was in
the likeness of a serpent,[2] and that he was brought by Nicagora
of Sicyon, the mother of Agasicles and the wife of Eche-
timus. . . . She who is (represented) on the serpent they say is
Aristodama, the mother of Aratus, and Aratus they hold to be
a son of Asclepius.[3]

[1] Cf. T. 709.　　　　　[2] Cf. T. 695.　　　　　[3] Cf. T. 288.

749. Pausanias, Descriptio Graeciae, II, 11, 5-7.

. . . Ἀλεξάνωρ ὁ Μαχάονος τοῦ Ἀσκληπιοῦ . . . ἐν Τιτάνῃ τὸ
Ἀσκληπιεῖον ἐποίησε. περιοικοῦσι μὲν δὴ καὶ ἄλλοι καὶ τὸ

πολὺ οἰκέται τοῦ θεοῦ, καὶ κυπαρίσσων ἐστὶν ἐντὸς τοῦ περιβόλου δένδρα ἀρχαῖα. . . . τῷ δὲ Ἀλεξάνορι καὶ Εὐαμερίωνι—καὶ γὰρ τούτοις ἀγάλματά ἐστι—τῷ μὲν ὡς ἥρωι μετὰ ἥλιον δύναντα ἐναγίζουσιν, Εὐαμερίωνι δὲ ὡς θεῷ θύουσιν. εἰ δὲ ὀρθῶς εἰκάζω, τὸν Εὐαμερίωνα τοῦτον Περγαμηνοὶ Τελεσφόρον ἐκ μαντεύματος, Ἐπιδαύριοι δὲ Ἄκεσιν ὀνομάζουσι.

. . . Alexanor, the son of Machaon, the son of Asclepius,[1] . . . built the sanctuary of Asclepius at Titane.[2] Those who live in the neighborhood are chiefly servants of the god, and within the enclosure are old cypress trees[3] . . . . There are images also of Alexanor and of Euamerion; to the former they give offerings as to a hero after the setting of the sun; to Euamerion, as being a god, they give burnt sacrifices. If I conjecture aright, the Pergamenes, in accordance with an oracle, call this Euamerion Telesphorus (Accomplisher),[4] while the Epidaurians call him Acesis (Cure).

[1] Cf. T. 188.   [2] Cf. T. 667; 753.   [3] Cf. T. 757.   [4] Cf. T. 287; 313; 808.

750. Pausanias, Descriptio Graeciae, II, 13, 5.

Κατιόντων δὲ ἐκ τῆς ἀκροπόλεώς ἐστιν Ἀσκληπιοῦ ναὸς ἐν δεξιᾷ . . .

As you go down from the citadel [sc., at Phlius] you see on the right a temple of Asclepius . . .

751. Pausanias, Descriptio Graeciae, II, 21, 1.

Κατελθοῦσι δὲ ἐντεῦθεν καὶ τραπεῖσιν αὖθις ἐπὶ τὴν ἀγοράν, ἔστι μὲν Κερδοῦς Φορωνέως γυναικὸς μνῆμα, ἔστι δὲ ναὸς Ἀσκληπιοῦ.

Having descended thence [i. e., from the temple of Aphrodite], and having turned again to the market-place [sc., of Argos], we come to the tomb of Cerdo, the wife of Phoroneus, and to a temple of Asclepius.

752. Pausanias, Descriptio Graeciae, II, 23, 2.

. . . ἱερὸν Ἀμφιαράου καὶ τοῦ ἱεροῦ πέραν Ἐριφύλης μνῆμα. ἑξῆς δὲ τούτων ἐστὶν Ἀσκληπιοῦ τέμενος . . .

. . . farther on a sanctuary of Amphiaraus, and opposite the sanctuary the tomb of Eriphyle. Next to these is a precinct of Asclepius [sc., in Argos] . . .

753. Pausanias, Descriptio Graeciae, II, 23, 4.

Τὸ δ' ἐπιφανέστατον Ἀργείοις τῶν Ἀσκληπιείων ἄγαλμα ... ἔχει
... ἐξ ἀρχῆς δὲ ἱδρύσατο Σφῦρος τὸ ἱερόν, Μαχάονος μὲν
υἱός, ἀδελφὸς δὲ Ἀλεξάνορος τοῦ παρὰ Σικυωνίοις ἐν Τιτάνῃ
τιμὰς ἔχοντος.

The most famous of all the Asclepius sanctuaries at Argos contains ... an image [1] ... The original founder of the sanctuary
was Sphyrus, son of Machaon [2] and brother of the Alexanor
who is honored among the Sicyonians in Titane. [3]

[1] Cf. T. 663.
[2] Cf. T. 188.
[3] Cf. T. 749. For the Asclepius sanctuary at Halieis, cf. T. 423, 33; at Troezen,
cf. T. 423, 23.

LACONIA

754. Pausanias, Descriptio Graeciae, III, 24, 5.

Ἱερὰ δὲ αὐτόθι τὸ μέν ἐστιν Ἀσκληπιοῦ, τὸ δὲ Ἀχιλλέως ...

There are temples here [sc., in Brasiae], one of Asclepius, the
other of Achilles ...

755. Pausanias, Descriptio Graeciae, III, 24, 2.

Προελθόντι δὲ ἀπὸ Ζάρακος παρὰ τὴν θάλασσαν ἑκατόν που
στάδια καὶ ἐπιστρέψαντι αὐτόθεν ἐς μεσόγαιαν καὶ ἐπαναβάντι
σταδίους ὡς δέκα, Κυφάντων καλουμένων ἐρείπιά ἐστιν, ἐν δὲ
αὐτοῖς σπήλαιον ἱερὸν Ἀσκληπιοῦ, λίθου δὲ τὸ ἄγαλμα. ἔστι δὲ
καὶ ὕδατος ψυχροῦ κρουνὸς ἐκβάλλων ἐκ πέτρας.

If one proceeds from Zarax along the coast for about a hundred stades, and from there turns inland, and ascends about
ten stades, there are the ruins of the so-called Cyphanta, among
which is a cave sacred to Asclepius; the image is of stone. [1]
There is also a fountain [2] of cold water springing from the
rock. (*)

[1] Cf. T. 668.   [2] Cf. T. 707.

756. Pausanias, Descriptio Graeciae, III, 23, 10.

... θέας δὲ αὐτόθι ἄξια τὸ μὲν Ἀφροδίτης ἐστὶν ἱερόν, τὸ δὲ
Ἀσκληπιοῦ ...

. . . here [sc., in Epidaurus Limera] the sanctuary of Artemis is worth seeing, also that of Asclepius . . .

757. Pausanias, Descriptio Graeciae, III, 23, 6-7.

Τῇ δὲ Βοιατῶν ὅμορος Ἐπίδαυρός ἐστιν ἡ Λιμηρά, σταδίους ὡς διακοσίους ἀπέχουσα Ἐπιδηλίου. φασὶ δὲ οὐ Λακεδαιμονίων, τῶν δὲ ἐν τῇ Ἀργολίδι Ἐπιδαυρίων εἶναι, πλέοντες δὲ ἐς Κῶν παρὰ τὸν Ἀσκληπιὸν ἀπὸ τοῦ κοινοῦ προσσχεῖν τῆς Λακωνικῆς ἐνταῦθα καὶ ἐξ ἐνυπνίων γενομένων σφίσι καταμείναντες οἰκῆσαι. λέγουσι δὲ καὶ ὡς οἴκοθεν ἐκ τῆς Ἐπιδαύρου δράκοντα ἐπαγομένοις αὐτοῖς ἐξέφυγεν ἐκ τῆς νεὼς ὁ δράκων, ἐκφυγὼν δὲ οὐ πόρρω κατέδυ θαλάσσης, καί σφισιν ὁμοῦ τῶν ὀνειράτων τῇ ὄψει καὶ ἀπὸ τοῦ σημείου τοῦ κατὰ τὸν δράκοντα ἔδοξεν αὐτόθι καταμείναντας οἰκῆσαι. καὶ ἔνθα ὁ δράκων κατέδυ, βωμοί τέ εἰσιν Ἀσκληπιοῦ καὶ ἐλαῖαι περὶ αὐτοὺς πεφύκασιν.

The country of the Boeatae is adjoined by Epidaurus Limera, distant some two hundred stades from Epidelium. The people say that they are not descended from the Lacedaemonians but from the Epidaurians of the Argolid, and that they touched at this point in Laconia when sailing on public business to Asclepius in Cos [1] and that, warned by dreams that appeared to them, they remained and settled here. They also say that the snake, which they were bringing from their home in Epidaurus, escaped from the ship, and disappeared into the ground not far from the sea, and that as a result of the portent of the snake [2] together with the vision in their dreams they resolved to remain and settle here. And where the snake disappeared, there are altars to Asclepius with olive trees [3] growing round them. (*)

[1] Cf. T. 794 ff.   [2] Cf. T. 846.   [3] Cf. T. 749.

758. Pausanias, Descriptio Graeciae, III, 22, 13.

Καὶ Ἀπόλλωνος ναὸς ἐν τῇ Βοιατῶν ἀγορᾷ ἐστι καὶ ἑτέρωθι Ἀσκληπιοῦ καὶ Σαράπιδός τε καὶ Ἴσιδος . . . ἰόντι δὲ ἐς αὐτὰ ἄγαλμα Ἑρμοῦ λίθινον . . . καὶ ἐν τοῖς ἐρειπίοις ἱερὸν Ἀσκληπιοῦ καὶ Ὑγείας ἐστὶν οὐκ ἀφανές.

In the market place of Boeae is a temple of Apollo, and in another part of the town are temples of Asclepius, of Sarapis, and of Isis . . . On the way to them [sc., the ruins of Etis]

there stands a stone image of Hermes . . . and among the ruins
is a not insignificant sanctuary of Asclepius and Health.

759. Pausanias, Descriptio Graeciae, III, 22, 9.

. . . καὶ ἀνωτέρω τῆς πόλεως ὅσον τε σταδίους δώδεκα ἱερόν
ἐστιν Ἀσκληπιοῦ· Φιλόλαον τὸν θεὸν ὀνομάζουσι.

. . . and about twelve stades inland from the city [sc., of
Asopus] is a sanctuary of Asclepius. They call the god
Philolaus (loving the people).[1]

[1] Cf. T. 626, n. 2.

760. Pausanias, Descriptio Graeciae, III, 22, 10.

Ἔστι δὲ ἐν τῇ γῇ ταύτῃ καὶ ἱερὸν Ἀσκληπιοῦ στάδια ἀπέχον ὡς
πεντήκοντα Ἀσωποῦ. τὸ δὲ χωρίον, ἔνθα τὸ Ἀσκληπιεῖον, Ὑπερ-
τελέατον ὀνομάζουσιν.

There is also in this district a sanctuary of Asclepius, about
fifty stades from Asopus; the place where the sanctuary is they
name Hypertcleatum.

761. Pausanias, Descriptio Graeciae, III, 21, 8.

Ἑτέρωθι δὲ Ἀπόλλων Κάρνειος καὶ ἱερὸν Ἄμμωνος καὶ Ἀσκληπιοῦ χαλκοῦν ἄγαλμά ἐστιν, οὐκ ἐπόντος ὀρόφου τῷ ναῷ. καὶ
πηγὴ τοῦ θεοῦ . . .

In another part of the town [sc., of Gythium] is Carnean
Apollo,[1] and a sanctuary of Ammon, and a bronze image[2] of
Asclepius—the temple is roofless—and a spring belonging to
the god[3] . . . (*)

[1] Cf. T. 747.   [2] Cf. T. 676.   [3] Cf. T. 707.

762. Pausanias, Descriptio Graeciae, III, 24, 8.

Τῶν δὲ ὀρῶν ἐπὶ μὲν τοῦ Ἰλίου Διονύσου τέ ἐστι καὶ ἐπ' ἄκρας
τῆς κορυφῆς Ἀσκληπιοῦ ναός, . . .

On Mount Ilius [sc., near Las] is a temple of Dionysus and
on the very summit[1] a sanctuary of Asclepius . . .

[1] Cf. T. 708.

763. Pausanias, Descriptio Graeciae, III, 19, 7.

Πρὶν δὲ ἢ διαβῆναι τὸν Εὐρώταν . . . ἱερὸν δείκνυται Διὸς Πλουσίου. διαβᾶσι δὲ Κοτυλέως ἐστὶν Ἀσκληπιοῦ ναός, ὃν ἐποίησεν Ἡρακλῆς· καὶ Ἀσκληπιὸν Κοτυλέα ὠνόμασεν ἀκεσθεὶς τὸ τραῦμα τὸ ἐς τὴν κοτύλην οἱ γενόμενον ἐν τῇ πρὸς Ἱπποκόωντα καὶ τοὺς παῖδας προτέρᾳ μάχῃ.

Before the Eurotas is crossed [sc., between Amyclae and Therapne], . . . is shown a sanctuary of Zeus Wealthy. Across the river is a temple of Asclepius Cotyleus (of the Hip-joint) ;[1] it was built by Heracles, and he named Asclepius Cotyleus because he was cured of the wound in the hip-joint that he received in the preceding battle with Hippocoon and his sons.[2] (*)

[1] Cf. T. 626, n. 2.   [2] Cf. T. 87.

764. Pausanias, Descriptio Graeciae, III, 14, 7.

Τὸ δὲ τοῦ Ἀγνίτα πεποίηται μὲν ἐν δεξιᾷ τοῦ Δρόμου, Ἀσκληπιοῦ δέ ἐστιν ἐπίκλησις ὁ Ἀγνίτας, . . .

The sanctuary of Agnitas has been made on the right of the Course [sc., in Sparta] ; Agnitas is a surname of Asclepius . . .[1]

[1] Cf. T. 626; cf. 764a. Inscriptiones Graecae, V, 1, no. 602 [3rd c. A. D.]:
. . . καὶ Ἀσκληπιοῦ Σχοινάτα ἐ[ν] | τῷ ἕλει . . . [sc., ἐν τῇ Σπάρτῃ].

765. Pausanias, Descriptio Graeciae, III, 14, 2.

Ἀσκληπιοῦ δὲ οὐ πόρρω τῆς λέσχης ἐστὶν ἱερόν, ἐν Ἀγιαδῶν καλούμενον.

Not far from the meeting hall [sc., in Sparta] is a sanctuary of Asclepius, called " in the place of the Agiadae." (*)

766. Pausanias, Descriptio Graeciae, III, 21, 2.

Θέας δὲ ἄξια αὐτόθι ἰδὼν Ἀσκληπιοῦ τε οἶδα ἱερὸν καὶ τὴν πηγὴν Πελλανίδα.

Remarkable sights I remember seeing here [sc., at Pellana] were the sanctuary of Asclepius and the spring Pellanis.

MESSENIA

767. Strabo, Geographica, VIII, 4, 4.[1]

Δείκνυται δ᾽ ἐν τῇ Γερηνίᾳ Τρικκαίου ἱερὸν Ἀσκληπιοῦ . . .

In Gerenia is to be seen a temple of Triccaean Asclepius . . .

[1] Cf. T. 715.

768. Pausanias, Descriptio Graeciae, IV, 30, 1.

Ἡράκλειον δὲ ἦν αὐτόθι ἐπιφανὲς καὶ Ἀσκληπιεῖον.

There was a conspicuous Heracleion here [*sc.*, in Abia] and an Asclepieion. (*)

769. Pausanias, Descriptio Graeciae, IV, 34, 6.

Θεῶν δέ ἐστιν ἐνταῦθα . . . καὶ Ἀσκληπιοῦ ναός.

The gods who have temples here [*sc.*, in Corone] are . . . and Asclepius.

770. Pausanias, Descriptio Graeciae, IV, 31, 10.

Πλεῖστα δέ σφισι καὶ θέας μάλιστα ἀγάλματα ἄξια τοῦ Ἀσκληπιοῦ παρέχεται τὸ ἱερόν.

The most numerous statues [1] and the most worth seeing are to be found in the sanctuary of Asclepius [*sc.*, in Messene].

[1] Cf. T. 636.

771. Pausanias, Descriptio Graeciae, IV, 36, 7.[1]

Ἐν δὲ Αὐλῶνι καλουμένῳ ναὸς Ἀσκληπιοῦ καὶ ἄγαλμά ἐστιν Αὐλωνίου.

In the so-called Aulon there is a temple and statue of Asclepius Aulonius[2] [*sc.*, near Cyparissiae]. (*)

[1] Cf. T. 637.   [2] Cf. T. 626, n. 2.

ARCADIA

772. Pausanias, Descriptio Graeciae, VIII, 54, 5.

Ἡ δὲ ἐς Ἄργος ἐκ Τεγέας ὀχήματι ἐπιτηδειοτάτη καὶ τὰ μάλιστά ἐστι λεωφόρος. ἔστι δὲ ἐπὶ τῆς ὁδοῦ πρῶτα μὲν ναὸς καὶ ἄγαλμα Ἀσκληπιοῦ.

The road from Tegea to Argos is very well suited for carriages and largely frequented by people. On the road[1] come first a temple and image[2] of Asclepius. (*)

[1] Cf. T. 776.   [2] Cf. T. 640.

773. Pausanias, Descriptio Graeciae, VIII, 32, 4-5.

> Ἐνταῦθα ἔστι μὲν ἱερὸν Ἀσκληπιοῦ . . . . ἔστι δὲ καὶ ἄλλο
> ὑπὸ τὸν λόφον τοῦτον Ἀσκληπιοῦ Παιδὸς ἱερόν . . .

> Here [sc., at Megalopolis] there is a sanctuary of Ascle-
> pius.¹ . . . Under this hill there is another sanctuary of the
> Child Asclepius.² . . . (*)

¹ Cf. T. 638.   ² Cf. T. 626, n. 2; 599.

774. Pausanias, Descriptio Graeciae, VIII, 28, 1.¹

> . . . Γόρτυς κώμη τὰ ἐπ᾽ ἐμοῦ, τὰ δὲ ἔτι ἀρχαιότερα πόλις.
> ἔστι δὲ αὐτόθι ναὸς Ἀσκληπιοῦ λίθου Πεντελησίου . . .

> . . . Gortys [sc., in Arcadia] which today is a village, but of
> old was a city. Here there is a temple of Asclepius, made of
> Pentelic marble. . . .

¹ Cf. T. 652.

775. Pausanias, Descriptio Graeciae, VIII, 26, 6.

> Ἀλιφηρεῦσι δὲ τὸ μὲν ὄνομα τῇ πόλει γέγονεν ἀπὸ Ἀλιφήρου
> Λυκάονος παιδός, ἱερὰ δὲ Ἀσκληπιοῦ τέ ἐστι καὶ Ἀθηνᾶς . . .

> The city of Aliphera has received its name from Alipherus,
> the son of Lycaon, and there are sanctuaries here of Asclepius
> and Athena . . .

776. Pausanias, Descriptio Graeciae, VIII, 25, 1-2.

> Ἐν δὲ τῇ γῇ τῇ Θελπουσίᾳ ποταμός ἐστιν Ἄρσην καλούμενος·
> τοῦτον οὖν διαβήσῃ καὶ ὅσον πέντε ἀπ᾽ αὐτοῦ σταδίοις ἀφίξῃ
> καὶ εἴκοσι ἐπὶ ἐρείπια Καούντος κώμης καὶ ἱερὸν Ἀσκληπιοῦ
> Καουσίου πεποιημένον ἐν τῇ ὁδῷ. τοῦ δὲ ἱεροῦ τούτου σταδίους
> τεσσαράκοντα μάλιστα ἀφέστηκεν ἡ πόλις.

> In the Thelpusian territory is a river called Arsen. Cross this
> and go on for about twenty-five stades, when you will arrive
> at the ruins of the village Caüs, with a sanctuary of Caüsian
> Asclepius,¹ built on the road.² The town (Thelpusa) is some
> forty stades distant³ from this sanctuary.

¹ Cf. T. 626, n. 2.   ² Cf. T. 772.   ³ Cf. T. 708, n. 3.

777. Pausanias, Descriptio Graeciae, VIII, 25, 3.

Ἔστι δὲ ἐν Θελπούσῃ ναὸς Ἀσκληπιοῦ καὶ θεῶν ἱερὸν τῶν δώδεκα . . .

Thelpusa[1] has a temple of Asclepius and a sanctuary of the twelve gods . . .

[1] Cf. T. 17.

778. Pausanias, Descriptio Graeciae, VIII, 21, 3.

Κλειτορίοις δὲ ἱερὰ τὰ ἐπιφανέστατα Δήμητρος τό τε Ἀσκληπιοῦ, τρίτον δέ ἐστιν Εἰλειθυίας . . .

The most celebrated sanctuaries of the Cleitorians are those of Demeter, Asclepius, and, thirdly, Eileithyia[1] . . .

[1] Cf. T. 783.

ELIS

779. Pausanias, Descriptio Graeciae, VI, 26, 5.

Θεῶν δὲ ἱερὰ ἐν Κυλλήνῃ Ἀσκληπιοῦ . . .

As for sanctuaries of the gods in Cyllene, there is one of Asclepius . . . (*)

780. Pausanias, Descriptio Graeciae, VI, 21, 4.

Τεσσαράκοντα δὲ ἀπὸ τῆς Σαύρου δειράδος προελθόντι στάδια ἔστιν Ἀσκληπιοῦ ναός, ἐπίκλησιν μὲν Δημαινέτου ἀπὸ τοῦ ἱδρυσαμένου, ἐρείπια δὲ καὶ αὐτός· ᾠκοδομήθη δὲ ἐπὶ τοῦ ὑψηλοῦ παρὰ τὸν Ἀλφειόν.

Forty stades beyond the ridge of Saurus is a temple of Asclepius, surnamed Demaenetus[1] after the founder. It too is in ruins. It was built on the height[2] beside the Alpheius.

[1] Cf. T. 626, n. 2.　　　　　　　　[2] Cf. T. 708, n. 2.

781. Pausanias, Descriptio Graeciae, V, 7, 1.

. . . παρὰ δὲ Γόρτυναν ἔνθα ἱερὸν Ἀσκληπιοῦ . . .

. . . past Gortyna [sc., near Olympia], where is a sanctuary of Asclepius . . .

ACHAIA

782. Pausanias, Descriptio Graeciae, VII, 27, 11.

Ἀπωτέρω δὲ οὐ πολὺ ἀπὸ τοῦ Μυσαίου ἱερόν ἐστιν Ἀσκληπιοῦ καλούμενον Κῦρος, καὶ ἰάματα ἀνθρώποις παρὰ τοῦ θεοῦ γίνεται. ὕδωρ δὲ καὶ ἐνταῦθα ἀνέδην ἐστί, καὶ ἐπὶ τῇ μεγίστῃ τῶν πηγῶν τοῦ Ἀσκληπιοῦ τὸ ἄγαλμα ἵδρυται.

At no great distance from the Mysaeum [near Pellene] is a sanctuary of Asclepius, called Cyrus, where cures of patients are effected by the god.[1] Here too there is a copious supply of water, and at the largest of the springs[2] stands the image of Asclepius.

[1] Cf. T. 391.
[2] Cf. T. 707 and Footnote.

783. Pausanias, Descriptio Graeciae, VII, 23, 7.

Τῆς δὲ Εἰλειθυίας οὐ μακρὰν Ἀσκληπιοῦ τέ ἐστι τέμενος . . .

Not far from Eileithyia[1] is a precinct of Asclepius [sc., in Aegium] . . .

[1] Cf. T. 778.

784. Pausanias, Descriptio Graeciae, VII, 21, 14.

Ἔστι δὲ καὶ ἱερὸν Πατρεῦσιν Ἀσκληπιοῦ· τοῦτο τὸ ἱερὸν ὑπὲρ τὴν ἀκρόπολιν τῶν πυλῶν ἐστιν ἐγγὺς αἱ ἐπὶ Μεσάτιν ἄγουσιν.

There is also at Patrae a sanctuary of Asclepius. This sanctuary is beyond the Acropolis near the gate leading to Mesatis.

785. Strabo, Geographica, VIII, 7, 4.

Δείκνυται δ' ἴχνη μεταξὺ Πατρῶν καὶ Δύμης τοῦ παλαιοῦ τῶν Ὠλενίων κτίσματος· αὐτοῦ δὲ καὶ τὸ τοῦ Ἀσκληπιοῦ ἱερὸν ἐπίσημον, ⟨ὃ⟩ Δύμης μὲν ἀπέχει τετταράκοντα σταδίους, Πατρῶν δὲ ὀγδοήκοντα.

Traces of the old settlement of the Olenians are shown between Patrae and Dyme; and there, too, is the notable temple of Asclepius, which is forty stades distant from Dyme and eighty from Patrae.

ISLES

Thasus

786. Inscriptiones Graecae, XII, 8, no. 265 [4th c. B. C.].

Ἐπὶ Λυσιστράτου [τοῦ Α]ἴσχρωνος ἄρχο[ντος· τύχηι ἀγαθῆι·]
ἐπὶ τοῖσδε ἐκδέδοται [ὁ κῆπος ὁ] Ἡρακλέος ὁ πρὸς [τῆι πύληι. ὁ ἀναι-]
ραιρημένος τὸν κῆ[πον τὸ χωρ]ίον καθαρὸν παρέξει [τὸ περὶ τὰς]
πύλας, ὅπου ἡ κόπρος [ἐξεβάλλ]ετο. ἢν δέ τις ἐγβάλλη[ι κόπρον ἐς]
τὸ χωρίον, εἶναι τὸ ἄγγος τοῦ ἀναιρερημένου τὸν κῆπο[ν, τὸν δὲ]
δοῦλον μαστιγώσαντα ἀθώϊον εἶναι. ὅπως δὲ τὸ χωρίον καθ[αρὸν]
παρέχηι, ἐπιμέλεσθαι τὸν ἀγορηνόμον καὶ τὸν ἱερέα τοῦ
Ἀσκληπιοῦ τοὺς ἑκάστοτε ἐόντας· ἢν δὲ μὴ ἐπιμέλωνται,
ὀφείλεν αὐτοὺς τῆς ἡμέρης ἑκάστης ἡμίεκτον ἱρὸν τῶι
Ἀσκληπιῶι· δικάζεσθαι δὲ τοὺς ἀπολόγους ἢ αὐτοὺς ὀφείλεν·
τὸν δὲ ἀναιρερημένον τῶ[ι ἱ]ρεῖ καὶ τῶι ἀγορανόμωι ἕκτην ὀφείλεν
τῆς ἡμέρης.

> In the archonship of Lysistratus, son of Aeschron; with good
> fortune. On these conditions the garden of Heracles near the
> gate has been leased. The lessee of the garden shall keep clean
> the area around the gates where the dung was dumped. More-
> over, if anyone throws dung into the area, the pail belongs to
> the lessee of the garden and he is not liable to prosecution if he
> flogs the slave. The *agoranomos* and the priest of Asclepius
> who are in office shall see that the lessee keeps the area clean.
> If they do not do so, they are liable to pay a sacred twelfth to
> Asclepius daily. The auditors must bring them to trial, or
> themselves be liable. The lessee must pay the priest and the
> *agoranomos* a sixth daily.

Euboea

787. Inscriptiones Graecae, XII, 9, no. 194 [4th or beginning of 3rd c.
B. C.].

－－－－－－－－－－－－－－－－－－－－－－－－

－－ ἐξέστω δὲ καὶ ἄλλω[ι τῶι βουλομένωι Ἐρετ]-
[ριε]ίων πένπειν τὴν θυγ[α]τέρα ἢ [τὸν παῖδα εἰς]
τὴμ πομπήν· ἐξέστω δὲ πένπειν κα[ὶ －－ το]-

ὺς νεωτέρους τῶν ἑ[π]τὰ ἐτῶν· συμπ[ένπειν δὲ καὶ τοὺς ἱ]-
ππεῖς τὴμ πομπὴν ἐν ἐσθῆτι ποικίλ[η]ι, ὅ[πως ἂν ὡς καλλίστη]
ἡ θυσία καὶ ἡ πομπὴ γίνηται τῶι Ἀσκληπιῶ[ι· καθιστᾶν δὲ τὰς]
παῖδας ἢ τοὺς παῖδας μ[ὴ] ἔ[χ]ειν καταγράψα – –
ὅπως δὲ εἰδῶσι[ν οἱ ἀεὶ] γ[ιν]όμενοι ἱεροποιοὶ τὰς [ἀεὶ συμπεν]-
[ψ]ούσας καὶ τ[οὺς ἀεὶ συνπεπομφ]ότας τῶμ παίδω[ν, ἀπογρ]-
[άψ]αι τὰ ὀνό[ματα αὐτῶν ἐν] λευκώματι τοὺς εἰ[ρημένους ἱερ]-
[οποι]οὺς τ[– καὶ στῆσαι ἐ]ν τῶι ἱερῶι τοῦ Ἀσκληπιο[ῦ]
. . . .

To any one of the Eretrians who wishes let permission be
granted for his daughter or for his son to take part in the
procession; and let permission be granted also for children
younger than seven years of age to take part in the procession;
and for the knights,[1] dressed in multi-colored raiment,[2] to help
in conducting the procession so as to make the sacrifice and
procession in honor of Asclepius most magnificent; and for the
girls or boys to be arrayed without enrollment . . . But in order
that the officiating *hieropoioi* may know who of the girls may
come and who of the boys has participated, the said *hieropoioi*
shall inscribe their names on a white tablet and place it in the
temple of Asclepius . . .

[1] Cf. T. 861.   [2] Cf. however T. 296; 513; 806.

DELOS

788. Inscriptiones Graecae, XI, 2, no. 161 A, 72-73 [279 B. C.].

Τοῦ νεὼ τοῦ Ἀσκληπιοῦ τῆς ὀροφῆς τὰ κρίνα λευκώ░░σαντι
κα[ὶ] | τὰς κάλχας χρυσώσαντι Ἀντιδότωι ἀρχιτέκτονος καὶ
ἐπιμελητῶν ἐγδόντων ἐδώκαμεν δραχμὰς· ΔΔΔΔ⊢⊢⊢⊢.

To Antidotus for whitening the architectural ornaments of the
roof of the temple of Asclepius and for gilding the rosettes on
the capitals of the columns in accordance with the contract
made by the architect and the *epimeletai*, we gave 40 drachmae.

PAROS

789. Inscriptiones Graecae, XII, 5, no. 119 [4th c. B. C.].

– – – – – –

[– – Ἀσκλ]ηπιῶ[ι]

－－－－τὰς θ[υ]-
[σίας －－ ἡ π]όλις
[－－－－－]ν
－－－－－

. . . to Asclepius . . . the sacrifices . . . the city . . .

ANAPHE

790. Inscriptiones Graecae, XII, 3, no. 248 [end of 2nd c. B. C.].

. . . .
Ἐπερωτᾶι Τιμόθεος ·[τὸ]ν θεὸν πότερον
αὐτῶι λῶιον καὶ ἄμει[νό]ν ἐστιν αἰτήσασθαι
τὰν πόλιν ἐν ⟨τ⟩ῶι ἐπινοεῖ τόπωι, ἐν τῶι τοῦ
Ἀπόλλωνος τοῦ Ἀσγελάτα, ὥστε ναὸν τᾶς
Ἀφροδίτας οἰκοδ[ο]μῆσαι καὶ ἦμεν δαμόσιον,
ἢ ἐν τῶι ἱερῶι τοῦ Ἀ[σκ]λαπιοῦ ἐν ὧι ἐπινοεῖ
τόπωι. Ὁ θεὸς ἔχρησε αἰτήσασθ[α]ι ἐν τῶι το[ῦ]
Ἀπόλλωνος· . . . .

Timotheus asks the god whether it would be more desirable for
him and better to ask permission of the city to build a temple
of Aphrodite and to make it public property in the place which
he has in mind in the sanctuary of Apollo Asgelatas, or in the
sanctuary of Asclepius in the place which he has in mind. The
god gave the response that he should apply for that in the
sanctuary of Apollo . . .

CRETE

791. Inscriptiones Creticae, I, XVII, no. 21 [1st c. A. D.].

Πράτωι μὲν [Σώσωι] γενέται Ἀσκληπί' ἔδειξας
　　ὕδατος εὐσήμ[ω]ς εἰς ναὸν ἀτραπιτόν,
φ]ανθεὶς μὲν καθ' ὕπνον, πένψας δ' ὕπαρ αὐτὸς ὁδαγὸ[ν
　　θεῖον ὄφιν, πᾶσιν θαῦμα βροτοῖσι μέγα,
τῶι Ἀριστωνύμω [[ι]] υἱῶι, ἐπεὶ κατὰ πάντα θεουδὴς
　　νακόρος εἰς ναὸν σαῖς μόλ' ἐφημοσύναις.
νῦν δὲ Σοάρχωι αὖθι φανεὶς κατὰ πᾶν κλυτῶι υἱῶι
　　ὡσαύτως ὅσιον νακόρον ἀγάγαο,

τεσσαρακοστῶι ἔτει τε καὶ ἑβδομάτωι ἵνα κράνας
λειπούσας πλήσηι νάματι τὰς πατέρος.
Παιάν, σοὶ δ' εἴη κεχαρισμένα, καὶ δόμον αὔξοις
τοῦδε καὶ ὑψίσταν πατρίδα Γόρτυν ἀεί.

First to Sosus, my father, the son of Aristonymus, O Asclepius,
you showed by favorable omens the water's path towards the
temple, appearing to him in his sleep, and yourself sending him
—while he was awake—as a guide the holy serpent, a great
marvel to all men, when as a most god-fearing sacristan he
went to the temple on your command. But now again you have
appeared to Soarchus, his widely known son, and guided in the
same way the devout sacristan in order that in the forty-seventh
year he might fill with running water the failing springs of his
father. Paean, may it be gratifying to you, and may you increase his house and his sublime home town, Gortys, for ever.

792. Philostratus, Vita Apollonii, IV, 34.

. . . ἐπορεύθη καὶ ἐς τὸ ἱερὸν τὸ Λεβηναῖον. ἔστι δὲ Ἀσκληπιοῦ,
καὶ ὥσπερ ἡ Ἀσία ἐς τὸ Πέργαμον, οὕτως ἐς τὸ ἱερὸν τοῦτο
ξυνεφοίτα ἡ Κρήτη, πολλοὶ δὲ καὶ Λιβύων ἐς αὐτὸ περαιοῦνται.
καὶ γὰρ τέτραπται πρὸς τὸ Λιβυκὸν πέλαγος κατὰ γοῦν τὴν
Φαιστόν, ἔνθα τὴν πολλὴν ἀνείργει θάλατταν ὁ μικρὸς λίθος.
Λεβηναῖον δὲ τὸ ἱερὸν ὠνομάσθαι φασίν, ἐπειδὴ ἀκρωτήριον ἐξ
αὐτοῦ κατατείνει λέοντι εἰκασμένον, οἷα πολλὰ αἱ ξυντυχίαι τῶν
πετρῶν ἀποφαίνουσι, μῦθόν τε ἐπὶ τῷ ἀκρωτηρίῳ ᾄδουσιν, ὡς
λέων εἷς οὗτος γένοιτο τῶν ὑποζυγίων ποτὲ τῇ Ῥέᾳ.

. . . he [sc., Apollonius] passed on to the shrine of Leben.[1]
This is a shrine of Asclepius, and just as the whole of Asia
flocks to Pergamum, so the whole of Crete flocked to this
shrine; and many Libyans also cross the sea to visit it. For it
faces towards the Libyan sea close to Phaestus, where the little
rock keeps out the mighty sea. And they say that this shrine
is named that of Leben, because a promontory juts out from it
which resembles a lion, as often rocks by chance suggest such
forms; and they tell a legend about this promontory, how it
was once one of the lions yoked in the chariot of Rhea. (*)

---

[1] Cf. T. 709.

RHODES

793. Diodorus, Bibliotheca Historica, XIX, 45, 4.

Τοῦ δ' ὕδατος παραδόξως ἀθροιζομένου πᾶς μὲν ὁ περὶ τὸ δεῖγμα
καὶ Διονύσιον τόπος ἐπεπλήρωτο, πρὸς δὲ τὸ Ἀσκληπιεῖον ἤδη
τοῦ λιμνάζοντος τόπου προσιόντος ἐκπλαγεῖς μὲν ἦσαν ἅπαντες,
πρὸς δὲ τὴν σωτηρίαν διαφόροις ἐχρῶντο κρίσεσιν.

When [in 316 B. C.] the water unexpectedly accumulated, the
whole region around the bazaar and the Dionysium was filled
[sc., in Rhodes], but when the flooded area reached even the
Asclepieion,[1] then everybody was frightened, but for their
rescue they used different devices.

[1] Evidently the Asclepieion was situated on a higher level, cf. T. 708, n. 2.

Cos

794. Strabo, Geographica, XIV, 2, 19.

Ἐν δὲ τῷ προαστείῳ τὸ Ἀσκληπιεῖόν ἐστι, σφόδρα ἔνδοξον καὶ
πολλῶν ἀναθημάτων μεστὸν [ἱερόν], ἐν οἷς ἐστι καὶ ὁ Ἀπελλοῦ
Ἀντίγονος. ἦν δὲ καὶ ἡ ἀναδυομένη Ἀφροδίτη . . .

In the suburb[1] [sc., of Cos] is the Asclepieion, a temple exceedingly famous and full of numerous votive offerings,[2]
among which is the Antigonus of Apelles. And also Aphrodite
Anadyomene used to be there. . . .

[1] Cf. T. 708, n. 3.
[2] Cf. T. 731; 841. For official documents set up in the temple, cf. **794a.** *Inscriptiones
Coae*, no. 14 [Paton-Hicks]: . . . ὅπως ἀναγραφῇ ἐς στάλαν καὶ τεθῇ[ι] ἐς τὸ ἱερὸν τοῦ
Ἀσκλαπιοῦ, ὅπως εἰδῶ[ν]τι Κῷο[ι] . . .

795. Plinius, Naturalis Historia, XXIX, 1 (2), 4.

Tunc eam revocavit in lucem Hippocrates, genitus in insula
Coo in primis clara ac valida et Aesculapio dicata. is, cum
fuisset mos, liberatos morbis scribere in templo eius dei quid
auxiliatum esset, ut postea similitudo proficeret, exscripsisse
ea traditur atque, ut Varro apud nos credit, templo cremato iis
instituisse medicinam hanc, quae clinice vocatur.

At that time [i. e. during the Peloponnesian war] Hippocrates
called it [sc., medicine] back to light;[1] he was born on the

island of Cos which was particularly famous and powerful and
dedicated to Asclepius. He, as it had been the custom for those
freed from disease to put down in the temple of that god what
had helped them, so that thereafter a similar case might have
the benefit of it,—he is said to have copied these (inscriptions),
and, as with us Varro believes, after having burnt the temple
he [*sc.*, Hippocrates] made use of them in instituting that
medicine which is called " bedside " medicine.[2] (**)

[1] Cf. T. 357.
[2] Cf. however T. 360.

796. Plinius, Naturalis Historia, XX, 24 (100), 264.

Et discessuri ab hortensiis unam conpositionem ex his clarissimam subteximus adversus venenata animalia incisam in lapide
versibus Coi in aede Aesculapi: serpylli duum denariorum
pondus, opopanacis et mei tantundem singulorum, trifolii
seminis pondus denarii, anesi et feniculi seminis et ammi et
apii denarium senum e singulis generibus, ervi farinae denarium
XII. haec tusa cribrataque vino quam possit excellenti digerun-
tur in pastillos victoriati ponderum. ex his singuli dantur ex
vini mixti cyathis ternis. hac theriace Magnus Antiochus rex
adversus omnia venenata usus traditur aspide excepta.

But as we are now about to leave the garden plants, we append
one most famous preparation extracted from them, an antidote
against venomous animals which is inscribed in verse upon a
stone in the temple of Asclepius at Cos : " Take two denarii of
wild thyme, and the same quantity of opopanax and meum
respectively; one denarius of trefoil seed; and of aniseed, fennel-
seed, Ammi, and parsley, six denarii respectively, with twelve
denarii of meal of fitches. After these ingredients have been
beaten up together, and passed through a sieve with the best
wine that can be had, they are made into lozenges of one vic-
toriatus each : these are given to the patient one at a time,
steeped in three cyathi of wine." King Antiochus the Great,
it is said, employed this antidote against all kinds of venomous
animals, the asp excepted.[1]

[1] For the same story cf. **796a.** Plinius Secundus, *De Medicina*, XXXVII; **796b.**
Gargilius Martialis, *Medicina*, XXXIX.

797. Valerius Maximus, *Facta et Dicta Memorabilia*, I, 1, 19.

Nec minus efficax ultor contemptaé religionis filius quoque
eius, Aesculapius, qui consecratum templo suo lucum a Turullio,
praefecto Antonii, ad naves ei faciendas magna ex parte
succisum ulturus,* . . . imperio Caesaris morti destinatum
Turullium manifestis numinis sui viribus in eum lucum, quem
violaverat, traxit, effecitque ut ibi potissimum a militibus Cae-
sarianis occisus, eodem exitio et eversis iam arboribus poenas
lueret et adhuc superantibus immunitatem consimilis iniuriae
pareret, suamque venerationem, quam apud colentes maximam
semper habuerat, sic ** multiplicavit.

* ulturus addidit Halm   dolens dett. lacunam supplentes
** sic Halm   dis B   deus dett.

A no less effective avenger of sacrilege was his [*i.e.*, Apollo's]
son, Asclepius. For after the grove which was consecrated to
his temple had to a great extent been cut down by Turullius,
the prefect of Antony, in order to build ships [after 42 B. C.],
Asclepius intending to take vengeance . . . drew Turullius,
condemned to death by Caesar, into the very grove which he
had violated, so revealing the power of his deity. And by
Turullius' execution at the hands of Caesar's soldiers [in 30
B. C.] on that very spot Asclepius caused him with one death
to pay the penalty for the trees already overthrown and to
provide immunity from similar injury for those which were
still extant. Thus Asclepius increased the great veneration
which he had always enjoyed with his worshippers.[1]

[1] Cf. **797a.** Cassius Dio, *Historia Romana*, LI, 8, 3: Καῖσαρ δὲ τὸν μὲν Τουρούλλιον
ἀπέκτεινε (καὶ ἔτυχε γὰρ ἐκ τῆς ἐν Κῷ τοῦ 'Ασκληπιοῦ ὕλης ξύλα ἐς ναυτικὸν κεκοφώς, δίκην
τινὰ καὶ τῷ θεῷ ὅτι ἐκεῖ ἐδικαιώθη, δοῦναι ἔδοξε) . . . ; **797b.** Lactantius, *Divinae
Institutiones*, II, 7, 17: Praefectus etiam Marci Antonii Turullius cum aput Coos
everso Aesculapii luco classem fecisset, eodem postea loco a militibus Caesaris
interfectus est.

798. Tacitus, *Annales*, IV, 14, 1-2.

Is quoque annus legationes Graecarum civitatium habuit, Samiis
Iunonis, Cois Aesculapii delubro vetustum asyli ius ut firma-
retur petentibus . . . neque dispar apud Coos antiquitas, et
accedebat meritum ex loco: nam cives Romanos templo Aesculapii induxerant, cum iussu regis Mithridatis apud cunctas
Asiae insulas et urbes trucidarentur.

This year [*i. e.*, 23 A. D.] also brought delegations from two Greek communities, the Samians and Coans, desiring the confirmation of an old right of asylum[1] to the temples of Juno and Asclepius respectively . . . The Coans had equal antiquity on their side, and, in addition, a claim associated with the place itself: for they had sheltered Roman citizens in the temple of Asclepius at a time [in 88 B. C.] when, by order of King Mithridates, they were being butchered in every island and town of Asia. (*)

[1] Cf. T. 811.

799.   Tacitus, Annales, XII, 61, 1-2.

Rettulit dein de inmunitate Cois tribuenda, multaque super antiquitate eorum memoravit: Argivos vel Coeum Latonae parentem vetustissimos insulae cultores; mox adventu Aesculapii artem medendi inlatam maximeque inter posteros eius celebrem fuisse, nomina singulorum referens et quibus quisque aetatibus viguissent. quin etiam dixit Xenophontem cuius scientia ipse uteretur eadem familia ortum, precibusque eius dandum, ut omni tributo vacui in posterum Coi sacram et tantum dei ministram insulam colerent.

He [*sc.*, Claudius, in 53 A. D.] next proposed to grant immunity to the inhabitants of Cos. Of their ancient history he had much to tell: "The earliest occupants of the island had," he said, "been Argives—or, possibly, Coeus, the father of Latona. Shortly the arrival of Asclepius had introduced the art of healing, which attained the highest celebrity among his descendants"—here he gave the names of the descendants and the epochs at which they had all flourished. "Xenophon," he observed again, "to whose knowledge he himself had recourse, derived his origin from the same family; and, as a concession to his prayers, the Coans ought to be exempted from all forms of tribute for the future and allowed to tenant their island as a sanctified place subservient only to its god."

ASIA MINOR

BITHYNIA

800. Pausanias, Descriptio Graeciae, III, 3, 8.

Βεβαιοῖ δὲ καὶ ἄλλως μοι τὸν λόγον ἐν Φασήλιδι ἀνακείμενον ἐν
Ἀθηνᾶς ἱερῷ τὸ δόρυ Ἀχιλλέως καὶ Νικομηδεῦσιν ἐν Ἀσκληπιοῦ
ναῷ μάχαιρα ἡ Μέμνονος.

My statement is likewise confirmed by the spear of Achilles
dedicated in the sanctuary of Athena at Phaselis, and by the
sword of Memnon[1] in the Nicomedian temple of Asclepius.

[1] Cf. T. 548.

MYSIA

801. Pausanias, Descriptio Graeciae, II, 26, 8.[1]

. . . Ἀρχίας ὁ Ἀρισταίχμου . . . τὸν θεὸν ἐπηγάγετο ἐς Πέργαμον.

. . . Archias, son of Aristaechmus . . . brought the god to
Pergamum.

[1] Cf. T. 709.

802. Polybius, Historiae, XXXII, 15, 1.[1]

Ὅτι Προυσίας μετὰ τὸ νικῆσαι τὸν Ἄτταλον καὶ τὸ παρελθεῖν
πρὸς τὸ Πέργαμον, παρασκευασάμενος θυσίαν πολυτελῆ προσ-
ήγαγε πρὸς τὸ τέμενος τ’ Ἀσκληπιοῦ . . .

Prusias on reaching Pergamum after his victory over Attalus
[in 155-4 B. C.] prepared a magnificent sacrifice which he
brought to the temple of Asclepius . . .[2]

[1] Cf. T. 546; 658.
[2] Cf. 802a. Philostratus, *Vita Apollonii*, IV, 11: Βαδίσας οὖν [*sc.*, ὁ Ἀπολλώνιος] ἐς
τὸ Πέργαμον καὶ ἡσθεὶς τῷ τοῦ Ἀσκληπιοῦ ἱερῷ . . .

803. Galenus, De Anatomicis Administrationibus, I, 2 [II, pp. 224-
25 K.].

Ἐγὼ δὲ ἐν τῇ πατρίδι κατ’ ἐκεῖνον ἔτι διέτριβον τὸν χρόνον, ὑπὸ
Σατύρῳ παιδευόμενος, ἔτος ἤδη τέταρτον ἐπιδημοῦντι τῇ Περγάμῳ

μετὰ Κοστουνίου Ῥουφίνου, κατασκευάζοντος ἡμῖν τὸν νεὼν τοῦ
Διὸς Ἀσκληπιοῦ.

I [sc., Galen] was still living in my native land at that time,
receiving instruction from Satyrus who was then spending his
fourth year in Pergamum with Costunius Rufinus who was
building for us the temple of Zeus Asclepius.

804. Aristides, Oratio XXXIX, 1-18.

ΕΙΣ ΤΟ ΦΡΕΑΡ ΤΟ ΕΝ ΑΣΚΛΗΠΙΟΥ

Τί δ' ἂν εἴποις τὸ φρέαρ τὸ ἱερόν; ἢ δῆλον ὅτι τὴν τοῦ λόγου
μέμφῃ φύσιν ὡς οὐχ ἱκανῶς ἐπὶ πάντα ἀφικνουμένην οὐδ' ἐνδείξασθαι δυναμένην ἐνίοτε ἀρκούντως τὸ ὄν; τοῦτο γοῦν οὐδ' ἂν εἷς
λόγος ὁποῖόν τι τὸ κάλλος καὶ τὴν ἡδονήν ἐστιν ἐνδείξαιτο, ἀλλ'
ἀμείνους ἐσμὲν πίνειν τούτου τοῦ ὕδατος καὶ λούεσθαι καὶ προσ-
ορᾶν ἡδέως ἢ ἔχειν εἰπεῖν τι περὶ αὐτοῦ, ὥσπερ οἱ τῶν καλῶν
ἐρῶντες, οἳ τῇ μὲν δυνάμει τοῦ κάλλους ἑαλώκασιν καὶ ἴσασιν οἵων
ἐρῶσιν, ἐὰν δέ τις αὐτοὺς ἔρηται, οὐκ ἂν ἔχοιεν εἰπεῖν περὶ ἀπόν-
2 των, παρόντας δὲ οἶμαι δείξαιεν ἄν. τὸ δ' αὐτὸ καὶ ἐπὶ τούτου τοῦ
φρέατος πεπόνθαμεν, καὶ ἐσμὲν ἐρασταὶ μὲν αὐτοῦ πάμπολλοι,
μᾶλλον δὲ πάντες σχεδόν, ὁποίου δὲ ὄντος οὐκ ἔχομεν εἰπεῖν. ἀλλ'
ἐάν τινα ἡμῶν ἀπολαβών τις ἐρωτᾷ, παραλαβόντες ἂν αὐτὸν ἄγειν
ἀξιοίημεν ἐπ' αὐτὸ καὶ δεικνύοιμεν. τῷ δὲ οὐδὲ τοῦτο ἀποχρήσει,
γευσάμενος δὲ καὶ πειραθεὶς τοῦ παρ' Ὁμήρῳ λωτοῦ γεγεῦσθαι
δόξει, μένειν ἐθέλων καὶ χαλεπῶς ἀξιῶν ἀποχωρεῖν ἀπ' αὐτοῦ.
3 ἀλλ' οὐ μὲν δὴ χρὴ τὸ τῶν διψώντων, φασί, πίνειν σιωπῇ, ἀλλ'
ἐπικοσμῆσαί τι καὶ λόγῳ, καὶ προσειπεῖν τόν τε σωτῆρα θεόν, οὗ
καὶ τόδ' ἐστὶν ἔργον τε καὶ ποίημα, καὶ τὰς ἐχούσας αὐτὸ Νύμφας
καὶ συνεργαζομένας καὶ ἡμῖν χρῆσθαι τῇ χάριτι τοῦ θεοῦ χαρι-
ζομένας τε καὶ συνυπηρετούσας.
4 Τίς οὖν δὴ γένοιτ' ἂν ἀρχή, ἢ ὥσπερ ἡνίκ' ἂν ἀπ' αὐτοῦ πίνωμεν,
προσθέντες τοῖς χείλεσι τὴν κύλικα οὐκέτι ἀφίσταμεν, ἀλλ'
ἀθρόον εἰσεχεάμεθα, οὕτως καὶ ὁ λόγος ἀθρόα πάνθ' ἕξει λεγό-
μενα; ἔστω δὲ ἀντὶ τῆς τῷ χείλει προσαγωγῆς ἐκεῖνο ἡμῖν, ὅτι
ἐν τῷ καλλίστῳ τῆς πάσης οἰκουμένης ἐστίν. ὃ γὰρ ἐξ ἁπάντων
χωρίων εἵλετο ὁ θεὸς ὡς ὑγιεινότατον καὶ καθαρώτατον καὶ ὃ ταῖς
εὐεργεσίαις ταῖς παρ' αὐτοῦ πεποίηκεν ἁπάντων ἐκφανέστατον,
5 ἦ που σφόδρα τοῦτο κάλλιστόν ἐστι τῶν ἐν γῇ πάντων. καὶ γὰρ
οὐχ ὥσπερ ἄλλοι ἄλλοις τόποις θεοὶ ἐξ ἀρχῆς ἐγγενεῖς εἰσιν,
καίτοι καὶ τούτους τιμίους, οἶμαι, χρὴ δοκεῖν κατ' αὐτὸ τὸ λαχεῖν

αὐτοὺς τοὺς θεούς, ἀλλὰ τούτῳ γε μεῖζόν ἐστιν ὅτι ἀπ᾽ αὐτῆς τῆς
Ἐπιδαύρου δεῦρο ὁρμηθεὶς ὁ θεὸς ἠράσθη δὴ μάλιστα, ὡς
δῆλόν ἐστιν ὅτι ἑλόμενός τε αὐτὸ ἐγκατέμεινεν τὸ λοιπὸν καὶ
προκρίνας τῶν ἄλλων. ὁ δὲ θεὸς καὶ θεῶν ὁ πραότατός τε καὶ
φιλανθρωπότατος προέχειν ἔκρινεν, πῶς ἡμῖν γε, καὶ ταῦτα τοῖς
τούτου θεράπουσιν, ἄλλο τι λέγειν ἔνεστιν ἢ ὡς τοῦτ᾽ ἔστι τὸ
6 βέλτιστον; ἐν καλλίστῳ μὲν δὴ τῆς οἰκουμένης οὕτως ἔστιν εἰπεῖν.
ἔτι δ᾽ αὐτοῦ τοῦ ἱεροῦ ὅσος ὑπαίθριος χῶρος καὶ βάσιμος ἐν τῷ
καλλίστῳ ἐστίν· μέσον γὰρ ἐν μέσῳ ἵδρυται. τὸ δὲ ὕδωρ εἰ μὲν
βούλει, ἀπὸ πλατάνου ῥεῖ—ὥσπερ γὰρ ἄλλο τι σύμβολον καὶ τοῦτο
παραπέφυκεν—, εἰ δὲ βούλει, τὸ ἔτι κάλλιόν τε καὶ ἱερώτερον,
ἀπ᾽ αὐτῶν τῶν βάθρων ἐκρεῖ, ἐφ᾽ ὧν ὁ νεὼς ἔστηκεν, ὥστε παντί
γε ταύτην τὴν δόξαν καὶ πίστιν ὑπεῖναι, ὅτι ἀπὸ ὑγιεινοῦ καὶ
ὑγιείας χορηγοῦ χωρίου φέρεται, ἀπό γε τοῦ ἱεροῦ καὶ τῶν ποδῶν
τοῦ Σωτῆρος ὁρμώμενον· οὐ γὰρ ἄν τι ἐξ ὑγιεινοτέρων ἢ καθαρω-
7 τέρων τόπων ὕδωρ ῥυείη ἢ τοῦτο ἐκ τούτων ῥέον. ἐν τοιούτῳ
δὴ φαινόμενον καὶ ἀπὸ τοιούτων ὁρμώμενον, ὡς τὸ εἰκὸς ἔχει,
κάλλιστόν ἐστι. πρῶτον μέν γε λεπτότατον ἐγγυτάτω ἀέρος,
ἔπειτα ὃ τούτῳ ἕπεται, κουφότατόν τε καὶ πραότατον, τρίτον γλυκύ-
τατόν τε καὶ ποτιμώτατον, [αὐτόχυτον], ὃ πίνων οὐκ ἂν οἴνου
προσδεηθείης. Ὅμηρος μὲν γὰρ ἔφη τὸν Τιταρήσιον ἐπιρρεῖν ἐπὶ
τοῦ Πηνειοῦ, ὥσπερ ἄνδρα ἐπινηχόμενον, ὑπὸ κουφότητος τοῦ
ὕδατος· τὸ δ᾽ ἐμοὶ δοκεῖν εἰ ἐπιρρήξαις αὐτῷ ὕδωρ ἕτερον, ἀντάν-
εισιν εἰς τὸ ἄνω, τὸ δὲ δύεται, ὥσπερ ὕφαλοι νέοντες** εἰς μυχὸν
ἐκ τοῦ μετεώρου· εἰ μὴ καὶ τοῦτ᾽ ἔστιν εἰπεῖν ὅτι μοι δοκεῖ κἂν τὸ
ἐπεγχυθὲν ὑψῶσαι τῇ ἑαυτοῦ κουφότητι. ὡς δὲ οὐ κομπάζομεν,
σταθμῷ κρίνεται· καίτοι τί ἂν εἴποι ὁ τῆς Στυγὸς ἀπορρώξ, ὅταν
8 ἀνθιστάμενος ῥέπῃ; πρόσεστι δὲ τούτῳ ὅτι οὔτε Στυγός ἐστιν
ἀπορρὼξ τόδε τὸ ῥεῦμα οὔτε ἄλλο ἔχει φρικῶδες οὐδέν, ἀλλ᾽
ὑγιείας ἂν αὐτὸ προσείποις ἢ νέκταρος ἤ τινος τῶν τοιούτων
9 ἀπορρῶγα. τεκμηριοῖ δὲ καὶ τούτῳ, χρόνος γοῦν αὐτοῦ οὐχ
ἅπτεται, ἀλλ᾽ ἀπαντληθὲν τὸ ὕδωρ καὶ ἔξω γενόμενον τὸ αὐτὸ
ποιεῖ οἷόνπερ τὸ αἰεὶ λειπόμενον ἐν τῷ φρέατι, ἄσηπτον καὶ ἀπαθὲς
10 μένει. πλῆθος δ᾽ αὖ τοῦ φρέατος τούτου τοσοῦτον ὅσον, ὡς ἔπος
εἰπεῖν, οὐδενὸς ἑτέρου φρέατος· μάλα ἀκμῆτας εἶναι δεῖ τοὺς
ἀρυτομένους καὶ νοῦν ἔχοντας, ἵνα μὴ φθάνῃ τὸ ἐγκαταλαμβάνον.
μόνον γοῦν τοῦτο πάντων φρεάτων ἀρυτομένων [καὶ ἑώντων]*** τὸ
ἴσον αἰεὶ μέτρον παρέχεται, τῷ τετρημένῳ πίθῳ τὸ ἀντίστροφον
ποιοῦν· ὁ μὲν γὰρ οὐδέποτε πληροῦται, τὸ δὲ αἰεὶ τοῦ χείλους
11 ἐγγύς ἐστιν. ἅτε γὰρ ὂν διάκονόν τε καὶ συνεργὸν τοῦ φιλανθρω-

ποτάτου τῶν θεῶν ἑτοιμότατον πρὸς τὴν ὑπηρεσίαν καὶ αἰεὶ
πλῆρές ἐστι, καὶ οὔτε ἐκεῖνος ἄγει σχολὴν ἄλλο τι πράττειν ἢ
σῴζειν ἀνθρώπους καὶ τοῦτο μιμούμενον τὸν δεσπότην αἰεὶ πληροῖ
τὴν τῶν δεομένων χρείαν, καὶ ἔστιν ὥσπερ ἄλλο τι θρέμμα ἢ
δῶρον Ἀσκληπιοῦ, ὥσπερ Ὅμηρος ἐποίησεν ὅπλα καὶ ἔργα

12 Ἡφαίστου πρὸς τὸ ἐκείνῳ δοκοῦν κινούμενα. ἐπεὶ δὲ ἐνταῦθα ἤδη
ἐγενόμην, ποῖον ὕδωρ ἂν τῶν κατ' ἀνθρώπους τούτῳ παραβληθείη
χρείας εἴνεκα; οὐ γὰρ μόνον πόμα, τὸ δ' αὐτὸ καὶ λουτρόν ἐστιν
ἥδιστον καὶ ἀβλαβέστατον· οὐδ' . . .* ἀντέστραπται πρὸς τὰς
ὥρας τοῦ ἔτους, θέρους μὲν ψυχρότατον ὂν αὐτὸ αὑτοῦ, χειμῶνος
δὲ ὡς ἠπιώτατον γιγνόμενον, τὰ τοῦ παρόντος αἰεὶ καιροῦ δυσχερῆ
λῦον καὶ παραμυθούμενον, οἵαν χρὴ τὴν Ἀσκληπιοῦ πηγὴν ἱερὰν

13 εἶναι. καλὰ μὲν γὰρ ταῦτα καὶ ἡδέα καὶ αὐτῷ χρωμένῳ καὶ
ἑτέρους ὁρῶντι τοῦτο μὲν θέρους ὥρᾳ περὶ τὰ χείλη τοῦ φρέατος
περιεστηκότας ἑξῆς, ὥσπερ ἑσμὸν μελιττῶν ἢ μυίας περὶ γάλα,
ἐξ ἕω ζητοῦντας τὸ πνῖγος προκαταλαβεῖν ἀντ' ἄλλου πόματος
τῶν κωλυόντων τὸ δίψος καὶ ἰσχόντων, τοῦτο δὲ ὅταν τις κρυστάλ-
λου πεπηγότος τὴν χεῖρα προτείνας ἀπονιψάμενος θερμότερος

14 αὐτὸς αὑτοῦ καὶ ἡδίων γένηται. ἀλλὰ καὶ τἆλλα ὁ θεὸς αὐτῷ
χρῆται ὥσπερ ἄλλῳ τῳ συνεργῷ, καὶ πολλοῖς ἤδη πολλάκις τὸ
φρέαρ τοῦτο συνεβάλετο εἰς τὸ τυχεῖν ὧν ἔχρῃζον παρὰ τοῦ θεοῦ·
ὥσπερ γὰρ οἱ παῖδες οἱ τῶν ἰατρῶν τε καὶ θαυματοποιῶν γεγυμνα-
σμένοι πρὸς τὰς διακονίας εἰσὶ καὶ συμπράττοντες ἐκπλήττουσι
τοὺς θεωμένους καὶ χρωμένους, οὕτω τοῦ μεγάλου θαυματοποιοῦ
καὶ πάντα ἐπὶ σωτηρίᾳ πράττοντος ἀνθρώπων εὕρημα τοῦτο καὶ
κτῆμά ἐστι· συμπράττει δὴ πρὸς ἅπαντα αὐτῷ καὶ γίγνεται

15 πολλοῖς ἀντὶ φαρμάκου. πολλοὶ μὲν γὰρ τούτῳ λουσάμενοι
ὀφθαλμοὺς ἐκομίσαντο, πολλοὶ δὲ πιόντες στέρνον ἰάθησαν καὶ τὸ
ἀναγκαῖον πνεῦμα ἀπέλαβον, τῶν δὲ πόδας ἐξώρθωσεν, τῶν δὲ
ἄλλο τι· ἤδη δέ τις πιὼν ἐξ ἀφώνου φωνὴν ἀφῆκεν, ὥσπερ οἱ τῶν
ἀπορρήτων ὑδάτων πιόντες μαντικοὶ γιγνόμενοι· τοῖς δὲ καὶ
αὐτὸ τὸ ἀρύτεσθαι ἀντ' ἄλλης σωτηρίας καθέστηκεν. καὶ τοῖς
τε δὴ νοσοῦσιν οὕτως ἀλεξιφάρμακον καὶ σωτήριόν ἐστιν καὶ τοῖς
ὑγιαίνουσιν ἐνδιαιτωμένοις παντὸς ἄλλου χρῆσιν ὕδατος οὐκ

16 ἄμεμπτον ποιεῖ. πάντα γὰρ ἤδη μετὰ τοῦτο τὸ ὕδωρ γίγνεται
πειρωμένοις, οἷον εἴ τις μετὰ ἀνθοσμίαν οἶνον τῶν ἐξεστηκότων
τινὰ πίνοι. μόνον δὲ τοῦτο τὸ αὐτὸ νοσοῦσι καὶ ὑγιαίνουσιν ὁμοίως
ἥδιστον καὶ λυσιτελέστατον ἑκατέροις τε καὶ συναμφοτέροις ἐστίν,
καὶ οὔτ' ἂν γάλα παραβάλοις οὔτ' ἂν οἶνον ποθήσαις, ἀλλ' ἔστιν
ὥσπερ Πίνδαρος τὸ νέκταρ ἐποίησεν αὐτόχυτον, πότιμον θείᾳ

τινὶ κράσει κεκραμένον ἀρκούντως. ὥστε εἰ δύο εἶεν κύλικες, ἡ
μὲν ἑτέρου του ὕδατος καὶ οἴνου τοῦ καλλίστου, ἡ δὲ τούτου τοῦ
17 ὕδατος, ἀπορήσαις ἂν πότερον λάβοις. ἔτι δὲ τὰ μὲν ἄλλα ἱερὰ
ὕδατα τὴν τῶν πολλῶν ἀνθρώπων χρῆσιν πέφευγεν, οἷον τὸ ἐπὶ
Δήλῳ καὶ εἴ τί που ἄλλοθι ἄλλο τοιοῦτόν ἐστι, τὸ δὲ τῷ σῴζειν
τοὺς χρωμένους, οὐ τῷ μηδένα αὐτοῦ ψαύειν, ἱερόν ἐστιν· καὶ
τὸ αὐτὸ καθαρσίοις τε ἐξαρκεῖ τοῖς περὶ τὸ ἱερὸν καὶ ἀνθρώποις
καὶ πίνειν καὶ λούεσθαι καὶ προσορῶσιν εὐφραίνεσθαι.

18 Ἐγὼ μὲν οὔτε Κύδνον οὔτε Εὐρυμέδοντα οὔτε Χοάσπην, ὅθεν
βασιλεὺς ἔπινεν περιφέρων, οὔτε ᾧ τοὺς καλλίστους στεφάνους
ἀνῆκεν ἡ γῆ περὶ τὴν ὄχθην ἑκατέραν, Πηνειόν, οὔτε . . .*
πηγὴν ἄβυσσον, οὔθ᾽ ὅ τι ἐρεῖς ἕτερον ὕδωρ παραβάλοιμ᾽ ἂν
τούτῳ τῷ πάντα ἱερῷ, ἀλλ᾽ εἶναι φαίην ἂν αὐτὸ ἐν ὕδασι τοσούτῳ
νικῶν ὅσονπερ τὸν προστάτην αὐτοῦ θεὸν ἐν θεοῖς. λοιπὸν ἐν
εἰπεῖν, ὅτι καὶ ὅσια ἂν ποιοῖμεν οὕτω κρίνοντες· ὁ γὰρ θεὸς
πρῶτος περὶ αὐτοῦ ταύτην τὴν ψῆφον ἤνεγκεν, ὥς φασιν.

* lacunam indicavit Keil.
** νεύοντες Friedländer   νεῦον MSS   ⟨οἱ⟩ ὕφαλοι, νεῦον Keil.
*** καὶ ἑώντων secl. Canter   καὶ κενούντων Keil.

To the Well in the Temple of Asclepius

What would you say about the holy well? Or do you clearly
find fault with the nature of the speech because it does not
adequately reach all aspects and cannot at times sufficiently
express the reality? In any case no single speech could express
the quality of this well, its beauty and taste; but we are better
at drinking of this water and at bathing in it and at beholding
it with pleasure than at having something to say about it, just
as the lovers of beautiful youths, who are caught by the power
of their beauty and know what they love—yet, if anyone should
question them, they would have nothing to say about their
absent loved ones, whom if they were present they could easily
point out, I am sure. This self same thing we have experienced
with regard to this well, and many—or rather almost all—
of us are lovers of it, though we cannot express of what nature
it is. But if someone should take one of us aside and question
him, we would not hesitate to lead him there [i. e., to the well]
and to show it to him. Not even this, however, will suffice him;
but having tasted and tried it, he will think he has tasted
Homer's lotus, and he will wish to remain and will find it diffi-

cult to leave. Yet, we should not drink in silence as the thirsty do, according to the saying,[1] but on the contrary we should pay it some honor also in words and should address both the savior god whose work and accomplishment it is and the Nymphs who are in charge of it and who work with him and kindly grant us to enjoy this favor of the god.

What, then, should be the beginning (of our speech), or, just as when we drink from the well, raising the cup to the lips we never stop again, but pour in the liquid all at once, so too should our speech say everything all at once? Instead of the approach to the lip, let this be our starting point: that the well is in the fairest spot on earth. For the place which the god selected from all places as the healthiest and purest, the place which he has made most notable of all by the services performed there by himself, surely this place is the fairest of all there are on earth. For other gods in other places are indigenous from the beginning—and I think we should esteem even these places for the very reason that they were allotted to the gods; but this place is unlike the others and even greater for this reason, that the god, having set out here from Epidaurus herself,[2] loved the place more than any other; this is clear from the fact that, having chosen the spot and having preferred it to all others, he remained thereafter. In regard, then, to that which the god who is the gentlest and most man-loving of the gods[3] decided to prefer, how is it possible for us, particularly those of us who are his servants, to say anything except that this is the best? In the fairest spot on earth, it is permissible, then, to say. Furthermore, of the portion of the sanctuary which is in the open and accessible, the well is in the fairest spot. For it is centered precisely in the middle (of the *temenos*). And, if you wish to put it in this way, the water flows from the plane tree—for as another symbol the plane tree has grown beside it—or, if you prefer, what is even more beautiful and holy, it flows from the very foundations on which the temple stands, so that this opinion and conviction forces itself upon everyone: that the water is brought from a spot that is healthful and promotes health, inasmuch as it comes forth from the shrine and the feet of the Savior. For no water could flow from more healthful and purer places than that which flows from these.

Appearing in such a place, then, and having its origin in such sources, naturally, it is the fairest. In the first place, the water is very thin, almost like air; then, in consequence of this, it is both very light and very gentle; in the third place, it is both very sweet and very fresh. Drinking it, one would not ask for wine to be added. For Homer said[4] that the Titaresius, because of the lightness of its water, flowed on top of the Peneius, just like a man floating. If you pour some other water into this, it seems to me, this (holy water) rises to the top, while the other sinks, just as divers do, dropping down from the surface to the innermost recesses; or I may say even this: that it seems to me to raise by its own lightness even the water that is poured in. That we do not boast, can be judged by its weight. And yet what would the branch of the Styx say when, being compared, it would turn the scale? In addition to that this current is neither a branch of the Styx nor does it have any other terrifying quality; rather you may call it a distillation of health or nectar or some other such thing. Proof of this is also the following fact: time does not affect it, but the water that has been drawn off and has come out has the same properties as that which is always left in the well; it remains unpolluted and unchanged. On the other hand, the supply of water in this well is certainly greater than that of any other well. Those who draw the water must be untiring and heedful lest the water that is being caught outstrip them. This alone, to be sure, of all wells that are constantly drawn always provides an equal measure of water, thus being the opposite of the perforated wine jar: the latter is never filled while the former is always near the brim. For, since the well is the servant and co-worker of the most man-loving of the gods, it is always full and ready for service. And as the god has no leisure to do anything except to save men, so the well, imitating its master, always fills the need of those who require it and is a sort of nursling or gift of Asclepius, just as Homer made the tools and works of Hephaestus act in accordance with his (the god's) intention.

Since I have now reached this point, what water of all that men have could be compared with this in usefulness? For it is not merely a drink, but it is also a most pleasant bathing water which prevents harm; nor . . . is it subject to changes accord-

ing to the seasons of the year, being (on the contrary) coldest
in the summertime and becoming mildest in the wintertime,
relieving and assuaging the discomforts of the respective season,
as it is proper for the holy spring of Asclepius to do. For this
is fair and pleasant to anyone whether he uses the water himself or watches others, either in summer, when like a swarm of
bees or like flies around milk they are standing crowded around
the rim of the well, seeking from daybreak to ward off the
stifling heat (with the help of the well) instead of with one of
the other drinks that check and restrain thirst; or when the ice
is frozen solid and someone stretches forth his hand, he becomes warmer upon being washed with the ice than he was
and feels more comfortable.

But even in other cases the god uses this well as a kind of
co-worker, and it has often been useful to many in obtaining
what they had asked from the god. For just as the servants of
physicians and miracle-workers are trained to ministrations,
and, working with their superiors, astonish those who behold
them and ask their advice, so is this well the discovery and
possession of the great miracle-worker who does everything
for the salvation of men: it works with him in all matters and
for many it comes to take the place of a drug. For when bathed
with it many recovered their eyesight, while many were cured
of ailments of the chest and regained their necessary breath by
drinking from it. In some cases it cured the feet, in others
something else. One man upon drinking from it straightway
recovered his voice after having been a mute, just as those who
drink sacred waters become prophetic. For some the drawing
of the water itself took the place of every other remedy.

Furthermore, not only is it remedial and beneficial to the
sick but even for those who enjoy health it makes the use of
any other water improper. For to those who try any other
water after this it is as if one were to drink some soured wine
on top of wine with a fine bouquet. This water alone is equally
pleasant to the sick and to the healthy, and it is most useful
for the one as for the other. And you would not compare
milk with it nor long for wine, but it is just as Pindar pictured
nectar:[5] self-flowing, a potion blended sufficiently by some
divine process of mixing, so that if there were two drinking-
cups, one containing some other water and the finest wine,

the other containing this water, you would be undecided which
to take. Besides, the other holy waters are removed from the
use of the many, as for example that in Delos and whatever
other similar water there is anywhere, while this water is sacred
because it saves those who use it and not because no one
touches it. And the same water suffices both for purifications
of what belongs to the shrine and for men to drink and be
bathed in and to rejoice over when they behold it.

To this water which is holy in all respects I would not compare Cydnus or Eurymedon or Choaspes, from which the king
carried water around and drank it,[6] or the Peneius, on both of
whose banks the earth brings forth the fairest garlands, or . . .
the bottomless spring [i. e., the Nile], or any other water you
name, but I would say that among waters this water is just as
superior as its patron god is eminent among gods. One thing
remains for me to say: that by so judging we seem to act in
accordance with the divine will. For the god was the first to
cast this vote concerning it (the well), so they say.

[1] Plato, *Symposium*, 214B.　　　　[4] *Iliad*, II, 754.
[2] Cf. T. 709; 801.　　　　　　　　　[5] *Olympians*, VII, 7.
[3] Cf. T. 422.　　　　　　　　　　　 [6] Cf. Herodotus, I, 188.

805. Aristides, Oratio LIII, 1–5.

ΠΑΝΗΓΥΡΙΚΟΣ ΕΠΙ ΤΩΙ ΥΔΑΤΙ ⟨ΤΩΙ⟩ ΕΝ ΠΕΡΓΑΜΩΙ

Ὁμήρῳ μὲν εἰς τὴν σύνοδον τῶν χειμάρρων εἴρηται· 'τῶν
δέ τε τηλόθι δοῦπον ἐν οὔρεσιν ἔκλυε ποιμήν,' καὶ φρίξαντα
δή φησιν αὐτὸν εἰς τὸ σπήλαιον εἰσελαύνειν τὰς οἶς· ἐγὼ δὲ
καίτοι τοσοῦτον ὑμῶν ἀπέχων τὸ νῦν, ἀκούσας τοῦ ὕδατος τὴν
εἰσβολὴν καὶ ὅσον τι κόσμου προσγέγονεν τῇ πόλει, οὐχ οἷός τε
ἦν ἡσυχάζειν ὑφ' ἡδονῆς, ἀλλ' ἐφθεγγόμην τε ἃ ἐφθεγγόμην καὶ
τοῦ σώματος ᾐσθανόμην ἐλαφροτέρου καὶ χαρᾶς ἀπῆν οὐδέν.
2　δυοῖν δ' ἡμέραιν πρότερον πρὶν ἀκοῦσαι—οὐ γὰρ χεῖρον ἴσως
πρὸς ὑμᾶς εἰπεῖν· ἀκούσεσθε γὰρ ἡδέως τοῦ θεοῦ χάριν τοῦ
προδείξαντος καὶ ἅμα τῆς εὐφημίας—ὄψις ὀνειράτων γενομένη μοι
ὡσπερεὶ διπλασίαν ἐδείκνυε τὴν πόλιν, χωρίου τε δή τινος
προσθήκῃ, πεπορισμένου συνεχῶς πρὸς αὐτήν, καὶ δημοσίων δὴ
κόσμων προσγενομένων παραπλησίων μάλιστά πως τοῖς περὶ
τὸν Φίλιον. διὰ ταῦτ' οὖν ὄναρ τε ἐγανύμην καί, ἐπειδὴ ἀνέστην,
3　ἐλάμβανον εἰς ἀγαθὸν τῇ τε πόλει καὶ ἐμαυτῷ. τριταία δὲ ἐπὶ

τούτοις ἀγγελία παρὰ ἀνδρὸς τῶν ἐπιτηδείων ἀφικνεῖται φράζουσα
καὶ δὴ πᾶσαν ὑμῖν τὴν Ἀσίαν συνεορτάζειν τῆς περὶ πάντα
ἀγαθῆς τύχης· εἶναι γὰρ τὸ ὕδωρ πλήθει τε πλεῖστον καὶ κάλλει
κάλλιστον ὅσων ἔλαχον πόλεις. ἦγον οὖν οὐχ ὅσον ἠρινὴν
ἡμέραν, ἀλλ᾽ οἴαν εἰκὸς ἄγειν Διός τε Εὐαγγελίου καὶ Ἀσκληπιοῦ
Σωτῆρος πανταχῇ τιμῶντος. καὶ συνέχαιρον δὴ τῇ πόλει μὲν τῶν
προσγεγονότων, ἐμαυτῷ δὲ ὡς ἠξιώθην προακοῦσαι, δῆλον ὅτι ὡς
4   οὐδενὸς ἧττον ἐμοὶ τῆς πόλεως προσῆκον. μετὰ δὲ τοῦτο ἐλογι-
ζόμην ὡς τὸ μὲν χαίρειν κοινὸν ἁπάντων καὶ ἀνδρῶν καὶ παιδαρίων
καὶ γυναικῶν, ἅτε τῆς ὄψεως προξενούσης τὴν ἡδονήν, λόγῳ δὲ
ἐπικοσμῆσαι τὴν τῶν Νυμφῶν δόσιν τάχα ἄν τινος εἴη τῶν περὶ
τὸν Παιῶνα διατριψάντων καὶ τῶν ἐπιταχθέντων ζῆν ἐν λόγοις.
ἀνεμιμνησκόμην δὲ τῶν ποιητῶν, ὅτι Νύμφας καὶ Μούσας αἰεί
πως συνάγουσι, καὶ τὸν Ἑρμῆν ὡς χορηγὸν αὖ προσαγορεύουσι
τῶν Νυμφῶν, καὶ πάλιν γε Ἀπόλλωνα χορηγὸν Μουσῶν· ὁ δ᾽
αὐτὸς οὗτος ὑμῖν θεὸς Καλλιτέκνου προσηγορίαν ἔχει τοῦ πατρὸς
εἵνεκα· ἀπανταχῇ δὴ πρέπον τε καὶ οὐκ ἄωρον ἐφαίνετο τῇ τῶν
Νυμφῶν χάριτι συγκεράσαι τὴν παρὰ τῆς μουσικῆς. πᾶσι γὰρ
ἂν προσήκοντα πράττειν οἷς εἶπον θεοῖς.
5   Ἐξ ἀρχῆς δ᾽, ὡς ἔοικε, τὰ κάλλιστα ἐδόθη τῇ πόλει καὶ παρὰ
θεῶν καὶ παρὰ ἀνθρώπων. τοῦτο μὲν πρεσβύτατοι δαιμόνων
ἐνταυθοῖ λέγονται γενέσθαι Κάβειροι καὶ τελεταὶ τούτοις καὶ
μυστήρια, ἃ τοσαύτην ἰσχὺν ἔχειν πεπίστευται ὥστε χειμώνων τε
ἐξαισίων . . . *

* The rest of this oration is lost.

Panegyric on the Water in Pergamum

In regard to the junction of swollen torrents it has been said
by Homer:[1] "The shepherd in the mountains heard the roar
of them from far away", and shuddering, he says, he drove
the sheep into the cave. Yet I, although I am now so far away
from you [sc., the people of Pergamum], when I heard about
the impact of the water and how much honor has accrued to
the city from it, could not keep silent in my happiness but sang
my praises and felt my body grow lighter and wanted for no
joy.

Two days before I heard (the news)—for it is probably not
amiss to say this to you, for you will gladly listen because of
the god who was foreshowing (the event), and because of the
auspiciousness (of the omen)—a vision came to me in dreams

which made the city appear twice as large by the addition not only of a certain district contiguous to it but also of public adornments in some ways very similar to those around (the temple of Zeus) Philius. For these reasons, then, I was happy at the dream, and when I got up I took it as a good portent for the city and for myself. On the third day after this a message comes from one of my friends saying that all Asia joins with you in the festival of good fortune in all things, since the water is the largest in quantity and the fairest in beauty of all that cities ever acquired. I spent this day not as an (ordinary) spring day, but as one should celebrate a feast day of Zeus the Bringer of good Tidings and Asclepius the Savior who bestows honor in every way. And I was jubilant for the city on account of what had been added to it and for myself because I had been deemed worthy to hear the tidings beforehand, although clearly that was less fitting for me than for anyone else in the city.

After this it occurred to me that to rejoice would be common to all men, children, and women since the vision procured happiness, but that to honor the gift of the Nymphs with words would be the right of one of those who spend their time in the service of Paeon and are enjoined to pass their lives in writing. I recalled that the poets always somehow join the Nymphs and the Muses and address Hermes as the leader of the chorus of the Nymphs and again Apollo as the leader of the Muses. This same god with you has the name *Kalliteknos* [*i. e.*, with beautiful children] because of his being their father. It appeared in every way proper and timely, therefore, to mingle the grace of the Nymphs with that of the Muses. For in that way one would do what was fitting for all the gods whom I have mentioned.

From the beginning, it seems, the fairest things were granted to the city by both gods and men. The oldest of the daemons, the Cabiri, are said to have been born there, and initiations and mysteries for them which one believes to have such power that of violent tempests . . .

¹ *Iliad*, IV, 455.

806. Aristides, Oratio XLVIII, 30.

Ἐδόκει ὁ μὲν Φιλάδελφος—τοσαῦτα γὰρ οὖν ἔχω διαμνημονεῦσαι —ἐν τῷ θεάτρῳ τῷ ἱερῷ πλῆθος ἀνθρώπων εἶναι λευχειμονούντων

καὶ συνεληλυθότων κατὰ τὸν θεόν· ἐν δ' αὐτοῖς ἑστῶτα ἐμὲ δημη-
γορεῖν τε καὶ ὑμνεῖν τὸν θεόν, . . .

Philadelphus [a friend of Aristides] dreamed — for so much
can I remember distinctly — that in the holy theater there was
a crowd of people clad in white,[1] gathering in honor of the god;
and standing among them I made a speech and sang the praises
of the god . . .

[1] Cf. T. 296; 513; cf. however T. 787.

807. Phrynichus, Ecloga Nominum, p. 421 [Lobeck].

Κατ' ὄναρ· Πολέμων ὁ Ἰωνικὸς σοφιστὴς Δημοσθένους τοῦ
ῥήτορος εἰκόνα χαλκῆν ἐν Ἀσκληπιοῦ τοῦ ἐν Περγάμῳ τῇ Μυσίᾳ
ἀναθεὶς ἐπέγραψεν ἐπίγραμμα τοιόνδε . . .

According to a dream: Polemo, the Ionian sophist [under
Trajan],[1] having set up a bronze statue of Demosthenes, the
orator, in the Asclepieion at Pergamum in Mysia, inscribed it
with the following epigram . . .

[1] Cf. T. 433.

808. Aristides, Oratio L, 46.

Μετὰ δὲ ταῦτα βουλευομένοις ἡμῖν κοινῇ περὶ τοῦ ἀναθήματος
συνεδόκει καὶ τῷ ἱερεῖ καὶ τοῖς νεωκόροις ἀναθεῖναι ἐν Διὸς
Ἀσκληπιοῦ, ταύτης γὰρ οὐκ εἶναι χώραν καλλίω· καὶ οὕτω δὴ
τοῦ ὀνείρατος ἡ φήμη ἐξέβη. καὶ ἔστιν ὁ τρίπους ὑπὸ τῇ δεξιᾷ
τοῦ θεοῦ, εἰκόνας χρυσᾶς ἔχων τρεῖς, μίαν καθ' ἕκαστον τὸν πόδα,
Ἀσκληπιοῦ, τὴν δὲ Ὑγιείας, τὴν δὲ Τελεσφόρου. καὶ τὸ ἐπίγραμμα ἐπιγέγραπται καὶ ὅτι ἐξ ὀνείρατος προσπαραγέγραπται.

After this we were taking counsel together concerning the
votive offering and it was resolved, in agreement with the
priest and the temple attendants, to place it in the temple of
Zeus Asclepius,[1] for it seemed to us that there was no fairer
place than this. In this way, then, the prophecy of the dream
was fulfilled. And the tripod is under the right hand of the
god, with three gold images—one on each foot—of Asclepius,
Hygieia, and Telesphorus[2] respectively. And the epigram has
been engraved on it and in addition the fact has been recorded
that it was written as I dreamed it.

[1] Cf. T. 317, 4.   [2] Cf. T. 287; 313; 504; 749.

809. Appianus, Historia Romana, XII, 23.

Περγαμηνοὶ τοὺς ἐς τὸ Ἀσκληπιεῖον συμφυγόντας, οὐκ ἀφισταμένους, ἐτόξευον τοῖς ξοάνοις συμπλεκομένους.

The Pergamenes [in 88 B. C.] shot with arrows those who had fled to the temple of Asclepius, while they were clinging to his statues and would not withdraw. (**)

810. Appianus, Historia Romana, XII, 60.

. . . ἐπανῆλθεν ἐς Πέργαμον, καὶ ἐς τὸ τοῦ Ἀσκληπιοῦ ἱερὸν παρελθὼν ἐχρήσατο τῷ ξίφει.

. . . he [sc., Fimbria, in 85 B. C.] returned to Pergamum, entered the temple of Asclepius, and stabbed himself with his sword.[1]

[1] Cf. **810a.** Orosius, *Historiae adv. Paganos*, VI, 2, 11: Idem Fimbria apud Thyatiram cum ab exercitu Sullae obsideretur, desperatione adactus in templo Aesculapii manu sua interfectus est.

811. Tacitus, Annales, III, 63, 2.

Consules super eas civitates, quas memoravi, apud Pergamum Aesculapii conpertum asylum rettulerunt, . . .

The consuls [in 22 A. D.] reported—apart from those communities which I have already mentioned—that at Pergamum (the right of) asylum of Asclepius had been ascertained[1] . . . (**)

[1] Cf. T. **798.**

812. Aristides, Oratio L, 3.

Ἔστι δὲ Ποιμανηνὸς χωρίον τῆς Μυσίας καὶ ἐν αὐτῷ ἱερὸν Ἀσκληπιοῦ ἅγιόν τε καὶ ὀνομαστόν.

Poemanenus is a place in Mysia and in it is a temple of Asclepius both holy and renowned.

LYDIA

813. Pausanias, Descriptio Graeciae, VII, 5, 9.

Ἐποιήθη δὲ καὶ κατ’ ἐμὲ Σμυρναίοις ἱερὸν Ἀσκληπιοῦ μεταξὺ Κορυφῆς τε ὄρους καὶ θαλάσσης ἀμιγοῦς ὕδατι ἀλλοίῳ.

There was made in my time by the Smyrnaeans also a sanctuary of Asclepius[1] between Mount Coryphe and a sea into which no other water flows. (*)

[1] Cf. T. 709.

814. Aristides, Oratio XLVII, 17.

Ἐνάτῃ ἐδόκουν ὡς ἐν Σμύρνῃ περὶ ἑσπέραν προσιέναι τῷ ἱερῷ τοῦ Ἀσκληπιοῦ τῷ ἐν τῷ γυμνασίῳ . . . καὶ εἶναι τὸν νεὼν μείζω τε καὶ ἐπειληφότα τῆς στοᾶς ὅσον ἐστὶν τὸ ἐστρωμένον.

On the ninth day I dreamed that toward the evening I approached the temple of Asclepius in the gymnasium at Smyrna . . . and that the temple both was larger and occupied all the levelled portion of the stoa.

CARIA

815. Vitruvius, De Architectura, VII, Praefatio, 12.

. . . item Arcesius de symmetriis Corinthiis et Ionico Trallibus Aesculapio, quod etiam ipse sua manu dicitur fecisse.

. . . Arcesius (published a work) on Corinthian proportions, and the Ionic temple of Asclepius at Tralles which he [sc., Arcesius, otherwise unknown] is said likewise to have built with his own hand. (**)

CILICIA

816. Philostratus, Vita Apollonii, I, 7.

Μεθίστησιν οὖν τὸν διδάσκαλον . . . ἐς Αἰγὰς . . . ἐν αἷς . . . καὶ ἱερὸν Ἀσκληπιοῦ, καὶ ὁ Ἀσκληπιὸς αὐτὸς ἐπίδηλος τοῖς ἀνθρώποις.

He [sc., Apollonius] therefore transferred his teacher [sc., Euthydemus] . . . to the town of Aegae . . . where he found . . . a temple of Asclepius,[1] and Asclepius reveals himself in person to men.[2] (*)

[1] Cf. T. 307.
[2] Cf. T. 389.

817. Libanius, Orationes, XXX, 39.

Καὶ νῦν οὓς ἄγει μὲν εἰς Κιλικίαν νοσήματα τῆς τοῦ Ἀσκληπιοῦ
χρήζοντα χειρός, αἱ δὲ περὶ τὸν τόπον ὕβρεις ἀπράκτους ἀπο-
πέμπουσι, πῶς ἔνεστι μὴ κακῶς τὸν τούτων αἴτιον λέγοντας
ἀναστρέφειν;

Now as for those whom illnesses requiring the hand of Asclepius bring to Cilicia, but outrages in the district send away
uncured, how is it possible for them to leave without speaking
evil of him who is to blame for this?

818. Eusebius Caesariensis, De Vita Constantini, III, 56.[1]

Ἐπειδὴ γὰρ πολὺς ἦν ὁ τῶν δοκησισόφων* περὶ τὸν τῶν Κιλίκων
δαίμονα πλάνος, μυρίων ἐπτοημένων ἐπ᾽ αὐτῷ ὡς ἂν ἐπὶ σωτῆρι
καὶ ἰατρῷ, ποτὲ μὲν ἐπιφαινομένῳ τοῖς ἐγκαθεύδουσι, ποτὲ δὲ τῶν
τὰ σώματα καμνόντων ἰωμένῳ τὰς νόσους· ψυχῶν δ᾽ ἦν ὀλετὴρ
ἄντικρυς οὗτος, τοῦ μὲν ἀληθοῦς ἀφέλκων Σωτῆρος, ἐπὶ δὲ τὴν
ἄθεον πλάνην κατασπῶν τοὺς πρὸς ἀπάτην εὐχερεῖς· εἰκότα δὴ βασιλεὺς** πράττων, Θεὸν ζηλωτὴν ἀληθῶς Σωτῆρα προβεβλημένος,
καὶ τοῦτον εἰς ἔδαφος τὸν νεὼν ἐκέλευσε καταβληθῆναι.*** ἑνὶ δὲ
νεύματι κατὰ γῆς ἡπλοῦτο, δεξιᾷ καταρριπτόμενον στρατιωτικῇ
τὸ τῶν γενναίων φιλοσόφων βοώμενον θαῦμα, καὶ ὁ τῇδε ἐνδο-
μυχῶν, οὐ δαίμων, οὐδέ γε θεός, πλάνος δέ τις ψυχῶν, μακροῖς
καὶ μυρίοις ἐξαπατήσας χρόνοις. εἶθ᾽ ὁ κακῶν ἑτέρους ἀπαλλάξειν
καὶ συμφορᾶς προϊσχόμενος, οὐδὲν αὐτὸς ἑαυτῷ πρὸς ἄμυναν
εὕρατο φάρμακον μᾶλλον, ἢ ὅτε κεραυνῷ βληθῆναι μυθεύεται.
ἀλλ᾽ οὐκ ἐν μύθοις ἦν τὰ τοῦ ἡμεδαποῦ βασιλέως Θεῷ κεχαρισμένα
κατορθώματα· δι᾽ ἐναργοῦς δέ γε ἀρετῆς τοῦ αὐτοῦ Σωτῆρος,
αὐτόρριζος καὶ ὁ τῇδε νεὼς ἀνετρέπετο, ὡς μηδὲ ἴχνος αὐτόθι τῆς
ἔμπροσθεν περιλελεῖφθαι μανίας.

* δοκησισόφων Wilamowitz   δοκήσει σοφῶν MSS.
** βασιλεὺς supplevit Valesius.
*** τὸν νεὼν ἐκέλευσε καταβληθῆναι A   φέρεσθαι τὸν νεὼν ἐκέλευσεν Heikel.

As to the god of the Cilicians, great was, indeed, the deception
of men seemingly wise, with thousands excited over him as if
over a savior and physician who now revealed himself to those
sleeping (in the temple) and again healed the diseases of those
ailing in body; of the souls, though, he was a downright destroyer,[2] drawing them away from the true Savior and leading
into godless imposture those who were susceptible to fraud;

the emperor [*sc.*, Constantine], therefore, acting fairly, holding
the true Savior a jealous god, commanded that this temple, too,
be razed to its foundations.[3] At one nod it was stretched out
on the ground—the celebrated marvel of the noble philosophers
overthrown by the hand of the soldier—and with it (fell) the
one lurking there, not a demon nor a god, but a kind of deceiver
of souls, who had practiced his deceit for a very long time.
Then, he who offered to rid others of evils and misfortunes
could find for his own defense no remedy any more than when
he is fabled to have been struck with the thunderbolt.[4] But
not in the sphere of fables was the righteous action of the
Roman emperor which found favor with God. Manifestly
through the miraculous power of the Savior Himself the temple
there was destroyed to the roots so that not even a trace remained here of the former madness.

[1] Cf. T. 390.
[2] Cf. **818a.** Eusebius Hieronymus, *Vita Hilarionis*, 2: . . . Aesculapii . . . non remedi-
antis animas sed perdentis; cf. T. 330.
[3] Cf. T. 819; 821.   [4] Cf. T. 105 ff.

819. Sozomenus, Historia Ecclesiastica, II, 5.

> Κατεσκάφησαν δὲ τότε καὶ ἄρδην ἠφανίσθησαν, ὁ ἐν Αἰγαῖς τῆς
> Κιλικίας Ἀσκληπιοῦ ναός, καὶ ὁ ἐν Ἀφάκοις τῆς Ἀφροδίτης . . .
> ἄμφω δὲ ἐπισημοτάτω νεὼ ἐγενέσθην καὶ σεβασμίω τοῖς πάλαι.
> καθότι Αἰγεᾶται μὲν ηὔχουν τοὺς κάμνοντας τὰ σώματα νόσων
> ἀπαλλάττεσθαι παρ' αὐτοῖς, ἐπιφαινομένου νύκτωρ καὶ ἰωμένου
> τοῦ δαίμονος.

At that time [in 331 A. D.] the temple of Asclepius at Aegae
in Cilicia, and that of Aphrodite at Aphaca . . . were razed to
the ground and disappeared completely.[1] Both these temples
had been most distinguished and venerable to the ancients.
As to the Aegeatae, they boasted that with them those sick in
body were freed of their diseases since at night the demon
appeared and healed.[2]

[1] Cf. T. 818; T. 821.   [2] Cf. T. 389.

820. Zonaras, Epitome Historiarum, XIII, 12 C-D.

> Κατὰ Περσῶν δὲ τὴν στρατιὰν κινήσας κατήντησεν εἰς Ταρσὸν
> τῆς Κιλικίας πόλιν ἐπιφανῆ· ἔνθα γενομένῳ Ἀρτέμιος προσῆλθεν

ὁ τοῦ Ἀσκληπιοῦ ἱερεύς. ἦν γὰρ ἐν Αἰγαῖς, πόλις δὲ καὶ αὗται
τῆς Κιλικίας, περίφημον Ἀσκληπιοῦ ἱερόν, καὶ ἤτησεν αὐτὸν τοὺς
κίονας, οὓς ἔτυχεν ἀφελόμενος ἐκ τούτου τοῦ ἱεροῦ ὁ ἀρχιερεὺς
τοῦ λαοῦ τῶν Χριστιανῶν καὶ ἐποικοδομήσας αὐτοῖς οἰκεῖον ναόν,
ἀποκαταστῆσαι αὖθις τῷ ἱερῷ τοῦ Ἀσκληπιοῦ. καὶ ὁ παραβάτης
αὐτίκα τοῦτο γενέσθαι προσέταξε δαπάναις τοῦ ἐπισκόπου. μόλις
οὖν οἱ Ἕλληνες καὶ πόνοις πολλοῖς καὶ ἀναλώμασι πλείστοις ἕνα
τῶν κιόνων καθελόντες καὶ μέχρι τῆς φλιᾶς τῆς πύλης τῆς
ἐκκλησίας σὺν μηχανήμασιν ἀγαγόντες, καὶ χρόνῳ συχνῷ περαι-
τέρω προενεγκεῖν ἐκεῖνον οὐκ ἠδυνήθησαν· καὶ καταλιπόντες
αὐτὸν ἀνεχώρησαν. τοῦ δὲ Ἰουλιανοῦ θανόντος, αὖθις αὐτὸν ὁ
ἐπίσκοπος ἀνορθώσας ῥᾷστα εἰς τὸν τόπον τὸν ἑαυτοῦ ἐπανήγαγε.

(Julian) moving the army against the Persians [in 363 A. D.],
came to Tarsus, a famous city of Cilicia. While he was there,
Artemius, the priest of Asclepius, came to see him; for in
Aegae, also a city of Cilicia, there was a very famous shrine
of Asclepius. The pillars which the high priest of the Christian
people happened to have carried off from the shrine and with
which he had built a temple for his own people he urged him
to restore again to the shrine of Asclepius. And the Apostate
straightway commanded that this be done at the expense of the
bishop. With great toil and expenditure, then, the Greeks
barely managed to take down one of the columns and bring it
to the lintel of the doorway of the church by means of mechanical devices; yet in a long period of time they were unable to
drag it further. Therefore they left it and went away. When
Julian died, the bishop again easily lifted it up and restored it
to its own place.

MEDIA

821. Arrianus, Alexandri Anabasis, VII, 14, 5-6.

. . . ἄλλοι δέ, ὅτι καὶ τοῦ Ἀσκληπιοῦ τὸ ἕδος ἐν Ἐκβατάνοις
κατασκάψαι ἐκέλευσε, βαρβαρικὸν τοῦτό γε, καὶ οὐδαμῇ Ἀλεξάνδρῳ πρόσφορον, ἀλλὰ τῇ Ξέρξου μᾶλλόν τι ἀτασθαλίᾳ τῇ
ἐς τὸ θεῖον καὶ ταῖς πέδαις ἃς λέγουσιν ἐς τὸν Ἑλλήσποντον
καθεῖναι Ξέρξην, τιμωρούμενον δῆθεν τὸν Ἑλλήσποντον. ἀλλὰ
καὶ ἐκεῖνο οὐ πάντη ἔξω τοῦ εἰκότος ἀναγεγράφθαι μοι δοκεῖ,
ὡς ἐπὶ Βαβυλῶνος ἤλαυνεν Ἀλέξανδρος, ἐντυχεῖν αὐτῷ κατὰ

τὴν ὁδὸν πολλὰς πρεσβείας ἀπὸ τῆς Ἑλλάδος, εἶναι δὲ δὴ ἐν
τούτοις Ἐπιδαυρίων πρέσβεις· καὶ τούτους ὧν τε ἐδέοντο ἐξ
Ἀλεξάνδρου τυχεῖν καὶ ἀνάθημα δοῦναι αὐτοῖς Ἀλέξανδρον
κομίζειν τῷ Ἀσκληπιῷ, ἐπειπόντα ὅτι· καίπερ οὐκ ἐπιεικῶς
κέχρηταί μοι ὁ Ἀσκληπιός, οὐ σώσας μοι τὸν ἑταῖρον ὅντινα
ἴσον τῇ ἐμαυτοῦ κεφαλῇ ἦγον.

. . . yet others tell us that he bade the temple of Asclepius at
Ecbatana be razed to the ground[1] — a barbaric order, and not
in Alexander's way at all; but rather suitable to Xerxes' insolence towards things divine and harmonizing with those fetters
which they say Xerxes let down into the Hellespont, with the
notion of punishing the Hellespont. But this also I think has
been recorded not wholly outside the bounds of likelihood,
that, when Alexander was going to Babylon, there met him on
the way several delegations from Greece, and that among these
were also Epidaurian envoys; these received from Alexander
what they sued for, and Alexander gave them an offering to
take back to Asclepius,[2] with the words: "Although Asclepius
has not been kind to me, for he did not save for me the comrade
whom I valued as highly as my own life." (*)

[1] Cf. T. 818; 819; cf. also T. 713.   [2] Cf. T. 548.

SCYTHIA

822. Stephanus Byzantius, Ethnica, s. v. Ἅγιον.

Ἅγιον, τόπος Σκυθίας, ἐν ᾧ Ἀσκληπιὸς ἐτιμᾶτο, ὡς Πολυίστωρ.

Hagium, a place in Scythia, where Asclepius was honored
according to the Polyhistor [sc., Alexander, 1st c. B. C.].

PHOENICIA

823. Strabo, Geographica, XVI, 2, 22.

. . . μεταξὺ δὲ ὁ Ταμύρας ποταμὸς καὶ τὸ τοῦ Ἀσκληπιοῦ
ἄλσος . . .

. . . but between the two places [i. e., Berytus and Sidon in
Phoenicia] are the Tamyras River and the grove of Asclepius . . .

824. Scholia in Caesaris Germanici Aratea, 173.

Is per pelagus Sidoniam venit ibique Europam inter aequales suas ludentem in templo Aesculapi conspexit . . .

He [*i. e.*, the Bull] came by sea to Sidon and there beheld Europa playing among her companions in the temple of Asclepius . . .

825. Philo Byblius, Fr. 2, 20 [Müller].

Κρόνῳ δὲ ἐγένοντο ἀπὸ ᾿Αστάρτης θυγατέρες ἑπτὰ Τιτανίδες ἢ ᾿Αρτέμιδες. . . . Συδύκῳ δέ, τῷ λεγομένῳ δικαίῳ, μία τῶν Τιτανίδων συνελθοῦσα γεννᾷ τὸν ᾿Ασκληπιόν.

To Cronus were born of Astarte seven daughters, Titanids or Artemids. . . . With Sydycus, called the just, one of the Titanids cohabited and bore Asclepius.[1]

[1] **825a.** Eusebius, *Praeparatio Evangelica*, I, 10, 25: Συδύκῳ—᾿Ασκληπιόν. Cf. **825b.** Philo Byblius, Fr. 2, 27 [Müller]: . . . οἱ ἑπτὰ Συδὲκ παῖδες Κάβειροι, καὶ ὁ ἴδιος αὐτῶν ἀδελφὸς ᾿Ασκληπιός.

826. Damascius, Vita Isidori, 302.

῞Οτι ὁ ἐν Βηρυτῷ . . . ᾿Ασκληπιὸς οὐκ ἔστιν ῞Ελλην, οὐδὲ Αἰγύπτιος, ἀλλά τις ἐπιχώριος Φοῖνιξ. Σαδύκῳ γὰρ ἐγένοντο παῖδες, οὓς Διοσκούρους ἑρμηνεύουσι καὶ Καβείρους. ὄγδοος δὲ ἐγένετο ἐπὶ τούτοις ὁ ῞Εσμουνος, ὃν ᾿Ασκληπιὸν ἑρμηνεύουσιν. οὗτος κάλλιστος ὢν θέαν καὶ νεανίας ἰδεῖν ἀξιάγαστος, ἐρώμενος γέγονεν, ὥς φησιν ὁ μῦθος, ᾿Αστρονόης θεοῦ Φοινίσσης, μητρὸς θεῶν. εἰωθώς τε κυνηγετεῖν ἐν ταῖσδε ταῖς νάπαις ἐπειδὴ ἐθεάσατο τὴν θεὸν αὐτὸν ἐκκυνηγετοῦσαν καὶ φεύγοντα ἐπιδιώκουσαν καὶ ἤδη καταληψομένην, ἀποτέμνει πελέκει τὴν αὐτὸς αὑτοῦ παιδοσπόρον φύσιν. ἡ δὲ τῷ πάθει περιαλγήσασα καὶ Παιῶνα καλέσασα τὸν νεανίσκον τῇ τε ζωογόνῳ θέρμῃ ἀναζωπυρήσασα θεὸν ἐποίησεν, ῞Εσμουνον ὑπὸ Φοινίκων ὠνομασμένον ἐπὶ τῇ θέρμῃ τῆς ζωῆς. οἱ δὲ τὸν ῞Εσμουνον ὄγδοον ἀξιοῦσιν ἑρμηνεύειν, ὅτι ὄγδοος ἦν τῷ Σαδύκῳ παῖς.

The Asclepius in Berytus . . . is neither Greek nor Egyptian, but rather a native Phoenician.[1] For to Sadycus were born the children whom they interpret as the Dioscuri and the Cabiri. Eighth, after these, was born Esmunus, whom they interpret as Asclepius. Being the fairest in appearance and a young man worthy to view with admiration, he was beloved, so the story goes, by the Phoenician goddess Astronoë, mother of the gods. He was accustomed to hunt in these woodland glens; but when he saw the goddess pursuing him, following hard after him when he fled, now almost overtaking him, with a sharp blade he emasculated himself. Distressed by his suffering, the goddess summoned Paeon and having rekindled the flame of life in the young man with life-giving warmth, she made him a god. He was named Esmunus by the Phoenicians with reference to the warmth of life. But some desire to make Esmunus mean the eighth, since he was the eighth child of Sadycus.

¹ Cf. **826a**. Marinus, *Vita Procli*, cp. 19: . . . δηλοῖ δὲ ἡ τῶν ὕμνων αὐτοῦ πραγματεία, οὐ τῶν παρὰ τοῖς Ἕλλησι μόνον τιμηθέντων ἐγκώμια παρέχουσα, ἀλλὰ . . . ὑμνοῦσα καὶ Ἀσκληπιὸν Λεοντοῦχον Ἀσκαλωνίτην . . .

AFRICA

Egypt

827. Ammianus Marcellinus, Res Gestae, XXII, 14, 7.

. . . ducitur Memphim, urbem frequentem praesentiaque numinis Aesculapii claram.

. . . (Apis) is taken to Memphis, a populous city, famed for the presence of the god Asclepius.

828. Epiphanius, De XII Gemmis, 32.

Arbitrati sunt autem eum de superiore esse Thebaida, proptei quod affirmant aliquos lapides ex lapide hoc in aede Asclepii, quae sic appellatur, in Memphi Aegyptia reperiri.

They are of the opinion, however, that this [stone, *i. e.*, the *ligyrium*] came from the upper Thebaïd; that is why, they claim, certain stones cut from this block are to be found in the so-called temple of Asclepius in Memphis in Egypt.¹

¹ Cf. T. **849c**.

829. Clemens Alexandrinus, Stromateis, I, 21, 134.

Ἀλλὰ καὶ τῶν παρ' Αἰγυπτίοις ἀνθρώπων ποτέ, γενομένων δὲ ἀνθρωπίνῃ δόξῃ θεῶν, Ἑρμῆς τε ὁ Θηβαῖος καὶ Ἀσκληπιὸς ὁ Μεμφίτης . . .

But of those who among the Egyptians at one time were men,
but according to human opinion became gods, were Hermes of
Thebes and Asclepius of Memphis . . .

830. Tacitus, Historiae, IV, 84, 5.

Deum ipsum multi Aesculapium, quod medeatur aegris corporibus . . . coniectant.

Many regard the god himself [*sc.*, Sarapis] as identical with
Asclepius, because he cures sick bodies . . . (*)

CYRENE

831. Tacitus, Annales, XIV, 18, 1.

Motus senatu et Pedius Blaesus, accusantibus Cyrenensibus
violatum ab eo thesaurum Aesculapii . . .

Pedius Blaesus also was removed from the senate [in 59
A. D.]: he was charged by the Cyreneans[1] with profaning
the treasury of Asclepius . . .

[1] Cf. T. 532; 709.

CARTHAGE

832. Livius, Ab Urbe Condita, XLI, 22, 2.

Conpertum tamen adfirmaverunt legatos ab rege Perseo venisse,
iisque noctu senatum in aede Aesculapi datum esse.

They [*sc.*, the Roman legates] affirmed that they had ascertained that ambassadors had come from King Perseus, and
that at night they had been permitted to appear before the
senate in the temple of Asclepius [*sc.*, at Carthage, in 174
B. C.]. (**)

833. Livius, Ab Urbe Condita, XLII, 24, 3.

In aede Aesculapi clandestinum eos per aliquot noctes consilium principum habuisse . . .

In the temple of Asclepius they [*sc.*, the Carthaginians] had
held secret consultation of their nobles through several nights
[in 172 B. C.] . . .**

834. Strabo, Geographica, XVII, 3, 14.

Κατὰ μέσην δὲ τὴν πόλιν ἡ ἀκρόπολις, ἣν ἐκάλουν Βύρσαν,
ὀφρὺς ἱκανῶς ὀρθία, κύκλῳ περιοικουμένη, κατὰ δὲ τὴν κορυφὴν
ἔχουσα Ἀσκληπιεῖον, ὅπερ κατὰ τὴν ἅλωσιν ἡ γυνὴ τοῦ
Ἀσδρούβα συνέπρησεν αὐτῇ.

Near the middle of the city [sc., of Carthage] was the acropolis, which they called Byrsa; it was a fairly steep height and
inhabited on all sides, and at the top[1] it had a temple of
Asclepius, which, at the time of the capture of the city [146
B. C.], the wife of Hasdrubal burnt along with herself.

[1] Cf. T. 708, n. 2.

835. Appianus, Historia Romana, VIII, 130-31.

Πολλῶν δ' ἔτι πορθουμένων, καὶ τοῦ κακοῦ μακροτάτου δοκοῦντος
ἔσεσθαι, προσέφυγον ἑβδόμης ἡμέρας αὐτῷ τινες ἐστεμμένοι
στέμματα Ἀσκληπίεια· τόδε γὰρ ἦν τὸ ἱερὸν ἐν ἀκροπόλει
μάλιστα τῶν ἄλλων ἐπιφανὲς καὶ πλούσιον, ὅθεν οἵδε τὰς ἱκετηρίας
λαβόντες ἐδέοντο τοῦ Σκιπίωνος περὶ μόνης συνθέσθαι σωτηρίας
τοῖς ἐθέλουσιν ἐπὶ τῷδε τῆς Βύρσης ἐξιέναι. ὁ δὲ ἐδίδου, χωρὶς
αὐτομόλων. καὶ ἐξῄεσαν αὐτίκα μυριάδες πέντε ἀνδρῶν ἅμα καὶ
γυναικῶν, ἀνοιχθέντος αὐτοῖς στενοῦ διατειχίσματος. καὶ οὗτοι
μὲν ἐφυλάσσοντο, ὅσοι δ' αὐτόμολοι Ῥωμαίων ἦσαν . . . ἀπο-
γνόντες αὐτῶν ἐς τὸ Ἀσκληπιεῖον ἀνέδραμον μετ' Ἀσδρούβα καὶ
τῆς γυναικὸς τῆς Ἀσδρούβα καὶ δύο παίδων ἀρρένων. ὅθεν
εὐμαρῶς ἀεὶ ἐμάχοντο, καίπερ ὄντες ὀλίγοι, διὰ τὸ ὕψος τοῦ
τεμένους καὶ τὸ ἀπόκρημνον, ἐς ὃ καὶ παρὰ τὴν εἰρήνην διὰ
βαθμῶν ἑξήκοντα ἀνέβαινον. ὡς δὲ ὅ τε λιμὸς αὐτοὺς καθῄρει καὶ
ἡ ἀγρυπνία καὶ ὁ φόβος καὶ ὁ πόνος, τοῦ κακοῦ προσπελάζοντος,
τὸ μὲν τέμενος ἐξέλιπον, ἐς δὲ τὸν νεὼν αὐτοῦ καὶ τὸ τέγος ἀνέτρε-
χον. κἀν τούτῳ λαθὼν ὁ Ἀσδρούβας ἔφυγε πρὸς τὸν Σκιπίωνα
μετὰ θαλλῶν. καὶ αὐτὸν ὁ Σκιπίων ἐκάθισε πρὸ ποδῶν ἑαυτοῦ καὶ
τοῖς αὐτομόλοις ἐπεδείκνυεν. οἳ δ', ὡς εἶδον, ᾔτησαν ἡσυχίαν
σφίσι γενέσθαι, καὶ γενομένης Ἀσδρούβᾳ μὲν ἐλοιδορήσαντο
πολλὰ καὶ ποικίλα, τὸν δὲ νεὼν ἐνέπρησάν τε καὶ κατεκαύθησαν.
τὴν δὲ γυναῖκα τοῦ Ἀσδρούβα λέγουσιν ἁπτομένου τοῦ πυρὸς . . .
παραστησαμένην τὰ τέκνα εἰπεῖν ἐς ἐπήκοον τοῦ Σκιπίωνος· ' σοὶ
μὲν οὐ νέμεσις ἐκ θεῶν, ὦ Ῥωμαῖε· ἐπὶ γὰρ πολεμίαν ἐστράτευσας.
Ἀσδρούβαν δὲ τόνδε πατρίδος τε καὶ ἱερῶν καὶ ἐμοῦ καὶ τέκνων

προδότην γενόμενον οἵ τε Καρχηδόνος δαίμονες ἀμύναιντο, καὶ
σὺ μετὰ τῶν δαιμόνων.’

While much was being destroyed already, and it seemed likely
that the disaster would become very great, on the seventh day
some suppliants presented themselves to Scipio [in 146 B. C.]
bearing the sacred garlands of Asclepius. For this temple was
on the citadel most conspicuous and richest of all, wherefrom
these (suppliants) took suppliant branches and besought Scipio
to make an agreement concerning the mere salvation of those
who under this covenant were willing to depart from Byrsa.
This he granted, except to the deserters. Forthwith there came
out 50,000 men and women, a narrow gate in the wall being
opened for them. And these were under guard, but the Roman
deserters . . . despairing of their lives, ran up to the temple
of Asclepius with Hasdrubal and his wife and their two boys.
Here they defended themselves a long time with ease, although
they were few in number, on account of the height and pre-
cipitous nature of the precinct, to which even in time of peace
people went up by sixty steps.[1] But when finally hunger, want
of sleep, fear, and weariness overcame them and the catas-
trophe was approaching, they abandoned the precinct and fled
to the temple and its roof. Thereupon Hasdrubal secretly fled
to Scipio, bearing a suppliant branch. Scipio commanded him
to sit at his feet and there showed him to the deserters. When
they saw him, they asked silence, and when it was granted,
they heaped all manners of reproaches upon Hasdrubal, then
set fire to the temple and were consumed in it. It is said that
as the fire was lighted the wife of Hasdrubal . . . setting her
children by her side, said, so as to be heard by Scipio, "For
you, Roman, the gods have no cause of indignation, since you
were waging war against a hostile city. But upon this Hasdrubal, betrayer of his country and of (its) temples and of
myself and his children, may the gods of Carthage take vengeance, and you with the gods." (**)

[1] Cf. T. 708, n. 2; cf. also T. 850, v. 685.

836. Cassius Dio, Historia Romana, XXI, Fr. 71 [Zonaras, 9, 30].

Ἐκεῖνος δὲ μετὰ τῶν αὐτομόλων (ὁ γὰρ Σκιπίων οὐκ ἐσπείσατο
αὐτοῖς) εἰς τὸ Ἀσκληπιεῖον ἀνειλήθη μετὰ τῆς γυναικὸς καὶ τῶν

παίδων, κἀντεῦθεν ἠμύνετο τοὺς προσβάλλοντας, μέχρις οὗ
ἐμπρήσαντες τὸν νεὼν οἱ αὐτόμολοι ἐπὶ τὸ τέγος αὐτοῦ ἀνέβησαν,
τὴν ἐσχάτην τοῦ πυρὸς ἀνάγκην ἀναμένοντες· τότε γὰρ ἡσσηθεὶς
πρὸς τὸν Σκιπίωνα ἦλθεν ἱκετηρίαν ἔχων. ἰδοῦσα δὲ αὐτὸν ἡ
γυνὴ ἀντιβολοῦντα ὀνομαστὶ ἀνεκάλεσεν, καὶ ἐξονειδίσασα ὅτι
ἑαυτῷ τὴν σωτηρίαν πράξας οὐκ ἐπέτρεψεν ἐκείνῃ σπείσασθαι,
τὰ τέκνα ἐνέβαλεν εἰς τὸ πῦρ καὶ ἑαυτὴν προσεπέρριψεν.

He [i. e., Hasdrubal], together with the deserters (for Scipio
would grant no truce to them), crowded into the temple of
Asclepius along with his wife and children; and there he defended himself against the assailants until the deserters, having
set fire to the temple, climbed to its roof to await the last
extremity of the flames.[1] Then, vanquished, he came to Scipio
holding the suppliant branch. His wife, seeing his entreaties,
called him by his name and, having cast reproaches on him for
securing safety for himself when he had not allowed her to
obtain terms, threw her children into the fire and then cast
herself in. (*)

[1] Cf. 836a. Orosius, *Historiae adv. Paganos*, IV, 23, 4: Transfugae, qui Aesculapii
templum occupaverant, voluntario praecipitio dati igne consumpti sunt.

837. Apuleius, Florida, 18.[1]

Nunc quoque igitur principium mihi apud vestras auris auspi-
catissimum ab Aesculapio deo capiam, qui arcem nostrae
Karthaginis indubitabili numine propitius tegit.*

* tegit Krueger   strepit MS.

Even now, therefore, I will make a beginning most auspicious
to your ears by starting with the god Asclepius who protects
the citadel of our Carthage propitiously with his indubitable
power. (*)

[1] Cf. T. 608.

SPAIN

838. Polybius, Historiae, X, 10, 8.

. . . ὧν ὁ μὲν μέγιστος ἀπὸ τῆς ἀνατολῆς αὐτῇ παράκειται,
προτείνων εἰς θάλατταν, ἐφ᾽ οὗ καθίδρυται νεὼς Ἀσκληπιοῦ.

. . . the biggest of (these hills) lies on the east side of the town [*i. e.*, New Carthage] and juts out into the sea, and on it is built a temple of Asclepius.[1]

[1] Cf. T. **708**, n. 2.

ITALY

Sicily

839. Polybius, Historiae, I, 18, 2.

. . . διελόντες οἱ στρατηγοὶ τῶν Ῥωμαίων εἰς δύο μέρη τὴν δύναμιν τῷ μὲν ἑνὶ περὶ τὸ πρὸ τῆς πόλεως Ἀσκληπιεῖον ἔμενον . . .

. . . the Roman generals [in 261 B. C.] divided their force into two bodies, remaining with one near the temple of Asclepius outside the walls[1] [*sc.*, of Acragas] . . .

[1] Cf. T. **708**, n. 3.

840. Cicero, Verrinae Orationes, IV, 43, 93.

Quid? Agrigento nonne eiusdem P. Scipionis monumentum, signum Apollinis pulcherrimum, cuius in femore litteris minutis argenteis nomen Myronis erat inscriptum, ex Aesculapi religiossimo fano sustulisti?

Next, did you not remove at Agrigentum another memorial of Scipio,—a magnificent statue of Apollo on whose thigh was inscribed in small silver letters the name of Myron,—from the much-venerated temple of Asclepius? (**)

841. Polyaenus, Strategemata, V, 2, 19.

Διονύσιος ἐν σπάνει χρημάτων παρὰ τῶν πολιτῶν εἰσφορὰς ᾔτει. τῶν δὲ πολλάκις δεδωκέναι φασκόντων βιάζεσθαι μὲν οὐκ ἔγνω, διαλιπὼν δὲ ὀλίγον ἐκέλευσε τοὺς ἄρχοντας, ὅσα ἦν ἀναθήματα ἐν Ἀσκληπιοῦ—πολλὰ δὲ ἦν ἀργύρου καὶ χρυσοῦ—ταῦτα ἔξω τοῦ ἱεροῦ κομίσαντας ὡς βέβηλα προκηρύττειν ἐπὶ τῆς ἀγορᾶς. οἱ Συρακούσιοι μετὰ πολλῆς σπουδῆς ἐπρίαντο, ὥστε ἠθροίσθη χρημάτων πλῆθος. ὧν κρατήσας Διονύσιος προσέταξε κηρῦξαι, εἴ τίς [τι] τῶν τοῦ Ἀσκληπιοῦ ἀναθημάτων ἠγόρασεν, αὐτίκα εἰς τὸ ἱερὸν ἀνακομίζειν, ἀποδιδόναι τῷ θεῷ, ἢ θάνατον εἶναι τὴν ζημίαν. οἱ μὲν ἀπέδωκαν τῷ θεῷ, Διονύσιος δὲ εἶχε τὰ χρήματα.

Dionysius [*sc.*, the tyrant of Sicily, 4th c. B. C.] being in want of money asked contributions from the citizens. When they reminded him that they had already paid many times, he decided not to compel them; but when a little time had passed, he ordered the authorities to take out of the sanctuary the offerings, so many as were in the temple of Asclepius,—there were a great many in silver and gold,[1]—and to advertise them for sale on the market place. The Syracusans bought them most eagerly, so that a large amount of money was collected. Dionysius having seized this now gave orders to proclaim that, if anybody had bought any of the offerings of Asclepius, he should immediately return them to the temple[2] and restore them to the god, or death should be his penalty. So they returned them to the god, while Dionysius kept the money.

[1] Cf. T. 545; 743.　　　　　[2] Cf. T. 554.

SOUTHERN ITALY

842. Iamblichus, De Vita Pythagorica, 27, 126.[1]

. . . ἐν Ἀσκληπιείῳ . . . καὶ τὸ μεταφερόμενον δὲ ὑπὸ τῶν ἀγνοούντων εἰς τόπους ἑτέρους ἐν Κρότωνι γενέσθαι λέγουσιν, . . .

. . . in the Asclepieion . . . and this story, which by the ignorant is referred to other places, happened in Croton, they say . . .

[1] Cf. T. 487.

843. Julianus, Contra Galilaeos, 200 B.[1]

Ἦλθεν εἰς Πέργαμον, εἰς Ἰωνίαν, εἰς Τάραντα μετὰ ταῦθ' . . .

He [*sc.*, Asclepius] came to Pergamum, to Ionia, to Tarentum afterwards . . .

[1] Cf. T. 307.

844. Livius, Ab Urbe Condita, XLIII, 4, 7.

. . . tabulis quoque pictis ex praeda fanum Aesculapi exornavit.

. . . he [*i. e.*, the tribune Lucretius, in 170 B. C.] also decorated the temple of Asclepius [*sc.*, at Antium] [1] with pictures taken from among the spoils.

[1] Cf. T. **848**; **849**; cf. however T. **850**, v. 722, where only an Apollo temple at Antium is mentioned.

ROME

845. Strabo, Geographica, XII, 5, 3.

Ἐπιφανὲς δ᾽ ἐποίησαν Ῥωμαῖοι τὸ ἱερόν, ἀφίδρυμα ἐνθένδε τῆς θεοῦ μεταπεμψάμενοι κατὰ τοὺς τῆς Σιβύλλης χρησμούς, καθάπερ καὶ τοῦ Ἀσκληπιοῦ τοῦ ἐν Ἐπιδαύρῳ.

The Romans made the temple [*sc.*, at Pessinus] famous when they sent for the statue of the goddess there, in accordance with the oracles of the Sibyl, just as they did in the case of that of Asclepius at Epidaurus. (*)

846. Livius, Periocha, XI.

Cum pestilentia civitas laboraret, missi legati, ut Aesculapi signum Romam ab Epidauro transferrent, anguem, qui se in navem eorum contulerat, in quo ipsum numen esse constabat, deportaverunt; eoque in insulam Tiberis egresso eodem loco aedis Aesculapio constituta est.

When the state was troubled with a pestilence [in 292 B. C.], the envoys dispatched to bring over the image of Asclepius from Epidaurus to Rome fetched away a serpent which had crawled into their ship and in which it was generally believed that the god himself was present. On the serpent's going ashore on the island of the Tiber, a temple was erected there to Asclepius.[1]

[1] Cf. T. **757**; T. **691**; cf. however, T. **583**.

847. Livius, Ab Urbe Condita, XXIX, 11, 1.

Nullasdum in Asia socias civitates habebat populus Romanus; tamen memores Aesculapium quoque ex Graecia quondam hauddum ullo foedere sociata valetudinis populi causa arcessitum, . . . decernunt . . .

As yet the Roman people had none of the states of Asia in alliance with them. Recollecting, however, that formerly Asclepius, on account of a sickness among the people, was fetched from Greece, which was not then united with them by any treaty . . . they resolved [in 205 B. C.] . . .

848. Valerius Maximus, Facta et Dicta Memorabilia, I, 8, 2.

Sed ut ceterorum quoque deorum propensum huic urbi numen exequamur, triennio continuo vexata pestilentia civitas nostra, cum finem tanti et tam diutini mali neque divina misericordia neque humano auxilio inponi videret, cura sacerdotum inspectis Sibyllinis libris animadvertit non aliter pristinam recuperari salubritatem posse quam si ab Epidauro Aesculapius esset accersitus. itaque eo legatis missis unicam fatalis remedii opem auctoritate sua, quae iam in terris erat amplissima, impetraturam se credidit. neque eam opinio decepit. pari namque studio petitum ac promissum est praesidium, e vestigioque Epidauri Romanorum legatos in templum Aesculapii, quod ab eorum urbe $\bar{V}$ passuum distat, perductos ut quidquid inde salubre patriae laturos se existimassent pro suo iure sumerent benignissime invitaverunt. quorum tam promptam indulgentiam numen ipsius dei subsecutum verba mortalium caelesti obsequio comprobavit: si quidem is anguis, quem Epidauri raro, sed numquam sine magno ipsorum bono visum in modum Aesculapii venerati fuerunt, per urbis celeberrimas partes mitibus oculis et leni tractu labi coepit, triduoque inter religiosam omnium admirationem conspectus haud dubiam prae se adpetitae clarioris sedis alacritatem ferens, ad triremem Romanam perrexit, paventibusque inusitato spectaculo nautis eo conscendit, ubi Q. Ogulni legati tabernaculum erat, inque multiplicem orbem per summam quietem est convolutus. tum legati perinde atque exoptatae rei conpotes, expleta gratiarum actione cultuque anguis a peritis excepto laeti inde solverunt, ac prosperam emensi navigationem postquam Antium appulerunt, anguis, qui ubique in navigio remanserat, prolapsus in vestibulo aedis Aesculapii murto frequentibus ramis diffusae superimminentem excelsae altitudinis palmam circumdedit, perque tres dies, positis quibus vesci solebat, non sine magno metu legatorum ne inde in triremem reverti nollet Antiensis templi hospitio usus, urbi se nostrae advehendum restituit, atque in ripam Tiberis egressis legatis in insulam, ubi templum dicatum

est, tranavit, adventuque suo tempestatem, cui remedio quae-
situs erat, dispulit.

But let me report also about the friendly spirit of the other
gods towards this city [*i. e.*, Rome]. When for three continuous years our community had been troubled by a pestilence,
and it seemed impossible either through divine grace or through
human aid to put an end to such a long-lasting evil, it was
discovered through the study of the priests who consulted the
Sibylline Books that the former state of health could not be
restored unless Asclepius were summoned from Epidaurus.
By sending a legation to this place our city believed, therefore,
that through its authority which was then already well established all over the world, it would be able to obtain the extraordinary help of the fateful remedy. And it was not far wrong
in this belief. For with equal eagerness protection was asked
for and promised. And instantly the Epidaurians escorted the
ambassadors of the Romans into the temple of Asclepius —
which was located at a distance of five thousand steps from
their city [1] — and most kindly invited them to feel entitled to
take and carry home whatever they might deem wholesome for
their country. This prompt indulgence of the citizens was
instantly followed by the divine will of the god himself who
confirmed the words of the mortals with heavenly complaisance :
if indeed the serpent, who was seen rarely, though never with
small benefit, and was worshipped by the Epidaurians as an
epiphany of Asclepius, began to glide through the most conspicuous parts of the town, with mild eyes and soft movements ;
and on the third day, under the pious admiration of all the
people, he was seen exhibiting definite eagerness to reach a
more renowned abode and turned toward the Roman trireme.
While the crew was trembling in view of the unusual spectacle,
he entered the boat at the very spot where the ambassador
Q. Ogulnius had his tent, and there curled himself up most
quietly in a circle of many folds. Whereupon the ambassadors,
as if they had attained what they had wished for, performed the
ceremony of thanksgiving and learned from the priests the rites
of the serpent. Then they gladly lifted anchor and left. At the end
of a happy trip, after they had reached Antium,[2] the serpent,
who all the time had remained in the vessel, escaped, and in

front of the temple of Asclepius he wound himself round a palm tree of tremendous height which was towering above the myrtle grove with its many widespread boughs. And for three days, while they gave him what he was used to being fed on, the ambassadors had great fear lest he might not return from there into the trireme, using the temple at Antium as a refuge. But again he betook himself to approaching our city, and when the ambassadors had disembarked at the bank of the Tiber, he swam across to the island where a temple has been dedicated; and through his arrival he dispelled the plague for the cure of which he had been summoned.[3]

[1] Cf. T. **708** and **708**, n. 3.
[2] Cf. T. **844**; **849**; **850**, v. 722.
[3] Cf. **848a**. Valerius Maximus, *Iulii Paridis Epitoma*, I, VIII, 2: Ob pestilentiam continuam ab Epidauro Aesculapius accersitus, cum immensae magnitudinis serpens legatorum navem intrasset, Romam pro deo advectus in ripam Tiberis egressus est, et in insula continenti dicatus illic numen suum posuit; cf. also Valerius Maximus, *Nepotiani Epitoma*, IX, 3.

849. Anonymus, De Viris Illustribus, 22, 1-3.

Romani ob pestilentiam responso monente ad Aesculapium Epidauro arcessendum decem legatos principe Q. Ogulnio miserunt. qui cum eo venissent et simulacrum ingens mirarentur, anguis e sedibus eius elapsus venerabilis, non horribilis, per mediam urbem cum admiratione omnium ad navem Romanam perrexit et se in Ogulnii tabernaculo conspiravit. legati deum vehentes Antium pervecti, ubi per mollitiem maris anguis proximum Aesculapii fanum petiit et post paucos dies ad navem rediit; et cum adverso Tiberi subveheretur, in proximam insulam desilivit, ubi templum ei constitutum et pestilentia mira celeritate sedata est.

The Romans sent ten ambassadors under the leadership of Q. Ogulnius to summon Asclepius from Epidaurus when the response of the oracle regarding a pestilence advised them (to do so). When they had arrived there and were admiring the enormous image[1] a serpent slipped out from under its base, venerable rather than horrible looking; to the amazement of all the people he crept through the middle of the town towards the Roman vessel and coiled himself up in the tent of Ogulnius. The ambassadors, carrying the god with them, reached Antium,[2] where the serpent swam through the gentle sea to the near-by

temple of Asclepius and after a few days returned to the boat;
and when he was taken further up the Tiber, he jumped out
toward the near-by island where a temple was erected for him
and the pestilence was stopped with miraculous celerity.[3]

[1] Cf. T. 630.
[2] Cf. T. 844; 848; 850, v. 722.
[3] Cf. 849a. Lactantius, *Divinae Institutiones*, II, 7, 13: Illut aeque mirum, quod lue
saeviente Aesculapius Epidauro accitus urbem Romam diuturna pestilentia liberasse
perhibetur; 849b. *ibid.*, II, 16, 11: . . . hinc quod serpens urbem Romam pestilentia
liberavit Epidauro accersitus. nam illuc daemoniarches ipse in figura sua sine ulla
dissimulatione perlatus est, siquidem legati ad eam rem missi draconem secum mirae
magnitudinis advexerunt. Cf. 849c. Orosius, *Historiae adv. Paganos*, III, 22, 5: Nam
tanta ac tam intolerabilis pestilentia tunc corripuit civitatem, ut propter eam quacumque
ratione sedandam libros Sibyllinos consulendos putarint horrendumque illum Epidaurium
colubrum cum ipso Aesculapi lapide advexerint: quasi vero pestilentia aut ante sedata
non sit aut post orta non fuerit.

850. Ovidius, Metamorphoses, XV, 622-744.

Pandite nunc, Musae, praesentia numina vatum,
(scitis enim, nec vos fallit spatiosa vetustas),
unde Coroniden circumflua Thybridis alti
625   insula Romuleae sacris adiecerit urbis.
Dira lues quondam Latias vitiaverat auras,
pallidaque exsangui squalebant corpora morbo.
funeribus fessi postquam mortalia cernunt
temptamenta nihil, nihil artes posse medentum,
630   auxilium caeleste petunt mediamque tenentis
orbis humum Delphos adeunt, oracula Phoebi,
utque salutifera miseris succurrere rebus
sorte velit tantaeque urbis mala finiat, orant:
et locus et laurus et, quas habet ipse pharetras,
635   intremuere simul, cortinaque reddidit imo
hanc adyto vocem pavefactaque pectora movit:
"quod petis hinc, propiore loco, Romane, petisses,
et pete nunc propiore loco: nec Apolline vobis,
qui minuat luctus, opus est, sed Apolline nato.
640   ite bonis avibus prolemque accersite nostram."
iussa dei prudens postquam accepere senatus,
quam colat, explorant, iuvenis Phoebeius urbem,
quique petant ventis Epidauria litora, mittunt;
quae simul incurva missi tetigere carina,
645   concilium Graiosque patres adiere, darentque,
oravere, deum, qui praesens funera gentis

finiat Ausoniae: certas ita dicere sortes.
dissidet et variat sententia, parsque negandum
non putat auxilium, multi retinere suamque
650 non emittere opem nec numina tradere suadent:
dum dubitant, seram pepulere crepuscula lucem;
umbraque telluris tenebras induxerat orbi,
cum deus in somnis opifer consistere visus
ante tuum, Romane, torum, sed qualis in aede
655 esse solet, baculumque tenens agreste sinistra
caesariem longae dextra deducere barbae
et placido tales emittere pectore voces:
" pone metus! veniam simulacraque nostra relinquam.
hunc modo serpentem, baculum qui nexibus ambit,
660 perspice et usque nota visu, ut cognoscere possis!
vertar in hunc: sed maior ero tantusque videbor,
in quantum debent caelestia corpora verti."
extemplo cum voce deus, cum voce deoque
somnus abit, somnique fugam lux alma secuta est.
665 postera sidereos aurora fugaverat ignes:
incerti, quid agant, proceres ad templa petiti
conveniunt operosa dei, quaque ipse morari
sede velit, signis caelestibus indicet, orant.
vix bene desierant, cum cristis aureus altis
670 in serpente deus praenuntia sibila misit
adventuque suo signumque arasque foresque
marmoreumque solum fastigiaque aurea movit
pectoribusque tenus media sublimis in aede
constitit atque oculos circumtulit igne micantes:
675 territa turba pavet, cognovit numina castos
evinctus vitta crines albente sacerdos;
"en deus est, deus est! animis linguisque favete,
quisquis adest!" dixit "sis, o pulcherrime, visus
utiliter populosque iuves tua sacra colentes!"
680 quisquis adest, iussum venerantur numen, et omnes
verba sacerdotis referunt geminata piumque
Aeneadae praestant et mente et voce favorem.
adnuit his motisque deus rata pignora cristis
et repetita dedit vibrata sibila lingua;
685 tum gradibus nitidis delabitur oraque retro
flectit et antiquas abiturus respicit aras

adsuetasque domos habitataque templa salutat.
inde per iniectis adopertam floribus ingens
serpit humum flectitque sinus mediamque per urbem
690   tendit ad incurvo munitos aggere portus.
restitit hic agmenque suum turbaeque sequentis
officium placido visus dimittere vultu
corpus in Ausonia posuit rate: numinis illa
sensit onus, pressa estque dei gravitate carina;
695   Aeneadae gaudent caesoque in litore tauro
torta coronatae solvunt retinacula navis.
inpulerat levis aura ratem: deus eminet alte
inpositaque premens puppim cervice recurvam
caeruleas despectat aquas modicisque per aequor
700   Ionium zephyris sextae Pallantidos ortu
Italiam tenuit praeterque Lacinia templo
nobilitata deae Scylaceaque litora fertur;
linquit Iapygiam laevisque Amphrisia remis
saxa fugit, dextra praerupta Cocinthia parte,
705   Romethiumque legit Caulonaque Naryciamque
evincitque fretum Siculique angusta Pelori
Hippotadaeque domos regis Temesesque metalla
Leucosiamque petit tepidique rosaria Paesti.
inde legit Capreas promunturiumque Minervae
710   et Surrentino generosos palmite colles
Herculeamque urbem Stabiasque et in otia natam
Parthenopen et ab hac Cumaeae templa Sibyllae.
hinc calidi fontes lentisciferumque tenetur
Liternum multamque trahens sub gurgite harenam
715   Volturnus niveisque frequens Sinuessa columbis
Minturnaeque graves et quam tumulavit alumnus
Antiphataeque domus Trachasque obsessa palude
et tellus Circaea et spissi litoris Antium.
huc ubi veliferam nautae advertere carinam,
720   (asper enim iam pontus erat), deus explicat orbes
perque sinus crebros et magna volumina labens
templa parentis init flavum tangentia litus.
aequore placato patrias Epidaurius aras
linquit et hospitio iuncti sibi numinis usus
725   litoream tractu squamae crepitantis harenam
sulcat et innixus moderamine navis in alta

puppe caput posuit, donec Castrumque sacrasque
Lavini sedes Tiberinaque ad ostia venit.
huc omnis populi passim matrumque patrumque
730   obvia turba ruit, quaeque ignes, Troica, servant,
Vesta, tuos, laetoque deum clamore salutant.
quaque per adversas navis cita ducitur undas,
tura super ripas aris ex ordine factis
parte ab utraque sonant et odorant aera fumis,
735   ictaque coniectos incalfacit hostia cultros.
iamque caput rerum, Romanam intraverat urbem:
erigitur serpens summoque acclinia malo
colla movet sedesque sibi circumspicit aptas.
scinditur in geminas partes circumfluus amnis
740   (Insula nomen habet) laterumque a parte duorum
porrigit aequales media tellure lacertos:
huc se de Latia pinu Phoebeius anguis
contulit et finem specie caeleste resumpta
luctibus inposuit venitque salutifer urbi.

Reveal to me now, O Muses, ye ever-helpful divinities of
bards (for you know, nor has far-stretching time dimmed your
memory), whence did the island bathed by the deep Tiber bring
Coronis' son and set him midst the deities of Rome.

In olden time a deadly pestilence had corrupted Latium's air,
and men's bodies lay wasting and pale with a ghastly disease.
When, weary with caring for the dead, men saw that their
human efforts were as nothing, and that the healers' arts were
of no avail, they sought the aid of heaven, and, coming to
Delphi, situate in the earth's central spot, the sacred oracle of
Phoebus, they begged that the god would vouchsafe with his
health-bringing lots to succour them in their wretchedness and
end the woes of their great city. Then did the shrine and the
laurel-tree and the quiver which the god himself bears quake
together, and the tripod from the inmost shrine gave forth
these words and stirred their hearts trembling with fear:
"What you seek from this place you should have sought, O
Roman, from a nearer place. And even now seek from that
nearer place. Nor have you any need of Apollo to abate your
troubles, but of Apollo's son. Go with kindly auspices and call
on my son." When the senate, rich in wisdom, heard the com-

mands of the god, they sought in what city the son of Phoebus
dwelt, and sent an embassy by ship to seek out the coast of
Epidaurus. When the embassy had beached their curved keel
upon that shore, they betook them to the council of the Grecian
elders and prayed that they would give the god who with his
present deity might end the deadly woes of the Ausonian race;
for thus the oracle distinctly bade. The elders disagreed and
sat with varying minds. Some thought that aid should not be
refused; but the many advised to keep their god and not let go
the source of their own wealth nor deliver up their deity. And
while they sat in doubt[1] the dusk of evening dispelled the
lingering day and the darkness spread its shadows over the
world. Then did the health-giving god seem in your dreams
to stand before your couch, O Roman, even as he is wont to
appear in his own temple,[2] holding his rustic staff[3] in his left
hand and with his right stroking his flowing beard,[4] and with
calm utterance to speak these words: "Fear not! I shall come
and leave my shrine. Only look upon this serpent which twines
about my staff, and fix it on your sight that you may know it.
I shall change myself to this, but shall be larger and shall seem
as great as celestial bodies should be when they change."
Straightway the god vanished as he spoke, and with the voice
and the god sleep vanished too, and the kindly day dawned as
sleep fled. The next morning had put the gleaming stars to
flight when the chiefs, still uncertain what to do, assembled at
the sumptuous temple of the sought-for god and begged him
by heavenly tokens to reveal where he himself wished to abide.
Scarce had they ceased to speak when the golden god,[5] in the
form of a serpent with high crest, uttered hissing warnings of
his presence, and at his coming the statue, altars, doors, the
marble pavement and gilded roof,[6] all rocked. Then, raised
breast-high in the temple's midst, he stood and gazed about
with eyes flashing fire. The terrified multitude quaked with
fear; but the priest, with his sacred locks bound with a white
fillet,[7] recognized the divinity and cried: "The god! behold
the god! Think holy thoughts and stand in reverent silence,
all ye who are in this presence. And, O thou most beautiful,
be this vision of thee expedient for us and bless thou this people
who worship at thy shrine." All in the divine presence worshipped the god as they were bid, repeating the priest's words

after him, and the Romans, too, performed their pious devotions with heart and lips. The god nodded graciously to them and, moving his crest, assured them of his favor and with darting tongue gave forth repeated hisses. Then he glided down the polished steps[8] and with backward gaze looked fixedly upon the ancient altars which he was about to leave, and saluted his well-known home and the shrine where he had dwelt so long. Thence the huge serpent wound his way along the ground covered with scattered flowers, bending and coiling as he went, and proceeded through the city's midst to the harbor guarded by a curving embankment. Here he halted and, seeming with kindly expression to dismiss his throng of pious followers, he took his place within the Ausonian ship. It felt the burden of the deity and the keel was forced deep down by the god's weight. The Romans were filled with joy and, after sacrificing a bull[9] upon the beach, they wreathed their ship with flowers and cast loose from the shore. A gentle breeze bore the vessel on, while the god, rising on high and reclining heavily with his neck resting upon the ship's curving stern, gazed down upon the azure waters. With fair winds he sailed through the Ionian sea and on the sixth morning he reached Italy, sailed past the shores of Lacinium, famed for Juno's temple, past Scylaceum, left Iapygia behind, and, avoiding the Amphrisian rocks upon the left and the Cocinthian crags upon the right, skirted Romethium and Caulon and Narycia; then passed the Sicilian sea and Pelorus' narrow strait, sailed by the home of King Hippotades, past the copper mines of Temesa, and headed for Leucosia and mild Paestum's rose-gardens. Thence he skirted Capreae, Minerva's promontory, and the hills of Surrentum rich in vines; thence sailed to Herculaneum and Stabiae and Parthenope, for soft pleasure founded, and from there to the temple of the Cumaean Sibyl. Next the hot pools [sc., of Baiae] were reached, and Liternum, thick grown with mastic-bearing trees, and the Volturnus, sweeping along vast quantities of sand beneath its whirling waters; Sinuessa, with its thronging flocks of snow-white doves; unwholesome Minturnae and the place [sc., Caieta] named for her whose foster-son [sc., Aeneas] entombed her there; the home of Antiphates, marsh-encompassed Trachas, Circe's land also, and Antium[10] with its hard-packed shore. When to this place the sailors turned their ship with sails full spread (for the sea was rough) the god unfolded his coils and, gliding on with many a sinuous curve and mighty fold, entered his father's temple set on the tawny strand. When the sea had calmed again, the Epidaurian god left his paternal altars and, having enjoyed the hospitality of his kindred deity, furrowed the sandy shore as he dragged his rasping scales along and, climbing up the rudder, reposed his head on the vessel's lofty stern, until he came to Castrum, the sacred seats of Lavinium and the Tiber's mouth. Hither the whole mass of the populace came thronging to meet him from every side, matrons and fathers and the maids who tend thy fires, O Trojan Vesta, and they saluted the god with joyful cries. And where the swift ship floated up the stream incense burned with a crackling sound on altars built in regular order on both the banks, the air was heavy with sweet perfumes, and the smitten victim warmed the sacrificial knife with his blood. And now the ship had entered Rome, the capital of the world. The serpent raised himself aloft and, resting his head upon the mast's top, moved it from side to side, viewing the places fit for his abode. The river, flowing around, separates at this point into two parts, forming the place called the Island; on each side it stretches out two equal arms with the land between. On this spot the serpent-son of Phoebus disembarked from the Latian ship and, resuming his heavenly form, put an end to the people's woes and came to them as health-bringer to their city.

[1] Cf. however, T. 848.
[2] Cf. T. 414 ff.
[3] Cf. T. 689; 630.
[4] Cf. T. 684.
[5] Cf. T. 630.
[6] Cf. T. 739.
[7] Cf. T. 493.
[8] Cf. T. 835.
[9] Cf. T. 517; 536.
[10] Cf. T. 844; 848; 849.

851. Claudianus, De Consulatu Stilichonis, III, 171-73.

Huc depulsurus morbos Epidaurius hospes
reptavit placido tractu, vectumque per undas
insula Paeonium texit Tiberina draconem.

Hither [sc., to Rome] to expel disease came, gliding with quiet motion, the Epidaurian friend, and Tiber's isle gave shelter to the Paeonian serpent[1] who had been carried across the sea. (*)

[1] Cf. T. 614: . . . qui [sc., Asclepius] quondam versus in anguem / templa Palatinae subiit sublimia Romae.

852. Arnobius, Adversus Nationes, VII, 44-48.

Post advectos, inquitis, transmarinis ex gentibus deos quosdam
postque condita his templa, post cumulatas sacrificiis aras male
habens sese recreatus convaluit populus et pestilentes morbi
inducta sinceritate fugerunt. qui, effamini, dii quaeso? Aesculapius, inquitis, Epidauro, bonis deus valetudinibus praesidens
et Tiberina in insula constitutus. si esset nobis animus scrupulosius ista tractare, vobis ipsis obtineremus auctoribus, minime
illum fuisse divum, qui conceptus et natus muliebri alvo esset,
qui annorum gradibus ad eum finem ascendisset aetatis in quo
illum vis fulminis, vestris quemadmodum litteris continetur, et
vita expulisset et lumine. sed quaestione ab ista discedimus,
Coronidis filius sit, ut vultis, ex immortalium numero et perpetua praeditus sublimitate caelesti. ex Epidauro tamen quid
est aliud adlatum nisi magni agminis coluber? fidem si an-
nalium sequimur et exploratam eis adtribuimus veritatem,
nihil, ut conscriptum est, aliud. quid ergo dicemus? Aesculapius iste quem praedicatis, deus praestans, sanctus deus,
salutis dator, valetudinum pessimarum propulsator, prohibitor
et extinctor, serpentis est forma et circumscriptione finitus, per
terram reptans, caeno natis ut vermiculis mos est, solum mento
radit et pectore, tortuosis voluminibus se trahens, atque ut
pergere prorsus possit, partem sui postremam conatibus prioris
adducit. et quoniam legitur usus cibis etiam, quibus vita in
corporibus inmoratur, habet patulas fauces, quibus cibos trans-
voret oris hiatibus adpetitos, habet receptaculum ventris et ubi
mansa et vorata decoquat viscera, sanguis detur ut corpori et
viribus redintegratio subrogetur, habet et extremos tramites,
per quos inmunda faex eat aversabili corpora foeditate deo-
nerans. si quando mutat loca et ab aliis transgredi in alias
regiones parat, non ut deus obscure per caeli evolat sidera
punctoque in temporis ubi causa postulaverit sistitur, sed velut
animal brutum vehiculum quo sustineatur petit, undas pelagi
vitat atque ut tutus possit incolumisque praestari, cum hominibus navem ascendit et ille publicae sanitatis deus fragili se ligno
et tabularum compagibus credit. non arbitramur evincere atque
obtinere vos posse, Aesculapium illum fuisse serpentem, nisi
hunc colorem volueritis inducere, ut in anguem dicatis con-
vertisse se deum, quominus curiose possent quisnam esset aut
qualis homines intueri. quod si fuerit a vobis dictum, quam

infirmiter invalideque dicatur, ipsa rerum inaequalitas indicabit.
si enim se deus videri ab hominibus evitabat, nec in forma serpentis velle debuit conspici, cum in qualibet forma non ab ipso
se alius, sed ipse esset futurus. sin autem intenderat cernendum
se dare — non enim debuit oculorum negare se conspectui —
cur non talem videri se praebuit, qualem sciebat se esse qua-
lemque se noverat sui numinis in potentia contineri? erat enim
hoc potius multoque praestantius augustaeque conveniens digni-
tati quam fieri beluam horrentisque animalis in similitudinem
verti et dare ambiguis contradictionibus locum, essetne verus
deus an nescio quid aliud longeque ab supera sublimitate
seiunctum. sed si deus, inquit, non erat, cur e navi postquam
extulit sese et Tiberinam ad insulam repsit, nusquam statim
conparuit et viderier ut ante desiit? possumus enim scire,
utrumne aliquod obstaculum fuerit, cuius sese obiectu atque
oppositione protexerit, an hiatus aliquis? vos pronuntiate, vos
dicite, quidnam illud fuerit aut cuinam rerum generi debuerit
applicari, si personarum officia sunt certa certarum. vestra cum
res ista sit deque vestro numine vestraque et religione tractetur,
vestrum est potius edocere vestrumque monstrare, quid illa res
fuerit, nostras velle quam exaudire sententias nostraque ex-
spectare decreta. nam nos quidem quid aliud possumus dicere,
nisi quod fuit et visum est, quod historiae prodidere omnes et
oculorum sensibus est conprehensum? hanc tamen scilicet
colubram validissimi corporis et prolixitatis inmensae aut,
si nomen hoc sordidum est, anguem dicimus, serpentem nomi-
nitamus, aut si quod aliud nobis usus vocamen obtulerit aut
ampliatio sermonis ecfinxerit. si enim repsit ut coluber non
pedibus se ferens neque suas subexplicans itiones sed ventre
nisus ac pectore, si ex materia formatus carnis longitudinem
porrigebatur in lubricam, caput si habuit atque caudam, si obsita
squamis terga, si macularum corium suffectionibus varium, si
os dentibus horridum et ad infligendos instructum morsus:
quid aliud possumus quam generis eum dicere fuisse terreni,
quamvis fuerit inmanis et nimius, quamvis illum ab Regulo
exercitus vi caesum longitudine corporis et robore anteierit?
sed aliud nos remur et labefactamus et destruimus veritatem.
ergo vestrum est explicare, quidnam ille fuerit vel cuius generis,
nominis et qualitatis cuius. nam deus esse qui potuit, cum
haberet ea quae diximus, quae dii habere non debent, si cogitant

dii esse et vocaminis huius eminentiam possidere? Tiberinam postquam ad insulam repsit, nusquam continuo apparuit: qua ex re numen fuisse monstratur. possumus enim scire, utrumne istic aliquod rei fuerit alicuius obstaculum, cuius sese obiectu atque oppositione protexerit, an hiatus aliquis aut ex molibus inaequaliter aggeratis successus quidam et fornices, in quos intulit se raptim circumscripto tuentium visu? quid enim, si flumen transilivit? quid, si transnatavit? quid, si silvarum densitatibus se dedit? argumentatio flaccida est, ea re suspicari deum illum fuisse serpentem, quod ab oculis sese properata agminis festinatione subtraxit, cum deum non fuisse eadem rursus possit argumentatione monstrari. sed si deus praesens anguis ille non fuit, cur post illius adventum pestilentiae vis fracta est et populo salus est reddita Romano? referimus et nos contra: si ex libris fatalibus vatumque responsis invitari ad urbem deus Aesculapius iussus est, ut ab luis contagio morbisque pestilentibus tutam eam incolumemque praestaret, et venit non aspernatus, ut dicitis, colubrarum in formam conversus: cur totiens Romana civitas mali huius afflicta est cladibus, totiens aliis aliisque temporibus dilacerata, vexata est et innumeris stragibus civium minor facta est millibus? cum enim deus adcitus in hoc esse dicatur, ut omnino omnes causas quibus pestilentia conflabatur averteret, sequebatur ut civitas intacta esse deberet flatuque a noxio inmunis semper innocuaque praestari. atquin videmus, ut superius dictum est, saepenumero his morbis cursus eam vitae habuisse funestos nec dispendiis levibus esse populi fractas debilitatasque virtutes. ubi ergo Aesculapius fuit, ubi ille promissus oraculis venerabilibus? cur templa post condita sibique exaedificata delubra diutius aditus habere perpessus est bene meritae civitatis luem, cum in id esset adcitus, ut et malis mederetur instantibus nec sineret in futurum tale aliquid quod metueretur inrepere? nisi forte aliquis dicet, minoribus et consequentibus saeculis idcirco dei talis defuisse custodiam, quod impiis iam moribus et inprobabilibus viveretur, opem autem contulisse maioribus, quod innoxii fuerint et ab omni scelerum contagione dimoti. quod ratione cum aliqua et audiri forsitam potuisset et dici, si aut in temporibus priscis omnes essent usque ad unum boni aut sequentia tempora malos omnes generarent et nulla diversitate discretos. cum vero res ita sit, ut in magnis populis, nationibus, quin immo et in civi-

tatibus cunctis mixtum sit humanum genus naturis voluntatibus
moribus tamque potuerint in prioribus saeculis quam in novellis
aetatibus boni simul malique existere, stultum satis est dicere,
propter malitias posteros auxilia numinum non meruisse mor-
tales. si enim propter malos sequentium saeculorum boni non
sunt protecti temporum novellorum, et propter antiquos malos
boni aeque maiores non debuerant mereri benivolentiam numinum: sin propter bonos priscos mali etiam conservati sunt
prisci, et propter bonos minores aetas debuit sequens, quamvis
esset inprobabilis, protegi. aut ergo iam fracta atque inminuta
vi morbi anguis ille perlatus famam conservatoris adsumpsit,
cum nihil omnino commoditatis attulisset, aut fatalia dicenda
sunt carmina multum veris aberravisse praesagiis, cum remedium ab his datum non deinceps cunctis sed auxilio fuisse uni
tantum reperiatur aetati.

After certain gods were brought from among nations dwelling
beyond the sea, you say, and after temples were built to them,
after their altars were heaped with sacrifices, the plague-stricken
(Roman) people grew strong and recovered, and the pestilence
fled before the soundness of health which arose. What gods,
I beseech, do you say? Asclepius, you say, from Epidaurus,
the god who supervises good health, and who is established on
the Tiberine island. If we were disposed to be very scrupulous
in dealing with your assertions, we might prove by your own
authority that he was by no means divine who had been conceived and born from a woman's womb, who had by yearly
stages reached that term of life at which, as is related in your
books, a thunderbolt drove him at once from life and light.
But we leave this question: let the son of Coronis be, as you
wish, one of the immortals, and possessed of the everlasting
blessedness of heaven. From Epidaurus, however, what was
brought except an enormous serpent? If we trust the annals,
and ascribe to them well-ascertained truth, nothing else, as
has been recorded. What shall we say then? That Asclepius,
whom you extol, an excellent, a venerable god, the giver of
health, the averter, preventer, destroyer of sickness, is contained
within the form and outline of a serpent, crawling along
the earth as worms are wont to do which spring from mud;
he rubs the ground with his chin and breast, dragging himself

in sinuous coils; and that he may be able to go forward, he draws on the last part of his body by the efforts of the first.

And as we read that he used food also, by which bodily existence is kept up, he has a large gullet, that he may gulp down the food sought for with gaping mouth; he has a belly to receive it, and bowels where to digest what he has eaten and devoured, that blood may be given to his body, and his strength recruited; he has also a draught, by which the filth is got rid of, freeing his body from disagreeable burden. Whenever he changes his place, and prepares to pass from one region to another, he does not as a god fly secretly through the stars of heaven, and stand in a moment where something requires his presence, but, just as a dull animal (of earth), he seeks a conveyance on which he may be borne; he avoids the waves of the sea; and that he may be safe and sound, he goes on board ship along with men; and that god of the common safety trusts himself to weak planks and to sheets of wood joined together. We do not think that you can prove and show that that serpent was Asclepius, unless you choose to bring forward this pretext, that you should say that the god changed himself into a snake, in order that men might not curiously observe who he was, or of what nature. But if you say this, the inconsistency of your own statements will show how weak and feeble such a defence is. For if the god shunned being seen by men, he should not have chosen to to be seen in the form of a serpent, since in any form whatever he was not to be other than himself, but (always) himself. But if, on the other hand, he had been intent on allowing himself to be seen—it was not necessary for him to withdraw from onlooking eyes—why did he not show himself such as he knew that he was and such as he considered himself in his own divine power? For this was preferable, and much better and more befitting his august majesty, than to become a beast, and be changed into the likeness of a terrible animal, and afford room for objections, which cannot be decided, as to whether he was a true god, or something different and far removed from the exalted nature of deity. But, says (my opponent), if he was not a god, why, after he left the ship, and crawled to the island in the Tiber, did he immediately become invisible, and cease to be seen as before? Can we indeed know whether there was anything in the way under cover of which he hid himself,

or any opening (in the earth)? Do you declare, say yourselves,
what that was, or to what race of beings it should be referred,
if your service of certain personages is (in itself) certain.
Since the case is thus, and the discussion deals with your deity,
and your religion also, it is your part to teach, and yours to
show what that was, rather than to wish to hear our opinions
and to await our decisions. For we, indeed, what else can we
say than that which took place and was seen, which has been
handed down in all the narratives, and has been observed by
means of the eyes? This, however, we call, of course, a colubra
of very powerful frame and immense length, or, if that name
is despicable, we say snake, we call it a serpent, or any other
name which usage has afforded to us, or the development of
language devised. For if it crawled as a serpent, not supporting itself and walking on feet, but resting upon its belly and
breast; if, being made of fleshly substance, it (lay) stretched
out in slippery length; if it had a head and tail, a back covered
with scales, a skin diversified by spots of various colors; if it
had a mouth bristling with fangs, and ready to bite, what else
can we say than that it was of earthly origin, although of immense and excessive size, although it exceeded in length of
body and (greatness) of strength that which was slain by
Regulus through the power of his army? But (if) we think
otherwise, we subvert and overthrow the truth. It is yours,
then, to explain what that was, or what was its origin, its name,
and nature. For how could it have been a god, seeing that it
had those things which we have mentioned, which gods should
not have if they intend to be gods, and to possess this exalted
title? After it crawled to the island in the Tiber, forthwith it
was nowhere to be seen, by which it is shown that it was a
deity. Can we, then, know whether there was anything in
the way under cover of which it hid itself, or some opening
(in the earth), or some caverns and vaults, caused by huge
masses being heaped up irregularly, into which it hurried,
evading the gaze of the beholders? For what if it leaped across
the river? what if it swam across it? what if it hid itself in
the dense forests? It is weak reasoning from this to suppose
that that serpent was a god because with all speed it withdrew
itself from the eyes (of the beholders), since by the same
reasoning, it can be proved, on the other hand, that it was
no god.

But if that snake was not a present deity, why, after its arrival, was the violence of the plague overcome, and health restored to the Roman people? We, too, ask in reply: If, according to the books of the fates and the responses of the seers, the god Asclepius was ordered to be invited to the city, that he might cause it to be safe and sound from the contagion of the plague and of pestilential diseases, and came without spurning (the proposal) contemptuously, as you say, changed into the form of serpents,—why has the Roman state been so often afflicted with the disaster of this evil, so often at one time and another been torn, harassed, and diminished by thousands, through the destruction of its citizens times without number? For since the god is said to have been summoned for this purpose, that he might drive away utterly all the causes by which pestilence was excited, it followed that the state should be safe, and should be always maintained free from pestilential blasts, and unharmed. But yet we see, as was said before, that it has over and over again had seasons made mournful by these diseases, and that the manly vigor of its people has been shattered and weakened by no slight losses. Where, then, was Asclepius? where that (deliverer) promised by venerable oracles? Why, after temples were built, and shrines reared to him, did he allow a state deserving his favor to be any longer plague-stricken, when he had been summoned there for the very purpose, that he should cure the diseases which were then raging, and not allow in the future that any such thing as they feared should creep in?

But some one will perhaps say that the care of such a god has been denied to later and following ages, because the ways in which men now live are impious and objectionable; that it brought help to our ancestors, on the contrary, because they were blameless and guiltless. Now this might perhaps have been listened to, and said with some reasonableness, either if in ancient times all were good without exception, or if later times produced only wicked people and no others. But since this is the case that in great peoples, in nations, nay, in all cities even, men have been of mixed natures, wishes, manners, and the good and bad have been able to exist at the same time in former ages, as well as in modern times, it is rather stupid to say that mortals of a later day have not obtained the aid of

the deities on account of their wickedness. For if on account
of the wicked of later generations the good men of modern
times have not been protected, on account of the ancient evil-
doers also the good of former times should in like manner not
have gained the favor of the deities. But if on account of the
good of ancient times the wicked of ancient times were preserved also, the following age, too, should have been protected,
although it was faulty, on account of the good of later times.
So, then, either that snake gained the reputation of (being)
a savior while he had been of no service at all, through his
being brought (to the city) when the violence of the disease
was already weakened and impaired, or the hymns of the fates
must be said to have been far from giving true indications
since the remedy given by them is found to have been useful,
not to all in succession, but to one age only. (*)

853. Augustinus, De Civitate Dei, III, 17.

. . . vel quando item alia intolerabili pestilentia Aesculapium
ab Epidauro quasi medicum deum Roma advocare atque adhi-
bere compulsa est, quoniam regem omnium Iovem, qui iam
diu in Capitolio sedebat, multa stupra, quibus adulescens vaca-
verat, non permiserant fortasse discere medicinam?

. . . or when Rome was driven by another intolerable plague,
to summon and invite Asclepius[1] from Epidaurus as a divine
physician, as it were; because the frequent adulteries with
which the king of all, Jupiter, who already had been residing
so long in the Capitol, had amused himself as a young man
had perhaps not left him any leisure to study medicine? (**)

[1] Cf. **853a.** Augustinus, *De Civitate Dei*, III, 12: Aesculapius autem ab Epidauro
ambivit ad Romam, ut peritissimus medicus in urbe nobilissima artem gloriosius
exerceret; cf. **853b.** *ibid.*, X, 16: . . . quod Epidaurius serpens Aesculapio naviganti
Romam comes adhaesit . . .

854. Augustinus, De Civitate Dei, III, 17.[1]

Quid? illa itidem ingens pestilentia, quamdiu saeviit. quam
multos peremit! quae cum in annum alium multo gravius
tenderetur frustra praesente Aesculapio, aditum est ad libros
Sibyllinos . . . tunc ergo dictum est eam esse causam pestilentiae, quod plurimas aedes sacras multi occupatas privatim

tenerent : sic interim a magno imperitiae vel desidiae crimine Aesculapius liberatus est.

And that other great pestilence, which raged so long and carried off so many; what shall I say of it? When in its second year it only grew worse, while Asclepius was present in vain, recourse was had to the Sibylline books . . . In this instance, the cause of the plague was said to be that so many temples had been used as private residences. And thus Asclepius for the present escaped the charge of either want of skill or ignominious negligence. (*)

[1] Cf. T. 363.

855. Ovidius, Fasti, I, 290-94.

. . . .

    sacravere patres hac duo templa die.
accepit Phoebo nymphaque Coronide natum
    insula, dividua quam premit amnis aqua.
Iuppiter in parte est; cepit locus unus utrumque
    iunctaque sunt magno templa nepotis avo.

. . . on this day [sc., the 1st of January, 291 B. C.] the senate dedicated two temples.[1] The island, which the river hems in with its parted waters, received him whom the nymph Coronis bore to Phoebus. Jupiter has his share of the site. One place found room for both, and the temples of the mighty grandsire and the grandson are joined together.

[1] Cf. **855a.** *Fasti Philocali, Ad III. Id. Sept.*: N(atalis) Asclepi.

856. Plinius, Naturalis Historia, XXIX, 1 (8), 16.

Non rem antiqui damnabant, sed artem, maxime vero quaestum esse manipretio vitae recusabant. ideo templum Aesculapii, etiam cum reciperetur is deus, extra urbem fecisse iterumque in insula traduntur . . .

The ancients did not condemn the thing itself [*i. e.*, healing] but the art [*sc.*, of medicine], and they rejected most of all that profit be made with life being the price. For this reason they are said to have built the temple of Asclepius, even after this god was received, outside the city,[1] and, again, on the island . . . (*)

[1] Cf. T. 708, n. 3.

857. Dionysius Halicarnasensis, Antiquitates Romanae, V, 13, 4.

Καὶ ἔστι νῦν μνημεῖον ἐμφανὲς τοῦ ποτε ἔργου νῆσος εὐμεγέθης
Ἀσκληπιοῦ ἱερά, περίκλυστος ἐκ τοῦ ποταμοῦ . . .

And there exists even now a conspicuous memorial of the former deed [*i. e.*, of throwing grain into the Tiber], the large island, sacred to Asclepius, which is washed all round by the river . . .

858. Suetonius, Claudius, 25, 2.

Cum quidam aegra et adfecta mancipia in insulam Aesculapi taedio medendi exponerent, omnes qui exponerentur liberos esse sanxit, nec redire in dicionem domini, si convaluissent; quod si quis necare quem mallet quam exponere, caedis crimine teneri.

When certain men were exposing their sick and worn-out slaves on the island of Asclepius because of the trouble of treating them, he [*i. e.*, Claudius] decreed that all who were exposed should be free,[1] and that if they recovered, they should not return to the control of their master; but if anyone preferred to kill such a slave rather than to expose him he should be liable to the charge of murder. (*)

[1] Cf. T. 718a.

859. Sidonius Apollinaris, Epistulae, I, 7, 12.

. . . capite multatus in insulam coniectus est serpentis Epidauri . . .

. . . he [*i. e.*, Arvandus, the Prefect of Gaul, in 469 A. D.] was sentenced to death and taken to the island of the serpent of Epidaurus . . . (*)

860. Cassius Dio, Historia Romana, XLVII, 2, 3.

> . . . καὶ ἐν τῷ Ἀσκληπιείῳ* μέλισσαι ἐς τὴν ἄκραν πολλαὶ
> συνεστράφησαν . . .

* τῷ Ἀσκληπιείῳ Boissevain τοῖς ἀσκληπείοις LM.

> . . . in the shrine of Asclepius [sc., in Rome, in 43 B. C.]
> bees gathered in swarms on the ceiling . . .

861. Varro, De Lingua Latina, VII, 57.

> Ferentarium a ferendo, . . . aut quod ferentarii equites hi dicti
> qui ea modo habebant arma quae ferrentur, ut iaculum. huius-
> cemodi equites pictos vidi in ⟨A⟩esculapii ⟨a⟩ede vetere et
> ferentarios ascriptos.

> 'Ferentarius,' from *ferre* 'to carry' . . . ; or because 'feren-
> tarii' were called those knights[1] who had only weapons which
> were carried (*ferrentur*), such as a javelin. Knights of this
> kind I have seen in a painting in the old temple of Asclepius,
> with the label 'ferentarii'. (**)

[1] Cf. T. **787.**

INDEX LOCORUM

MENANDER—*continued*
Papyrus Didotiana, b, 1-15: T. 419
Perikeiromene, 146: T. 618a
Samia, 94-95: T. 619
MINUCIUS FELIX, 3rd c. A. D.
Octavius, ed. J. P. Waltzing, 1926
[Teub.]; tr. G. H. Rendall, Tertullian
and Minucius Felix [Loeb]
VI, 1: T. 737
XXIII, 5: T. 685
XXIII, 7: T. 236
NICANDER, 2nd c. B. C.
Nicandrea, ed. O. Schneider, 1856; F. S.
Lehrs, Poetae Bucolici et Didactici,
1862
Theriaca
438-57: T. 697
685-88: T. 89

SCHOLIA IN NICANDRUM
Nicandrea, ed. O. Schneider, 1856
Ad Theriaca
438: T. 698
685: T. 98
687: T. 90

NICEPHORUS GREGORAS, 14th c. A. D.
Patrologia Graeca, 149 [Migne]
Byzantinae Historiae
XXVI, 20, p. 85 B: T. 479
OLYMPIODORUS, 6th c. A. D.
Commentaria in Aristotelem Graeca,
XII, 1, ed. A. Busse, 1902
Prolegomena, p. 8, 21-24: T. 223
In Platonis Phaedonem Commentaria,
ed. W. Norvin, 1913 [Teub.]
p. 205, 24: T. 526
p. 244, 17: T. 527
Vita Platonis, ed. A. Westermann in
Diogenes Laertius, ed. C. G. Cobet, 1878
p. 4, 39 ff.: T. 322a
ORIBASIUS, 4th c. A. D.
ed. J. Raeder, 1929-31 [C. M. G., VI, 1,
2; VI, 2, 1]; 1926 [C. M. G., VI, 3]
Collectiones Medicae
XIV, 14, 6: T. 370a
XLV, 30, 10-14: T. 425
Synopsis ad Eustathium
III, 162: T. 373
ORIGENES, *ca.* 185-255 A. D.
Contra Celsum, Patrologia Graeca, 11
[Migne]; Opera, I-II, ed. P. Koetschau,
1899; tr. F. Crombie, Origen, II, 1872
[Ante-Nicene Chr. Libr., XXIII]
III, 3, p. 924 B: T. 292; 711
III, 22, p. 944 C: T. 249
III, 23, p. 945 C: T. 112
III, 24, p. 948 A: T. 293
III, 25, p. 948 C: T. 332a

V, 2, p. 1184 A-C: T. 261
Contra Haereses, Patrologia Graeca, 16
[Migne]
IV, 32, p. 3095 B-C: T. 328
In Jeremiam Homilia, Patrologia Graeca,
13 [Migne]
V, 3, p. 300 C: T. 242
OROSIUS, 5th c. A. D.
Historiae Adversum Paganos, ed. C.
Zangemeister, 1882 [Corp. Script. Eccl.
Lat., V]
III, 22, 5: T. 849c
IV, 23, 4: T. 836a
VI, 2, 11: T. 810a
ORPHICA
ed. E. Abel, 1885
Hymni, 3rd c. A. D. (?)
LXVII: T. 601
LXVII, 7: T. 281e
Lithica, 4th c. A. D. (?)
346-54: T. 176
ORPHICORUM FRAGMENTA
ed. O. Kern, 1922
Fr. 202, see PROCLUS: T. 314
Fr. 297a: T. 315
OVIDIUS, 43 B. C.-18 A. D.
Fasti, ed. and tr. Sir J. G. Frazer, Ovid,
Fasti [Loeb]
I, 290-94: T. 855
VI, 743-62: T. 75
Ibis, ed. R. Ehwald-F. W. Levy,
Opera, III, 1, 1922 [Teub.]
403-04: T. 736a
Metamorphoses, ed. R. Merkel–R. Ehwald,
Opera, II, 1928 [Teub.]; ed. and tr. F. J.
Miller, Ovid, Metamorphoses, I-II [Loeb]
II, 534-41: T. 45
II, 542-648: T. 2
XV, 531-36: T. 76
XV, 622-744: T. 850
XV, 654-59: T. 689
OXYRHYNCHUS PAPYRUS, 2nd c. A. D.
The Oxyrhynchus Papyri, XI, ed. and
tr. B. P. Grenfell–A. S. Hunt, 1915
1381: T. 331
PAULUS AEGINETA, 7th c. A. D.
Opera, ed. J. L. Heiberg, 1921-24
[C. M. G., IX, 1-2]
III, 24, 7: T. 373c
IV, 1, 7: T. 373a
VII, 13, 17: T. 373b
PAUSANIAS, *ca.* 170 A. D.
Descriptio Graeciae, I-III, ed. F. Spiro,
1903 [Teub.]; ed. and tr. W. H. S.
Jones, I-IV [Loeb] (II by W. H. S.
Jones and H. A. Ormerod)
I, 21, 4: T. 628
I, 21, 4-5: T. 725

II
INTERPRETATION OF THE TESTIMONIES

PREFACE

There is no generally accepted pattern that one is able to follow in studying any of the ancient deities, nor can the aims and methods of such an investigation be established on general grounds. The divine figures differ so essentially in their nature that the discussion of each of them raises different issues. Besides, the material available in every instance, by its quality and its peculiarity, enforces ever varying rules of analysis. The questions to be asked and the approach to be pursued, therefore, have to be derived from the examination of the object of inquiry. They cannot even be made explicit otherwise than by interpreting the data; their only basis is that afforded by the display of the evidence.

Under these circumstances it would be futile by way of a prefatory statement to formulate or to justify the task which has been evolved for this study of Asclepius, and the manner in which its solution has been sought. The interpretation must be left to speak for itself; it must find vindication or refutation by its results. I should point out in advance only those features which are due not to any singularity of the problem dealt with but rather to the subjective point of view of the interpreter.

No endeavor has been made to assign to Asclepius his proper place among the many deities of antiquity. Although his relation to other gods, and especially that to other healing deities, has occasionally been touched upon, Asclepius himself has remained the focus of attention. Furthermore, there has been no attempt to compare Greek views with primitive notions, or to relate the religious attitude of the Greeks and Romans to that characteristic of other periods or peoples; not even the Oriental transformations of the Asclepius figure have been studied. Finally, all theorizing regarding the origin of religion in general or of Greek religion in particular has been avoided. Human belief in gods, in divine and supernatural powers, has been taken simply as a fact to be acknowledged by the historian.

However, equal emphasis has been laid on the elucidation of realistic details and on the understanding of religious ideas. That is why account has been taken not merely of the character of Asclepius' godhead, of the sacred ritual of his worship, of the dissension between the Asclepius religion and Christianity; the medical activity of the god, the parapher-

nalia of his cult, the chance circumstances of the artistic representation
of the deity, the architecture and history of his temples have been discussed as well. Thus the accent of the interpretation lies as much on
the " how it was " as on the " what it meant."

In other words, although this book aims at an insight into the
essence of the Asclepius figure and of the cult attached to him, it also
intends to provide a commentary, as it were, on the whole ancient testimony on Asclepius preserved in books and on stone.[1]

A few words concerning the composition of the second volume must
be added. To give unity of thought and expression to the interpretation,
it seemed advisable that only one of us should undertake the actual
writing. Yet no solution was definitely accepted, no sentence received
its final form, until we had together examined and weighed it. Nor is
it only in this sense that the interpretation remains our common work.
It would be difficult indeed for me to say in any one instance how far
my views were shaped by these discussions, or how far I followed suggestions made by the partner in a dialogue that continued through
several years.

As in the preparation of the first volume, Paul A. Clement assisted
untiringly in the redaction of the second. He read the manuscript
and improved it by proposing changes in language and by pointing to
those problems that had not found adequate clarification. Owsei Temkin
also went over the typescript. He cautioned me where I had been too
hasty in my judgment, and he encouraged me to look further where I
had been inclined to rest my case. The galley proofs were read by F.
Michael Krouse who subjected the text to a close scrutiny and thus
enabled me to remove deficiencies of the final wording. I ask all of them
to accept my thanks for their help which they extended to me as friends
and scholars.

L. E.

Baltimore,
March, 1945.

---

[1] For a discussion of the principles according to which the material was collected,
cf. the Preface to Vol. I. Every statement made in the analysis refers to the passage
on which it is based by citing the number under which the testimony is listed in the
volume of texts.

TABLE OF CONTENTS

I. THE HERO ASCLEPIUS

The process by which the great deities of antiquity came into being is in most cases hidden from the eyes of the modern interpreter, and it is with conjectures alone that he can penetrate the mystery of their existence. For such gods as Zeus or Apollo assumed power long before the dawn of history, in ages from which no records have survived. Like aboriginal inhabitants of the divine abode they occupied their seats throughout the span of ancient civilization. With Asclepius it stands otherwise. As one of the immortals, as god of medicine, he was revered from the classical period to the end of antiquity. To earlier generations, however, he was known only as a hero who had lived on this earth. For Homer, for Hesiod, even for Pindar, Asclepius was a mortal. To be sure, most scholars nowadays contend that this hero is but one of the "decayed gods" or "gods in disguise," deities, that is, who lost the divine prerogatives which they once possessed, and appeared in the world of the Epic as human beings.[1] Even if this claim proved to be correct, it would still be true that the god Asclepius, who held sway over the mind and the imagination of the ancients, was considered a hero for centuries before his divine dignity was generally recognized. This fact cannot have been without significance for the history of the Asclepius worship, nor can it have been without influence on the formation of the concept of Asclepius' godhead. The career of the hero Asclepius must have constituted at least part of the matter that went into the making of the god. At any rate, there are innumerable testimonies dealing with the hero Asclepius, and testimonies that in point of age if not in value precede those which speak of the god. If justice is to be done to the data transmitted by tradition, the investigator must start with an interpretation of the heroic figure; it is impossible for him to turn his attention to Asclepius, the god, unless he has previously reached an opinion on Asclepius, the hero.

1. THE HOMERIC ASCLEPIUS

It is in the *Iliad* that the hero Asclepius is mentioned for the first time. Here he is represented as one of the aristocrats of old, as a king like Agamemnon or Menelaus, though of lesser might than is given to them.

[1] Cf. *e. g.* Pfister, *Bursian*, 1930, p. 128, and in general, below, p. 65.

1

Does this mean that the saga which Homer followed was that of a warrior, a chieftain of the Greeks? Is the tale of Asclepius and his sons the reflection of historical events, or is it the product of poetical fantasy? And assuming that the Asclepius myth is older than the Epic, in which region of Greece did it originate? Modern views are sharply divided on these fundamental issues, and it is difficult indeed to reach a decision in regard to them. For the Homeric Asclepius is a rather shadowy figure. Reference is made to him only as the father of Machaon and Podalirius who fought before Troy. He himself has no active share in the deeds that are recounted, nor is his personality portrayed in detail even where the mention of his name would have permitted a more elaborate description. It seems necessary, therefore, to consider carefully even such data as are furnished by the Homeric and post-Homeric tradition concerning Asclepius' children. In this way—despite the scantiness of the material—one may perhaps succeed in descrying Asclepius himself as he was seen in the beginning.

i. Asclepius' Heroic Rank

Asclepius, as Homer says in the *Catalogue of Ships* (T. 135), is the father of Machaon and Podalirius, leaders of men from Tricca, Ithome and Oechalia. In another passage, where Machaon alone is mentioned, the poet calls Asclepius the father of Machaon, who is in command of the soldiers from Trice (T. 164, *vv. 201 f.*). Unmistakably, then, the *Iliad* delineates Asclepius as a king whose sons and subjects participated in the common venture of the Greeks.[2]

The ancients did not hesitate to accept the Homeric claim as to Asclepius' rank and time of life. He was generally held to have belonged to that generation of great heroes who of yore had been the pride of the Greeks (T. 5; 127). There was dissension only concerning the question whether he had been a contemporary of the happenings described by Homer, or whether he had died too early to witness them. Some believed that Asclepius had passed away while Podalirius was only a young child (T. 184), that is, before the Trojan War. A few chroniclers consequently fixed the time of Asclepius' demise in the thirty-eighth year

---

[2] Cf. Wilamowitz, *Isyllos*, p. 44: ". . . nach der Ilias ist Asklepios . . . ein . . . König"; Kjellberg, *Asklepios*, pp. 20 f.: "irdischer Herrscher"; W. Leaf, *The Iliad*, I², 1900, *ad* II, 729: "a mortal chieftain." Farnell, *Hero Cults*, pp. 234 f., states that in the view of the Homeric poem Asclepius is a physician. Such a claim is untenable and warranted only by the fact that Farnell does not discuss at all the implications contained in Homer's characteristic of Machaon and Podalirius as princes. But cf. also below, n. 30. For the Homeric Asclepius in general, cf. Walton, *Cult of Asklepios*, Ch. I.

after Heracles became King of Argos (T. 129).[3] Others thought that he was still alive when the Greeks battled the Trojans (T. 92), and this assumption is the earlier one attested.[4] That Asclepius lived in the Heroic Age no one ever questioned.

The modern historian will have little use for the ancients' attempts to determine with such exactness the year of the death of Asclepius. But on the basis of the Homeric testimony he, too, may be inclined to assume that the Asclepius legend, like other heroic sagas, reflects the memory of a warrior or a tribal chief.[5] However, if he scrutinizes the evidence under this supposition he immediately encounters difficulties. For Homer's assertion that Asclepius was a king is strangely at variance with the few data otherwise furnished by the Epic in regard to this hero.

To begin with, the only characterization given of Asclepius bears on his medical knowledge rather than on his authority and dignity as a king. Twice Machaon is called the son of Asclepius, the blameless physician (T. 164, *v. 194*; T. 165, *v. 518*).[6] It is true, other Homeric kings and heroes, too, know medicine and do not recoil from treating wounds. But none of them is ever designated as a physician.[7] Such an epithet, in the world of the *Iliad*, is singular and very different indeed from those usually attached to an epic hero. It is hard to believe that it would have been given to one who really was the peer of princes.

Additional reasons for suspecting Asclepius' royal rank seem to be

[3] Concerning this testimony = Apollodorus, *Fr.* 87 [Jacoby], cf. Jacoby's commentary, *F. Gr. H.*, II, p. 752, 22 ff. Contrary to his earlier opinion he now holds that it is not genuine. If the year here assumed for the fall of Troy were that accepted also by Apollodorus, Asclepius' death would fall in the year 1237 B. C.; if, on the other hand, one of the earlier datings of Troy's destruction is presupposed, Asclepius would have died even sooner. Cf. also below, pp. 91 f.

[4] Sophocles proffers this opinion; cf. also below, pp. 41 f.

[5] Such an interpretation of the heroic saga which prevailed for a long time is now beginning to be discredited; cf. *e. g.* Wilamowitz, *Die griechische Heldensage*, I, *S. B. Berl.*, phil.-hist. Kl., 1925, pp. 38 ff. Still, it must be reckoned with as a possibility, and it has been applied also to the Asclepius myth; for such theories cf. Thraemer, Roscher, *Lexikon, s. v.* Asklepios, I, 1, p. 622, 43 ff., and Wilamowitz, *Isyllos*, p. 44. Of course, for the moment I am disregarding the opinion of those who believe that Asclepius originally was a god.

[6] T. 164 has been emended by Wilamowitz, *Isyllos*, p. 46, n. 3 (cf. also *Glaube*, II, p. 228, n. 2) to show Homer calling Machaon, not Asclepius, a blameless physician (ἀμύμονα ἰητῆρα instead of ἀμύμονος ἰητῆρος). I do not think that this change of the text is permissible; cf. also L. Weber, *Philologus*, LXXXVII, 1932, p. 389, n. 1 a. Even if Wilamowitz' emendation of T. 164 were accepted, T. 165 would still denote Asclepius as a physician. In passing I wish to stress that the epithet ἀμύμων which is here given to Asclepius is never attributed to any god in the Homeric Epic; cf. Panofka, *Asklepios und die Asklepiaden, Berl. Phil.-Hist. Abh.*, 1845, p. 271, n. 2; cf. also T. 257a.

[7] The one exception to this rule is found in connection with the sons of Asclepius (T. 135 f.). In regard to them, cf. below, pp. 5 ff. For the medical knowledge of the Homeric heroes in general, cf. *e. g.* Wilamowitz, *Isyllos*, p. 45, n. 2.

provided by the only incident in the hero's life that is recounted by
Homer. The poet notes that Machaon heals Menelaus with remedies
which his father had once received from Chiron. The Centaur, being
amicably disposed toward Asclepius, gave him some drugs which Asclepius in turn handed over to his son (T. 164, *v. 219*). The few words
referring to Chiron's gift to Asclepius hardly suffice to prove that the
two were placed on equal footing, as has sometimes been maintained.[8]
Such a conclusion might be warranted if an exchange of presents between Chiron and Asclepius were related. The statement as it stands
seems to imply that Chiron was thought to be superior to Asclepius, be
it in rank or knowledge. On the other hand, Homer's report does not
authorize the claim that the Centaur was Asclepius' teacher. While the
poet explicitly states that Achilles, the only other hero whose association
with Chiron he attests, was educated by the sage and among other things
was taught by him to prepare remedies, he says of Asclepius simply that
Chiron presented him with certain drugs.[9] Not until later on, when it
became customary to suppose that every hero of some standing had been
Chiron's pupil, was it thought that Asclepius, too, had been instructed
by him in medicine as well as in the other arts. But in the *Iliad*, the
connection between the two is not pictured in this conventional form of
the heroic saga.[10]

What, then, is the significance of the Homeric story? Is it not
perhaps indicative of the fact that in the legend taken up by Homer
Asclepius was represented as a physician? Would Chiron have donated
healing herbs to Asclepius, had he not trusted that Asclepius, being
especially interested in medicine, would appreciate such a gift? Had he
considered him primarily a warrior, Chiron could have presented him
with a weapon, a token that he found appropriate for a hero like Peleus.
Moreover, Asclepius apparently did value Chiron's gift higher than anything else and thought it most helpful. Otherwise he would hardly
have passed it on to his son, when Machaon went to war. Peleus, when
his son, Achilles, sailed against Troy, gave to him the sword which he
had received from Chiron.[11] On such an occasion to choose the one gift

---

[8] Cf., *e. g.*, Weber, *op. cit.*, p. 391.

[9] Kern, *Religion*, II, pp. 304-5, goes too far, therefore, in asserting that, according to
Homer, Chiron was Asclepius' teacher; cf. Weber, *op. cit.*, p. 391, n. 3. For Achilles
and Chiron, cf. *Iliad*, XI, 832. Achilles (Peleus) and Machaon (Asclepius) are the
only figures in Homer who are connected with Chiron, cf. Leaf, *op. cit.*, *ad* IV, 219.

[10] Concerning the later interpretation of Chiron as the mentor of Asclepius, cf. below,
p. 38.

[11] Concerning Chiron, Peleus and Achilles, cf. *Iliad*, XVI, 143; XIX, 390.

seems suitable for a physician, just as to choose the other is worthy of a prince. Finally, Chiron was not merely the friend and patron of heroes; he was also the inventor of herbal medicine whose descendants were known even to Hellenistic writers as experts in drugs which they had inherited from their father. He was considered the first physician and was worshipped in this capacity by the Magnesians.[12] Throughout antiquity Chiron enjoyed great and lasting fame on account of his medical accomplishments. To Pindar, he was the one who first comes to the poet's mind as a helper for his ailing friend (T. 1, *vv. 1 ff.*), and down to late Roman centuries Chiron was respected as the instigator of the medical art.[13] Surely, it makes good sense that this medical hero should have presented Asclepius with drugs, and it is quite plausible that Asclepius, even if he was a physician rather than a king, should have been his friend.

But the assumption that Asclepius was a physician, a craftsman, seems to be contradicted by Homer's representation of his children. Are they not princes, aristocrats like all the other heroes? Unquestionably, this is the impression conveyed by the Epic. Nevertheless, a closer examination of the rôle played by Machaon and Podalirius in the *Iliad* reveals, I think, that the two are by no means the equals of their fellows-in-arms.

To begin the inquiry with Machaon, who in the *Iliad* is more clearly described than his brother Podalirius: Machaon is said to be in command of the soldiers from Trice (T. 164, *vv. 202 f.*). Yet, does he really act as becomes a leader of men, the son of a king? In the first instance (T. 164), in which Machaon is one of the principal figures of the story unfolded, Agamemnon, in deadly fear for his wounded brother, sends his herald to summon Machaon. The messenger quickly succeeds in finding him on the battle field among his troops, and delivers his message. Immediately the Asclepiad complies with his call. When they arrive at the place where Menelaus lies hurt, Machaon draws forth the arrow, loosens Menelaus' armor, and, having inspected the injury, sucks out the blood and spreads upon the wound the simples which he had received from his father.

Now it is not astonishing—as has been stated before—that a hero should be able to treat wounds. Achilles, Patroclus, and all the other

---

[12] For the Chironidae, cf. *F. H. G.*, II, p. 263 (Dicaearchus, *Fr.* 61); for the Magnesians, cf. Plutarch, *Conv.*, III, 1, 3 (647a) and below, p. 96.

[13] As regards Chiron, cf. Welcker, *Kleine Schriften*, III, pp. 3 ff.; W. A. Jayne, *The Healing Gods of Ancient Civilizations*, 1925, pp. 359-61. For Chiron in the Roman period, cf. *e. g.* Pliny, *Nat. Hist.*, VII, 196.

princes know how to do that.[14] Nor would it be particularly strange that Agamemnon should be so anxious to have Machaon take care of his brother. Machaon may have had more experience in dealing with such cases, and the leader of all Greeks, the commander-in-chief, naturally must secure the services of the man who is most capable of doing the job. However, it is peculiar that Machaon apparently expects such calls. He does not express any astonishment at Agamemnon's summons; he is not too busy fighting to act in accordance with the king's wishes. Moreover, although he is on the field of battle in the midst of his soldiers, he has his simples with him; he does not rush to his tent to get them; he proceeds straightway to Menelaus and renders assistance. Is this not the behavior of a regular physician rather than of a warrior or a hero who if need arises may also dress wounds?

In the second scene in which Machaon is portrayed (T. 165), he is shown in actual fight and is said to be the foremost warrior. Nevertheless his exploits are not recited; that Paris' arrow puts a stop to them is all that the poet has to report.[15] But he adds with meticulous care that the Greeks, who are battling by Machaon's side, are greatly disturbed by his being wounded, and Idomeneus asks Nestor to transport Machaon to the ships, for " a leech is of the worth of many other men " (v. 514). Not only is Machaon here called a physician—the demand to save him is motivated not by his excellence as a warrior but as a doctor. Nestor then takes Machaon back, not to Machaon's tent, but to his own (T. 166). Here Machaon is given wine, here a bath is prepared for him (T. 168), here Patroclus comes to inquire about his condition (T. 167). Why is Machaon not brought to his own people? When Eurypylus is wounded and asks for Patroclus' help he is conducted to his own tent, as is only natural.[16] Does Machaon, the leader of the contingent from Trice, have no servants, no tent, and is he therefore left to Nestor's compassion?

In both scenes, then, in which Machaon appears he behaves like a physician. His actions are not those of a leader of men, nor is he esteemed by the others as a fighter. To be sure, the *Iliad* states that he is in command of the Tricaeans, he is mentioned as a participant in the battle that is raging, but his rank as a chieftain, his soldierly achievements remain mere accessories. I suggest, then, that the heroic character

---

[14] Concerning Achilles and Patroclus, cf. *e. g. Iliad*, XI, 828 ff.

[15] T. 165, *vv. 505-6*: . . . Ἀλέξανδρος . . ./ παῦσεν ἀριστεύοντα Μαχάονα. Scholion T on these lines (T. 165a) also notes the discrepancy between Homer's use of the term ἀριστεία and the content of the verses. Cf. below, n. 53.

[16] Cf. *Iliad*, XI, 828; 843.

of the Machaon figure is an invention of the epic poet. Machaon, the physician, was elevated to the dignity of a hero in order to be acceptable to the circle of heroes in which he was supposed to move. On one occasion he has to play a prominent part in the *Iliad*: he heals the king, Menelaus, and his treatment is described in great detail. Only a hero can perform so important a task, can act so conspicuously on the epic stage.[17] Therefore the physician Machaon must be endowed with the qualities of a prince; he must have men to command. Yet all this heroic attire is nothing but artistic ornament.[18]

It is true, the post-Homeric saga seems to take Machaon, the hero, more seriously than did the *Iliad*. Leaving aside those feats which he accomplishes together with his brother,[19] he is now represented as a fierce fighter: he dies as a soldier, be it through Penthesilea (T. 179), or through Eurypylus (T. 180 ff.), or upon entering Troy (T. 177; 178). Yet it must be stressed that even in the later Epic the actual description of Machaon's deeds is meager and indistinct. Although his very name is supposed to indicate his warlike disposition (T. 139), Eurypylus, the slayer of Machaon, is full of disdain for his opponent, who, he says, hoped to escape his doom through his knowledge of " sooth-

[17] For Machaon's rôle, cf. also below, p. 16. The aristocratic bias of the epic poet hardly needs to be elaborated. But it is interesting in this connection to compare *Iliad*, XIV, 110 ff. In these verses the Greeks are represented as being in difficulties; good advice is sought. Diomedes thinks that he can offer it, and since he is afraid that the leaders will not listen to him on account of his youth, he begins his speech with a long diatribe on his noble descent. High birth legitimates the epic figure and its actions. Physicians, in the Homeric world, are of inferior standing. With the exception of Machaon and Podalirius they have not even individual names, cf. *Iliad*, XIII, 213; XVI, 28. In the *Odyssey*, XVII, 383-4, they are classed together with other craftsmen.

[18] Such "heroizations" are not uncommon in the Epic. It is now generally recognized that the heroic saga, as it is transmitted by Homer, may be the vehicle of stories of quite a different character, cf. *e. g.* Wilamowitz, *S. B. Berl.*, 1925, pp. 59 f.; O. Weinreich in L. Friedlaender, *Sittengeschichte Roms*[9-10], ed. G. Wissowa, IV, 1921, pp. 101 ff. In the case of Diomedes, on the other hand, one can still observe how the dignity of a king is bestowed upon a figure in accordance with the poet's designs. In *Iliad*, XIV, 110 ff., Diomedes is characterized as the son of a wealthy aristocrat. In another passage, *Iliad*, XXIII, 471, he is named a prince; while in *Iliad*, II, 559 ff., he is described as almost the mightiest among the Argive kings, cf. M. P. Nilsson, *Homer and Mycenae*, 1933, p. 259. One might be tempted to assume that the "King" Machaon, who at the same time is a great healer, is one of those ancient kings who are sometimes likened to the "medicine-man." Cf. *e. g.* G. Murray, *The Rise of the Greek Epic*[3], 1924, pp. 135 ff.; *Four Stages of Greek Religion*, 1912, p. 39. Yet, apart from the fact that it seems impossible to trace Asclepius or his children to Mycenaean times—Nilsson, *The Mycenaean Origin of Greek Mythology, Sather Classical Lectures*, VIII, 1932, at any rate has not attempted to do so—the concept of divine kingship, if applicable at all to the interpretation of Greek history, in the Homeric Epic has receded into the background; cf. Nilsson, *Homer and Mycenae*, pp. 223; 220. Medicine, in the *Iliad* and in the *Odyssey*, is considered to be rational knowledge. Cf. in general Ch. Daremberg, *La Médecine dans Homère*, 1865.

[19] Cf. below, pp. 13 f.

ing salves " (T. 181), and the Greeks mourn the death of Machaon, a mighty warrior, as he is called (T. 180, *v. 406*), because he knew " skillful arts " (T. 183, *v. 15*).[20] Moreover, Machaon's share in medical achievements is increased along with his soldierly deeds. He is named as the one who restored the health of Philoctetes (T. 173 ff.).[21] Finally, when Machaon's life previous to the war is set forth, it is again his medical knowledge that is celebrated. After the early death of their father, Machaon educated Podalirius and taught him the art of healing (T. 184).[22] There is hardly any doubt that even in the post-Homeric tradition, just as in the Homeric Epic, Machaon is a physician dressed in the trappings of a hero.

The same holds good of Podalirius. In the *Iliad* this son of Asclepius remains a mere name; Homer is satisfied with taking note of his existence.[23] Like his brother Machaon, he calls him a physician (T. 135 f.), he implies that Podalirius and Machaon are regular army physicians (T. 136).[24] These few data intimate at least that the Homeric Podalirius is hardly a genuine hero.

In the *post-Homerica*, where Podalirius assumes an individual character and is given a more active rôle, the fact that he is a doctor is unmistakable. Leaving aside again the deeds accomplished by the two brothers together, it is Podalirius who treats Ajax (T. 141 f.) and Philoctetes (T. 201);[25] in addition, he heals Epeius and Acamas, who were injured in the games held in honor of Achilles (T. 199), and he also cures Thoas and Eurypylus (T. 200). Like Machaon, Podalirius, too, had received salves from his father with which to assist the wounded

[20] Quintus Smyrnaeus gives the story in detail. He believes that Machaon was killed by Eurypylus—the version of the *Ilias Parva* (T. 186), in which Machaon heals Philoctetes (T. 173). It is probable, therefore, that Quintus' report reflects the tradition of the Cyclic Epic, even if he did not directly copy from this source, as is generally believed, cf. W. Schmid-O. Stählin, *Geschichte d. griech. Literatur*, I, 1, 1929, p. 198, n. 5, and below, n. 72. For Machaon in Quintus' poem, cf. also Pauly-Wissowa, *s. v.* Machaon, XIV, p. 146, 20.

[21] For the rôle which this treatment plays in the Epic, cf. also below, p. 13.

[22] This, too, is attested by Quintus, cf. above, n. 20.

[23] For the difference in importance which the two brothers have in the *Iliad*, cf. also below, pp. 11 ff.

[24] For a detailed interpretation of T. 136, cf. below, p. 11.

[25] The exploits of Machaon and Podalirius are discussed below, pp. 13 f. Incidentally, the same cases are ascribed both to Machaon and to Podalirius respectively. It is impossible to inquire here into all the variants of the late saga concerning Asclepius' children. It may suffice to say that the divergences in regard to their medical achievements are due in part to the differences in dating the death of Machaon. If it was assumed that he was killed by Penthesilea (T. 179), he could not act in those healings which occurred after the death of Achilles; if he was thought to have died through Eurypylus (T. 180), or during the conquest of Troy (T. 177), Machaon lived long enough to treat Ajax and Philoctetes.

(T. 199).[26] Certainly, Podalirius is also pictured as a fighter. He is named among those who man the Trojan horse (T. 204).[27] Still, where he is shown in action, his unheroic character becomes evident. He is able to contend in combat, but he has a special reason for doing so: while busy tending the wounded near the ships, he learns of Machaon's death. Enraged at the murder of his brother he clothes himself in his armor and rushes forth to the battlefield where he kills many a man (T. 182). Were it not in order to take revenge, he would heal men instead of killing them.[28]

Behind the array of the Epic, then, it can still be recognized, I think, that Machaon and Podalirius are physicians rather than warriors, craftsmen rather than kings. The Epic, to be sure, makes them appear in the garb of heroes. Yet it is only for the purposes of the Epic that the heroic mask is superimposed upon the physicians Machaon and Podalirius; for otherwise they would not be suitable companions of the noblemen in whose society they are to live and act.[29] As the princely character of the Asclepiads vanishes, not much remains of Asclepius, the king. Homer's representation of Asclepius' children creates the impresssion that their father, too, should be considered a leader of men. But this is the result of poetical fiction. It does not contradict the other data related by Homer concerning Asclepius himself: he may well have been a physician.[30]

[26] This testimony derives from Quintus; cf. above, n. 20. That here the drugs are a magic cure-all (T. 199)—for this detail, cf. also below, p. 43—and that Asclepius' name is invoked during the treatment (T. 201), are certainly late inventions; for another healing of Podalirius, cf. below, p. 19. It is interesting that the story of Podalirius assumes features earlier attested for that of Machaon. Compare how later Podalirius, to whom the death of his brother is almost unbearable (T. 184), is consoled by Nestor (T. 185), apparently a reminiscence of Machaon's relation to Nestor, as it is described in the *Iliad* (T. 165 ff.).

[27] This testimony has been overlooked by Wilamowitz, *Isyllos*, p. 48, who states that Podalirius is never mentioned in connection with the fall of Troy.

[28] Note that the discussion of his actual exploits in battle (T. 203) is very vague, indeed even more indistinct than in the case of Machaon (cf. above, pp. 6 ff.); the better part of this testimony contains the description of a cave! It seems symbolic that in the opinion of the ancients Podalirius' name indicated that he was more peaceful than Machaon (T. 197; cf. 139). The saga of his return from Troy is in keeping with this character, cf. below, p. 19.

[29] As far as I am aware, Türk in Roscher, *Lexikon, s. v.* Podalirius, III, 2, p. 2587, 16 ff., is the only one to have proposed a similar thesis. He too holds that Machaon and Podalirius originally were physicians and were ranged among the heroes only by the Epic. But his claim is based solely on the fact that the two, although called physicians, are depicted as warriors and leaders of men. He has failed to notice, as have all those who considered Asclepius a king, the inconsistency throughout the epic description of the Asclepiads. The complicated history of the Podalirius figure can be fully evaluated only in section ii, below.

[30] For the originally unheroic character of certain heroic figures, cf. above, n. 18.

As a matter of fact, such a contention is also in accord with the ancients' esteem of the Homeric Asclepius. While everybody accepted the claim of the Epic that he was the father of Machaon and Podalirius, the leaders of men from Tricca, Oechalia and Ithome, no one actually had anything to tell about the king or chieftain Asclepius, his reign and deeds. Describing his character and his accomplishments, they spoke of the physician alone.[31] That simple medical knowledge was ascribed to him with which his time was generally credited, that is, medicine as it was cultivated before the rise of scientific studies (T. 125; cf. 145). Or he was said to have brought about a slight improvement in the art of his day (T. 354). Plato, projecting his own designs into the past, dared to see in Asclepius the physician who had realized the ideal of political medicine as conceived by the philosopher (T. 124).[32] If Asclepius had really been a king, if as a king he had had an existence outside the poetical realm of the Epic, it would be strange indeed that his memory left no trace in ancient tradition.[33]

ii. Machaon and Podalirius

As Homer tells the story of Asclepius and his sons, it seems to deal with real or historical characters. Even after it has been recognized that Asclepius was not a chieftain and therefore cannot be considered a tribal hero in the strict sense of the word, one might still maintain that he actually existed. If the memory of great warriors could be preserved by posterity, the memory of great physicians could be retained as well.[34] Asclepius, Machaon and Podalirius, then, one might suggest, were human beings who once lived on this earth; their story is transformed by Homer only in so far as he has elevated them to the dignity of princes. Yet, it is hardly probable that this is so. Granted that Asclepius,

Fundamentally, then, I agree with Farnell, who claims without giving any reasons for his assumption that Asclepius, according to Homer and the saga which he followed, was a physician, cf. above, n. 2.

[31] Concerning Asclepius' participation in the Voyage of the Argonauts and in the Calydonian Hunt, usually called his heroic deeds, cf. below, pp. 38 f.

[32] For the interpretation of this and other statements concerning Asclepius' " political medicine," cf. below, pp. 178 ff.

[33] Of course, no strictly historical sources were extant in regard to any of the epic heroes; yet there were certain sectional traditions by which the Homeric picture usually was supplemented, or information was gathered from the Cyclic Epic. With regard to Asclepius, the king, these sources seem to have yielded nothing, in spite of the fact that the Asclepiads were favorites of the post-Homeric saga; cf. also below, n. 44.

[34] Farnell, *Hero Cults*, pp. 236 ff., has strongly insisted on this possibility; for earlier proponents of such a theory, cf. Roscher, *Lexikon*, I, 1, p. 621, 52 ff.

Machaon and Podalirius were historical figures—Asclepius, as father of Machaon and Podalirius, seems to be a fictitious character. For he was not always known in this capacity, not even to Homer. The references to Podalirius are obviously later additions which were made in accordance with the post-Homeric saga as it was codified in the Cyclic Epic.[35]

Podalirius in the *Iliad* is mentioned only twice, in both cases in connection with Machaon. The first statement is to the effect that neither Machaon nor Podalirius can be called upon to help the wounded Eurypylus because Machaon is wounded himself, while Podalirius is engaged in battle (T. 136). The other information furnished concerns the names of the cities over which the two brothers were ruling, and the number of ships commanded jointly by Podalirius and Machaon is added (T. 135). Now as for the first of these passages, it is strange indeed. For it presupposes that Machaon and Podalirius usually take care of the wounded and that it is extraordinary that Eurypylus should appeal to Patroclus for help. But though it is true that in some books regular army physicians are referred to, in others the heroes themselves are wont to treat their wounds and those of their comrades.[36] Thus Nestor, whom Patroclus has just left when he meets Eurypylus, has taken charge of Machaon, and nobody has thought of calling in Podalirius or any other physician. The explanation given by Eurypylus therefore reads like an addition made by someone who knew that regular physicians were attached to the Greek army and that Machaon and Podalirius were famed for their medical skill. The words in question can hardly be original in their place.[37] Considering the fact that the only other mention of Podalirius (T. 135) is found in the *Catalogue of Ships*, which represents a tradition in many respects at variance with that of the *Iliad*, it seems safe to conclude that Podalirius' name in the beginning did not occur in this poem.[38]

[35] Cf. Wilamowitz, *Isyllos*, pp. 45 ff., esp. p. 51. In my interpretation I follow Wilamowitz for whose conclusions I shall propose some additional arguments. Moreover, the question why Podalirius was included in the *Iliad* has not yet been satisfactorily solved, cf. Weber, *op. cit.*, p. 390, n. 2.

[36] Cf. above, pp. 3; 5.

[37] Note that Ennius (T. 146) in his description of the scene under discussion makes Eurypylus ask for Patroclus' assistance, because "the wounded crowd the entrance ways of the sons of Asclepius; there is no access." Here the fiction that Machaon and Podalirius are fighters is discarded entirely. Ennius' source is supposed to be Aeschylus or Homer, cf. M. Schanz-C. Hosius, *Geschichte d. römischen Literatur* I⁴, 1927, p. 89.

[38] This is not the place to enter into a detailed discussion of the difficult question of the composition of the Homeric Epic. That the text of the *Iliad* shows traces of interpolation even the "unitarians" admit, cf. *e. g.* H. J. Rose, *A Handbook of Greek Litera-*

This inference, based on literary criteria alone, can be corroborated by an evaluation of the medical knowledge ascribed to Machaon and Podalirius respectively in the post-Homeric saga. Here, medicine is divided between the two physicians in such a way that the one is the representative of surgery, that is, of the treatment of wounds, while the other is in charge of the cure of internal diseases. Arctinus, who mentions this partition which he regards as ordained by Poseidon, also attests that it was Podalirius who through his great skill was able to diagnose the illness of Ajax (T. 141 f.).[39] Arctinus' statement in revealing the specific task of Podalirius makes it possible to surmise why another physician was needed in the later saga. While in the *Iliad* only wounds and their treatment are depicted,[40] diseases like that of Ajax and Philoctetes are described in the post-Homeric Epic, in fact, they are of great conse-

---

ture[2], 1942, p. 46. That the *Catalogue of Ships* deviates from the main text even T. W. Allen, *The Homeric Catalogue of Ships*, 1921, pp. 168 f., concedes. It is quite a different problem whether the books in which Machaon and Podalirius appear belong to the older or to the younger layer of the *Iliad*. Wilamowitz (*Isyllos*, pp. 45 ff.; *Die Ilias und Homer*[2], 1920, p. 200) is of the opinion that *Iliad*, IV and XI, from which the testimonies concerning Machaon and Podalirius are taken, are relatively late songs; in this assumption he is in agreement with Leaf in his introduction to Books IV and XI and *Prolegomena*, XXII-XXIII. For the late date of the *Catalogue of Ships* from which the third reference to the Asclepiads stems, cf. below, n. 78. In spite of the more recent pleas for the "unity of Homer" (cf. *e. g.* J. A. Scott, *Sather Classical Lectures*, I, 1921), it seems still permissible to believe that not all the books of the *Iliad* or all their parts were written at the same time; the extreme unitarian view is as unconvincing as the extreme separatist position. On these questions cf. Nilsson's discriminating survey, *Homer and Mycenae*, pp. 1 ff. The Machaon and Podalirius episodes are of too little consequence to allow a decision concerning the date of the books under discussion, nor is this essential for my interpretation. One can say only this: the fact that Machaon and Podalirius are represented as physicians argues for a later rather than for an earlier date; for treatment of wounds by physicians is a sign of later specialization of knowledge, whereas treatment by the warriors themselves is characteristic of more primitive conditions. Of course, the unitarians claim that such variations are not mutually exclusive and that the poet's picture of life is no more varied than was life itself; cf. *e. g.* Scott, *op. cit.*, pp. 117 ff. At least in this instance I think it is safer to side with those who interpret the divergencies to be observed as distinct changes of customs which betray the reworking of the Epic by succeeding generations, cf. *e. g.* Murray, *The Rise of the Greek Epic*, pp. 146 ff. It is not likely that, except in emergencies, anybody took the risk of meddling with wounds when regular physicians were attached to the army.

[39] T. 141, according to tradition, is ascribed to Arctinus and his poem, the *Iliupersis*. Some scholars, however, attribute it to the *Aethiopis*, a change that is hardly permissible, cf. Wilamowitz, *Isyllos*, pp. 47 f.; E. Bethe, *Homer*, II, 1922, p. 179. In this connection, the important fact is that the testimony goes back to the Cyclic Epic, whatever the name of the poem and whoever its author; for these complicated problems, cf. Schmid-Stählin, *op. cit.*, I, 1, p. 212; Bethe, *op. cit.*, pp. 207 ff. For the relation of the Asclepiads to Poseidon, cf. also below, p. 15.

[40] The pestilence sent by Apollo, *Iliad*, I, 43 ff., is the only exception. The ancients already noticed that references to the treatment of diseases by diet were missing in the *Iliad* (T. 140 ff.), and consequently discussed the question whether dietetics, that is, internal medicine, had existed in Homer's time, or whether it had been invented only afterwards.

quence for its plot: Ajax' madness forms the climax of the Ajax drama; the healing of Philoctetes is decisive for the outcome of the Trojan War.[41] Such cases, however, complicated as they were, could hardly be taken care of by Machaon, a surgeon whose routine concerned the tending of injuries.[42] In the more sophisticated times of the Cyclic Epic they seemed to require a specialist's knowledge, and such a specialist was found in Podalirius, who is always referred to as healer of internal diseases.[43]

But why should Podalirius' name have been interpolated in the *Iliad* if he was a figure of the late saga and if his presence in the Homeric Epic was not essential to the actions there described? It seems that through the later poems on the Trojan War Podalirius had become so closely associated with Machaon that the one could not be thought of without the other. The Cyclic Epic was wont to represent the two together and to give them a common share, and not an unimportant one, in the events in which they were involved.[44] Thus Machaon and Podalirius, now pupils of Chiron like their father (T. 56; 148), are named among the wooers of Helen (T. 150). When the Greeks sail against Troy they invite the two heroes whom they respect as good doctors (T. 151) to join the army,[45] and their confidence in Machaon and Podalirius is fully justified by the subsequent behavior of the two Asclepiads. Both prove capable of fighting fiercely (T. 178). In the games held in honor of Patroclus they win the fourth prize after Agamemnon, Nestor and Ajax (T. 155). In their medical function they excel in the healing of Telephus (T. 153-4) and Philoctetes (T. 152 a-b).[46] The sick flock to their tent, and so busy are they with tending

[41] For Ajax in the post-Homeric saga and particularly for Arctinus' representation, cf. Roscher, *Lexikon*, *s. v.* Aias, I, 1, pp. 126 ff. For Philoctetes, cf. *e. g.* T. 282, 10.

[42] Homer does not call Machaon a surgeon but a physician. Yet he defines the physician's task thus: "to cut out arrows and to spread soothing simples" (T. 165, *v. 515*). This verse was considered a later addition by some ancient critics, because it restricts the realm of medicine to one of its provinces and thus degrades the value and dignity of the art. Modern commentators usually follow suit; cf. G. M. Bolling, *The Athetized Lines of the Iliad*, 1944, p. 127. But Leaf, *ad loc.*, rightly insists that this argumentation is not convincing and that the words in question are genuine because the line "fairly represents the primitive stage of Homeric medicine"; cf. also above, n. 40.

[43] Cf. T. 140-142. The late variants concerning Machaon as the healer of diseases or Podalirius as the healer of wounds should not obscure the original character of Podalirius as a specialist for internal illnesses, cf. also Preller-Robert, *Mythologie*, I, p. 524.

[44] Wilamowitz, *Isyllos*, p. 56, rightly says that the Asclepiads play a subordinate rôle in the Homeric Epic. The situation changes considerably in the later saga.

[45] Concerning the ships of which the Asclepiads were in command and later variants in regard to their number, cf. T. 137; 137a ff.

[46] Here and in the following characterization of Machaon and Podalirius I have made

the wounded that many a warrior must seek assistance from his comrades in order to receive quick relief (T. 146); their value to the Greek
cause is such that in appreciation of their merits exemption from war
services and taxes is granted to them later on (T. 147).[47] Podalirius and
Machaon, then, are constantly mentioned together, Machaon, the " brave,
great, sure, wise, patient, compassionate " (T. 149), a " man of great
beauty " (T. 180a), and Podalirius, the " stout, strong, haughty, and
stern " (T. 149).[48] Under these circumstances, is it astonishing that
Podalirius made his way from the post-Homeric saga into the *Iliad*, that
his name was inserted into the older poem? That such changes in favor
of Podalirius were actually made in the Homeric text is certain. For in
that edition of the *Iliad* which Plato read, Podalirius was named even as
the healer of Menelaus along with Machaon (T. 143).[49] Like the
Dioscuri, with whom they were compared in late centuries (T. 282, *24*),
the two, it seems, could not exist separately.

use also of testimonies taken from Dictys (fourth century A.D.) and Dares (fifth
century A.D.), *e.g.* T. 153 ff.; 149. The Greek original which these late authors
translated into Latin in part undoubtedly depended on good sources. As for Dictys,
T. W. Allen, *The Journal of Philology*, LXII, 1910, pp. 207 ff., especially p. 223, has
tried to show that in those incidents which do not occur in the *Iliad*, his exemplar
copied from the Cyclic Epic. Even if this cannot be proved conclusively in regard to
every detail, it must be admitted that Dictys' statements need not be mere inventions,
but sometimes do reproduce old material, cf. Wilamowitz, *S.B.Berl.*, 1925, p. 47.
Concerning Dares, cf. O. Rossbach, Pauly-Wissowa, IV, p. 2212, *s.v.*, and N. E. Griffin,
Dares and Dictys, Diss. Baltimore, 1907, p. 13, n. 2. At any rate, it seems permissible
to quote a few items from Dares and Dictys in order to round out the picture of the
Asclepiads as it was presented in the late Epic.

[47] The medical knowledge ascribed to the Asclepiads is outlined in T. 143 ff., 123 ff.
In T. 145 it is said to have been as great as could be expected in their time. At any
rate, it exceeded that of their fellow-comrades, as is shown by the passages just referred
to. Wilamowitz, *Isyllos*, p. 45, n. 2, claims that the Asclepiads did not know more than
Patroclus; but this is certainly not true as far as the post-Homeric saga is concerned,
and it seems questionable whether such a conclusion can be drawn even from the *Iliad*;
cf. below, p. 16.

[48] The connection between the two heroes is limited to the time before the fall of Troy.
No author, except Aristides (T. 282, *11*), speaks of their common return. Generally
it is supposed that Machaon died in the war, and Aristides admits that he is thinking
of Machaon and Podalirius as if they were one (T. 282, *13*; cf. also *ibid.*, *1-2*). It is
possible that this version represents the Coan tradition, cf. Wilamowitz, *Isyllos*, p. 49,
but in that case it would be much later than the saga of the Cyclic Epic; cf. below,
n. 52. That the divergencies concerning the various deeds of the heroes are due to the
diverse assumptions concerning Machaon's death has been stated above, n. 25.

[49] There is no reason to assume with A. Ludwich, *Die Homervulgata*, 1898, p. 135,
that Plato should have composed the line himself, nor is it satisfactory to say with the
commentators on Plato that he is loosely quoting. That in the fourth century B.C.
texts of Homer were current which were different from that to be found in the preserved manuscripts is a well established fact; cf. in general Murray, *The Rise of the
Greek Epic*, pp. 282 ff., and concerning Plato in particular, *ibid.*, pp. 293 ff. Note moreover
that Plato claims (T. 143) that the Asclepiads took care of Eurypylus, while according
to *Iliad*, XI, 842 ff., it was Patroclus who healed him.

At any rate, originally Podalirius was not mentioned in the *Iliad*.[50] Is it possible to assume that Homer simply had no occasion to refer to both sons of Asclepius, while Podalirius was remembered by the later epic poets as soon as his special knowledge was needed? Was Asclepius from the beginning known as the father of the two Asclepiads in spite of the fact that in the *Iliad* at first Machaon alone was referred to as his son? Such a thesis, improbable in itself, is refuted also by another testimony. Arctinus, who is the first to differentiate between the specific tasks assigned to Machaon and Podalirius respectively, is of the opinion that they were the sons of Poseidon (T. 141-142).[51] There is no way of telling whether this claim is original with Arctinus, or whether it goes back to older sources and was more widely accepted. This much, however, follows with necessity from Arctinus' statement: the post-Homeric saga was still undecided as to the genealogy of the two physicians. Homer's version of the descent of Machaon and Podalirius was only one of two, if not of many possibilities. Certainly, this circumstance is not in favor of the assumption that Homer's claim concerning the Asclepiadic origin of Machaon—to say nothing about that of Podalirius—represents a genuine historical tradition. One cannot escape the conclusion that the two physicians were associated with Asclepius only in an artificial way.[52]

But why and by whom was their connection with Asclepius established? This question one must try to answer first in regard to Machaon, and for this purpose the moment of Machaon's main action, the occasion on which

[50] Thraemer, Pauly-Wissowa, II, p. 1658, 30 ff., suggests that Podalirius' name was inserted first in the *Catalogue of Ships* (T. 135) and that subsequently a reference to Podalirius was included in the other book (T. 136). This is not impossible since the *Catalogue* contains other names which do not occur elsewhere in the *Iliad*, cf. Thraemer, *ibid.*, and below, n. 78.

[51] The testimony in question is much debated. Concerning its attribution to the *Iliupersis* or the *Aethiopis*, cf. above, n. 39. The text of line 1, where Poseidon is named (T. 141), is corrupt, but the name itself ('Εννοσίγαιος) is assured and an emendation is not permissible; cf. Wilamowitz, *Isyllos*, pp. 47-8, where he refutes Welcker, *Kleine Schriften*, III, p. 47; cf. also Bethe, *op. cit.*, p. 179. Th. Lefort, *Musée Belge*, IX, 1905, pp. 215 ff., has only shown that the line in question is metrically incorrect; but this fact does not justify a rejection of the authenticity of the genealogy therein proposed (contrary to Thraemer, *E. R. E.*, VI, p. 546a). That Poseidon is said to have been the father of Machaon and Podalirius follows also from Eustathius' paraphrase of the Arctinus verses (T. 142). Eustathius does not quote lines 1-2 because he is interested only in the task given to the heroes by Poseidon (lines 3 ff.), but in paraphrasing them he says expressly that Machaon and Podalirius were the children of Poseidon. For Poseidon's relation to medicine, cf. below, n. 57.

[52] Wilamowitz, *Isyllos*, p. 51, claimed that Machaon and Podalirius, the representatives of Cos in the Epic, were introduced as sons of Asclepius because in the old Coan saga they appeared as his children. This suggestion he had to retract later, when the Coan Asclepius was found to be of relatively late date, cf. *Glaube*, II, pp. 228-9, and below, p. 243. There is no tradition older than that of the Epic on the basis of which the two could be proved to be Asclepiads.

he proves his real prowess assumes particular significance (T. 164): [53] Menelaus is wounded in the fighting that takes place after Achilles in his wrath has retired from the battle. However, Achilles, and, through him, Patroclus are the two heroes among the Greeks who are best versed in the preparation of especially efficacious simples. Chiron had taught Achilles this art.[54] Now Agamemnon could not possibly ask either Achilles or Patroclus to help his brother. On the other hand, the dangerous injury of the king, the concern of all, requires an outstanding physician. Here, Machaon is introduced into the fable. He is not one of the nobles; he is a craftsman, who, since he must appear on the Homeric stage in the costume of a hero, is dubbed a knight and invested with a train of vassals. Naturally, he needs a pedigree, too. Therefore, he is made the son of Asclepius, an excellent choice indeed. For Asclepius was the friend of Chiron and gave to his son a share in the benefit of the Centaur's gifts. Thus Machaon becomes a worthy competitor of Achilles whose place he is supposed to take, so that in spite of Achilles' wrath, the wound of Menelaus can be treated properly.[55]

Such an hypothesis has at least some probability and is consistent with the results drawn from the analysis of Homer's representation of Asclepius. It merely presupposes that in Homeric times Asclepius was renowned as a great physician whose name was apt to add to the glory of those who could trace their descent to him. This much may safely be assumed. For Homer certainly did not invent Asclepius' relation to Chiron, a feature not even quite in line with the rôle ascribed to Asclepius in the *Iliad*. There must have been a legend of Asclepius, the physician, before the time of the Homeric Epic.[56]

Later on, when Podalirius was introduced because a specialist for internal diseases was needed, he was pictured as the brother of Machaon, the famous physician of the *Iliad*. At the same time the two were made children of Poseidon, because this origin gave more dignity to their achievements, and medicine now was held in higher esteem.[57] But

---

[53] The poet speaks of Machaon's ἀριστεία when he mentions his war deeds (T. 165); cf. above, p. 6. But actually Machaon shows himself the best and is most useful in his treatment of Menelaus, not in his military achievements which are never detailed and are irreconcilable with the true character of this hero.

[54] Cf. above, p. 4.

[55] For the parallelism in the representation of Machaon and Achilles, cf. above, p. 4.

[56] This fact, though of great importance for the appreciation of the Asclepius figure, as far as I can see, has never been explicitly stated in the modern discussion. Yet, it is tacitly presupposed by all who believe with Farnell that Homer affiliated Machaon and Podalirius with Asclepius "merely perhaps because of their medical character" (*Hero Cults*, p. 237). Otherwise it would be inexplicable why Asclepius should have been made the father of Machaon and Podalirius.

[57] This has been intimated already by Wilamowitz, *Isyllos*, p. 51. Poseidon in Greek

Podalirius was also drawn into the story of the *Iliad*, and once Homer's authority had superseded all other epic versions, Asclepius' relation to Machaon and Podalirius became predominant to the exclusion of all others. Gradually the two Asclepiads were made the helpers of Asclepius, the administrators of their father's might.[58] Yet, it should be remembered that they had not always been his sons, that in the oldest saga which can still be detected Asclepius was not the father of Machaon and Podalirius.

iii. Origin of the Asclepiads and of Asclepius

The last question that arises in connection with Homer's representation of Asclepius is that concerning the regional origin of the saga, as it was known to the poet. The *Iliad* does not state where Asclepius lived, just as it is silent about his forebears.[59] But from the data given in regard to Machaon and Podalirius it should be easy to infer which was the country of the Asclepiads and consequently that of their father. Even if one holds that their heroic attire is wholly fictitious, that Machaon and Podalirius were only the "adopted children" of Asclepius, one must suppose that the poet made them reside in that region of Greece where their father was believed to have lived.

For the ancients themselves, however, any attempt to derive Homer's opinion concerning Asclepius' homeland from his statements in regard to Machaon and Podalirius led into difficulties. Once Machaon is said to have come from Trice (T. 164, *v. 202*); another time Machaon and Podalirius are reported to be in charge of the troops from Tricca, Ithome and Oechalia (T. 135). In historical times Tricca was a famous town in Thessaly, while Ithome and Oechalia were well known cities in Messenia.[60] But obviously the poet did not believe that Asclepius ruled over these two regions or that his sons came from places so far apart. It was necessary, therefore, to assume that Homeric geography and later geography were at variance and to reinterpret historical reality in the light of the myth.[61]

---

mythology is sometimes associated with medicine. Not only is he himself called ἰατρός (cf. Wilamowitz, *Glaube*, I, p. 215)—Chiron, the inventor of medicine, is also named as son of Poseidon, *Scholia in Homerum* [Nicole], IV, 219.

[58] Concerning the late concept of the Asclepiads, cf. Aristides (T. 282), and below, p. 78.

[59] Of course, Homer may well have been familiar with more facts than he cared to tell; Asclepius is not one of his principal characters but sketched only in a superficial way, cf. above, p. 2.

[60] Cf. Wilamowitz, *Glaube*, II, p. 228: "Den Homererklärern ist es nicht gelungen, die Städte Ithome and Oichalia nachzuweisen [*sc.*, in Thessaly], die der Schiffskatalog mit Trikka verbindet; es gab sie dagegen in Messenien (Trikka aber nicht) . . ." Cf. also Kjellberg, *Asklepios*, pp. 26 ff.

[61] Kjellberg, *loc. cit.*, thinks that the general context in which T. 135 is to be found

The Messenians accomplished this task in a very simple way. Ithome and Oechalia were theirs. Tricca, which did not exist in Messenia, they believed to have disappeared in post-Homeric centuries, and as proof of this assertion they called a desolate spot in their country by this name (T. 38). Such a theory was advanced at least as early as the fourth century B. C. Ingenious as it may be, it is certainly far from being convincing.[62]

On the other hand, the reasoning of those, who tried to make Homer's account consistent by placing all three towns in Thessaly, though somewhat subtler, was fundamentally not sounder. Apollodorus denied that there had been such a town as Oechalia in Messenia in Homer's time; he contended that in those days only Thessaly had had a city by that name.[63] Ithome was identified with old Thessalian Thome, which in turn had become part of Metropolis through joint settlement; for *Ithome*, the Thessalians held, was an incorrect pronunciation of *Thome*.[64] Tricca, moreover, the surest warrant of the Thessalian claim, was declared to be the birthplace of Asclepius. Strabo attests this fact (T. 11), and so does Hyginus (T. 12). Since Strabo's authority for his statement is Apollodorus' book *On the Homeric Catalogue of Ships* and Hyginus' remark is taken from the Aristotelian *Peplos*, a treatise also dealing with the Homeric Epic, the Thessalian theory like the Messenian can be traced at least to the end of the fourth century B. C.[65]

No final agreement was ever reached among the ancients concerning Homer's opinion. The Messenian, like the Thessalian, interpretation encountered considerable obstacles. Any decision in favor of one or the other was due solely to the bias of the respective critics.[66] For the

indicates that the poet of the *Catalogue* placed the cities of the Asclepiads in Thessaly. Even if this were correct, the ancients at any rate paid no regard to this contention of the poet.

[62] It seems safe to ascribe this Messenian argumentation to the fourth century B. C., since the Messenian claims to Asclepius were brought forward immediately after the restoration of Messenia in 369 B. C., cf. below, p. 68. But the Messenian saga itself is much older; cf. below, pp. 32 f.

[63] Cf. Strabo, VIII, 3, 6; IX, 5, 17, and in general, F. Bölte, Pauly-Wissowa, *s. v.* Oichalia, XVII, pp. 2097 ff.

[64] Cf. Strabo, *loc. cit.*, and in general, Stählin, Pauly-Wissowa, *s. v.* Ithome, IX, p. 2307.

[65] For Apollodorus as Strabo's authority, cf. B. Niese, *Rh. M.*, XXXII, 1877, pp. 267 ff.; E. Schwartz, Pauly-Wissowa, *s. v.* Apollodorus, I, p. 2866; for Aristotle as the source of Hyginus, cf. A. Wendling, *De Peplo Aristotelico*, Diss. Bonn, 1901, pp. 35 ff. Why Farnell, *Hero Cults*, p. 243 a, says that Strabo gives the genuine tradition of Tricca, I cannot tell. The Thessalian origin of Asclepius apparently was accepted in the scholarly literature. This fact was first recognized by Kjellberg, *Asklepios*, p. 31.

[66] Kjellberg, *loc. cit.*, again was the first to insist that the Triccaean or Thessalian claim can hardly be called "the best" or "the oldest" one (contrary to the almost unanimous assumption of earlier scholars). This is true, at least as long as the argument

modern scholar, there seems to be a slight balance in favor of Tricca as Asclepius' or rather Machaon's home; for this particular town, though in a metrically different form, is attested by the earlier authority (T. 164, *v. 202*), where it occurs in connection with the "older" son of Asclepius.[67] And since this place is characterized as "the pastureland of horses," an epithet well fitting Thessaly, one should conclude that Homer meant to place Asclepius in Thessaly—if indeed Tricca was situated only in Thessaly, and if the Messenian claim to a municipality of this name was wholly unfounded.[68]

But could it not be that the Asclepius story is one of those myths that were current in two different sections of Greece? Twofold localizations of one and the same saga were not uncommon.[69] It is true, the Messenians quoted Nestor's assistance to Machaon (T. 165 ff.) as proof of the Messenian origin of Machaon (T. 38). But those who believed that he was a Thessalian referred to Achilles' interest in him (T. 166 ff.) as evidence that their opinion was supported by Homer (T. 166b). These arguments may well be of equal value.[70] If so, why should Thessaly have had a better title to Asclepius than Messenia? Why should Tricca be a Thessalian place rather than a Messenian, especially since Oechalia and Ithome most probably were Messenian localities?

As a way out of these difficulties one might for a moment be inclined to believe that a new regional claim arose when the Podalirius figure was inserted into the Homeric Epic. But Podalirius has no original connection with either Messenia or Thessaly. In late centuries, he was revered together with Machaon in both countries (T. 657; T. 161). All other data that concern him alone point to his affiliation with Caria. As the late saga has it, Podalirius, the physician, survived the Trojan War, and on his way home was driven to Caria. When he landed there, the only daughter of the king was ill as a result of an accident. Podalirius restored her health and thus ingratiated himself with her father to such a degree that he was given her hand and part of the country (T. 209).

is based on geographical data; but cf. below, p. 22. Thraemer, Pauly-Wissowa, II, p. 1649, 64 ff., gives the impression that only the Messenian interpretation of Homer is not borne out by the facts.

[67] This has been pointed out by Wilamowitz, *Glaube*, II, p. 228 (T. 164, *v. 202*: Τρίκης; T. 135: Τρίκκην).

[68] For Thessaly as the land of horses, cf. H. D. Westlake, *Thessaly*, 1935, p. 4, n. 3. For Tricca in general, cf. E. Kirsten, Pauly-Wissowa, *s. v.* Trikka, VII A, pp. 146 ff.; 1273 ff.; in my opinion he underrates the difficulties with which the interpretation of the ancient testimony concerning Tricca is faced.

[69] Cf. *e. g.* the Thamyris saga, *Iliad*, II, 595, and Leaf's commentary, *ad loc.*

[70] Wilamowitz, *Isyllos*, p. 54, n. 26, who has drawn attention to these claims and counterclaims, is of the opinion that the weight of the Messenian interpretation is neutralized by the Thessalian one. But this is not *a priori* certain.

Having acquired a wife and a home in this none-too-heroic manner, he settled down in Caria and founded two cities, Syrna (T. 209) and Bybassus (T. 209a).[71] He was said to have been buried near the tomb of Calchas in Daunia, that is, in Apulia (T. 205 f.; 158 ff.). Even this Italian legend may easily be understood in connection with the Carian saga, for certain regions of Apulia were colonized by Rhodians and Coans.[72] Evidently, then, Podalirius was linked up with Caria. That in this province he was an indigenous hero, however, is not likely.[73] According to Theopompus (T. 212), the children of Podalirius went from Syrna—the city which supposedly he had founded—to Cos and Cnidus, the famous medical centers of the sixth and fifth centuries B. C. Here in Asia Minor he was believed to be the ancestor of a family of physicians, among them even Hippocrates (T. 213 ff.), who through Podalirius and his wife traced their descent to Asclepius and to royal blood. This they did most probably under the influence of the epic representation of Podalirius, the Asclepiad.[74] At any rate, if to any country, Podalirius belonged to Caria rather than to Thessaly or Messenia.

The twofold localization of the Asclepius saga cannot be understood,

[71] Another version of the saga presupposes that Podalirius went to Delphi and asked the Oracle where he should settle; the Oracle answered: in Caria (T. 207 f.). This, no doubt, is a late sanction of the Carian Podalirius through Delphi according to the usual concept that colonizations were under the tutelage of the Oracle, cf. Wilamowitz, *Glaube*, II, p. 40. Wilamowitz, *Isyllos*, p. 48, assumes that Podalirius was mentioned in the *Nostoi*, yet, as he himself admits later on, the fable of Podalirius' return to Caria can be traced only to Timaeus, cf. *ibid.*, p. 50, n. 14.

[72] T. 205 goes back to Timaeus, cf. Wilamowitz, *Isyllos*, p. 50, n. 15. Calchas travelled together with Podalirius to Caria (T. 207). For the Coan-Rhodian colonization of Apulia, cf. Farnell, *Hero Cults*, p. 237. Incidentally, the Apulian worship of Podalirius, who was honored here as healer of men and animals (T. 206), may account for the preference given to Podalirius by Latin poets. Thus the *Homerus Latinus* (first century A. D., cf. M. Schanz, *Geschichte d. römischen Litteratur*, II, 2, 1913, p. 120) names Podalirius as the physician who treated Menelaus (T. 198); for purposeful deviations of this poem from the older saga, cf. *ibid.*, p. 119, n. 2. Quintus Smyrnaeus, in whose poem Podalirius plays so important a rôle (cf. above, pp. 8 f.), is perhaps also influenced by Roman sources, cf. W. Christ-W. Schmid-O. Stählin, *Geschichte d. griechischen Literatur*, II, 2⁶, 1924, p. 963.

[73] Most interpreters think that Podalirius was an old Carian hero, cf. Pauly-Wissowa, II, p. 1659, 34 ff. But in Caria no tomb or any independent worship of Podalirius is attested. Wilamowitz, *Isyllos*, p. 51, believes that the name Podalirius is Carian; cf. also Türk in Roscher, *Lexikon*, III, 2, p. 2589. Farnell, *Hero Cults*, p. 237, holds that the names of the Asclepiads are real names, "culled from Mynian-Thessalian tradition." I do not think that the evidence is sufficient to decide the issue either way, although it seems more likely that the name Podalirius is not Greek. For Machaon's name, cf. below, n. 76. On the other hand, Podalirius' name does not provide evidence for the assumption that he was a medical deity, cf. Usener, *Götternamen*, p. 170; for the ancient etymologies, cf. above, n. 28.

[74] For the concept of "epic heroes," heroes, that is, who were adopted from the Epic, cf. Farnell, *Hero Cults*, ch. XI. Whether Podalirius was an historical personality or an invented figure, it is impossible to decide.

then, as a consequence of the introduction of the Podalirius figure into the Epic. But perhaps it may be explained by the later history of Machaon. For this son of Asclepius, even if originally he was a Thessalian, continued to lead an existence of his own only in Messenia. Here he was worshipped as a healing hero. He had a Messenian wife, Anticleia (T. 191), he had sons, Nicomachus and Gorgasus (T. 170), apparently two older healing heroes who were conquered by Machaon.[75] His influence extended even to the Argolid where Polemocrates (T. 189), Alexanor (T. 187) and Sphyrus (T. 188) were named as his children, these again most probably healing deities of old.

That in Messenia Machaon was revered as a hero who was taken over from the Epic seems certain.[76] According to the Messenians, Nestor, on his return from Troy, had brought Machaon's corpse to Gerenia (T. 156). His worship was said to date from the time of the Messenian kings (T. 186a). The Messenian saga made even Asclepius a Messenian, the child of a Messenian mother (T. 37), whereas Machaon was believed to be the son of Asclepius and Xanthe (T. 169). And in this birth myth Asclepius was described as the hero of whom Homer spoke, for he was called " the leader of men," the " blameless and strong " (T. 37).[77] The story was codified by Hesiod as early as the end of the seventh century B. C. The *Catalogue of Ships*, therefore, could well include a reference to the Messenian home of the Asclepiads. The addition of names of cities in accordance with the legendary tradition of certain Greek provinces is a feature not uncommon in this enumeration of the troops before Troy.[78]

[75] It is the Messenian Machaon to whom Aristotle traced his origin (T. 192 f.; cf. Wilamowitz, *Aristoteles und Athen*, I, 1893, p. 311). Aristotle's ancestors are the only known human descendants of this hero. The epic legend asserted that he had no children (T. 157). No tale was told about his fate after the fall of Troy; he was generally believed to have died before that event. Gruppe, *Griechische Mythologie*, p. 638, is therefore hardly justified in saying that the stories about his return were lost at an early date.

[76] Machaon, the Thessalian, could well be taken over by the Messenians. Concerning the worship of epic heroes in countries other than their own, cf. Wilamowitz, *Glaube*, II, pp. 9 ff. I should mention that most interpreters consider Machaon a healing hero; cf. *e. g.* Pauly-Wissowa, *s. v.* Machaon, XIV, pp. 144 ff. Yet, even the name does not furnish conclusive proof for such a theory, cf. *ibid.*, p. 150, 17 ff. Usener, *Götternamen*, pp. 170 ff., tries to explain Machaon as " Kneter " (kneader), and Wilamowitz, *Glaube*, II, p. 228, n. 4, suggests that Machaon is related to μάχεσθαι. Again, as with Podalirius, the historicity of the figure remains uncertain.

[77] The Greek words used in T. 37 are ὄρχαμος ἀνδρῶν, ἀμύμων and κρατερός; these are typical epithets of epic heroes, cf. *e. g. Iliad*, II, 837; XXI, 546.

[78] For such interpolations in the *Catalogue of Ships*, cf. *e. g. Iliad*, II, 557-8 (Salamis is added to Athens), for this passage in particular, G. M. Bolling, *The External Evidence for Interpolation in Homer*, 1925, pp. 16; 73; cf. also Wilamowitz, *S. B. Berl.*, 1925, p. 45, n. 2. The time of the *Catalogue of Ships* is difficult to determine. The lists are usually considered an addition, cf. *e. g.* Rose, *A Handbook of Greek Literature*, p. 46, n. 77, and are dated around 600 B. C.; cf. *e. g.* Schmid-Stählin, *op. cit.*, I, 1, p. 151,

From all that has been said it is unlikely that the Asclepius saga originally was localized in two different regions. Homer's testimony seems to place Machaon, and thereby Asclepius, in Thessaly. This is well in agreement even with that report on Asclepius which Homer himself apparently followed. For the fact that Asclepius was associated with Chiron, the wise Centaur, skilled in the knowledge of drugs,[79] suggests that Asclepius was a well-known figure in Thessalian mythology. Chiron was a Thessalian hero; he was living on Mount Pelion which was renowned for its herbs.[80] Asclepius, therefore, should likewise be a Thessalian. Of course, one might argue that Asclepius, as a physician, became related to Chiron, the inventor of medicine, and on account of this connection was subsequently made a Thessalian.[81] The data reviewed so far do not admit of a choice between these two possibilities of interpreting the Chiron episode, which seems to form the nucleus of the Asclepius saga. The decision depends upon the character of the legend as a whole, and in order to make such an evaluation, it is necessary to question other witnesses. Later writers, not much inferior to Homer in their claim to authority, have more to tell about the physician Asclepius.

2. THE HEROIC LEGEND

The proper legend of a hero must give his ancestry, the tale of his birth and education, his deeds and his death. Concerning Asclepius, a poem of Pindar's is the earliest testimony preserved in which all these data are set forth coherently and in detail.[1] Asclepius, as Pindar relates

and F. Jacoby, *Die Einschaltung des Schiffskatalogs in die Ilias*, S. B. Berl., phil.-hist. Klasse, 1932, pp. 572 ff. Allen, *op. cit.*, p. 168, seems the only one to take the lines in question as composed by Homer whom he places at around 900 B. C. I cannot concur with his argument. That the *Catalogue of Ships*, even if relatively late, reflects earlier traditions seems certain; it is in the main intended to give a picture of the state of affairs in the eighth century, cf. H. Berve, *Griechische Geschichte*, I, 1931, p. 50. But supposing the enumeration were older, this would still not invalidate the interpretation given. The Messenian saga itself may have originated in the eighth century, cf. below, p. 33, n. 40. Whether the poet consciously combined two regional sagas, whether for him all these cities were situated either in Messenia or in Thessaly, I do not presume to decide. The inconsistencies which even the ancients noticed (cf. above, p. 17) cannot be resolved. They point to the fact that here intrinsically different traditions were amalgamated without too much regard for historical accuracy.

[79] Cf. above, pp. 4 f.

[80] For Chiron as a Thessalian and the inhabitant of the Pelion Mountain, cf. Preller-Robert, *Mythologie*, II, 1, pp. 19 ff.

[81] Many sagas, in the beginning, had no specific localizations whatever; they were locally fixed only when through the Epic they were transformed into heroic legends. Cf. Wilamowitz, *S. B. Berl.*, 1925, p. 60, and below, p. 63, n. 36.

[1] That Pindar recounts the heroic saga is certain: he himself refers to Asclepius as "hero" (T. 1, *v.* 7).

(T. 1, *vv. 8 ff.*), was the son of Apollo and Coronis. His mother, when with child by the god, fell in love with a mortal. The god, having discovered her unfaithfulness, sent Artemis to kill Coronis. His child, however, he himself saved and put it under Chiron's care. Educated by this friendly sage Asclepius grew up and became a great physician, skillful in everything that pertains to medicine, a helper of men in their distress. Yet he was given to unlawful greed. For money's sake he dared to heal those who were doomed to die. Therefore, Zeus, the king of gods and men, the administrator of justice, slew Asclepius with his thunderbolt, thus rightly punishing the crime of the licentious son of a licentious mother. So Pindar. His poem has all the features of an ancient tragedy. Generation after generation falls into crime; men are at the mercy of their passions; they are irreverent toward the gods and thereby bring about their own fall.

It was not Pindar's version of the saga, most familiar to modern men, that in antiquity was generally accepted. Rather was the Asclepius myth handed down from one century to the other in the form attested by Apollodorus, the mythographer (T. 3). He speaks of two genealogies of Asclepius. In both, Apollo was the father, but the mother in the one was Arsinoë, in the other—and according to Apollodorus this was held true only by some—it was Coronis. Summarizing, however, only the Coronis story, Apollodorus tells about Coronis' having been killed by Apollo because she intended to marry Ischys, about Asclepius, her child, who was saved by the god and entrusted to Chiron from whom he learned the arts of hunting and of medicine. He became an especially good surgeon; he healed the sick and he revived the dead. But besides being a physician, he was a sorcerer as well. Athena had shared with him the blood of the Gorgon which he used for the benefit of men no less than for their destruction. Finally Zeus killed Asclepius because he was afraid that the privileges of the gods would be infringed upon if men no longer were prey to death and thus were led to irreverence. Apollo resented the punishment imposed upon his son and took vengeance on the Cyclops who had furnished the thunderbolt with which Zeus slew Asclepius. Whereupon Apollo in turn was punished by the father of the gods for his uprising against the highest authority and barely escaped utter destruction. Thus the common saga about Asclepius' life and death, composed no doubt in a spirit quite other than that exhibited in Pindar's verses.[2]

---

[2] That the Coronis story in antiquity was more popular than that of Arsinoë follows from the testimonies and is also in agreement with the fact that Apollodorus' work is

Whatever the reasons for the differences in the two accounts just referred to, he who reads these stories after having examined Homer's representation of Asclepius will immediately ask whether this Asclepius, the son of Apollo, is the Asclepius whom the epic poet had in mind, whether this is the physician whom Homer introduced into his world, clad in the armor of a prince. Apollodorus, the Hellenistic scholar, did not hesitate to take this for granted. To him, Asclepius, the son of Coronis or Arsinoë by Apollo, was somewhat older than Achilles, as he concludes from the fact that Asclepius participated in the enterprise of the Argonauts, whereas his son Machaon was one of the fellow soldiers of Achilles in the siege of Troy (T. 5).[3] Apollodorus, it seems, saw no difficulty in identifying the Asclepius of the later legend with the Homeric Asclepius. No one will dare to follow him without having found out first when the saga of Asclepius, Apollo's son, originated, and whether Homer could have been familiar with this legend.

i. Reconstruction of the Hesiodic Version

Asclepius, who descended from Coronis or Arsinoë, was undoubtedly mentioned in the Hesiodic *Catalogue*, written around 600 B. C.[4] Both genealogies were recounted in this work. Of both only fragments are extant, but the main content of the stories told by Hesiod has long been reconstructed with a fair degree of certainty.[5] In view of the importance of this reconstruction for the history of the Asclepius legend it would hardly do here simply to recount its results. Moreover, essential details are still debated, and the tendency of the whole myth by which alone its date can be determined must, I think, be understood otherwise than is usually done.[6]

"an accurate record of what the Greeks in general believed," cf. *Apollodorus*, ed. J. G. Frazer, Introduction, p. XVII [Loeb]. Why Apollodorus himself states that only some believed in Asclepius' descent from Coronis, and yet recounts this genealogy alone, I cannot explain. For the Arsinoë myth, cf. below, pp. 32 f.

[3] For an evaluation of Apollodorus' method and thought as represented in T. 5, cf. F. Jacoby, *F. Gr. H.*, II, Kommentar, p. 773, 30 ff.

[4] Niese and Kirchhoff date the *Catalogue* between 620 and 580 B. C.; Wilamowitz says: "not long before 600." Cf. Schmid-Stählin, *Geschichte d. griech. Literatur*, I, 1, pp. 267; 269; 269, n. 2.

[5] Cf. Wilamowitz, *Isyllos*, pp. 57 ff., an admirable analysis indeed.

[6] In my discussion of the evidence I shall not, as does Wilamowitz, begin with a comparison of Hesiod and Pindar who deviates greatly from the former, but rather reconstruct the content of the Hesiodic poem from the testimony of those writers who are known to have followed the *Catalogue* closely. I should add that if it seems possible now to understand certain details of the myth as well as its general meaning in a new way, this is due to a large extent to the later studies of Wilamowitz on Greek mythology and religion.

To start with the Coronis saga which in antiquity was the more popular one: Coronis, according to Hesiod, was a young maiden living in Thessaly, near the Boebaean lake, in the Dotian plain (T. 21); she married Ischys, the son of Elatus; a raven reported her marriage to Apollo, who at that time happened to be in Pytho and was unaware of what had taken place (T. 22). That is all that the Hesiodic fragments tell expressly about Coronis. Although none of the lines extant from Hesiod designate this Coronis as the mother of Asclepius by Apollo or allude to her having been with child when she became the bride of Ischys, unquestionably it was she about whom Hesiod meant to speak. The ancients always quoted the Hesiodic verses as referring to Coronis, the mother of Asclepius by Apollo.[7] Moreover, the crow's report to Apollo, to which Hesiod himself makes reference (T. 22), indicates that in the poet's opinion the god had reason to be concerned with Coronis and her marriage to Ischys.

As for the events which occurred after Apollo had received the bad tidings, Pherecydes, in a relatively early paraphrase of the Hesiodic tale, says (T. 24) that, on Apollo's bidding, Artemis killed Coronis and many other women; Apollo himself, however, slew Ischys, while he delivered Asclepius to Chiron.[8] If Hesiod narrated so much, he must also have assumed what all the later representations of the heroic saga maintain: that the unborn child was saved from his mother's womb.[9] Hesiod, furthermore, must have recounted some of the accomplishments of Asclepius, the skillful physician, who even revived the dead and consequently was killed by Zeus; this can be concluded from Acusilaus, who was the first to transpose the Hesiodic genealogies into prose and who mentions the Coronis myth for the first time after Hesiod.[10] Acusilaus attests that Zeus slew Asclepius (T. 106) and that Apollo subsequently killed the Cyclops who had furnished the thunderbolt with which Asclepius was destroyed; that as punishment for this deed Zeus intended to throw

---

[7] Cf. e. g. Scholia in Pindarum, Ad Pythias, III, 52 a.

[8] T. 24 speaks only of the death of many women, but it is clear from the context that Coronis too was killed. That Pherecydes told the Coronis myth in agreement with Hesiod follows from his reference to the crow; cf. in general F. Jacoby, F. Gr. H., I, Kommentar, p. 389, 1 ff., and below, pp. 29 f. Concerning the localization of the saga by Hesiod, Pherecydes and others, cf. below, n. 45. The Hesiodic Fragment 124 [Rzach] may perhaps be part of the Coronis poem. In that case, it would afford direct evidence for a connection between Asclepius and Chiron, but the content of the lines is so vague that such an identification necessarily remains uncertain. It is also possible that the fragment belonged to the Jason saga, cf. Hesiod, Fr. 19 [Rzach].

[9] Concerning this detail of the birth legend, cf. below, pp. 36 f.

[10] For Acusilaus as a transcriber of Hesiod, cf. Jacoby, F. Gr. H., I, Kommentar, p. 375, 42; that he copied the Coronis saga follows from T. 23.

Apollo into Tartarus, but on the intercession of Leto was satisfied with making him serve as a thrall to a mortal (T. 107b).[11] One might claim even direct evidence for the fact that the latter events were recounted in the Coronis poem. For those testimonies which state that Hesiod spoke about Asclepius' death (T. 105 f.)[12] and Leto's mediation (T. 107b) may well refer to the Coronis *Eoee*.[13]

Like all the other poems that formed part of the Hesiodic *Catalogue*, the Coronis song, the outlines of which have now become clear, cannot be dated earlier than the end of the seventh century B. C.[14] Does this imply that the myth which is told here was also composed at that time, be it by the poet himself, or by somebody else? Or is it more probable that the legend was invented at an earlier period? If it were correct that the story as it stands was written down as propaganda for the Delphic Oracle, as is generally believed, one should conclude that the myth itself cannot have been earlier than the seventh century. For the great reform movement which brought about the rise of the typically Delphic religion did not set in before 700 B. C.[15] Now, it is true that the Apollo of whom Hesiod speaks resided in Delphi (T. 22). But this fact does not necessarily imply that he was the " Delphic Apollo " in the strict sense of the

[11] Acusilaus himself gives only the bare facts (Asclepius' death, if the fragment is rightly restored [cf. Jacoby, *op. cit.*, p. 378, 14 ff.]; Zeus' decision; Leto's plea; Apollo's servitude); connection and motivation of the events I have supplied in accordance with the usual form of the saga (cf. *e. g.* T. 3 and T. 108; that Pherecydes here speaks of the sons of the Cyclops instead of the Cyclops themselves, is a later rationalization of the myth). Since this part of the saga is always given in an identical fashion, such a restoration seems unobjectionable.

[12] As regards the expression Λητοΐδην for Asclepius, cf. Wilamowitz, *Isyllos*, p. 64, n. 37, but cf. also L. Weber, *Philologus*, LXXXVII, 1932, p. 394.

[13] It must be stressed, however, that all these Hesiodic fragments at first glance would seem to belong to the Arsinoë story (cf. below, pp. 32 f.) with which they were usually connected until Wilamowitz restored the Coronis poem. For the killing of the Cyclops is expressly attested as part of the Arsinoë *Eoee* (T. 107a; cf. Wilamowitz, *Isyllos*, p. 79). Consequently, as far as the direct evidence goes, T. 105, 106, 107b, which enlarge on the circumstances following this event, should be considered as belonging to the Arsinoë fragments, and it is strange that Wilamowitz, *ibid.*, and all other scholars should speak of the Cyclops and their death as " also " mentioned in the Arsinoë poem. As a matter of fact, the assumption that these testimonies can be claimed for the Coronis *Eoee* rests only on the authority of Acusilaus and on the agreement of all later witnesses concerning this feature.

[14] For the date of the Hesiodic *Catalogue*, cf. above, n. 4.

[15] For the common interpretation of the tendency of the Coronis story, cf. Wilamowitz, *Isyllos*, p. 72: " Der Gott von Delphi steht überall im Mittelpunkt der Handlung. . . . Ich möchte es gradezu ein delphisches Gedicht nennen." Cf. *ibid.*, p. 76: " Die Eoee benutzt Asklepios nur für eine Episode zum Ruhme des Apollon." Cf. also L. Malten, *Kyrene, Philologische Untersuchungen*, XX, 1911, p. 160, n. 1; Pauly-Wissowa, II, p. 1647, 14 ff.; Kern, *Religion*, II, p. 75; Weber, *op. cit.*, p. 395. For the date of the Delphic reform, cf. Wilamowitz, *Glaube*, II, p. 34, and for the concept of the Delphic Apollo, *ibid.*, pp. 26 ff.

word. Homer, too, knows of the Apollo in Delphi. Nevertheless, his Apollo is of a character quite different from that of the later god of Delphi.[16] The same must be said of the Apollo of whom Hesiod sings. For while the " Delphic Apollo " is the interpreter of Zeus' will, the preserver of law and order, the persecutor of any blood guilt, the Apollo of the Coronis myth dares to oppose Zeus, to revolt against his will; he avenges Asclepius' death by slaying the Cyclops. Under these circumstances, is it permissible to claim that Hesiod's Apollo is the god, as he was later venerated in Delphi?

Yet, the Delphic tendency of the poem, it is maintained, is definitely established by Apollo's repentance recounted at the end of the Coronis story. The god became a thrall to a mortal (T. 107b), to Admetus as another testimony (T. 107c) specifies. Consequently, scholars have even gone so far as to assert that in the *Eoee* Apollo helped Admetus in assuaging Artemis' wrath and in regaining Alcestis. Both Apollo and his sister thus finally appear as the holy and benign protectors of men; here, a conversion of the gods is represented in the true spirit of the Delphic Oracle.[17] Recent research, however, has shown that such a claim certainly is exaggerated. It is impossible to prove that the Admetus-Alcestis episode was referred to at all in the Coronis song.[18] To be sure, Apollo is made to pay penalty for his misdeeds. Still, the god's servitude to a mortal, to Admetus, may have been introduced in an attempt to reconcile the new, the Delphic, concept of Apollo with that of an earlier period.[19] The god, as he is represented throughout the legend with the exception of the final episode, is a wrathful god rather than a mild and lawful deity.

This becomes abundantly clear if one takes into consideration not only what Apollo does, but also the motives which are ascribed to him, especially his intentions in killing Coronis. The god punishes her after he has heard that she married Ischys (T. 22). Why does he do so? The fact that Coronis, in spite of her being with child by Apollo, becomes

[16] For Delphi as the seat of Apollo, cf. *Iliad*, IX, 404-5; for the Homeric Apollo in general, Wilamowitz, *Glaube*, I, pp. 324 ff.

[17] Cf. especially Wilamowitz, *Isyllos*, pp. 72-3.

[18] Cf. A. Lesky, *Alkestis, Der Mythus und das Drama, Wiener Sitzb.*, phil.-hist. Klasse, CCIII, 1927, Abh. 2, pp. 50-54; cf. also H. Drexler, *Gnomon*, III, 1927, p. 443; L. Weber, *op. cit.*, p. 403. In this connection it should also be remembered that Acusilaus who follows the Coronis story does not mention Admetus, but speaks only of " a mortal " with whom Apollo served (T. 107b). The Hesiodic fragment in which Admetus' name is given may have been taken from the Arsinoë story for which even the Admetus-Alcestis saga would make a fitting end indeed, cf. below, n. 39.

[19] Thus Anaxandrides, the Delphian, made Apollo serve with a mortal because he killed the Pythian dragon, cf. Wilamowitz, *Glaube*, II, p. 38, n. 4.

the wife of another is nothing unheard-of in the world of ancient mythology; on the contrary, it is quite usual for a god's love to marry a
mortal.[20] Acusilaus claims that Coronis accepted Ischys as her husband
" for fear of contempt " (T. 23). This may be a rational explanation for
what the old saga did not think worth explaining because it was self-
evident; or it may be the reason adduced by Hesiod.[21] At any rate, it
is quite certain that Coronis could not marry in solemn ceremony against
the wishes of her father who had to give the bride away. Nor had the
father any reason for opposing the wedlock of his daughter with Ischys,
her kinsman who, according to ancient custom, was the appropriate
husband for her.[22] No, Phlegyas could only wish for the union between
Ischys and Coronis, and Coronis when marrying Ischys did so with her
father's consent.[23] Coronis, then, acted as she was allowed or even expected to do. What right had Apollo to punish her, to take her life?

[20] That Hesiod speaks of a legal marriage between Coronis and Ischys is now
generally recognized, cf. Wilamowitz, *Isyllos*, p. 59; Lesky, *op. cit.*, pp. 45-6; Weber,
op. cit., p. 393 (note in T. 22 especially the words γῆμε; ἱερῆς ἀπὸ δαιτός). For the naive
attitude of the ancient saga, cf. the stories of Europa (Hesiod, *Fr.* 30 [Rzach]) and of
Leda and Alcmene, and in general Wilamowitz, *Isyllos*, p. 59.

[21] Cf. Jacoby, *F. Gr. H.*, I, Kommentar, p. 378, 20-21: "(Acusilaus) fügt . . . einen
rationalistischen Einzelzug ein." The Greek term κατὰ δέος ὑπεροψίας is ambiguous.
Is Coronis afraid of the contempt of men, or of the negligence of the god, as is
Marpessa (cf. Bacchylides, 20 [Kenyon])? The insistence on Coronis' preferring a
mortal seems to point to the latter interpretation.

[22] In general, cf. Wilamowitz, *Isyllos*, p. 59. Not even Pindar dares to impute that
Coronis' marriage would be objectionable; that is why he changes this feature of the
old saga, cf. below, p. 31. For Phlegyas, the enemy of Apollo, cf. below, n. 30 and
p. 35.

[23] That the Hesiodic poem viewed the events in this light I find also confirmed by
Apollodorus' words (T. 3): καὶ φασιν ἐρασθῆναι ταύτης Ἀπόλλωνα [*sc.*, τῆς Κορωνίδος]
καὶ εὐθέως συνελθεῖν· τὴν δὲ παρὰ τὴν τοῦ πατρὸς γνώμην ἑλομένην Ἰσχυϊ τῷ Καινέως ἀδελφῷ
συνοικεῖν. The text of the passage is corrupt, but certain changes are more or less agreed
upon. All editors and interpreters read τὴν δὲ instead of τῆς δὲ (R?), τοῦ δὲ A. On the
other hand, ἑλομένην is read instead of ἐλωμν᾽ R, ἐλωμένου Rᵃ, ἐλομένου A by Wagner;
it is bracketed by Frazier [Loeb]. Lesky, *op. cit.*, p. 48, keeps ἐλομένου, but his explanation and the addition of τὸν θεῖον γαμβρὸν which it necessitates is rightly refuted by
Drexler, *op. cit.*, p. 444, who himself proposes ἐλέσθαι, a very plausible emendation;
Wilamowitz' restitution of the text, *Isyllos*, pp. 62 f., is now generally rejected as going
too far. What Apollodorus wishes to say is in the main clear; Coronis chooses to marry
Ischys (for συνοικεῖν = marry, cf. Apollodorus, I, 7, 9: ἐπέτρεψεν αὐτῇ τῇ παρθένῳ ἐλέσθαι
ὁποτέρῳ βούλεται συνοικεῖν). But what does he mean by παρὰ τὴν τοῦ πατρὸς γνώμην?
This phrase usually is translated " against her father's will," and Lesky and Drexler,
loc. cit., consider it equivalent to Pindar's words κρύβδαν πατρός (T. 1, *v. 13*). Yet
certainly " without her father's knowledge " and " against her father's will " are not
identical expressions. Moreover, Apollodorus in his account obviously follows Hesiod,
not Pindar, for he speaks of Coronis' marriage and he alludes to the story of the raven
which Pindar suppresses; cf. Wilamowitz, *Isyllos*, p. 63, and below, p. 31. The expression παρὰ τὴν τοῦ πατρὸς γνώμην therefore should, I think, not be taken to mean " against
her father's will," but rather " in accordance with her father's will." For παρά in this
sense, cf. Kühner-Gerth, *Griech. Grammatik*, II, 1, 1898, p. 513, and J. Bernays, *Die
Dialoge des Aristoteles*, 1863, p. 138.

Perhaps Apollo had demanded of Coronis that she should not wed, hence in not following his wishes she violated no human law, but rather the personal command of the god. Later versions of the legend intimate such an attitude on the part of the god. They claim that the raven from whom Apollo heard about Coronis (T. 22) was assigned to her by the god as her guardian; some even say that the raven was expressly charged with protecting Coronis from any violation (T. 46; 48; 49). Yet not only does such an assumption presuppose that Coronis becomes the love of Ischys, not his wife, as she does in Hesiod[24]—the raven as protector of Coronis certainly is a later transformation of the myth. The guardian is a typical figure of the love poetry of the decadence, in which the lover is always suspicious of his beloved and by all possible means tries to guard her and himself against the interference of a rival.[25] In Hesiod, the raven cannot have made his report to Apollo because the god had charged him with watching Coronis; the raven must have acted on his own account, as other sources aver (T. 44; 45). Neither by human nor by divine law was Coronis bound to renounce marriage.

If nevertheless the god punished her, he did so because he has a bad and quick temper. His anger is easily roused, and if it is, he vents it without paying attention to reason or justice. Thus, when the raven told him about Coronis, the god was infuriated and, in return for the report, punished the bird by changing his white plumage into black (T. 2, vv. 631 f.; T. 45).[26] Now although Apollo had not ordered the bird to watch Coronis, the fowl by informing Apollo of what had happened thought to do a good service to his master, and rightly hoped for a reward.[27] But only too true were the warnings which on his way he received from

---

[24] In other words, these reports adapt a feature of Pindar's version, cf. below, p. 31.

[25] Cf. e. g. Ovid, Amores, II, 2, and in general M. Schanz, Gesch. d. röm. Litteratur, II, 1, 1911, p. 301. At best one might regard this feature as a Hellenistic transposition of the old saga, which may have seemed the more plausible since the raven was considered the exemplar of chastity among the birds, cf. Aristotle, Hist. Animal., 488 b 5.

[26] The story is most charmingly told by Ovid (T. 45) who follows Callimachus (T. 44); cf. M. Haupt-R. Ehwaldt, Die Metamorphosen des P. Ovidius Naso, I, 1915, ad II, 531; M. de Cola, Callimaco e Ovidio, Studi Palermitani di Filologia Classica, II, 1937, pp. 44 f.; Wilamowitz' hypothesis that Ovid is dependent on Nicander (Isyllos, p. 60, n. 33) was refuted by the discovery of the Callimachus fragment in 1896. Contrary to Wilamowitz' claim, Isyllos, pp. 57 f., it is not attested that Hesiod mentioned the change of the raven's plumage. T. 41 speaks of this story only as one generally told. Yet one may safely ascribe it also to Hesiod since Artemidorus' assertion is borne out by all testimonies, except that of Pindar who omits the raven episode altogether.

[27] The raven is the sacred bird of Apollo. Wilamowitz' denial of this fact (Glaube, I, p. 146) is refuted by the testimony of Plutarch, De Iside et Osiride, 379 d, and Aelian, Historia Animalium, I, 48. For other stories concerning Apollo and the raven, cf. Preller-Robert, Mythologie, I, 1, p. 516, n. 3. In Homer, the bird of Apollo is the ἴρηξ (Iliad, XV, 237), or the κίρκος (Odyssey, XV, 526); the word κόραξ does not occur in the Epic.

a fellow bird who on account of his sad experiences with Athena had urged
him not to meddle in affairs that were not of his concern, for the gods
do not distinguish between the messenger reporting an offence and the
culprit himself who committed it (T. 44; 45). Apollo, irascible and unreasonable as he was, felt hurt (T. 41) and retaliated on whoever was
nearest at hand.[28] In the same way, the god acted in slaying the Cyclops,
who certainly had no responsibility for what Zeus had done to Asclepius
but, in furnishing the thunderbolt with which to fell him, had obeyed the
command of the king of gods, as was their duty. In the same way, Apollo
acted in killing Coronis. She was in her right, but the god did not ask
whether she was justified, or whether he himself had wronged her and
was now inflicting cruel punishment upon one who was innocent. He was
enraged at Coronis, and just as he was quick to take possession of her
whom he loved (T. 3), so he was quick to take revenge on her whom he
hated.[29] Coronis died and with her many others, they too, without guilt,
her maidens, her bridegroom (T. 24).[30] Certainly, the god who imposes
such a fate upon men is the Homeric Apollo, the rash and wrathful deity,
not the " Delphic Apollo," the guardian of law and order.[31]

If further proof for this claim were needed, it could easily be found in
Pindar's refashioning of the old saga. The prophet of the Delphic god

[28] The raven story is the typical tale of the bearer of ill tidings (cf. Wilamowitz,
Glaube, I, p. 146, n. 1), which at the same time explains why the plumage of the bird
is black. The myth is told of *the* raven rather than *a* raven; cf. also W. Aly, Pauly-
Wissowa, *s. v.* Mythos, XVI, p. 1396, 55 ff.; *ibid.*, p. 1407, 1 ff. No one will believe that
in Hesiod, Apollo was pictured as the sentimental lover, as he was in Ovid, the lover
who hates the raven " by which he had been compelled to know the offence that brought
his grief " (T. 2, *vv. 614 f.*). Such an attitude fits the spirit of Hellenistic and Roman
love poetry rather than that of the Hesiodic saga. It is significant that Ovid makes the
god immediately repent of what he has done and dispute his right to any action whatsoever (T. 2, *vv. 612 ff.*).

[29] That Apollodorus' words ἐρασθῆναι ταύτης 'Απόλλωνα καὶ εὐθέως συνελθεῖν (T. 3),
characterizing Apollo's rash love, are probably taken from Hesiod is a most plausible
suggestion of Wilamowitz, *Isyllos*, p. 70, n. 48. Concerning the punishment of Coronis
it is stated by Servius, who also has preserved other details of the old saga (cf. n. 30),
that the god acts *iratus* (T. 46); cf. also T. 2, *v. 613*.

[30] Servius reports (T. 27) that Phlegyas, enraged at the violator of his daughter, set
Apollo's temple aflame and therefore was killed by the god. This account is quite
singular; in its rationalizing form it may indicate that in the old saga Phlegyas,
too, suffered death from the hands of Apollo. Cf. Weber, *op. cit.*, p. 395, n. 8.

[31] There are other features of the Coronis story which can be understood only under
the presupposition that the myth reflects the Homeric spirit rather than that of Delphi.
Artemis, here, is still the goddess who brings sudden death upon women, as she does
in Homer, cf. Wilamowitz, *Glaube*, I, p. 179; Apollo and Artemis together inflict death
upon Coronis and Ischys, just as they do upon the children of Niobe, cf. Homer, *Iliad*,
XXIV, 602 ff. The Delphic Apollo, on the other hand, has nothing whatever to do
with death, cf. Wilamowitz, *ibid.*, II, p. 42 (curiously enough, Wilamowitz himself,
ibid., I, p. 315, says: " Dass Apollon und Artemis Menschen bei Homer erschiessen,
später höchstens in Nachahmung Homers wie bei Koronis . . .").

could not help altering the legend completely so as to make it agree with his concept of Apollo. The ancients themselves noticed his changes and attributed them to his religious criticism (T. 41; 42). He could take over only the bare facts: Coronis' death, Asclepius' salvation, his education through Chiron, his career as a physician and finally his destruction through Zeus. Yet, out of these data, Pindar of necessity created an entirely new myth (T. 1).[82] Coronis now does not marry Ischys, her kinsman, with the consent of her father; although she is the love of Apollo she consorts with Ischys, a stranger, concealing her shameless indulgence from her father. However, her crime does not remain hidden from the god who, not through a bird's report, but through his own omniscience, detects Coronis' misdeed.[83] Justly, then, he punishes her for what she has done, not because he is angry with her, but because she has failed to observe the obligation imposed upon her by human and divine law: to wait for the day appointed for her marriage feast. It is for this reason and for no other, that Artemis kills her, rightly, as the poet emphasizes, who is himself not remiss in heaping blame upon Coronis.[84] Thus Pindar's Delphic version presents the moral drama in which the transgressions of sinful men are visited upon them by divine justice, whereas the old legend described the fateful tragedy in which men, the playthings of the gods, were innocently destroyed to please the divine whims.

It stands to reason, then, that the original Coronis myth was independent of Delphic influence. It may have been invented at any time after Apollo had taken possession of Delphi, yet before the reform move-

[82] For Pindar's changes, cf. Wilamowitz, *Isyllos*, pp. 58 ff.; *id.*, *Pindaros*, 1922, p. 281. Weber, *op. cit.*, pp. 397 ff., tries to prove that Pindar's representation of the saga is dependent upon earlier sources, among them formulations of Delphic priests. I do not think that this is probable. Pindar's version of the Asclepius myth seems original with him. For the attitude of the Delphic Oracle towards the saga, cf. below, n. 35.

[83] It is Pindar's omission of the raven episode which the ancients themselves expressly explained by his religious scruples (T. 41; 42). Omniscience is characteristic of the god of the Delphic Oracle, cf. Wilamowitz, *Glaube*, II, p. 27. It is hardly necessary to emphasize the fact that the reference to the raven, while being in opposition to the concept of the Delphic God and his prophecy, is in agreement with older religious concepts. In Homer, Apollo like all the other gods has a bird as a servant or messenger, cf. above, n. 27; he even appears himself in the disguise of a bird, cf. *Iliad*, XV, 237. For the general significance of birds in early Greek religion, cf. Wilamowitz, *Glaube*, I, p. 148. The raven episode, therefore, is one of the strongest indications of the antiquity of the original Coronis myth.

[84] It should be noted that Pindar suppresses the killing of Ischys and the death of the Cyclops. The latter fact in particular, which involves Apollo's opposition to Zeus, under no circumstances could be accommodated to Pindar's concept of Apollo. I have dealt here only with those features which serve to clarify the difference in the Hesiodic and Pindaric representations of Coronis and Apollo. For other discrepancies in their judgment concerning Asclepius, cf below, n. 93.

ment of the seventh century B. C. Certainly it was not composed as
Delphic propaganda or in praise of the Delphic Apollo. Delphi and the
Coronis saga belong to worlds wide apart. When the story of Coronis
and Apollo was conceived, the thought of men was still dominated by the
early amoral concepts concerning gods and mortals and their relation to
each other, concerning divine prerogatives and human strength.[35]

Moreover, it may safely be assumed that the saga, according to which
Asclepius was the son of Coronis and Apollo, constitutes the oldest
version of the Asclepius legend. At any rate, no other tale known from
the testimonies has an earlier or better-founded claim to this title. Even
that story which makes Asclepius the son of Apollo and Arsinoë and
which is also related in the Hesiodic *Catalogue* must be of later origin.
Its real aim is not that of tracing the genealogy of Asclepius, but rather
that of establishing the descent of his son Machaon, who was worshipped
as a national hero in Messenia (T. 38).[36] Arsinoë was a Messenian of
royal blood, the daughter of Leucippus, who was widely renowned among
the Messenians (T. 36). Besides being the mother of Asclepius Arsinoë
had also a daughter, Eriopis, by Apollo (T. 37). Obviously her relation
to the god is seen here as the happy union of a mortal woman with a
divine partner.[37] The raven episode is suppressed. There is no trace of

[35] This result implies that the serfdom imposed upon Apollo at the end of the *Eoee*
must be understood as an adaptation of the old fable to the spirit of the time in which
the Hesiodic poem was written, cf. above, p. 27. This feature does not provide the clue
to the meaning of the story itself. Once the essentially Homeric character of the old
legend is recognized, one might even hold that the punishment meted out to Apollo is
not different from that which the Homeric Zeus inflicts upon other gods or threatens
against them in retaliation for their misdeeds; cf. *e. g. Iliad*, I, 580 ff., and VIII, 13 ff.
That Zeus intends to throw his disobedient son into Tartarus (T. 107b) reads like a
reminiscence of the second of the epic passages just referred to. It is therefore not
impossible that the Coronis song may follow the original version throughout. In my
opinion no decision in favor of one or the other of these interpretations can be reached.
It is noteworthy that the Delphic Oracle did not approve of the Asclepius myth as it
was recounted by Hesiod. Still further expurgations were needed before the saga was
sanctioned by Delphi, cf. below, p. 68. The necessity of such moralistic changes of the
early legends is not taken into account by Wilamowitz, *Glaube*, II, p. 35, where he
points out that the love stories of Apollo were retained as part of the old-inherited
myth even by Delphi.

[36] Hesiod mentions Asclepius and Xanthe as Machaon's parents (T. 169). Since it is
known that the Arsinoë myth was intended to make Asclepius a compatriot of the
Messenians—Pausanias even charges that it was invented in their interest (T. 36)—
this testimony must be part of the Arsinoë myth, cf. Wilamowitz, *Isyllos*, pp. 77 f.

[37] That Arsinoë had a daughter Eriopis is attested by Asclepiades (T. 37), but it
cannot be decided with certainty which of the lines preserved in the testimony are
Hesiodic, cf. Wilamowitz, *Isyllos*, p. 79. Concerning Arsinoë in general, cf. Wilamowitz,
Isyllos, pp. 77ff.; Pauly-Wissowa, II, pp. 1279 ff.; p. 1648, 64 ff. Wilamowitz, *op. cit.*,
p. 77, believes that Arsinoë originally was not a member of the Leucippus family; for
conjectures as regards her earlier relationship to other families, cf. Pauly-Wissowa,
II, p. 1649, 41 ff. Leucippus is sometimes connected with Laconia (T. 34). The Homeric

lawlessness or violence in this story, as is fitting for a myth which is destined to serve religious and political aspirations. Nevertheless it is affirmed that Apollo killed the Cyclops (T. 107 a).[38] In view of the general tendency of the poem, this can be understood only under the presupposition that at the time when the Arsinoë genealogy was put in verse, the slaying of the Cyclops was a feature already so closely connected with the Asclepius saga that it could not be omitted.[39] The Arsinoë poem, like the Coronis song, was composed probably around 600 B. C. Yet, the Arsinoë myth must have been preceded by the Coronis saga.[40] It is the latter which, according to the evidence available, represents the archetype of the Asclepius legend, a story that originated in a world where the gods were still viewed as the epic poet had depicted them.

Does this result imply that Homer knew of Asclepius, the son of Apollo and Coronis? Was it this hero whom he made the father of Machaon? As long as the date of the myth alone serves as a touchstone for settling these questions, one might be inclined to say that this may have been the case, for the Coronis saga originated before 700 B. C. On the other hand, when those books were written in which Machaon's treatment of Menelaus is described, a physician was hardly esteemed so highly that he would have been considered the descendant of a god. Such an assumption seems more fitting for the age of the Cyclic Epic, when medicine became increasingly important and when Machaon and Podalirius, too, were said to be of divine origin.[41] It is more likely, therefore, that a saga, later than the one which Homer followed, elevated Asclepius to the kinship with Apollo. Asclepius, Chiron's friend, through Homer became a king; the Epic endowed him with greater human dignity than he had formerly possessed. The later transformation of the original tale, by making As-

---

Hymn to Apollo possibly refers to a fight between Apollo and another of Arsinoë's wooers, cf. Wilamowitz, *Isyllos*, pp. 80 ff.

[38] Cf. above, n. 13, where it was pointed out that in fact the Cyclops incident is directly attested only for the Arsinoë poem.

[39] On the other hand, the Admetus-Alcestis myth would be an appropriate conclusion of the Arsinoë story, cf. above, n. 18. For in both sagas the gods are seen as the friends of men. An interesting hypothesis concerning the connection of Arsinoë and Alcestis is proposed by Wilamowitz, *Isyllos*, p. 79, n. 52.

[40] This is also the opinion of Wilamowitz, though his argument is different from mine, cf. *Isyllos*, p. 80. At the end of the seventh century Messenia fought her second war against Sparta. At this time of national uprising an interest in the Messenian saga is well understandable. But one should believe that the myth itself originated before 700, that is, in the period of Messenian independence, cf. J. B. Bury, *A History of Greece*, 1937, p. 120. On the other hand, the Messenian saga seems later than the Homeric Epic, for as has been shown above, p. 21, Asclepius in the Arsinoë poem appears as a truly epic hero.

[41] Cf. above, p. 15.

clepius the son of a god, formed the first link in the chain which was to bind Asclepius to the immortals.[42]

ii.   The Meaning of the Legend

The main content and the date of the Asclepius legend have now been determined. But the data reviewed so far do not give more than the broad outlines of the saga. Its purport has not yet been clarified. What kind of myth is the one attached to Asclepius? Does it echo certain facts in the lives of real individuals or tribes? Is it designed to explain a cult or a ritual? Or if it is neither a historical nor a hieratic story, does it symbolize in images of human fantasy the hopes and fears of men, their estimation of life, of the destiny and fate of mortal beings? One cannot hope to grasp the significance of the Asclepius figure without deciding which one of these meanings applies to the Asclepius myth.[43] A close scrutiny of the details reported about Asclepius, of the local and temporal transformations of the saga, should make it possible to ferret out the gist of the oldest draft of the legend and thus to determine its original import.

A hero, a being half human, half divine, must be a member of a human family; he must also be assigned to some native land. Asclepius' mother, according to Hesiod, was Coronis (T. 22), the daughter of Phlegyas and the sister of Ixion (T. 27). She thus belonged to a race which played an important rôle in the Thessalian saga. The Phlegyans probably were a conquering people who of old held the regions of Thrace and Thessaly under their sway.[44] That Coronis lived in the Dotian plain, as Hesiod said (T. 21), was generally agreed upon in antiquity. Phere-

---

[42] Such " promotions " of human beings are not unusual. Meleager for instance in the Homeric Epic is a mortal, while in the Hesiodic *Catalogue* he appears as the son of Ares; cf. Wilamowitz, *Die griechische Heldensage*, II, *S. B. Berl.*, phil.-hist. Klasse, 1925, p. 216.

[43] For a general appreciation of ancient mythology, cf. the studies of Wilamowitz and Weinreich, referred to above, p. 7, n. 18. Cf. also A. Lang, *Encyclopaedia Britannica*[11], 1913 *s. v.* Myth; Farnell, *The Value and the Methods of Mythologic Study*, in *Proc. Brit. Acad.*, 1919-20, pp. 37 ff.; H. J. Rose, *Modern Methods in Classical Mythology*, 1930; Pfister, *Bursian*, 1930, pp. 146 ff.; W. Aly, Pauly-Wissowa, *s. v.* Mythos, XVI, pp. 1374 ff.

[44] Concerning the Phlegyans who for a long time were considered merely a mythical tribe, cf. Pauly-Wissowa, II, p. 1643, 65 ff.; Wilamowitz, *Glaube*, I, p. 52, n. 1; for Coronis, cf. Pauly-Wissowa, *s. v.* Koronis, XI, p. 1431, 57 ff. Whether she was the genuine mother of Asclepius (Pauly-Wissowa, II, p. 1644, 53), or whether her name is only poetic fiction (Wilamowitz, *Isyllos*, p. 79, n. 52), it is impossible to say before the myth has been interpreted as a whole, cf. below, p. 64, n. 37. The name Coronis often appears in the heroic saga; it may signify a crow, the symbol of longevity, cf. Preller-Robert, *Mythologie*, I, p. 515, n. 3.

cydes, Pindar and almost all the poets and mythographers concurred in this assumption of the *Eoee*. There were only slight variations as to the exact location of Coronis' home within that region.[45] The hero Asclepius, then, was born in the Dotian plain. To Hesiod, he was a Thessalian, just as he probably was in the eyes of the poet of the Homeric Epic.[46]

Within the framework of the whole legend one might understand Asclepius' descent on his mother's side as an indication that in his physical make-up there were ingredients of daring enterprise, of bold venture. Phlegyas and his kin were notorious for their ruthlessness and their hostile attitude towards the gods.[47] Beyond this, the reports concerning Coronis yield hardly anything, nor can they be expected to be of great consequence. They concern Asclepius' human existence alone, and in the legend of a hero the mother's task is merely that of bearing the child who is destined for fame.[48] It is her divine partner through whom her son shares in superhuman power and through whom he acquires the strength which he needs for his exploits.

Now in all versions of the myth, Asclepius is said to be the son of Apollo; with this god and no other is he connected throughout antiquity.[49] To explain the relationship between the two, it would not do to reason on racial or geographical grounds. Apollo was revered everywhere; any

[45] Hesiod adds that Coronis lived near the Boebaean Lake (T. 21); cf. also Pindar, T. 1, *v. 34*. Pherecydes speaks of Lacereia near the Amyrus river (T. 24); cf. also Apollonius Rhodius, T. 15; 28. Incidentally, Lacereia may be a mythical place, cf. L. Weber, *op. cit.*, p. 401, and since the bird from which Coronis' name may be derived is once called λακέρυγα κορώνη, "a cawing crow" (Hesiod, *Opera*, 747), etymological play may have been responsible for the connection of Coronis and Lacereia (cf. Wilamowitz, *Isyllos*, p. 60, n. 32).

[46] Cf. above, p. 22. Note moreover that the Dotian plain, the Amyrus river and Lacereia, the places where Coronis lived, are all located in the neighborhood of Mount Pelion where Chiron resided, cf. Strabo, IX, p. 442; XIV, p. 647.

[47] In general cf. Preller-Robert, *Mythologie*, II, pp. 26 ff. Concerning Phlegyas and Apollo, cf. also above, n. 30.

[48] Cf. Aeschylus, *Eumenidis*, 658 ff. Apollo himself here advocates the patrilineal theory, and even the poems of the *Catalogue*, although they glorify the mothers, do not profess matrilineal beliefs, cf. Schmid-Stählin, *op. cit.*, I, 1, p. 267.

[49] This fact is the more remarkable, since in Greek mythology variations as to a hero's father are nothing unheard of. W. A. Jayne, *The Healing Gods of Ancient Civilizations*, 1925, p. 254, and L. Dyer, *Studies of the Gods in Greece* etc., 1891, p. 219, speak also of Zeus as the father of Asclepius, adducing as evidence P. Gardner, *New Chapters in Greek History*, 1892. Gardner, p. 358, however, says only that Zeus and Asclepius later were identified. Walton, *Cult of Asklepios*, p. 85, lists Aristetes, Arsippus, Hephaestus, Ischys, Elatus, Lapithas, Sydykus as "fathers" of Asclepius. Sydykus is the father of the Phoenician god Asclepius (T. 825). Lapithas is not mentioned at all by Eustathius, *ad* B 732, the passage to which Walton refers (at another place, T. 30, Eustathius merely says that Asclepius sprang from the Lapithian race). All the other names given by Walton are connected with the genealogy of the three Asclepii (T. 379 f.), distinguished by the "sondernden Theologen," and these *stemmata* do not concern the son of Coronis.

Greek tribe could consider him the father of a hero. Nor is it permissible
to assume, as is usually done, that the Delphic Apollo adopted Asclepius,
delegating to him the realm of medicine of which he wanted to dispose,
because his medical functions did not fit his newly acquired character.
The Coronis poem, the oldest attestation of Asclepius' kinship with
Apollo, is not a tribute to the god of Delphi.[50] Rather is it Apollo, as he
is known from the Epic, who became the father of Asclepius in the tale
told by the Hesiodic poet. But what is the significance of Asclepius' being
treated as the son of Apollo? Is this divine genealogy merely a title to
higher rank and greater dignity? Or is he related to Apollo, the representative of one particular task, one of the many skills over which this
god presided?

From the rôle which in the old saga was attributed to Apollo in connection with Asclepius' birth it seems to follow that Apollo was represented here as the god of medicine, the sender of disease and the liberator
from illness.[51] Pindar, the oldest witness (T. 1, *vv. 43 f.*), relates that
when Coronis was burning on the pyre, Apollo snatched the unborn
child from her corpse. The god seized the babe from the flames and from
its mother's womb, says Ovid (T. 2, *vv. 629 f.*). Having cut open
Coronis' womb, Apollo extracted the child from it, says Hyginus (T. 48;
cf. also T. 47). Much as the language of these testimonies differs—
Pindar conceals the details, Ovid is less reticent, Hyginus is quite outspoken and technical—all of them agree on one fact: Asclepius was
brought forth to light by Apollo. This description of Asclepius' birth is
a very singular account indeed, not paralleled in the birth saga of any
other hero or god. Besides, Apollo here is made to play the part of a
physician, whether the hand of the god, miraculously penetrating Coronis'
body, achieved the saving deed, or whether, in rationalization of the
inscrutable ways of the deity, he is believed to have performed a regular
operation on Coronis.[52] That the god's action, ascribed to him by the

---

[50] Cf. above, pp. 31 f. For the generally accepted view, cf. *e. g.* Wilamowitz, *Glaube*,
II, p. 39: "Die Eoee von Koronis machte den Asklepios, einerlei was er gewessen war,
zu seinem sterblichen Sohne. So war die Heilkunst zwar von Apollon gelehrt, aber
er brauchte sich nicht mehr selbst mit ihr zu befassen."

[51] For Apollo as the god of medicine, cf. especially Nilsson, *Griech. Feste*, p. 97;
Griechische Religion, I, pp. 507 ff. The pertinent material is collected in Pauly-Wissowa,
s. v. Apollo, II, p. 15, 52 ff. Cf. also below, p. 56.

[52] Weber, *op. cit.*, p. 400, n. 15, finds a contradiction between Pindar's report (T. 1)
and that of Servius (T. 47); only according to the latter, he says "verfährt . . .
Apollon wie ein erfahrener Arzt." It is true that Servius like Hyginus, whom Weber
does not mention, deviates from Pindar in his words, but this does not imply a difference
in fact, cf. the commentary on the Pindaric verses given by Ch. G. Heyne, *Pindari
Carmina*, I-III, 1798, p. 238: "Simpliciter poeta dicit, etsi splendide, utpote de deo,
uno gressu facto Apollinem eripuisse flamma puerum, quem mater utero gestabat." Cf.

earliest source, also formed an integral part of the original story seems certain. Otherwise Pindar would hardly have referred to it. Moreover the help of the god was needed. When Coronis died, Asclepius was not yet born.[53] If Apollodorus simply says that Apollo snatched Asclepius from the pyre (T. 3), if another unknown authority asserts that Hermes rescued the babe from the flames (T. 26), these reports are only later transformations of the old saga, changes made in agreement with more refined religious concepts, or accommodations to other famous myths.[54] At any rate, it is the original and unique features that provide the clue to the character of Apollo, as it was viewed when the god was made the father of Asclepius: it was Apollo, the physician, the medical deity of old.

Certainly, this choice was not fortuitous, nor merely an eccentric play of poetic or mythical imagination; it was made on purpose, meant to have symbolic significance. Through his origin from the god of medicine, Asclepius inherited the natural gifts required for the task which he had

also E. C. J. von Siebold, *Versuch einer Geschichte der Geburtshülfe*, I, 1839, pp. 64 ff. In passing it should be pointed out that the Romans called the operation performed by Apollo *sectio Caesarea* and attributed to the god the tutelage over all children who were born in this way (T. 47).

[53] Cf. Wilamowitz, *Isyllos*, p. 70, in his recapitulation of the old legend: "Als er die Leiche der Geliebten auf dem Scheiterhaufen sah, da erbarmte sich Apollon wenigstens des ungeborenen Sohnes. Der den Tod gesendet, gab Leben. . . ."

[54] Note that even in T. 3 and T. 26 the child is saved from the pyre where Coronis is burnt; she did not give birth to her son before she died. In this respect the saga remains absolutely consistent. The Delphic Apollo renounced medicine; his medical deeds, therefore, tended to be forgotten, and this might explain Apollodorus' report (T. 3), if he did not simply omit a few words. The source of T. 26 cannot be determined; the version recounted seems to be a contamination of the Hesiodic tale with Pindar's poem (Ischys, the lover of Coronis). Wilamowitz (*Isyllos*, p. 77; *Glaube*, I, p. 132, n. 2) finds in the story an influence of the Dionysus myth. This is quite likely as far as the reference to Hermes is concerned, for Hermes is fixed in the birth saga of Dionysus, while in that of Asclepius he is nowhere else mentioned. Farnell's objection (*Hero Cults*, p. 243) to Wilamowitz' claim is not justified. But how did it come about that Hermes was made to play the rôle of Apollo? Some versions of the Dionysus myth are not satisfied with Hermes' carrying the prematurely born babe to Zeus, they make Hermes do with Dionysus what Apollo does with Asclepius: he brings the child forth by a *sectio Caesarea*; cf. Lucian, *Dialogi Deorum*, IX; also Ovid, *Metam.*, III, 310-11. But Hermes is not a physician, as is Apollo. Moreover, the far greater number of authorities, and the older ones at that, assert that Semele gave birth to the child in the moment of death, cf. Preller-Robert, *Mythologie*, I, p. 661. Lucian's story, therefore, can only be an adaptation of the Dionysus saga to the Asclepius story. It is impossible to speak of a general and basic similarity of the two tales, as has sometimes been done, e. g. Pauly-Wissowa, *s. v.* Koronidae, XI, p. 1434, 38 ff. Once Hermes had usurped Apollo's art, he could also usurp Apollo's place in the Asclepius story. Beyond this connection between the late myth of Asclepius and that of Dionysus, however, there are no relations of the Asclepius tale to any other birth saga. Farnell, *loc. cit.*, compares it with the story of the Coronidae. But here, stars or twins are born from the ashes of their mothers. It is only the name Coronidae that this late tale consciously derives from the Asclepius myth, and any dependence of the latter on the former seems out of the question. The original story of Asclepius' birth is unique.

to fulfill and for which his father himself had designated him: the pursuit
of medicine. For Apollo, as Pherecydes asserts in his transcription of
the Hesiodic poem (T. 24), gave the child into Chiron's custody.
Obviously he wanted the Centaur to take care of Asclepius and to educate
him in the medical art. Pindar says so expressly: it was Apollo's intention that Asclepius should learn " how to heal mortal men of painful
maladies " (T. 1, vv. 45-6).[55] Thus the hero was not merely descended
from Apollo, the physician; he was expected by his father to follow the
paternal calling. Indeed, no better recommendation of Asclepius, the
physician, could be imagined. Sharing through his birth in the power of
the god of medicine, expressly chosen by him to be a healer, educated on
Apollo's instigation by the sage who was preëminently skilled in medicine,
how could Asclepius fail to be a good physician himself? Here, foresight
was at work, divine providence, or wise planning on the part of him,
whoever he was, who designed the genealogy of Asclepius.

But at this point one might ask: was it really Asclepius' destiny to
be a physician and nothing else? Such a career, it seems, is incompatible
with the prerogatives of a great hero; such an occupation could hardly
have engrossed all his thoughts and activities. Even granted that Asclepius practiced medicine— being a hero, he must have been notable also
for his exploits in war and in other ventures. Yet, in fact Asclepius'
record as a hero among heroes is not too impressive, as the ancients
themselves noticed (T. 265).[56] Of all the enterprises, famous in Greek
mythology, he is said to have participated only in the Voyage of the
Argonauts (T. 5; 63 f.) and in the Calydonian Hunt (T. 65), and even
these two heroic deeds of his are attested by late authorities alone, that is,
by Alexandrian scholars, and have the air of rather stereotyped
inventions.[57]

[55] In general, cf. T. 51 ff. Concerning the subjects of Chiron's instruction, cf. T. 57;
as regards those who were educated by him together with Asclepius, cf. T. 51; 56; 58.
Only a few sources eliminate Chiron as the teacher of Asclepius and mention Apollo
himself as his instructor in medicine (T. 355; 356), whereas others make both, Apollo
and Asclepius, students of Chiron (T. 62). In one version (T. 55) Chiron and Asclepius
are even more closely connected by the fact that Chiron's death is linked up with the
Asclepius story. Since in Homer Chiron was the friend of Asclepius, not his mentor
(cf. above, p. 4), it seems most probable that this ancient relationship was changed into
that of pupil and teacher when Asclepius became Apollo's son and was entrusted to
Chiron by his father. Incidentally, in the eyes of the ancients who were accustomed to
ask with whom a physician had studied in order to check on his abilities (cf. Edelstein,
Problemata, IV, p. 92 and p. 109, n. 2) Asclepius' education through Chiron must have
been of great significance.
[56] Modern scholars, too, are unanimous in this verdict, cf. *e. g.* Thraemer, Pauly-
Wissowa, II, p. 1652, 66 ff.; Farnell, *Hero Cults*, p. 244.
[57] Thraemer and Farnell (see preceding note) are going too far in discrediting the
value of the three references to the heroic activities of Asclepius that are preserved.

As for the Argonauts, Asclepius seems to have been included among them as a famous Thessalian figure. The Voyage of the Argonauts was the foremost Thessalian saga. Moreover, in this myth Chiron was an important character, and Asclepius was thought to have been his pupil after Jason (T. 51), whose son, as Homer says, gave aid to the Greeks before Troy.[58] How could Asclepius, the Thessalian, the father of Machaon, fail to be included in the list of the Argonauts which the fantasy of the ancients constantly changed in accordance with their varying wishes? As for the Calydonian Hunt, Asclepius may have been mentioned along with others who had participated in the enterprise of the Argonauts and were later engaged in this new venture.[59] On the other hand, Asclepius' association with the Argonauts and the Calydonian Hunt in some way also fits the Messenian Asclepius. Nestor, the national hero of the Messenians and the friend of Machaon, took part in these two exploits of the heroic age, in fact they are the only ones in which he appears, with the exception of the Trojan War.[60] Asclepius, then, became a partner in the most famous sagas of those regions of Greece from which he was supposed to have come. It is fair to say that such claims represent fictitious adornments rather than old-inherited and " genuine " beliefs. When Asclepius had risen to a place of primary importance, it seemed impossible to admit that he should not have been connected with these outstanding events.[61]

The evidence for Asclepius' heroic exploits being so slight and of such artificial character, modern scholars have been wont to affirm that Asclepius has no real place in the heroic saga, and they have been inclined to add that this is so because Asclepius originally was either a god or a human being.[62] Yet, such an assertion takes account merely of what

Theodoretus' statement (T. 5) is taken from Apollodorus; that of Hyginus (T. 65) goes back at least to Alexandrian sources (cf. Pauly-Wissowa, s. v. Mythographie, XVI, p. 1366, 23 ff.) ; Clement (T. 63) names Apollonius Rhodius, Argonautica, where Asclepius, however, is not mentioned in the text preserved, cf. Pauly-Wissowa, II, p. 1653, 2. For the participants in the Voyage, cf. ibid., s. v. Argonauten, pp. 751-53.

[58] Concerning the Argonauts in general and Chiron's rôle, cf. Preller-Robert, Mythologie, II, pp. 760 ff.; 770. It is Pindar, who says that Asclepius came to Chiron after Jason (T. 51). This statement may have been one of the sources of the later claim that Asclepius was one of the Argonauts. For Jason's son, cf. Iliad, VII, 468-9; XXI, 41. W. Heidel's assertion (The Heroic Age of Science, 1933, p. 134, n. 15) that the Greeks generally dated the Argonautic expedition two generations before the Trojan era is incorrect, at least as far as Homer is concerned.

[59] Cf. Preller-Robert, Mythologie, II, esp. pp. 96-97.

[60] Cf. Ovid, Metam., VIII, 313; Valerius Flaccus, I, 380.

[61] Cf. e. g. Gruppe, Griechische Mythologie, p. 1452, n. 1: " Dass Asklepios Argonaut und Teilnehmer an der kalydonischen Jagd heisst, scheint später frei hinzugefügt zu sein."

[62] Thraemer, Pauly-Wissowa, II, p. 1653, 4 ff., reasons that Asclepius must have

might properly be called " heroic deeds "; it disregards all those statements which testify to Asclepius' curing or reviving of other heroes.
Undoubtedly, however, the fact that Asclepius is said to have healed the
wounds of Heracles (T. 87) gives him a share in the heroic saga, as well
deserved as does his active fighting or enduring of danger in the cause
of Jason—if not indeed even in the Argonautic expedition Asclepius was
a partner merely as a physician, or at best in his twofold capacity as a
warrior and a medical man. At least one of his mythical cures must have
taken place during the Voyage of the Argonauts; [63] the same double
activity may have been represented in the saga of the Calydonian Hunt.
Whether or not this was so, for an appreciation of Asclepius' rôle in
the heroic myth it is indispensable to take cognizance also of his medical
performances, the more so since the reports on these achievements are
by far the older ones and the better attested. As early as about 600 B. C.
Asclepius is mentioned as the one who revived Capaneus, Lycurgus, and
Hippolytus, and in the fifth century he is said to have healed Philoctetes
and Heracles.[64] Asclepius' part in the heroic saga, then, is considerably
more extensive than is generally supposed. The fact that he is not
credited with many heroic deeds, or more correctly with none at all, only
proves, therefore, that the saga was always consistent in viewing
Asclepius not as a warrior or adventurer, but as a physician, who by
his very nature was a man of peace rather than a fighter.

Under these circumstances, should one assume that Asclepius' medical
task consisted in reviving and healing heroes? Such a contention would
hold good only for that branch of the tradition which accepted as true the
Homeric representation of Asclepius; for all that is told about the
physician of heroes seems derived in the main from those data which the
Iliad furnished in regard to Asclepius and the age in which he lived.[65]

been a god; Farnell, *Hero Cults*, p. 244, advocates that he was a human being. The
value of these conclusions is not enhanced by their contradictory character.

[63] Cf. below, n. 72 (the healing of the Phinidae).

[64] Stesichorus mentioned Capaneus and Lycurgus (T. 69; 70); the *Naupactica* referred to Hippolytus (*ibid.*). According to Sophocles, Philoctetes was healed by
Asclepius (T. 92). Since the poet makes Heracles recommend Asclepius to Philoctetes,
it is most probable that he knew of Asclepius' treatment of Heracles which is attested
by later sources (T. 87; 88); cf. below, n. 71. It should be emphasized that the
authority of these reports, in part at least, is not inferior to that of all the other data
known about Asclepius. The oldest statements (Stesichorus, *Naupactica*) go back to
a few decades after the time in which the Hesiodic *Catalogue* was written and are
contemporaneous with the Homeric *Catalogue of Ships*, cf. above, n. 4; p. 21, n. 78.
The usual contention (cf. *e. g.* Jacoby, *Fr. H. G.*, I, Kommentar, p. 519, 29) that the
names of those revived by Asclepius are given by late tradition only, is therefore not
correct.

[65] Wilamowitz, *Glaube*, II, p. 227, n. 2, suggests that the names of the heroes revived
by Asclepius are derived from speculations concerning the Hesiodic *Eoee*, where

Capaneus and Lycurgus, whose revivals are the first to be attested, were characters of the Argive saga. With this province, however, Asclepius is not connected by any other early testimony; it is therefore unlikely that the revival of the two heroes was a genuine feature of the Argive tradition.[66] But Capaneus was one of the protagonists of the expedition against Thebes in which he was killed, and since the participants in this campaign were the fathers of the fighters before Troy, Asclepius' mention on the occasion of Capaneus' death is probably due to the circumstance that as the father of Machaon and Podalirius, who were soldiers in the Trojan War, he seemed a fitting partner in the Theban venture.[67] Lycurgus, too, was linked up with the enterprise against Thebes. Thus, he was a contemporary of Asclepius, the rescuer of Capaneus, whose sons were of one age with Achilles.[68] The same may be said of Hippolytus, Theseus' son, whom Asclepius restored to life. Other children of Theseus had joined the Asclepiads in the conquest of Troy.[69]

What is true of the revivals of heroes brought about by Asclepius is equally valid with regard to his mythical healings. Heracles, who was supposed to have been treated by Asclepius (T. 87, 88),[70] was a famous contemporary of the Homeric Asclepius. Besides, like Asclepius, Heracles had also been one of the Argonauts. Finally, the report that Asclepius was called upon to heal Philoctetes (T. 92) obviously represents an

Asclepius was also said to have revived men, but where no names of these "patients" were mentioned, cf. below, n. 74. This may well have been the case. Nevertheless all these identifications may depend on the Homeric picture of Asclepius.

[66] For the Argive background of the two heroes, cf. Preller-Robert, *Mythologie* II, pp. 914; 922; 935.

[67] The death of Capaneus is discussed also below, p. 47. The Theban enterprise was the greatest undertaking before the Trojan War, cf. Hesiod, *Opera*, 162 ff.

[68] For Lycurgus and the heroes who attacked Thebes, cf. Preller-Robert, *Mythologie*, II, pp. 934 f.

[69] Hippolytus, too, was an Argive figure, cf. Preller-Robert, *Mythologie*, II, pp. 738 ff.; but at an early date he was taken over by the Athenians and made a son of Theseus, cf. Preller-Robert, *ibid.*, pp. 741 ff., and even the Epidaurian god was said to have revived Hippolytus, the son of Theseus (T. 739, 4; cf. below, p. 74, n. 24). Granted that the reanimation of Hippolytus was also told in a Troezenian saga (cf. Preller-Robert, *ibid.*, p. 739), and that there, Hippolytus was not the son of Theseus—originally the story must have referred to Theseus' descendant. For the Thesidae in the Trojan War, cf. Preller-Robert, *ibid.*, p. 1286. Note that Hippolytus and Asclepius were brought together in the *Naupactica*, a poem that follows the Hesiodic pattern, cf. Schmid-Stählin, *op. cit.*, I, 1, p. 292. This fact confirms Wilamowitz' suggestion, referred to above, n. 65.

[70] The healing of Heracles, attested by Pausanias, was believed to have taken place in Laconia. The story, then, could be a Laconian invention. As such it would be relatively late, for in Laconia even the god Asclepius did not enjoy popularity in early times, cf. below, p. 241. On the other hand, Nicander (T. 89 ff.) already reports that Asclepius treated Iphicles when he was fighting the Hydra together with his brother, Heracles. This sounds as if Nicander had known of a relation between Heracles and Asclepius, which then in turn led to a connection between Iphicles and Asclepius, cf. also n. 71.

attempt in some way to associate Asclepius, the greatest doctor, with the Trojan War in which, according to Homer, he had no part. It seemed inconceivable that he should not have helped his countrymen in their dire plight. And thus he was made responsible for that medical achievement which decided the fate of Troy.[71]

Doubtless, the heroic saga gradually and systematically extended the reach of Asclepius' practice among his fellow heroes. The Asclepius of whom Homer spoke turned into the famous healer of wounded fighters, the reviver of those who had died in battle. Just as his children took care of their contemporaries, so he was believed to have treated the heroes of his own time and generation.[72] But Asclepius, the father of Machaon and Podalirius, was a figure invented by Homer; there must have been another notion of a physician Asclepius who was not related to these two medical men of the Trojan War.[73] All the stories concerning the

[71] Concerning the healing of Philoctetes, its importance and its original connection with the sons of Asclepius, cf. above, p. 13. T. 92 is in seeming contradiction with another passage of the same play (T. 152), where Sophocles says that the Asclepiads will heal Philoctetes. The difficulty has been resolved by R. C. Jebb, Sophocles, *The Plays and Fragments*, IV, 1890, *The Philoctetes*, ad 1437. In T. 152, Neoptolemus, the mortal, is speaking who cannot foresee the intervention of Heracles, the god (T. 92). Jebb assumes that Sophocles thinks " of Asclepius as a god, whom Heracles is to send from heaven." It is probable that Sophocles here wants to pay his respect to the god whose devotee he was, cf. below, p. 117. But this does not necessarily imply that in his opinion the divine Asclepius will take care of Philoctetes. Moreover, the connection between Asclepius and Philoctetes may be older than the Sophoclean play and may have originated in a time in which Asclepius was not yet considered a god. Heracles may well promise to procure the aid of the great physician whose merits he knows from his own experiences on this earth.

[72] In the text I have dealt only with the oldest and most important testimonies concerning Asclepius' reviving and healing of heroes. The later statements, in part, support the interpretation given, or at least they do not contradict it. T. 69 and 70 name as those whom Asclepius called back to life also Tyndareus (Panyassis), Glaucus (Amelesagoras), Orion (Telesarchus), Hymenaeus (Orphics). Tyndareus, according to the Cyclic Epic, was the father of Helen, cf. Preller-Robert, *Mythologie*, II, pp. 331; 1066. Glaucus, the son of Minos, was a contemporary of Theseus, but his revival is usually attributed to Polyides, cf. Preller-Robert, II, pp. 199-201; later, it was apparently ascribed to Asclepius, the most famous physician. Propertius even claims that Androgeon, another son of Minos, was restored by Asclepius (T. 86); but this, it seems, is simply a mistake. For the Androgeon saga, cf. Preller-Robert, II, p. 689. Orion and Hymenaeus, mentioned in Hellenistic times, are gods rather than men (cf. Wilamowitz, *Glaube*, H, p. 199, n. 2; Preller-Robert, I, pp. 448 ff.), and their connection with Asclepius seems the mere play of poetic fancy—it was never generally accepted; but cf. also below, n. 87. As for patients whom Asclepius cured of grave diseases, only two more are mentioned: the sons of Phineus and the daughters of Proetus (T. 69; 71). The story of the Phinidae is of Thessalian origin, and Asclepius gets involved in it as a participant in the Argonautic Voyage, cf. Preller-Robert, II, pp. 810 ff.; 815; Jacoby, *Fr. G. H.*, II, Kommentar, p. 136, 44, who assumes also Athenian influence on the version of Phylarchus. The healing of the Proetidae, on the other hand, is usually attributed to Melampus, cf. Preller-Robert, II, p. 251. Again, a famous case is transferred to Asclepius, the famous physician. For the significance of these healings, cf. also below, p. 48.

[73] No children of Asclepius appear in the *Eoee*, cf. below, p. 54.

heroic clientele of Asclepius unquestionably are of a relatively late date and hardly reveal the original meaning of the legend. In the world of true heroes, Asclepius, the physician, seems to be a stranger, or at best a denizen. Is it possible that once he had been viewed as the healer and helper of common people?

Indeed, Asclepius' accomplishments were usually summed up in the statement that " he healed the sick and revived the dead " (T. 56).[74] He was educated, Pindar says, to give medical help to men (T. 1, vv. 45 f.). In these instances no names of patients are listed, no cures are described, no revivings are recounted. Nor does it become apparent where Asclepius followed his profession. To be sure, in late centuries an attempt was made by some to localize his practice in Delphi (T. 98).[75] But on the whole, the saga is as vague in regard to the place where Asclepius practiced, as it is in regard to other details of his life. The general assertion that he was able to heal and to revive is all that the myth avers.

The medical procedure of Asclepius, a healer of mortal men, was thought to have been no different from that of any human physician. True, while, according to Homer, Asclepius was in possession of certain drugs which apparently had an extraordinary efficacy (T. 164, vv. 218 f.), the common legend was not entirely satisfied with Asclepius' having only simple, though potent, remedies at his disposal. He was believed to own a share in the blood of the Gorgon which he received from Athena and which he used to heal or to destroy at will (T. 3).[76] But

---

[74] T. 56 is taken from Xenophon; the same phrasing, however, occurs over and over again, cf. e. g. T. 3, and it can safely be attributed already to the Hesiodic poem, for it is in agreement with the tenor of the whole story; this is also the assumption of Wilamowitz, cf. above, n. 65. Pindar, in his description of Asclepius' cures (T. 1, vv. 47 ff.), likewise does not mention any names of patients.

[75] The source is a Nicander Scholion (T. 98). Yet, even Pherecydes spoke of Asclepius' death in Delphi (T. 71). This is a unique version of the saga, cf. below, n. 95; it may be connected with the worship of Asclepius Archegetes, cf. below, pp. 96 f. Incidentally, Farnell, Hero Cults, p. 248, suggests that the localization of Asclepius' healing in Delphi " may have arisen from the identification of the Delphians with the Hyperboreans and from the story that the life of the latter people was miraculously prolonged."

[76] Thraemer, following O. Müller, suggests that the Gorgon-Asclepius saga was Argive, cf. Pauly-Wissowa, II, p. 1654, 32 ff. Gruppe, Griech. Mythologie, p. 1452, n. 2,, proposes an Arcadian origin, because in Tegea, Athena and Asclepius were worshipped together. That the myth was told also in Arcadia is not impossible, but an Argive origin of the fable is more probable since it is above all Argive heroes whom Asclepius is said to have revived (cf. above, p. 41) and since the Gorgon was represented in the main temple of Asclepius at Epidaurus (T. 739, 2). At any rate, it is hardly permissible to explain the passage in Apollodorus as an " unorganisches Einschiebsel," as Wilamowitz does (Isyllos, p. 63, n. 35), though the story is certainly late like all other Argive tales. It seems that the Gorgon, who is terrifying and at the same time beautiful (cf. Preller-Robert, Mythologie, I, p. 192), symbolizes the ambiguous character of the

such occasional contentions are refuted by the ever-repeated claim that Asclepius' success was simply due to his knowledge of the healing art. Through his noble education, through his excellent training he had become a physician of consummate skill (T. 5).[77] Pindar, who most exactly describes Asclepius' medical procedures, says (T. 1, *vv. 47 ff.*) : " And those whosoever came suffering from the sores of nature, or with their limbs wounded either by gray bronze or by far-hurled stone, or with bodies wasting away with summer's heat or winter's cold, he loosed and delivered divers of them from diverse pains, tending some of them with kindly incantations, giving to others a soothing potion, or, haply, swathing their limbs with simples, or restoring others by the knife." [78] There is nothing mysterious about Asclepius' treatment, he healed with the means which every physician had at his command. Nor do Asclepius' motives for practicing medicine seem at variance with those of ordinary human physicians. Pindar intimates that in treating his patients Asclepius was seduced by love of gain (T. 1, *vv. 54 ff.*). The tragedians make the same charge (T. 99).[79] Plato cannot believe that Asclepius, the son of a god, should have been given to unlawful lust for gold; if he really was avaricious, the philosopher says, he could not have been of divine origin (T. 99). But neither Pindar, nor Plato seem to object to the fact that Asclepius should have accepted money, as long

φάρμακον, its healing and destructive power, which was so clearly realized in antiquity, cf. W. Artelt, *Studien z. Geschichte d. Medizin*, XXIII, 1937, pp. 38 ff. Apollodorus' version of the Gorgon story (T. 3) must therefore be considered more correct than that of Tatian (T. 3a), contrary to Pauly-Wissowa, *loc. cit.* Ovid's claim that Asclepius came into possession of the miraculous drugs of Glaucus (T. 75) constitutes another turn of the tradition toward the supernatural. Cf. also Nicander's story about the herb Panacea (T. 89). Such reports probably are transformations of Homer's data concerning Chiron's gift to Asclepius.

[77] For the explanation of Asclepius' reviving of the dead, cf. below, p. 45.

[78] That the kindly incantations which Pindar mentions were not miraculous words, but rather songs and music, I have tried to show, *Bulletin of the Institute of the History of Medicine*, V, 1937, p. 224; p. 235, n. 102. However, even if μαλακαὶ ἐπαοιδαί were understood as incantations in the usual sense of the word, they would still be nothing divine, but rather a human method of treatment.

[79] Who these poets are, whom Plato mentions, is unknown. Certainly he does not refer to Aeschylus or Euripides, as is usually suggested in the commentaries on the *Republic*; for in these writers, Asclepius' death is not motivated in this way, cf. below, p. 46 and n. 84. Wilamowitz, *Isyllos*, pp. 60 ff., considers the whole story a Pindaric invention, and it is undoubtedly conceived in the spirit of Pindar whose contempt for the φιλοκερδής is well illustrated in *Isthmiae*, II. O. Schroeder, *Pindars Pythien*, 1922, p. 30, opposes the opinion of Wilamowitz with reference to T. 100, which according to him gives a similar tradition and is independent of Pindar; cf. however below, n. 86. Herzog, *Wunderheilungen*, p. 142, n. 7, rightly suggests that the story told by Pindar, whoever may have invented it, was meant as censure of medicine as it was practiced by human doctors. That they often cared more for their fees than for their patients, that their aim was nothing other than gain, is admitted even in the medical literature of the 5th and 4th centuries; cf. Edelstein, *Problemata*, IV, pp. 89 ff.; 103.

as he did so in a proper way. As a doctor he was entitled to receive a remuneration for his services. Wandering around from one place to another, and offering his help to those who were in need of it (T. 126), he practiced medicine like all his fellow craftsmen in order to make a living (T. 104).

While in the pursuit of his medical profession Asclepius apparently was not unlike other ordinary physicians, he differed from them in one respect: he was more successful than they were: he possessed the skill to raise from the dead (T. 66). Through his craftsmanship he was able to revive those who had died (T. 121; 75, v. 748; 97; 109), through his wisdom he succeeded in saving those who were already the lawful prey of Hades (T. 1, vv. 55-6).[80] To be rescued from death was the highest expectation, the greatest hope which the ancients cherished in regard to medicine. Its fulfillment was held to be the most adroit feat of the physician. Some human doctors pretended, or were reputed, to have achieved this task.[81] But for the average physician, it remained a goal that was inaccessible to him; in vain, he aspired to reach it. The son of Apollo the god of medicine, the pupil of Chiron the wise Centaur, however, attained that perfection of which the others fell short through the inadequacy of their art.[82]

Yet, although the achievements of Asclepius were incomparable, he still remained a physician. He was no equal of the other heroes with whom some of the poets brought him into contact. A craftsman, a medical practitioner, doing his daily work and earning a livelihood, was not on the same footing with those aristocrats whose lives consisted in warfare and hunting. He was nearer to the life of the common people,

---

[80] Note that Asclepius, unlike Heracles, Orpheus, or Dionysus, does not save from death by a κατάβασις εἰς Ἅιδου (for such descents into the nether world, cf. Pauly-Wissowa, s. v. Katabasis, X, pp. 2397; 2398; 2400), but rather through his medical art.

[81] Leaving aside the deeds of magicians and miracle workers (cf. Weinreich, Heilungswunder, pp. 171 ff.), Empedocles was believed to have revived the dead; cf. Diogenes Laertius, VIII, 67, and below, n. 82. The same was told about Asclepiades, cf. Pliny, Nat. Hist., XXVI, 15. Plato, Charm., 156 D, speaks of τῶν Ζαλμόξιδος ἰατρῶν, οἳ λέγονται καὶ ἀποθανατίζειν. Moreover, the successful physician was commonly viewed as the savior from death which was threatening his patients. Thus Herodotus, III, 130, says of Democedes of Croton that he gave back life to the dying: ὃς τὴν ψυχὴν ἀπέδωκε. Cf. also Callimachus, Hymnus in Apollinem, 46: ἰητροὶ δεδάασι ἀνάβλησιν θανάτοιο, or Eunapius, Vitae philosophorum [Oribasius], 499: προσεκυνεῖτο καθάπερ τις θεός, τοὺς μὲν ἐκ νοσημάτων χρονίων ἀνασώζων, τοὺς δὲ ἀπὸ τῆς τοῦ θανάτου πύλης διακλέπτων; cf. also T. 468. It is against the background of such testimonies that the reports concerning Asclepius must be considered.

[82] Welcker, Götterlehre, II, p. 738, n. 17, has seen that this is the meaning of the revivals ascribed to Asclepius, and he rightly compares Empedocles, Fr. 111 [Diels-Kranz]: ἄξεις δ' ἐξ ἀΐδαο καταφθιμένου μένος ἀνδρός, words which picture the achievements of him who possesses perfect knowledge. Cf. also T. 227.

closer to men; indeed, he was one of them. In the legend, as it was current throughout the centuries, the Homeric Asclepius, the king and peer of princes, had taken off his fancy costume. The pretender appeared in his simple and natural dress of old.

That it was Asclepius, the craftsman, the physician of men, who was the subject of the old myth, is finally confirmed by what was told about his death. All versions of the saga unanimously relate that Zeus killed Asclepius because he restored the dead to life. Asclepius, then, suffered death on account of his deeds as a physician, and modern interpreters usually explain his destruction as a divine judgment visited upon a sinner against the eternal rules of the cosmos. Asclepius, they say, in reviving the dead, in making men immortal, transgressed the limits set for mankind, he infringed upon the rights which death has over mortals. Zeus, the lord of men and the preserver of the laws of nature, could not but punish him because he had violated the order of the universe.[83] Such a moralistic interpretation, to be sure, agrees with the view found in some of the ancient sources. The story went that Hades complained to Zeus that as a consequence of Asclepius' activity his power was diminishing; the dead, he charged, were continually growing fewer. Whereupon Zeus, enraged at Asclepius and in defence of Hades' rights, struck the perpetrator with his thunderbolt (T. 4; cf. also T. 75, vv. 757 ff.). This tale, although it is known only from Diodorus and Ovid, must have been current even in the fifth century B. C., for it fits the temper of that period and the same reasoning seems to be presupposed by Aeschylus (T. 66).[84] But does it give that motivation for Asclepius' death which was proposed in the oldest version of the saga?

[83] Wilamowitz, *Isyllos*, pp. 70-1, says: "Ein helfender Arzt, vielen zum Segen, die siech waren von Wunden und Krankheit. Aber seine Kunst verführte ihn, die Schranken der Menschheit zu durchbrechen, er erweckte Gestorbene. Dafür zerschmetterte ihn Zeus mit dem Donnerkeil, und dem Tode verfiel, der des Todes Rechte gekürzt hatte." Cf. also *Glaube*, II, p. 227. Farnell, *Hero Cults*, p. 235a, remarks: "It is interesting to note here [in regard to Asclepius whom Zeus killed "for the *sin* of raising the dead to life"] the germ of a higher thought, worthy of the Hellenic scientific spirit, that the divine powers are concerned with the maintenance of the normal order rather than with miracles." The earlier assumption that Asclepius was actually killed by a stroke of lightning and that the saga echoes this event (cf. Pauly-Wissowa, II, p. 1654, 8 ff.) is now generally discarded, and that with good reason. Such a contention is beyond reasonable proof.

[84] Aeschylus' words, in their brevity, are somewhat obscure; besides, there are textual difficulties. With Smyth and others I read: οὐδὲ τὸν ὀρθοδαῆ / τῶν φθιμένων ἀνάγειν Ζεὺς ἀπέπαυσεν (Hartung αὖτ' ἔπαυσ' MSS) ἐπ' εὐλαβείᾳ. N reads ἐπ' ἀβλαβείᾳ, F reads ἐπ' αὐλαβείᾳ, the latter probably being a mere misspelling of ἐπ' ἀβλαβείᾳ in accordance with the identical value of αὐ and ἀβ in Byzantine pronunciation, cf. G. Thomson, *The Oresteia of Aeschylus*, II, 1938, p. 108. If ἐπ' ἀβλαβείᾳ is accepted, the meaning of the words would be that Zeus killed Asclepius in order "to prevent his arrangements being

Around 600 B. C., at any rate, when Asclepius was believed to have suffered destruction not because he had raised men, but because he had restored certain heroes to life, death came upon him on account of his opposition to the gods rather than to the laws of nature. For Capaneus, according to the ancient saga, was a great evil-doer, and it was Zeus who killed him when he attempted to storm Thebes.[85] If Asclepius revived the victim of divine judgment and was himself destroyed afterwards, he hardly perished because he had violated the order of the universe. His crime rather lay in that he helped those whom the gods had punished; this offence and no other drew the wrath of Zeus upon him. The same is true of the reviving of Hippolytus. He had met death at the hands of Aphrodite; as a favor to Artemis, Asclepius restored him to life (T. 77; 79).[86] Yet, she who had killed Hippolytus was a goddess, too. Asclepius had no right to revoke the divine decree, to act " against the will of Dis " (T. 76). Naturally, Zeus was indignant; he annihilated the mortal who meddled with the affairs of the gods.[87]

thwarted—or ἐπ᾽ εὐλαβείᾳ as a precautionary measure to that end"; cf. W. Headlam, *Agamemnon*, 1910, p. 233. In both cases the intent of the statement is the same, but in the whole context ἐπ᾽ εὐλαβείᾳ seems to me more probable. Weber, *op. cit.*, pp. 397 f., claims that according to Aeschylus, Zeus forbade Asclepius to revive men, yet did not kill him. This seems to be a misinterpretation of the passage in question. Wilamowitz, *Glaube*, II, p. 227, on the other hand, gives the impression that not only Aeschylus but also Euripides and Pindar adopted the moralistic version of the saga. Yet the former (T. 67; 107) simply states the fact of Asclepius' death and does not dwell on Zeus' motives. According to Pindar (T. 1, *vv. 54 ff.*), Asclepius was killed because he was too greedy (Wilamowitz' assertion: " Pindar Pyth. 3, der die Eöe nacherzählt . . ." is a *lapsus calami*).

[85] For Capaneus, cf. Preller-Robert, *Mythologie*, II, p. 937. Nothing is known about Lycurgus whose revival Stesichorus mentions together with that of Capaneus, cf. above, n. 64. His participation in the Theban expedition is attested only by this writer, cf. Preller-Robert, *ibid.*, p. 914.

[86] In the interpretation of the Hippolytus healing I follow in part the report of Ovid (T. 75). It is Phaedra, on whose account Hippolytus dies, but Aphrodite is responsible for Phaedra's falling in love with the son of Theseus; for she has kindled this passion so as to take vengeance on Hippolytus (cf. also Euripides, *Hippolytus*). The claim of the Pindar Scholion (T. 100) that in the case of Hippolytus, too, Asclepius was enticed by gold, sounds like a reinterpretation of the old myth in the Pindaric spirit (contrary to Schroeder, cf. above, n. 79). Incidentally, Ovid's poem proves that the Hippolytus whom Asclepius revived was believed to have been the son of Theseus, cf. above, p. 41. For the identity of Hippolytus with Virbius whom Ovid mentions, cf. Roscher, *Lexikon*, *s. v.* Virbius, VI, esp. p. 330.

[87] Ovid, who in many respects takes up the old tradition (*e. g.* his report on the raven, above, n. 26), has neatly expressed Asclepius' position (T. 2, *vv. 642 ff.*): it was his privilege to revive men, but when he interfered with the wishes of the gods, he met death. Of other figures whom, according to late evidence (cf. above, n. 72), Asclepius restored to life, Orion fits into the same scheme: he had been destroyed by Artemis, cf. *Odyssey*, V, 121 ff. Glaucus originally was not a patient of Asclepius; the revival of the god Hymenaeus is an isolated case. Concerning the death of Tyndareus no details are known. All these names may have replaced older ones without regard for the meaning of the earlier attributions.

The same reasoning is reflected even in those testimonies which connect the death of Asclepius with certain healings which he supposedly had performed. In these cases Asclepius certainly was not violating any human or divine laws, he was doing only what was his duty as a physician. But his choice of patients again was rather incautious. He healed the daughters of Proetus who had been maddened through the wrath of Hera (T. 69).[88] Certainly, it is the thwarting of the divine will which here brings about Asclepius' death. His destruction is not the punishment of a sinner, it is the annihilation of one who has trespassed against the personal wishes of the gods.[89]

In a similar vein Asclepius' death is explained by some authors who believe that he died because he had restored his human patients to life, and not only one but many of them. He was wont to raise the dead, Euripides avers (T. 67). He was killed because Zeus was afraid lest men, no longer liable to death, would become too mighty and therefore would come to the rescue of each other (T. 3).[90] It is, then, fear for himself which determines the action of the supreme god, anxiety for his own might and that of the other deities, hatred against men who shall not become the equals of their celestial masters. To put it differently: the attitude of the lord of the world toward Asclepius is the same which he takes toward Prometheus. Both he hates, and apparently for the same reason: they take sides with the mortals; they help men; they try to better the lot of the human race. That is what Zeus resents. But whereas he is unable to destroy the divine Prometheus and must be satisfied with tormenting him, he can annihilate the hero Asclepius, and he has no qualms in doing so. To be sure, he would not have troubled to interfere, had Asclepius revived just one or two individuals. Yet he who makes it a practice to raise the dead sets Zeus trembling. The reign of the gods is

---

[88] This healing usually is attributed to Melampus, cf. Preller-Robert, *Mythologie*, II, pp. 246 ff.; it is ascribed to Asclepius only once, and that by a late authority, cf. *ibid.*, p. 251, n. 4, and above, n. 72. But the testimony in which the motive of divine wrath is especially evident is in keeping with older tradition and not a later rationalization, as Jacoby, *Fr. G. H.*, I, Kommentar, p. 519, 27, suggests. Divine wrath is hardly a category characteristic of Greek enlightenment. Nor can the healing of madness be called a miraculous cure, contrary to Jacoby, *ibid.*; it is nothing uncommon in Greek medicine.

[89] Of all the other healings ascribed to Asclepius, only that of the sons of Phineus is supposed by Phylarchus to have led to his death (T. 69; cf. above, n. 72). Sophocles (T. 421a), however, saw in it a well-deserving deed of Asclepius (cf. Preller-Robert, *Mythologie*, II, p. 819) and apparently did not hold it the reason for Asclepius' destruction. Did Phylarchus identify the sons of Phineus with Phineus himself who was blinded by Helius in punishment for his misdeeds (cf. Preller-Robert, *ibid.*, p. 811-812)?

[90] T. 3 is taken from Apollodorus the mythographer. Cf. also T. 121. Even Ovid who mentions Hades' complaints to Zeus (cf. above, p. 46) characterizes Zeus as fearing the example set by Asclepius (T. 75, *v.* 759).

imperiled; action is necessary in order to preserve their power. Envious of the overskillful craftsman, Zeus kills the healer of the sick and the reviver of the dead (T. 233; 102).[91] As for Asclepius, however, his death is the final proof of the greatness of his achievements. He had reached the goal, he had become all-powerful in his craftsmanship. Like others whose art attained perfection, he was on that account destroyed by the gods.[92]

There is hardly any doubt that this version of the saga echoes the explanation for Asclepius' fate, as it was given in the Hesiodic poem and in the original pre-Delphic myth. For it is in agreement with those ancient Greek beliefs in regard to the nature of the deity and the destiny of men which are characteristic of the old legend in general.[93] All other stories that were told must be considered attempts at reconciling the objectionable content of the original fable with the changing ideas of later generations. In the course of this process, Asclepius' death at first was justified by his opposing the will of the gods in a specific case in which death had been decreed by Zeus, or disease inflicted by divine wrath; later, the raising of the dead was viewed as a violation of the cosmic laws which Zeus was bound to defend.[94] But in the beginning,

[91] K. Ph. Moritz, *Götterlehre*, 1795, p. 251, is the first to have drawn a comparison between Asclepius and Prometheus. Another casual reference to the relationship between the Asclepius saga and the Prometheus myth is to be found in Dyer, *Greek Gods*, p. 239, and Herzog, *Wunderheilungen*, p. 142, n. 7. But even these scholars do not fully appreciate the importance of the parallel which seems to have escaped the notice of all other interpreters. The resemblance of the motives attributed by the two tales to the main actors is close indeed. Prometheus makes men independent of the gods so that they may help each other; the same is said of Asclepius (T. 3). Zeus is envious of Prometheus as he is of Asclepius (T. 233). For the Prometheus saga, cf. K. Lehrs, *Vorstellung der Griechen über den Neid der Götter und die Ueberhebung, Populäre Aufsätze*, 1875, pp. 35 ff.

[92] Cf. Welcker, *Götterlehre*, II, p. 738: " Nicht minder wird Asklepios verherrlicht durch den Blitz des Zeus . . . der ihn tödet. Dies bedeuten die vielen Mythen seit dem Dichter der Naupaktika . . . von durch ihn auf erweckten Todten. . . . Wie die Blendung des Sängers Thamyris durch die Musen die Bedeutung hat dass er den Gipfel seiner Kunst einnahm, so ist auch das Leiden des Asklepios nur aus der Vorstellung entsprungen dass die Kunst des Heros ins Überheroische gegangen sei." For the Thamyris story, cf. Preller-Robert, *Mythologie*, II, pp. 413 ff.

[93] In this connection it is important to remember that Apollodorus (T. 3) who in many details follows the Hesiodic poem (cf. above, n. 23), ascribes Asclepius' destruction to the fear of Zeus. Note also Pindar's effort to make Asclepius appear guilty by insinuating that he was seduced by love of gain. Although this motive is meant to satirize the avarice of physicians (cf. above, n. 79), it is also needed in order to justify Zeus' action which, to Pindar, otherwise would remain inexplicable.

[94] Still later, Asclepius' death through a thunderbolt was taken to mean that he was consumed by a fiery disease (T. 115), an explanation in the truly euhemeristic spirit; or his being killed by a stroke of lightning was understood as the symbol of his heroization or deification (T. 236). Rohde, *Psyche*, I, pp. 320 ff., believes this to be the correct interpretation of the myth and claims that such stories are much older than is generally assumed. Cf. however below, pp. 50 f.; 75.

Asclepius was thought to have suffered destruction in return for his benevolence toward mankind; his fate was the inevitable outcome of his Promethean endeavor, it was the climax of his life, the last act of the drama.

That the saga saw Asclepius' death in this light follows also from the fact that the heroic myth is not at all concerned with the fate of Asclepius, once Zeus had destroyed him. Just as the locality where his death took place is not identified,[95] so his grave is never mentioned. Although usually the tomb of a hero was a sacred place, where he was believed to continue his existence and where he could still be found by men who wanted his help, it was apparently of no consequence where Asclepius was buried. True, in later centuries a few regions claimed to be in possession of the grave of an Asclepius: a spot near the river Lusius in Arcadia (T. 118), Cynosura (T. 101; 116), and Epidaurus (T. 119 f.).[96] But the Asclepius who supposedly was entombed in these places was not the character of the heroic saga whose mother, Coronis, was killed by Apollo and who himself was slain by Zeus. It was not the Asclepius whom the Hesiodic poem celebrated. At the river Lusius or in Cynosura it was the "second" Asclepius, the son of Arsippus and Arsinoë, who was buried there, and in Epidaurus, if the late testimony can be trusted at all, the god Asclepius had his sepulchre.[97] The hero Asclepius, however, the helper of men, the healer of the sick, the raiser of the dead, was nowhere to be found on this earth.[98]

Nor did he ascend to heaven or live on in the firmament, visible to

[95] Pherecydes is the only one to assert that Asclepius died in Delphi (T. 71), cf. above, n. 75. Wilamowitz, *Isyllos*, p. 64, n. 38, believes that Hesiod made the same claim. Yet, had this been the case, it would be hard to understand why all the other versions which go back to the Hesiodic poem, without exception, leave the place of Asclepius' death undetermined; for Pherecydes, cf. also below, p. 97, n. 21. Firmicus (T. 114), in reviewing the places where gods and heroes died and were buried, says of Asclepius: "He was destroyed somewhere else." Besides, had Hesiod spoken of Asclepius' death in Delphi, one would certainly expect to find a Delphic tomb of Asclepius attested by the tradition.

[96] Cf. in general, Pauly-Wissowa, II, p. 1654, 47 ff.; Farnell, *Hero Cults*, p. 239. The river Lusius and Cynosura are mentioned by Cicero, Clement, Ampelius and Lydus. They are dependent on the so-called "sondernden Theologen" (around 100 B. C., cf. Pauly-Wissowa, *s. v.* Mythographie, XVI, p. 1371, 21 ff.), who, in turn, seem to have borrowed from the Peripatetic *Peplos*; cf. Pauly-Wissowa, *ibid.*, and Wilamowitz, *Glaube*, II, p. 420. Epidaurus occurs in Ps. Clementine treatises whose source is not known, but which may also have used the tradition of the "sondernden Theologen." Cynosura, according to the ancient interpreters, lies in Lacedaemon; S. Wide, *Lakonische Kulte*, 1893, pp. 196 f., seems to have proved that Cynosura in Arcadia must have been meant.

[97] Concerning the various Asclepius figures, cf. also below, p. 93. For the possible explanation of the reports regarding Asclepius' tombs, especially that in Epidaurus, cf. below, p. 191, n. 2. Note that for Tricca no grave of Asclepius is attested.

[98] Cf. Wilamowitz, *Glaube*, II, p. 12. For the significance of this fact in regard to the later history of Asclepius' worship, cf. below, p. 93.

men for all eternity, as did so many other heroes. To be sure, Eratosthenes asserts that Asclepius, the son of Apollo and Coronis, he whom Zeus killed with his thunderbolt, eventually became a star (T. 704a).[99] Yet, such a dénouement of the saga cannot have formed part of the Coronis poem, still less of the myth which was taken up by Hesiod. For in the star saga Asclepius was identified with Ophiuchus. This presupposes that the serpent was the servant and symbol of Asclepius; only after their association had been firmly established could Ophiuchus be taken for Asclepius. In the old heroic myth, however, no mention was made of the serpent at all.[100] Moreover, in the star saga Apollo intercedes with Zeus on behalf of his son Asclepius, and the king of gods, at the instigation of Apollo and as a favor to him, transports Asclepius to the stars. He takes pity on him whom he had first destroyed (T. 121; 121a-b). Neither Zeus' action, nor that of Apollo, as they are represented here, is in line with the Hesiodic poem. Their characters are changed completely; the relation between Apollo and Zeus is altered; the enemies have become friends. Such a reversal of the main features of the story being the indispensable presupposition of Asclepius' translation to the stars, his elevation to heaven must be the invention of a later period.[101] According to the original myth, Asclepius in dying as he

[99] F. Boll and W. Gundel, Roscher, *Lexikon, s. v.* Sternbilder, VI, p. 921, 66, have stressed the fact that in T. 704a Eratosthenes himself is quoted as authority. The other testimonies for the star legend are Ps. Eratosthenes, *Catasterismi* (T. 121) and Servius (T. 122). T. 121a is taken from Hyginus who himself seems dependent on Eratosthenes, cf. Pauly-Wissowa, XVI, p. 1369, 31 f. T. 121b comes from the commentary on the *Aratea* of Caesar Germanicus, who wrote under Augustus or Tiberius. Cf. also T. 704.

[100] The healing of Glaucus, through which the hero Asclepius acquired the serpent as his servant, is attested only in the fifth century B. C. (Amelesagoras, T. 70, cf. above, n. 72). During the same period the serpent became generally known as the symbol of the god Asclepius, cf. below, pp. 229 ff. It is therefore likely that the connection between the hero Asclepius and the serpent is a reflection of the divine worship. At any rate, the Glaucus story has no original relation with Asclepius, cf. Pauly-Wissowa, II, p. 1654, 27 ff.

[101] At what time it became customary in Greece to place heroes among the stars, it is impossible to say with certainty. The earliest star saga is that of Perseus, and even this can at best be traced to Sophocles; cf. Wilamowitz, *Glaube*, I, p. 261; Preller-Robert, *Mythologie*, II, p. 239. In the fifth century there seems to have been a general belief in "translations of heroes to the stars," cf. F. Cumont, *Astrology and Religion among the Greeks and Romans*, 1912, p. 175. At that time one may also have begun to consider Asclepius a star. Though Eratosthenes is the oldest authority for such a claim, his sources go back to the fifth century B. C. (cf. Pauly-Wissowa, XIV, p. 1369, 31 f.). On the other hand, Aratus, in describing Ophiuchus, does not mention Asclepius; in his time, that is a few decades before Eratosthenes, the star legend can hardly have been commonly accepted or even commonly known. A Hellenistic origin of the story, therefore, is not entirely to be excluded, the less so since in the Hellenistic period star sagas were the favorites of mythological fancy (cf. F. Boll-G. Bezold, *Sternglaube u. Sterndeutung*, 1926, p. 53). In late centuries the legend of the Asclepius star was very popular

did had fulfilled his destiny, " he was dead and done with." [102]

Now that the data concerning Asclepius' origin, his life and deeds, his death have been analyzed in detail, it should be possible to form an opinion about the character of the original legend and thus to grasp the meaning of the Asclepius figure. Invented long before the time in which it was put into verse by the poet of the Hesiodic *Catalogue*, the story underwent the usual transformations; it was adapted to the purposes and ambitions of various sections of Greece; religious developments left their imprint upon it; changes were made in accordance with new ideals. Yet the gist of the ancient fable which made Asclepius, Chiron's friend, the son of a god, is still discernible through all later distortions; it is roughly this:

Coronis, the daughter of Phlegyas who was famous for his violence and daring, became the love of Apollo. When she gave her hand to the man whom her father had chosen for her, the god heard about this through the raven, his messenger. Apollo's anger was roused against Coronis and she died at his command. But the divine physician snatched his child from the dying mother's womb, thus saving it from death which threatened it even before it was born. The god himself then entrusted Asclepius to the care of Chiron, and the Centaur educated Asclepius and taught him medicine. As the son of Apollo, the physician, as the pupil of Chiron, the medical hero, Asclepius himself grew to be a great doctor. He chose to live the simple life of a craftsman, not that of a hero, engaged in daring ventures. The enemies whom he, as a friend of men, conquered with indomitable spirit were disease and death. For he knew his art to perfection. Nay, such was his wisdom that he not only healed the sick and prevented them from dying, he also restored those who had perished. He who himself saw the light of the world through the saving hand of his divine father was to save all mankind from the most dreaded darkness of Hades. He achieved the greatest triumph which the medical art can gain and bestowed upon mortal beings the greatest boon which can be granted to them: eternal existence. But Zeus feared that men, no longer subject to death, would unseat the Olympians. He could not let it come to pass that the barriers which separate mortals from immortals be laid low. The king of gods, in defence of the divine prerogatives, put a stop to Asclepius. Like Prometheus, Asclepius fell victim to the jealousy of the supreme deity.

and it was known even during the Middle Ages (cf. A. Warburg, *Journal of the Warburg Institute*, II, 1938-39, p. 289). For the importance of the star Asclepius in astrology, cf. Boll-Gundel, *op. cit.*, p. 922, 44 ff.

[102] Cf. Wilamowitz, *Glaube*, II, p. 227: " Damit ist er tot und abgetan."

Such a story, told in a grave and melancholy mood, obviously is symbolic of bold heroes and hateful gods, of human endeavor against the unsurmountable boundaries of fate, of a conflict which cannot be avoided or solved. More specifically, it is the tale of the physician, of his chances and his limitations, a simile of eternal truth. That explains why the facts recounted are left in the vagueness and uncertainty of the " once upon a time " instead of being placed within the realistic setting of the " here and now." To be sure, Asclepius' birth and death are described with exactness, but only as far as they are of symbolic significance. Moreover, what takes place between the beginning and the end of his life remains a blank, except for the assertion that he was a medical man. No love affairs are attached to him, no tales of friendship or hatred, no accounts of rivalry or quarrel with men or gods. In this regard, Asclepius is one of the least humanized of the Greek heroes. All mythological figures of the Greeks, at least those who are of some importance in the heroic myth, have taken on a definite appearance; their actions are known in minute detail, their characters are clearly visualized and not to be confounded one with another. It would be hard indeed to mistake Achilles for Odysseus, Agamemnon for Menelaus, or even Machaon for Podalirius. Asclepius, however, has no such sharply delineated individuality, nor has he a distinct life story in which his personality could unfold. He is a shadowy inhabitant of this earth. All his being is integrated into one function, that of healing and helping mankind. His story is neither historical nor hieratic, it is the expression of an idea, a compelling inspiration or a warning example. Asclepius himself is the embodiment of the ideal ancient physician.

3. ASCLEPIUS THE HERO OF PHYSICIANS

The legend of Asclepius is that of a culture hero. It celebrates the achievements of the artist and his art.[1] Stories of this type, in Greek mythology, are very common. Again and again the life and death of great craftsmen were recounted, their contributions to human knowledge were praised. In all sections of Greece such tales were current; people apparently took delight in them.[2] But why did Hesiod put the story

---

[1] Farnell, *Hero Cults*, pp. 243-4, seems the only scholar to have characterized the Asclepius saga in such a way. It is fair, however, to add that Welcker's interpretation of the myth (cf. above, p. 45, n. 82; p. 49, n. 92) presupposes an opinion similar to that which is here expressed.

[2] Such sagas concerning poets, craftsmen and the like, important and interesting as they are, so far have not received much attention in the literature on ancient mythology. Some data have been collected by L. Preller, *Griechische Mythologie*[3], II, 1875, pp. 470 ff.; cf. also Roscher, *Lexikon, s. v.* Heros, I, 2, p. 2486.

into verse? Why did it become part of the Hesiodic *Catalogue*? All the poems included in that collection serve the purpose of establishing the genealogies of noble families, of connecting mortal beings with a divine or semidivine ancestor.[3] Asclepius, too, therefore, should have been dealt with by Hesiod as the father of famous offspring. Yet, no children of his are mentioned here. The preserved fragments, at any rate, do not speak of the descendants of Asclepius, and scholars have always agreed that this is so not because of the chance of tradition: the myth of Coronis and of the physician Asclepius was hardly the proper genealogy of a proud family.[4] Why, then, is the story to be found in a collection of genealogies?

That this is due to fortuitous circumstances is the less likely since there were sons of Asclepius who certainly had an interest in the history of their father. To be sure, they were not related to him by blood but were his children only in a spiritual sense; just the same, they were called his children. From early times physicians were known under the name of "Asclepiads," sons of Asclepius. The term was used generically and applied to those who by no means maintained that they were real descendants of Asclepius. As the hero-father of physicians, Asclepius naturally must have had a legend which told who he was and what he had accomplished. And while the story recounted by Hesiod may contain features which would not suit the aspirations of an aristocratic family, it could hardly infringe upon the feelings of those who were Asclepius' children only in so far as they followed his calling. His boldness, his Promethean endeavor was in line with their own ideals; his suffering, in their eyes, could only be merit and distinction. I suggest, then, that it was Hesiod's intention to celebrate Asclepius as the father of physicians; that the genealogical claims made in the *Catalogue* concerned not Asclepius' sons in the flesh and their offspring, but all those who, like their ancestor, undertook to lighten human suffering; that the culture hero, in an appropriate manner indeed, was extolled as the progenitor of all his fellow craftsmen.

As far as chronological arguments are concerned, this hypothesis seems quite justifiable. The name "Asclepiads" is attested for the first time in one of the elegies of Theognis (T. 219). It is uncertain whether the

---

[3] Cf. *e. g.* Schmid-Stählin, *op. cit.*, I, 1, p. 267.

[4] For the generally accepted view, cf. *e. g.* Wilamowitz, *Isyllos*, p. 76; Weber, *Philologus, loc. cit.*, p. 405 (*Rh. Mus.*, LXXXI, 1932, p. 1). Cf. also above, pp. 32 f., where it has been pointed out that in the Arsinoë poem, which gives the genealogy of the Messenian hero Machaon and his family, the original myth is changed so as to adapt it to the intended purpose.

verses in which the word appears were written by the poet of the sixth century—they may constitute a later addition—, but there is no reason for assuming that they were inserted long after 500 B. C.[5] In the fifth century, at any rate, the name " Asclepiads " in its generic sense was well known and generally accepted (*e. g.* T. 220).[6] Under these circumstances it is hard to believe that the designation of physicians as " Asclepiads " should not have been current even earlier; if the term was commonly applied during the fifth century, it is safe to assume that its use goes back to the sixth century. Now the poems of the Hesiodic *Catalogue* are usually dated around 600 B. C., or even as late as the first two decades of the sixth century.[7] Granted that there still remains a slight margin between the time in which the Coronis *Eoee* was composed and that for which the application of the term " Asclepiads " to physicians in general can be proved, the suggestion as to the purpose of the Hesiodic poem hardly loses in plausibility.

Whatever Hesiod's intention, this much is undoubtedly true: physicians were called " Asclepiads," they were considered Asclepius' children, and it was the hero Asclepius who came to be recognized as their father, their patron saint, as the one whose name they carried. It was not Asclepius, the god of medicine, to whom the physicians were related. As a medical deity, Asclepius was not recognized even in the fifth century B. C., except in a few isolated areas of Greece.[8] The generic use of the term " As-

[5] Farnell, *Hero Cults*, pp. 267-8, says: " 'Ασκληπιάδης, 'a son of Asklepios,' occurs as a generic name for all physicians as early as the poetry of Theognis . . ."; cf. also R. Herzog, *S. B. Berl.*, 1935, p. 992. Yet the verses do not have the " seal " or " signature," the most important criterion of genuineness in Jacoby's opinion (F. Jacoby, *Theognis*, *S. B. Berl.*, phil.-hist. Klasse, 1931, pp. 90 ff.), and an important one also according to Edmonds (J. M. Edmonds, *Elegy and Iambus*, I, 1931, p. 20 [Loeb]). I therefore hesitate to date the lines in the 6th century. On the other hand, since they are preserved in A (cf. Jacoby, *op. cit.*, p. 155), they can hardly be of a very late date, and the tenor of the poem, its insistence on good breeding in contrast to good education, fits the Megarian Theognis rather than the Athenian poet. Edmonds, *op. cit.*, p. 13, suspects the verses because some of their metrical characteristics do not occur before Hellenistic times; but the lines are quoted already by Aristotle, *N. E.*, 1179 b 5 ff.

[6] That Asclepius was considered the ancestor of physicians and that they were called after him in the early fifth century B. C. would be attested in T. 105a, provided that Weber's interpretation of the words in question (*op. cit.*, p. 394) were correct. He understands πατέρος as referring to Asclepius, " der auch für Pindaros sterbliche Ahnherr der Ärzte." Usually, however, πατέρος is taken to refer to Apollo, cf. Farnell, *The Works of Pindar*, II, 1932, pp. 140 f., a much more plausible interpretation. Besides, Pindar's words seem to mean no more than " a second Asclepius or Apollo." The expressions used are too ambiguous to support any conclusive argument.

[7] Cf. above, p. 24, n. 4.

[8] For the rise of the worship of the god Asclepius, cf. below, pp. 111 ff. Apart from the fact that the use of the term " Asclepiads " apparently is older than the divine cult of Asclepius, no one will share any longer the view accepted even in the beginning of the 20th century that lay physicians were called " Asclepiads," because ancient medi-

clepiads," however, implies a concept of Asclepius that was generally known and accepted, and this was true only of the hero Asclepius, whether the legend attached to him was that told by Hesiod or not. Moreover, that Asclepius, the craftsman, half human, half divine, should have become the patron saint of physicians seems well in accord with the development of Greek concepts concerning the origin of arts and sciences in general.

As Homer attests, in his time doctors were called "Paeonii," the sons of Paeon.[9] Paeon was the ancient god of medicine, the physician of the Olympians. Since in early days it was common belief that knowledge was bestowed upon men by the gods, that they had obtained their skill from the deity—the artisans revered Athena and Hephaestus, the poets worshipped the Muses as their patron gods—it was natural that the human doctors were considered members of the family of Paeon.[10] Gradually Paeon was replaced by Apollo or identified with him who, even in the *Iliad*, had some medical functions and finally conquered the whole province of medicine. Consequently one would expect that physicians should have been called sons of Apollo; yet, although it is sometimes asserted that it was Apollo to whom they owed their art, physicians were never named after him.[11] Instead, from the sixth century on they were called Asclepiads, sons of Asclepius. For at that time it was no longer taken for granted that the wisdom of the craftsman was a gift of the gods. What men knew and accomplished, they believed to owe to themselves; by

cine originated in the temples of Asclepius, a widespread prejudice already at the end of antiquity (T. 414; 231, where the father of the famous physician Democedes, himself a physician, is said to have been a priest of Asclepius). Cf. in general, E. T. Withington, *The Asclepiadae and the Priests of Asclepius, Studies in the History and Method of Science,* ed. C. Singer, II, 1921, pp. 192 ff.; cf. also below, n. 32. As far as I am aware, J. Rosenbaum in K. Sprengel, *Geschichte der Arzneikunde,* I⁴, 1846, p. 189, n. 59, was the only one to realize clearly that the Asclepiads are related not to the god, but rather to the hero Asclepius.

[9] Cf. *Odyssey,* IV, 232; that here all physicians are named "Paeonii," not only those from Egypt, has been shown by Welcker, *Kleine Schriften,* III, p. 49. Cf. also Wilamowitz, *Isyllos,* p. 83, n. 58; R. Pohl, *De Graecorum medicis publicis,* Diss. Berl., 1905, p. 11.

[10] Cf. *Odyssey, loc. cit.:* γενέθλη. Solon, *Fr.* 1, 57 [Diehl] speaks of ἔργον Παιῶνος. For the early attitude towards the crafts, cf. C. F. Nägelsbach, *Die Homerische Theologie,* 1840, pp. 54-56; W. Graf Uxkull-Gyllenband, *Griechische Kulturentstehungslehren, Bibliothek f. Philosophie,* XXVI, Beilage zu Heft 3-4 des *Archiv f. Gesch. d. Philosophie,* XXXVI, 1924, p. 3, no. 16.

[11] For Apollo as the god of medicine, cf. above, p. 36, n. 51. For Paeon and Apollo, cf. Usener, *Götternamen,* pp. 153 ff.; the distinction between the two is still upheld by Hesiod, *Fr.* 194 [Rzach]; cf. also Scholion, *Odyssey,* IV, 231. Euripides (T. 220) claims that Apollo gave to the Asclepiads the knowledge of drugs. Callimachus, *Hymnus in Apollinem,* 45 f., states that they know from Apollo how to save from death. But these assertions are rather isolated. At any rate, physicians were never known as "Apollonidae."

slow and painful toil, progress was supposed to have been brought about through human endeavor. " Forsooth the gods have not revealed everything to men from the beginning, rather men by themselves, with time going on, have found out the better through their own investigations," philosophers asserted.[12] In their opinion, those who through valuable inventions and artful skills had remedied human deficiencies and bettered the miserable lot of mortals remained anonymous. In the imagination of the many, however, these individuals were given names and became definite personalities, beings of unusual gifts, of a superior character, of an origin half human, half divine. Such figures, then, usurped the place which had formerly been held by the gods and became the patronymic heroes of the various crafts. Thus the artists who of old had known and followed the art of Athena or Hephaestus were put under the patronage of Daedalus and were called " Daedalidae "; thus the rhapsodists who had pursued the art of the Muses now considered Homer their patron and named themselves " Homeridae." In the same way, physicians who, as long as the arts were under divine tutelage, had been considered the sons of Paeon, became the sons of the hero Asclepius when it was assumed that the crafts were the accomplishment of human exertion.[18]

To be sure, in late antiquity some people believed that Asclepius was the real forefather or ancestor of all physicians. He had invented medicine, they thought, and had imparted his knowledge to his sons, Machaon and Podalirius, who in turn had transmitted it to their children and their children's children. Medicine, then, for quite some time had been the privilege of one family, the Asclepiads, the descendants of Asclepius in the literal sense of the word. But at a certain point in the historical process the medical art had ceased to be their property. Other people, too, in some way or other had acquired medical knowledge and dared to practice; medicine had slipped out of the hands of the sons of Asclepius (T. 229 f.). Those who assumed such a development probably believed that the Asclepiads had first tried to remedy the situation by admitting

---

[12] Xenophanes, *Fr.* 18 [Diels-Kranz]; in general cf. Uxkull, *op. cit.*, p. 3; also A. Kleingünther, Πρῶτος Εὑρετής, *Philol.*, Suppl. XXVI, 1, 1933, pp. 26 ff.

[18] The parallel between Asclepiads, Homerids and Daedalids has often been drawn, though for purposes other than that of explaining the name Asclepiads, cf. *e. g.* Welcker, *Götterlehre*, III, p. 102; Wilamowitz, *Glaube*, II, p. 231. That in early centuries the term Homerids was regarded as derived ἀπὸ ὀνόματος, not ἀπὸ γένους, follows from Acusilaus, *Fr.* 2, and Hellanicus, *Fr.* 20 [Jacoby]; cf. also Jacoby's comment on the latter fragment, I, p. 438, 35 ff. T. W. Allen, *Classical Quarterly*, I, 1907, pp. 135 ff., maintains that originally the Homerids were a real family (cf. also Pauly-Wissowa, *s. v.* Homeridai, VIII, pp. 2145 ff.). But he admits that in Pindar the expression is not used in such a sense and that the Asclepiads and Daedalids at any rate were only a fictitious family organization, cf. also below, n. 22.

through adoption outsiders who were not blood relations, in this way
clinging at least to the semblance of a family craft.[14] But this attempt
proved to be futile. Every doctor began to be called an Asclepiad. The
name which had once denoted the true children of Asclepius was finally
given to all medical men by misuse of the term (T. 228). Thus, the real
ancestor of the first physicians, the founder of the first family of doctors,
came to be considered the ancestral hero of the craft in a merely figurative
sense.

Now regardless of the question whether Asclepius was a historical
personality or not, whether Machaon and Podalirius were in reality his
sons or were named his sons only by the epic poets, it is safe to say that
such a view, even if taken less literally than it had been intended, is
untenable and utterly at variance with the facts. In Greek history there
is no indication that knowledge of any kind was ever restricted to one
particular family, nor was it believed in early centuries that craftsmanship
and skill were secrets to be guarded by the initiated.[15] In the Homeric
period physicians were included among the itinerant craftsmen who were
generally welcomed because of their unusual experience and training.
Nothing intimates that at that time castes of craftsmen existed; those
who had to earn their livelihood apparently were free to choose whatever
work they wanted.[16] Later it was precisely the fact that everybody was
allowed to practice without any restriction, even without examination,
that was constantly discussed by medical writers and was held responsible
for the low standard of the art of medicine.[17] Certainly, in every age and
in all civilizations, especially in their early stages, it was customary for the
son to follow his father's trade. It is not impossible that in this way some
medical knowledge was handed on from generation to generation and
thus remained restricted to one particular family, at least for a certain
time;[18] but this can have been true only in a limited sense, not of

[14] Such adoptions are presupposed by the Hippocratic *Oath, C. M. G.,* I, 1, p. 4, 2 ff.,
though without specific reference to the family of the Asclepiads. For the explanation
of the passage in question, cf. L. Edelstein, *The Hippocratic Oath, Supplements to the
Bull. of the Hist. of Med.,* I, 1943, pp. 39 ff.

[15] This statement of course concerns only historical times. Whether all crafts were
originally considered "magisches Charisma" (cf. J. Hasebroek, *Griech. Wirtschafts- u.
Gesellschaftsgesch. bis zur Perserzeit,* 1931, pp. 24 f.) is quite a different problem.
Asclepius, the ancestor of physicians, was a figure of the Homeric Epic, and at that time
the concept of charisma was not known (contrary to Hasebroek, *loc. cit.*). Cf.
Wilamowitz, *Staat u. Gesellschaft d. Griechen, Die Kultur d. Gegenwart,* Teil II, Abt.
IV, 1, 1923², p. 55.

[16] Cf. *Odyssey,* XVII, 383 ff., and Wilamowitz, *op. cit.,* p. 40.

[17] In general cf. Edelstein, *Problemata,* IV, p. 89; cf. also the Hippocratic *Law,
C. M. G.,* I, 1, p. 7, 2 ff.

[18] Cf. below, p. 59.

medicine in general. The concept of Asclepius as the real ancestor of the doctors, of the Asclepiads as the one family of physicians—such notions resulted from late theories, invented to explain the historical process in accordance with certain ideas concerning the development of society as a whole, or in conformity with what was known of other civilizations.[19]

Would it be any more plausible to assume that in post-Homeric centuries the family of the Asclepiads was one of those in which medicine was hereditary, and that later, when its fame eclipsed that of all others, its name became equivalent to " physicians " ? It is at least questionable whether in Homeric or post-Homeric centuries there was any such family of Asclepiads in the literal sense of the word.[20] It was Machaon and Podalirius who were considered the forefathers of certain families whose members gloried in their descent from these heroes, whereas Asclepius played only a secondary rôle in their family history.[21] Moreover, the term Asclepiads was used rather vaguely, not in connection with any definite skill or achievement. " Sons of Asclepius " were simply those who practiced medicine—good and scientific medicine to be sure—but no specific knowledge was ever ascribed to them (T. 219 ff.), while other similar families, such as the Chironidae, were famous for instance for certain drugs which they had received from Chiron.[22] There is then no certainty that a real family of Asclepiads ever existed, still less is there any indication that it conquered all its competitors in the field of medical practice.

In short, the assumption that Asclepius was the " father " of physicians was mere fiction. But it was a fiction that had great practical significance.

[19] Aristotle assumed a gradual development of social life from the household to the family stage and city formation, cf. Wilamowitz, op. cit., pp. 33; 43. Besides, the example of Egypt, where there were castes and family crafts, must have impressed people who found the origin of culture in Egyptian civilization.

[20] In the Epic, the term " Asclepiad " means only that Machaon was regarded as son of Asclepius (T. 164, v. 204) ; for the strictly patronymic use of such names in the early Epic, cf. Wilamowitz, Aristoteles und Athen, II, 1893, pp. 180 ff. Later, Sophocles seems the only one to refer to Machaon and Podalirius as Asclepiads in the sense that they were Asclepius' real sons (T. 152). Plato speaks of them as παῖδες τοῦ Ἀσκληπιοῦ (T. 143), or ἔκγονοι (T. 123), and reserves the term Ἀσκληπιάδαι for physicians in general, e. g. Republic, III, 405 D.

[21] Theopompus mentions Podalirius and his descendants who settled in Caria (T. 212). Hippocrates, though lastly descended from Asclepius, was a member of the Nebridae family (T. 214). Aristotle traced his origin to Machaon (T. 192).

[22] Allen, who claims that the Homerids were a real family (Classical Quarterly, I, 1907, pp. 135 ff.), presupposes that they were in possession of the Homeric books as their distinctive privilege marking them off from all other rhapsodists. I do not think that this assumption is correct, for the name Homerids is not a family name but rather a generic one, cf. above, n. 13. But it may be true of some families of priests and seers that they had a knowledge of rites which they did not divulge to outsiders (the names are collected by Allen, p. 138), just as the Chironidae had drugs unknown to others.

By calling themselves Asclepiads, by considering themselves a unity, a family though not in blood, the physicians formed one of the first corporations of craftsmen. In order to understand why they did so, one must take into account the position of the itinerant craftsman in early Greek society.

Generally speaking the travelling trades had an important share in the formation of juridical concepts. To protect those who by leaving their homesteads had lost all inherited protection, the right of hospitality was evolved. On the basis of its assurance the craftsman came to stay with the lord of the house; through him he was safe from wanton attacks; through him he also took part in the religious rites of the house and the family, a privilege that he needed since with his home he had given up his gods too.[23] Naturally in the beginning the number of such strolling minstrels and travelling craftsmen was small; their opportunities were few indeed. Later on, when society progressed, when the number of artisans increased because there was greater need for them, it was not so much safety which was desired; life had become more civilized and more peaceful than before. But there were still two things wanting for him who had migrated from his home: he had no standing among men and he had no gods of his own. He acquired both when, as a physician, he became a member of the Asclepius " family," or, as a minstrel, one of the Homerids. For the hero after whom he was called provided him with a pedigree even in foreign communities where his own family was unknown; at the same time the patron saint of his art was the divinity to whom he could pray and on whose tutelage he could rely. He was no longer dependent upon being admitted to the religious ceremonies of strangers, an admission which became the more difficult and unlikely, the more he travelled and the shorter the time he stayed at the various places. That is why these craftsmen were the first to feel the necessity of uniting themselves under the patron of their art, however loose and devoid of formal organization such an association may have been in the beginning. That is why physicians so early rallied around the hero Asclepius.[24]

[23] For the right of hospitality, cf. Wilamowitz, *Gesellschaft*, p. 40, who also points out how dangerous the life of these men remained. Wilamowitz does not emphasize the religious aspect of the problem; cf. however Nilsson, *Popular Religion*, pp. 77-8.

[24] Recent investigations have shown that in pre-Solonic or at least in Solonic times corporations (ὀργεῶνες) existed which were intended to assume the rôle of the family for ritualistic and genealogical purposes, cf. E. Ziebarth, Pauly-Wissowa, *s. v.* Orgeones, XVIII, p. 1024; cf. also Nilsson, *Popular Religion*, p. 82. The hypothesis proposed here, which suggests a similar meaning of the association of the Asclepiads at exactly that time, was intimated by Wilamowitz, *Aristoteles und Athen*, II, p. 182: "Asklepiadai . . . sind vielleicht schon eher Gilden als Geschlechter, aber sie fingieren den Geschlechtsverband"; cf. also *S. B. Berl.*, 1901, p. 22, n. 1. For the development of

But why was it Asclepius, who was chosen as the patron of physicians, not any other mythological figure? There were many more sagas dealing with physicians, with helpers in bodily pain. Almost every ancient city had its " hero doctor "; almost every section of Greece had its worship of a physician who supposedly had been the first to attempt the cure of diseases or who was a master of the healing art.[25] It would be presumptuous to believe that Asclepius' selection among his numerous competitors could ever be fully explained by modern inquiry. The predilection for one religious or legendary character, the rejection of another, fundamentally remains a riddle. It is only after one particular figure has proved its power of attraction, its hold over the imagination of men, that one can surmise wherein the specific appeal of the chosen hero may have consisted. As far as Asclepius is concerned, one might venture to suggest that he outranked all his rivals as a protector of physicians because he, and he alone together with his " real " sons—they, too, great physicians—had found a place in the *Iliad* and in the Cyclic saga. The Epic was the constant source of inspiration for the creation of new heroes. Moreover, even if Hesiod did not expressly praise the hero of physicians, his poem should have contributed at least to enhancing Asclepius' fame.[26] Around 600 B. C. the name of Asclepius must have been better known in all Greece than that of any other physician of old. Furthermore, physicians in their activity were not restricted to one place. Their " surname " had to be one which was full of significance for everybody. Thus the name of Asclepius naturally suggested itself to the poet or to the physicians themselves, or to him, whoever he was, who assigned the term " Asclepiads " in its generic sense to all doctors, once Homer had used it at least for Machaon: it was understandable and full of meaning to every Greek, while the fact that their hero had found favor with Homer added honor and dignity to the professional standing of the doctors.[27]

gentilitial organizations in Greece in general, cf. Wilamowitz, *Gesellschaft*, pp. 43 ff. Hasebroeck, *op. cit.*, pp. 96 f., has stressed the fact that on the whole the " Geschlecht" was a relatively late form of organization. There can be no doubt that most of these family groups were artificial creations. On the other hand, such gentilitial fictions formed the basis for the organization of Greek society. Every association was held together through the relationship of its members to a patronymic hero, who figured as the " Rechtsperson," cf. Wilamowitz, *op. cit.*, p. 50. Note moreover that even in later centuries the itinerant craftsmen, who came from foreign lands, were leading in the establishment of guilds, cf. E. Ziebarth, *Das griechische Vereinswesen*, 1896, pp. 192 ff.

[25] Cf. in general, Wilamowitz, *Glaube*, II, p. 12; Farnell, *Hero Cults*, p. 236, and below, pp. 96 f.

[26] Wilamowitz, *Glaube*, II, p. 227, n. 2, expresses the opinion that Asclepius became known only through the Hesiodic poem. He underestimates the importance of the Homeric mention.

[27] That considerations such as these were of importance in the choice of collective names, seems to follow again from the parallel cases of the Homerids and the Daedalids.

But were Asclepius' achievements sufficient to make him acceptable as the patron of physicians? According to the Hesiodic poem—and no other legend of the hero Asclepius is known—he had healed the sick and revived the dead. The tale, however, is not preserved in its entirety. It is possible that he was also pictured as the first to cure certain diseases or as the first to invent certain remedies or instruments. The later saga is full of such instances.[28] Or the invention of medicine may have been ascribed to him, as was done afterwards. Eryximachus in the Platonic *Symposium*, speaking of Asclepius, calls him the forefather of physicians, the founder of the medical art (T. 348).[29] Yet, it would have sufficed had Asclepius been celebrated only as the healer of the sick and the restorer of the dead. These two achievements epitomize and symbolize all the ideals and aims that instigated human physicians.[30] As Pindar expresses it, Asclepius was the " hero who gave aid in all manner of maladies " (T. 1, *v. 7*). In exactly the same terminology, a Hesiodic poem characterizes Heracles as the " preserver of good and enterprising men from ruin." Perhaps it was thought that Asclepius had been begotten by Apollo, the " savior from death," to be a helper in disease, just as Heracles had been begotten by Zeus to be a liberator from the distress that is the lot of mortals.[31] Whether the saga went that far or not, physicians could easily venerate in Asclepius not so much the giver of medicine, the founder of the art, but the model physician, the ideal representative of their calling, the practitioner who spent all his life in healing the sick and in raising the dead even at the risk of his own life. The hero Asclepius, the friend of Chiron, the son of Apollo, was indeed well qualified to become the patron of human healers even long before he came to be venerated as a god.[32]

It is hardly by chance that the bards, who travelled around like the physicians, were named after Homer with whom everybody was familiar. Daedalus' name, at first most popular in Athens, had become generally known through the fame of the Athenian products, cf. Preller, *op. cit.*, II, pp. 497 ff. Thus all artisans could be called after him. Note that Asclepius was the only physician referred to by name in the Epic who was "available" as the hero of all doctors; Machaon and Podalirius were fixed in Caria and Messenia respectively, cf. above, pp. 19 ff.

[28] Cf. below, pp. 140 f.

[29] Cf. Farnell, *Hero Cults*, p. 268 d, who corrects Walton, *Cult of Asklepios*, p. 27, where this passage is understood as referring to the Athenians in general. In passing it should be stressed that Eryximachus, when acknowledging that Asclepius is "our forefather," adds: "as these poets here say and I believe it." His statement seems to indicate how clearly the artificial character of the term Asclepiads was remembered.

[30] Cf. above, p. 45.

[31] Cf. the Ps. Hesiodic *Shield of Heracles*, 28-9, and for Apollo as savior from death, Hesiod, *Fr.* 194 [Rzach].

[32] Wilamowitz, *Glaube*, II, p. 231, says: " Den Hippokrates mögen unsere Ärzte als ihren ἥρως κτίστης verehren; eine Asklepiosbüste würde ich mir nicht ins Zimmer stellen,

Now that the significance of the early Asclepius figure has been sketched as far as it seems still possible to glean its meaning from the testimonies, two questions may at last find an answer which have repeatedly been referred to, although judgment has always been suspended: was Asclepius originally connected with any regional saga or not? Was he an historical person, did he actually live on this earth? The fact that Homer and Hesiod localize the Asclepius saga in Thessaly can be taken as an indication that he really was a hero of that country.[33] But Thessaly in antiquity was considered the home of many a legendary character who had no genuine relation to that section of Greece. The mythical fantasy of the ancients showed a certain predilection for placing the lives of heroes in Thessaly.[34] Moreover, this region was a particularly appropriate country for Asclepius to live in. Throughout antiquity, it was famous for its healing plants. Medea, so the story went, while fleeing through the Thessalian plain, spilled her drugs and thus made the soil abound in herbs (T. 282, 15). Mount Pelion, near which Asclepius was reputed to have been born and where Chiron resided, was especially rich in remedial plants, the "hands of the gods," as they were called even by physicians.[35] Asclepius' localization in Thessaly, therefore, like other features of the saga, seems to be a purposeful invention, well in accord with Homer's intentions no less than with the symbolic character of the whole legend.[36]

As regards the second problem, all the evidence points to the conclusion that Asclepius was not a historical personality. In every respect his story is that of the physician rather than that of a definite individual. Against the modern proponents of the historicity of the Asclepius figure it must be maintained that if Asclepius was a real man, tradition has obliterated all

wenn ich Mediziner wäre, denn er ist der Gott von Epidauros." But ancient doctors did have an Asclepius statue in their homes (T. 501); they were called sons of Asclepius, not sons of Hippocrates. It is hardly permissible to see their attitude as an "historical error." Wilamowitz' judgment is determined by the prejudice that Asclepius was only the god of medicine, as is also Farnell's interpretation of the relation between Asclepius and the Asclepiads (*Hero Cults*, pp. 265 ff.).

[33] For Homer, cf. above, p. 22. For Hesiod, cf. above, p. 34.

[34] Cf. Preller-Robert, *Mythologie*, II, p. 4; Kjellberg, *Asklepios*, p. 14, n. 1.

[35] For Thessaly and its wealth in herbs, cf. Kern, *Religion*, II, p. 304; for Mount Pelion, cf. above, p. 22. The expression χεῖρες θεῶν, used in regard to remedial plants, occurs *e. g.* in Scribonius Largus, p. 1, 1-3 [Helmreich]; in general, cf. L. Edelstein, *Bulletin of the Institute of the History of Medicine*, V, 1937, p. 229.

[36] I see no way of determining with certainty whether Homer was the first to place Asclepius in Thessaly, or whether the saga taken up by Homer already knew of the Thessalian hero. It would be nothing uncommon if in the beginning the story had been told without any local detail. Even in the *Iliad* there appear heroes who have no country, *e. g.* Ajax (cf. Wilamowitz, *S. B. Berl.*, 1925, p. 242; *ibid.*, p. 60).

detail concerning his existence to such a degree that it can no longer be proved.[87] Such an assertion, however, does not yet mean that Asclepius must have been a wholly fictitious character of the saga. After all, there was a god Asclepius. Is it not possible that the hero was only a " decayed god" who was adorned with all the features of a heroic legend only after he had lost his original divinity? A decision on this issue will depend upon the analysis of those testimonies which deal with Asclepius, the god of medicine.

[87] Cf. Farnell, *Hero Cults*, pp. 235 ff. He holds that the legend " somewhat confirms our impression of his humanity as the primary fact " (p. 242). But even though Asclepius' humanity, according to the legend, is the very core of his existence, this fact does not imply his historicity at the same time. It is hardly necessary to enumerate again all those features of the saga which seem to have symbolic meaning. Only so much I wish to point out, that the result of the interpretation given also implies that Wilamowitz, not Thraemer, is correct in his explanation of the name Coronis, cf. above, p. 34, n. 44.

II. THE GOD ASCLEPIUS

Unlike the hero who had vanished and lived on only in the memory of men full of admiration for the feats which he had once accomplished, the god Asclepius was one of the immortals and ever-present in the holy shrines at Epidaurus, Tricca, Athens, Cos, Pergamum and wherever else he had taken up his residence. Leaning on his staff around which the serpent was coiled, he revealed himself to the race of mortals unremittingly and without envy. In their dreams he healed the sick who came to sleep in his magnificent temples. He was the helper to whom almost everybody prayed in supplication, imploring his succor in every kind of distress. He was the savior, mightier and stronger than any other deity; he was the lover of man, kinder and more benevolent than any other god, lending his hand to help mankind.

Who was this deity who, when the god of a new Gospel appeared, became perhaps his most significant and most powerful antagonist in the spiritual struggle that ensued between paganism and Christianity? How is this deity related to the hero Asclepius of whom Homer and Hesiod knew? These are the first questions which one will naturally ask if, after having studied the history of the hero of physicians, one turns his attention to the god of medicine.

When in the nineteenth century the first attempts were made to explain the nature of Greek divinities within the limits of the scientific categories which result from modern theories concerning the origin of human belief in superior beings, the god Asclepius was characterized as the representative of a cosmic element, or as a totem, or as the ancestral hero of one or the other of the Greek tribes.[1] Such hypotheses are no longer credited. Today the majority of scholars agree in calling Asclepius a chthonian deity. The hero Asclepius is viewed as a " decayed god," that is, as the younger of the two Asclepius figures that are known. And it is maintained that the god was revered from time immemorial.[2]

---

[1] The oldest theories on Asclepius are summarized by Thraemer, Roscher, *Lexikon*, I, pp. 621 ff. For Asclepius as the god of the Phlegyans, cf. Pauly-Wissowa, II, p. 1643, 65 ff.; as the deification of air, *e. g.* Preller-Robert, *Mythologie*, I, p. 514; as (dog) totem, S. Reinach, *Rev. Arch.*, IV, 1884, pp. 129 ff. H. Usener, *Rh. Mus.*, XLIX, 1894, p. 470, interprets Asclepius as a god of light.

[2] For Asclepius as a chthonian god, cf. *e. g.* Kern, *Religion*, II, p. 305; for the prevalence of this explanation in modern literature, cf. also Farnell, *Hero Cults*, p. 235. The consequence that the hero Asclepius is a god in disguise, is expressly drawn by Pfister, *Bursian*, 1930, p. 128, and is silently presupposed by all who believe in the aboriginal divinity of Asclepius.

Yet there can be no doubt that the god Asclepius, whose fame resounded throughout the ancient world, was received by all the Greeks at a rather late date. The first testimony which unambiguously refers to Asclepius as a god is furnished by an Athenian inscription recording his arrival at Athens in 420 B. C. (T. 720).[3] Of course, Asclepius' divinity must have ben recognized before, at any rate in Epidaurus from where he came to Athens. On the other hand, Pindar in 475 B. C., when writing about Asclepius, still speaks of him as a hero.[4] In the beginning of the fifth century B. C., then, the god Asclepius can hardly have been acknowledged or even have been known everywhere. It is therefore assumed that the chthonian Asclepius at first was venerated only in an isolated part of Greece, in Thessaly, where Tricca most probably was the oldest place of his cult. From there he gradually extended his influence over the neighboring countries, even down to Epidaurus, until in the fifth century the god suddenly emerged from the darkness of his provincial existence into the full light of Panhellenic fame.[5]

As far as arguments have been brought forward in support of the theory that Asclepius is a chthonian god, they concern mainly the name of the deity, his cult and his myth. Yet, from the very same instances it has also been concluded by some that Asclepius must originally have been a heroized mortal who attained godhead in prehistoric centuries.[6] Again, speculation on the etymology of the name alone has prompted the assumption that Asclepius was a pre-Greek deity while, curiously enough, it is admitted at the same time that according to the other evidence the god Asclepius seems to be an Epidaurian " invention " of relatively late date.[7]

The modern debate, then, has not led to any generally accepted solution of the problem. It has proved, I think, only this much: the name of the god cannot be understood at all and therefore must be disregarded in the argument, while the data on the cult may at best be called indecisive evidence.[8] Of all the tests usually applied in investigations of this kind,

---

[3] The inscription itself was set up in the early fourth century B. C. That the events mentioned belong in the year 420 has been shown by A. Koerte, *Ath. Mitt.*, XVIII, 1893, pp. 247 ff.; XXI, 1896, pp. 313 f. Cf. also Kutsch, *Attische Heilgötter*, pp. 16 ff.

[4] Cf. above, p. 22. The year 475 is the date of the composition of *Pythiae*, III.

[5] Cf. *e. g.* Kern, *Religion*, II, p. 305.

[6] Cf. Farnell, *Hero Cults*, pp. 238 ff. That the hero was revered as a god very early is assumed *ibid.*, pp. 247 ff.

[7] Cf. Wilamowitz, *Glaube*, II, pp. 229 ff., who candidly states that his interpretation ends in an insoluble *aporia*, *ibid.*, p. 223.

[8] Cf. in general Farnell, *Hero Cults*, pp. 238 ff. The various opinions concerning the name are summarized below, pp. 80 ff. For the cult, cf. Ch. IV, *passim*. Frequently the chthonian character of Asclepius is deduced from the fact that one of his attributes is

there remains only the myth as a criterion of the correctness of the two main theories according to which Asclepius is a chthonian deity or a deified mortal. It therefore appears necessary to scrutinize more closely the myth attached to the god Asclepius.

In doing so, one has to face the fact that during recent decades it has become questionable whether mythology has any bearing on problems of religion. It is held by not a few that the divine figures, at least the true gods, are older than the myths told about them, and that myths therefore are irrelevant for the interpretation of the nature of a deity. However, the issue which has not yet been clarified is precisely whether Asclepius was one of those true gods for whom myths were invented only in later times, or whether he was a hero who gained divinity through deeds and whose myth expresses adequately the essence of his divine being.[9] In the case of Asclepius, then, a study of the myth seems indispensable.

1. THE DIVINE MYTH

In the preserved literature the divine myth of Asclepius is nowhere recounted at great length. Only that form of the saga which was current in the second century A. D. is attested in some detail. Pausanias (T. 7) relates that on his visit to Epidaurus he learned from the people there that in their belief Asclepius' mother, Coronis, when with child by Apollo, came to the Epidaurian country. Near the city, where later the most famous sanctuary of the god was to rise, Coronis gave birth to Asclepius and exposed her child immediately after it was born. A goat that pastured about the mountains nourished the infant, and the watchdog of the flock guarded it until it was found by the herdsman. When he beheld the child, lightning flared from its face, thus revealing the divinity of the babe. At the same time it became known all over the world that the great discoverer of medical help, the savior of the sick, the reviver of the dead, had made his appearance on this earth at Epidaurus. Thus Pausanias. At first glance his account seems worlds apart from the myth of Hesiod. Yet there are certainly also traits distinctly reminiscent of the heroic saga. The god Asclepius, too, is the son of Apollo and Coronis; he has come into this world to heal the sick and to raise the dead. But why is he now

a snake; cf. *e. g.* Kern, *Religion*, II, p. 305; Pauly-Wissowa, II, p. 1655, 25 ff. For the meaning of this symbol, cf. below, pp. 229 f.

[9] For the modern point of view concerning the relationship between mythology and religion, cf. especially Pauly-Wissowa, *s. v.* Mythos, XVI, pp. 1375-6, and Wilamowitz, *Glaube*, I, p. 7. In regard to heroes or saints who became gods, the value of the myth, as far as I am aware, has never been doubted. Cf. also Preller-Robert, *Mythologie*, II, p. 2.

regarded an Epidaurian? Why is he now believed to have been exposed by his mother rather than to have been brought forth to life through the saving hand of his father? The Epidaurian myth apparently is a late version of the divine saga,[1] and Pausanias' report is incomplete; it concerns only the god's birth. Did other tales circulate, and how were they related to each other? What were the outlines of the whole story?

The oldest divine birth myth attested is the one to be found in the Homeric *Hymn to Asclepius* (T. 14; 31), a document that probably goes back to the end of the fifth century B. C.[2] The god, it says, was the son of Apollo and Coronis, the daughter of Phlegyas; he was born in the Dotian plain. The hymn then is in perfect agreement with the data given by Hesiod in regard to Asclepius' ancestry and place of birth. Only one difference can be detected: all traits of violence and unusualness are eradicated. The simple assertion that Coronis gave birth to Asclepius has replaced the unique story of his being born by means of an operation. It is the myth of a god that is told here; it is Delphi whose influence pervades this account; consequently all offensive features are repressed.[3] But otherwise the narratives are identical.

Throughout antiquity the tale of the Homeric hymn basically remained the pattern of the Asclepius saga. Apollo was always considered Asclepius' father. Tradition never varied in this respect.[4] Coronis, after some dissension, came to be generally recognized as the mother who was found worthy to bear the future god. Only one detail was changed completely: Asclepius lost his affiliation with Thessaly.

As for Coronis, the divine hymn of the early fourth century still commends her, the daughter of Phlegyas, as the mother of Asclepius who was born in the Phlegyan country (T. 592).[5] But shortly after the restoration of Messenia, that is after 370 B. C., the Messenians renewed their demands on Asclepius. His mother, they said, was Arsinoë (T. 36).[6] In making this claim they came in conflict with those raised by

---

[1] Cf. Wilamowitz, *Glaube*, I, p. 132, n. 2, and below, pp. 72 f.

[2] For the date of the Homeric Hymn, cf. R. Pohl, *De Graecorum medicis publicis*, Diss. Berl., 1905, p. 12; Schmid-Stählin, *op. cit.*, I, 1, p. 244.

[3] Cf. A. D. Nock, *Conversion*, 1933, p. 23, who points out that the Homeric hymns are written in the interest of those cults and traditions with which Delphi identified itself; cf. also Pauly-Wissowa, *s. v.* Orakel, XVIII, p. 844, 30 ff.

[4] Cf. above, p. 35, where it has been shown that the same is true of the father of the hero Asclepius.

[5] The importance of this testimony has rightly been stressed by Kern, *Religion*, II, p. 309; it shows how strong the belief in the Thessalian origin of Asclepius was in the beginning. A later copy of the old hymn (T. 592a) omits the reference to Thessaly and speaks only of Coronis, the daughter of Phlegyas.

[6] For the Messenian saga, cf. above, pp. 32 f.; for the restoration of the Asclepius

the Arcadians in the same century. They gave to the god an Arcadian mother whom they probably called Arsinoë, just like the Messenians. On what grounds the people of Arcadia founded their title to Asclepius is not clear from the evidence preserved; their saga is the one least known.[7] Although the disagreement between the two contestants probably was no more than a local argument—beyond Arcadia and Messenia Arsinoë, the mother of Asclepius, never had any importance—the discussion became so embittered that the Delphic Oracle was asked to settle the conflict. But when Apollophanes, the Arcadian, put before the Delphic priestess the question whether Asclepius was a Messenian, obviously wishing for a decision in favor of Arcadia, the Oracle answered that Asclepius was an Epidaurian (T. 36).[8] For Delphi in the meantime had abandoned the Thessalian myth which it had formerly approved, it had shifted its allegiance to the Epidaurian saga.

The Epidaurians at first seem to have accepted the version of the Homeric hymn; it was probably one of their festivals for which this song was composed or at which it was recited.[9] But around 300 B. C. Isyllus

cult in connection with the rise of the new Messenian state, cf. Wilamowitz, *Isyllos*, p. 78, and above, p. 33, n. 40.

[7] The claims of the Arcadians seem strange indeed, for the original legend of the hero has no connection whatsoever with Arcadia. Nor had the Arcadian worship of the god Asclepius more than regional significance, although his temples rose relatively early in that region. On the other hand, it is again in Arcadia that the three Asclepii, distinguished by later tradition (T. 379 f.), were supposed to have lived. The occurrence of the Asclepius saga in this province, which up to the fifth century was somewhat isolated from the rest of Greece, could perhaps be explained by the fact that part of the Messenian population after the second conquest of their country by the Lacedaemonians migrated to Arcadia (cf. Polybius, IV, 33, and K. J. Beloch, *Griechische Geschichte*, I, 1, 1912, p. 334). The Messenian emigrants doubtless did not forget their old sagas, and from them the Arcadians may have received their hero Asclepius. In the Arcadian birth myth of the god (T. 17) which, as it is told by Pausanias, seems of late origin (cf. below, n. 21), Asclepius' mother is not referred to by name. But she is identified by modern scholars with Arsinoë who is the mother of one of the Arcadian heroes (T. 379); O. Immisch, *Jahrb. f. klass. Philologie*, Suppl. XVII, 1890, p. 201, also suggests that this Arsinoë is the daughter of Poseidon and Demeter. Immisch's opinion has been critically reviewed by Thraemer, Pauly-Wissowa, II, p. 1652, 13 ff. The data are too scanty and too vague to warrant any definite conclusions. It seems significant, however, that the name Arsinoë, which occurs in the Messenian saga, is to be found also in that of Arcadia.

[8] The date of the oracle is not attested, but the question asked would have had little justification before the restoration of Messenia in 370 B. C.; cf. H. W. Parke, *A History of the Delphic Oracle*, 1939, p. 353. On the other hand, the decision made shows that Delphi had already recognized the Epidaurian version of the birth saga, an action that was taken around 300 B. C., but this cannot yet have been common knowledge; otherwise Apollophanes' inquiry would have had no sense. Incidentally, Wilamowitz, *Isyllos*, p. 90, believes that the words of the Oracle are directly reminiscent of those used by Isyllus (T. 594). But they also seem rather similar to the phraseology of the Homeric *Hymn to Asclepius* (T. 31); cf. T. W. Allen-W. R. Halliday-E. E. Sikes, *The Homeric Hymns*, 1936, ad XVI, 4.

[9] For the connection of the Homeric hymn with Epidaurus, cf. below, p. 210.

(T. 594) dared to assert that both the mother and the maternal grand-father of Asclepius were inhabitants of Epidaurus, thus for the first time connecting the god's birth with that city, and he added that Coronis' name in reality had been Aigle, although on account of her beauty she was later named Coronis.[10] The Delphic Oracle encouraged Isyllus to engrave on stone his paean in which the new story was recited, and to dedicate it in the temple (T. 594); Delphi thereby threw its influence on the side of the Epidaurian aspirations. Great as its authority was, the Delphic sanction could not insure the acceptance of so arbitrary a claim as the one made by Isyllus. The old-inherited beliefs proved stronger, at least in regard to Asclepius' mother. The name Aigle was never accepted by the worshippers of the god; she is not mentioned in any testimony other than that of Isyllus. The discussion whether the god was the son of Arsinoë or of Coronis went on for some time. Reconciliations of the divergent views were attempted: either it was claimed that Arsinoë had originally been called Coronis (T. 34), or Asclepius, while remaining the son of Arsinoë, became the adopted child of Coronis (T. 35).[11] Even in these accommodations it is apparent that Coronis was gaining ascendancy; it was she who took precedence as the mother of Asclepius. At last she emerged as the uncontested claimant.[12] In this respect, the Hesiodic poem had won the day.

[10] The date of Isyllus' poem is variously given. Wilamowitz, *Isyllos*, p. 39: around 280 B.C.; F. Hiller v. Gaertringen, *IG*, IV², 1, Prolegomena, p. XXVI, 29: "certo definiri non potest, quando Isyllus, qui a. 338 puer fuerat, senex rem publicam Epidauriorum ad Lycurgi Spartani mores revocaverit"; cf. also below, p. 103, n. 9. I have called Isyllus' poem the first claim on the Epidaurian origin of Coronis. K. Latte, *Gnomon*, VII, 1931, p. 115, on the evidence of a late inscription takes Ἀξαντίδα κούρην in the Homeric *Hymn to Apollo* (T. 33) to mean "a girl from Epidaurus." These words are usually assumed to refer to Coronis, though this identification is by no means certain (cf. Allen-Halliday-Sikes, *op. cit., ad loc.*). Coronis, then, would have been called an Epidaurian not later than 600 B.C., the *terminus post quem* for the Homeric hymn. But in such an interpretation too many doubtful factors are involved to make it acceptable. Moreover, it would be difficult to understand, why in the Homeric *Hymn to Asclepius*, which is of later date than that to Apollo, Coronis is said to live in Thessaly (cf. above, p. 68). Finally, Isyllus' claim that Coronis was an Epidaurian would certainly have been successful had it been supported by so old an authority. Incidentally, Latte seems to identify Ἀξαντίδα κούρην with Arsinoë, as does Wilamowitz, *Isyllos*, p. 80, to whom Latte refers. This interpretation is entirely unfounded. That the words in question can not be emended so as to attest an Arcadian mother of Asclepius (Ἀξανίδα Martin) has rightly been pointed out by Wilamowitz, *ibid.*, p. 80, n. 53.

[11] These theories evidently were local transformations of the saga, for they occur in books on Argos and Cnidos; the authority for T. 35 is Socrates (cf. Pauly-Wissowa, III [2. Reihe], p. 806; 808 f.), that for T. 34 is Aristides, whose date is unknown, cf. *op. cit.*, II, p. 886, 10 ff. Yet the divergence of the saga regarding Asclepius' mother is also referred to by the historian Apollodorus (T. 5), as well as by the mythographer (T. 3).

[12] In T. 8 for instance the Coronis version is called "the current story" (ὁ παρὰ πᾶσι λόγος).

Yet Asclepius no longer was considered a Thessalian. Coronis had borne him in Epidaurus—this came to be the belief commonly adopted throughout antiquity. No other place, whatever its fame as a cult center of Asclepius, neither Cos nor Pergamum, could lay any claim upon Asclepius as a native of their cities. Scholars advocated Tricca as the birthplace of the god (T. 11),[13] but no attention was paid by the pious to their deductions. In the eyes of Julian, the great devotee of the god, Asclepius was an Epidaurian (T. 307); Proclus held the same belief (T. 446), and for the many, Epidaurus unquestionably was the holy city of Asclepius (e. g. T. 20).[14] Unfounded as the Epidaurian demands were, they remained victorious and they were sanctioned by Delphi— for the sole reason, no doubt, that Epidaurus was the center and bulwark of the Asclepius religion.[15] Delphi first gave preference to the Thessalian version over that of the Messenians despite the fact that the latter was of an almost equally early date and even more in conformity with the Delphic ideals. It was probably political considerations that were primarily responsible for this decision. Delphi, which up to the seventh century B. C. had very close relations with the Dorian tribes, later came more and more under Thessalian influence; around 600 B. C. the Thessalians held the hegemony over the Delphic amphictyony. The Messenians, on the other hand, from the beginning of the seventh century ceased to exert political influence and power. Moreover, the Asclepius whose fame resounded through the sixth and the beginning of the fifth centuries, was the hero Asclepius, who was reputed to have come from Thessaly.[16] When in the fourth century the independence of Messenia

[13] Strabo places the oldest sanctuary of the god Asclepius in Tricca (T. 714). Like his authority, Apollodorus, he seems to have identified the Homeric hero with the god, cf. above, pp. 18; 24, and below, p. 93. That is why I have not hesitated to quote his testimony for both Asclepius figures. For the scholars' decision in favor of Tricca as the birthplace of the hero Asclepius, cf. above, p. 18. At the end of antiquity the demands of Tricca were revived (cf. e. g. T. 13); this, one can only understand as an archaistic predilection for Homeric opinions, in keeping with the whole attitude of late centuries.

[14] Panofka (Berl. Phil.-Hist. Abh., 1845, pp. 276 ff.; 320; cf. Welcker, Götterlehre, II, p. 737), starting from the presupposition that any cult of Asclepius παῖς proves a birth saga, attempted to ascertain a number of local traditions concerning the birth of Asclepius beyond those discussed here. Mainly on the evidence of archaeological material he names twelve such places. His conclusions were rejected by Thraemer (Pauly-Wissowa, II, s. v. Asklepios, passim) as unfounded, or at least as too far-fetched. In my opinion, no certain results can be arrived at by Panofka's method, the less so since the cult devoted to a child can be interpreted in a very different way; cf. Farnell, Hero Cults, p. 276, and Wilamowitz, Glaube, I, p. 132, who considers this cult an old one which became associated with Asclepius only later.

[15] For the rôle of Epidaurus in connection with the Asclepius worship, cf. below, pp. 238 ff.

[16] For Thessaly and Delphi, cf. Parke, op. cit., pp. 120-21; for Messenia, cf. Berve, op. cit., I, p. 73; for the hero Asclepius, cf. above, p. 63.

was finally restored, when the Messenian cult of Asclepius was revived, when Arcadia announced its title to Asclepius, the supremacy of the Coronis myth was too well established to be challenged successfully by either Messenia or Arcadia. The Oracle expressly rejected the Messenian pretensions and, indirectly, those of Arcadia. It was forced to support the claims which Epidaurus had vindicated for herself. Where else could the god have been born if not in that city where his greatest sanctuary was to be found, where everybody went to seek his help; the place to which most of the other temples traced their origin?

While Asclepius' birthplace was thus changed, partly for political and partly for religious reasons, Hesiod's account of Asclepius' birth was not retained merely because it offended the religious conscience of later generations. As is evidenced already by the Homeric hymn, the Hesiodic tale was irreconcilable with the feelings of the pious. They could not believe that Asclepius' mother married a mortal before the divine child was born, that Apollo killed her whom he had chosen as the mother of the savior, and finally that he snatched the child from its mother's womb.[17] Isyllus alone, audacious as he was in proposing another name for Coronis, remained faithful to the old tradition, at least in so far as he made Apollo act as an obstetrician, though not as a surgeon: together with the Fates and Lachesis he assisted Aigle in the pangs of birth (T. 32).[18] But usually the divine myth is satisfied with stating that Asclepius was the child of Coronis (T. 14; 587; 592 f.). Awe and reverence make it no longer permissible to say more. Instead, religious fantasy begins to picture and amplify other details that were connected with the god's birth. Extraordinary circumstances, indicative of the future greatness of the child, are embroidered into the design. Isyllus recounted that Asclepius was born in the sacred temple of Epidaurus (T. 32), an unheard-of event, since for mortals it was unlawful that any birth should take place within the precincts of the sanctuary (T. 488).[19] In the later legend, Asclepius' mother comes from abroad; thus the god, though born in Epidaurus, is a foreigner; the child is exposed; it is nourished by an animal; a stepfather rears it.[20] All these data, just as the light that

---

[17] Kern, *Religion*, II, p. 308, rightly states the fact that in the religious tradition no punishment can be inflicted upon Coronis.

[18] Isyllus' version of the birth saga may also be influenced by Dorian religious beliefs, according to which male deities help the parturient; cf. Wilamowitz, *Isyllos*, p. 194; *Glaube*, I, p. 99, n. 1; p. 232.

[19] For the Epidaurian law, cf. below, p. 192, n. 3.

[20] Both the Epidaurian and the Arcadian myths assume that Asclepius was exposed, and nursed by an animal, and reared by a stepfather. In the Arcadian saga (T. 17) Τρύγων, a dove, is mentioned; the stepfather is the illegitimate son of the eponymous

radiates from the child and the immediate proclamation of its birth to all mankind, are typical of the birth saga of a " divine man," a θεῖος ἀνήρ.[21] In short, the birth myth of the god gradually changed from a unique report into a common pattern that was in accord with the religious tenets of late centuries. Yet this gradual transformation cannot obscure the fact that it is still the characters of the old heroic saga with whom the story deals, that basically it is the Hesiodic tale which is here retold.

Nor does the divine myth exhibit substantially new features in regard to Asclepius' biography. Explicit and definite as the legend was concerning the ancestry and the birth of the god, it seems to have been strangely reticent about the particulars of his ensuing life. There are no testimonies reporting the experiences of the god when he was a youth and when he grew up; nor is it reported whether he remained in Epidaurus all his life or went also to other places. All realistic details seem to be missing in the divine legend, just as they were in that of the hero.[22] As for his deeds, the fact that Asclepius was a healer of the sick, a raiser of the dead, was emphasized by the divine myth (T. 7), as it was by the heroic legend. Asclepius, so the Delphic Oracle affirmed, was born " a great joy to men, a soother of cruel pangs " (T. 31; cf. 36).[23] Healing,

Arcadian hero; in the Epidaurian saga (T. 7) the stepfather is the shepherd Aresthanas, the animal, according to Pausanias (T. 7), is a goat, according to others, among them Apollodorus (T. 5; 9), a dog. The latter statement is usually considered erroneous and a misrepresentation of the original myth. Yet the cult statue of the Epidaurian god (T. 630) showed him with a dog. It seems, then, that this feature is part of old tradition. The mention of the goat may have a hieratic meaning, for goats were not sacrificed in Epidaurus (T. 532). For a more detailed discussion of these topics, cf. below, p. 227.

[21] In general, cf. L. Bieler, ΘΕΙΟΣ ΑΝΗΡ, I-II, 1935-36. His analysis of the Asclepius story (II, p. 106) does not cover all pertinent data. Concerning the general significance of the additional facts to which I have referred, cf. Bieler, op. cit., I, pp. 29; 46. Other details which would fit into the legend of the θεῖος ἀνήρ-type are the change of Asclepius' name (T. 271 ff.), his birth in the temple (T. 32), the prophecy of his future greatness and of his resurrection through the gods (T. 2, vv. 642 ff.); cf. Bieler, op. cit., I, pp. 28; 37; 48. The Arcadian and the Epidaurian sagas, on account of these typically late features, hardly reflect early traditions. Wilamowitz, Isyllos, p. 84, n. 61, traced Pausanias' account to Ister, the pupil of Callimachus (Wendel, Pauly-Wissowa, s. v. Mythographie, XVI, p. 1366, 54 ff., rightly stresses the impossibility of deciding whom Pausanias follows). An early date was also assumed by Thraemer, Pauly-Wissowa, II, p. 1650, 36 ff.; Farnell, Hero Cults, p. 253; Kern, Religion, II, p. 308. Later, however, Wilamowitz changed his opinion and declared the myth to be a very late invention, though he failed to give any reasons (Glaube, I, p. 132, n. 2). Note that Imperial Epidaurian coins represent the birth myth as attested by Pausanias, cf. Imhoof-Gardner, Numismatic Commentary on Pausanias, ad loc. That the θεῖος ἀνήρ-version bears striking resemblance with the legend of the Christ Child has been stressed by Kern, Religion, II, p. 308, n. 3.

[22] Cf. above, pp. 52 f.; 63 f. One late hymn (T. 593) mentions Asclepius' education by Chiron; for the family of the god, cf. below, pp. 85 ff.

[23] The hymn of Sophocles (T. 587) calls Coronis the mother of the god " who wards off pains." The hymn of Macedonius (T. 593) holds that Asclepius was born " as a

then, was the task which was set before the god, the mission for which he was destined. Yet concerning the cures which the god performed while living on earth, the data are scanty again. Hippolytus must have been among Asclepius' patients; his thank-offerings were recorded in the Epidaurian temple (T. 739, 4). One of the sanctuaries, moreover, was believed to have been dedicated by Heracles after his wound had been treated by Asclepius; obviously, then, it was the god who was thought to have taken care of this case.[24] Only one healing is attributed specifically to the god, and to the god alone: that of Askles, the tyrant of Epidaurus. From him, so the Epidaurians believed, the name of the god was derived (T. 271 ff.).[25] Beyond these instances, however, the god's practice cannot be inferred with any degree of certainty. Nor can the medical discoveries be specified which were referred to in the Epidaurian account of his birth (T. 7). Only a few data may be surmised from dispersed references dealing with certain improvements in medicine with which the god was credited.[26]

One thing seems certain: there was nothing in the Asclepius myth that was in the least reminiscent of other divine legends which ascribed to the deities " all of the acts which are counted by men disgraceful and shameful, thieving and wenching and dealing deceitfully one with another."[27] Granted that the tradition is fragmentary, that stories may have been current which are not preserved, there can have been no stories of love affairs or of dissension, tales amoral in tone or character. Otherwise it would be incomprehensible that the Christian polemic, eager as it was to find fault with the outrageous behavior of the pagan gods, does not refer to any derogatory incident in the life of Asclepius, the most dangerous enemy of Christ.[28] The record of the god must have been spotless indeed, just as was that of the hero.

---

defender from illness and a healer of pain." The hymn of Erythrae (T. 592) proclaims that Apollo " begat great joy for man when he mingled in love with Coronis."

[24] One can hardly assume that the divine Asclepius restored to life those who were the enemies of the gods, cf. above, p. 47. But the old sagas may have been changed in regard to motivations or persons, or less offensive variants of the same story may have been selected. In this connection it is interesting to note that according to the tradition of the Epidaurian temple, Hippolytus met his death through the curse of his father, Theseus, not through a goddess; cf. above, p. 41.

[25] Cf. above, n. 21; below, p. 80.

[26] Cf. below, pp. 141 ff.

[27] Xenophanes, *Fr.* 11 [Diels-Kranz] = Sextus Empiricus, *Adv. Math.*, IX, 193 (transl. by R. G. Bury, *Sextus Empiricus*, III [Loeb]).

[28] The only reproach which the Church Fathers were able to bring forth against Asclepius is the accusation that he was " a bastard " (T. 8; 9; 233)—a claim that had been made already in Cicero's time (T. 9)—and that he was in love with Hippolytus (T. 83). Both these charges are commonly found in the Christian slander of

Born like a human being, having practiced medicine like a physician, the god Asclepius finally suffered the death of a mortal. That this was so is certified by no testimony which can unambiguously be referred to the divine myth, as it was told by Asclepius' priests and held true by his worshippers. Yet the arguments of the Christian Fathers warrant the conclusion that in the opinion of the many death had been the appointed lot of the god. Time and again the Apologists insist that to believe in the godhead of Jesus, who died the death of a mortal, is by no means stranger than to acclaim the god Asclepius who suffered the same fate (e. g. T. 235; cf. 112). They could hardly have made use of this analogy, had not the death of the god Asclepius been a fact commonly agreed upon.[29]

But if Asclepius the god had died, he had to come to life again, he had to become immortal, for he was ever-present in his temples. At the end of antiquity some seem to have assumed that Zeus struck Asclepius with his thunderbolt in order to make him a god (T. 236).[30] There were also other stories: Asclepius was supposed to have come back with the permission of the Fates from the nether world to which he had fallen prey just like any other human being (T. 237; cf. also T. 2, vv. 647 f.). Or it was the gods who were believed to have revived Asclepius, and their motives in making him one of their kind are variously given; it was either Zeus' wish to gratify Apollo (T. 75),[31] or it was the compas-

Greek gods. Concerning the importance which the lack of objectionable features in the Asclepius myth had for the rise of the cult and for the rôle which Asclepius assumed in the discussion between Christians and heathens, cf. below, pp. 113; 135.

[29] How the divine myth explained Asclepius' death remains an open question. The Christian polemic tends to give the impression that Asclepius was killed by Zeus "deservedly" (T. 233). Lucian too asserts that Asclepius died because he did what he ought not to have done (T. 265). In the same way Ovid (T. 75) explains Asclepius' death as caused by his having infringed upon the rights of Hades. On the other hand, Origenes maintains that men do obeisance to Asclepius "as to one who changed from mortal to god through his virtue" (T. 242). This interpretation seems to imply that Asclepius instead of being punished by Zeus died a natural death. Wilamowitz, Isyllos, p. 76, holds that the worshippers of Asclepius cannot possibly have believed in his destruction by Zeus on account of sinful deeds. In this sharply accentuated form the statement may be correct. But is it really incompatible with ancient religious feelings that Zeus should have been supposed to have destroyed the helper of men, the raiser of the dead, and that nevertheless people later revered him as a god? The ancient testimonies often mention these two facts together without any further comment. Cf. also below, n. 30.

[30] Thraemer, Pauly-Wissowa, II, p. 1654, 40, says: "An den Blitztod des Asklepios schliessen spätere Schriftsteller seine Apotheose. . . ." This statement is hardly correct in its generality. Rohde, Psyche, I, p. 320 ff., suggests that such an interpretation of death through lightning, though it is attested only by late authors, is very ancient. If this is so, it would have been easy indeed to reinterpret the heroic saga and from the beginning to give to Asclepius' death a meaning that fitted his new dignity.

[31] In reviving Asclepius, Zeus himself, as Ovid says (T. 75), does for the sake of

sion of the gods on Asclepius that brought about his resurrection, an act of mercy for which no further reasons are adduced (T. 265). Which of these versions was the canonized one, cannot be determined.

While, unlike the hero, the god was resurrected by the intervention of the gods, it is nevertheless clear that the divine legend is nothing but a transformation of the heroic saga. To be sure, its tendency was completely reversed; its main content, however, remained the same. Old features were omitted only in so far as they were incompatible with the divinity now ascribed to Asclepius; new facts were introduced only in so far as they were indispensable for the story of a god.[32]

To express it differently: according to the divine myth, one must conclude, the god is only the deified heroic figure. The myth attached to the god lacks all individual characteristics; in no way is the existence of a divine saga indicated that essentially deviates from the heroic legend. This would be strange indeed if the god Asclepius, as most modern interpreters assume, were a chthonian deity and one who was revered from time immemorial in certain sections of Greece from which he finally conquered the ancient world. Such a god one should expect to have possessed a legend of his own and one very different in character.[33] It should show at least certain traces of his one-time existence as an independent being. As things stand, it is hard to believe that a god Asclepius was ever worshipped anywhere irrespective of the heroic figure.

2. THE NATURE OF ASCLEPIUS' GODHEAD

The modern assumption that Asclepius was an aboriginal deity is not borne out by the evidence available. It seems necessary to seek for another explanation of his divinity. How did the ancients judge about the nature of this god? Their opinion cannot be taken lightly. Asclepius apparently was a late deity. Many of those who wrote about him were

Apollo, the father, what he forbade Asclepius, the son, to do. The star saga also gave Zeus' desire to do a favor to Apollo as the reason for Asclepius' elevation to heaven, cf. above, p. 51.

[32] An analysis of the divine legend in its entirety has never been undertaken. Even Wilamowitz in his attempt to reconstruct the content of the ἱεροὶ λόγοι (*Isyllos*, pp. 84 ff.) restricts himself to an investigation of the genealogy and the birth legend which, as he admits, are dependent on the heroic saga. It seems to me that this judgment must be extended to the whole myth.

[33] Farnell, *Hero Cults*, p. 244, has pointed out that Asclepius' death through Zeus' thunderbolt is inconsistent with the assumption that he was a "chthonian daimon." His thesis is correct, I think, but his argument is invalidated by the fact that as evidence he simply quotes Hesiod who speaks about the hero, while he has failed to show that the story of Asclepius' death in one form or another must have been part of the divine myth.

themselves witnesses of the rise of his worship. What they report, therefore, should have more weight than what Greek and Roman writers have to tell about those gods who were venerated from earliest times.

That in his cult Asclepius was thought to be a man-god follows with certainty from the myth attached to him. His worshippers saw in him one of the many human beings of half divine origin who had risen to the dignity of a god.[1] And it was by no means only the pious who saw in Asclepius a deified mortal. A like view of his godhead was generally held by the ancients, educated and uneducated alike (T. 248; 262). The only deviation from the more dogmatic belief was that, in the opinion of the less reverent, Asclepius was a human physician, deified by men after his death in grateful recognition of the help that he had bestowed upon them. Xenophon states that Asclepius was revered as a god because he had healed the sick and revived the dead (T. 243).[2] Other authors declare that Asclepius was regarded as a god because he perfected medicine (T. 244), or because he was the founder of the medical art (T. 245; 246). Again, the fact that he had endured so many hardships is proffered as reason for his deification (T. 241), or, in a more general way, his elevation to the rank of a god is attributed to his merits (T. 240), his benefactions (T. 239), his virtue (T. 242).[3] The motivations adduced in these explanations differ, but there is agreement in one respect: Asclepius owed his new dignity to men who honored him as a god because of his accomplishments; originally he had been a mortal.

It is true, from the fifth century B. C. onward, there was a tendency to interpret all gods as human beings, as benefactors of men, as inventors of arts and crafts who later were considered gods by their grateful fellow men. Yet, in regard to Asclepius it is not the usual euhemeristic explanation that is also applied in this particular case; rather is Asclepius together with such figures as the Dioscuri, or Heracles, or Romulus taken as an indisputable example which by analogy would prove beyond doubt that all gods had originally been human beings; for even those philosophers who were principally opposed to the euhemeristic argument could

---

[1] In the Epidaurian saga, Asclepius was nothing but a θεῖος ἀνήρ, cf. above, p. 73. Even those who, like Pausanias, claimed that Asclepius was a god from the beginning (T. 255), hardly intended to say more than that he was a θεῖος ἀνήρ or a god in the shape of man. Pausanias himself accepted the Epidaurian birth of Asclepius; his statement is directed against those who claimed that the deification of Asclepius had taken place very late, and is intended to support the belief that he was recognized as a god even in Homer's time; cf. below, n. 6.

[2] The text of the passage is corrupt; but, whether the reading θεὸς ὤν or θεός ὡς is accepted, the meaning is the same as far as the problem under discussion is concerned.

[3] This testimony is interesting in regard to the moral character of the god, as it was indicated by the divine legend, cf. above, p. 74.

not, and did not deny that Asclepius was a man-god (T. 251).[4] Nay, so strong was this conviction, that he was never made subject to the physical interpretation of the nature of the divine, so widely current in antiquity. Even the Stoics, eager as they were to understand the gods as elements of the universe, did not dare to identify Asclepius with any of them and declared him to be the divine protector of medicine (T. 6).[5]

It is easily understandable that this was so. Homer had sung of Asclepius as the human father of human sons who were among the fighters before Troy (T. 135-36). Such a testimony could hardly be quoted in regard to any other god, and Apollodorus, the scholar, basing his conclusions on this historical evidence, had endorsed the assumption that Asclepius in Homeric times had not yet been a god, but simply a mortal (T. 257).[6] Nor did the pious deny that their god was identical with the Homeric hero. To be sure, the divine myth represented the life of Asclepius as it had been pictured by Hesiod. But Asclepius, the god, retained as his sons Machaon and Podalirius, who had been celebrated by Homer and by the epic saga. The earliest known hymn in honor of the deity refers to them as his children (T. 592).[7] Another one expressly calls Machaon and Podalirius " two healers of Greek spearsmen " (T. 593; cf. also T. 482, *vv. 7 ff.*), thus showing that the two were viewed as the sons of the epic Asclepius. And at the end of antiquity Aristides in his speech in honor of the Asclepiads stated that just as Triptolemus distributed the gifts of Demeter over the earth, thus Machaon and Podalirius, who had grown up in the garden of health and had studied medicine with their father, after the Trojan War imparted medical knowledge, the gift of Asclepius, to all parts of the inhabited world (T. 282, *7; 11; 15*). Not only in a vague sense was Asclepius a deified mortal: his life, as everybody admitted, could be exactly dated. The ancients were quite justified, indeed, in assuming that Asclepius, if any of the generally acknowledged deities, was a man-god.

Nevertheless, one will hardly venture to follow their supposition in all

---

[4] It is noteworthy that the mythological explanations of Asclepius' deification through the gods (cf. above, p. 75) are partly identical with those offered by the saga for the deification of the other figures with whom Asclepius is paralleled in the euhemeristic argument (Heracles, the Dioscuri, Romulus). Incidentally not only the Christians, but even the Greeks themselves sometimes refrained from calling Asclepius a god at all (T. 249).

[5] Cf. below, pp. 105 f.

[6] Pausanias' forced interpretation of the Homer passage (T. 255; cf. below, p. 91, n. 3) indicates the importance which the statement must have had in the ancient discussion of Asclepius' nature.

[7] Even in late centuries, the two had temples wherever the doors were open for Asclepius; wherever the name of the father was invoked, his sons, too, were commemorated, as Aristides says (T. 282, *21*).

its implications. To ascertain the historical reality of the Homeric or of the Hesiodic hero proved impossible. He seemed to be an invention of mythical fantasy, the character of a fairy tale which in eternal symbols expresses the mission of the physician.[8] Now, if the god was no more than the deified hero, as one is led to believe by the testimony, then historicity cannot be attributed to him either. His human nature can be granted only in the sense that he was one of those saints who through their deeds acquired divine prerogatives.[9] But this can be asserted: over a long period of history Asclepius, before being elevated to the dignity of a god, had held only the inferior rank of a hero, of the patron of physicians.[10] As such a hero of artisans, who made his way into the pantheon, Asclepius would not be an isolated figure among the Greek deities. Hephaestus, too, was originally the saint of craftsmen.[11] If in contrast to Hephaestus Asclepius became a god revered by everybody, this was due to the fact that his task, the healing of diseases, concerned all men. Nor is it astonishing that the greatest divine physician should have come from the realm of secular medicine. In antiquity, unlike modern times, divine healing and human medicine, though strictly separated, were not hostile or in irreconcilable opposition. Moreover, the help expected from Asclepius, though superhuman in range and efficacy, in the opinion of the ancients, was nevertheless rational and fashioned after the human art of medicine.[12]

Yet, for an appreciation of the nature of Asclepius' godhead it is not enough to state that the god was the deified hero. What was the power that he acquired with his new standing? Did he receive the unlimited

[8] Cf. above, pp. 52 f.; 63 f.

[9] In this respect, then, I differ from Farnell and all those who believe that the god Asclepius grew out of a real personality. In my opinion the case stands otherwise with the Greek god of medicine than it does with the Egyptian one, Imhotep, with whom Asclepius in late antiquity was identified (T. 331), and who, according to modern research, was a historical figure, cf. K. Sethe, *Imhotep, Unters. z. Gesch. u. Altertumsk. Aegyptens*, II, 4, 1902; R. S. Forbes, *Proc. of the Royal Society of Medicine*, Sec. of the Hist. of Med., XXXIII, 4, 1940, p. 733; a parallel between Asclepius and Imhotep is drawn *e. g.* by Farnell, *Hero Cults*, p. 236. If it were objected that originally all heroes, as all gods for that matter, were human beings, this would introduce into the debate a general philosophical problem which I am not prepared to discuss. I wish to state only this much, that Asclepius cannot be proved to have been a definite historical personality. On the other hand, I need hardly defend the theory proposed against the claim that all ancient heroes originally were gods. This assumption, much favored in the nineteenth century, is now beginning to be discarded; cf. Wilamowitz, *S. B. Berl.*, 1925, p. 62. Some heroes certainly are "decayed gods," but this is by no means true of all of them; cf. in general, Farnell, *Hero Cults, passim.*

[10] For the interpretation of the process of Asclepius' deification, cf. below, pp. 91 ff.

[11] Cf. Wilamowitz, *G. G. N.*, 1895, pp. 217 ff.; Nilsson, *Griechische Religion*, I, pp. 495 ff.

[12] For an elaboration on these theses, cf. below, pp. 139 f.; 152 ff.

might of the great gods, but in self-restriction choose to heal diseases?
Or did he assume from the beginning only one special duty, namely that
of tending the sick? Besides, in the world of polytheism, the gods were
not equals; they formed a society in which every member had his own
rank. What was Asclepius' station?

Even the name Asclepius, as it was understood in antiquity, points
to the fact that he was thought of as a god who was in charge of
one limited domain. The word Asclepius, as all ancient interpreters
agreed, was a compound.[18] Its second part was not difficult to explain:
no one doubted that it consisted of the expression ἤπιος, *mild* (T. 267).
There was a difference of opinion only in so far as the concept of mildness
was referred to various subjects, to the gentleness of Asclepius' hands, to
his way of healing in general, to his art, or to his character (T. 268 ff.).[14]

It is the explanation of the first syllable which proved to be a rather
baffling problem. Some people said that Asclepius was originally called
Ἤπιος, *the Mild*, but was surnamed Asclepius after he had healed Askles,
the tyrant of Epidaurus (*e. g.* T. 273).[15] Older perhaps than this
interpretation, though unique, is the claim of Isyllus that Apollo named
his son Asclepius after his mother Aigle; for thus Coronis was first called
in Epidaurus (T. 594).[16] Both these etymologies apparently were determined by the wish to connect Asclepius as closely as possible with Epidaurus. What better proof could be found for the city's importance in
regard to Asclepius than the fact that even his name was derived from

[18] For such a theory the accentuation of the word provides a difficulty; Demosthenes
was quite right in saying Ἀσκλήπιος instead of Ἀσκληπιός, since the name was supposed
to be a compound (T. 266). At the same time this testimony proves that the interpretation of the name as a compound was current at least as early as the fourth century
B. C. Dialectical modifications of the name are collected in Pauly-Wissowa, II, p. 1642,
52 ff. Concerning the Latin transcription, cf. T. 267a-b. Asclepius is one of the few
Greek gods who even after migrating to Rome kept their Greek names (cf. Deubner,
Die Römer, p. 462; K. Latte, *Gnomon*, VII, 1931, p. 121, n. 2).

[14] For the importance of the word ἤπιος, cf. below, p. 81. It should be noted that the
character and the attitude of the god as outlined by the saga (cf. above, pp. 73 f.) are in
many respects reminiscent of the ideal physician who is pictured in the Hippocratic
Oath (*C. M. G.*, I, 1, pp. 4, 18; 5, 1 ff.).

[15] Askles or Askletos (T. 272) is otherwise unknown and may have been invented for
etymological purposes. It is interesting to note that this etymology would imply that
Asclepius came to Epidaurus at the end of the seventh century B. C.; for at that time
the city was under the domination of tyrants; cf. Hiller v. Gaertringen, *I. G.*, IV², 1,
p. xi. Yet undoubtedly this is an intentional attempt at archaizing; for the Epidaurian
god is much younger, cf. below, p. 98.

[16] It is this etymology that has found favor with Wilamowitz (*Isyllos*, pp. 91 ff.;
Glaube, II, p. 229); it is in fact the corner-stone of his belief that Asclepius is a pre-
Hellenic god, in spite of all the qualms which he feels himself ("befremdender Laut-
wandel," "sehr seltsame Ableitung mit noch seltsamerer Betonung," *Isyllos*, p. 93);
cf. below, n. 18.

the Epidaurian legend? Such explanations were hardly anything but tendentious inventions and they were never commonly accepted.[17] In general the ancients did not derive the meaning of the name Asclepius from a combination of the word ἤπιος with a proper noun. Taking as a fact that the second syllable meant mild, they referred the other, Ἀσκλ to ἀσκ with the superfluous addition of the λ, thus understanding Asclepius as " the one who gently takes pains with the sick, that is, considers them deserving of watchful care " (T. 272); or they referred it to σκέλλω, to make harsh, with an alpha privative, that is " he who by the agency of the medical art does not permit dryness " (T. 269). What are the implications of these etymologies? [18]

The explanation of Asclepius as " the one who gently takes pains with the sick " (T. 272) seems to convey the notion that the god exercised his art as, in the opinion of the Greeks, a good physician should do, that is, kindly. Not only was the word ἤπιος always connected with drugs, " the healing hands of the gods," not only does it occur as surname of Apollo, the physician; it was also characteristic of the attitude of the human doctor, and more specifically of the Greek doctor in contrast to the barbarian physician.[19] Mildness and friendliness are inherent qualities of the ideal physician. To be sure, the doctor must sometimes hurt his patient; he must burn or cut, but, just as he does so only in order to help, he is in general the friend of his patient; he is thought to be mild rather than rough, kind rather than harsh or severe.[20] The name Ascle-

[17] The obvious Epidaurian tendency of these etymologies has never been pointed out. Weinreich, Heilungswunder, p. 38, n. 3, emphasizes that treatment of eye diseases, as in the case of Askles, was very common among the Epidaurian temple cures. Concerning the importance of such a change of the name in the legend of the θεῖος ἀνήρ, cf. above, p. 73, n. 21.

[18] I wish to stress that I am not concerned with the correctness of the ancient explanations. I think it necessary, however, to find out their significance, the more so since the fact that the name Asclepius was considered transparent allows of certain conclusions regarding the nature of Asclepius' godhead, as it was understood in antiquity. For modern etymologies, cf. the survey in Roscher, Lexikon, I, 1, p. 616, 4 ff.; Pauly-Wissowa, II, p. 1643, 26 ff.; D. Detschew, Mitt. Bulg. Arch. Inst., III, pp. 131 ff. (Thracian god); cf. P. Kretschmer, Glotta, XVI, 1928, p. 193. No agreement has been reached among the modern critics, and the more cautious interpreters believe that the problem has not yet been solved; cf. the most emphatic statement of Farnell, Hero Cults, p. 238: " Now the name ' Asklepios ' has hitherto defied all attempts to explain it and remains an unsolved mystery, at least for those who possess a philological conscience." Cf. also Kern, Religion, II, p. 305, n. 2.

[19] For the connection of the word ἤπιος with drugs and all aspects of healing, cf. Usener, Götternamen, p. 165, n. 49. For the contrast between Egyptian and Greek procedure, cf. Herodotus, III, 130: Ἑλληνικοῖσι ἰήμασι χρεώμενος καὶ ἤπια μετὰ τὰ ἰσχυρὰ προσάγων (sc., Democedes). Cf. also T. 355a.

[20] Hippocrates (C. M. G., I, 1, p. 22, 1 ff.) insists upon the necessity of a mild procedure in surgery; another passage (Hippocrates, V, 308 L.) speaks of αἱ τοῖσι

pius, then, was understood to express a certain standard; it symbolized the ethical demands of the craft which Asclepius represented. Taken as functional, the name delineates him as the prototype of the good doctor.

The other etymology expounded in antiquity is even more revealing. By saying that Asclepius is the one "who makes the harsh in illnesses mild" (T. 271), or "who does not permit dryness" (T. 269), the ancients intimated that Asclepius is the one who, by making the body soft again, "does not allow men to be parched and dried up and mortified by diseases" (T. 271), that is, he gives them their earlier health and lively freshness.[21] In other words: Asclepius is visualized here as the protector from "the withering that comes with death" (T. 268), from the mortification of the body; he is the one who defers "the cessation which comes with death" (T. 273). To accomplish this feat was the aim of every physician; it was his highest ambition; it was the sum total of his endeavor as an artisan and a scientist, the divine mission of his craft.[22] Thus understood Asclepius in providing for the sick what they most needed, help in deadly diseases, in acting as the savior from their most dreaded enemy, death, is again the impersonation of the ideal physician.[23]

Whatever the correct etymology of the word Asclepius, the name of the god, the "cult statue of human speech" as Democritus calls it,[24] to the ancients had a definite and perspicuous meaning; it reflected the hopes and expectations with which they entered the temples of the deity, it symbolized the perfect application of the medical art, the liberation from utter destruction. If one remembers that those who came to visit the sanctuaries in most cases did not suffer from trifling ailments, but were desperately ill and took refuge in the Asclepieia when given up by human physicians as "hopeless cases,"[25] one immediately realizes how appro-

---

κάμνουσι χάριτες which the specialist in internal medicine should observe; Galen too (XVII B, pp. 145 ff. K.) discusses the topic of kindness at the bedside. Cf. Edelstein, *Problemata*, IV, p. 102.

[21] The Greek etymology was curiously interpreted by the Methodist Caelius Aurelianus in his translation of the works of Soranus into Latin. He claims that Asclepius was the first to cure chronic diseases that are hard to overcome, and that from this accomplishment he received his name (T. 364). But, of course, originally the etymology cannot have referred to any specific healing.

[22] Cf. above, p. 45, and especially the Callimachus verses referred to p. 45, n. 81. The etymology proposed by the ancients summarizes the gist of the Asclepius legend.

[23] Cf. also below, p. 112, n. 4. It is well to remember that σκέλετον and σκέλλω contain the same root and that Hippocrates was said to have dedicated to the god in Delphi μίμημα χαλκοῦν χρονιωτέρου κατερρυηκότος τε ἤδη τὰς σάρκας καὶ τὰ ὀστᾶ ὑπολειπομένου μόνα (Paus. X, 2, 6). The dedication obviously symbolized Hippocrates' victory over death. For the interpretation of this passage, cf. H. Pomtow, *Klio*, XV, 1918, pp. 306 ff.

[24] 68 B 142 [Diels-Kranz]; cf. also M. Warburg, *N. Philol. Unters.*, V, 1929, pp. 72 ff.

[25] Cf. below, p. 169.

priate the ancient etymology was in connection with the god's activity.
Moreover, it seems most unlikely that a deity, whose name was so definitely connected with medicine and the task of the physician, had at any
time been thought of as being more than the incarnation of the divine
power of healing.[26]

The limitation of Asclepius' power is also expressed in the divine
appellation usually accorded to him by his worshippers, and in the position which he held in comparison with the more ancient gods. If the
sceptics of Cicero's time speak about him as " one of the newly enrolled
citizens in heaven " (T. 262), or if Lucian pokes fun at Asclepius as
" a naturalized alien " among the gods (T. 264), if some call him " a
newcomer to heaven " (T. 263), or designate him as one of those " who
are named gods " (T. 262a), all such statements may first of all refer to
the fact that he was a late deity who rose to godhead long after the other
inhabitants of the pantheon. Besides, these assertions are indicative of
the inferior rank of the new god. Even the oldest hymns invariably
address Asclepius not as god, but as " daimon " (T. 592 f.). It has been
claimed that in poetry the terms θεός and δαίμων are used interchange-
ably.[27] This may be true in some instances. But the constant use of the
word " daimon " in regard to Asclepius can hardly be fortuitous.[28] Moreover, in the language of later theology Asclepius, in contradistinction to
the full-fledged gods, is unambiguously characterized as " daimon " (T.
260), as demigod, " and it is doubtless through the possession of emotion
that the demigod is inferior to the god. For what their natural characteristics were when they dwelt on earth, these they do not desire
wholly to relinquish." In the procession of the pre-Christian era As-

[26] Even Wilamowitz who believes that the word Asclepius is pre-Greek (cf. above, n.
16) grants " das ἤπιον . . . , das man in seinem Namen hörte " (*Glaube*, II, p. 232). Of
the name Asclepius, I think, the same is true as of the words Eleusis or Artemis, concerning which Kern (*Religion*, I, p. 102) says: " Wie die Hellenen bei dem nach moderner
Ansicht ursprünglich ungriechischen Namen Eleusis zweifelsohne stets die Bedeutung
der 'Ankunft' empfunden haben, werden sie auch bei der Namensform Artemis immer
an die grosse Schlächterin, die Herrin über alles Getier des Waldes und der Berge,
gedacht haben."

[27] Cf. Roscher, *Lexikon*, I, 1, p. 938, 15 ff. Wilamowitz too, *Glaube*, I, p. 363, n. 2,
holds that in some cases the term " daimon " is used only for grammatical reasons.

[28] Note also that it was only in the Homeric religion that the concepts of θεός and
δαίμων were not clearly distinguished (cf. Wilamowitz, *Glaube*, I, pp. 362 ff.; II, pp.
15 f.; K. Lehrs, *Populäre Aufsätze*, 1875, pp. 143 ff.). But even here the gods are
preferably called " daimones " when they are considered in their active interference in
human affairs. Popular opinion at least from the time of Hesiod accepted the belief in
" daimones " who were different from the gods (cf. Preller-Robert, *Mythologie*, I, p.
112). This distinction should be valid, therefore, also in the case of the Asclepius hymns,
the oldest of which dates from the fourth century B. C. The term δαίμων also occurs
in T. 595.

clepius like all the other demigods (ἡμίθεοι), who had left their bodies and whose souls had risen to heaven, followed behind the great Olympian gods and received honors similar to those paid to the gods (T. 258). He never became one of the Olympians.[29] In fact, he chose for all times to live on earth; he was a fugitive from the regions of the gods, as it were (T. 261).[30] He was a terrestrial god rather than one of the Olympians, or one of the celestial deities, or one of the chthonian divinities (T. 259). In short, Asclepius was considered one of those " in-between existences," whom Hesiod calls ἐπιχθόνιοι δαίμονες.[31] This was appropriate indeed for the son of a divine father and a mortal mother who even after his death continued to heal men in his divine temples, for a divinity who was thought to be in charge of one specific function.

I have now marshalled the evidence concerning the ancients' opinion in regard to Asclepius' nature and dignity. Translated into modern categories it suggests that Asclepius was a " special god," although not one of the genuine gods of this type, for he became a " special god " only after he had been a hero.[32] With such a character it is concordant that he should have been regarded a " daimon." " Special gods " being minor deities most frequently were classified in this way.[33] With such a character the name too seems to agree, as it was understood in antiquity.

---

[29] This statement is correct in spite of the fact that Aristides (T. 317, 4) and Quintus (T. 181) speak of Zeus Asclepius or the Olympian Asclepius. Such characterizations, to be found only at the end of antiquity, may be understood in connection with the Neo-Platonic speculation, according to which all gods were emanations of one identical divine power, cf. below, pp. 107 f. Or else they may be superlative forms of worship as they also occurred in the cases of Heracles, Agamemnon, Trophonius (cf. L. Preller, *Griechische Mythologie*, II³, 1875, p. 361, n. 3). At any rate, these expressions do not place Asclepius on the same level with the other Olympians (cf. also Wilamowitz, *Glaube*, II, p. 223).

[30] Here I am borrowing from Origenes who with these words characterizes the activities of all gods who appear to men on earth.

[31] The importance of this concept in Greek religion has been stressed by M. P. Nilsson, *Gercke-Norden, Einl. in d. Altertumswissenschaft*, II³, 1922, p. 280.

[32] The term " special god " (" Sondergott "), which Usener introduced into the interpretation of Greek religion, is foreign to Greek speculation and therefore somewhat dangerous to use. Yet the phenomenon itself, the personification of certain limited powers, is to be found in Greek no less than in Roman religious life. Wilamowitz, *Glaube*, I, p. 11, proceeds much too summarily in rejecting Usener's theory altogether; cf. Kern, *Religion*, I, pp. 113 ff., who discusses the pro and con of the question and vindicates the indisputable importance of the concept of " special gods " against undue exaggerations. He says, p. 114: "Echte Sondergötter sind göttliche Gestalten, die lediglich zu dem Zweck geschaffen sind, Schutzheilige einer ganz bestimmten und meist eng begrenzten Naturkraft oder menschlichen Beschäftigung zu sein." From these, Kern distinguishes " die falschen Sondergötter," that is, " die *numina specialia*, die aus Heroen dazu geworden sind." In this differentiation of genuine and non-genuine " special gods " Kern follows G. Wissowa, *Gesammelte Abhandlungen zur röm. Religions- und Stadtgeschichte*, 1904, pp. 304 ff. Cf. also Nilsson, *Griechische Religion*, I, p. 363.

[33] Cf. Usener, *Götternamen*, pp. 247 ff.; 291 ff. Cf. also Eitrem, Pauly-Wissowa, *s. v.* Heros, VIII, p. 1112, 49.

A functional name was usual for " special gods." [34] Can the suggestion implicit in the testimony reviewed so far be verified by additional arguments?

First of all it is only under the presupposition that Asclepius was a " special god " that the peculiarity of the myth can be fully appreciated. In his divine legend Asclepius is represented as a man-god, and yet how indistinct does he remain as an individual, how dimly is he pictured as a personality compared with other deities of the Greeks and Romans; how barren of facts is the account, how incomparable to the myths attached to other gods of equally high renown. [35] The beginning and end of his life, cardinal points on which the exposition of the story and explanation of the god's nature hinge, are clearly marked. What happens in between the boundaries of his earthly existence, however, is left vague and not adorned with the concreteness of a definite setting. As far as his " life " becomes visible at all, it centers on his medical activities, on his achievements as a healer of the sick, a raiser of the dead. If Asclepius was a god invented or created for the sole purpose of taking care of the province of medicine, it is quite understandable that his myth remained as hazy as that of the hero. The mythological embroidering of such a figure was of necessity restricted, because the original concept of the god was functional and as such demanded certain features and scenes for its elucidation, but they were few indeed. The legend, viewed from this angle, cannot be considered the product of mere chance. [36]

Even the god's relation to his family now becomes meaningful. Asclepius was surrounded by a divine retinue composed of a number of gods and goddesses. None of them, except Machaon and Podalirius, who were taken over from the Epic, [37] are mentioned in the myth; their con-

---

[34] Farnell, *Hero Cults*, p. 238, says: " The original significance, then, of the name Ἀσκληπιός remains unknown; but one cannot help feeling that it has more the fashion and sound of a personal-human than of a divine name." This feeling, at any rate, is not that of the Greeks, as is evidenced by the ancient etymologies which Farnell does not discuss at all. When on the other hand he claims that a functional name is typical of a chthonian god (*ibid.*, p. 239), he does not take into account the " special gods," of whom the same can be said; cf. Kern, *Religion*, I, p. 115.

[35] Cf. above, pp. 73 f.

[36] Wilamowitz seems to be the only one to have noticed the peculiarity of the Asclepius myth, but he explains it quite differently (*Isyllos*, p. 103): " . . . er ist zu spät in die Reihen der Götter getreten, zu denen die edelsten der hellenischen Volksstämme beteten, als dass Dichter sein göttliches Wirken und Walten so verschönt hätten, wie sie es an Apollon oder Athene getan haben, denen er freilich auch an innerem Gehalte nicht zu vergleichen ist." Yet, the phenomenon in question is rather the typical one in connection with " special gods," cf. Usener, *Götternamen*, esp. pp. 330 f., also p. 304 (Helius as contrasted with Apollo); pp. 314 f. (Uranos).

[37] Cf. above, p. 78, where it has also been shown that Machaon and Podalirius were

nection with Asclepius is in no way elaborated. All his divine children as well as his divine wife are personifications of abstract concepts, of medical functions, a fact recognized by ancient and modern critics alike.[38] Asclepius was indeed a god who took charge of medicine alone; even his family followed his calling.

One must go still farther. The association with these deities had a special significance for Asclepius. To have a family, for him, was more important than it was for most other deities.[39] When the Athenian peasants invoked Asclepius, they called him " Sire in his offspring blest " (T. 277). The hymns carefully enumerated all the sons and daughters of the god; they asked them to appear and to lend their help (T. 592-93). Obviously the father, the mother and their offspring, in the eyes of the worshippers, were a unity. It is not by accident that Aristophanes, when picturing one of the cures of Asclepius, made him perform his task supported by his daughters (T. 421, *vv. 701-02*). Rather is it true that the god of medicine, at least at the beginning of his divine practice, rarely acted without his children; [40] they were part of his own nature, as it were. What does this fact mean in regard to the character of Asclepius' godhead?

Not much can be concluded from his constant association with Epione, who from the end of the fifth century B. C. was generally considered his wife. Her name is derived from the word ἤπιος, *mild*, the same root that all the ancients found in the name of Asclepius himself; she is nothing but the personification of mildness, as the testimonies assert (T. 281).[41] There is no word about her origin, no tale concerning her marriage to Asclepius; nor is it ever told where and when she gave birth to her children.[42] Epione lives only in her union with her divine husband; she

considered physicians. Here, I am concerned only with those members of Asclepius' family who were associated with him when he became a god.

[38] Cf. *e. g.* T. 278, and Pauly-Wissowa, II, p. 1656, 62 ff.; Usener, *Götternamen*, pp. 163 ff.

[39] The only gods comparable to Asclepius in this respect seem to be Ares and Aphrodite. Cf. below, n. 49.

[40] The early monuments, too, picture Asclepius with one or another of his children assisting him; cf. *e. g.* the relief made around 405 B. C., now in the Piraeus Museum (L. Curtius, *Die antike Kunst*, II, *Handb. d. Kunstwissenschaft*, p. 236). The great temple inscriptions of Epidaurus (T. 423), dating from the second half of the fourth century, naturally put the emphasis on the work performed by the god alone. But even in the third century another Epidaurian inscription (T. 431) is dedicated to the god and his children. Cf. also below, p. 102, n. 4.

[41] This etymology is also accepted by modern scholars, cf. *e. g.* Usener, *Götternamen*, p. 165; Kern, *Religion*, II, p. 311; as for Asclepius and the importance of the term ἤπιος in medicine, cf. above, n. 19.

[42] Once Epione is called the daughter of Merops (T. 281b), the hero of Cos (cf.

is the woman whom Asclepius needs so that he may have offspring. Apparently she is a mere double of the god.[43]

It stands differently with Asclepius' and Epione's children. In addition to Machaon and Podalirius, there were Aceso, Iaso, Panacea, Hygieia and Aegle. At any rate from the fourth century on, when they are mentioned in the hymns (T. 592), they constituted the divine family of Asclepius and were recognized everywhere.[44] Among them the iatric children,

Preller-Robert, *Mythologie*, II, p. 562, n. 2; Thraemer, Pauly-Wissowa, *s. v.* Epione, VI, p. 187, 41 ff., reads Μεροπίς instead of Μέροπος, thus making Epione a Coan by birth) ; in the Hippocratic legend she is called the daughter of Heracles (T. 281a). Both versions apparently are Coan transformations of the saga (Heracles was famous in Cos, cf. Thraemer, *loc. cit.*; for the Coan origin of the Hippocrates legend, cf. Pauly-Wissowa, Supplement VI, pp. 1301, 26 ff.). Since the Coan Asclepius, according to the recent excavations, is very late (cf. below, p. 243), these genealogies cannot be considered original (contrary to Thraemer, *loc. cit.*, p. 189, 3 ff.). At any rate, the tradition is purely local and has never had any general significance. Usener, *Götternamen*, p. 163, suggests the Epidaurian origin of Epione; this can neither be proved nor disproved.

[43] It should be remembered that Epione, though uncontested in the cult, is not the only wife of Asclepius mentioned by tradition. One testimony, the oldest one that speaks about Asclepius' family, names her *Lampetia*, the daughter of Helius (T. 281c) ; she may have had some connection with Rhodes where Helius was the most important god (cf. Wilamowitz, *Glaube*, I, p. 116) and where an old family of Asclepiads existed (thus Thraemer, Pauly-Wissowa, II, pp. 1656, 38 ff.) ; or Lampetia may have been given as the name of his wife because of the importance of the sun for the healing of diseases, or because the light of day represented health and life (cf. Usener, *Götternamen*, p. 178), or because the Heliads were symbols of knowledge (cf. Preller-Robert, *Mythologie*, I, p. 433) ; Asclepius' wisdom is often stressed (*e. g.*, T. 593). One other source names *Hipponoë* as the wife of Asclepius (T. 281f). She is probably one of the Nereids (Hesiod, *Theog.*, 251; cf. Pauly-Wissowa, II, p. 1656, 48), but is otherwise unknown; she may have been connected with Asclepius because of the healing power of wells and rivers (cf. Preller-Robert, *op. cit.*, I, pp. 547; 551; concerning wells in the Asclepieia, cf. below, p. 237). The Orphics gave to Asclepius as his wife *Hygieia* (T. 281e) ; concerning her, cf. below, pp. 89 f. These are all the women ever named as mother of all the children of Asclepius, none of them being universally recognized. Machaon's mother, in the Messenian legend, is *Xanthe* (T. 281d, cf. above, p. 32, n. 36), perhaps the daughter of Oceanus and Tethys (Hesiod, *Theog.*, 356)—she again would point to the influence of water on healing. That Arsinoë and Coronis (in T. 169) are mentioned as Asclepius' wives only by mistake, has been shown by Wilamowitz, *Isyllos*, p. 49, n. 12. Aristodama is the mother of the Sicyonian hero Aratus (T. 281g; cf. T. 288).

[44] Some of the ancients added *Ianiscus* and *Alexenor* (T. 279). But they were at no time generally accepted in the cult; cf. Usener, *Götternamen*, p. 170, n. 61. *Ianiscus* originally came from Attica and was king of Sicyon. As such he probably received heroic worship after his death, but was later superseded by Asclepius, a fact which is reflected in the local tradition that he was the son of the god of medicine (cf. Pausanias, II, 6, 6). Usener, *op. cit.*, p. 170, connects the name Ianiscus with Ion, the healer; cf. Gruppe, *Griechische Mythologie*, p. 739, 7. *Alexenor* apparently is the Alexanor honored at Titane (T. 188) who was thought to have founded the Asclepius temple at that place (T. 187). There is hardly any doubt that Alexanor was made the son of Asclepius when he was superseded by Asclepius, just as Machaon had once ousted Alexanor and had adopted him as his son; cf. above, p. 21. Alexanor is " der den Männern Helfende, Abwehr Bringende," Usener, *op. cit.*, p. 170. At any rate, Ianiscus and Alexanor are local deities, connected with Asclepius only in the tradition of certain sanctuaries; they are no genuine members of his divine retinue. Another local variation is the Coan claim that *Epio* was a daughter of Asclepius (T. 482, *v. 6*) ; cf. Usener,

Aceso, Iaso, and Panacea are clearly personifications of functions. As indicated by their names, they represent the healing power itself.[45] When and where they became connected with Asclepius one cannot decide with certainty. Aceso is always referred to as daughter of Asclepius. She may have originated in Epidaurus.[46] Iaso even in the fifth century B. C. appears also as the child of Amphiaraus (T. 285). She is, then, a healing spirit in her own right, originally associated with various gods, but finally fixed in the Asclepius family.[47] The same is true of Panacea, the "universal remedy." She too originally was an independent goddess and was linked together with the god of medicine only later.[48]

For what reasons were these figures believed to be the children of Asclepius? Were they thought of as hypostases of his power? In that case it would be difficult to understand why they never made use of their own faculties, why they never undertook to heal by themselves. Nor would it be meaningful under this presupposition that the god acted in unison with them. As emanations of his own strength they would be mere shadows of their father. The close cooperation between Asclepius and his iatric daughters in the beginning of his career seems to indicate that his children were not hypostases of his nature but rather an enlargement upon it, and such a relationship is well in accord with Asclepius' position. Since originally he was only a hero, it was necessary that as a newly

---

op. cit., p. 165; that Epio is identical with Epione, as suggested in Pauly-Wissowa, II, p. 1656, 53, cannot be proved. As regards Acesis, Euamerion, and Telesphorus, cf. below, n. 50.

[45] In the oldest hymn to Asclepius, Aceso is not mentioned (T. 592), while her name appears in a later copy of the hymn (T. 592a) and in the hymn of Macedonius (T. 593). On the other hand, she is referred to in a sacrificial ritual of the fourth century B. C. (T. 515). Generally speaking one has the impression that she was somewhat neglected and less important than her sisters (cf. Wilamowitz, *Griech. Verskunst*, 1921, p. 353, n. 3, who explains the epithet πολύλλιτος in T. 592a as "Entschädigung für frühere Übergehung"). Maybe in certain cases she was omitted because she was identical with Iaso, as Usener, *Götternamen*, p. 165, suggests. It is noteworthy that the Athenians apparently were not very fond of Aceso: no ship was named after her, whereas Iaso, Panacea, and Hygieia occur as names of Athenian ships; cf. Kutsch, *Attische Heilgötter*, p. 33.

[46] Nothing is known about Aceso except her relation to Asclepius. Names similar to hers in connection with medicine have been collected by Usener, *Götternamen*, pp. 158 f.; 164. An Epidaurian descent seems indicated by the fact that a god Acesis was worshipped there later; cf. below, n. 50.

[47] Concerning Iaso in general, cf. Pauly-Wissowa, *s. v.* Iaso, IX, p. 758.

[48] Concerning Panacea (T. 284), cf. Thraemer, Roscher, *Lexikon, s. v.* Panakeia, III, 1, pp. 1482 ff.; here the centers of her cult and her possibly Rhodian origin are discussed. In the temple of Amphiaraus she had an altar like Iaso, Amphiaraus' daughter (cf. Pausanias, I, 34, 3). Thraemer, *E. R. E.*, VI, p. 551, says rightly that she is "the outstanding iatric figure of the group . . . a personification of the popular notion of the miraculous all-healing herbs already mentioned in connection with Cheiron and Herakles. . . . In the ancient oath of the physicians she alone—as a healer—is contrasted with Hygieia."

created special god he should possess the divine " healing power " and the " universal remedy." [49] Iaso, Aceso, and Panacea symbolized this extension and intensification of the might of the hero of old. That is why they accompanied their father, as the apprentices accompanied the human doctor. Later, when people had become accustomed to the fact that Asclepius, the god, was ever-present in his temples, his descendants were gradually forgotten. Asclepius alone, then, represented the healing power; in fact he finally presided over other minor deities who mediated between him and men.[50]

That the connection between the god and his children originally signified an extension of his power is most clearly brought out by another instance, his unchanging association with Hygieia. She is the one daughter who is " worth as much as all the others " (T. 282, *22*) and later she precedes all her brothers and sisters; sometimes she is even called the wife of Asclepius (T. 601). She seems to have been connected with the god around 400 B. C., for she is not mentioned in the oldest testimony about Asclepius' family which dates from the middle of the fifth century B. C.[51]

[49] In the same way, I think, the fact that Iaso is connected with Amphiaraus must be explained. Originally he was a warrior and seer; his daughter Iaso signifies that he had become a healer. Or again, Ares, " the one who does damage " (cf. Kern, *Religion*, I, p. 118), according to literary as well as archaeological testimonies, is surrounded by a number of deities, sometimes called his children, by Fear and Terror, and others, personifications which apparently complement the original concept of this " special god " (cf. Kern, *Religion*, I, pp. 120 f.); the retinue of Aphrodite also consists of personifications that enlarge on the goddess' own power.

[50] Cf. below, n. 54. *Euamerion, Acesis* and *Telesphorus* are usually counted among the children of Asclepius (cf. *e. g.* Pauly-Wissowa, II, p. 1656, 62 ff.; Usener, *Götternamen*, pp. 170-71, judges differently). However, *Telesphorus* is the only one who in ancient tradition is once expressly referred to as son of Asclepius on an inscription (T. 287; cf. F. Schwenn, Pauly-Wissowa, *s. v.* Telesphorus, V(2), p. 389, 30 ff.). The three who are attested only in late centuries seem to have been but one deity, worshipped under different names in different temples. *Euamerion* is known only from Pausanias (T. 749). Venerated at Titane in the temple of Asclepius along with many other deities he was an originally independent healing deity with a functional name, probably meaning " the giver of good and healthy days " (cf. Usener, *op. cit.*, p. 170; Wilamowitz, *Glaube*, II, p. 472, n. 1). According to Welcker, *Götterlehre*, II, p. 739, the name indicates that Asclepius himself was ἥμερος (cf. Pauly-Wissowa, VI, p. 838, 18 ff.). *Telesphorus*, created perhaps at Pergamum and later adopted also by other cities (cf. Schwenn, *loc. cit.*, p. 387, 39 ff.; Wilamowitz, *Glaube*, II, p. 471, without reasons and against Pausanias' express statement [T. 749] claims Telesphorus for Epidaurus), was a god in his own right, he too a " special god " (Usener, *op. cit.*, p. 171: " der die Heilung bringende Gott "; Boeckh, however, according to Welcker, *Götterlehre*, II, p. 740, connected the name with τελεταί, a not improbable interpretation in view of the late mysteries; cf. below, p. 213, n. 21). Of *Acesis* nothing is known except his name, which apparently is functional, and the fact that he was worshipped at Epidaurus (T. 749; cf. Pauly-Wissowa, *s. v.* Akesis, I, p. 1164). Since he was identical with the other two, as Pausanias says, he too must have been an independent deity, an inferior healing god, later related to Asclepius, the supreme medical divinity.

[51] That Hygieia is an Athenian creation has been suggested by Usener, *Götternamen*,

Why was she associated with the god? Asclepius, certainly from the fourth century B. C., was venerated not only as the healer of diseases but also as the giver and preserver of health, as the hymns written in his honor clearly show. The worship paid to him by the inhabitants of whole cities at regular intervals, by those who enjoyed perfect health, was directed toward the god who keeps men healthy rather than toward him who restores them to health.[52] The twofold activity which was thus attributed to the god reflects the twofold activity of the human physician who was supposed to take care not only of the sick, but of the healthy as well; health no less than disease, in the opinion of these generations, required medical supervision.[53] The god of medicine, the deified physician, therefore, was obliged likewise to take care of both health and sickness. Since originally he had been only a healer of diseases, a savior from deadly illness, Hygieia was made his daughter, and he retained her as his associate. Only in this way could the concept of the preserver of health become part of his own godhead.[54]

p. 167; his theory is upheld and supported with additional arguments by Tambornino, Pauly-Wissowa, s. v. Hygieia, IX, pp. 93 ff. Thraemer, Roscher, Lexikon, I, 2, p. 2776, maintains that Hygieia is an emanation of Asclepius and had her origin at Titane, although she was worshipped in Epidaurus at a relatively early time; cf. E. R. E., VI, p. 551. Yet Thraemer's reasoning is not conclusive and does not disprove Usener's thesis. Wilamowitz, Glaube, II, pp. 223 f., is satisfied with stating that she too, of course, came from Epidaurus. Farnell, Hero Cults, p. 260, assumes that Hygieia was connected with Asclepius in Athens, though he renounces his earlier opinion (Cults, I, pp. 316 ff.) that she arose as an emanation of Athena Hygieia. Cf. also K. Neugebauer, Die Hygieia von Athen, Forschungen u. Fortschritte, X, 1936, pp. 109 ff.

[52] Cf. below, p. 102.

[53] Wilamowitz' opinion that "die Gesundheit . . . den Arzt überflüssig macht" (Glaube, II, p. 224, n. 1) is erroneous. Concerning the importance of the dietetics prescribed for the healthy, cf. L. Edelstein, Die Antike, VII, 1931, pp. 255 ff. Human physicians were instigated by the belief that health is a balance of humors and activities which is always disturbed and therefore has to be restored constantly. This concept finally led to a division of the medical art into two branches: medicine proper and dietetics. Concerning the influence of the ideal of health on the Asclepius cult, cf. below, pp. 122 f.

[54] Hygieia, I take it, superseded Aegle who is called the youngest of Asclepius' children in an enumeration of the fifth century B. C. which does not yet know of Hygieia (T. 279). That Aegle was added later is obvious from the testimony itself; that she was added as symbol of health before the reception of Hygieia is at least probable. Aegle, who is mentioned only in connection with Asclepius, is a goddess of light, as her name indicates; cf. Usener, Götternamen, pp. 164-65. As such she may personify the light of day which the Asclepius worshippers asked him to let them see forever when invoking him as the giver of health in solemn prayer (T. 592). For the connection between light and health, cf. above, n. 43. Thraemer, Pauly-Wissowa, II, p. 1657, 50 ff., also places Hygieia and Aegle on the same level. On the other hand, one must admit that as goddess of light Aegle may also symbolize Asclepius' wisdom; cf. again above, n. 43. That Asclepius' mother is called Aigle by the Epidaurian Isyllus seems to indicate that the name Aegle had an Epidaurian connotation; cf. Usener, op. cit., p. 164, and Roscher, Lexikon, I, 2, pp. 2776, 20 ff., where reference is made to Asclepius Αἰγλάηρ (T. 623).

To sum up: the god Asclepius is surrounded by a number of other gods and goddesses who are personifications of abstract concepts or symbols of the medical art. No myth which would transpose them into a more personal existence is attached to any of these members of his family, nor is their connection with the god elaborated by mythological fancy; it is a purely abstract relation that exists between them. Just as the god is the embodiment of the ideal physician, so are his children representations of the various aspects of medicine. Moreover, the myth attached to Asclepius himself is indistinct; his name, in the opinion of the ancients, was functional. All these features point to the hypothesis that Asclepius was a " special god," created by men to be in charge of one task alone, that of healing.

3. TIME AND PLACE OF ASCLEPIUS' DEIFICATION

Now that the nature of Asclepius' godhead has been evaluated, the question remains where and at what time the god of medicine was first admitted to the Greek pantheon. Of course, such a problem cannot be decided with certainty, but it should be possible on the basis of the known data to ferret out the place from which the god originated and to suggest an approximate date of his deification.

The ancients themselves, agreeing as they did that Asclepius had been elevated to the rank of a deity either by the gods or by his human admirers, had very definite opinions at least concerning the time at which his deification had taken place. Those who believed that he had ascended to heaven immediately upon his death computed that he had become a god shortly before the Trojan War. Some gave as the precise date the year 1237 B. C. and held Asclepius' apotheosis to be contemporaneous with that of Heracles (T. 129).[1] Others made Asclepius the senior of Heracles in his elevation to heaven (T. 265).[2] A few authorities were less explicit. They simply maintained that Asclepius had been a god already in the time of Homer (T. 255).[3] Those, on the other hand, who

---

[1] For the interpretation of T. 129, cf. above, p. 3, n. 3.

[2] As a curiosity it should be noted that according to one Christian writer Asclepius proclaimed himself god during his lifetime (T. 8). This statement is perhaps made in analogy to what was reported about the physician Menecrates who styled himself Zeus (Athenaeus, VII, 289a ff.). But Empedocles too (*Fr.* 112 [Diels-Kranz]) had spoken of himself as a god. There may have been many variations concerning the time of Asclepius' apotheosis, depending on the lifetime ascribed to him, cf. above, pp. 2 f. The Christian Fathers, in their wish to prove that the Jewish Prophets were older than the Greek sages and gods and that consequently Asclepius was of a very recent date, put him long after Moses (T. 130 ff.).

[3] Pausanias (T. 255) quotes *Iliad*, IV (T. 164) and takes the words φῶτ' Ἀσκληπιοῦ υἱόν as meaning "mortal son of Asclepius," thus implying that the father was divine.

saw in Asclepius' deification the work of men, naturally allowed some decades, if not centuries to have elapsed in which his reputation could gradually spread and the greatness of his deeds could become apparent to all, until eventually men began to revere him as a deity. Thus Apollodorus insisted that in Homer's time Asclepius was not yet recognized as a god and that he was regarded as such only afterwards (T. 257).[4] It seems that the adherents of such a theory did not try to define more precisely the period in which Asclepius became a god. The testimonies preserved, at least, furnish no data in this respect.

Modern scholars, as far as they believe that Asclepius was an aboriginal deity, have no reason to discuss the date of his elevation to godhead. For them, the problem is rather when the god Asclepius came to be known to all Greeks.[5] The few, however, who admit that Asclepius became a god only after having first been a hero, and who consequently must face the issue of his deification, are very vague in their pronouncements in regard to the manner in which this deification was brought about, and to the time at which it took place. It is held that the divine worship goes back to prehistoric centuries, that most likely it originated in Thessaly.[6] Moreover, some interpreters maintain that the Asclepius who was elevated to godhead was a healing hero bound to the earth in which he was buried and helping those who came to his tomb. Others speak of him as a hero who was believed to be dwelling at a certain place to which he had been removed as an immortal spirit while still alive, and where he took care of the sick.[7]

This interpretation undoubtedly is erroneous (cf. Leaf's commentary ad Iliadem, IV, 194; II, 729). Interestingly enough, Pausanias omits the words ἀμύμονος ἰητῆρος (T. 164, v. 194) ; the epithet ἀμύμων is given by Homer only to mortals (cf. above, p. 3, n. 6). That Pausanias cannot have intended to claim original divinity for Asclepius has been suggested above, p. 77, n. 1. The word θεός used by him and also in a similar statement of Galen (T. 245), in late centuries could well be applied to "human" gods. Even the Athenians of the fourth century had no qualms in addressing Demetrius Poliorcetes as θεός (Athenaeus, VI, 253 d). Jacoby, F. Gr. H., II, Kommentar, p. 773, 35, asserts that Pausanias, or rather his authority, was not familiar with any attempt to prove from Homer that Asclepius was mortal. This assumption he seems to make because he thinks that Pausanias' account is derived from a source older than Apollodorus. But this is hardly correct, for Pausanias' version is late, cf. above, p. 73, n. 21. Granted even that Pausanias should have copied a third-century writer (Ister), it does not seem likely that Apollodorus (T. 257) was the first to make use of Homer's testimony in order to show that Asclepius was a human being. The euhemeristic interpretation is much older, cf. e. g. T. 243.

    [4] Note that Apollodorus quotes the words which Pausanias omitted, cf. previous note.
    [5] Cf. above, p. 66.
    [6] Cf. Farnell, Hero Cults, pp. 247 ff.
    [7] Rohde, Psyche, I, pp. 159 ff., was the first to emphasize the difference between those genuine heroes for whose worship a tomb is an indispensable prerequisite (cf. Nilsson, Popular Religion, pp. 18-19; Wilamowitz, Glaube, II, p. 12; Roscher, Lexikon, s. v.

To disregard for the moment the chronological and regional argumentation, is it permissible to assume that the hero Asclepius was a healer of diseases? The analysis of the testimonies has shown that he was the hero of physicians, the patron of their craft. Is there any evidence that he ever had any other function or capacity? As for the first claim that there was an Asclepius whose tomb was approached by the sick, such a grave was not known in early centuries. According to the ancient legend, the hero Asclepius met death, yet his burial place was not mentioned.[8] The few late testimonies that refer to the grave of the Arcadian Asclepius do not concern the son of Apollo and Coronis who became a god. If the Christians speak of the tomb of the Epidaurian Asclepius, one can understand this assertion only as the usual attempt to prove that the heathen deities were but mortals.[9] Otherwise tradition is silent about the burial place of Asclepius. One must conclude, therefore, that Asclepius was never worshipped in the typical fashion of a healing hero, enshrined in his tomb.[10]

Nor is it possible to suppose that Asclepius, like Amphiaraus or Trophonius, with whom he is often compared, was believed to have been removed from human sight, and to live at a certain place where he was healing. Nothing indicates that the ancients ever thought that Asclepius had not actually died but had disappeared while still alive. Nay, the death of the hero is an essential feature of the Asclepius saga; it forms the climax of the legend.[11] Asclepius, then, in spite of all the traits which in

Heros, I, 2, pp. 2482 ff.) and those "pseudo heroes" who were originally thought to have been removed from sight, but were later ranked among the common heroes.

[8] Cf. above, p. 50.

[9] Cf. T. 118 f., and above, p. 50, n. 96. One of the well known enumerations by Christian writers of the graves of pagan gods is given e. g. in T. 117. I should also stress the fact that Epidaurus is referred to only in Ps. Clementine treatises, while Clement himself speaks of the tomb in Cynosura (T. 101), although it is true, he identifies the hero there buried with the son of Coronis. Incidentally, it is not certain even of the Arcadian Asclepius that he was a healing hero.

[10] Wilamowitz, Glaube, II, p. 12, says: "Die Kranken werden sich an dem Grabe [sc. of the heroes] zum Schlafe niedergelegt und den Heros im Traume oder auch ὕπαρ gesehen und gehört haben, wie später den Asklepios, der zwar kein Grab hatte, aber auch gestorben war." This rather cryptic statement can hardly invalidate the reasoning that Asclepius, if he had no grave, could not be worshipped in the same way as those who were revered at their tombs. Later, however, Wilamowitz says (op. cit., p. 231): "Dann hatten Heroen aus ihren Gräbern aufsteigend die Schlafenden beraten, Amphiaraos z. B.; es kann sehr wohl sein, dass Asklepios in Tricca es als Heros ebenso getan hat, der pythische Gott durch den Dichter der Eöe ihn zurückwies." This, as Wilamowitz admits, is a mere hypothesis, and not a probable one according to the testimonies; the judgment about the intention of the Eoee is certainly not correct.

[11] Rohde who likens Asclepius to such gods as Amphiaraus (Psyche, I, pp. 141 ff.; cf. also Wilamowitz, Isyllos, pp. 95; 101) tries to prove (op. cit., p. 320) that from the beginning Asclepius' death through Zeus' thunderbolt meant nothing else but his ele-

regard to his healing power he may have had in common with Amphiaraus
or Trophonius, cannot be compared with them in respect to his original
character. Just as he was not one of the healing heroes, he was not one
of the removed heroes either.

This result is confirmed by one consideration which vitally affects the
two theories under discussion: Asclepius came to be a god who was
worshipped throughout the ancient world. Had he ever been restricted
to his tomb or to a certain dwelling place, why and how should he, and
he alone, eventually have become separated from his regional fixation?
None of those who were revered at their graves ever found worship in
foreign cities or countries. The ancients, too, noticed certain similarities
between Amphiaraus and Trophonius on the one hand, and Asclepius on
the other, but they rightly insisted on the difference that Asclepius was
revered everywhere, whereas the other two were restricted to certain
localities (T. 254; cf. also T. 282, 21).[12] It seems safe to say: if Asclepius had been a healing hero or a removed god, he could not have become
the deity that he was later on. Apart from all the other objections
furnished by the evidence preserved, the simple fact of Asclepius' mobility
militates against the modern views.[13]

Any solution of the problem of Asclepius' deification, then, has to start
from the supposition that the hero Asclepius was nothing but the patron
of physicians. With such a thesis it is consistent that Asclepius had no
grave. The hero worship, paid to him by his fellow craftsmen, did not
center around a tomb, it was not fixed at a certain locality. The ancient
doctors were strolling craftsmen, travelling from one place to another;
they could not revere their hero where he was buried; they prayed to
him wherever they were practicing.[14] Moreover, Asclepius' function as
the hero of physicians implies that he himself had no concern with the
sick. He was thought to be the protector of the physicians; he was
invoked by them to guard their daily welfare. They may also have asked

---

vation to godhead. That this is not the meaning of the original legend has been shown
above, pp. 50 f.

[12] The cult of Amphiaraus, to be sure, was transferred to one other region, but the
legends concerning his dwelling place varied accordingly; it became a matter of debate
where his removal had actually taken place; cf. Roscher, *Lexikon*, *s.v.* Amphiaraos,
I, 1, pp. 298, 52 ff. Needless to say, there was no difference of opinion concerning
the place of Asclepius' removal: there were no reports concerning his disappearance.

[13] Rohde, *Psyche*, I, p.121, n. 1, admits that under the presupposition that Asclepius
was a removed hero or god the later development of the cult would be singular. For
even granted that Machaon and Podalirius, his only parallels, were removed heroes,
their worship in late antiquity cannot be separated from that of Asclepius.

[14] For ancient physicians as wandering craftsmen, cf. above, p. 60. On the worship
paid to heroes of other crafts, *e. g.* Homer or Daedalus, cf. in general Roscher, *Lexikon*,
s. v. Heros, I, 2, pp. 2474 ff.

his support for their difficult task, but it was they who healed the patients, while their hero, leading a vague and shadowy existence in a Beyond that men could not penetrate, did not himself heal or perform miracles. He had no supernatural healing gifts to carry over into his new status as a deity, he did not continue his work only on a higher level, as it were, or in a different form. Although at one time he had been a physician on this earth, his divine power of healing he must have acquired through his deification. Not only did the hero of physicians grow in dignity, he also gained in capacity.

Now there are two factors which must be regarded as ferments that instigated a development which finally brought about the complete reversal of the old concept of the hero Asclepius. The first is the fact that gradually a general change of thought took place with regard to the valuation of hero cults. When the hero Asclepius was first created, heroization was the form in which men expressed their reverence for those whom they considered great and whom they regarded as their benefactors. The Homeric and post-Homeric centuries had no higher honor to bestow than that of a hero. Later, however, the dignity of the hero began to lose importance. Men of ever so small merit were heroized after their death.[15] On the other hand, the belief that great individuals, persons of singular achievements, were able to attain godhead gradually gained momentum. The barriers between god and man became less distinct.[16] Under these circumstances it is not astonishing that Asclepius, the hero of physicians, should have tended to become a god. The only presupposition necessary for such a change in his status was that his accomplishments were considered highly important. Of course, the patron of the cooks, the saint of the heralds were not deified. But the patrons of other arts were elevated to godhead. Homer, the son of a god, as the later saga has it, was himself made a god, because poetry is such a great achievement.[17] The same is true of medicine. The author of the Hippocratic treatise *On Ancient Medicine* states that the invention of the medical art must rightly be attributed to a god because it involves so much skill, and he adds that this is indeed generally done.[18] Thus Asclepius in the course of time was bound to be regarded a god, not merely a hero.

[15] Cf. in general Wilamowitz, *Glaube*, II, pp. 8 ff. In Hellenistic times the word hero no longer carried any distinction, the term "the heroes" eventually came to be an equivalent of "the dead"; cf. Wilamowitz, *ibid.*, p. 19.

[16] Cf. in general O. Weinreich, *Antikes Gottmenschentum, N. Jahrb.*, II, 1926, pp. 633 ff.; Kern, *Religion*, III, pp. 111 ff.

[17] Homer was considered the son of a god as early as in the sixth century, cf. Kern, *Religion*, II, p. 31.

[18] Cf. Hippocrates, *C. M. G.*, I, 1, p. 45, 16 ff. This writer, who believes that medical

On the other hand, it is quite plausible that the hero of physicians soon found worship even outside the circle of doctors. It was customary for the ancients to pay heroic or divine honors to those who were credited with the invention of medicine. The various Greek tribes revered various such figures: in Magnesia, Chiron was worshipped by all the inhabitants of that country as the one who had first practiced the medical art; in Tyrus, the same achievement was attributed to Agenoridas who accordingly was revered as a hero.[19] That Asclepius, too, in the same capacity was venerated in certain sections of Greece is indicated at least by one testimony. The people of Tithorea are said to have honored the Asclepius Archegetes, and it is added that all the Phocians concurred with them in paying tribute to this particular Asclepius (T. 719). The epithet Archegetes can hardly mean anything except that Asclepius, the hero or the god, was revered as the leader of medicine and in Phocis had the same position which Chiron had in Magnesia.[20] The Arcadian Asclepius, however distantly related to the son of Apollo, was likewise praised for

knowledge is the product of right reasoning and observation (*ibid.*, p. 39, 6 ff.), evidently means one of those human beings whose superior wisdom made them equals of the gods.

[19] Cf. Plutarch, *Conv.*, III, 1, 3 (647a). Compare also the hero worship paid to Hippocrates by all Coans (Soranus, *C. M. G.*, IV, 1929, p. 175, 14 f.). The names of Chiron and Agenoridas are the only ones preserved by tradition. Yet it is probable that there were many more such figures. Perhaps the ἥρως ἰατρός, who in many Greek cities was revered as helper in diseases (cf. Usener, *Götternamen*, pp. 147 ff.; Kern, *Religion*, II, pp. 317 f.), was also believed to have been the first to practice medicine, at least at the particular place where he was venerated. Nilsson, *Griechische Religion*, I, p. 507, rightly points out that in all matters connected with medicine people preferred to have recourse to local deities.

[20] The interpretation of "Archegetes" as "founder or leader of medicine" was first proposed by J. Rosenbaum, in K. Sprengel, *Versuch einer pragmatischen Geschichte d. Heilkunde*, I, 1846, p. 148, who as proof referred to Diodorus' characterization of Asclepius (T. 355) as ἀρχηγός (τῆς ἰατρικῆς) καὶ κτίστης; cf. also T. 341. Usually the Asclepius Archegetes is understood as the tribal god of the Phocians, and Farnell (*Hero Cults*, p. 247) even claims that the epithet "Archegetes" in Greek religion was attached only "to god or hero who by some received tradition was accepted as the ancestor of the tribe or stock or who was the leader and settler of the colony either actually or in the popular belief." This is certainly not correct. Archegetes may be any founder or head of a family; cf. Pauly-Wissowa, *s. v.* Ἀρχηγέτης, II, p. 441, 37 ff. Similar to the term Asclepius Archegetes seem such expressions as Διόνυσος ἀρχηγέτης and Διόνυσος καθηγεμών (with reference to the god as leader of the actors; cf. Wilamowitz, *Glaube*, II, p. 375; Preller-Robert, *Mythologie*, I, p. 697, n. 4), or ἀρχηγέτης τῶν μυστηρίων (with reference to the genius of Demeter; Strabo, X, 468). Apollo too is called ἀρχηγός as well as ἀρχηγέτης τῆς φιλοσοφίας (Julian, *Oratio* VI, 188 a-b). Note moreover that no other testimony indicates a tribal function of Asclepius (cf. Kern, *Religion*, II, p. 307), and that according to Pausanias, who mentions the Asclepius Archegetes, the ancestral hero of the Phocians, worshipped by the whole tribe, was Xanthippus or Phocus (X, 4, 10). It may also be of some importance that in Phocis Pausanias saw a statue which some believed to be that of Asclepius, others that of Prometheus (T. 671). Asclepius must have been represented in some way similar to that other culture hero, cf. below, p. 214, n. 2.

his medical inventions (T. 379), and not even the Epidaurians had forgotten that their great healer was also the discoverer of many medical improvements (T. 7). Thus it is likely that the patron of physicians came to have also a regional worship in certain provinces of Greece, that his cult was no longer restricted to the craft alone.[21]

Everything, then, pointed in the direction of Asclepius' becoming a god and being worshipped by the people at large. There was an inner tendency toward his elevation to the god of medicine. Yet, the saint of the craftsmen, even the Asclepius Archegetes, was still not the deity who was revered in Epidaurus, in Tricca, finally everywhere.[22] The latent power of the Asclepius figure needed actualization. Where and when was Asclepius, the god of medicine, created? What was it that eventually brought about a complete change in the position of the hero of old?

As for the place from which the god came—and this in ancient terminology is equivalent to his first place of worship—only two cities can enter into serious competition: Tricca and Epidaurus. The opinion of the testimonies is divided. Common belief was in favor of the Epidaurian origin of the god, while ancient scholars advocated Tricca as his birthplace. His oldest sanctuary was supposed to have been in Tricca (T. 714).[23]

In spite of the latter assertions which certainly cannot be dismissed lightly, it seems that Epidaurus' rights are better founded. The diffusion of the Asclepius cult is closely connected with the temple of Epidaurus and the god, as he was revered later, is the Epidaurian rather than the Triccaean Asclepius.[24] Moreover, there is one other fact which indicates that Epidaurus must have had some importance for Asclepius which no other city had. Delphi approved of the Epidaurian claim that Epidaurus

[21] The Phocian worship of Asclepius Archegetes cannot be dated with certainty. But it is probable that it goes back at least to the sixth century B. C. For Pherecydes claims that Asclepius practiced and died in Delphi (cf. above, p. 50, n. 95). This statement sounds like the rationalization of an older Phocian saga, and certainly not of a saga celebrating the settler of a colony or a tribal god and hero. Pherecydes may well have known of the Asclepius Archegetes and may have been influenced in his account of the hero Asclepius by the myth attached to the Phocian hero, just as Apollodorus (T. 5) in his report on Asclepius' practice in Tricca and Epidaurus obviously is influenced by the divine myth. The Phocian worship, then, seems older indeed than that of the god of medicine and should probably be placed between the original hero cult and the divine cult of later centuries.

[22] In late centuries, of course, the patron of physicians and the Asclepius Archegetes were identified with the god of medicine, cf. below, pp. 139 f.; 248.

[23] For the interpretation of the scholarly tradition concerning Asclepius, cf. above, p. 71, n. 13. That the assertion of Strabo concerning Tricca's rôle (T. 714) is not borne out by the evidence will be shown in the discussion of the history of the temples and of their affiliation, cf. below, pp. 238 ff.

[24] Cf. below, pp. 238 ff.

was the town in which the god was born, whereas the Oracle is not known to have approved of the Triccaean contention, or even to have considered it.[25] The judgment of the ancient scholars, then, seems to be the result of historical pedantry. From Homer they concluded that Machaon and Podalirius, the sons of Asclepius, came from Tricca. The Triccaeans too said this and showed the cenotaph of their heroes (T. 161).[26] Once the god had been established they probably were not slow in claiming that he was rightly theirs. And if the data of the religious development were not in accord with their pretensions, historical inquiry decided that they, certainly not the Epidaurians, had Homer on their side.[27] It was a kind of compromise when Apollodorus made Asclepius practice at first in Tricca and Epidaurus (T. 5), as if the demands of the two cities on Asclepius were equally well founded. But taking the whole evidence into consideration, it appears more likely that the god Asclepius was created in Epidaurus.[28]

While the problem of the god's place of origin can be solved only with a high degree of probability, the time of Asclepius' deification can be determined with certainty, I think. For it is safe to say that Asclepius did not become a god until the end of the sixth century B. C. The excavations at Epidaurus have shown that the buildings dedicated to Asclepius are not to be dated earlier. It is impossible to trace the Thessalian cult beyond the fifth century B. C. Nowhere else is there any sign of an older worship of the god Asclepius.[29] That this is so squares well with the fact

[25] Cf. above, pp. 69 f. The lack of any Delphic statement concerning Tricca is the more astounding since the Oracle at first had been in agreement with the claim that Asclepius was born in the Dotian plain; cf. above, p. 68.

[26] Cf. above, p. 18.

[27] Wilamowitz too explains the decision in favor of Tricca as a consequence of the Homer exegesis (*Glaube*, II, p. 228).

[28] Cf. also below, n. 32. I side, then, with Wilamowitz who characterizes Asclepius as the god of Epidaurus (*Glaube*, II, pp. 227 ff.). Farnell (*Hero Cults*, pp. 247 ff.) believes that Tricca is the " original home " of the Asclepius cult, and admits Epidaurus' importance only from the fifth century on (*ibid.*, p. 254). But nothing definite is known about any earlier divine cult, as will be shown immediately. If the question were asked why the Epidaurians should have chosen a Thessalian hero—for in the beginning they acknowledged that Asclepius was born in the Dotian plain (cf. above, p. 69)—one might well answer that the hero Asclepius was known to everybody. Moreover, in the sixth century the saga had already established a connection between Asclepius and Argos: many of his " patients " were of Argive descent, cf. above, p. 41.

[29] As regards Epidaurus, cf. Wilamowitz, *Glaube*, II, p. 227; Kern, *Religion*, III, p. 155, and below, p. 243. Even Farnell admits that nothing is known that would prove a Triccaean cult before the fifth century B. C. (*Hero Cults*, p. 247), and yet he claims (*ibid.*, p. 254) that the Asclepius cult came to Epidaurus from Tricca " at some period after the date of Hesiod." It is true, the Triccaean sanctuary has not yet been excavated. The possibility of surprising new finds cannot be excluded. But the extant evidence unanimously points in the same direction. For the history of the sanctuaries in general, cf. below, pp. 244 ff.

that around 600 B. C. Asclepius had become the patron of physicians who from that time onward were called after him. Such a rôle must have increased the interest in Asclepius, and it was an easy step to elevate him, who had attained so much fame, to even greater honors, to the status of a god.

But how did it come about that the hero Asclepius gained divine prerogatives? Any solution of such a problem must necessarily be conjectural; yet a conjecture at least can be attempted; it is almost urged upon the interpreter by one striking fact characteristic of the beginnings of Asclepius' career as a medical god. In Tricca no less than in Epidaurus, Asclepius at first appeared together with Apollo Maleatas, one of the many Apollo figures worshipped in antiquity.[30] On him Asclepius was so dependent that even the famous Epidaurian miracles (T. 423) are still recorded in the name of Apollo and Asclepius. In Tricca the worshippers sacrificed to Apollo Maleatas before entering the sanctuary of Asclepius (T. 516).[31] It seems to follow, therefore, that Asclepius, the hero of doctors, the son of Apollo the physician, became himself a god capable of healing sickness through his connection with Apollo Maleatas, a medical deity who healed the sick in their dreams.[32]

In more technical terms: Asclepius began his divine practice as a coadjutor or assessor of another god. With such a hypothesis the fact that Asclepius was a demigod or a " daimon " is well in accord; for thus it was customary to address the coadjutors of great gods.[33] Moreover,

[30] Apollo Maleatas, whether he was a fusion of an originally independent Maleatas with Apollo (Wilamowitz, *Glaube*, I, pp. 393 ff., especially p. 397), or whether he was only a specific form of Apollo (Farnell, *Cults*, IV, pp. 235 ff., especially pp. 238 f.), certainly was an iatric god. This is generally admitted.

[31] In Athens too initial sacrifices were made to Apollo Maleatas (T. 515). For the rôle of this god in Epidaurus, where the association between Asclepius and Apollo Maleatas lasted throughout antiquity (T. 424), cf. Herzog, *Wunderheilungen*, p. 46. Note also that many other Asclepius temples were built over the foundations of Apollo sanctuaries; cf. below, p. 243, n. 3; p. 246, n. 16.

[32] In Epidaurus in the fifth century B. C. Asclepius had only an altar and an *abaton* in the temple of Apollo (cf. G. Karo, *Weihgeschenke in Epidaurus*, 1937, p. 1). The fact that Asclepius, when he first started out, was connected with Apollo Maleatas is one more argument in favor of the Epidaurian origin of the god. Nowhere was the relationship between the two divinities closer than in Epidaurus (Farnell, *Cults*, IV, p. 239; *Hero Cults*, p. 254). Incidentally, Isyllus' statement (T. 516) that *even* in Tricca Apollo Maleatas received sacrifices, in my opinion, supports the belief that Asclepius was created in Epidaurus. Pointing to the importance of Maleatas even for Tricca, Isyllus seems to indicate that Asclepius' connection with Maleatas was of decisive importance and that also those who contested Asclepius' Epidaurian origin admitted that much. Religious arguments are here opposed to historical ones as they were derived from Homer.

[33] For Asclepius as a demigod or "daimon," cf. above, pp. 83 f. Wilamowitz, *Glaube*, I, p. 366, says: "Im Anhange der Theogonie 991 wird Phaëton als Tempelwächter

such a hypothesis explains how the new god could become a divinity revered everywhere. The connection between Apollo and Asclepius, being only a cult-association, could be repeated wherever Apollo Maleatas, or Apollo, was worshipped. Asclepius, as soon as he had risen to equal rank with the one who originally was his superior, could likewise go everywhere, even by himself. He may have had favorite temples, but so had other gods. Nothing impeded his mobility; he could extend his power over the whole earth.[34]

In the religious language of the ancients such a connection as that between Apollo Maleatas and Asclepius might have been understood as one made by the superior deity himself. Thus Athena was believed to have established Erechtheus in her temple, thus Aphrodite established Phaëton in her sanctuary.[35] The modern interpreter is interested in finding a reason why the two, Apollo Maleatas and Asclepius, should have been related, and such a reason is not far to seek. Apollo Maleatas was an iatric Apollo. As such he must have grown obsolete in the course of time, for the various local Apollo cults could not remain aloof from the Delphic movement which gradually changed the former concept of Apollo with the result that the god lost his interest in medicine, at any rate in his personal function as healer of diseases.[36] Yet the god, or his priests, naturally had the tendency to retain the domain over which he once reigned, even if he himself could no longer be in charge of it. Therefore he needed an associate to do the job, while he remained the titular head of the sanctuary. He sought a helper who in his stead would take over the task of healing diseases. Is it so astonishing that he should have chosen Asclepius, his son, the great physician, the patron of doctors? Asclepius, to be sure, had died, no one could tell where his remains were to be found. But should the greatest of all doctors be really dead, should

Aphrodites zu einem δῖος δαίμων. Das setzt voraus, dass πάρεδροι grosser Götter als Dämonen bezeichnet werden. Die Absonderung der Olympier drängte auf eine solche Unterscheidung," and he adds (p. 367) : " Es mag Zufall sein, dass ich aus alter Zeit die Bezeichnung Dämonen für die πάρεδροι eines Gottes nicht belegen kann." The figure of Asclepius is perhaps such an example of relatively early time; for he was called " daimon " from the beginning. For πάρεδροι in general, cf. Roscher, Lexikon, s. v., III, 1, especially p. 1573, 11 ff.

[34] This explanation, then, does not encounter the difficulties of the older theories. Concerning the introduction of healing cults in various cities, cf. below, p. 119.

[35] Iliad, II, 546 ff.; Hesiod, Theogony, 987 ff.; cf. Rohde, Psyche, I, pp. 135 ff. Phaëton is expressly said to have been removed by the goddess when still alive. But after the fourth century B. C. even mortals, immediately after their death, were worshipped as πάρεδροι, cf. Roscher, loc. cit., p. 1577, 40 ff. It is not improbable, therefore, that in earlier centuries heroes were chosen by the gods as coadjutors.

[36] Concerning the renunciation of medicine by the Delphic Apollo, cf. Wilamowitz, Glaube, II, p. 39, and above, p. 36.

he no longer be able to help, should he be approachable no more, should he not be revived and restored to men by the gods?[37] Certainly, it is quite understandable that it was Asclepius who became an associate of his father, his coadjutor.

4. THE RANGE OF ASCLEPIUS' POWER

The hero of physicians, having been elevated to divinity, assumed the task of healing the sick. In his sanctuaries he cured those who were willing to be his patients. As to the manner in which he showed his power, there can be no doubt: the divine healer performed miracles. The diseases which he treated vanished overnight; the sick, having slept in the temple, awoke free from illness. Yet miraculous as the effect of the god's cures may have been, his procedure was not supernatural at all; it resembled that of a human physician. In dream-apparitions the god cut veins, he operated on his patients, he applied remedies. That he did at least in the beginning. Later on he restricted himself more and more to the rôle of a consulting physician, a physician of superhuman knowledge to be sure, but a physician nevertheless. In oracles he advised the treatment which had to be carried out by his patients; in dreams he prescribed drugs which the patient himself or his doctor was supposed to administer.[1]

The cures of the god, then, though miraculous, his precepts, though beyond human wisdom, followed the pattern of rational and empirical medicine. The power of Asclepius was now divine, but in its nature his ability was not different from that which he had had when still on this earth. This fact distinguished the treatment of Asclepius from that provided by all the other healing deities. They, too, cured the sick, but nobody knew how they did it. Their accomplishments were outright witchcraft which could not be checked. Asclepius' medicine differed from human medicine only in so far as the god was infinitely better versed in his art and more successful than any human doctor, he was the god of "lofty art" (T. 593a).[2]

In the beginning Asclepius' power had two important limitations. For quite some time the new god was still so deficient that he could not perform his task all by himself, he depended on the might of his father,

---

[37] In this connection the assertion of the divine myth that it was the gods who revived Asclepius, and more specifically that it was Zeus who did this as a favor to Apollo (cf. above, pp. 75 f.), takes on a pregnant meaning.

[1] For a detailed interpretation of temple medicine, cf. below, Ch. III.

[2] What the hymn expresses by this solemn epithet (ὑψιτέχνης), Aristophanes brings out by calling Asclepius an "ordure-taster" (σκατοφάγον, T. 421, v. 706), an appellation often used for the human practitioner. Cf. also below, p. 135, n. 10.

although the father was participating in the deeds of his son only in name. It was relatively late that Asclepius grew into manhood and took his place alone, acting by himself without the mediation of his father.[3] Moreover, at first the power of the divine Asclepius seems to have been such that he could not treat at two different places simultaneously. When he was in Epidaurus, the patients in Troizen had to wait; so at least the Epidaurian tablets say (T. 423, 23). In such instances his sons sometimes tried to handle the cases in his absence; yet they failed completely. Though the god could not be everywhere at the same time, he was the only one who could do the job right, and therefore he had to do it himself.[4] Fortunately, as time progressed, he learned how to take care of his many patients at widely separate localities and to be able to satisfy the demands made upon him, or even stimulated by him, for he was wont to call his patients to his temples.[5]

While accomplishing these feats the healer of diseases also rose to the position of the giver and preserver of health. Wherever he came, heart-felt prayers were offered to him as the one who granted men to live free from sickness. The husband prayed for his wife, the parents for their children, the friend for his friend. But also the city as a whole revered Asclepius as the god who gave health. Solemn hymns were sung to him who kept the people sound in body; festivals were celebrated in honor of him who did not allow diseases to overcome men.[6]

It was in this twofold capacity as healer of diseases and giver of health that Asclepius finally extended his practice all over Greece and over the whole ancient world. The many other healing deities bowed before his might; he became the unchallenged master of the province of medicine. Once Asclepius had gained this position, the range of his activity did not basically change. Up to the end of antiquity he was mainly revered as the god of healing dreams and oracles.[7] But it is only natural that in the long course of his reign different aspects of his power were accentuated. Changing concepts of the divine and of the world modified the understanding of the gods in general and thus also of Asclepius' influence. If

[3] Cf. above, p. 99. But there remained traces of his original dependence on Apollo, cf. below, p. 183, n. 6.

[4] As has been shown above, p. 86, in the beginning Asclepius used to act together with his children. T. 423, *23* is interesting because it indicates that the power of the god was nevertheless thought to be supreme. Note that the historian, Hippys, rationalizes the story and speaks of the servants of Asclepius (T. 422); cf. also below, p. 160, n. 12.

[5] Cf. below, p. 112.

[6] For Asclepius, the giver of health, cf. below, pp. 182 ff.

[7] It should be remembered, however, that this same god throughout antiquity was also the protector of physicians and of laical medicine, cf. below, pp. 139 f.

the final form, in which the Asclepius religion emerged, is to be grasped, it is necessary to follow up this historical process in which the full import of the Asclepius figure was unfolded.

The first consequence of the new position which Asclepius occupied was, of course, that he was brought into contact with the life of the various cities in which he dwelt. Soon after the god had been received by all Greeks, the Epidaurian Isyllus attempted to make Asclepius together with Apollo the guarantor of the new constitution, of the peace and welfare of his city (T. 296).[8] Isyllus, the prophet of Asclepius, also saw his god going in arms to the aid of the Lacedaemonians, the admired friends of Isyllus. Overjoyed by his vision, Isyllus told the people of Sparta that Asclepius had come to save them, and trusting in his message they forthwith revered Asclepius as their Savior (T. 295).[9] In this instance Asclepius acted as the ancient gods and heroes used to do in behalf of those who were under their protection. But only once did he trouble to play such a rôle, to accomplish such deeds, and then only for the sake of his beloved Epidaurians and of his friend Isyllus. In general he partook only in small matters of civic life. He was satisfied if documents of importance were deposited in his shrine (T. 794a), if he contributed to the dignity of the community through the right of asylum that was connected with almost all the great Asclepieia (e. g. T. 798), if he was in charge of the manumission of slaves (T. 718a).[10] It was the character of his godhead that excluded him from political strife. His was the work not of war, but of peace. Heracles was quite right in saying of Asclepius contemptuously that he did not show "any bravery" (T. 265).[11]

[8] For a detailed interpretation of Isyllus' political aims, cf. Wilamowitz, Isyllos, pp. 38 ff.

[9] Concerning the event referred to by Isyllus (Philip's attack on Sparta in 338 B. C.) and the historical situation in general, cf. Wilamowitz, Isyllos, pp. 24; 30 ff. S. Schebelev, according to E. Bickermann, Philol. Wochenschrift, 1930, p. 242, dates the inscription 218 B. C., but his arguments are not convincing. Manifestations of gods or heroes as saviors of their peoples were again experienced in Isyllus' time (cf. Nilsson, Die Griechen, p. 381).

[10] Cf. Farnell, Hero Cults, p. 248; such manumissions are known from Tithorea, Orchomenos, Thespiae, Stiris, Elateia, Amphissa, Naupactus. The practice followed was identical with that applied by Apollo in Delphi (cf. Farnell, ibid.), and it is by no means necessary to see in this function of the god "the tribal memory of ancestral kinship with Asklepios" (ibid., pp. 248-9; concerning the Asclepius Archegetes in Tithorea, cf. above, pp. 96 f.). The importance of the Asclepius worship in general is quite sufficient for an explanation of the phenomenon in question. It is noteworthy that on the Tiber island sick slaves were freed when their masters did not support them (T. 858). In the Greek temples too there may have been a connection between the medical activity of the god and manumission.

[11] I wish to stress once more that the account of Asclepius' activities as given by Isyllus is singular. It is well understandable that he should have tried to ascribe in-

While thus the political functions of Asclepius remained limited indeed, what he did for the family was all-important. He was the patron and savior of the household. It is Aristides who most eloquently praises him as such (T. 317, 5), but his words, no doubt, express a feeling that was generally prevalent among men. And although this aspect of the Asclepius worship is especially characteristic of late centuries, it must have been of importance from relatively early times, as is evident from the prayer of the Coan woman in Herondas' mime (T. 482, *vv. 83 ff.*). The helper in disease has a natural affinity to the life of the family, for the family no less than the individual suffers from the burden and dangers of diseases, is troubled over their outcome and rejoices over their healing. And still another consideration played its part: since Asclepius gave health, he also gave healthy children, and thus immortality on this earth (T. 317, 5). It is the continuation of life that he guarded. Generally speaking, it seems permissible to state that Asclepius became the protector of the family. Gradually he usurped the place which in the classical period Zeus held, but relinquished when in Hellenistic and Roman centuries he was left in charge only of the royal household and of the empire at large.[12] As the god of the family, Asclepius naturally was concerned with every little detail of daily life. It is in this sense that Aristides can say that Asclepius, if any of the gods, was the one to whom one could pray, and did pray, in regard to everything (T. 316).[13]

In the position which the god had gradually acquired for himself, he could not afford to neglect one domain to which as a physician he had a birthright anyhow: that of giving oracles. Men are curious to know the future. For Greeks and Romans in particular foretelling and fore-knowledge had great attraction. He who was to be their helper had to be able to satisfy their demands even in this respect. On the other hand, to the ancients, medicine and mantic were closely connected (*e. g.* T. 366); the physician could best prove his superiority by giving a

terest in political affairs to Asclepius, for Isyllus was a statesman and Asclepius the god in whom he believed. Moreover Epidaurus was the town sacred to the god above all others (T. 7). And yet even here such endeavors at reinterpreting the nature of the god were doomed to failure. Only in matters pertaining to medicine did Asclepius show if not political, at least social consciousness, cf. below, pp. 173 ff.

[12] For this change in Zeus' position, cf. K. Ziegler, Roscher, *Lexikon, s. v.* Zeus, VI, pp. 696, 51 ff. This book was already in print when I found that Nilsson, *Griechische Religion*, I, p. 378, has drawn attention to an inscription of Panamara (*Bull. Corresp. Hell.*, XII, 1888, p. 269, n. 54), according to which Zeus Ktesios, Tyche and Asclepius were revered as ἐνοικίδιοι θεοί.

[13] Some examples of the god's helpfulness are his support in distress at sea (T. 317, *10*; cf. also the inscriptions of Syros [Pauly-Wissowa, II, p. 1672, 47] and Thasos [Gruppe, *Mythologie*, p. 1442, n. 11]), and his protection during travels on land, cf. below, p. 190. Yet Asclepius did not disdain even to give advice to a boxer (T. 317, *11*).

prognosis.[14] It is therefore not surprising that the divine physician should have predicted the future or revealed that which was hidden, especially after he had changed his cures from the kind in which he actually performed the treatment to the kind in which, often in the form of an oracle, he only advised the right means of proceeding in a given case. Yet, some of the Epidaurian inscriptions attest even oracles of a non-medical character (T. 423, 24).[15] In the second century B. C. it is expressly noted that Asclepius superintended divination and augury (T. 301).[16] Later, the fact that Asclepius was the interpreter of the future, seemed a matter of course to Aristides (T. 317, 10). According to Celsus, the oracular activity of the god extended even to whole cities, especially to those which were dedicated to him, like Tricca, Epidaurus, Cos, and Pergamum (T. 292; cf. T. 294). The evidence, then, is quite sufficient to warrant Lucian's assertion (T. 291) concerning an oracular Asclepius who was consulted for general purposes.[17] Nor were the gods justified in enjoining upon the giver of oracles, as they did according to Lucian, that he should not overstep the realm allotted to him. He did only what by his very nature he was privileged to do.

In the same way, finally, the cosmic significance which Asclepius took on was strictly within the boundaries of his own province.[18] Not that great importance attaches to Pausanias' assertion, according to which Asclepius was identified with the air because of its usefulness (T. 297). True, Pausanias claims that this was no less a Greek conviction than it was a Phoenician one, but it is obvious from his own words that he only wishes to refute the reproach that the notions of the

[14] In general, cf. Edelstein, *Problemata*, IV, p. 85, and P. Friedländer, *G. G. A.*, 1931, pp. 251 f.

[15] In general, cf. Herzog, *Wunderheilungen*, pp. 112 ff. and *ibid.*, nos. 46; 63.

[16] That Apollodorus' statement does not concern medical oracles alone has rightly been stressed by Jacoby, *F. Gr. Hist.*, Kommentar, II, p. 770, 1 ff. (*ad* 244 F 116). Thraemer, Pauly-Wissowa, II, p. 1656, 13 ff., is wrong in giving the impression that Apollo restricted his son to medical oracles.

[17] Farnell, *Hero Cults*, p. 246 b, says: "The vague phrase in Lucian, *Div. Concil.* 16, suggests an oracular Asklepios consulted for general purposes; but I am not aware of any recorded example except the Epidaurian inscription." But, as has been shown, there are many more testimonies of this kind preserved. Rohde's evaluation of the oracular Asclepius is more in agreement with the facts (*Psyche*, I, p. 141, n. 3). Incidentally, the Asclepius of Memphis too gave oracles, cf. L. Deubner, *De Incubatione*, 1900, pp. 36 f.

[18] As far as I am aware the cosmic influence of Asclepius is the last aggrandizement of his power to be considered. Thraemer (*E. R. E.*, VI, pp. 551-52) supposes Asclepius also to have been in charge of agriculture and gymnastics; but such a claim is based on a misunderstanding of Plato, *Symposium*, 186 E. Eryximachus finds in gymnastics and agriculture the same principles that hold good of medicine; he does not say that these departments are under the care of the god of medicine. For Asclepius, the god of the literati, cf. below, pp. 206 ff.

Phoenicians in regard to the nature of the gods were superior to those of the Greeks. Such an opinion was not typical and never generally accepted; it is indicative of the religious syncretism of late centuries.[19] Nor is it of any consequence that in testimonies of the fourth and fifth centuries A. D. Asclepius is equated with the sun (T. 298 ff.) or, in union with Hygieia, with the sun and the moon (T. 301). Although even here the influence of the two heavenly bodies on health as well as on sickness is stressed, these statements are conventional and determined by the common solar theology of the dying pagan world.[20] It was from his position as the protector of medicine that Asclepius rose to a peculiar rôle, that he became a vitally essential cosmic factor. For he was thought now to be in charge of the health of the universe, he was believed to be the physician of the whole world. He was acclaimed as the one " who guides and rules the universe, the savior of the whole and the guardian of the immortals, . . . he who saves that which always exists and that which is in the state of becoming " (T. 303).

Such an interpretation reflects a belief, a philosophical attitude, which was very common at least from the early Christian period on. At that time there was a general tendency to remove the highest god from this world; he was supposed to have no direct influence on it; the lesser gods, created by the supreme Being, were those who acted and who protected the world of phenomena. On the other hand, the perceptible world, separated from the true god, was not self-sufficient; it could not subsist without permanently being guarded and supervised. Otherwise it would grow old, become ill, even be destroyed. The cosmos, then, like the human body was held to be subject to disease; its health was apprehended as a harmony that has to be restored constantly, just as the health of the human body which consists in the harmony of the humors has to be reëstablished ever anew. In such a system the divine physician naturally had an essential function. Of necessity he, rather than any other god,

---

[19] The Pausanias passage is corrupt so that the details of his argument remain obscure, but its general meaning is quite clear: in proof of his assumption he can refer only to the temple at Titane where the same statue, it seems, was supposed to represent the sun and Hygieia. In the nineteenth century Pausanias' interpretation was favored as the true explication of Asclepius' nature; cf. e. g. Panofka, *Berl. Phil.-Hist. Abh.*, 1845, p. 271. In Preller-Robert, *Mythologie*, I, p. 514, Asclepius is still characterized as the divine representative of the healing power of nature, though Welcker, *Götterlehre*, II, pp. 746 f., long before had correctly evaluated the philosophy behind Pausanias' declamations. For the common ancient interpretation of Asclepius as a deified mortal, cf. above, pp. 77 f.

[20] Deubner, *Die Römer*, II, pp. 502 f., speaking of the late solar theology finds it especially characteristic of Macrobius from whom T. 301 is taken. For the Neo-Platonic interpretation of Asclepius which in spite of certain similarities still differs from that of Macrobius, cf. below, n. 24.

became the patron of the world, the guardian deity of the universe; for he alone was able to preserve the cosmic organism intact.[21]

Moreover, the god himself, of his own free will, had volunteered the information that he was the soul of the cosmos, the one whom Plato called " the soul of the universe " (T. 302). There could be no doubt that this revelation, communicated to Aristides, the most devoted servant of Asclepius, referred to the scheme of the world, as outlined in the Platonic *Timaeus*.[22] Consequently the god, after having taken his place in the Platonic world and after having determined his rôle in his own words, came to be of great importance for the Neo-Platonists, who in their attributions of the various powers to the individual deities paid close attention to what the gods themselves had revealed about their own nature.[23] In the opinion of the Neo-Platonists, Asclepius was in fact the soul of the world by which the creation is held together and filled with symmetry and balanced union (T. 304).[24] Through Asclepius, the

[21] Farnell, *Hero Cults*, pp. 278-9, takes Aristides' interpretation of Asclepius (T. 303) more or less as the rhetor's personal opinion. But this can hardly be correct. Aristides was not an original metaphysician, or rather he was no metaphysician at all. He refers to philosophical speculations only occasionally, cf. A. Boulanger, *Aelius Aristide*, 1923, p. 199. That he talks about Asclepius as he does, immediately raises the suspicion that he is echoing a more general belief. Aristides' views seem reminiscent of the Neo-Pythagorean philosophy and of the Platonism of the first and second centuries A. D.; cf. K. Praechter, *Die Philosophie des Altertums*, 1926[12], pp. 521; 542 f.; 556.

[22] The reference made by the god is to *Timaeus* 34 B. Aristides, in his dream, is accompanied by Pyrallianus, an otherwise unknown rhetor who lived in the temple of the god and was well versed in Platonic philosophy. Concerning the self-revelation of Asclepius to Aristides, cf. Boulanger, *op. cit.*, p. 202. It is noteworthy that, while the Stoics identified Zeus with the cosmos and therefore called him the soul of the world, Aristides names Asclepius thus. For him, Zeus was the creator of the world and of the gods, but he himself was beyond the world. On the other hand, all the lesser gods were emanations of Zeus, and it is in this sense that Aristides' concept of Zeus-Asclepius must be understood (T. 317, 4), not as an identification of Zeus and Asclepius (contrary to Farnell, *Hero Cults*, p. 278). In regard to Aristides' concept of Zeus, cf. J. Amann, *Die Zeusrede d. Ailios Aristeides, Tüb. Beitr. z. Altertumswiss.*, XII, 1931; K. Ziegler in Roscher, *Lexikon, s. v.* Zeus, VI, pp. 701-2. Only in late Orphic theology (T. 315) are the various gods no longer differentiated. The significance of the Pergamene cult of Zeus-Asclepius (T. 317, 4), to which Aristides' concept of the god may be related (cf. Farnell, *ibid.*, and Pauly-Wissowa, II, p. 1661, 59 ff.), cannot be determined; no details about this worship are known. Maybe the people of Pergamum shared Aristides' views, or they may have celebrated their god as Zeus in order to express the high regard in which they held him, cf. above, p. 84, n. 29. For the attestation of Zeus Asclepius in literature and inscriptions, cf. A. B. Cook, *Zeus*, II, 2, 1925, pp. 1076 ff.

[23] The Neo-Platonists collected such oracles, cf. Porphyry, *De Philosophia ex Oraculis haurienda*, ed. G. Wolff, 1856 (T. 13 is another example taken from the same book, cf. *ibid.*, p. 127). Boulanger, *op. cit.*, p. 208, n. 1, was the first to suggest a dependence of the Neo-Platonists on Aristides.

[24] Iamblichus here, in contrast to Porphyry, interprets Asclepius not as lunar, but as solar intellect. These Neo-Platonic identifications of the gods with heavenly bodies are of course different from the physical interpretations referred to above, pp. 105 f.

savior of the whole world, the health and safety of all is guaranteed (T. 306). Through him, the elements do not relax their indestructible bonds; through him, the universe remains young and healthy (T. 309).[25]

Since the Neo-Platonists, progressing along the lines of earlier speculation, stressed even more than did their predecessors the remoteness of the highest god, the frailty and weakness of the world which requires supervision and salvation at every moment, Asclepius' task, to them, was all-important. In their opinion, therefore, Asclepius was a god even before the beginning of existence, a transcendental deity (T. 305; cf. also T. 259), although he was ruling over the phenomenal world, although he was within it. Zeus had engendered Asclepius from himself; but through the sun, through Apollo, he had revealed him to the mundane regions (T. 307). In his earthly appearance Asclepius was the third from Zeus (T. 303). Thus the god of medicine took his place in the pagan trinity.[26] He had risen high indeed. But despite all changes in his influence and in his position, he did not change his nature: he remained the healer of diseases and the giver of health.

5. THE RISE OF THE ASCLEPIUS RELIGION

The cult of Asclepius, who was created a god relatively late, is first to be found in such cities of Greece as were of small influence in the religious or political affairs of the ancient world. Nevertheless the god soon lost his merely provincial character. His worship spread everywhere; it became one of the most renowned among the many ancient cults; it outlasted most of them; it was the hated enemy and dreaded competitor of Christianity. What were the forces that caused this astounding development?

In modern discussions the rise of the Asclepius religion is given scant attention. If an attempt is made at all to explain this puzzling phenomenon, it is restricted to casual remarks; no systematic investigation has ever been made. Among the reasons proposed, that which attributes the ascendancy of the cult to the propagandistic efforts of Asclepius' priests is the most favored. At first migrating with their tribes, later from Epidaurus, the metropolis of the cult, they are supposed to have

[25] The discussion of Asclepius' part in Neo-Platonic philosophy as given here is by no means exhaustive; the problem is too complex to be solved in this context. Note T. 308: relation of Apollo to Asclepius; T. 310; 311: relation of Asclepius to Asclepii; T. 312: Asclepius surrounded by a retinue of associated and subordinate gods; T. 313: Asclepius-Telesphorus; T. 314: demiurgic and Asclepiadic health.

[26] For the importance which this fact assumed in the discussion between pagans and Christians, cf. below, p. 136.

colonized the ancient world for their god.[1] The great success of their propaganda is traced to the fact that the lower classes, those who at all times believe in miracles which are nothing but the deceit of priests, showed a special sympathy and regard for Asclepius. The rise of the new god is thus linked together with the less respectable trends of ancient religious life. Moreover, the belief in Asclepius is thought to be a relapse into old superstitions which more enlightened generations had discarded long ago; it is considered " the beginning of the end," the death of truly Hellenic religion.[2] Some critics grant that the Asclepius worship was capable of " deep and genuine sentiment," but they hasten to add that it lacked " all higher religious values and a really ethical foundation." [3] A very few connect certain tendencies in the Asclepius cult with the general development of late religious thought, thereby giving to the worship of the god a share in higher aspirations; yet, on the whole, these scholars, too, find the god one of the less attractive religious figures, whose power was acclaimed by the vulgar, not by the best.[4] Even the fact that Christianity had to fight against Asclepius more than against any other ancient god is taken as a proof of the imperfection of the Asclepius cult rather than of its dignity. For it was only early Christianity that had to fear Asclepius, only that Christianity which was a kind of " medicinal religion," not yet imbued with those lofty ideals which later were integrated into the Christian faith.[5] Generally speaking, the verdict on the cult and the influence which it undoubtedly had, is one of repri-

[1] Cf. *e. g.* Rohde, *Psyche*, I, pp. 142 f.: " Eine unternehmende Priesterschaft hatte, mit ihren Stammesgenossen wandernd, seinen, unter diesen alt begründeten Dienst weit verbreitet und damit den Asklepios selbst an vielen Orten heimisch gemacht." Or Farnell, *Hero Cults*, p. 256: " The later Panhellenic position of Asklepios owes much, then, indirectly to his early development at Epidauros; and directly also, in so far as she propagated it by missionary effort at least when she was invited to do so"; cf. *ibid.*, p. 258. Cf. also Kern, *Religion*, I, p. 178.

[2] Cf. *e. g.* Wilamowitz, *Isyllos*, p. 102: " Asklepios . . . dessen Walten immer höher geschätzt wird, je kränker die nationale Religion, ja die Nation selber wird"; cf. also *Glaube*, II, pp. 231; Kern, *Religion*, II, pp. 318-19: " Aber sein Siegeszug, der noch während der Verbreitung von Christus- und Mithrasglauben anhält, zeigt den Niedergang echter Religiosität . . . Asklepios wurde . . . der Totengräber der althellenischen Religion. Für die griechische Volksreligion aber bedeutet die hohe Blüte des Asklepioskults den Anfang vom Ende, viel mehr als die sogenannte Aufklärung."

[3] Cf. Nilsson, *Die Griechen*, p. 411; he, too, speaks of " gröbste Mirakelkuren, die dem massiven Glauben zusagten." S. Wide, *Gercke-Norden*, II, p. 226, is one of the few to acknowledge that there is " auch eine ethisch vertiefte Religiosität."

[4] Cf. *e. g.* V. Ehrenberg, *Die Antike*, VII, 1931, p. 289 (relation to the worship of kings); Welcker, *Kleine Schriften*, III, p. 102 (relation to the assessment of values); Farnell, *Hero Cults*, pp. 275 ff. (relation to medical healing); Nilsson, *Popular Religion*, p. 95 (relation to individualism). Yet it is fair to say that such statements concern only details and do not modify the modern disapproval of the cult as such.

[5] Cf. A. Harnack, *The Expansion of Christianity in the first three Centuries*, transl. by J. Moffatt, I, 1904, pp. 121 ff.

manding rejection: propaganda, not belief; superstition, not religion; selfishness, not devotion.

Such views regarding the reasons for Asclepius' ascendancy indicate an attitude toward the understanding of religious phenomena in general which from the outset does not seem propitious. Granted that the priests of Asclepius propagandized their god, how did they succeed in persuading the people to accept him? Greek priests had no worldly authority or power on which they could rely in furthering their aims; there were many more miracle workers among the ancient gods. Why then did Asclepius win the day? Only because his priests were more adroit than those of the other gods?[6] Furthermore, supposing that Asclepius' healings were trickery, apt to please and convince the simpletons, can mere deceit create pious reverence, and such reverence as was paid to Asclepius for many centuries? Could fools alone have supported and successfully carried on a cult which gained the upper hand over the cults of the wise and educated? Of all the Greek gods it was Asclepius who kept his hold over the people by far the longest. When the Oriental religions had almost completely superseded the Greek and Roman cults, Asclepius was still a powerful deity.[7] Would all this have been possible if he had not had the attractive power and the charm of a great and a true god?

The appearance of Asclepius and the rise of his worship must first of all be interpreted as a religious phenomenon; it must be related to the religious history of the ancients and can be understood only in this connection. There were miracles to be sure, but what did these miracles mean to the ancients? There were external factors, no doubt, outward means which came into play in this religious movement as in any other historical process, and they must be investigated carefully. But it seems safe to contend that even religious propaganda becomes effective only in so far as it meets with religious needs, that it remains sterile unless men feel that the new doctrine accords with the conscious or unconscious hopes and wishes within themselves. Certainly, the belief in Asclepius presupposes an attitude different from that which instigated the worshippers of the Olympians, an experience other than that of men of the Homeric or classical centuries—Asclepius preëminently was the god of a later period

---

[6] As a matter of fact, the assumption that Greek priests made propaganda in the interest of their gods is highly debatable. Delphi supported certain cults; the Orphics seem to have made missionary efforts (Kern, *Religion*, I, pp. 162; 178; II, pp. 121 ff.; 148); beyond that, there is hardly any evidence of a propagandatory expansion in regard to any one of the ancient cults. For Delphi and Asclepius, cf. below, pp. 121 f.

[7] Cf. *e. g.* Farnell, *Hero Cults*, p. 275: "We discern that above all the genuinely Hellenic deities he was endeared to the hearts of the people"; cf. also Deubner, *Die Römer*, II, p. 490; Harnack, *op. cit.*, p. 129.

—and the respect paid to him was based on values different from those cherished before. The historian cannot allow himself to view them with aversion. He must acknowledge the god in Asclepius, as did the ancients. Otherwise he cannot hope to achieve an understanding of the Asclepius religion.

In approaching the subject with such an attitude, moreover, one will not only try to explain why Asclepius was accepted at all. One will also endeavor to explain why he, rather than any other healing deity, was chosen as the god of medicine by all the ancients; to ascertain how it came about that such a late god found recognition in spite of all his older and strongly entrenched competitors; to determine for what reasons the god of medicine became the foremost savior; to account for the fact that he, of all the ancient gods, was the leading deity in the struggle between the dying world of the pagans and the rising world of the Christians.

i. Religious Reasons

If one wishes to understand the significance of the Asclepius cult and the reasons for its ascendancy, he must first of all try to analyze the motifs that instigated the worshippers to approach the god, and to evaluate the god's attitude toward the pious as they saw it. Next, it is necessary to confront the religious experience of the followers of Asclepius with those characteristic of the centuries in which his cult was rising, and to ask how this experience fits into the temper of the times.[1]

People came to Asclepius above all to invoke his help in diseases. They came, then, not as members of a family or a community but as individuals. For in diseases men are left to themselves; their fellows, much as they may sympathize with the fate of the sick, cannot really share in it. Again the god, if he gave his succor, removed the most personal griefs and sorrows of his worshippers. That is, Asclepius was concerned primarily with the individual, as is every physician.[2] But it is not enough to describe the relationship between the god and his suppliants in such general terms. The ancients understood the rôle of the physician not only as that of a personal advisor and helper. They believed that the doctor in taking care of the patient assumed the responsibility which the superior

---

[1] In the following analysis I shall make reference to those factors that seem to have had the most general importance for the acceptance of the god of dream healings, the performance of which was Asclepius' specific function. Other aspects of his godhead and the influence of non-religious values on the rise of the god will be discussed below, pp. 119 ff.; 125 ff.

[2] This fact has long been recognized, cf. *e. g.* Nilsson, *Die Griechen*, II, p. 411; Deubner, *Die Römer*, p. 490; Wilamowitz, *Glaube*, II, pp. 471; 505.

has for his subordinates, the king for his subjects; he was supposed to act as the guardian of the patient who gave himself over into the tutelage of the doctor.[3] Thus the connection between the healer and the sick became even closer and more intimate. Certainly, what was true of the good human practitioner and his patients applied no less to the divine physician and those under his care: it was a kind of patronage, of guardianship that the god assumed over his worshippers. Under his hands they felt safe; his temples were justly compared to a most secure and steadfast port receiving the greatest number of people and affording the most in tranquility (T. 402). On the other hand, so conscious was Asclepius of his responsibility toward his followers that he imparted his assistance not only when they approached him; he summoned his patients, he deigned to seek them instead of being sought by them (T. 400; 402; 432; 436).[4] Truly, he was "looking after man" (T. 317, 5), truly, he showed providence (T. 317, 7) in pursuing his art day and night, in healing the limbs of the body that had been destroyed by nature.

Now, ever since the end of the fifth century B. C., when the Asclepius cult began to make its way, a desire was felt for a more personal contact with the divinity than was afforded by the ancient state cults, for a relation that was not shared by all the others.[5] The collective worship of the citizens was supplemented by individual communion with the godhead. As time went on, the state cults became petrified and turned into a merely formalistic ritualism.[6] Men, freer than they had been to choose their own gods, naturally wanted them to care most of all for the fate of the individual, his sorrows and joys, his sufferings and his happiness. There was a craving for a personal relationship to the deity, and the belief in divine providence progressed steadily.[7] In such a world it was natural that

---

[3] Cf. *e. g.* Plato, *Politicus*, 293 C; *Laws*, 720 A ff.; Hippocrates, *The Physician*, *C. M. G.*, I, 1, p. 20, 18 ff.; Galen, X, p. 4 K.: ['Ασκληπιάδαις] οἳ τῶν νοσούντων ἠξίουν ἄρχειν ὡς στρατηγοὶ στρατηγουμένων καὶ βασιλεῖς ὑπηκόων.

[4] To look out for patients was customary for human physicians, cf. *e. g.* T. 126, and in general Edelstein, *Problemata*, IV, p. 91. It is interesting to observe again and again how closely the concept of the god resembles that of the medical practitioner.

[5] Cf. Nilsson, *Reflexe von dem Durchbruch des Individualismus in der griech. Religion um die Wende des 5. u. 4. Jahrh. v. Chr. (Mélanges F. Cumont*, pp. 365 ff.) ; *Griechische Religion*, I, pp. 760 ff. Nilsson refers to the Asclepius religion as one of the examples of the new trend.

[6] For the state cults in later centuries and the attitude of people toward them, cf. Wilamowitz, *Glaube*, II, p. 271; Deubner, *Die Römer*, p. 490.

[7] I am aware of the fact that in the Hellenistic period *Tyche*, the personification of blind fate, held sway over men. But *Pronoia*, Providence, was considered a goddess too, and found worship even among the common people, together with *Tyche* (cf. Wilamowitz, *Glaube*, II, p. 390; for *Pronoia*, cf. also Kern, *Religion*, III, p. 84). Moreover, philosophers succeeded in destroying the belief in the rule of irrational Fate (cf. Wilamowitz,

Asclepius found favor, for if any god was interested in the private needs of men, in their most personal affairs, if any god showed providence, it was Asclepius.

Moreover, the deified physician gave help freely and without envy. He refused assistance only on moral grounds; those who were themselves not virtuous he would not heal (T. 394). But this was the only request which he made of his patients. Otherwise he did not expect anything, no reward (T. 320), not even belief in his power; he cured his patient, whether he was a devotee or a disbeliever (T. 423, 3).[8] As the testimonies aver, he was the most man-loving of the gods (T. 422; 804, 5). This attitude of philanthropy was, so to say, native to the divine physician. Even a human doctor who lacked love of man was not considered a good physician; to practice the art of healing out of concern for man, not for money's sake or for glory, was held characteristic of good craftsmanship.[9] No matter how often human beings may have failed in exhibiting the moral quality demanded of them, the god unswervingly acted in accordance with the ideal set before the physician. But in the god, philanthropy was more than a professional virtue. It constituted the essence of his divinity. His life on earth had been pure, given to healing, to helping mankind. His entire divine existence consisted in fulfilling this function of being beneficent to mankind (T. 320).[10]

Likewise from the end of the fifth century it was no longer thought that the divinities were the enemies of man. The sinister belief in the envy of the gods, the terrifying conviction that they delight in harassing

ibid., p. 309). And the growing influence of the mystery religions indicates the kind of religious experience that was sought: an affinity as close as possible between god and man.

[8] Farnell, Hero Cults, pp. 271 f., rightly stresses that Asclepius "was not vindictive like the God of the Old Testament." To be sure, he does not heal those who cheat him but rather punishes them (T. 423, 7), yet he is easily ready to forgive any offence (T. 423, 36; also T. 423, 11). It should be added that Asclepius, unlike the God of the New Testament, did not demand faith. Latte's claim to the contrary (Gnomon, VII, 1931, p. 120) is unfounded; cf. below, pp. 161 f.

[9] The earliest testimonies to this effect are to be found in Hippocrates (Precepts and Physician, C. M. G., I, 1, pp. 32, 9; 20, 12). The Oath does not yet mention philanthropy, but its doctrine was interpreted as implying this virtue (cf. Scribonius Largus, p. 2, 17 ff. [Helmreich]). In Roman times the view that medicine was a calling that required philanthropy became ever more widespread; cf. e. g. Galen, De Placitis Hippocratis et Platonis, pp. 764 ff. [Mueller].

[10] Farnell, who has emphasized the relation between the god's philanthropy and the ancients' evaluation of medicine, the most philanthropic of all social activities (Hero Cults, p. 276), says in regard to Julian's statement (T. 320): "Julian's phrase concerning the φιλανθρώπευμα of Asklepios, who heals not for reward, but to gratify his own loving heart, not inaptly expresses the popular conviction, and marks off this man-god from the older Hellenic divinities." Asclepius seems to have assumed the rôle of Hermes who in early centuries was considered the most philanthropic of the gods (cf. Kern, Religion, II, p. 19).

and destroying men, gradually disappeared. Plato's demand that the divinity be revered as free from grudge and hatred became generally recognized.[11] The opinion was abandoned that the gods were subject to affections, just like men; only the belief in their goodness seemed acceptable. For a god was now visualized as the one " who rises superior to inordinate desire, to envy and jealousy, and is generous, beneficent, a lover of man." [12] Did Asclepius not answer all the wishes that men could have in regard to divine perfection?

Finally, Asclepius was ever-present in his temples; every cure that he performed was an epiphany of his divine person, a proof of his divine existence. His healings, his miracles, as they are called by the modern, to the ancients were " his glorious deeds," attesting that divine might was still alive, that there were gods and that they were powerful.[13] This experience, in fact, at least from the third century B. C., provided the most convincing evidence for the belief in gods. Just as the Athenians acknowledged the divinity of Demetrius because he had proved his power through his deeds, because he had achieved what they were not capable of doing themselves, so the people at large were ready to believe in the deified Asclepius because they experienced his might.[14] Like all deified

[11] Plato, *Phaedrus*, 247 A; *Timaeus*, 29 E; cf. Kern, *Religion*, III, p. 15.

[12] In later centuries the Stoics were the main exponents of such doctrines—the definition of the divine referred to here is taken from Musonius (p. 90 [Hense]). But the Epicureans too rejected the belief that the gods were passionate, or haters of men. The Stoics again were foremost in the " accommodations " of the old myths through which the objectionable features were interpreted away. Unquestionably, what in earlier periods had been denounced only by some individuals as incompatible with the character of a god, was now discarded by the majority. For what does the worship paid to *Aidos* and *Nemesis* mean, if not the victory of moral concepts (Wilamowitz, *Glaube*, I, p. 45)? Even a writer like Pliny treated the mythological fables of old with contempt.

[13] The word ἀρετή, " glorious deed " or " power," seems first to have been used by Isyllus in describing Asclepius' help for Sparta (T. 295). To the god's medical assistance it was applied by Hermodicus in the third century B. C. (T. 431; the genuineness of the inscription has been reaffirmed by G. Karo, [*Weihgeschenke in Epidaurus*, 1937, p. 3] against the objections raised by Herzog, *Wunderheilungen*, pp. 100 ff.); the Epidaurian tablets (T. 423) still speak merely of cures. Later, the word ἀρεταί occurs in T. 438 and T. 441. Ἐπιφάνεια commonly denotes Asclepius' appearance in dream healings, *e. g.* T. 382. For the history of the terms ἀρεταί and ἐπιφάνεια, cf. Herzog, *op. cit.*, pp. 49 ff., who also points out that the word miracle (θαῦμα) begins to become customary usage only in Christian collections of divine healings. Yet, Aristides who usually speaks of the δύναμις of Asclepius (T. 317, 4) once calls him θαυματοποιός (T. 409).

[14] V. Ehrenberg, *Die Antike*, VII, 1931, p. 289, in his interpretation of the Athenian hymn on Demetrius (Athenaeus, VI. 253 d ff.) rightly says: " Die Folgerung, die hier gezogen wird, ist nicht, dass die Götter im Weltbild überhaupt überflüssig seien oder ein völlig anderes, rein geistiges Wesen hätten, sondern dass an die Stelle der 'hölzernen' oder 'steinernen' ein anderer 'wahrer,' das heisst lebender Gott tritt." And he adds: " Es ist eine durchaus verwandte Erscheinung, wenn der Glaube an den Heilgott Asklepios, der im Traume, natürlich in menschlicher Gestalt, erschien und so heilte, ungeachtet aller Fortschritte der rational-wissenschaftlichen Medizin im gleichen 3.

kings of that century and of later times he was a man-god. While he was not a warrior, he was an artisan of superior knowledge and skill, of incomparable wisdom, and many a great genius was held to be divine.[15] On the other hand, the belief in the might of the Olympians was rapidly declining. It was not so much the sceptical criticism that destroyed their old-inherited position, it was not so much the philosophical and scientific explanation of nature that led to their downfall. Though still worshipped, they vanished from the scene of real life, because they no longer seemed able, or willing, to help. Their place was largely usurped by other deities —for men want gods—by lesser ones, be they man-gods who had risen to godhead through their achievements, like deified kings, or be they personifications of abstract concepts.[16] This was another reason why Asclepius, the "daimon," who never became one of the Olympians, the hero, who had become a god in charge of one special domain, that of healing, was acceptable to men.

The Asclepius cult, then, was most closely akin to the trends of religious thought prevailing in Hellenistic and Roman centuries. In fact, the god seems to have represented some of the new ideals with more determination than any of his peers.[17] Small wonder that the reverence paid him increased steadily, that his worship gradually spread over the

Jahrhundert stärkste Intensivierung erlebte"; for the concepts of the παρουσία and ἐπιφάνεια of these gods, cf. ibid., p. 292. Note that the religious experience here presupposed is similar to the "Beweis des Geistes und der Kraft" of the Christians (Hebrews, 2, 4; cf. G. E. Lessing, Sämmtliche Schriften, ed. K. Lachmann, X, 1839, pp. 33 ff.). Cf. also below, pp. 133 f.

[15] For the nature of Asclepius' godhead, cf. above, pp. 84 f.; 91. His wisdom is praised e. g. T. 593. It is important to remember that the worship of kings was only one instance of the deification accorded to great men. The founders of philosophical schools also received divine honors. Even Lucretius can speak of Epicurus as a god (De Rer. Nat., V, 8; III, 15). The words of the Oracle concerning Lycurgus' divinity (Herodotus, I, 65) correspond exactly to Galen's statement about Asclepius (T. 245). In general cf. Kern, Religion, III, pp. 111 ff.

[16] In general, cf. Kern, Religion, III, pp. 74 ff. Ehrenberg, loc. cit., pp. 279 ff.

[17] I need hardly add that the evaluation I have given of the new concepts that from the end of the fifth century B. C. fundamentally reshaped ancient religious life is in no way intended to be complete. I have selected only those features which seem to have had the greatest bearing on the Asclepius religion. Moreover, the historical development naturally was not as consistent as I have sketched it. During the many centuries through which Asclepius' reign lasted religious thought fluctuated, the various tendencies gained or lost momentum. But the evidence preserved does not suffice to give a precise picture of the position of Asclepius at every moment. One must take recourse to generalizations. Finally, I have made no attempt to distinguish strictly between that which is commonly called philosophical religion and that which figures as common religious consciousness. It is quite true that some aspects of the cult may have had less weight with the many than with the intelligentsia, and vice versa. Only all aspects taken together, in my opinion, account for the ascendancy of Asclepius. The cults of foreign gods, on the other hand, despite their general importance, I have not taken into consideration, because the worshippers of Asclepius were those who still adhered to the truly Greek deities.

entire ancient world, that he acquired a position of pre-eminence. Asclepius truly was a living deity, near to men's hearts. It was he who now gave shelter to the more ancient gods and supported their failing might; it was his festivals that were still celebrated when those of other deities began to be forgotten—in Athens, the *agon* in honor of Asclepius was one of the last to be commemorated; it was to Asclepius that hymns were dedicated in ever-growing number when poems had ceased to be written for the rest of the gods; it was Asclepius who remained rich in offerings at a time when other sanctuaries decayed and who, while hardly any money or other gifts were presented anywhere else, tried to get rid of the overflow of dedications in his temples.[18]

But here one might object: it is just this popularity of Asclepius which serves to prove that he was a god of the masses rather than of the few that count, of the lower strata rather than of the upper classes, of the uneducated not of the cultivated, of the egotists not of the moralists. Superficially the god may have conformed even to higher ideals, but fundamentally it was the downright fraud of his miracles that attracted men, and certainly not the best among them.[19]

Now, to disregard for the moment the question whether Asclepius' miracles were frauds, it is true that they impressed people. But to say it once more, they impressed everybody because they seemed to reveal the reality of divine power and providence.[20] One may also admit that a large number of Asclepius' patients consisted of the poor. The god prided himself on healing the paupers (T. 321), on being satisfied with slight thanksgiving if no more could be presented to him (T. 482, *vv. 11 ff.; 79 ff.*). There can be no doubt that those who had no money or very little availed themselves of the opportunity thus offered, and consequently were among the staunchest adherents of Asclepius.[21] That this was so, does not mean, however, that the rich were not also in favor of the new god. Without the support of wealthy patients who paid hand-

[18] In this short characterization of the god's position I have followed Wilamowitz, *Glaube*, II, pp. 347; 359; 468; 474; 482. For an elaboration of the details, cf. the discussion of the cult and of the history of the temples, below, *passim*.

[19] Cf. *e. g.* Wilamowitz, *Glaube*, II, p. 505: "Der Gott, an dessen Wunder die besten Männer nicht glaubten"; cf. also above, p. 109; Deubner, *Die Römer*, p. 490: "Gottheiten wie Asklepios und Serapis, die damals wirklich lebendige Geltung im Volk hatten . . ."; Kern, *Religion*, II, p. 319: "Es ist oft in den Religionen der Menschheit wiederkehrende Erscheinung, dass der hellsten Einsicht hoher Geister der abergläubische Wahn der Menge entgegensteht. Der Asklepioswelt fehlt das θεῖον, das längst gefunden war."

[20] Cf. above, pp. 114 f. The whole problem of temple medicine and especially that of the cures of Asclepius is discussed in detail below, pp. 162 ff.

[21] In this respect the Asclepieia had a very important social function in the ancient world, cf. below, pp. 175 ff.

somely, the god could not possibly have taken care of his poor clientele; nor could he have built his splendid temples, had there not been many who gave him more than their share. That the upper strata of society frequented his temples even in early centuries is proved by the fact that the old Epidaurian inscriptions mention so many patients from out of town; poor people could not afford to travel. Moreover, those who in grateful recognition of the god's help introduced the cult in their home-cities were certainly members of the nobility; no ordinary citizen could attempt to propagate a new god.[22] In later times, of course, rich and poor were mingled among his clients, and peasants no less than senators, shoemakers no less than emperors asked the advice of Asclepius. It is, then, well nigh impossible to draw any distinction as to the social standing of the god's visitors.[23]

But one might still insist that poor and rich alike, people of low and high rank, may be fools, and that they were attracted by the miracles of Asclepius in which they believed with the superstitious credulity and naïveté of the uneducated. It has been claimed, in fact, that the great spirits, those who had any scientific or philosophical training, those who had culture, and those who had true religion, were certainly not swayed by this god of temple dreams and oracles that were nothing but the deceit of clever priests. It is strange, indeed, that such an argument should ever have been put forward. As for the great men, the best, the erudite, one need not be satisfied with Aelian's general statement to the effect that Asclepius, as he was the helper of the poor, was also the protector of the cultured (T. 456; 321). Was it not Sophocles who wrote the most famous paean on Asclepius (T. 587)? Did he not receive the god into his home (T. 590)?[24] Did not other great poets of the fifth century B. C. write tragedies dedicated to Asclepius (T. 455)? Did not the most renowned artists create the cult statues of the god? Did not the foremost

---

[22] Cf. below, p. 119.

[23] For more particulars on Asclepius' clientele, cf. below, p. 120.

[24] Wilamowitz' interpretation of this testimony is strange indeed. He says (Glaube, II, pp. 232-33): "Am stärksten musste das Auftreten des Sophokles wirken" (namely in favor of the Asclepius cult), but then he adds: "und für uns ist es am wichtigsten, dass dieser fromme Mann für einen solchen Gott enpfänglich sein konnte, so sehr, dass er dessen Erscheinung in seinem Hause erlebte. Das scheint zu dem Bilde schlecht zu stimmen, das wir uns von dem Dichter der Tragödien machen. In der Tat würden weder Aischylos noch Euripides dazu fähig gewesen sein. Es stimmt aber dazu, dass er allein die eleusinischen Weihen verherrlicht und offenbar selbst empfangen hat." It is more adequate, I think, to turn the argument around and to say that if such a man as Sophocles favored Asclepius, the god must have been a true representative of the divine. Again, at another place, Wilamowitz adds (ibid., p. 235): "Für Sophokles war es keine Erscheinung, die ihn zur Umkehr von seinem alten Glauben zwang, sondern eine Bestätigung seiner Hingabe an den Glauben der Väter." This may have been the experience of many.

architects build his sanctuaries? [25] Did not Socrates in his last words ask his friends to make a sacrifice to Asclepius (T. 524)? [26] Was Theophrastus (T. 318), who saw in Asclepius' ethics a model of what men should do, a poor judge of morals and human behavior? [27] Was Epictetus (T. 406), who speaks with reverence of Asclepius, a worse philosopher than Dion and Plutarch who do not mention him? Was Marcus Aurelius, in whose opinion Asclepius was a great healer (T. 407), a man of smaller stature than Lucian who belittled Asclepius (T. 265)? [28] Were the Neo-Platonists, who abound in praise of Asclepius, not the great philosophers of their time? Were not Rufus and Galen great physicians, and did they not believe in the cures of Asclepius, as did nearly all ancient doctors? [29]

Yet it is neither necessary, nor helpful, to collect names—the god did not keep a file of his patients and his friends, or at any rate not every great man who asked his help left a note for posterity attesting his visit. The forces behind the Asclepius cult were the same that were operative throughout the religious life of later centuries, the beliefs and convictions of the Asclepius worshippers were in accord with all that was sacred to the ancients. This fact alone would fully justify the assumption that the best no less than the lowest found in this god their savior. Otherwise his worship would never have gained the importance that it actually had throughout the ancient world.

[25] Cf. the chapters on the artistic representation of Asclepius and on the history of the temples.

[26] In this case too Wilamowitz' comment can hardly be accepted. He claims (*Glaube*, II, p. 236): "Sokrates zeigt sich auch hier als guter Bürger, der den Göttern des Staates das Ihre gibt. Er wird freilich die ärztliche Hilfe des neuen Gottes nicht für sich in Anspruch genommen haben, aber er hatte eine Frau und kleine Kinder und hinderte Xanthippe oder auch Kriton nicht, wenn sie zu Asklepios gingen." For the interpretation of Socrates' words and the importance which they assumed in late antiquity, cf. below, pp. 130 f. Incidentally, Kern goes so far as to say that Plato's criticism of Pindar's poem (T. 99) shows "seinen Glauben an die Göttlichkeit des Asklepios, der die Hellenenwelt im Sturm erobert hat" (*Religion*, III, p. 21). This, it seems to me, is exaggerated, for Plato speaks about the hero Asclepius whose medical art he praises as the prototype of good medical care (T. 124).

[27] For the interpretation of T. 318 and its relation to Theophrastus, cf. below, p. 126, n. 6.

[28] Wilamowitz claims (*Glaube*, II, p. 471) that for a great many people, yet not for Plutarch, Asclepius was a living deity, and he adds (p. 505) that Plutarch, Dion and Epictetus, to be sure, do not attack the belief in miracles, but that their very silence is betraying and should be acknowledged by modern critics with all its implications. Conclusions *ex silentio* always are precarious. Moreover, Epictetus is not silent about Asclepius, and the Emperor Hadrian would hardly have dedicated a bust of the philosopher in the temple of Epidaurus (cf. below, p. 254, n. 13), had Epictetus abhorred the god. Nor is Wilamowitz justified in refusing to believe that Marcus Aurelius saw in Asclepius the sender of his healing dreams (*Glaube*, II, p. 494, n. 1). Marcus Aurelius' respect for Asclepius is clearly expressed in his letter to Fronto (T. 577), and his court physician, Galen, thanked Asclepius for the help bestowed upon him in his dreams (T. 458). The Emperor was hardly more enlightened than his doctor.

[29] For the Neo-Platonists, cf. above, pp. 107 f.; for Rufus and Galen, and the attitude of ancient doctors toward temple medicine, cf. below, pp. 139; 147.

ii. Non-religious Influences

Certain as it is that the god Asclepius exercised his influence because, in the eyes of the ancients, he was a true god, one should not overlook the subsidiary causes which helped in bringing about the rise of his cult. Generally speaking, the post-classical centuries were favorably inclined toward the introduction and acceptance of new worships. Once the influence of the old Greek gods had been weakened, even Oriental deities found recognition. The conquest of new territories, the building of new cities brought a re-orientation of life which allowed people to select their own gods for their homesteads. It was relatively easy then, in a world not as inflexibly joined as of old, to win proselytes for a new deity. But why did Asclepius outdo all other gods in the extension of his might?

To begin with the most obvious reasons: cults in antiquity were often spread through private individuals, and it was especially the healing cults that were disseminated in such a way. The Asclepius cult in this respect formed no exception. Patients who had visited one of the great temples of the god, in grateful recognition of the help bestowed upon them, established Asclepius sanctuaries in their home towns.[1] Thus Naupactus was founded (T. 444), thus Pergamum (T. 457); the custom remained in vogue throughout the centuries. And though such a cult in the beginning was a private worship, not officially recognized, it tended to become a state cult and in fact succeeded in most cases.[2] Moreover, very often a city, threatened by disease, felt the need of Asclepius' help and asked the god to take up his abode within its boundaries. It was thus that Asclepius came to Rome (T. 846).

Beyond this power of expansion inherent in the Asclepius worship as a healing cult, there must have been many personal influences, some of them no longer to be determined, some at least to be conjectured, which

---

[1] Herzog (*Wunderheilungen*, pp. 37 ff.; 106) has rightly stressed this kind of "self-propaganda" which furthered the extension of the cult. He has also drawn attention to a passage where Plato, adversely criticizing this custom, describes it quite vividly (*Laws*, 909 E-910 A, transl. by R. G. Bury, *Laws*, II, 1926 [Loeb]): "It is no easy task to found temples and gods, and to do this rightly needs much deliberation; yet it is customary for all women especially, and for sick folk everywhere, and those in peril or in distress (whatever the nature of the distress), and conversely for those who have had a slice of good fortune, to dedicate whatever happens to be at hand at the moment, and to vow sacrifices and promise the founding of shrines to gods and demi-gods (δαίμοσι) and children of gods; and through terrors caused by waking visions or by dreams, and in like manner as they recall many visions and try to provide remedies for each of them, they are wont to found altars and shrines, and to fill with them every house and every village, and open places too, and every spot which was the scene of such experiences."

[2] This happened for instance in Athens, cf. below, n. 4.

contributed their share to the astonishing process of Asclepius' rise: the skill and shrewdness of certain priests in charge of the main centers of the cult, their ability to cooperate and to remain in contact with the changing temperament of the centuries; the interest of famous men in the god, be they powerful political individuals like Alexander the Great or Marcus Aurelius, or at least influential statesmen like the Epidaurian aristocrat, Isyllus, or the Roman Senator, Antoninus, or be they great literary personalities like Apuleius or Aristides.[3] There is no reason to deny that Asclepius was not only a great god but also a lucky one; apparently it was the right people who were interested in him, those who would further his ambitions.

Moreover, two events within the historical process of the acceptance of the cult were of outstanding importance, and seem to have been decisive in accelerating the ascendancy of the god: his introduction into Athens and his sanction through the Delphic Oracle. The Athenians received Asclepius into their city relatively early, at the end of the fifth century B. C., and soon he was recognized as one of their official gods.[4] Many factors may have contributed to the Athenian interest in him. The Athenians considered themselves the originators of culture and civilization, and Asclepius was a culture hero who had become a god, the protector of medicine, an art which was regarded as one of the greatest human accomplishments.[5] Besides, Athena, the goddess of the city, had shown a very distinct sympathy for Asclepius; she had given him a share in the blood of the Gorgon, as she had done to Erechtheus (T. 3). This instance, too, must have endeared Asclepius to the Athenians.[6] But, what-

[3] For the names referred to here, cf. e. g. T. 821, 407, 295, 739, 608, 317, and the chapters on the cult and the history of the temples where more details concerning these persons are given and additional names are mentioned.
[4] It was Telemachus who brought Asclepius to Athens in 420 (T. 720); cf. above, p. 66. When the state officially accepted the cult, is not known (cf. Wilamowitz, Glaube, II, p. 232). Apollophanes still calls Asclepius a ξενικὸς θεός (T. 720a). But around 350 B. C., according to inscriptions, the recognition had already taken place, cf. Kutsch, Attische Heilgötter, p. 26. The reason for the original introduction was most probably a successful healing (cf. Herzog, Wunderheilungen, p. 38). It is hardly correct to assume with Kern (Religion, II, p. 312) that the worship was brought to Athens because Asclepius had helped the city during the great pestilence; some ten years had passed since, and in that case one would not expect private, but rather official action, as later in Rome (cf. below, p. 252); besides, it was Apollo Alexikakos who was credited with saving the city from the dangerous epidemic (cf. Pausanias, I, 3, 4). Nor can one believe with Farnell (Hero Cults, p. 258) that the Athenians submitted to Epidaurian propaganda in 420, for in that year Athens was on the brink of war with Epidaurus; cf. J. B. Bury, A History of Greece, p. 442. (Pericles had also attacked Epidaurus in 430; cf. Bury, op. cit., p. 390).
[5] Concerning the Athenian claims, cf. Nilsson, Popular Religion, pp. 56 ff.
[6] For the interpretation of T. 3, cf. above, p. 43, n. 76. At first sight one might think that the saga connects Asclepius with Athena because he was revered in Athens. But

ever the reasons for Asclepius' recognition, his position was certainly strengthened by the fact that he was accepted in that city whose power was almost supreme in matters spiritual and religious. The attitude of the Athenians toward him was bound to make an impression on other regions and cities. With his entry into Athens, the " world-city," the Epidaurian Asclepius lost his provincial character; he became a Pan-hellenic god.[7]

The other conspicuous element in the rise of the Asclepius worship, the support of the Delphic Oracle, began to assert itself at almost the same time at which the god arrived in Athens. Near the end of the fifth century B. C. Asclepius was honored with a sacred precinct in Delphi.[8] The Oracle also assisted in the expansion of the cult to other places. In a few instances it recommended the founding of Asclepius temples.[9] But doubtless the most important and most effective assistance which Delphi gave to Asclepius was that it legitimated him as the son of Apollo and Coronis (T. 31) and put an end to the dissension about the place of his birth, throwing its whole influence behind one city, Epidaurus (T. 36).[10] The Oracle also sanctioned Asclepius' position as healer of diseases. The Homeric hymn expressly referred to this, his medical function (T. 31), and Isyllus, with the approval of Delphi, addressed Asclepius as " the reliever of illness, the granter of health, great boon to mankind " (T. 594). Thereby Asclepius was designated by Apollo as the medical deity to whom he had entrusted the domain over which he

---

Athena is the typical helper of such figures as Heracles and Perseus, men who through their own endeavor accomplished great feats, and she is the patroness of the arts as well; cf. Kern, *Religion*, II, pp. 22-23. The connection between Athena and Asclepius, then, is hardly derived from historical facts, but rather is a purposeful invention of the saga (Aristides still speaks of the Ἀσκληπιοῦ καὶ Ἀθηνᾶς συμφωνία (p. 310, 3 ff. [Keil]) ; cf. also T. 449a ; 732).

[7] Concerning Athens' influence on religious life in general, cf. Kern, *Religion*, II, pp. 265; 78 f. The importance of the Athenians' acceptance of Asclepius has always been felt, cf. *e. g.* Farnell, *Hero Cults*, p. 258; Latte, *Gnomon*, VII, 1931, pp. 118 ff. Wilamowitz, *Isyllos*, p. 94, stresses the Athenian "Gegengabe für die Stiftung des attischen Asklepieion," that is, he assumes that Athenian artists created the likeness of the god. If it were true that the Asclepius type originated in Athens, that the god was seen by men through the eyes of Athenian talent, this would certainly have been an important factor in the history of his ascendancy. But modern attempts to trace the origin of the cult statue to the circle of Phidias have failed, cf. below, p. 218, n. 12.

[8] Cf. Kern, *Religion*, III, pp. 156 f. Farnell's statement that Asclepius was never admitted within the Delphic area (*Hero Cults*, p. 248) is not correct. Cf. below, p. 246.

[9] The temples of Halieis and Rome are said to have been founded on the instigation of Delphi (T. 423, *33*; 850, *vv. 631 ff.*). Kern, *Religion*, III, p. 156, therefore is quite right in saying that the testimonies for the active support of the cult through Delphi are not numerous. But the Oracle had other means of furthering Asclepius' ambitions, and it made use of them.

[10] Cf. above, pp. 71 f.

himself had once ruled. Thus confirmed by the highest authority, the new god was even more likely to find the acclamation of all men.[11]

Yet not even with all these factors to his advantage, with all this array of supporting powers would the temples of Asclepius have arisen everywhere, had there not been a general willingness on the part of the people to receive him and to abandon in his favor those divinities with whom they were accustomed to take refuge in sickness and distress. Long before Asclepius had become the accessory of Apollo, long before the god had been established in Epidaurus, his name had been familiar to all Greeks. Homer had sung of him; Hesiod had recounted his story. Physicians had taken him wherever they went; with them, he had entered villages and towns, the great cities, all sections of Greece. Whenever the people had used the name Asclepiads, they had been reminded of the patron of their doctors, of Asclepius.[12] In short, the hero had been received by the Greeks long before the god began to aspire to their adherence. When the god finally came, he found the ground well prepared for his worship by the fame of the hero. Moreover, the fact that the physicians revered Asclepius must have inclined people favorably towards him. If even the doctors acknowledged his medical abilities by regarding him as their patron, how could he fail to be a good physician, an efficient healer of diseases? The extension of the Asclepius worship, therefore, once it had been established, was a development that started from within rather than from without. The god hardly forced his followers into subjection, nor did he impose his reign after hard battles; he found the doors wide open.

Finally, when Asclepius at last deigned to come into a world which was predisposed to receive him, he came as the dispenser of a good which men learned to estimate ever more highly: health and freedom from disease. This change in the appreciation of values was a development almost contemporaneous with the rise of the Asclepius cult.[13] In Homeric centuries the warrior, the hero, found his sole aim in glory; he was intent only on battle, on victory; health to him was no value in itself. In post-Homeric centuries, however, physical well-being became more

---

[11] It must be stressed, however, that the attitude of Delphi towards the Asclepius religion presupposes a strong general interest in the god. Fundamentally the Oracle was opposed to new gods, and it did not favor " incubations." Yet, just as it finally allied itself with the worship of Heracles and Dionysus because the people approved of them (cf. Wilamowitz, *Glaube*, II, pp. 37; 41-42; H. W. Parke, *A History of the Delphic Oracle*, 1939, p. 431), it seems to have recognized Asclepius because he had become popular. Concerning Delphi's influence on the ethics of the Asclepius religion, cf. below, p. 126, n. 6.

[12] Cf. above, pp. 54 ff.

[13] In general, cf. Edelstein, *Die Antike*, VII, 1931, pp. 267 ff.

important and began to assume a place among the goods for which men were striving. In consequence of a rather pessimistic evaluation of the outcome of all human endeavor, of the balance between happiness and suffering in the life of the individual, the estimation of health grew increasingly strong. It was health even more than beauty or riches that constituted the *summum bonum* of later generations; to be healthy seemed the indispensable pre-requisite of all joy and happiness. If this is true of the classical period, it is even more valid for the Hellenistic era and later centuries. With the destruction of political freedom, the interest in health, the appreciation of bodily fitness became more intense. Merchants and businessmen, scholars and artists, who were no longer free citizens but subjects of a king, and who left war and politics, except the petty politics of their own communities, to professional soldiers and royal servants—this bourgeois society naturally delighted in the worship of the god. " When health is absent, wisdom cannot reveal itself, art cannot become manifest, strength cannot fight, wealth becomes useless, and intelligence cannot be applied "; thus a physician epitomized the value of health.[14] And men in general agreed with him and put soundness of body before all other goods pertaining to human happiness. Not even philosophers, not even the Stoics could dare to minimize the value of health. It is true, the new masters of the world, the Romans, in the beginning of their reign revolted against such a philosophy. But at that time Asclepius already had entrenched himself too strongly to be shaken in his power, and soon the Romans, too, submitted to the Greek idolatrous belief in health as a value in itself. No wonder that among people of this disposition the god of medicine, the giver of health, was a great god.[15]

One more fact is to be considered: Asclepius was the god who would postpone the day of death; he intervened when the sick were despaired of,

---

[14] Herophilus *apud* Sextum Empiricum, *Adv. Math.*, XI, 50.

[15] Welcker, *Kleine Schriften*, III, p. 102, was the first to see the importance which the value ascribed to health had for the rise of the Asclepius religion. He says: " Eine so grosse und allgemeine Angelegenheit ist die Gesundheit dass darum der ihnen (*sc.*, Machaon and Podalirius) zum Vater gegebene Asklepios späterhin in grossen und reichen Tempeln verehrt wurde, während einem Homeros, einem Dädalos nur hier und da bescheidene Opfer von ihren Kunstgeschlechtern gebracht wurden. Nur nach dem Masstab wonach die Lebensgüter im Allgemeinen gemessen werden, ist die Heiligkeit des Asklepios wie anderer Götter gestiegen." Farnell, too (*Hero Cults*, pp. 275-6), stresses the significance of health: " No physical need of man is more imperious than the need to escape from physical suffering; no divine ministration is more vehemently craved, more ardently besought than that which provides this escape; and none other awakens so warmly the gratitude of the worshipper." Yet, both scholars have failed to notice the special significance of health for the ancients. Compared to this most essential factor it is no sufficient explanation for the later expansion of the new cult to say that civilized people need recreation in fresh air, as it was to be found in the Asclepieia, cf. *e. g.* Wilamowitz, *Isyllos*, p. 86, n. 64. Cf. also below, p. 158.

when all other help was failing.[16] He saved men from Hades, the god who to the ancients was the most hateful of all. Hades alone had no altar among them; no hymn was sung in his honor; no sacrifices were offered to him. For he was not fond of gifts, nor could he be bought by persuasion; he was adamant and implacable.[17] Through Asclepius those who had approached the doors of the nether world were given back to life. Through him they became alive again, as they said (T. 419).[18] They lived not only once but twice (T. 317, 6), and sickness seemed almost advantageous on account of this experience (T. 402). For who loves life more dearly than do those who were at the point of losing it? And who was fonder of life than were the ancients? Most human beings have a desire to live, but the Greeks delighted in the beauty of life and in the brightness of light more perhaps than any other people. Even Achilles did not refrain from exclaiming: "Nay, seek not to speak soothingly to me of death, glorious Odysseus. I should choose, so I might live on earth, to serve as a hireling of another, of some portionless man whose livelihood was but small, rather than to be lord over all the dead that have perished." [19]

To men who thus judged about the value of life, Asclepius was indeed the savior (T. 463 ff.), he who granted " that we may see the sunlight in joy, acceptable with bright Hygieia, the glorious " (T. 592).[20] It was

[16] Cf. above, p. 82.

[17] Cf. Homer, *Iliad*, IX, 158-9; Aeschylus, *Fr.* 156 [Nauck]; Philostratus, *Vita Apollonii*, V, 4; and in general Pauly-Wissowa, *s. v.* Thanatos, IX (2), pp. 1252; 1257 f.

[18] T. 419: ἀναβεβίωκα. The testimony is taken from one of Menander's plays. That a comedy represents the reaction of the patients of Asclepius in this manner, warrants the conclusion that this was what people in general felt. It was not only Aristides who used such language (*e. g.* T. 402, but cf. also 317, 6) or any other devotee whom one might accuse of a bias in favor of the god.

[19] Homer, *Odyssey*, XI, 488-91 [tr. by A. T. Murray, Loeb]. Plato in his criticism of Homer considers these verses the first to be deleted in the poem (*Republic*, III, 386 C). His polemic proves how strongly they reflected a common feeling. To be sure, fear of death was not universally the rule. The pessimism of classical centuries praised as the highest, the greatest blessing not to be born, and as the second to hasten back from whence one came (Sophocles, *Oedipus Colonus*, 1224-27). Stoic and Epicurean philosophers proudly asserted man's mastery over death. Still, even under the cover of their debates the horror of death is noticeable, and perhaps what Bacon says of the Stoics was true of all those who defied death by glorifying its nothingness: they bestowed "too much cost on death, and by their preparations made it more fearful " (*Essay on Death*, cf. W. E. H. Lecky, *History of European Morals*, I, pp. 202 f. [Appleton]). Needless to say, even those who most strongly emphasize the pessimistic component in Greek thought do not go so far as to deny that also among the Greeks the majority loved life and feared death (cf. J. Burckhardt, *Griechische Kulturgeschichte*, II⁵, pp. 401 ff.; 412). Asclepius himself, it seems, was well aware of this fact and tried to assuage the hearts of his worshippers like a good Platonist rather than a Stoic or an Epicurean, cf. below, pp. 127 ff.

[20] The expression " to see the sunlight " is equivalent to the expression " to live "; cf. *Iliad*, XVIII, 442, and for more parallels, Pauly-Wissowa, *s. v.* Naturgefühl, XVI, p. 1829, 4 ff.

with good reason that his temples became a new center of religious life for the ancients. One might well say of all of them what Aristides says of the sanctuary of Pergamum: " Here fire-brands friendly to all men were raised on high "; here " the stern-cable of salvation for all was anchored in Asclepius " (T. 402). And it was probably not only Aristides who felt that " neither belonging to a chorus nor sailing together nor having the same teacher is as great a thing as the boon and profit of being a fellow pilgrim to the temple of Asclepius and being initiated into the first of the holy rites by the fairest and most perfect torchbearer and leader of the mysteries, to whom every rule of necessity yields " (T. 402).

6. THE TEACHING OF ASCLEPIUS

Asclepius was the impersonation of the divine healing power; his function was that of giving and preserving health, of relieving from disease. At first glance this fact seems certain proof that his worship, in spite of all its truly religious aspects, was a materialistic one, that he was a god of the body rather than of the soul. Asclepius played a dominant rôle in antiquity, one might say, because of the Greeks' this-worldliness; like them he preferred the visible to the invisible, the life here to the Beyond.

To be sure, Asclepius was concerned primarily with men's bodies. It is the physician's duty to look after the bodily well-being of his patients, and Asclepius in doing so only fulfilled his appointed task. Nor can the fact that the god prolonged the lives of the sick be looked upon as objectionable, while at the same time human doctors are highly praised for similar achievements. Moreover, it is necessary to stress that the god or his worshippers in their concern for the body in no way forgot that health was only a means, not an end in itself, that men must be well in order to be able to do their daily work. What other than the realization of this truth is voiced in the fervent words of his priest (T. 428) : " How shall I come to your golden abode, O blessed, longed-for, divine head, since I do not have the feet with which I formerly came to the shrine, unless by healing me you graciously wish to lead me there again so that I may look upon you, my god, brighter than the earth in springtime? "[1]

---

[1] The passage has been strangely misrepresented by Farnell, *Hero Cults*, p. 277, who prints it with the omission of all reference to the disease from which Diophantus suffered (podagra) and then says: " the tone approaches the warmth of some of the best and most personal of our Psalms." Such a statement, even granted that the poem shows a great strength of emotion, hardly seems appropriate. An attitude similar to that of Diophantus is evident from the words of Aristides (T. 317, 5) : " [Asclepius granted

Nor did the god disregard spiritual values. He recognized that mind and body should be harmonized, that the latter must even be subordinated if the true goal of life is to be achieved. What other than the awareness of such an interdependence of soul and body found expression in the prayer (T. 323): " Relieve me of my disease and grant me as much health as is necessary in order that the body may obey that which the soul wishes, and, to say it in one word, a life lived with ease "?[2] Or in the words of Julian (T. 324): " The Muses train our souls with the aid of Asclepius and Apollo and Hermes." Such utterances show that the god was not too great a materialist. He had remained in contact with the progress of laical medicine which from the Hellenistic centuries on acknowledged that the mind cannot work properly if the body fails, or impedes the soul, that health is a balance of corporeal and intellectual factors.[3] Nor had he stood aloof from the development of ethical thought that put more and more emphasis upon the values of the spirit. Certainly the ancients never felt that the body was detestable, that sickness was a grace bestowed by God.[4] Still, this does not mean that their concept of health was wholly somatic.

What is more, the god respected and enforced moral ideals. Not only did he sift his patients by refusing to help those whom he did not deem worthy of help, not only did he himself follow the highest moral standards,[5] he even expected those who came to him to strive for perfection as it was envisaged by philosophers. For in contrast to the outward purity of ceremonies and rites he desired for them the inner purity of the heart. It was Asclepius on whose temple at Epidaurus the words were inscribed: " Pure must be he who enters the fragrant temple; purity means to think nothing but holy thoughts " (T. 318).[6] Besides, the god expected his

men] other things individually, such as the arts and professions and all the various modes of life, using one common remedy for all pains and troubles, namely health."

[2] The significance of this prayer has been clearly recognized by L. Dyer (*Studies of the Gods in Greece*, p. 236), and W. Pater (*Marius the Epicurean*, Ch. III) uses it to illustrate the ancients' attitude toward health. The fact that the words occur in Aristides' *Oration on the Asclepiads* is hardly an objection to their being quoted for Asclepius.

[3] Cf. Edelstein, *Die Antike*, VII, 1931, pp. 258 ff. There is a fundamental difference between the theory of the fourth century B. C. (Diocles) and that of the first century A. D. (Athenaeus). While in the beginning physical exercises alone are mentioned, later on they are counterbalanced by reading and studying, especially as regards elderly people. Galen expressly mentions Asclepius' " psychic gymnastics " (T. 413).

[4] In this connection Aristides' testimony (T. 402) is of interest. Only because the god restores health to the sick, can disease be called almost advantageous; cf. above, p. 124.

[5] Cf. above, p. 113.

[6] T. 318 is taken from Porphyry. That he is copying this passage from Theophrastus has been shown by J. Bernays (*Theophrastos' Schrift über Frömmigkeit*, 1866, pp. 62 ff.). Theophrastus concurred with Asclepius' views, and they are in agreement also with the opinion of Plato (*Phaedrus* 279 B; cf. also Epicharmus, *Fr.* 26 [Diels-Kranz]).

patients, after they had been healed by him, to lead a better life in the future than they had done before, he urged them: " Enter a good man, leave a better one " (T. 319). So great was the responsibility which the divine physician felt for those whom he treated, so firmly was he convinced that health and freedom from disease must be regarded as the means to some higher attainment. Asclepius was a true son of Apollo, he was a teacher as well as a healer, and it is symbolic that some testimonies name in one breath as children of the god of Delphi Asclepius, the physician of the body, and Plato, the physician of the soul (T. 322).[7]

It seems, moreover, that the god was concerned not only with the life here but also with the Beyond. He himself was a *mystes*, one of those initiated into the sacred revelation of Eleusis (T. 564 f.). The evidence to this effect is late, to be sure, it goes back to the second century A. D. But a connection between Asclepius and Demeter is attested much earlier, at the end of the fifth century B. C. When the god came to Athens, he resided first in the temple of the goddesses of Eleusis (T. 720).

Strange as it is that Asclepius should have been received by Demeter rather than by his father Apollo,[8] it would be rash to draw any conclusion from this fact concerning Asclepius' particular interest in the Eleusinian cult. Perhaps Demeter opened her sanctuary to Asclepius because he arrived at the time of her festival, the Great Eleusinia. On the other hand, there was a natural affinity between the goddess of agriculture and the god of medicine. Both husbandry and medicine preserve life. Even later medical writers parallel the two: the one gives nourishment to the healthy, the other to the sick.[9] Besides, the Demeter cult, in the eyes of the Athenians of the fifth century B. C., was the foundation of civilization; agriculture, they thought, was the starting point of a refined life. Medicine too, in the opinion of the ancients, was a product of civilization; barbarians and primitive people, if they were not believed to live free of disease, were at least supposed to be unfamiliar with the medical art.[10]

Asclepius' attitude may have been determined by Delphic teaching, cf. Kern, *Religion*, II, pp. 125, n. 2; 126 f.

[7] Bernays (*op. cit.*, p. 76), quite rightly says: "Allmählich hatte sich Asklepios im griechischen Glauben aus seiner ursprünglichen Stellung eines nur leibliche Gesundheit schaffenden Heros zu der Würde eines höchsten Gottes, eines Spenders allseitigen geistigen wie leiblichen Heils (σωτήρ) emporgehoben." Cf. below, p. 133.

[8] Kern, *Religion*, II, p. 312, has rightly stressed this point. Latte, *Gnomon*, VII, 1931, p. 118, suggests that Telemachus, who introduced Asclepius into Athens, was a priest of Eleusis.

[9] Celsus, *De Medicina, init.*

[10] Concerning Demeter, cf. Nilsson, *Popular Religion*, pp. 56 ff. In 418 B. C. the Athenians invited all Greeks to send sacrifices to Demeter, a demand that Nilsson explains by their conviction that Eleusis was the cradle of agriculture and civilization.

Thus the peaceful god who relieved suffering naturally was a favorite of Demeter, the goddess of peace-loving people.

But even if the relation between Asclepius and Demeter originally was brought about by chance, or did not signify more than a harmonious friendship; even if the report that Asclepius was a *mystes* in the beginning simply was an elaboration on his stay at the sanctuary of Demeter, instigated by the feeling that Asclepius, like other heroes or demigods, needed the initiation into the Mysteries in order to become more pious and just, to become in every respect better than he was before[11]—still, the fact remains that Asclepius became a *mystes*, and that he was revered as such in Eleusis (T. 565a).[12] Moreover, Asclepius' worshippers, at any rate in the Christian era, were also devotees of Demeter, thus imitating their god's friendship with her. Many dedications to Demeter were found in Epidaurus.[13] Asclepius asked Apellas, his patient, during the period of the treatment to make offerings to himself and to Epione, as well as to Demeter and Kore (T. 519). It is hard to believe that such a sacrifice should have been without meaning.

The Demeter cult, apart from its importance for civilization, was supposed to give its worshippers an inkling of a higher existence, of the life after death, and consequently it was thought to convey also the true knowledge of life in this world. Demeter, as she was revered by men from the fourth century B. C. onward, was in ever increasing measure the goddess of other-worldliness.[14] Those whom the god treated, sick with grave illnesses, were on the borderline between life and death. Asclepius cured them; he saved them from Hades. In urging them to pray to the Eleusinian goddess, did he not imply that, after their torments and anguish, their narrow escape from death, he expected them to rectify their way of living and henceforth to act in accordance with the knowledge of the initiated, to see the things here in the true light of the things

That medicine is unknown to barbarians and to primitive races is stated by Hippocrates, *On Ancient Medicine* (*C. M. G.*, I, 1, p. 39, 11). The similarity between agriculture and medicine, between Demeter and Asclepius, serves also to explain why Machaon and Podalirius, in their medical activity, could be compared with Triptolemus (T. 282, *15*), or why Hippocrates, the servant of Asclepius, could be compared with Triptolemus, the dispenser of Demeter's gifts (T. 467).

[11] Diodorus, V, 49, 6.

[12] Cf. Kern, *Religion*, II, pp. 314 f., and Pauly-Wissowa, *s. v.* Mystes, XVI, p. 1350, 25 ff. That Asclepius was a *protomystes*, as has sometimes been claimed, is hardly correct; cf. Kern, *ibid.*, p. 1350, 60 ff.

[13] For the close relation between Epidaurus and Eleusis, cf. Latte, *Gnomon*, VII, 1931, pp. 118 f., and Kern, *Religion*, II, p. 315.

[14] Cf. Nilsson, *Popular Religion*, pp. 53; 59; 63 f. The ethical and metaphysical message of the Eleusinian cult is clearly established by such statements as that of Isocrates (*Panegyricus*, 28).

there? Did he not intimate that, although he had saved them this time, it was their lot as mortals eventually to suffer death, and that he wanted them to accept their destiny with that hopeful confidence which was accorded to men by divine revelation?

The god of medicine himself had nothing to tell about the life to come: as a physician it was not his task to concern himself with the Beyond; in saving men, he was concerned with this world.[15] Yet Asclepius seems to have felt that those who came to him, who were in the extreme peril of life, ought to be initiated into the mystery of eternity by those who were privileged to reveal it. This feature of the Asclepius cult is remarkable indeed, for the Greek gods in general showed little interest in transcendental problems. None of the Olympians had anything to teach about the other world; even Apollo was intent on forming the ethical attitude of men here and now, without regard for metaphysical doctrines, without appeal to a Hereafter.[16] The only ancient worship characterized by an experience other than that of this world was the cult of Eleusis. Asclepius, by allying himself with Demeter, by joining together this world and the other, gave to his worship a significance far beyond that of a merely materialistic healing cult.[17]

[15] It should be noted that only Asclepius Imouthes, the Egyptian Asclepius, was supposed to reveal the mysteries of creation (T. 331, *vv. 170 ff.*). Aristides, in a vision in which the wonders above and below were shown to him, and in which Sarapis was the main actor, saw this god together with his great protector, Asclepius (T. 325). Yet, the true mysteries of Asclepius, in the opinion of Aristides, were his healings; as the one through whom men live twice, he calls him the fairest and most perfect torch-bearer and leader of the mysteries (T. 402; cf. above, p. 125). It seems that the Greek Asclepius, the god to whom this inquiry is restricted, did not reveal mysteries. That he was believed to have been a *mystes* himself may explain why a grandson of his, he too named Asclepius, was made the pupil of Hermes Trismegistos (T. 329) whose fame was founded on his doctrine of salvation; cf. F. Ueberweg-K. Prächter, *Die Philosophie des Altertums*, 1926[12], p. 514, and W. Scott, *Hermetica*, I, 1924, pp. 286 ff. Naturally, the advocates of magic likewise tried to use Asclepius for their purposes. Those who sought a short cut to knowledge would conjure up the inventor of medicine and ask him for a share in his wisdom (T. 326). More modest people wanted him to enhance the efficacy of love potions (T 327), and it was as helper in love affairs that Asclepius was reproached by the Christians (T. 330). A magic formula by which the leader of magicians induced the god to appear is preserved (T. 328), and here, in a complete reversal of the old tradition (cf. above, p. 45, n. 80), Asclepius seems to have been credited with a descent into the nether world. All these features, however, are late reflections of the growing influence of mysticism and magic and have little to do with the genuine Asclepius religion.

[16] Cf. *e. g.* Kern, *Religion*, II, p. 206.

[17] The strong connection between Asclepius and Demeter is generally recognized by modern scholars. Yet they are wont to minimize its importance (cf. *e. g.* Farnell, *Hero Cults*, p. 244: " His connexion with the Eleusinian mysteries is merely that of an outsider, he is admitted to them as a new denizen of Athens "; Wilamowitz, *Glaube*, II, p. 474: " Eine innerliche Verbindung . . . bestand nicht "). It seems to me that a relationship between two deities that lasted through centuries can hardly have failed to become meaningful in the opinion of the pious.

Finally for men who believed in Asclepius, the god had a message not only when he deigned to restore them to life. Aristides avers his conviction that if the god were unwilling to help him, this would mean that he thought it time for him to die (T. 464b). In the opinion of the many, Asclepius was the judge of life and death (T. 453). As an umpire may disqualify a competitor from a contest, so Asclepius could disqualify an individual from life. In the worship of the healer of diseases, meaning was given to death; the decision of man's ultimate fate was not in the hands of blind fortune, but lay with divine providence. The god was all-powerful. If his assistance was not forthcoming, he was believed to have reasons for not acting in behalf of those who besought him, and men reconciled themselves to their destiny by acknowledging the superior judgment of Asclepius.[18]

Nor did the divine physician, it seems, entirely abandon his devotees in their last hour. To be sure, Asclepius barred from his sanctuaries those who were about to die (T. 488); for like all the other gods, he was afraid lest death might alloy his divine purity. None of the ancient deities remained near the dying. Hermes escorted the dead and Hades received them, but in the dire moments that precede death man was alone. One votary of Asclepius, however, and no less a one than Socrates, thought of this god when he was dying. His last words were: " Crito, we owe a cock to Asclepius. Pay it and do not neglect it " (T. 524).

The ancients were puzzled by the fact that Socrates upon taking leave of life should have felt himself in the debt of the god of medicine, of the healer of disease, and some people tried to laugh away the enigma. " Poor fellow," says Lucian (T. 525), " he drank hemlock before he had even paid the cock to Asclepius," as if Socrates had forgotten to offer a sacrifice which he had once promised to Asclepius, and in the face of death remembering his pledge had asked his friends to act in his stead. The Neo-Platonists, however, answered that Socrates, being a watchful and pious man, could not possibly have failed to give to the god what he had vowed to him on a previous occasion. It must have been in the moment of death, they rightly felt, that Socrates incurred his obligation to Asclepius (T. 527).[19] Socrates, they therefore decided, was suppli-

---

[18] The testimonies concerning Asclepius, the judge of life and death, are of late origin. That Artemidorus characterizes him in this way (T. 453) proves at least that this interpretation was quite common at the end of antiquity. On the other hand, it seems natural that death that was not averted by the god of medicine should have been considered as unavoidable. One will therefore not hesitate to ascribe this concept even to earlier periods.

[19] The interpretation rejected by the Neo-Platonists is still considered probable by A. E. Taylor, *Socrates*, 1933, p. 119. However, one need only remember the beginning

cating the healer of disease because the soul in the process of generation becomes sick, and at the point of death needs help so as to become sound again (T. 526). Or it was believed that Socrates, in accordance with the advice of the Oracle, desired to return to his native land singing the paean, since the soul dismissing its cares requires Asclepius' foresight (T. 527).[20]

Thus Socrates' last words became fraught with deep meaning. The god who for people at large was the one who granted life and happiness in this world, for those initiated into philosophy was the one who opened the path into the Beyond, the one on whom they relied when the soul escaped from the prison of the body. Asclepius, says the Orphic Hymn (T. 601), the giver of gentle gifts, the savior, grants a good end to life. It may be that this Neo-Platonic or Orphic interpretation of Socrates' cryptic remark is correct. It may be that Socrates was not so much thinking of the healer of disease, that in his subtle irony he was invoking the reviver of the dead. For Socrates, life was death rather than illness, and death for him marked the beginning of real life.[21] Thus he might well have thanked Asclepius because he was redeeming him from that death which is called life, and because he was restoring him to that life which is called death. Whatever its true intent, Socrates' vow to Asclepius became a watchword of later generations. When a new god asked for man's allegiance, philosophers recalled the last words of Socrates. They did not feel that they were in need of a new revelation. Asclepius led them to immortality and to the true life, just as Christ promised to do for his adherents.[22]

of the *Phaedo* and Socrates' punctiliousness in fulfilling his obligations to Apollo, in order to agree with the Neo-Platonists that Socrates' vow could not possibly have concerned an event of the past.

[20] Modern commentators in general follow the Neo-Platonic lead, although they try to remove the tinge of Neo-Platonic philosophy. According to them, Socrates considered life a fitful fever from which he hoped to recover with the help of Asclepius; or he trusted to awaken, cured from the disease of life like those who recovered from their illness in the Asclepieia, cf. *e. g.* R. D. Archer-Hind, *The Phaedo of Plato*, 1894[2], p. 146; M. Wohlrab, *Platons Phaidon*, 1895, p. 158; J. Burnet, *Plato's Phaedo*, 1937, p. 118. Other interpretations are given by Wilamowitz, *Platon*, II, 1919, pp. 57 f.; J. Heiberg, *Danske Videnskab Selskabs Forhandl.*, 1902, p. 106.

[21] *Phaedo*, 64 A ff.

[22] T. R. Glover, *The Conflict of Religions in the Early Roman Empire*, 1923[10], p. 337, has stressed the importance which Socrates' vow to Asclepius assumed in the discussion between the Christians and the pagans. T. 528 ff. show how much trouble the Church Fathers took in belittling Socrates' last words (only after the victory of Christianity did Prudentius give a fairer interpretation of Socrates' intent, though one colored by his Christian ideals; the philosopher, he says [T. 531], appealed to Asclepius so that "the god who attends patients may deign to be just to them when they are dying"). It is perhaps not amiss to point out that Socrates' sacrifice to Asclepius was obviously intended to be made when the jailor brought him the drink of hemlock; at that moment

7. ASCLEPIUS AND CHRIST

Asclepius was firmly entrenched in his position as one of the great and most popular ancient gods, when at last the struggle between paganism and Christianity ensued. That he possessed real power the Christians denied as little as they refused to acknowledge the reality and strength of the other heathen deities. Asclepius was able to heal, they admitted (T. 113; 332a). But, as was true of all those gods and goddesses whom they dared to oppose, they believed that Asclepius was weaker than their own Lord,[1] and they delighted in pointing out that unlike Christ, he could not " command the wicked spirit of one out of his senses to come out of the man " (T. 333). Nor did they have any doubts that Asclepius, the performer of miracles, used his might for evil purposes, that he was an evil spirit, one " who does not cure souls but destroys them " (T. 330). This too was one of the usual charges brought forward by Apologists and Church Fathers alike in regard to the hated divinities in which the Greeks and Romans believed.

Yet such common protestations, found over and over again in the preserved evidence, by no means exhaust the Christian polemics against Asclepius, nor do they touch upon the essentials of the argument over Asclepius that, especially in the beginnings of Christianity, raged between the propagators of the new creed and the defenders of the old religion. If one studies the testimony more carefully, one soon discerns behind the familiar accusations and invectives a tone of uneasiness, an apprehension which is not apparent in the Christian censure of Zeus or Apollo, of Hera or Athena. There are outbursts of singular bitterness, there is unequalled concern when the criticism turns to Asclepius. He is, says Lactantius, the archdemon (T. 849b). Tertullian goes so far as to call him " a beast, so dangerous to the world " (T. 103).[2] And Eusebius still claims that

Socrates wished to make an offering to " some god " (*Phaedo*, 117 B ἀποσπεῖσαί τινι). But as the warden tells him, the whole drink is required to effect death. Socrates therefore omits the libation. But is it not likely that he is already thinking of Asclepius, to whom as he later says " we owe a cock," and that this most common gift to the god is meant to take the place of the libation which Socrates himself could not offer? Socrates' sacrifice, then, would be connected with his " last supper "; he certainly considers the drink a meal of which it was customary to dedicate a share to the gods, especially to Zeus the Savior, or to the Good Daemon. These circumstances would make the offering seem even more offensive to the Christians.

[1] For this attitude of the Christians toward the pagan deities, cf. A. D. Nock, *Sallustius*, 1926, p. LXXIX, n. 177.

[2] For the interpretation of this passage and the attitude of the Latin Apologists toward Asclepius, cf. Th. Lefort, *Musée Belge*, IX, 1905, pp. 212 ff. Incidentally, the fact that the serpent was the typical symbol of Asclepius, that he himself appeared in the form of a snake, as Arnobius points out at length (T. 852), may have contributed

he is the one who draws men away from their true Savior (T. 818). It seems that the Christians themselves realized that in Asclepius they faced their strongest enemy, the most dangerous antagonist of their Master.[3]

Why should they have been of this opinion? Simply because the Asclepius worship in the Christian era had spread all over the world and was a living force, whereas other cults, even if still respected, had lost much of their former hold over the imagination and the hearts of men? This fact may have contributed to their rancor against Asclepius, but it hardly explains their preoccupation with him. Nor is it likely that they feared Asclepius so intensely because like Christ he was a savior—most of the gods of late antiquity were called saviors, though it may be true that Asclepius was honored with this title more often than any of his divine associates. The decisive reason for the disquiet aroused in Christian writers by Asclepius lay in the fact that in the early Gospel Jesus appeared as a physician, as a healer of diseases. It was such an interpretation of the new god's mission that made him resemble Asclepius, the god of medicine, more than any other pagan divinity. It was the similarity between the deeds of Christ and of Asclepius that was bound to heighten the controversy between the Christian faith and the Asclepius religion.[4]

---

to the Christians' dislike of this god and may have made it easy for them to see in him the incarnation of the Devil. But the true reasons for their opposition to the god of medicine lie much deeper.

[3] With this statement I do not intend to take sides in the discussion regarding the influence which Oriental religions had in halting or obstructing the progress of Christianity. Their resistance against the new faith is usually seen as the stongest counterforce, e. g. Deubner, *Die Römer*, p. 504. But whatever the power of the cults of Mithras and Isis in the third century and later, in the first two centuries of our era the Greek cults were uppermost in the consideration of Christian writers; cf. *The Cambridge Mediaeval History*, I, 1924, p. 92. Moreover, the actual balance of power is only one of the factors to be taken into account by the interpreter. There was also a spiritual fight. The new creed had to defend its tenets against those of the other sects, and it is worth asking to what extent any of the earlier religious concepts were dangerous to the novel dogma because of their similarity to the Christian teaching, or because in certain points they seemed even superior. It is this clash of ideas with which I am mainly concerned, and it is in this struggle over values that Asclepius, as will become clear from the intepretation of the testimonies, appeared as *the* antagonist of Christ. It is impossible, of course, to discuss the subject in full here, and every attempt at giving a general picture of the situation necessarily involves a certain arbitrariness, the more so since in the first centuries there was no unified Christian dogma to which one could refer—yet the main facts can certainly be established.

[4] The fact that Asclepius stood in the foreground of the discussion between pagans and Christians has been noted by some modern scholars, cf. e. g. Thraemer, Pauly-Wissowa, II, p. 1662, 11 ff. But it was only through A. v. Harnack's study *Medicinisches aus der ältesten Kirchengeschichte*, 1892, pp. 89 ff., that the debate was set in its proper perspective. Harnack showed that Christ at first was seen as a physician and as such compared to Asclepius; he attempted to illustrate " the extent to which the Gospel in the earliest Christendom was preached as medicine and Jesus as a Physician, and how

For there was no question: Jesus' achievements when healing the lame and the paralytic and when resurrecting the dead were similar to those of Asclepius, even identical with them (T. 94). Therefore the heathens could compare the blessings bestowed by Christ to those bestowed by Asclepius (T. 584), they could even claim: what Jesus does, he does "in the name of the god Asclepius" (T. 334). The only answer left to the Christians beset by such contentions was the affirmation: "When they learned about the prophecies to the effect that He would heal every disease and would raise the dead, they brought forward Asclepius" (T. 332), or, "when the Devil brings forward Asclepius as the raiser of the dead and the healer of the other diseases, he has imitated the prophecies about Christ" (T. 95).[5] However, if from the Christian point of view Asclepius had come into the world to cheat men out of their salvation by his miracles (T. 818), to the heathens Christ naturally seemed but another Asclepius.[6]

To be sure, despite the similarity of the two physicians, fundamental differences existed between them. That which was revolutionary in Christ, that which was novel and unheard-of did not find any correspondence in Asclepius: Jesus came to heal not only the sick in body and soul; he extended his help to "the sinners and publicans." Asclepius, as a Greek god, had rejected those who were impure, those who did not think holy thoughts. In this respect—and the distinction was a crucial one, for it set Christianity apart from paganism— Asclepius and Christ were at variance.[7] Yet, just because the heathens were reluctant to accept the new

the Christian Message was really comprehended by the Gentiles as a medicinal religion"; cf. *id.*, *History of Dogma*, transl. by N. Buchanan, I, 1901, p. 147, n. 1; cf. also *Expansion of Christianity*, pp. 121 ff., where the matter is discussed in a "fresh version." Harnack has also pointed out that reference to Asclepius as the "Soter κατ' ἐξοχήν" (cf. *e. g.* Thraemer, *ibid.*) is not enough to explain the controversy between the followers of Asclepius and of Christ, rightly stressing the fact that Zeus and Apollo at that time were also addressed as saviors. Among the few who have considered or elaborated Harnack's thesis is S. Angus, *The Religious Quests of the Graeco-Roman World*, 1929, pp. 414 ff.

[5] Justinus' mode of reasoning is typical of the Christian conviction that whatever the prophecies of the Bible, the Devil will try to match them. Consequently the deeds of Asclepius were understood as an attempt to nullify the deeds of Christ.

[6] The significance of Asclepius' "glorious deeds" has been analyzed above, pp. 114 f., where it has also been pointed out that such a demonstration of the existence of the divine is not unlike the one sanctioned by the New Testament. It is interesting to note for how long a time the parallel Christ-Asclepius was remembered. Alcuin still calls Asclepius the false Christ, the "Scolapius falsator" (T. 336a). And when the purism of renaissance humanism forbade the use of words foreign to classical Latin, the name of Asclepius, or that of Apollo, was used to designate Christ; cf. E. Norden, *Die antike Kunstprosa vom VI. Jahrhundert v. Chr. bis in die Zeit der Renaissance*, II, 1923, p. 776, n. 1.

[7] K. Holl, *Die Antike*, I, 1925, p. 166, rightly insists on this point which was perhaps not sufficiently emphasized in Harnack's analysis.

dogma that God would commune with sinners and would heal them, they clung the more steadfastly to Asclepius, who had achieved the same as Christ without adopting an attitude so repulsive to the ancients, without even demanding faith from his adherents.[8] The Christians, on the other hand, had to attack Asclepius all the more violently.

Still, to find fault with him was difficult indeed. The usual incriminations against the pagan divinities were of no avail in regard to Asclepius. In his case, it was impossible to brand the absurd notion that God should be one of the cosmic elements; Asclepius had never been identified with any of them. Nor could one dwell on the immorality of the myth attached to Asclepius; the tales recounted about him were unobjectionable.[9] On the contrary, in addition to the similarity of the deeds of the two saviors, which even the later Christians seem to have found disconcerting,[10] there was a disturbing resemblance in their way of life and in their characters. Christ did not perform heroic or worldly exploits; he fought no battles; he concerned himself solely with assisting those who were in need of succor. So did Asclepius. Christ, like Asclepius, was sent into the world as a helper of men. Christ's life on earth was blameless, as was that of Asclepius.[11] Christ in his love of men invited his patients to come to him, or else he wandered about to meet them. This, too, could be said of Asclepius.[12] All in all, it is not astonishing that Apologists and Church

---

[8] The matter is summed up in the debate between Celsus and Origenes (e. g. T. 292 f.; 332a), but apparently was also discussed in Porphyry's book Against the Christians, e. g. T. 506; cf. in general Harnack, Expansion, p. 130; Holl, op. cit., p. 168; T. R. Glover, The Conflict of Religions in the Early Roman Empire, 1923[10], pp. 239 ff.

[9] Concerning the physical interpretation of Asclepius' godhead, cf. above, p. 78 and p. 105. Only once is Asclepius referred to as a sun-god when the belief in such deities is discussed in general by Eusebius (T. 298). For the dislike which Christians and "true" pagans felt in regard to the equation of gods with elements, cf. Nock, op. cit., pp. XLIII; XLVII. That the traditional mythology, to the Christian Fathers, seemed the most vulnerable point of paganism has been stressed by G. Murray, Four Stages of Greek Religion, 1912, pp. 167-8.

[10] To be sure, after the Christian dogma had been more fully elaborated, the figure of Christ assumed significance beyond that of a mere healer. Yet Eusebius (Hist. Ecc., X, 4, 11) still said of Jesus that "like the medical man He sees terrible sights, touches unpleasant things, and the misfortunes of others bring a harvest of sorrows that are peculiarly His." This characterization is taken from Hippocrates, Περὶ φυσῶν (C. M. G., I, 1, p. 91, 5-7); cf. J. Ilberg, Berl. Phil. Woch., 1893, pp. 402 f. In exactly the same words the healing of the god Asclepius had been depicted (T. 392). I should also draw attention to the fact that the name Iason in the Hellenizing period was often taken as the Greek analogon of Jesus and that the word Jesus was derived from the same root as Iason, namely ἰᾶσθαι, even by Eusebius; cf. The Catholic Encyclopedia, s. v. Jesus, VIII, p. 374.

[11] For Asclepius' character and "heroic" deeds, cf. above, pp. 38 ff.; 73 f. Concerning his mission as a helper of men, cf. above, p. 121. Just as Asclepius had been sent by Apollo, so Christ had been sent by God in his φιλανθρωπία (Ep. ad Titum, 3, 4; cf. also John, 3, 16).

[12] Cf. above, p. 112. Compare moreover T. 406 with Matthew, 11, 28: "Come unto

Fathers had a hard stand in their fight against Asclepius, in proving the superiority of Jesus, if moral reasoning alone was to be relied upon.

Nor were the metaphysical or theological problems fewer or less cumbersome for the followers of Christ than they were for the supporters of Asclepius. The nature of the godhead of the two saviors was indisputably identical: both were man-gods. Like Asclepius, Christ was the son of God and of a mortal woman. The story of Christ's birth in many details resembled the birth saga of the divine Asclepius.[13] Jesus had died, not through God to be sure, but through men, as the Christians rightly insisted (T. 112). Yet through God he had risen to heaven, as had Asclepius. Some of the early Christians even thought that their Master had been made immortal on account of his virtue. The same had been asserted of Asclepius.[14] Justinus certainly was most convincing to the heathens when he urged them with reference to Asclepius to believe in Christ because they had believed in the divinity of sons of their own gods (T. 335). Nor was it easy for the Christians to refuse assent to the godhead of Asclepius when they were asked by the pagans to admit his divinity (T. 8; 234). Moreover, Asclepius acted as son of Apollo; Christ came in the name of his Father.[15] The speculations concerning the Christian Trinity were dangerously near to the speculations concerning Asclepius, the third from Zeus (T. 303). While Christ was human and divine at the same time, Asclepius was called a terrestrial and intelligible god (T. 259).[16]

me all ye that labour and are heavy laden, and I will give you rest." Concerning Christ's seeking men and not waiting for them, cf. also Holl, *loc. cit.*, p. 167. C. Bonner (*Harvard Theological Review*, XXX, 1937, p. 123), has rightly pointed to the fact that the belief in Asclepius' providence is similar to that in the providence of Christ.

[13] For Asclepius as a man-god, cf. above, p. 77. For Christian parallels with the Asclepius birth legend, cf. Kern, *Religion*, II, p. 308, n. 3.

[14] Cf. above, pp. 75 ff. Paulus of Samosata (third century A. D.) said of Christ: ἐκ προκοπῆς τεθεοποιῆσθαι; cf. J. H. Kurtz, *Lehrb. d. Kirchengesch.*, I, 1892, p. 135; cf. also Porphyry's statement (Augustine, *De Civitate Dei*, XIX, 23): "The gods have declared Christ to have been most pious; he has become immortal, and by them his memory is cherished. Whereas the Christians are a polluted sect, contaminated and enmeshed in error"; cf. *The Cambridge Mediaeval History*, I, p. 94.

[15] Dyer, *Greek Gods*, p. 240, claims that the Romans were wont to address Asclepius as *filius dei*. I am unable to substantiate this claim. But this much is certain: Asclepius even in Epidaurus acted together with Apollo, his father, cf. above, p. 99, and T. 428, where the priest Diophantus invokes Asclepius "in the name of your father," a passage also referred to by Dyer, *ibid.*

[16] It is Aristides who speaks of Asclepius as the third from Zeus. Discussions of three kinds of providence or of three different gods were current among the Neo-Pythagoreans and the Platonists of the first and second centuries A. D.; cf. F. Ueberweg-K. Prächter, *Die Philosophie des Altertums*, 1926, pp. 556; 521; .542. The influence of these schools on patristic literature is certain, cf. *ibid.*, p. 556. It is usually overlooked that Asclepius had taken his place in the pagan "trinity" at such an early date. On the other hand, the characterization of Asclepius as a terrestrial and intelligible god is already to be found in Artemidorus. Again this concept must have been widely current in the second century A. D.

It may be that the Christian dogma was to a certain extent responsible for the claim of the later Neo-Platonists that Asclepius had been begotten by his father in the world, but that nevertheless he had been eternally existing by his father's side before the universe was made (T. 305). The important fact is that with regard to Asclepius such a reconciliation was in line with earlier beliefs and not merely an artifice.[17] Heathens and Christians were faced with the same difficulties in defining the position of their gods. They had to be Homoousians or Homoiousians, Monophysites or Nestorians. In short, the similarity between Asclepius and Christ was deep-rooted; it was founded in the very essence of the two figures. In the historical process that had shaped and reshaped the concept of Asclepius this god had become an anticipation, as it were, of Christ who was to be proclaimed to men.[18]

This truth the Christians themselves, it seems, were willing to recognize as soon as they learned to admit that the pagans had not been entirely deprived of the light of knowledge, that in fact an inkling of truth was given even to them, that the natural revelation had foreshadowed the clearer and truer Revelation of the Bible. Some of the great Christian teachers, at least, did no longer regard Asclepius as an evil spirit, invented by the Devil to be the false Christ; rather did they find in him and in his teaching the enigmatic prediction of Christ and of his message, as they now began to understand it. For as Clement says: " And we must adapt ourselves to Him through the divine charity, so that we may see the like through the like, listening to the truth of the word without fraud and purely after the manner of the children who obey us. And that is what he referred to as in a riddle, whoever he was, who inscribed over the entrance of the temple of Epidaurus: ' Pure must be he who enters the fragrant temple; purity means to think nothing but holy thoughts.' And ' except ye become as little children ye shall not enter in the kingdom of heaven.' For there is the sanctuary of God, set up over the three foundation-stones, Faith, Hope, Charity " (T. 336).[19] It is true, Christianity

[17] Farnell, *Hero Cults*, p. 279, has claimed a dependence of the Neo-Platonists (Julian) on the Christian dogma. On the other hand, he asserts (*ibid.*, p. 234) : ". . . and when at last vanquished by Christianity it [the Asclepius cult] left its impress on the vanquisher." Farnell does not give an account of those similarities which he has in mind.

[18] In this connection I should also draw attention to Asclepius' ethical teachings and to his relation to the Mysteries, cf. above, pp. 125 ff.

[19] Bernays, *op. cit.*, pp. 76 f., has noted Clement's praise of the Epidaurian inscription, but has failed to draw any conclusion as to the relation between Christianity and Asclepius religion. Clement was deeply influenced by Musonius, cf. H. Lietzmann, *Christl. Literatur, Gercke-Norden*, I, 5, 1923, p. 7; and it was Musonius' definition of the nature of God that seemed characteristic of the religious beliefs of the post-classical centuries, cf. above, p. 114.

and Asclepius religion, despite the many features which they had in common, were not identical. That which formed the essence of the new faith was unique.[20] And yet, as far as Christianity can be compared at all to any of the Greek and Roman cults, the Asclepius ideal seems nearest to the ideal of Christ. The Greek god of medicine was the most accomplished precursor of the god of a higher Gospel that paganism had brought forth.[21]

From his humble position as a friend of Chiron Asclepius had risen to be a symbol of human daring; he had become the patron of physicians; he had assumed divine prerogatives and in an unparalleled triumph secured for himself a foremost place among the deities of old. When Christianity was made the religion of the state, Asclepius, so the Christians charged, was still revered in secret corners, and libations and prayers were still offered to him (T. 5). Of all the Greek gods he persisted longest in exercising his full and undiminished power.[22] For he had been the embodiment of the highest expectations which men cherished, of the highest values which they had known before the ancient world was shattered and a new world was built in which men lived for other hopes and other virtues.

---

[20] Holl, *op. cit.*, pp. 162 ff., especially 166, has rightly shown that it is impossible to explain Christianity by comparing it to other religions, by adducing historical parallels which, however close, fall short of making the new message understandable.

[21] In recent discussions the similarity of the Heracles figure and that of Christ has been underlined, cf. F. Pfister, *Arch. Rel. Wiss.*, XXXIV, 1937, pp. 42 ff. But the resemblance between the Savior Heracles and the Savior Christ, between the man-god Heracles and the man-god Christ, concerns external facts, while that between Christ and Asclepius covers essential features, the spiritual meaning of the teaching. Of course, there were many more religions against which Christianity had to fight and by which it may have been influenced in turn (cf. above, n. 3), yet, it would be difficult, I think, to adduce testimonies from any other religion that show such comprehensive similarities as are to be found in the evidence concerning Asclepius. For the supposed identity of early artistic representations of Christ with those of Asclepius, cf. below, p. 224, n. 18.

[22] Cf. M. Besnier, *L'Ile Tibérine*, 1902, pp. 240 ff., who cites various Christian authors attesting the popularity of Asclepius in late centuries. Cf. also below, pp. 255 ff.

III. TEMPLE MEDICINE

Thus far, in the discussion of Asclepius and of the history of his worship mythological and religious problems have stood in the foreground. Only occasionally has reference been made to the practical aspect of the cult, to the miracle cures which the god performed in behalf of those who implored his help. Important as Asclepius' function may have been in giving oracles, in safeguarding the family, in preserving the order of the cosmos,[1] it was his dream healings which in antiquity constituted his greatest claim to fame. In recent times, on the other hand, the fact that Asclepius indulged in what seems to have been mere trickery forms the cornerstone of the criticism heaped upon the god himself and upon those who believed in him. The condemnation of divine healing, of temple medicine, is usually accompanied by high praise for the achievement of Greek physicians; a sharp dividing line is drawn between the religious beliefs, or delusions, of the ancients and their scientific accomplishments.[2]

Such an antagonism of science and religion may be self-evident to the modern mind, yet it was foreign to ancient thought. Especially in medicine, divine and human knowledge and action, though of a different nature, did not exclude each other. To be sure, in sickness people first took recourse to human physicians. The scope of their competence, however, was limited. There are cases so desperate, diseases so grave, the doctors themselves admitted, that human endeavor fights them in vain. Still, the divinity may be able and willing to help. It is not only the right, but almost the duty of everyone to ask for divine assistance where human power fails.[3] Greek physicians never opposed religious medicine, much as they abhorred magic. Consequently, the Asclepius cult was a friendly ally rather than a hated enemy of the medical art.

Moreover, one must not forget that the god of incubations continued to be the patron of physicians, the guardian of their craft. Just as of old

---

[1] Cf. above, pp. 104 ff.

[2] Cf. *e. g.* Wilamowitz, *Glaube*, II, p. 231, and above, pp. 109; 117.

[3] Cf. *e. g.* Hippocrates, *C. M. G.*, I, 1, p. 27, 13 ff.; II, p. 289 [Loeb]: "Physicians have given place to the gods. For in medicine that which is powerful is not in excess." Cf. Edelstein, *Bull. of the Inst. of the Hist. of Med.*, V, 1937, pp. 243 ff. That the attitude of later physicians did not differ from that of early doctors is shown by statements of Rufus (T. 425) and Galen (*e. g.* T. 436). In general, cf. also Herzog, *Wunderheilungen*, pp. 149 ff.

doctors had venerated the hero Asclepius, so later they honored the god of Epidaurus (T. 338 f.). To him, they offered sacrifices according to their ancient custom (T. 552) ; for him, they staged processions (T. 568) and games (T. 573). He was the protector of their guilds.[4] He remained the founder of their art (T. 341), the inspirer of their science (T. 468). Although Asclepius was now a healer in his own right, he still presided over the medical art (T. 342 ff.) ; he still was the leader of medicine (T. 340) ; physicians still were called Asclepiads, and nobody who was not a good doctor, could properly carry this name (T. 346 f.).[5]

That this was so, also explains why men, despite their preoccupation with the deeds of the god who was sought in his temples, did not cease to speculate on his share in the development of scientific medicine. The old saga saw in Asclepius the savior from death.[6] Later, his attainments were specified more in detail. Many were of the opinion that Asclepius had really been the first physician, the first to study medicine (T. 350 ff.). Or the view was held that Apollo had invented medicine and had imparted his art to his son who, as his earthly representative, had introduced it among men (T. 344; 356 f.).[7] The more scholarly assumption was that Asclepius had improved upon medicine, but had done so in such a measure that he was regarded as its inventor (T. 355). Medical knowledge was supposed to have no absolute beginning; men had always been familiar with some drugs, some methods of healing, and medicine as taught by Asclepius was certainly not a creation out of nothing.[8] It was a refinement of a crude art already existing that Asclepius had brought about (T. 354) ; his knowledge, though great, was in no way complete,

[4] For medical guilds in later times, cf. E. Ziebarth, *Das griechische Vereinswesen*, 1896, pp. 96 ff. ; Poland, *Geschichte d. griech. Vereinswesens*, pp. 514 ff. ; Pauly-Wissowa, *s. v.* Berufsvereine, Suppl. IV, pp. 176 f. Concerning medical schools and Asclepius, cf. also R. Herzog, *S. B. Berl.*, 1935, p. 1006. Guilds were based on a religious foundation, as were all associations in antiquity (cf. Wilamowitz, *Gesellschaft*, pp. 52 f.), even though this fact may have been of little practical consequence; cf. Pauly-Wissowa, *ibid.*, p. 202, 24 ff.

[5] As the leader of medicine, Asclepius was also the patron of certain cities which were famous for their medical art (T. 345). For Asclepius as symbol of medicine, cf. T. 467-81.

[6] Cf. above, p. 45.

[7] Asclepius is given here a task which, according to Aristides, was that of his sons: the promulgation of medicine all over the earth (T. 282, *11 ff.*). In late speculation, of course, Apollo as the originator of Asclepius' knowledge was replaced by Oriental gods (T. 358). Proclus ascribed divine medicine to Asclepius, while he regarded mortals as the representatives of medicine resulting from theory and experience (T. 312; cf. 381).

[8] Such a theory is probably influenced by Peripatetic speculations. The Peripatetics held that the beginnings of all arts and crafts were unknown. Their so-called inventors were men whose names happened to be the first to be remembered, but who actually had done no more than bring about a decisive progress, cf. Aristotle, *Poetics*, 1448 b 27 ff. (Homer) ; Theophrastus, *Doxographi Graeci* [Diels], p. 475, 10 ff. (Thales).

it could be extended by later generations (T. 349).[9] Likewise, it was maintained that Asclepius had transformed the experience of men before him into a rational system (T. 356). Or it was believed that he had established scientific medicine by teaching that the opposing qualities which constitute the human body should be balanced and reconciled (T. 348).[10]

More specifically, Asclepius was reputed to have invented surgery and pharmacology (T. 355), in fact the oldest branches of medicine.[11] Or the discoveries of empirical medicine (T. 359) and of bedside medicine (T. 360) were ascribed to him. He was even considered the representative of dietetic medicine (T. 361).[12] Certain cures, too, were traced to him: that of chronic diseases (T. 364), such as gout and consumption. Certain remedies and drugs were believed to be his findings (T. 365 ff.). And if not Asclepius, the god of Epidaurus, at least one of the three Asclepii distinguished in antiquity was credited with inventing the extraction of teeth, with the first use of purges, with the introduction of instruments and other medical means, such as the probe and surgical bandages (T. 379).[13]

The medical achievements of Asclepius, then, did not consist solely in his performance of miracles. He was also a scientist; his contribution extended from the discovery of slight medical details to that of whole departments of medicine, even to that of the art itself. This is the typical picture of the patron of a craft. In the same way, Daedalus was viewed as the inventor of tools needed by the artist, but he was also considered the discoverer of sculpture and architecture, the founder of the fine arts in general.[14] In turning to an investigation of the divine healing, as it was practiced by Asclepius throughout the centuries, one must keep in mind that this god was himself a physician, the protector of medicine, as it was developed by Hippocrates, by Herophilus, by Galen, and by many others famous in the history of science.

[9] Similarly the medical knowledge of Machaon and Podalirius is sometimes called imperfect in the sense that it was adequate only for their own age (T. 145).

[10] In passing it should be noted that Asclepius was also considered the healer of animals (T. 378).

[11] Cf. T. 142 f. According to Plautus (T. 362), the doctor sets a broken leg for Asclepius—apparently Asclepius, the surgeon.

[12] Curiously enough, Asclepius does not seem to have understood anything of obstetrics (T. 363). This may have some connection with the fact that birth could not take place in his temples (T. 488).

[13] Wilamowitz (*Isyllos*, p. 79, n. 52) is not correct in branding such statements as T. 379 as " Abgeschmacktheit der Theologen." Probably the various achievements attributed to Asclepius were treated in the literature περὶ εὑρημάτων; on this topic, cf. A. Kleingünther, *Philologus*, Suppl., XXVI, 1, 1933, pp. 143 ff.

[14] Cf. Pauly-Wissowa, *s. v.* Daidalos, IV, pp. 2004, 26 ff.; 2005, 31 ff.; 2006, 8 ff.

1. MODERN EXPLANATIONS OF TEMPLE CURES

When in the seventeenth century the testimonies concerning Asclepius' healings were for the first time critically examined by Meibom, a doctor who was a philologist and historian as well, he did not hesitate to accept the reality of Asclepius' cures at their face value.[1] To him, they were miracles; they were not, in a rational sense, medicine from which physicians could learn anything worth their while. Moreover, in Meibom's opinion, these miracles were deeds of the Devil, and being a Protestant, he could not desist from sneering at the Catholics in whose churches, as he charged, the saints repeated the deception of the pagan demons.[2]

Such a naive attitude toward the miraculous, reminiscent of that of the early Christians, was still almost universal in Meibom's time. It began to change slowly as Descartes' philosophy permeated natural sciences.[3] Around the middle of the eighteenth century, scholars in general no longer accepted the assumption that God's omnipotence, or the interference of demons, would make itself felt in supernatural actions contrary to common human experience, and this destruction of the belief in the self-evidence of miracles necessarily gave rise to the question how the healings of Asclepius had been brought about. That neither God nor Satan had performed them seemed a foregone conclusion. Men alone were responsible for the treatment given. The issue was: what procedure had they followed?

At first, the success of the cures was explained by the application of natural remedies, no matter what the pretenses made by patients and priests might have been.[4] But soon, a different interpretation was proffered. The miracles of the ancient god of medicine were taken as the first evidence of the efficacy of animal magnetism and somnambulism. Asclepius, or rather his priests, were considered forerunners of Mesmer, whose theories they seemed to have anticipated in their practices. Against such a claim, which made Asclepius a representative of the occult sciences, rationalistic philologists and theologians protested violently. They asserted that animal magnetism was unknown in antiquity, and at

---

[1] H. Meibom, *Exercitatio philologico-medica de incubatione*, etc., 1659.

[2] Cf. S. Herrlich, *Antike Wunderkuren, Wiss. Beilage z. Jahresber. d. Humboldt-Gymnasiums zu Berlin*, 1911, pp. 3-5; from Herrlich's survey of the early interpretations of the cures of Asclepius (pp. 3 ff.) I have drawn most of the material for my discussion of the subject.

[3] Cf. W. Dilthey, *Gesammelte Schriften*, II, 1923, p. 132.

[4] Cf. *e. g.* K. F. Hundertmark, *De incrementis artis medicae*, etc., 1749; Daniel Le Clerc, *Histoire de la Médecine*, I, 1729, pp. 64 ff.

the same time condemned incubations as impostures, as fantastic as somnambulism.[5]

The spiritists were not abashed by these strictures. Even in the nineteenth century they continued to regard Asclepius as one of their own kind.[6] Moreover, some scientists, accepting the correctness of the ancient reports, attempted to account for the healings performed by the neurotic disposition of the patients; they pointed to the influence which the soul has over the body. Humanists stressed the power of religious experience as a curative factor. The sick were healed, they contended, since they believed in divine help.[7] On the other hand, the number of those was increasing who thought that the cures of Asclepius had been without real success, mere charlatanry, the machination of clever priests who pretended that their ignorant contrivances were the work of a god, because they hoped in this way to raise the revenues of their temples.[8]

No unanimity of interpretation had been reached when in 1883 the scanty material on which the interpretation of the cures had been based so far—it had been composed in the main of Roman inscriptions and of the orations of Aristides—was greatly increased by archaeological discoveries. The Epidaurian tablets were published.[9] They seemed unrestrictedly to confirm the verdict of the sceptical scientists. It was now obvious to all that these so-called cures had been trickery and mere fraud.[10]

But again a reaction set in. First it was stated that the Epidaurian inscriptions could not be taken as evidence of the therapy actually performed in the temples of Asclepius; they were considered typical miracle stories, such as are to be found in all religions. The real cures, it was claimed, at all times consisted of the interpretation of dreams, which was provided by priests who were trained in medicine and who

[5] F. A. Wolff, *Beitr. z. Gesch. d. Somnambulismus aus dem Altertum*, 1787 (*Miscell. Halae*, 1802, pp. 283-430); J. F. A. Kinderling, *Der Somnambulismus unserer Zeit mit der Inkubation . . . in Vergleich gestellt . . .*, 1788. Cf. also Herrlich, *op. cit.*, p. 8.

[6] Literature of this type has been collected by Welcker (*Kleine Schriften*, III, p. 152). A late echo of the theory of somnambulism is to be found in C. du Prel, *Die Mystik der alten Griechen*, 1888, pp. 1 ff.

[7] For the scientific theory, cf. A. Gauthier, *Recherches historiques sur l'exercice de la médecine dans les temples chez les peuples de l'antiquité*, 1844; for the humanistic one, cf. Welcker, *op. cit.*, III, p. 124; he finds in Aristides an attitude similar to that of pietism.

[8] *E. g.* Ritter von Rittershain, *Der medizinische Wunderglaube und die Inkubation im Altertum*, 1878.

[9] P. Kabbadias, *Ephem. Arch.*, 1883, pp. 197 ff.

[10] Most emphatic was the verdict of H. Diels (*Nord und Süd*, 1888, pp. 29 ff.); cf. also S. Reinach, Daremberg-Saglio, *s. v.* medicus, III, 2, p. 1671; J. Heiberg, *Gesch. d. Mathematik u. Naturwissenschaften im Altertum*, p. 97 (*Aus Natur u. Geisteswelt*); cf. also Wilamowitz, *Glaube*, II, p. 231.

determined the treatment in accordance with rational procedures.[11] Later it was emphasized that, even though the temple of Epidaurus was given to deceitful miracles, other temples, like those of Cos and Pergamum, relied mainly on the help of physicians and on the benefit of a watering-place, together with a sound interpretation of the dreams of the patients. The miracle, as far as it was admitted to have come into play at all, was explained in accordance with the laws of hypnosis and suggestion or autosuggestion.[12] Finally the attempt was made to distinguish different periods of temple medicine which were characterized by different methods of healing: in pre-Roman centuries the god was supposed to have worked as the typical performer of miracles; but in the course of time he was believed to have become more rational, " he learned medicine "; the temples themselves became sanatoria; the dreams were cleverly interpreted so as to fit in with the other prescriptions; thus at least the cures brought about at the end of antiquity could be likened to modern scientific treatment.[13]

In all these more recent theories the priests of Asclepius—and there is hardly any doubt expressed that they were the main actors of the play—were no longer conceived of as tricksters; the deceitful priests had become good and benevolent physicians themselves. While it had once been assumed that laical medicine had been derived from clerical medicine, the latter now seemed dependent upon the former. The miracle, a dish so distasteful to the modern palate, had been cooked until it became acceptable and digestible; the irrational had been worked on until it finally evaporated into nothingness; the amoral had been sifted until it had cleared off a useful and orderly substance. Only a few scholars still dared to insist that from beginning to end the cures of Asclepius to the ancients had been a religious experience beyond all human understanding, and must be interpreted as such.[14] They were silenced, however, by their opponents' clamoring for a natural explanation of the phenomena in question.

---

[11] Cf. Thraemer, Pauly-Wissowa, II, p. 1686, 5 ff.; especially p. 1690, 8 ff.

[12] Wilamowitz, *Isyllos*, p. 37, first contrasted the " schwindelhafte Kurpfuscherei " of Epidaurus with the treatment given at Cos and Cnidus. R. Herzog, *Koische Forschungen und Funde*, 1899, pp. 202 ff., elaborated this view; cf. also *Wunderheilungen*, pp. 139 ff. The explanation of the miracles along the lines of modern psychological theories is outlined by Herzog (*Wunderheilungen*, pp. 67 ff.). O. Hirschberg, *Geschichte der Augenheilkunde*, 1899, p. 56, speaks of a " Vereinigung von Lourdes und Karlsbad."

[13] This distinction of different periods is implied by Weinreich, *Heilungswunder*, pp. 110 ff.; cf. also J. Ilberg, *N. Jahrb.*, XV, 1905, p. 8, 2 (*N. Jahrb. f. Pädagogik*, IV, 1901, pp. 309; 312) and the same, *Abh. Akad. Leipz.*, XLI, 1, 1930, p. 32: " Der Gott hat offenbar Medizin studiert, man sieht den Einfluss der Wissenschaft auf die Tempelpraxis um 100 nach Chr."

[14] Cf. Th. Lefort, *Musée Belge*, IX, 1905, pp. 197 ff.; Herrlich, *op. cit.*, p. 18.

From the modern point of view this demand is quite justified. Yet, it would not be improper to say that the healings of Asclepius are attested by witnesses of the events; that even though today miracles of this kind no longer occur, they may have occurred in antiquity, that they are historical facts just like many others which have to be accepted as true because their truth cannot be disproved. This was the attitude of the greatest among the rationalists of the Enlightenment in regard to the miracle cures of Christ.[15] It is the attitude of the pragmatic philosopher toward those who in our time claim that they are healed by their god. He acknowledges the reality of their experience, for it is proved by its practical fruits.[16] This should be the attitude of everybody who recognizes experience as the only criterion of truth, for he has no standard by which to contradict other people's experience. That the cures of Asclepius need any explanation in natural terms can be upheld only by the assumption on which the question was first founded in the eighteenth century: that God does not exist, or if He exists, that He never interferes in the affairs of this world. It is only on the basis of this hypothesis that one can and must ask how the patients of Asclepius got well if it was not the god who cured them.

2. ANCIENT REPORTS ON INCUBATION

Before attempting to pronounce judgment, before choosing between the various solutions proposed since the first critical examination of the testimony was undertaken, or suggesting another explanation if the older theories should prove inadequate, it is necessary to find out as exactly as possible what actually happened in the Asclepieia. This, to be sure, is not an easy task, or one which can be approached without hesitation. Not that there is any lack of material. Reports referring to healings from the fifth century down to the end of antiquity are preserved. But they are hardly of a kind that would seem satisfactory.[1]

Among the literary sources two comic poets have most to tell: Ari-

---

[15] Cf. Lessing, *Sämmtliche Schriften*, X, 1839, p. 36 [Lachmann].

[16] Cf. William James, *The Varieties of Religious Experience*, Lectures IV and V.

[1] Concerning temple medicine in general and incubations in particular, cf. Welcker, *Kleine Schriften*, III, pp. 89 ff.; Pauly-Wissowa, II, pp. 1686 ff.; cf. also *ibid.*, the articles Incubatio and Mantik; R. Caton, *Two Lectures on the Temples and Rituals of Asklepios* (*Otia Merceiana*, I, 1899); Walton, *Asklepios*, Ch. 5; L. Deubner, *De Incubatione*, Diss. Berl., 1900; M. Hamilton, *Incubation*, 1906; Weinreich, *Heilungswunder*, pp. 76 ff.; Herzog, *Wunderheilungen, passim*; C. R. Simboli, *Disease spirits and divine cures among the Greeks and Romans*, Diss. New York, 1921, pp. 57 ff.; cf. also W. R. Halliday, *On the Treatment of Disease in Antiquity*, in *Greek Poetry and Life*, Essays presented to Gilbert Murray, 1936, pp. 277 ff.

stophanes, the oldest witness of the cures of Asclepius (T. 420 f.), and Plautus, who is following a Greek original of the third century B. C. (T. 429 f.) ; [2] the one deals with what was going on in Athens, the other with the happenings in the temple at Epidaurus. It can hardly be expected that these authors should give a picture of the situation satisfying to the curiosity of the scholar or to be accepted without making allowance for poetic fantasy and comic license. On the other hand, Aristides' orations on Asclepius (e. g. T. 317) and his many casual remarks about the god's power, as it manifested itself mainly in Pergamum, are the utterances of a man who was a staunch believer, a partisan of Asclepius. He is pre-occupied with his own troubles. Hundreds of details which he gives, important as they may have been for himself, are of little interest for others, and naturally he is inclined to praise the god who helped him, rather than to investigate the divine doings.[3] Aelian knows of many cures performed by the god (T. 455 f.) ; but he is a bigot, if not a hypo-crite, who pretends to see the hand of divine providence in everything, in the sublime no less than in the trivial, and he finds it sufficient to proclaim the achievements of the god without expounding their particu-lars.[4] The statements of the Neo-Platonists, of men like Julian and Marinus (e. g. T. 462; 445 f.), are certainly honest, but they are written during a period not distinguished by its sense of criticism, nor are these philosophers given to fact-finding.[5]

As regards the inscriptions, they abound in details and tell at length about Asclepius' actions, but they are open to suspicion as to their objectivity. The most important among them are the ones found in Epidaurus: a whole set of stories (T. 423), which were inscribed on columns placed in such a way that everybody who came to the temple was attracted by them. They constitute almost an official record of the

[2] For the Aristophanes passage, cf. Weinreich, Heilungswunder, pp. 95 ff.; for Plautus, cf. Thraemer, Pauly-Wissowa, II, p. 1688.

[3] The literature on Aristides is very extensive. Of older interpretations that of Welcker, Kleine Schriften, III, pp. 114 ff., is still most valuable. Of more recent critics, cf. Weinreich, N. Jahrb., XXXIII, 1914, pp. 597 ff.; A. Boulanger, Aelius Aristide, 1923; Wilamowitz, S. B. Berl., 1925, pp. 333 ff.; C. Bonner, Harvard Theol. Review, XXX, 1937, pp. 124 ff. I should draw attention to a newly found statement of Galen in which he characterizes Aristides as a man whose soul was strong by nature, while his body was weak (Galeni in Platonis Timaeum Commentarii Fragmenta, ed. H. O. Schröder, C. M. G., Suppl. I, 1934, p. 33 [83]). To give all of Aristides' statements concerning Asclepius in the collection of the testimonies would almost have amounted to reprinting all of his works. A selection had to be made; some passages are occa-sionally referred to in the notes of the interpretation.

[4] For Aelian as a writer περὶ προνοίας, cf. Weinreich, Heilungswunder, pp. 120 ff.

[5] For the various stories told by the Neo-Platonists, cf. Weinreich, op. cit., pp. 35; 74; 90, n. 3; 94.

god's merits and virtues.[6] The many other individual Epidaurian dedications (*e. g.* T. 432) likewise, if not actually written by the priests, were certainly composed under their supervision. The same holds true of the Athenian inscriptions (T. 428), or those of Rome (T. 438), or of Lebena which outline the happenings at a place which was not one of the great world-centers of the cult but a second-rate, provincial sanctuary (T. 439 ff.).[7] Certainly, nothing detrimental to the dignity of the god was allowed to be mentioned in these temple reports, and on the whole they tend to glorify the greatness of the divine rather than to testify to it in simple and unadorned language.

The main sources, then, on which the modern student must rely do not constitute the best material that could be hoped for. There is no critical report of a detached observer who made it his business to study Asclepius' cures. Euarestos, the philosopher who came from Egypt in order to learn about the god and his deeds (T. 403), did not leave his notes to posterity. The records preserved are tinged with partisan zeal, and they are statements of Greeks who delighted in the mythical and who even in regard to human affairs showed little concern for the exactness and accuracy of minutiae.[8] But objectivity without prejudice for or against, in the matter under discussion, is extremely rare in all ages. Moreover, despite the shortcomings inherent in the character of the available evidence, one essential point is indubitable: people went to the Asclepieia, they had dream visions and awoke healed, or at least informed what to do in order to heal themselves. What the physicians Rufus (T. 425) and Galen (T. 459) report, is ample proof of the actuality of the dreams and of the effectiveness of the cures. These men were good scientists, keen observers, and interested in the results achieved rather than in any religious controversies or beliefs.[9] Their testimony gives assurance that one is not dealing with fiction only, but with facts. Having this in mind and guarding oneself against blind trust in every word to be found in

---

[6] For the Epidaurian inscriptions, cf. esp. Hamilton, *op. cit.*, pp. 8 ff.; Herzog, *Wunderheilungen, passim*; cf. also R. Nehrbass, *Philologus*, Suppl. XXVII, 4, 1935.

[7] The Athenian tradition can be supplemented by literary evidence, cf. Wilamowitz, *Isyllos*, p. 36, n. 6. For Rome, cf. M. Besnier, *L'Ile Tibérine*, 1902, pp. 203 ff.; Deubner, *De Incubatione*, pp. 44 ff.; Hamilton, *op. cit.*, pp. 63 ff. For Lebena, cf. Hamilton, *op. cit.*, pp. 69 ff.; M. Guarducci, *I miracoli di Asclepio a Lebena*, *Historia*, VIII, 1934, no. 3, pp. 410 ff.

[8] I am thinking of Burckhardt's characterization of the Greek temperament (*Griech. Kulturgeschichte*, III, pp. 426 ff.; cf. also *ibid.*, I, pp. 15 ff.).

[9] The sober judgment of Rufus is generally recognized; cf. J. Ilberg, *Abh. Akad. Leipzig*, 1930, p. 1. Wilamowitz' spiteful verdict on Galen (*Isyllos*, p. 122, n. 12), which has often been repeated, should not blind one to the fact that Galen was one of the greatest scientists of antiquity; for a more adequate evaluation of Galen's work, cf. *e. g.* J. Ilberg, *N. Jahrb.*, XV, 1905, pp. 276 ff.

the sources, one should be able to draw, in broad outlines at least, a picture of Asclepius' treatment.

In making this attempt the aim must needs be a general representation of temple medicine.[10] For none of the places is the information sufficient to sketch the whole procedure; the various data must be combined, unless the task of reconstruction is to be abandoned entirely. Local and temporal differences can only be hinted at. This involves a certain arbitrariness in representing the facts which can hardly be avoided. On the other hand, although the practice of one temple need not in every detail have been that of all the others, there is no reason to suppose that any basic divergencies existed.[11]

Now, like any other deity Asclepius had his favorite spots where he was invoked with the greatest probability of success. The god of Epidaurus, of Cos, of Tricca (T. 382), of Pergamum (T. 386), of Lebena (T. 426), of Aegae (T. 387) was especially famous, and the sick flocked to these sanctuaries with a certain preference. Yet there were many more shrines in which Asclepius appeared and took away diseases (e. g. T. 391), and it was customary to seek him wherever he was nearest (T. 385).[12]

In all these places incubations were usually made by the patients themselves, but it was also possible, though it happened rarely, that another person acted in behalf of the sick (T. 423, 21; 447).[13] Whoever came in quest of the god's help had to take some preparatory steps before he

[10] The testimonies refer almost exclusively to temple sleep (incubations). But the god also appeared in the daytime, outside of the temple, on the roadside (T. 423, 25); he came to the home of his friends and took away their illness (T. 446). Besides, people offered prayer to him asking for health and for release from disease; cf. below, pp. 182 ff. If they remained well, or felt relieved of their sufferings, the god had heeded their demands, he had performed a miracle no less impressive than when healing those who slept in his temples. These manifestations of Asclepius' power cannot be rated properly in any modern evaluation of his medicine, for they are rarely mentioned in literary or epigraphical sources; they vanished with the day that saw them. But incubation doubtless was the favorite, the typical method of insuring the assistance of the god, and the investigation therefore can safely be restricted to an analysis of this phenomenon.

[11] Herzog has claimed that the practice in the various Asclepius temples differed considerably in method (cf. above, p. 144). One of his arguments derives from the fact that in Cos and Pergamum no tablets similar to those of Epidaurus have been found (Wunderheilungen, p. 148). Yet votive tablets were erected in Epidaurus, Cos, and Tricca (T. 382). Farnell, Hero Cults, p. 274, therefore rightly insists that the lack of epigraphical findings does not prove that the cures in these places were basically different from those in others; cf. also Th. Wiegand, Abh. Berl. Akad., 1932, no. 5, p. 31.

[12] For the sanctuaries where incubations are known to have taken place, cf. Pauly-Wissowa, II, p. 1690, 26 ff. The popularity of the temples varied at various times, cf. below, pp. 253 ff. When Herondas speaks of Epidaurus and Cos as the cities which Asclepius inhabits (T. 482, vv. 1-2), such an expression should not be taken too literally.

[13] Cf. Weinreich, Heilungswunder, p. 85.

could approach him; he had to bathe and to offer sacrifices. That is all that is ever mentioned concerning the ritual preliminary to the incubations.[14] There is no evidence that the supplicants refrained from certain food, or wines, or fasted altogether for several days, as they did in the sanctuaries of other gods. Nor is there any reference to purifying rites or solemn ceremonies, which were so common in most places of oracular revelation.[15] What those who wanted to enter into communication with Asclepius were asked to do was little indeed and devoid of any strangeness. By cleansing themselves and by sacrificing to the deity they paid their proper respect to him before whom they were to appear. But since this god had proclaimed that purity consisted of thinking holy thoughts (T. 318), no elaborate ritual was deemed necessary.[16]

At night, then, without paying any fee of admission, the patients went to the place where they were supposed to wait for the god.[17] Their way led them through the temple district. No secret roads were to be followed; no horrifying or mysterious signs were to be seen. It was the sanctuary that they knew and that they had visited in daylight through which they now walked.[18] They were dressed in their usual apparel,

[14] Bathing is known from Athens (T. 421, v. 656 f.). It is justifiable to presuppose that it was demanded everywhere, for purification with water was important in Greek cults (cf. Kern, Religion, I, p. 174). Sacrifices are mentioned for Epidaurus (T. 511) and Athens (T. 421, vv. 660 ff.) ; for details, cf. below, pp. 186 f.

[15] From Philostratus (T. 397) it does not follow that wine was forbidden (contrary to Deubner, De incubatione, p. 17). Philostratus blames the Assyrian stripling for his debauchery only on moral grounds, and in the same way T. 415 must also be interpreted. Fasting was demanded for instance by Amphiaraus, cf. P. R. Arbesmann, R. G. V. V., XXI, 1, 1929, pp. 101 f. For the general ritualistic rules prescribed in other temples, cf. e. g. Pausanias, IX, 39, 5 ff. (Trophonius).

[16] Cf. above, p. 126. Even bathing was considered to purify body and soul, cf. Plato, Cratylus, 405 B, and one may also think of the bath of the μύσται; cf. Kern, Religion, II, p. 198. J. Ilberg, N. Jahrb., Sect. Pädagogik, IV, 1901, p. 298, seems inclined to explain Asclepius' concept of inner purity by the fact that he had to deal with sick people who were impure in a physical sense. But in antiquity, disease was never understood as a defilement of the body. Nor would it have discouraged people from coming to the sanctuaries had Asclepius imposed an elaborate ritual, as Ilberg thinks. Other healing deities demanded purifications. How singular the lack of complicated ritualistic rules in the Asclepieia was, becomes evident from the modern surveys on incubation rites in general, cf. Deubner, op. cit., pp. 14 ff.; Pauly-Wissowa, s. v. Incubatio, IX, p. 1258, 37 ff. The testimonies adduced there are almost exclusively taken from other cults.

[17] W. H. D. Rouse, Greek Votive Offerings, 1902, p. 200, n. 1, speaking of the entrance fees paid at Oropus in the temple of Amphiaraus, says: "this is not certain for the shrines of Asclepius." No such charges are mentioned in connection with the Asclepieia. W. Otto's interpretation of T. 513a (Priester und Tempel im hellenist. Ägypten, I, 1905, p. 395), according to which in Memphis entrance fees were paid, is erroneous; cf. R. Herzog, Arch. Rel. Wiss., 1907, p. 217, n. 4.

[18] Concerning the strange happenings in the temple of Trophonius, cf. again Pausanias, IX, 39, 5 ff.

rather than clothed in hieratic garments; nor did they wear crowns, as supplicants often did.[19] Having entered the temple or the halls especially built for incubation, they lay down on the floor, on a pallet (T. 421, *v. 663*). There were no animals' skins with which they had to cover themselves.[20] Lights were burning where the patients assembled, and were extinguished only shortly before the god himself was supposed to make his appearance (T. 421, *vv. 668 f.*).[21]

In these sober, almost trivial surroundings the god Asclepius revealed himself directly to everyone who needed his help, and he did so nightly. Whereas the other gods manifested themselves on their own initiative or, if approached by men and asked to appear, determined the time at which they would deign to enter into communication with their worshippers, Asclepius was always willing to come and to lend his assistance.[22] To be sure, sometimes he was "out of town" (T. 423, *25*), or he refused to be seen on moral grounds (T. 397). Otherwise he showed himself to everybody; he gave his succor in the trials of any individual, whenever he was invoked.

The god was seen by the incubant in his sleep (*e. g.* T. 423, *3*) or in a strange state between sleep and waking (T. 445).[23] Everything in his

---

[19] Cf. Pausanias, *ibid.* It is true, in Pergamum (T. 513) it seems to have occurred that people wore white garments, that rings and belts were forbidden, etc. The inscription is badly mutilated, cf. O. Deubner, *Das Asklepieion von Pergamon*, 1938, p. 11. Even if correctly restored, the many statements of Aristides do not give any indication of such rules; they can hardly have been commonly applied. Aristophanes, moreover (T. 421), though mentioning the bath, does not refer to a change of the garment, a wonderful topic indeed for the comic poet, had he been able to make use of it without doing too much injustice to the actual facts.

[20] In Epidaurus there was an ἄβατον (T. 423, *7*) opposite the temple (T. 739, *2*), where the patients slept; the same is probably true of Cos, cf. Herzog, *Wunderheilungen*, p. 152, n. 22. In Athens, they slept opposite the statue of the god (T. 427). As regards sleeping on animals' skins, cf. T. 206 (with reference to Podalirius); Amphiaraus, too, expected his visitors to sleep on the skin of the sacrificial animal, cf. Pausanias, I, 34, 5.

[21] Note that here the priests did not play an important rôle in the preparation of the epiphany, as they did for instance in the temple of Trophonius, cf. Pausanias, *loc. cit.*

[22] For epiphanies of gods in general, cf. below, n. 37. The great gods appeared to their favorites in order to assist them in their difficulties, or to their people during times of political danger in order to save them. In the case of Trophonius, it was hard to ascertain when he could be seen, cf. again Pausanias, *loc. cit.* Whether Amphiaraus gave oracles regularly or only at certain times, does not become clear from Pausanias' description, nor is it sure at all that Amphiaraus actually appeared to the incubants.

[23] It is important to emphasize that the epiphanies of Asclepius were dream visions. Even the protegés of the god, like Proclus, did not see him when they were awake, cf. T. 445 f.; cf. also T. 444; 447; 448. As far as I am aware, Asclepius only once appeared to one of his patients while she was awake (T. 423, *25*), but this happened outside the temple, by the roadside. I need hardly add that Aristophanes' statement, according to which the witness of an incubation scene is not asleep (T. 421, *vv. 714 ff.*), is poetical license. In situations other than temple healings, Asclepius could of course be seen in a waking vision, cf. *e. g.* T. 260.

appearance announced his divinity.[24] He came in the form in which he was portrayed in his sacred statues (T. 850, *v. 654*), that is, as a bearded man, the expression of his face gentle and calm (T. 448), holding in his hand a rustic staff (T. 850, *v. 655*), or as a youth of beautiful and fine appearance (T. 423, *17*; cf. *25*). He spoke with a harmonious voice (T. 427; 850, *v. 657*). Sometimes he laughed (T. 423, *35*); he had a sense of humor (T. 423, *8*). There was nothing terrifying in his appearance.[25]

Having approached the patient (*e. g.* T. 423, *4*) and entered into personal contact with him (T. 425), the god then immediately proceeded to heal the disease brought to his attention, or he advised a treatment to be followed; the former procedure is the older one attested. It is celebrated in the play of Aristophanes, where Asclepius restores the eyesight of Plutus in a flash (T. 421, *vv. 737 f.*). The Epidaurian inscriptions glory in the god whose patients, whatever the disease that plagued them, woke up the next morning well and healthy (T. 423).[26] The Coan Asclepius wiped off illness with his divine hand (T. 482, *17 ff.*). The Asclepius of Pergamum took away illness overnight (T. 425). The god

[24] The description of dream visions given here is restricted to those of the god alone. Aristophanes makes Asclepius appear together with Iaso and Panacea (T. 421, *vv. 701 ff.*); the monuments picture him together with his family, cf. B. Stark, *Die Epiphanien des Asklepios u. ihre Darstellung durch die Kunst, Arch. Zeitschr.*, 1851, pp. 341 ff. The Epidaurian inscriptions sometimes name the sons of Asclepius as his companions in the treatment (T. 423, *23*), probably referring to Machaon and Podalirius (cf. also *e. g.* T. 431), or they speak of slaves who follow the god (T. 423, *27*; cf. also T. 421, *v. 710*).

[25] Trophonius' terrifying sight was proverbial (Aristophanes, *Clouds*, 508 f.); it was hard to look upon the gods in their manifest presence (cf. *Iliad*, XX, 131). Only once, in a literary passage, is Asclepius' appearance described as rather frightening or awe-inspiring, and the difference of his features from those of the cult statue expressly mentioned (T. 448). (For the artistic representation of Asclepius, cf. below, pp. 218 ff.). Not even a mystic light radiated from the god; at least not when he appeared to his patients. Deubner (*De incubatione*, p. 11) quotes T. 295 as evidence for a light coming forth from the god; but this is the light shining forth from his armor when he manifests himself as Savior of Sparta. Aristides, who in mystic revelations sees Asclepius as a man of wondrous beauty and size (T. 325), once beholds him with three heads, surrounded by light (p. 438, 25 K), in a vision in which the god assures him of his patronage, and the god of magic appears in a firebrand (T. 328). Similar as the appearance of Asclepius may have been to that of other gods (cf. Deubner, *ibid.*, pp. 9 f.), still it was different in that it lacked all "superhuman" traits.

[26] Herzog, *Wunderheilungen*, p. 67, says that the Epidaurian inscriptions give the impression that the healings were performed hurriedly, but that in a few instances it is still possible to look behind the scene and to detect that actually time was required. Yet, the healings quoted by Herzog (pp. 67; 79) are not "nachträgliche Heilungen," but miracles outside the temple, or the recovery is supposed to occur later, yet spontaneously. In the case of Aeschines (T. 404), who stays over a longer period (Herzog, *ibid.*, pp. 39 ff.) it is not known whether this testimony refers to Epidaurus at all; nor is it certain that this Aeschines is the rhetor of the fourth century B. C., cf. Weinreich, *Heilungswunder*, p. 6, n. 4.

healed at Lebena in a trice (*e. g.* T. 426). At all times and in all places Asclepius proved to be a speedy healer.[27]

Yet he may also have acted as a consulting physician from the beginning of his career, although the oldest testimonies do not speak of him in this rôle.[28] At any rate, from Hellenistic times on the god proposed means of healing to his patients, and he seems to have given a certain preference to this method, although he did not cease to perform miracle cures and to intervene with his divine hand even at a time when in most cases he restricted himself to prescribing remedies (T. 432). Apparently he took increasing delight in working out a complicated schedule for his supplicants and in suggesting baths and diet to them (T. 407).[29] His advice in most instances was clear and direct. It rarely needed clarification through a professional interpreter (T. 430, *vv. 246 ff.*), or through a priest (T. 497), or through the patient himself and his friends (T. 417).[30]

What did the god actually do when immediately healing his patients, or what did he ordain when proposing a treatment? As healer of diseases he proved to be an excellent and daring surgeon. He did not abhor cutting the whole body open; he made incisions into the innermost parts of the human frame (T. 423, *13; 23; 25*). Nevertheless his patients, when awakening the following morning, could walk about and did not feel any bad aftereffects. Few details are given by the testimonies in regard to the god's technique: he used a knife (T. 423, *13*), and other instruments (T. 426). But on the whole Asclepius' procedure remains vague (T. 423, *4; 12; 27; 32*).[31]

[27] Herrlich (*op. cit.*, pp. 30 f.) was the first of the more recent students to insist again on this point. Herzog (*op. cit.*, p. 148) wants to exclude the miracle, at least as far as Cos is concerned. But Herrlich (*op. cit.*, p. 31, n. 1) was right in stressing the importance of the Herondas passage (T. 482). Cf. also above, n. 11.

[28] Thraemer (Pauly-Wissowa, II, p. 1688, 17 ff.) holds the view that the *Curculio* of Plautus (T. 429-30) proves that even in the Epidaurus of the fourth century B. C. dream oracles were given, and that this was the usual practice. But the *Curculio* relates only that Asclepius did not take care of his patient, although he appeared to him, and that the patient asked the interpreter what the apparition may have meant. Nothing certain can be concluded from this testimony as to the healings themselves. On the other hand, the tablets naturally celebrated the greatest achievements of the god, his miraculous cures; that does not necessarily mean that he did not also prescribe remedies, a slighter miracle indeed. Moreover, even the Epidaurian inscriptions mention oracles given by Asclepius, cf. above, p. 105.

[29] Weinreich, *Heilungswunder*, pp. 111 ff., has emphasized this change in the god's practice.

[30] Welcker, *Kleine Schriften*, III, pp. 130 ff., has collected those passages in Aristides which refer to the various persons with whom the rhetor takes counsel in interpreting his dreams. In very dubious cases he even asks the god himself for clarification.

[31] It must be stressed that the god continued his surgical practice throughout the centuries. The generation before Aristides witnessed especially astonishing examples

Besides, the god relied on his pharmacological knowledge. Salves and drugs are mentioned (T. 423, *9; 17; 41*), which were taken from Asclepius' medicine chest (T. 421, *v. 710*), but they are not specified. Moreover, he had certain animals to help him: just as it actually happened that his sacred dogs or serpents cured the patients by licking them (T. 423, *17; 26*), so serpents appeared to the sick in their dreams (T. 423, *39*; 421, *vv. 732 ff.*). In one case, the god, jokingly perhaps, used a brush to take away the disease (T. 423, *28*), but he could also banish an illness into a cloth (T. 423, *6*; cf. 421, *vv. 730 ff.*). Finally, he applied his hand, touching his patient's body with it (T. 423, *31*), or stretching it forth to him (T. 456),[82] and in Epidaurus (T. 423, *41*) and Athens (T. 446) the divine kiss was one of the means by which the sickness was removed instantaneously.

If the god advised a treatment, he recommended natural remedies— what else would his clients have been able to use? He gave drugs (T. 432), either the usual ones or those of his own invention (T. 410); he ordered the eating of partridge with frankincense (T. 434); he counselled his patients to apply ashes from his altar (T. 438), to take rides or any other kind of exercise (T. 407 f.; 435); to go swimming in rivers or in the sea, to bathe in the bathhouse or in springs (T. 408a; 432); he also favored mental gymnastics ranging from the composition of odes to that of comical mimes (T. 413). In fact he was very definite in his prescriptions (T. 439) and varied them considerably, according to the type of patients with whom he was concerned (T. 427). Natural as these remedies were, they had of course unusual effects. Water from the well in Pergamum, whether one drank of it or bathed in it, could make the blind see again, the lame walk about (T. 409). It was apparently also of some consequence whether a drug was taken in the home of the patient, or at a certain sacred spot in the temple (T. 411). In the latter case the efficacy of the remedy was enhanced. Nor did the god neglect to take a personal part in what was going on: his divine hand touched the body of the patient and thus imparted his blessings (T. 432). In addition, the medical treatment which he gave was in many instances contrary to

---

of his skill (T. 412). Weinreich (*Heilungswunder*, p. 30) interprets χειρουργία in T. 412 as "Handauflegen." There is no reason to take the word in such an unusual sense. Operations are attested also in an inscription of Lebena dating from the second century A. D. (T. 426). Incidentally, the operation of which Aristides dreams (T. 325) is performed by Sarapis, not by Asclepius, contrary to Weinreich, *op. cit.*, p. 86.

[82] Cf. also T. 317, *10*. For archaeological material, cf. K. Sudhoff, *Handanlegung des Heilgottes* etc., *Arch. f. Gesch. d. Medizin*, XVIII, 1926, pp. 235 ff., and Herzog, *Wunderheilungen*, p. 55, n. 17. For the symbolism of the gesture, cf. Weinreich, *Heilungswunder*, pp. 1 ff.

all human experience and theory. He directed the sick to exercise, when in their opinion rest was necessary (T. 408). What the god demanded, astonished patient and doctor (T. 417); it seemed strange to the human physician (T. 405). In short, Asclepius' precepts very often were unique or paradoxical (T. 317, *8*; 408).[88]

This much can still be ferreted out about the apparition of the god and about his doings. No doubt, the outward aspect of incubations, as they were practiced in the Asclepieia, was simple and sober. The god's cures were medical cures. Asclepius was not a wizard whose hocus pocus perplexes men, nor was he one of those gods who nod fulfillment of the wishes of their suppliants.[84] He acted as a physician; his healings were miracles—for his success was beyond all human reach—but they were strictly medical miracles. On the other hand, this rationality of Asclepius' treatment seems strangely interwoven with the fantastic and the unreal. There were dreams, there were oracles, there was a divinity who was believed to appear and " I seemed almost to touch him and to perceive that he himself was coming, and to be halfway between sleep and waking and to want to get the power of vision and to be anxious lest he depart beforehand, and to have turned my ears to listen, sometimes as in a dream, sometimes as in a waking vision, and my hair was standing on end and tears of joy (came forth), and the weight of knowledge was no burden— what man could even set these things forth in words? But if he is one of the initiates, then he knows and understands " (T. 417).

Yet one must not overemphasize the irrational element in these cures, at least not as far as their historical understanding is concerned. The belief in the existence of gods was never seriously challenged in Graeco-Roman centuries. Even the Epicureans admitted that there were gods; the Sceptics were satisfied with expressing their uncertainty in this matter, just as in all others; atheism in its true sense, meaning the absolute denial of gods, was an isolated creed of no consequence. The point under debate was rather, whether the gods actually interfered in human affairs. The many, of course, never had any qualms about this problem. But in the centuries in which the Asclepius cult gained ascendancy, even the predominant philosophy, that of the Stoics, asserted that God's providence did become apparent in His direct help to men more than in any other of His acts; the Epicureans meekly objected that God, although

---

[88] Cf. also Weinreich, *Heilungswunder*, pp. 198 f.; Welcker, *Kleine Schriften*, III, pp. 137 f.

[84] The latter statement, of course, is true only of the god of incubations. When Asclepius was approached in prayers and granted his help, he acted like all the other deities, cf. below, pp. 182; 184.

He existed, lived in blessed indifference far from this world.[35] On the whole, it was taken for granted that the deities showed concern for man and would come to his assistance.

As regards the particular form in which Asclepius administered his help, his epiphany, his active healing or his prophetic advice—it was generally admitted that epiphanies of gods actually happened. Plato, to be sure, had rejected this belief; he thought it beneath the dignity of a god to appear in human shape.[36] But his contention did not impress the people, not even the philosophers. Epiphanies continued to be experienced, though less frequently than in ancient times, and they never ceased entirely.[37] Moreover it should be borne in mind that every religious festival, every religious ceremony presupposed the presence of the god. Plutarch, the Academic philosopher and priest of Delphi, neatly summarizes the general belief by saying: " For it is not the abundance of wine or the roasting of meat that makes the joy of festivals, but the good hope and the belief that the god is present in his kindness and graciously accepts what is offered." [38] In short, Asclepius' epiphanies occurred in a world in which such events were usual and were accepted as a matter of course, as a necessary consequence of the existence of the gods. The only feature that distinguished his appearances from those of other deities was the regularity of his nightly revelations. Yet Asclepius was a demigod; it was in accord with his nature to live on earth, to be in constant touch with men.[39]

[35] Cf. *e. g.* the discussion between the Stoic, the Sceptic, and the Epicurean in Cicero, *De Natura Deorum.* That the belief in god's participation in this world increased with the rise of Neo-Platonic philosophy, is self-evident.

[36] Cf. *Republic*, III, 381, especially D; in general, cf. F. Pfister, Pauly-Wissowa, *s. v.* Epiphanie, Suppl. IV, pp. 277 ff.

[37] Such epiphanies are to be found as early as in the *Iliad* and *Odyssey.* After they had occurred less often for some centuries, they became again more frequent in the third century B. C. The attitude of the Lycaonians toward St. Paul and Barnabas, *Acts*, 14, 11-12, exemplifies the ever-ready willingness of the heathens to believe in the epiphanies even of the Olympians. Cf. in general Wilamowitz, *Glaube*, I, p. 23; Kern, *Religion*, III, pp. 256 ff.; K. Steinhauser, *Der Prodigienglaube u. das Prodigienwesen der Griechen*, Diss. Tüb., 1911, pp. 36 ff. The classic literary passage concerning epiphanies is the statement of Dionysius Halicarnasensis, *Antiquitates Romanae*, II, 68.

[38] *Non posse suaviter vivi*, 1102A; cf. Wilamowitz, *Glaube*, I, pp. 301 f., who rightly emphasizes that "(die Urzeit) den Gott leibhaftig erscheinend und am Opfermahl teilnehmend dachte"; cf. also the *Apollo Hymn* of Callimachus and Weinreich's interpretation, *Tüb. Beitr.*, V, 1929, pp. 229 ff.; even if Weinreich is right in saying that the god in this case is not "visibly" present (*loc. cit.*, p. 409), the god is present just the same; the miracle presages his coming.

[39] Cf. above, p. 84. The great gods were seen by the few who were their favorites; or during times of political danger they appeared to their people in order to save them. Their epiphanies remained isolated happenings, otherwise they would not have been so carefully recorded in later times (for ancient books on epiphanies, cf. Herzog, *Wunderheilungen*, pp. 49 ff.). Asclepius' epiphanies in consequence of his appointed task were

Asclepius not only appeared to the sick, he took away illness, he did what was beyond human power. It seems fitting for a god to be able to accomplish more than human beings, to achieve things which are beyond the reach of man. This fact in itself, to the ancients, was not a miracle in the modern sense; it was a wonder, a marvel; men were aghast at the power of the deity, they were deeply impressed by it. As far as the god was concerned, he only proved his divine virtue, for him it was not extraordinary to act as he did; his superhuman deeds were the genuine outgrowth of his divine nature.[40] Not even ancient scientists and philosophers had any clear-cut proof to offer that such divine actions were impossible; though generally insisting on a causal explanation of the phenomena, they could not exclude divine interference, as long as the possibility of divine action was recognized at all.[41] Moreover the ancient concept of natural laws was not as rigid as is the modern; it applied only to such phenomena as were within the range of the ordinary; extra-ordinary events and circumstances, however, were on a different level. There were certain facts which had to be recognized as long as they were accredited by good testimonies, even though a rational explanation could not be found for them, and miracles, or divine actions, formed part of these inexplicable though well-attested occurrences.[42] In other words: the ancients, even scientists and philosophers, did not debate so much

so numerous that they were never collected. The Epidaurian tablets (T. 423) are the only known "official" record, otherwise there are only the individual attestations of the various inscriptions set up in the temples.

[40] Again Plutarch expresses the general feeling when, after rejecting such petty miracles as the sweating or bleeding or speaking of statues, phenomena which he tries to explain on a natural basis, he adds, *Coriolanus*, 38: "But where history forces our assent with numerous and credible witnesses, we must conclude that an experience different from that of sensation arises in the imaginative part of the soul, and persuades men to think it sensation; as, for instance, in sleep, when we think we see and hear, although we neither see nor hear. However, those who cherish strong feelings of goodwill and affection for the Deity, and are therefore unable to reject or deny anything of this kind, have a strong argument for their faith in the wonderful and transcendent character of the divine power. For the Deity has no resemblance whatever to man, either in nature, activity, skill, or strength; nor, if He does something that we cannot do, or contrives something that we cannot contrive, is this contrary to reason; but rather, since He differs from us in all points, in His works most of all is He unlike us and far removed from us. But most of the Deity's powers, as Heracleitus says, 'escape our knowledge through incredulity'" (*Plutarch's Lives*, transl. by B. Perrin, IV, pp. 213 f. [Loeb]).

[41] The Epicureans tried to give the impression that they invented the principle *ex nihilo nihil fieri*, but in reality everybody took it for granted that each phenomenon had its cause. The non-Epicureans found one of the causes in God whom the Epicureans had banned from the world by making Him live in the *intermundia*, and thus arrived at a "natural explanation" of the phenomena.

[42] Cf. in general R. Hirzel, *Themis, Dike*, etc., 1907, pp. 387 ff., and Edelstein, *Bulletin*, V, pp. 240 ff.

the possibility of miracles, as their actuality. Asclepius, since he really healed his patients by performing operations or by applying drugs, as these patients confirmed, was certain to be recognized as a performer of divine deeds.[48] It is true, some wondered why the god should be a better physician than any human doctor, and ascribed greater accomplishments to Hippocrates than to Asclepius (T. 416a). A few sceptics even went so far as to claim that Asclepius could not possibly prescribe a treatment since, to their knowledge, Poseidon did not advise about shipping (T. 416). But in view of the cures related such an argument by analogy was not convincing. Generally speaking all agreed that the sick were healed by the god in his temples.

Equally positive was the judgment of the ancients concerning the reality of dreams which were supposed to give men a share in divine wisdom. Nobility and plebs, townsfolk and farmers believed in such revelations. Philosophers and scientists admitted that dreams were sent by the gods. Only the Epicureans and the adherents of the New Academy objected to such a belief, but even they did not necessarily reject the prophetic and revealing character of dreams.[44] Asclepius, then, as a giver of dream oracles only made use of that means by which God and men were supposed to communicate. In dreams the soul came into contact with those divine powers surrounding men and the world which it could not apprehend while it was awake.[45]

The fact, then, that Asclepius appeared to his worshippers, that he himself cured them, or that he told them how to take care of their illnesses, was nothing peculiar in the ancient world. In doing all this the god did not act contrary to any of the established scientific or philosophical

[48] The problem of miracles, as they were understood in antiquity, has never been studied in detail. Most books dealing with the questions involved are satisfied with stating that the ancients believed in miracles and that ancient and modern concepts differ in this respect. Cf. e. g. R. Lembert, *Das Wunder bei den Römischen Historikern,* 1905, pp. 5 f.; Weinreich, *Heilungswunder,* pp. VII f. Not even Weinreich's new investigations bring a general clarification of the matter (*Tüb. Beitr.,* V, pp. 204 ff.).

[44] Cf. in general A. Bouché-Leclercq, *Histoire de la divination dans l'antiquité,* I, 1879, pp. 277 ff.; Herzog, *Wunderheilungen,* pp. 61 f. A. Palm, *Studien z. hippokr. Schrift περὶ διαίτης,* Diss. Tübingen, 1933, pp. 47 ff. Aristotle, who denies the divine character of dreams (462 b20), admits their prophetic nature and recognizes that especially physicians considered the dreams of their patients a great help (463a 5). Cf. also the Hippocratic treatise *On Regimen,* Book IV. It is conceivable that even the Epicureans and Academics acknowledged the possibility that dreams could reveal something about diseases and their treatment, although they denied that dreams were sent by God. If this assumption is correct, they were unable to refute the contention that the dream healings in the Asclepieia were successful.

[45] Cf. especially Posidonius' concept of communication with the gods through dreams (Cicero, *De Divinatione,* I, 30, 64), rightly referred to in this context by Weinreich, *Heilungswunder,* p. 76, n. 1.

theories, nor did he assume any extraordinary position. He simply acted like a god. From every point of view, Asclepius' cures, performed continually in the Asclepieia, were well within the limits of that world which the ancients recognized as real.

3. HISTORICAL EVALUATION OF TEMPLE MEDICINE

While it is quite understandable that to the ancients Asclepius' cures were acceptable, that their belief in his divine deeds was never shaken, it is a different problem how the modern interpreter should explain them, how he should account for their success—if they were successful. Since for him it cannot have been the god who treated or advised treatment, he will be first of all inclined to assume that it was men who did the job. Moreover, on the basis of the medical character of Asclepius' miracles, it seems reasonable to suppose that the therapy applied in the Asclepieia actually was a medical therapy, that it was administered by physicians and priests, that the Asclepieia were sanatoria, that the dreams were only a means of enhancing the confidence of the patient, thus corroborating his will-power, a factor of great consequence for the restoration of his health.[1]

But such a thesis is untenable in view of the facts. There is no evidence whatever that physicians participated in the temple healings. Besides, the treatment given by the god, the true physician (T. 418), in many cases was contrary to all ancient medical theory. One would have to postulate a *medicina altera* different from that known from Hippocrates and Galen, an invention of the priests and the physicians who presumably worked together in the Asclepieia.[2] Last but not least, that the temples of Asclepius were health-resorts, distinguished through their especially favorable location, is a concept not to be substantiated. It has long been pointed out that the temple near Epidaurus had no better climate than the city proper, that the Tiber Island was certainly not an ideal place for the sick to live, and that the Laconian temple of Asclepius, which was situated in the swamps, could hardly be called a watering

[1] Cf. above, pp. 143 f.

[2] Cf. above, pp. 153 f. It should also be pointed out that most people came to the god when human help failed. Apart from the many statements to this effect, the cases brought to Asclepius' attention varied with the times. After physicians had learned to treat chronic diseases, such ailments are less frequently mentioned in the inscriptions, cf. Edelstein, *Bull. of the Inst. of the Hist. of Med.*, V, 1937, pp. 244 f.; cf. also below, p. 169. It is difficult to understand how Herzog (*Wunderheilungen*, p. 147) could infer from the testimonies of Aristides, Galen and Rufus a participation in the cures at Pergamum on the part of "scientific medicine." These witnesses speak solely of cures that were performed by the god.

place.[8] Occasionally the good effect of the place where the Asclepieion was situated may have contributed to the success of cures that extended over a long period, but more can hardly be admitted. For even the most magnificent wells, whatever their mineral qualities, cannot have done what the ancients believed them to do: make the blind see and the lame walk (T. 409). Certainly the god made use of natural remedies—but in what way and for what purpose![4] It will not do, then, to refer to medicine and good air alone in explanation of Asclepius' healings.

Now is it conceivable that the cures were brought about by somnambulism?[5] This might account at least for the dreams that people had in the Asclepieia. Yet it is highly questionable whether somnambulism could ever have been a mass-phenomenon. Moreover, people who go through this experience usually do not recall the content of their dreams, whereas the worshippers of Asclepius did.[6] Nor is there any indication that in preparation of the temple sleep artificial means were used to influence the supplicants. What they had to do before communing with the god, was restricted to the simplest and most common acts, demanded of everybody who approached a deity.[7] Finally, to attribute to the priests of Asclepius reliance on such devices as somnambulism means to place

[8] Herrlich, op. cit., pp. 29 f., insisted on the fact that many of the Asclepieia, even Epidaurus, were not built in healthful places. He omits a reference to the Laconian temple (T. 764a), about which Kern, Religion, I, p. 75, says: "Bei Asklepios überrascht (!), dass er als Heilgott ἐν ἕλει bei Sparta verehrt wurde." The modern discussion on the whole is determined not by the actual findings of the excavations, but rather by Vitruvius' theoretical demands for the establishment of temples in general, and of Asclepieia in particular (T. 707). Vitruvius urges that all sanctuaries be erected (constituantur) in healthful spots, because such a situation is an adornment to the temple, and as regards the sanctuaries of Asclepius and Salus, permits in times of epidemic fevers that people may live for some time in a salubrious climate. Vitruvius' plea is probably the result of the conditions which were prevalent in Italy due to malaria.

[4] Cf. above, p. 153. I need not discuss here the various scientific analyses of the waters in the Asclepieia (some of the results are summarized by Herzog, Wunderheilungen, p. 155, n. 31). These investigations have no bearing on the problem at hand. Herrlich, op. cit., p. 30, who also rejects the attempt to explain the cures through the mineral waters that were used, has pointed out that ancient physicians did not think highly of mineral waters. If the priests had applied water for its mineral qualities, they would have done so contrary to the opinion of ancient scientists.

[5] Cf. above, p. 142.

[6] This important point has been stressed already by Herrlich, op. cit., p. 9. That the patients remembered their dreams is the presupposition for their subsequent recovery, at least where the god gave only his advice. Needless to say, I am not concerned here with the problem of somnambulism itself. Nor do I wish to argue that somnambulism was known to the ancients, cf. above, p. 142. My only purpose is to raise the question, whether Asclepius' cures could possibly have been connected with this phenomenon.

[7] Cf. above, pp. 148 ff. Welcker (Kleine Schriften, III, pp. 110 ff.) in the same way has refuted the adherents of somnambulism, and he adds: "Mit Bestimmtheit darf behauptet werden dass . . . die alten Autoren nichts enthalten was uns berechtigte den Somnambulismus der neueren Zeit daran anzuknüpfen. . . ."

them on the same level with magicians. In antiquity, however, religion and magic were entirely separated. No priest of a true deity would have promoted magic, which was an unholy art, not worthy of those who believe in gods.[8]

But perhaps it is pointless to search for an elaborate explanation: the priests impersonated the god, they employed some means or other, the success of their contrivances was merely accidental, based on deceit and trickery.[9] To be sure there were deceitful priests, nor were the ancients unaware of them. They often condemned miracles as fakes, and much as they believed in manifestations of the divine, it was a commonplace that epiphanies of gods were often staged by men so as to mislead the credulous.[10] Once, even a propagator of the Asclepius cult was unmasked as charlatan, and the worship which he had founded in a provincial town in the second century A. D., and which had soon become renowned, was indicted as deception of the people. It was Lucian who thus exposed Alexander, the false prophet, as he calls him.[11] Otherwise the Asclepieia remained free from all charges of fraud. The priests of Epidaurus, of Cos, of Pergamum were never accused of being impostors, of having themselves fabricated the miracles ascribed to their god; not even those who were sceptical of the healings went so far.[12] Occasional deceit granted,

---

[8] K. H. E. de Jong, *Das antike Mysterienwesen*, 1919, p. 428, in a different context, has emphasized this essential distinction between magic and religion, characteristic of classical antiquity. The situation in Greece was different from that in Egypt or in India, where the priests were also magicians. What was true of Greek and Roman priests was also true of laymen. Physicians, for instance, abided by religion, but rejected magic. Cf. above, p. 139.

[9] Cf. above, p. 143.

[10] Even the pious Xenophon (*Hellenica*, VI, 4, 7) does not refrain from noting that a certain miracle was by some people held a fraud. Nor does Herodotus lack judgment when he speaks about the apparition of Athena in the time of Pisistratus (I, 60). For later times cf. *e. g.* Livy, XXXIX, 13, 13; Pausanias, III, 19, 10; Servius, *in Aeneidem*, XI, 787. For apparitions of gods used as ruses, cf. Pauly-Wissowa, *s. v.* Epiphanie, Suppl. IV, p. 294, 55; concerning mechanical devices for the performance of miracles, cf. Weinreich, *Tüb. Beitr.*, pp. 407 ff.; in general de Jong, *op. cit.*, pp. 122 ff. The fraudulent impersonation of gods through men must be distinguished of course from those instances in which it was customary for a human being to represent the deity, as in processions, cf. Kern, *Religion*, I, pp. 165 ff.

[11] For Lucian's *Alexander* or *Pseudomantis,* cf. especially O. Weinreich, *N. Jahrb.*, XLVII, 1921, pp. 129 ff.; A. D. Nock, *Conversion*, 1933, pp. 93 ff.; Kern, *Religion*, III, pp. 263 f. It hardly needs justification that Lucian's dialogue was not included in the collection of testimonies; only two passages (T. 210; 29 b) which seem to have some bearing on the cult in general have been printed; cf. below, p. 213, n. 20.

[12] Herzog (*Wunderheilungen*, p. 78) calls the report of Hippys (T. 422) " eine sehr bösartige Geschichte über das epidaurische Hieron." Here the priests in the absence of the god try to perform an operation but fail utterly, and the god chides them for their undertaking. I do not think that the story is told with any malicious intent, otherwise Aelian would not have repeated it; and Hippys, too, apparently was a believer in divine miracle stories which later authors copied from his book, cf. F. Jacoby, Pauly-

the ancients were hardly so easy to fool that a mummery performed daily in hundreds of places throughout the centuries should never have been detected or suspected, or at least hinted at. The fact remains that in antiquity the majority of the people, rich and poor, educated and uneducated, scientists and poets alike, had no doubt that Asclepius actually appeared to his worshippers and personally treated their ailments. The Christians too admitted that in the temples of Asclepius miracles were brought about. They differed from the heathens only in so far as they considered them the work of Satan rather than of the true God.

The theories which assume an interference on the part of men, be they physicians or priests, cannot be substantiated. Does this not enhance the probability that the cures of Asclepius resulted from the soul's mastery over the body, that it was the belief in God which healed the sick? It is difficult to argue for or against the influence of the soul on the body. Those who believe in it, admit themselves that they have not yet succeeded in clearly tracing the power whose manifestation seems apparent.[18] It is likely that some of Asclepius' patients were healed through autosuggestion. All cases can hardly be explained in this way. And what about the dreams which people acknowledged that they had? What about the medical treatment, which in the opinion of the ancients was responsible for the success attained? The modern thesis is too vague and explains too little to be satisfactory. On the other hand, although Asclepius' cures in their outcome were similar to those of Christ, they cannot be unrestrictedly paralleled with them: Christ's followers regained health through their trust in the Savior. Asclepius' help did not involve faith. Even those who did not believe were healed by the god (T. 423, 3; 4); being a heathen deity and a Greek he even delighted in treating people who doubted his power, although at times he was also inclined to punish those who did not acknowledge his might (T. 423, 7).[14] It seems almost

Wissowa, s. v. Hippys, VIII, p. 1930, 7 ff. The whole event serves to emphasize the might of the god which cannot be matched by human endeavor. At any rate, not even here are the priests accused of fraud, and their action is accounted for by the god's absence. Incidentally I cannot believe that Hippys gives the original tradition. This is represented rather by the report of the Epidaurian tablets (T. 423, 23) which in turn is rationalized by Hippys who substitutes the priests for the children of the god, cf. also Jacoby, ibid., p. 1928, 37 ff.

[13] Cf. above, p. 144.

[14] The difference between the miracles of Asclepius and those of Christ in this respect has rightly been pointed out by Herzog (Wunderheilungen, p. 125), who refutes Latte's attempt to explain Asclepius' deeds in terms of Christian faith (Gnomon, VII, 1931, p. 120). But Herzog himself sometimes overemphasizes the motif of faith (ibid., pp. 60; 64 f.; 71). It is only Aelian who knows that Asclepius refused to heal a pupil of Epicurus (T. 399), and even here the god seems more concerned with his patient's

characteristic of the god's patients that the firmness of their belief did not match the greatness of the god's deeds; even Aristides, a devotee if ever there was one, time and again showed distrust in what he was told by his patron saint (T. 417; cf. 447). Most of Asclepius' patients, to be sure, were hopeful that the god could and would help (T. 444); the god asked them to have confidence (T. 440), he despised cowards (T. 423, *35; 37*). In the same way, the human doctor expected his patient to be hopeful, to believe that he could be cured, to help him fight the disease; otherwise nothing could be achieved.[15] The patient's confidence, then, was a factor which came into play in human medicine no less than in divine, the only distinctive feature of Asclepius' healings being that they eclipsed all human hopes and expectations. Sometimes the patient's belief may have contributed to his recovery, but this disposition of the mind was not a conscious factor in the process of healing, nor can it have been a common one that would make the phenomenon of temple medicine as a whole understandable.[16]

All the explanations proposed in modern debate therefore in some way seem to fall short of their goal. Another interpretation of the facts related by the ancient testimonies is necessary. I suggest that an attempt be made to explain the cures of Asclepius historically. Even miracles, one should think, can be accounted for only against the background of the society in which they happen and with particular regard to the subject matter which they concern. Asclepius' healings, then, being the deeds of a Greek god, must be interpreted in their relation to Greek life and Greek medicine.[17] Thereby, one has to presuppose that cures were actually achieved in the

---

general lack of religiosity than with his disregard for his own power. Concerning the so-called "Strafwunder," cf. also Weinreich, *Heilungswunder*, pp. 189 ff.; Herzog, *op. cit.*, pp. 123 ff.

[15] Cf. Hippocrates, *Epidemiae*, I, xi: "The art has three factors, the disease, the patient, the physician. The physician is the servant of the art. The patient must co-operate with the physician in combating the disease" (transl. by Jones, *Hippocrates*, I, p. 165 [Loeb]).

[16] To express the same argument in more general terms: the ancient religious experience differed from that of later periods, and especially from that of the Christians. On the other hand, it would be rash to exclude the factor of belief entirely. Aristides, for whom the god was his special benefactor, his patron, has a relationship to the deity which one might well liken to Pietism (cf. Welcker, *Kleine Schriften*, III, p. 124). But even with him it is not yet the attitude to be found in modern faith-healing, cf. above, p. 145. The treatment given remains of paramount importance, and all other factors can have only secondary influence.

[17] It seems to me that the general tendency to interpret the miracles in the light of historical parallels is rather dangerous. Certainly, what happened in Greece is not an isolated phenomenon and can be clarified to a certain extent by references to the events taking place in Catholic churches or even among uncivilized tribes; yet for the interpreter it is first of all the singular and individual, the Greek solution of the problem, with which he should be concerned.

way in which they are described, that many of the healings were successful. This assumption can safely be made, for had all cures been failures,
the Asclepieia would certainly not have existed for so many centuries.
To be sure, one might well say of the dedications to Asclepius expressing
the patients' thanks for divine help in disease, what Diogenes said of the
votives at Samothrace commemorating the gratitude of the worshippers
for their salvation from shipwreck: " There would have been far more,
if those who were not saved had set up offerings." [18] Nevertheless, a
sufficient number of successful cures are known which would justify,
even require, an explanation.

Now two problems really need clarification: how could people have
such dreams as they claimed they had? How could they feel cured afterwards or get better by applying the devices proposed in their dreams or
concluded from them by the interpretation of their content? For under
the presupposition that the god did not act and that the priests did not
interfere, there remain only the dreams themselves, and the healings
accomplished instantaneously, or brought about in course of time through
some medical treatment resulting from dreamed prescriptions.[19]

That people who came to the Asclepieia dreamed in the sanctuaries
seems understandable. They came with this aim in mind, and if some
did not succeed in having visions, as the testimonies candidly admit (T.
423, 25; 395; 397), they must have been the minority. It is equally well
comprehensible that the sick dreamed of Asclepius and of their diseases,
that they saw the god, as he was represented in his statues, assisting them
or giving them counsel. Coming in quest of the god's help, excited by
the long journey which sometimes they had undertaken for this selfsame
purpose, preoccupied with their suffering as they must have been, having
seen the sights of the sanctuary and having stayed in these surroundings
at least for a number of hours, having read the tablets on which the
reports on portentous dreams and successful cures were inscribed—how
could the supplicants fail to dream as they did? It is hardly exaggerated
to say that anybody who in a world in which the gods were still alive
should visit a temple and wait for a divine vision would have such
dreams. In the given circumstances these visions were quite natural.[20]

[18] Diogenes Laertius, VI, 59 (transl. by R. D. Hicks, II, 1925, p. 61 [Loeb]).
[19] Cf. above, pp. 150 ff.
[20] Cf. Herzog's short statement to the same effect, *Wunderheilungen*, p. 67: ". . . die
Erzählungen der Geheilten und die Lektüre der Stelen mussten demnach die Träume . . .
in eine bestimmte Richtung lenken." Such an assumption, I think, does not need any
confirmation beyond that which everybody's personal experience provides; it is not necessary to go into the details of modern doctrines concerning dreams, a task which I am
not prepared to undertake. But it should be pointed out that an explanation of dreams

As regards the specific content of the dreams, it is first of all necessary to remember that such refined visions as the one reported by Hippys (T. 422), or that seen by Apellas (T. 432) were so to say individual achievements of good dreamers. Even the Epidaurian inscriptions give instances of very simple dreams (e. g. T. 423, 17). The examples which Artemidorus cites (T. 449 ff.) are mostly devoid of any complicated succession of actions. If a dream in which Asclepius and Athena appeared together meant that an enema of Attic honey was to be administered to the patient (T. 449a), it was really not difficult to receive healing oracles, and the typical dream of the worshipper certainly was of this plain and unsophisticated character.[21]

But many people with a special disposition for vivid dreams had long and detailed visions, full of medical meaning; they saw operations performed, remedies applied; they heard of devices which were fraught with medical implications. How is this to be understood? The framework of such visions was given by the fact that the god who was approached was himself a physician.[22] He could act therefore only in accordance with the established rules of his craft, and these rules were well known to the dreamers. For in antiquity, medicine was not yet a science working behind the closed doors of hospitals and universities, unintelligible in its technical detail to the untrained layman. Surgical treatment, as well as that by remedies or diet, was a matter of common knowledge; everybody could see how drugs were prepared and given, how bandages were applied. People were used to watching the performance of operations, or

---

as a reflection of daytime experiences is well in accord with the only rational explanation of dreams proposed in antiquity (Herodotus, VII, 16, 2 [Loeb]): "Those visions that rove about us in dreams are for the most part the thoughts of the day" (πεπλανῆσθαι αὗται μάλιστα ἐώθασι αἱ ὄψιες τῶν ὀνειράτων, τά τις ἡμέρης φροντίζει).

[21] T. 449a serves to prove that even a man like Aristides was not always successful in having magnificent dreams.

[22] I need hardly point out that the dreams of the worshippers of Asclepius provide one of the strongest proofs for the explanation of the nature of the god as it was given above, pp. 79; 84, just as this explanation, it seems to me, is the only one that makes the dreams themselves understandable. Usually modern scholars are inclined to take the medical character of the dreams as self-evident. Such a view is hardly justified. For any god, except one who is a physician himself, the dignified way of healing would be that of making the disease disappear without any technical intervention. If in other cults of later times, down to those of the Christian saints, medical dreams are attested, this, in my opinion, is due to the influence of the Asclepius cult. Herzog seems the only one to have noticed that the content of the dreams poses a problem. Since for him Asclepius is an old chthonian god, he believes that the epiphany of Asclepius and his medical procedure are later additions made after the god had become the patron of the Asclepiads (*Wunderheilungen*, pp. 143 f.). Apart from the objections to the basic assumptions of Herzog, how could the priests of Asclepius—or anyone else—have succeeded in making the worshippers dream of medical subjects, had not everybody believed that this god was a physician and the hero of doctors?

of any other kind of treatment. They were familiar even with medical theories. Physicians were wont to discuss their cases with the patients themselves and with the families and friends of their clients; they gave public lectures on problems relating to their art. The upper classes were always well versed in medicine; in late antiquity, medical instruction constituted part of general education. People of lower standing, on the other hand, were in the habit of doing a good deal of their own doctoring; many of the country districts never had any physicians in residence. Simple folk exchanged their experiences with their fellow men, especially with those who were well reputed for their wisdom; they tried to learn as much as possible, so as to be prepared for emergencies. Given such a situation, given such a relationship of the layman to health and its preservation and to disease and its treatment, it is not astonishing to find that some of the dreams, even a great many of them, were so technical in content.[23]

Indeed all the main features of these visions are clear reflections of the patients' every-day experiences. Not only did people see the god operating upon themselves, as they had seen the human doctor operating upon their friends, not only did they hear him give advice, as they had received prescriptions from their physicians, they argued with the god when they were displeased with his procedure(T. 427), as they were wont to argue with their doctors. They appealed to him for speedy recovery (T. 432), as they demanded quick relief from those who treated them in ordinary life. They were reminded by the god to pay his fees (T. 423, 25), just as it used to happen when they were under the care of human beings.[24] A child would offer ten dice to the god as an honorarium (T. 423, 8). In addition to being healed, a boxer would learn boxing tricks from him (T. 423, 29; cf. 317, 11).

To be sure, the dreams made the god appear to be doing the impossible. But was he not a god? They made him appear to be prescribing remedies which the human doctor would not have given. But why should the patients otherwise have consulted the god? Moreover, although in all these dreams Asclepius seemed to perform the impossible and to prescribe even the paradoxical, he did nothing that did not follow with perfect

[23] As regards Hippocratic medicine, cf. Edelstein, *Problemata*, IV, Chap. 3. The ancients were familiar with medical problems also on account of the importance of dietetics for the healthy, cf. *Die Antike*, VII, 1931, pp. 255 ff.

[24] Cf. Edelstein, *Problemata, loc. cit.* It is this realistic reflection of daily life which explains the many amusing features of the dreams, the business-like attitude of the god. To refer to the avidity of the priests is quite unnecessary, certain as it is that they were intent on getting their money.

logic from what people had seen or heard. His deeds kept abreast with the changing knowledge of the centuries, but at the same time they were superior and fulfilled the most daring wishes of men.

The god made incisions into the body of his patient, he used cupping glasses on him (T. 426), he bound him to a door knocker (T. 423, *27*), he suspended him head down (T. 423, *21*). All this was done also by the average doctor.[25] In the fourth century B. C. surgery was very much in vogue, great progress was accomplished; in the same century the god acted as a daring surgeon (T. 423, *13*). In the first century A. D. surgery again advanced considerably; consequently in that period the god of Pergamum was especially good at operating (T. 412). But the god was able to cut more deeply into the human frame than any human doctor could do (T. 423, *21; 23*). The Hippocratic physician exclaimed: " So, then, the impossibility of reducing such a dislocation either by succussion or any other method is obvious, unless after cutting open the patient, one inserted the hand into the body cavity and made pressure from within outwards. One might do this with a corpse, but hardly with a living patient. Why then am I writing this? " Only the god was capable of operating on a living man as if he were a dead body.[26]

[25] For the binding of patients to a door, cf. Hippocrates, IV, p. 92 L.; for suspension head down cf. *e. g.* Hippocrates, VIII, p. 318 L.; Soranus, IV, 36, 7; and Herzog, *Wunderheilungen*, p. 77. Unfortunately the later surgery of the god is known only from T. 426. Here Asclepius uses a cupping instrument on a woman who comes to him for the sake of offspring. Hippocrates recommends the use of this instrument for the stoppage of the menses (IV, p. 550 L.); cf. Herzog, *ibid.*, p. 74.

[26] Hippocrates, *On Joints*, ch. 46 (Hippocrates, III, p. 293 [Loeb]). For surgery in the fourth century B. C. (Praxagoras), cf. K. Sudhoff, *Quellen u. Stud. z. Gesch. d. Medizin*, III, 1932, pp. 347 ff.; for surgery in the first century A. D., cf. Celsus, *De Medicina*, VII, praef. Herzog denies that the operations outlined on the Epidaurian inscriptions are reflections of the contemporary surgery and considers them the products of a fantasy stimulated by the doings of butchers and embalmers, as well as by fairy tales (*Wunderheilungen*, pp. 83 f.). His opinion is based on the fact that it was only Praxagoras who dared to make incisions into the abdomen similar to those described in T. 423, *25; 27* (*op. cit.*, pp. 75 f.), while the cutting off of the head described in T. 423, *21; 23*, is of course a plain impossibility. But the Hippocratic passage referred to above shows that even physicians, and before the time of Praxagoras at that, ventilated the idea of what might be accomplished if one were able to cut open a human being like a corpse. That the god should be able to achieve what man cannot do is his divine prerogative. And how is one to understand the similarities which Herzog himself has proved to exist between real operations and those dreamed of, unless the latter are reflections of the former? In Herzog's opinion, who believes that the priests surely went over the content of the miracles, if they did not actually compose them (*op. cit.*, p. 57), it was the intention of the tales to show that the therapy of the god was not identical with that of human physicians (p. 66). This assumption cannot be proved. For if in no. 48 of the Epidaurian tablets the god advises a patient not to submit to the surgery of the physician and then proceeds to heal him without an operation (cf. Herzog, *op. cit.*, pp. 84 ff.), this dream seems to mean nothing but the well-understandable wish on the part of the sick to avoid painful treatment. Moreover, the important point about these

Asclepius gave remedies. Nothing astonishing about that,—his human colleagues did so, too. If the divine drugs were specified at all, they were the same as people were used to taking. Only their effect was infinitely better and more far-reaching than that of common drugs. When the limited pharmacopoeia of the Hippocratics in course of time had been extended, the god, too, knew a greater variety of remedies. He gave compound prescriptions (T. 410), as was customary among doctors. When diet, bathing and exercise had become fashionable means of human treatment, the god, too, made use of them (T. 407). Why was that so, why did the dreams change? Not because the god or the priests had learned medicine, but rather because those who dreamed knew of a more elaborate therapy than they had known before.

The god relied on the help of animals: serpents licked the patients' bodies (T. 421, *vv. 732 ff.*). He also prescribed the blood of a cock (T. 438). In popular medicine, being licked by animals was always considered helpful. The poison extracted from snakes, even in Galen's opinion, was very advantageous for the treatment of certain diseases. Animals were used as a sympathetic remedy by many physicians.[27] The touching of the body with the hand (T. 423, *31*), or the foot (T. 423, *3*), even the divine kiss (T. 423, *41*) reëcho the medical belief or imagination of the people. The god was not choosy in his means of healing; he integrated into his medicine all that was ever regarded as efficacious.[28] For men in their dreams made him trust in everything on which they themselves relied.

miracles is that although they are performed without pain, the dreamer afterwards remembers the god as operating. Herzog's position is determined by the belief that Epidaurus was hostile to laical medicine (pp. 152 f.), while other temples, where physicians cooperated (pp. 148 f.), were favorably inclined toward it. Nor does he take into account the fact that the pattern of medicine which the dreams follow was well known to the people at large.

[27] For the licking by animals, cf. H. Scholz, *Der Hund in der griech.-röm. Magie u. Religion*, Diss. Berl., 1937, pp. 12 f.; 23. For Galen's opinion on snake poison, cf. J. Ilberg, *N. Jahrb.*, XV, 1905, p. 298. Sympathetic remedies were applied by human physicians relatively late, cf. Edelstein, *Bull. of the Inst. of the Hist. of Med.*, V, 1937, pp. 229 ff. It is in accordance with this fact that after 300 B. C. such drugs were used more frequently by the god, too.

[28] For healing with the hand, the foot, or through a kiss, cf. the parallels gathered by Weinreich, *Heilungswunder*, pp. 67 ff.; 73 ff.; cf. also T. C. Albutt, *Greek Medicine in Rome*, 1921, pp. 33 ff. For the "divine tablet" (T. 444), cf. O. Weinreich, *Arch. Rel. Wiss.*, X, 1907, pp. 566 f.; for the hair miracle (T. 423, *19*), cf. Weinreich, *Abh. Heid. Akad.*, 1924-25, no. 7. Those dreams in which serpents were seen having intercourse with women (T. 423, *39*; *42*) presuppose the belief in the serpent as the symbol of the phallus, cf. Küster, *R. G. V. V.*, XIII, 2, 1913, p. 149. Note moreover that in other cases the god begets offspring by touching the body of the patient with his own hand (T. 423, *31*). Then again the apparition of a serpent is taken as the revelation of the god's healing power (T. 445; cf. also T. 423, *33*), because the serpent was sacred to Asclepius, cf. below, pp. 227 ff. For the healing power ascribed to the red color (T. 421, *v. 731*), cf. Weinreich, *Heilungswunder*, p. 97.

The dreams, then, which people claimed to have had, and which they doubtless did have, are easy to understand despite their technical content. But how did the patients get well if they only dreamed? How could they feel instant relief after having had merely a vision? How could they recover by means of a chance treatment that was based on dreamed remedies, or on a fantastic interpretation of dream pictures?

As for instantaneous healings, there is little doubt that in some cases nervous disturbances were involved, which were cured through the exaltation of the soul, through the confidence in the miracle that was expected to happen. It was not for nothing that the god in the dream vision asked his patient to have courage, to be no coward, to walk even though he had been unable to walk before (T. 423, *35; 37*; also *16; 38*), to bring to the temple as large a rock as he could carry, although he was paralyzed in body (T. 423, *15*). It actually happens that people suddenly regain the movement of their limbs through a shock, through any kind of excitement, be it joy or sadness, that they become able to move their fingers again (T. 423, *3*), even though they may have been stiff for a long time. Nor is it impossible that a boy should recover his voice (T. 423, *5*) and speak once more. Eyesight too occasionally returns all of a sudden (T. 423, *11; 20; 22*) ; blindness may be but a passing disorder.[29] On the other hand, in many instances spontaneous healing may have come about well timed by fate. A fever could supervene and thus heal epilepsy (T. 425). An abscess could open by itself and blood then would cover the floor (T. 423, *27*). An arrow point could come out of a festering wound (T. 423, *12; 30; 32*) through the healing power of nature. It is also conceivable that a stone was ejected together with a seminal discharge (T. 423, *14*; also *8*). Nor can one raise much objection against the assertion that the sick vomited, that the peccant matter was expelled, and when day came was found all over their clothes, whereupon they became well (T. 423, *41*).[30] This god did not turn only

[29] True, it takes a believer in the divine to become convinced of the fact that the god healed also those who had not even one eyeball left (T. 423, *9*). But then, the Epidaurian inscriptions, medical as their content may be, are not the diary of a physician, and the further advanced the decay of the body, the greater the power of the god who can restore it ! This much can be said even if one holds—as there are good . reasons to do—that generally speaking the Epidaurian inscriptions are realistic and do not exaggerate unduly. I have dealt only with the most remarkable cases which may possibly be accounted for by nervous disorders and which constitute the famous topics of all divine healing that has at all times made the blind see, the lame walk, the mute speak. Such small matters as the cure of headaches and insomnia (T. 423, *29*), or of baldness (T. 423, *19*) need not detain anyone.

[30] T. 423, *13* is a parallel example, although the fact that the patient upon awaking finds leeches in his hands, hardly enhances the likelihood of the story. One should

to the soul of his patient to help it in the recovery of the body. Pestered by the sick like every human physician, he was willing " to see dire sights, and to touch unpleasant things, and in the woes of others to reap sorrow for himself" (T. 392) ; he allowed his sacred *abaton* to become a hospital ward; he had no aversion to the natural, and therefore nature itself may have proved unusually cooperative.[81]

Yet not all diseases, and particularly not the more serious ones, are known to heal spontaneously or through an effort of the soul. If it is assumed, as surely it must be, that throughout the centuries people recovered in the Asclepieia in the twinkling of an eye, whatever their ailment—then temple medicine really seems but a miracle, the more so since men took refuge in the sanctuaries when human help no longer availed, when they had been given up by their physicians, when they were considered hopeless cases (T. 404; 438; 582).[82] However, one must not be swayed by such statements. In antiquity, the doctor's assistance failed much sooner than it does today : the knowledge which he had at his disposal was infinitely smaller than is that of his modern colleague ; he could actually help in infinitely fewer instances than is possible nowadays. Besides, there was reluctance on the part of the Greek physician to assume responsibility whenever he was not certain of his success. He had to guard his reputation, which would have been seriously threatened by failure; it was also considered the duty and the proper attitude of a good doctor not to undertake what he was unable to carry through.[83] Consequently, the number of hopeless cases, of patients given up by physicians, was much larger than it is under modern conditions, and accordingly the chances were much greater that among these so-called hopeless cases there were some who could still recover spontaneously, or at least get better. At any rate, this much is certain: Asclepius' patients were easily satisfied, they were grateful even for small benefits. Libanius acknowledges that in three dreams the god took away no small part of his sickness, and he adds that thus it was brought to a stage " which he

remember, however, that ancient physicians were of the opinion that polluted fluids may contain leeches which, if swallowed, may cause sickness; cf. Hippocrates, IX, pp. 42 f. L.; Galen, XIX, p. 88 K.; Herzog, *Wunderheilungen*, p. 82. T. 423, *21* and *25* are perhaps to be understood in the same way as T. 423, *41*.

[81] In some of the Christian churches where the saints performed their healings there was a similar atmosphere of the operating room (parallels with the Asclepius cures have been collected by Herzog, *Wunderheilungen*, pp. 75 ff.). But here, such naturalistic traits are only accidental, and some of the reports obviously were written in competition with the miracles of Asclepius.

[82] Cf. Weinreich, *Heilungswunder*, pp. 195 ff.; Edelstein, *Bulletin of the Inst. of the History of Medicine*, V, 1937, pp. 244 f.

[83] Cf. Edelstein, *ibid.*, pp. 224 ff.; 244.

may never take away" (T. 447). Even subsidence of the illness was a miracle (T. 445).[84] In many cases it was probably the symptoms alone that disappeared, pains ceased to excruciate the sick. No testimony avers how long the healing effect lasted, just as the diseases are seldom specified. And yet, ever so often the disappearance of the symptoms, the relaxation of pain may have coincided with the recovery brought about by nature; many of the patients whom the god healed at the spur of the moment, scoring a victory over his human competitors, were not quite so sick after all.[85]

The fact that some of the suppliants suffered from relatively slight disorders, though in antiquity they were diagnosed as very ill indeed, and the other that many people came to the sanctuaries who had not consulted a physician before, because they could find no doctor or because they preferred to doctor themselves—these facts should be kept in mind if one now turns to those cases in which the god prescribed a definite treatment, thus curing his patients by means of a lesser miracle, and one not so much to his credit, yet one that from Hellenistic times seems to have been preferable to him.[86] Sometimes when giving drugs or advising a diet and physical or mental exercise, the god did exactly what the human doctor would have done. For people dreamed in accordance with their own knowledge, or instigated by their dreams they simply felt encouraged to apply their usual drugs, as Aristides relates that at the behest of Asclepius he was able to endure a medicine that he had previously

---

[84] Cf. also T. 425. The Epidaurian tablets are cautious enough not to mention what the god did in a case of consumption, and how he succeeded in his treatment; they prefer to relate instead that the patient went home and founded a shrine of Asclepius in his town (T. 423, *33*).

[85] I need hardly say that I have no intention to show that all the healings ascribed to Asclepius can be explained rationally; they certainly cannot. Leaving aside the non-medical miracles recorded on the Epidaurian tablets (T. 423, *10*; cf. Herzog, *Wunder-heilungen*, pp. 123 ff.)—such stories as T. 423, *1* and *2* are plain miracles, although it is true, as Herzog has shown (p. 71), that Hippocratic physicians did not think a pregnancy of many years impossible (even in the *Journal des Savans* of 1678, pp. 305; 348, the possibility of a pregnancy of 26 years was seriously discussed—the child being extracted from the dead body of the mother, and Lessing, *Sämmtliche Schriften*, XI, 1839, p. 401, found this discussion worth noting). The significant point even in regard to the most incredible stories is again that at least the basic presuppositions are in agreement with the scientific beliefs of the times. It is also well to remember that some of the miracles hardly are miraculous (birth of children after one year from the visit in Epidaurus, T. 423, *34*; *42*; also *31*, and T. 426), and that some do not even pretend to be miracles, as in the case of a man who was bitten by a goose and was cured because this made him bleed (T. 423, *43*; cf. *26*; *17*). Incidentally, I have selected only the most characteristic cases and types. For an analysis of the entire Epidaurian material Herzog's commentary (*Wunderheilungen*) remains indispensable. Herzog has printed also those inscriptions which have been omitted from the collection of testimonies because they are too badly mutilated.

[86] Cf. above, p. 152.

abhorred (T. 410). Moreover, many people, having been told by the god what to do, nursed their ailments, while otherwise they would scarcely have done so at all, or at any rate not so painstakingly, had the regimen been prescribed by a human doctor; they were willing to do things that they would never have done at the request of their regular physicians, as Galen says (T. 401). Asclepius' success in these cases is not much more astounding than the success of Galen himself, or of Hippocrates, or of any other ancient physician.

Still, following the suggestion of the god the sick would take exercise when they themselves or their physicians would have been in favor of rest; they applied warm water where the practitioner of the day would have used cold water, or vice versa. In these instances, the procedure was paradoxical, that is, contrary to the established principles of Greek medicine.[87] Furthermore, people took fantastic remedies, they followed chance precepts resulting from their interpretation of the dreams. How, then, could they become healthy? As regards rest or exercise, warm or cold ablutions in the treatment of certain diseases, have there been no divergencies of opinion between the various generations of doctors of the nineteenth and twentieth centuries, and even among the doctors of one and the same generation? And have patients not recovered, or died, whatever procedure was adopted? Is not the same more or less true of dieting or fasting, of drinking wine or abstaining from it, and of other things? As for drugs and their success, again one must not tacitly parallel ancient and modern medicine and judge the prescriptions ordained by the god in the light of the science of today. Ashes of burnt sacrifices, or other remedies of that type administered by the god (e. g. T. 438) were not necessarily of less avail than excrements of doves or goats applied by ancient physicians.[88] Moreover, Hippocrates and Galen, when giving drugs, made their choice according to experience, the latter even according to fanciful theories concerning the qualities inherent in plants or minerals or animals. The dosage of remedies, although based on observation, remained arbitrary. Those sciences which for modern treatment guarantee a certain accuracy, by studying the reactions of the drugs on the one hand, and of the tissues of the human body on the other, thus allowing precise correlation of the effect of the remedy to the needs of the organism—these two sciences, chemistry and experimental pharmacology, were unknown in antiquity. There was no objective standard by

[87] Cf. above, pp. 153 f.

[88] Such remedies were applied by Galen, especially in his country practice; cf. XIII, p. 633 K.; XII, p. 298 K. In general, cf. Ilberg, N. Jahrb., loc. cit., p. 306. On the whole, the " Dreckapotheke " plays an important part in ancient pharmacology; cf. also T. 369.

which men could be guided, a fact that ancient physicians never tired of conceding and lamenting. The pharmacology of Asclepius was also based on experience, and his patients too in applying their dream prescriptions, which in most cases were made up of the usual healing matter, guessed sometimes rightly, sometimes wrongly; perhaps more often wrongly, for despite their considerable knowledge of medicine, they lacked the life-long experience of the physician. On the other hand, it may well be that the simple prescriptions, which laymen derived from their dreams, were more effective than those complicated drugs or salves which ancient physicians loved to concoct; [39] that the sojourn at the Asclepieion had in fact as good an influence on the patient as the practice of being carried around in litters, or whatever else Greek and Roman doctors imagined to contribute to health and recovery from disease. In short, if ancient medicine is seen as it really was, if the treatment which it could provide is taken at its true value, the miracles of the god seem less miraculous—or else, I think, it must be admitted that in the given circumstances it was almost as miraculous that those who were taken care of by human doctors so often recovered. [40]

Therefore, even if the patients of Asclepius only dreamed, even if their cures were only a rationalization of what they believed they had seen, it is not astonishing that the Asclepieia could claim so many real healings. The god himself, or his priests never asserted that he could do everything, nor did they promise immortality; they promised assistance. [41] The

[39] Wilamowitz (*Isyllos*, p. 122), in analyzing the Apellas inscription, says: "Im ganzen stimmt Apellas, wie mich dünkt, die abenteuerlichen Vorstellungen von den Asklepioskuren, die man aus der Lektüre des Aristides gewinnt, auf ein menschliches Mass herab. Und es ist wohl möglich, dass diese 'natürlichen' Heilmethoden, wie wir sagen, 'göttlichen,' wie das Altertum sagt, gegenüber den ungeheuren Mixturen, die Galen und seine methodischen wie empirischen Collegen verschrieben, ihre Berechtigung hatten." Wilamowitz, here, is nearer the truth than he is in his later condemnation of temple cures (*Glaube*, II, pp. 229; 231). Cf. also Farnell, *Hero Cults*, p. 273: ". . . the treatment is more sensible than an eighteenth-century physician would probably have prescribed; it consists chiefly in diet and exercise, and an appeal to the will-power of the patient: the god is eminently rational and considerate; he advises Apellas not to irritate himself. . . ."

[40] Recognition of the achievements of Greek medicine must not blind one to its short-comings. The greatness of the ancient medical art consisted in the observation of phe-nomena, especially of the individuality of the patient, in reliance on experience, on the intuition of the doctor, in rational elucidation of the possibilities of medicine, in the courage to explore methods, to outline programs, to start investigations in almost every field. The treatment, too, sometimes seems surprisingly adequate, and often great success may have been attained. Yet there was no scientific system of therapy in the modern sense of the term. While in their general attitude the ancients still are the masters of the moderns, in practical matters,—to modify a famous saying of Darwin—they were mere school boys compared to the moderns.

[41] In T. 739 it is frankly told that the Epidaurian temple had a building where people could die without desecrating the holiness of the sacred district. One should also re-

ancients, believing in gods as they did, were wont to turn to the deity in quest of help; they did so in every matter, however small and trivial; they did so particularly in circumstances which were serious and threatening. Disease, the fear of death, was a condition which made even the sceptic a devotee of the divine; [42] and people found help, or at least relief from suffering in their communication with him who among their deities was best versed in medicine.

Is it not a matter of common experience that even grave illnesses sometimes heal contrary to the expectations of the most experienced and best trained physician of the twentieth century? Does it not frequently happen even today that, when all has been done for the patient that is humanly possible, the doctor has to admit that only a miracle can save him? And the miracle sometimes happens. Men of today ascribe it to nature; men of antiquity ascribed it to the deity, though sometimes, if they were asked by the god to pay their fees, they remembered that diseases heal by themselves (T. 540)! Unless it is simply true that the gods in ancient times performed miracles and only left this world of ours as of old "did Justice loathe that race of men and fly heavenward," [43] there are good reasons for the success which nature and the experience of the suppliants brought about in the temples of Asclepius.

4. SOCIAL AND POLITICAL IMPLICATIONS OF DIVINE HEALING

Disease is a serious calamity, not only on account of the physical dangers that it involves, but also because it carries with it a heavy economic burden. The ancients were well aware of this fact; even the idealist Plato stressed it to the utmost.[1] Nor was the civic importance of medicine overlooked. Nay, in late centuries it was a favorite topic of discussion, whether the rhetor or the philosopher or the physician was the most valuable citizen, and whereas the philosopher was decried as a subversive character, the rhetor as a revolutionary, and neither therefore of any good for the state, the physician was praised as most useful for the community since he gives assistance to everybody where the others

member that men like Aristides felt that, if the god was not willing to help, it was time for them to die, cf. above, p. 130. Herzog (*Wunderheilungen*, p. 142) has drawn attention to the fact that for the Asclepius cult the miracle of reviving the dead is never attested. This is significant and indicative of the caution and rationality of priests and worshippers alike. But I should add that the healings themselves were understood by the patients as revivals, they were reborn through Asclepius; cf. above, p. 124.

[42] Cf. Edelstein, *Bulletin of the Inst. of the History of Medicine*, V, p. 244.

[43] Aratus, *Phaenomena*, 133-34 [Loeb].

[1] Cf. *Republic*, III, 406 D-E.

fail.[2] The social and political aspects of the problems connected with medicine and medical care, then, were clearly recognized in antiquity, and it is pertinent to ask how far the divine physician too took an interest in the economic difficulties of his suppliants and how far he was conscious of his obligations towards the commonwealth.

To evaluate Asclepius' position properly, however, one must first point out that the social risks of disease were great indeed in the ancient world due to its political structure and to the almost complete lack of any feeling of social responsibility on the part of the individual.[3] Of course, it was only in the lower strata of society that these risks prevailed. The rich had money to take care of themselves, to " nurture " their disease; to them, sickness was something that they dreaded and detested, because it was incompatible with their concept of a perfect human being.[4] Although such a reaction was certainly not entirely foreign to those who did not belong to the well-to-do—every natural and unsophisticated person has an instinctive aversion to disease and the disqualification which it entails —for the middle classes and the poor it must have been the economic damage resulting from sickness of which they were most afraid. If the small businessmen, the craftsmen, could not work, they lost their income; they had hardly any other means of paying for their livelihood; no insurance system provided money for the time during which they were incapacitated; consequently they themselves and their families faced hunger, if not starvation.[5] If their illness led to any permanent disablement, their lot was even more horrible. Some of the cities, though reluctantly, assisted their citizens when they had become unable to work. Generally speaking, however, those who could no longer earn what they needed were utterly lost; they had to live in the most pitiful circumstances. Whether the proletarians were taken care of when ill or found

---

[2] Cf. Ps. Quintilian, *Declamationes*, no. 268, p. 92 f. [Ritter]; cf. also Quintilian, *Institut.*, VII, 1, 38; 4, 39. Herzog, *S.B.Berl.*, 1935, p. 982, has discussed these passages in their significance for the competition between the various university faculties; their social implications are even more important. That in many cases the decision may have been made in favor of the rhetor or of the philosopher need hardly be stressed.

[3] For the problem of social responsibility in antiquity, cf. W. J. Woodhouse, *E. R. E.*, III, pp. 386 f., *s. v.* Charity (Greek); J. S. Reid, *ibid.*, pp. 391 f. (Roman); especially H. Bolkestein, *Wohltätigkeit u. Armenfürsorge im vorchristlichen Altertum*, 1939; cf. his article in *Mededeelingen der Koninklijke Nederlandsche Akademie van Wetenschappen*, Afd. Letterkunde, Nieuwe Reeks, Deel 12, No. 2, 1939.

[4] For the difference between this concept and that prevalent in other societies, where disease was sin or, as in Christianity, a certain distinction, cf. H. E. Sigerist, *Medicine and Human Welfare*, 1941, pp. 16 ff.

[5] All that has been said here is implied by Plato, *Republic*, III, 406 D-E. That even the guilds of later centuries did not provide any medical insurance for their members, has been shown by H. E. Sigerist, *Mitt. z. Gesch. d. Med. u. d. Naturwiss.*, XXV, 1926, pp. 65 ff.

any assistance at all, no state, no private individual was anxious to inquire.[6]

On the other hand, medical care was expensive; physicians wanted to make money and were not ashamed of admitting it. In medical literature there are innumerable admonitions to the effect that the doctor should be more interested in his patients than in his fees; Galen prided himself on treating senators and slaves with the same scrupulousness.[7] Philanthropy was the ideal set before the physician. Yet, with the necessary allowance for individual kindness and willingness to help even without adequate remuneration physicians on the whole were businessmen, and no moral or professional code enforced by the state or by an association or by religious organizations obliged them to give their service to the needy.[8] Some of the great cities had doctors in residence who were paid by the community, but even such an official status did not exclude their receiving or even demanding remuneration from their patients.[9] The hospitals of which these public physicians were in charge and to whose equipment the city contributed were run on a business basis. Patients who could not pay for their stay had no claim to be treated free of charge. Again, although help was certainly given in some of these cases without recompense, there were no vested rights of receiving assistance in sickness, no claims secured by legislation. Everything was left to the discretion of the individual.[10]

The economic threat of disease, then, to the lower and middle classes was grave, graver perhaps than at any later time. It is with this fact in mind that one must consider the tribute which is paid by the testimonies to the god Asclepius who gave help cheaply. He was well known for being satisfied with small thank-offerings (T. 482, *vv. 79 ff.*); it was one of his claims to fame and admiration that he took care of the poor (T. 405),

[6] Cf. Woodhouse, *op. cit.*, p. 387: "The ruthlessness [*sc.* in matters of charity] of ancient society, in which one must be hammer or anvil, is largely concealed from us by the fact that, with few exceptions, it is only the class which enjoys wealth and power that is articulate ; that is to say, ancient literature is mainly aristocratic in origin." It is only fair to add that the lack of social responsibility was partly compensated by a stronger family feeling, cf. Woodhouse, *loc. cit.*, p. 386. For the later change in the attitude of the individual, cf. below, pp. 177 f.

[7] Cf. *e. g. C. M. G.*, I, 1, p. 32, 4 ff.; Galen, X, 613 ff. K.; Ilberg, *N. Jahrb.*, 1905, p. 309.

[8] Concerning Hippocratic medicine, cf. Edelstein, *Problemata*, IV, Ch. III. The situation did not change essentially later on, as can easily be concluded from the income of physicians in the Roman empire, cf. L. Friedlaender, *Sittengeschichte*, I, 1919, pp. 195 f.

[9] R. Pohl, *De Graecorum medicis publicis*, pp. 57 ff., rightly concludes from the inscriptions preserved that the public physicians were allowed to accept money, since it is especially stressed in honorary titles if and when they did not make any distinction between rich and poor patients.

[10] Concerning private and public hospitals in antiquity, cf. Th. Meyer-Steineg, *Kranken-Anstalten im griechisch-römischen Altertum, Jenaer medizin-hist. Beiträge*, III, 1912, especially pp. 10 ff. Hospitals for slaves (*ibid.*, pp. 31 ff.) and soldiers (*ibid.*, pp. 34 ff.) are of course a different matter.

and he had done so from the very beginning of his career.[11] This charitable attitude was, so to speak, part of his being. The ancients thought that, as a terrestrial god, he was by nature the god of the destitute (T. 259).[12]

While it is certain from the evidence available that Asclepius' cures were of great importance for the paupers, it is very difficult to ascertain in detail in what way and to what extent the god showed his generosity. Whenever he deigned to relieve his patients immediately through a miracle, it is clear why such a cure was not expensive; naturally, in these cases he did not find the offering of a cock less attractive or less acceptable than that of hecatombs of oxen. But what was the situation if the god prescribed a treatment, if it was necessary for the sick to remain in his sanctuary for quite some time, if they had to apply remedies which they could not afford? Evidence regarding the god's reaction, or rather that of his priests, in such circumstances is rather scanty. Yet it seems permissible to combine the few data known with those testimonies which speak of the care of the poor in other ancient temples and thus to surmise how the god's help was administered to the needy.

First of all, the Asclepieia, like other sanctuaries, had special buildings where the worshippers could live. Such houses are known from Pergamum and other places.[18] Modern scholars used to take it for granted that the priests charged a certain amount for lodging, and that these houses were used by pilgrims only on festive occasions. It has now become increasingly clear that these hostels in fact represent the first form of the Christian hospitals or poorhouses, that these institutions were taken over by the Church along with the pagan temples. In late antiquity, at any rate, most temples provided refuge for the poor without exacting any fee from them, and it is improbable that the Asclepius temples formed an exception from this general development.[14]

---

[11] Cf. above, p. 84. That the statement of Aelian (T. 405) goes back to the old comedy, has been shown by F. V. Fritsche in his edition of the *Thesmophoriazousae,* 1838, pp. 380 f. Herondas' testimony (T. 482) confirms the fact that the god's generosity was recognized by everybody at an early date. The Epidaurian inscriptions tell the same story, cf. *e. g.* T. 423, 8.

[12] As has been pointed out above, pp. 116 f., Asclepius' clientele was not restricted to the poor, though they formed a considerable section of his patients. Here I only wish to draw attention to one aspect of temple medicine which must have had a special attraction for the lower classes.

[18] Cf. T. 719 and T. 499; Herzog, *Wunderheilungen,* p. 155; for Pergamum, cf. H. Hepding, *Philologus,* LXXXVIII, 1933, p. 102; in general, Bolkestein, Ξένων, *Mededeel. d. Koningkl. Akad. van Wetens.,* Afd. Letterkunde, Deel 84, Ser. B. No. 3, 1937, p. 23. A corresponding institution were the καταλύματα in the temples of Sarapis; cf. Bolkestein, *ibid.,* p. 24.

[14] Bolkestein (*op. cit.,* p. 39) has shown that Christian charity, although unique in its scope and its motif, was preceded by a gradual change in the attitude of the pagans.

But where did the means come from if the god advised a lengthy cure and remedies which the patient could not afford? Did the priests in that case say: " Go home, it is impossible for you to follow the god's counsel "? This is, indeed, unlikely. Even if it is assumed that the primary interest of the priesthood consisted in securing great revenues for their sanctuaries, they must have realized that sometimes it is wise to be generous and that money spent on the poor was a profitable investment. For whoever had been helped free of charge certainly did not minimize in his reports the greatness of the deeds of the god. Nor can one believe that the priests of Asclepius, the god who was called the most philanthropic of all deities (T. 422; 804, 5), should have entirely lacked the prime virtue of their master. Other priests, in conformity with the philanthropy of their gods, did not only offer feasts and banquets to those who at the festival of the deity came to the sanctuary, but sheltered them during their visit and made a present of money to everybody, rich or poor, free or slave.[15] In the Asclepieia, the treatment of the patients was the highest exultation, the perpetual manifestation of the god. Out of kindness the priests of Asclepius, at least occasionally, must have provided the resources for the therapy prescribed by the god. Actually not much was needed as far as expense was concerned. Baths could certainly be offered free of charge; exercises in the gymnasium hardly required admission fees. There remained then only the cost of drugs and purgatives and similar things. Weighing the whole evidence carefully, it is no exaggeration to claim that in many instances the temples, especially the great sanctuaries, put whatever was necessary at the disposal of those patients who were unable to pay.

Moreover, as time went on, the temples were certainly relieved of much of the burden laid upon them by needy patients through the munificence of their rich clientele. No doubt, in the truly Greek society of old the sick could not expect any assistance from their more opulent fellow sufferers, though disease sometimes may have opened the heart of the greatest sinner to the claims made upon him by the misery of his neighbor.[16] But in Hellenistic centuries, and in the Roman period, the

[15] Cf. H. Oppermann, *Zeus Panamaros*, R. G. V. V., XIX, 3, 1924, pp. 61 f., and Bolkestein, *op. cit.*, 1937, pp. 28 f.

[16] Cf. Woodhouse, *op. cit.*, p. 386: "[Almsgiving] was not inculcated as an item of a national ideal of conduct, reflected back upon the individual as a command of religious or philosophical sanctity. So much is true, at any rate, of the Greeks of the great age, in whom the instinct of generosity existed only in rudimentary form." Of course, exceptions to the general rule can never be excluded. When in the fourth century B. C. a private citizen donated a house to Asclepius (T. 729), he probably stipulated that it should be used for the benefit of the public. For the interpretation of T. 729, cf. R. Schlaifer, *Class. Phil.*, XXXVIII, 1943, pp. 39 ff.

situation changed a great deal. Stoic philosophy instigated a reversal of the attitude of the wealthy and infused the virtue of generosity toward the paupers into the ideal of humaneness.[17] Most important, for the aristocracy generosity became part of their devotion to the divine. Gifts, says Plutarch, " should be made at the proper time and for good and worthy reasons, that is, in connection with the reverence paid to the deity which kindles piety in everyone. For whenever the many see that men whom they respect and consider great show zeal toward the divine by being generous and liberal, they feel a strong conviction and belief that the godhead is great and holy." [18] It is in this spirit that Aristides was told by the god to distribute money among his fellow pilgrims (T. 504). It is in the same spirit that Antoninus erected buildings in Epidaurus for the use of women in childbirth and of those who were near death (T. 739).[19] From every point of view, then, it is most unlikely that those patients who had no sufficient means or no means whatever, had to leave the temples without being able to follow the advice of the god. In late centuries, private charity, if not the charity of the priests themselves, must have allowed them to stay on, to defray their personal expenses, and even to give a tip to the bath attendant, an action which the god did not refrain from urging upon his rich patients either (T. 432).

Under these circumstances it seems safe to say that in a world in which the poor were left to their fate by the state or by the communities at large, where laical medicine did not know of any special provisions for the needy, the Asclepieia and religious medicine were of the greatest importance for the medical welfare of the lower classes. To be sure, the heathens could never compete with the Christians in their efforts on behalf of the destitute. But when the Church made it its express aim to take care of the sick, when Christ invited to His temples all those who were ill and troubled by diseases as well as by poverty, He was in this respect, too, the fulfiller of a task which Asclepius had at least dimly visualized.[20]

Considering the fact that the ancient god of medicine was well aware of his social responsibility, it is the more astonishing that he never showed any interest in the political aspect of medical care, that he did

---

[17] Cf. Reid, op. cit., p. 391.

[18] Praecepta gerendae reipublicae, 30 (822 B). Bolkestein has stressed the importance of this passage, op. cit., 1937, p. 29, n. 2, though in a different connection. In general, cf. also Friedlaender, Sittengeschichte, II, p. 377.

[19] Antoninus probably was a character not unlike Herodes Atticus whose activities have been pictured by K. A. Neugebauer, Die Antike, X, 1934, pp. 92 ff.

[20] Cf. above, p. 137.

not realize his obligation toward the state. Plato had claimed (T. 124), to be sure, that even the hero Asclepius, in just recognition of the needs of the city, had restricted himself to a quick treatment of diseases. If he could not restore the health of his patients by some simple prescription, Plato said, he let them die; for men must go about their work and cannot spend their time in curing themselves, a truism which people are wont to overlook in the case of the rich, but which they well remember as concerns the poor. The hero Asclepius, then, in Plato's eyes, was " a most politic physician," who saw men as citizens and who asked whether his actions as a doctor agreed with the requirements of public life and with the demands of the state.[21] Moreover, for Aristophanes, Asclepius was " a clever, patriotic god " (T. 421, v. 726). For he restores the sight of Plutus whom Zeus has blinded; thus riches will be distributed among men more justly than before. On the other hand, he blinds Neoclides, the base extortioner, in order to make it impossible for him to do more harm.[22] The god, however, followed Plato's ideals no more than he assumed the rôle which Aristophanes ascribed to him. Although he delighted in healing his patients instantaneously, he also applied treatment that meant that for weeks or months they could not work; he protracted their lives with all the means of dietetic medicine which Plato detested.[23] Nor did he care to eliminate the enemies of society, to punish those who were unworthy of the light of day. Although he did not appear to Alexander Severus (T. 395), apparently because that emperor was too great a villain, he allowed the tyrant Dionysius to die in his bed peacefully. The ancient sceptics were not slow in pointing out that Asclepius, if he were a true deity, should have done otherwise (T. 396; cf. 437), and the Christians too blamed him for having failed to do his part in matters of public concern (T. 507).

The censure brought against him did not impress the divine healer.

[21] There is a strange resemblance between Plato's political interpretation of Asclepius and the interpretation of Christianity by modern Romanticists, such as Adam Müller, who claimed " dass Christus nicht für die Menschen, sondern für die Staaten gestorben sei," cf. G. A. Briefs, *Journal of the History of Ideas*, II, 1941, p. 285.

[22] It is noteworthy that in Aristophanes the god sides with men and in doing so acts against the intentions of the supreme deity. This feature the poet probably derives from the old myth, cf. above, p. 48. Whether Aristophanes' whole description should be taken as " spöttische Schilderung " (Herzog, *Wunderheilungen*, p. 61; Wilamowitz, *Glaube*, II, p. 232), or whether it represents a " Huldigung für Asklepios " (Kern, *Religion*, II, pp. 316 ff.) it is difficult to decide. Perhaps it is best to say with Nilsson (*Griechische Religion*, I, p. 739) that Asclepius just like the Eleusinian goddesses is treated with more respect because these deities were favored by the people. That Asclepius is made the butt of jokes is only natural in a comedy, and no conclusions can be drawn from this fact in regard to the poet's personal views.

[23] Cf. *Republic*, III, 406 A-B.

As a true physician, Asclepius was interested in individuals rather than in politics. His only concern was the disease and the patient who came to him in quest of help. Certainly, the god refused treatment on moral grounds; illness due to luxury and folly he did not cure (T. 398); he could not be bought (T. 394); he disliked the irreverent.[24] Yet, if his supplicant approached him in the hour of tribulation and honored him in the right way, the god did not ask what he had done before, what his usefulness was as a citizen, whether he was rich or poor—with his divine hand he took away sickness and suffering.

[24] Cf. above, pp. 113; 161.

IV. CULT

The interpretation of the temple medicine has shown what the god Asclepius did for men. How did men appeal to him for his support? "The gods of their own free will are prone to beneficence, but when invoked they are still better" (T. 538)—this statement fairly summarizes the attitude of the ancients. One cannot hope for divine succor if he does not ask for it, and he must do so in the right way. A rigid ritual dictates the forms of supplication and governs the relationship between man and gods.

Moreover, the deity demands continuous worship from his devotees. In antiquity a religion without rites, without sacrifices, processions, festivals, was inconceivable. To be sure, even for the ancients it was belief that created the ritual and found in it tangible expression. But if man did not perform the sacred ceremonies, the gods did not fulfill their task either, nor were they expected to do so. No wonder, says Porphyry, that our city is beset by diseases; since Christ is honored, Asclepius has not enjoyed any public service (T. 506). Such an assertion does not imply that rites were supposed to have any magic power over the gods. If this had been the conviction in prehistoric centuries, in historical times the ritual was rather the expression of the respect owed to the gods. One cannot even speak of an attitude of "I give that you may give." Men were bound to revere the deity who was their superior; those too who rejected the assumption that the help of the gods could be bought by prayers and sacrifices worshipped them devoutly. The cult was considered an intrinsic part of religion; it was the indispensable actualization of belief.[1]

An analysis of the ritual seems requisite, therefore, for a proper representation of the Asclepius religion. It should also throw light on the peculiar character of the service paid to Asclepius. It should show whether new forms of religious observances were created in the worship of the new god. Of course one cannot hope to give an adequate picture of what took place in the temples, to represent fully the performances, to enumerate completely the accessories, to characterize the whole personnel

---

[1] Wilamowitz, *S. B. Berl.*, 1921, pp. 959 f., and *Glaube*, I, p. 137, tries to differentiate between ritual and religion. Pfister, *Bursian*, 1930, p. 54, opposes Wilamowitz' views. He is right in so far as an attitude like that of Wilamowitz might lead to a depreciation of the actual importance of the rites. S. Wide, *Gercke-Norden*, II, 1922, p. 235, in my opinion, overemphasizes the utilitarian aspect of the worship.

who shared in the service of the god. In regard to almost every detail, the data are scanty and incomplete. Moreover, the rules and ceremonies differed in the various sanctuaries. There was no generally accepted pattern of veneration in the Asclepius cult or in that of any other ancient deity. It is only a faint likeness of the old splendor and diversity that the modern interpreter can still conjure up by putting together the scattered statements that chance has preserved for him.[2]

1. RITUAL FOR THE HEALTHY

It was first of all Asclepius, the preserver of health, whom people approached. Not to fall into disease is the greatest boon to man. How frequently prayer and sacrifices were offered to this end, the testimony does not reveal. But there can be little doubt that this kind of " preventive medicine " based on religion was much in vogue among the ancients. Even Marcus Aurelius did not disdain the thought of climbing the citadel of Pergamum where the temple of Asclepius stood and of praying for the preservation of his teacher's health (T. 577). Innumerable others must have gone beyond thought, actually imploring the god in his sanctuary or in the privacy of their homes.[3] Or again sacrifices were offered after a dream which was taken as a bad omen. By this means it was tried to ward off the threatening evil (T. 502; 454).[4]

Nor were such procedures restricted to individuals. In regular observances the Athenian priest asked the god's protection for the city (T. 553). With the same intent the Syracusans made an annual joint offering to Asclepius and Paean (T. 554).[5] In Pergamum, hecatombs were dedicated to Asclepius, just as in Delphi they were offered to Apollo

---

[2] In this inquiry it will often be necessary to repeat what is a matter of common knowledge. But only if the data preserved are interpreted within the framework of ancient rituals, as they have become known through the research done during the past few decades, can the specific meaning of the testimonies concerning the Asclepius worship be established. On the other hand, in some instances, the reports in regard to Asclepius give the only or the best evidence for certain procedures that are known to have been customary also in the cults of other deities.

[3] Wissowa, *Religion u. Kultus*, p. 308, has noticed the fact that " Aesculapius namentlich *pro salute alicuius* angerufen zu werden pflegte " (*C. I. L.*, VI, 13; 19; XI, 2092 f.). Greek inscriptions tell the same story, *e. g. I. G.*, II², 4372; cf. also T. 577a; 522; 537.

[4] Prayers subsequent to a bad dream are also mentioned by medical writers; these authors, however, took such visions as indications of bodily disorders which should be treated by rational means, cf. Hippocrates, *On Regimen*, IV (VI, pp. 640 ff. L.).

[5] Nilsson, *Griech. Feste*, p. 443, suggests that here Paean is identical with Apollo " da Paian ganz zu einer Epiklesis der grossen Götter herabgesunken ist." For similar offerings in Delos, cf. *ibid.*, p. 412; for Ceos, p. 421, n. 7; for Tamynae, p. 412; for Elaea, cf. O. Deubner, *Das Asklepieion von Pergamon*, 1938, p. 18 (sacrifice to Asclepius and Attalus). As regards Athens and a sacrifice for Asclepius and Hygieia on the day of the *Dissoteria*, cf. L. Deubner, *Attische Feste*, pp. 174 f.

(T. 569). Since the sacrifices to the latter were made to him in his capacity as savior from pestilence, it seems probable that the Pergamene sacrifice to Asclepius had the same meaning.[6] When on the other hand it was feared that disease might befall the town, the citizens invoked the god with apotropaeic prayers and sacrifices. In Rome, the simultaneous deaths of a praetor and a consul in one instance were interpreted as an indication of impending illness. The chief pontiff and the decemvirs, after due consultation of the sacred books, decided to vow gifts and gilded statues to the god. For two days supplications were held to secure the health of the city; processions went through the fora; all citizens over twelve years of age participated; they wore crowns and held laurel shoots in their hands (T. 505). And the god graciously listened to their humble petition; Rome was saved from disease.[7]

In addition, at decisive turns of the year the blessings of Asclepius were sought. In Titane, once every twelve months the god received a *Suovetaurilia*; during this service his mother's cult statue was taken to the temple of Athena; sacrificial animals were burnt on the ground, with the exception of birds which were burnt on the altar (T. 555). The whole ceremony seems to have been a rather old one, meant to celebrate the beginning of the year; it was dedicated to Asclepius, apparently with the wish of incurring the god's benevolence for the coming year.[8] Sacred

[6] Cf. Nilsson, *Griech. Feste*, p. 174: "Vor allen anderen Göttern ist Apollo der Hekatombaios, derjenige, der die grössten und glänzendsten Opfer empfängt. . . . Die Ursache ist wohl die Auffassung von ihm als Sender und Abwehrer von Seuchen und grossen Landplagen, der Schutz, den er gewährte, musste durch grosse Opfer erkauft werden." The fact that T. 569 parallels the hecatombs offered at Delphi with those offered at Pergamum is in favor of the interpretation here proposed. Nilsson, *loc. cit.*, points out that sometimes the name hecatomb promised more than was actually given. Lucian's words seem to emphasize the magnitude of the Pergamene sacrifice. A. Bouché-Leclercq (*Histoire de la Divination dans l'Antiquité*, III, 1880, p. 273) apparently assumes that it is Apollo who was approached in cases of pestilence, whereas Asclepius was restricted to individual healings; this is incorrect, cf. *e. g.* T. 451. Interestingly enough it is said in one passage that in great danger Apollo and Asclepius were invoked together (T. 586), a clear reminiscence of the original dependence of Asclepius on Apollo, cf. above, p. 99. Cf. also T. 460.

[7] These processions were modelled after the Greek pattern (cf. W. Warde Fowler, *E. R. E.*, *s. v.* Roman Religion, X, p. 831); the Roman cult of Asclepius in general was practiced in the Greek fashion (T. 509). The Christian polemic of course knows only of supplications that remained unheeded, *e. g.* T. 585.

[8] Cf. Nilsson, *Griech. Feste*, pp. 410 f., who equates the ceremonies held at Titane with the *Laphriae* and similar rites commemorated elsewhere. He says, *ibid.*, p. 54: "Alle diese 'Opfer' sind gleichartig und entsprechen den modernen Jahresfeuern. Der sogenannte Altar ist eigentlich ein Scheiterhaufen, oft als ein Gebäude ausgestattet, und wird mitsamt den Tieren ganz verbrannt." Cf. *ibid.*, p. 411: ". . . [in Titane] χαμαί, also auf einem direkt auf dem Erdboden aufgeschichteten Scheiterhaufen." For the *Suovetaurilia*, cf. L. Weber, *Philologus*, LXXXVII, 1932, pp. 409 ff. T. 553 also seems to attest sacrifices at the beginning of the year.

ablutions with clay were performed in Pergamum at the time of the spring equinox (T. 570). In late Roman centuries a vintage sacrifice was sacred to Asclepius (T. 574). On occasions such as these it was customary for people to say prayers for health and to herald the grapes as a good remedy for future ills.[9]

Finally, there was one group that revered the giver of health daily as well as on special occasions: the physicians who saw in him the patron of their art. Many of them probably had a house cult of Asclepius, as had Nicias who every day honored with incense the Asclepius statue in his possession (T. 501).[10] Then again, the physicians of a whole community, as those of Athens, sacrificed to the god twice a year in behalf of themselves and of those whom they had healed (T. 552). Among the devotees of Asclepius who zealously worshipped the dispenser of health his human colleagues were not the least noteworthy.[11]

2. RITUAL FOR THE SICK

Great as Asclepius' power was in protecting man from diseases, it was in the dark hours of sickness that his might shone forth most clearly. None of the other deities knew as well as he did how to bring back from Hades' innermost shrine, how to restore to light and life, those who were in peril.

The simplest way of securing the god's succor was that of approaching him with prayers. Proclus, lying on his sickbed, implored Asclepius and

---

[9] In earlier times, it was mainly Jupiter who was invoked as the god of vintage; cf. Wissowa, *Religion u. Kultus*, p. 115; F. Bömer, *Rhein. Mus.*, XL, 1941, pp. 30 ff. The attribution of the vintage festival to Asclepius is the more plausible as the *Meditrinalia*, the closing vintage rites, were sacred to Meditrina, a goddess of medicine, cf. Paulus *ex Festo*, p. 123, 16. Varro, *De Lingua Latina*, VI, 21, gives the formula recited: "novum vetus vinum bibo, novo veteri morbo medeor"; for the meaning of this prayer, cf. L. Deubner, *N. Jahrb.*, XXVII, 1911, p. 329, n. 5. The prayers offered at the Athenian *Pithoigia* were similar, and here the wine is directly called a remedy, cf. Plutarch, *Quaest. Conv.*, III, 7, 1. It is not impossible that Greek vintage festivals too became associated with Asclepius. One must remember that wine, in Greek and Roman medicine, was considered one of the most important remedies, cf. T. C. Allbutt, *Greek Medicine in Rome*, 1921, pp. 329 ff.

[10] Cf. Wilamowitz' interpretation of this poem, *Die Textgesch. d. griech. Bukoliker, Phil. Unters.*, XVIII, 1906, p. 118; he is right in emphasizing the fact that Nicias paid tribute to Asclepius as the saint of doctors.

[11] T. 552 relates that the Athenian physicians of the third century B. C. sacrificed to Asclepius according to their ancestral custom. Does this imply that they revered Asclepius still as a hero? An Athenian inscription, enumerating the sacrifices which the priest of Asclepius performed, mentions *Heroa* (T. 553); maybe they were *Heroa* in honor of Asclepius. Finally, Tertullian tells (T. 103) that the Athenians sacrificed to Asclepius and Coronis among their dead. A hero worship on the part of the physicians would be quite understandable, cf. also below, pp. 191, n. 1; 193, n. 7.

besought his assistance; the god came and took away his disease (T. 446). Many a man must have had this experience: he prayed, and afterwards he felt relief; the god had helped. Or, if it was not the patient himself who addressed the god, it may have been a friend who, moved by the sorrows of the sick, sent a heartfelt cry to heaven: " I sadly fear that the disease has become more serious; to you, Asclepius, and you, Health, I pray that there may be none of this " (T. 578).[1] If the patient got better, the deity had performed a miracle. Sometimes it seemed safer to go to the temple of the god and to approach him there. Thus a father sent his son to the sanctuary to offer prayers and vows for his recovery (T. 538).[2] In other cases the patient's relatives asked those who were known as the favorites of the god to pray to him and to insure his assistance for the sufferer (T. 582). It was felt that the god would not reject the plea of his friends if they took upon themselves the office of mediation.

The content and form of the prayers differed. Some wearied the god with long and elaborate expositions, they importuned him with most piteous vows (T. 584). Probably they believed that Asclepius would more easily listen if they talked at great length and showed him their subservience. Proclus, however, made his address to the god in the true old fashion (T. 582), he prayed in the simple, noble way of the ancient Athenians.[3] Many hoped that they might influence the god even through their posture while praying, and thought prostration before him more helpful than the upright position taken of old before the divine statue. Diogenes the Cynic, enraged at this blasphemy, dedicated to Asclepius a bruiser who, whenever the supplicants fell on their faces, would run up and beat them up (T. 580).[4] On the other hand, the intimacy, the personal fervor of the entreaties is well indicated by the fact that the Romans prayed to Asclepius with their heads covered (T. 579), in the Roman fashion, although in general the rites of the Roman cult of Asclepius were borrowed from the Greek ritual.[5]

---

[1] T. 578 is taken from Terence who follows either Menander or Apollodorus; cf. M. Schanz and C. Hosius, *Gesch. d. römischen Literatur*, I, 1927, p. 110.

[2] More sensible people were of the opinion that it was unnecessary to address the god in his temple. Asked to go there they answered that they were sure that Asclepius was not deaf and could hear their prayer wherever they happened to be (T. 577a). I should note that cities too offered prayers when disease gripped the town, cf. *e. g.* T. 585.

[3] For the old form of prayers, cf. *e. g.* Marcus Aurelius, V, 7, and in general A. W. Mair, *E. R. E.*, *s. v.* Prayer, X, pp. 182 ff.; L. R. Farnell, *The Evolution of Religion*, 1905, pp. 163 ff.

[4] Concerning the Greek attitude in prayer, cf. Wilamowitz, *Glaube*, I, p. 301; P. Stengel, *Die griech. Kultusaltertümer*, 1920[3], pp. 79 f.

[5] Cf. above, p. 183, n. 7, and for the Roman praying rites, Deubner, *Die Römer*, p. 462. It is true, Plautus generally changes the Greek original from which he copies, and

Whatever the attitude of the devout, the god expected from those who prayed to him the same sincerity that he demanded from those who came to his temple for the sake of incubations. When a barbarian prince once asked Apollonius to recommend him to Asclepius, the answer of the god's servant was: " You do not need any recommendation. If you care for good qualities, go boldly up to the god and tender what prayer you will " (T. 581).[6]

Yet, although Asclepius often gave help when merely invoked by the sick, it was in their temple dreams that he most effectively imparted his assistance. In order to receive healing dreams people came to the sanctuaries and lay down in the *abaton*. Before approaching the god, however, they had to perform certain rites which consisted of purifications of the body,[7] and of sacrifices (T. 511). Every religious act, in antiquity, demanded offerings to the deity, and temple cures were considered religious ceremonies.

In the beginning, when Asclepius had not yet become a great god, all-powerful by himself, the sacrifice brought to him before the act of incubation was composed of an initial offering to other deities and a main one dedicated to Asclepius alone. In Tricca and Epidaurus, around 300 B. C., Apollo Maleatas was the god who received the primary gifts of the supplicant (T. 516). Still earlier, around 400 B. C., in the Piraeus it was Maleas, Apollo, Hermes, Iaso, Aceso, Panacea, the dogs and the huntsmen or dog-guardians who were given the first share (T. 515).[8] Nothing is known about the kind of sacrifices made in Epidaurus or Tricca. In Athens, cakes were given to the deities who were invoked (T. 515).[9] Generally speaking, it is permissible to assume that as time

introduces Roman names and features (for the *Curculio*, cf. F. Leo, *Gesch. d. römischen Literatur*, I, 1913, pp. 145 ff.). Yet that he could do so in this particular instance seems to warrant the conclusion that Asclepius' worshippers prayed to their god *more Romano*.

[6] I fail to understand the interpretation of this passage given by R. Reitzenstein, *Die hellenistischen Mysterienreligionen*, 1927, p. 252, who quotes it in connection with the mystery of a ἱερὸς γάμος. Even if in the scene recorded there were a reminiscence of such a mystic union, the important point is that the god, through the mouth of Apollonius, rejects such concepts.

[7] Cf. above, p. 149.

[8] For the importance of Apollo Maleatas, cf. above, pp. 99 ff.; for Iaso, Aceso, Panacea, cf. above, pp. 87 ff. Hermes, as Wilamowitz has pointed out (*Glaube*, II, p. 227), is probably mentioned here as the giver of sleep; cf. Homer, *Odyssey*, XXIV, 1 ff. For dogs and dog-guardians, cf. next note.

[9] For a detailed interpretation of T. 515, cf. Farnell, *Hero Cults*, pp. 261 ff.; *Class. Quart.*, XIV, 1920, pp. 143 f.; Wilamowitz, *Glaube*, II, pp. 226 f. The reference made in this ritual of the fourth century B. C. to sacrifices to hunters and dogs is somewhat puzzling. The hunters must be heroes, rather than living human beings; sacrifices to mortals, and of this profession, are unknown at that time, as Farnell has shown; cf. also F. R. Walton, *Harvard Studies*, XLVI, 1935, p. 184. H. Scholz (*Der Hund in d.*

went on such sacrifices to other gods grew obsolete, and that, when at the height of his power, Asclepius alone demanded the whole attention of his worshippers.[10]

The main offering to Asclepius, as performed in early centuries, again is attested only for Athens. Honey-cakes, cheese-cakes, bakemeats, and figs were laid upon the holy table of the god, as Aristophanes says (T. 514; 490). The sacrifices prescribed in other sanctuaries may have been similar, or they may have changed in the course of time—, at any rate, there is no indication that the initial offerings required much expense. This fact is well in accord with Asclepius' general attitude towards gifts. He was not one of those gods who enjoyed excessive luxury.[11] On the contrary, he was suspicious of expensive offerings, especially if they were made before the supplicant had received any favors from him. When a wealthy Cilician sacrificed on a very lavish scale immediately after his arrival at the temple and without ever having been helped by Asclepius on previous occasions, Apollonius at once recognized the man as one who wanted to beg himself off from the penalty of some horrible and cruel deed, and the god confirmed the correctness of his servant's judgment: he refused to help this man and bade him leave the sanctuary (T. 517).

After the preparatory steps were taken, the patients could enter the *abaton* and wait for the apparition of the god. If Asclepius decided in favor of treatment, religious ceremonies often formed an integral part of the cure prescribed. Just as the god chose the drugs and determined all other measures of natural therapy, he also designated the sacrifices to be made. Thus an inscription records in one breath that the god imposed upon his patient: " to take a walk in the upper portico, to take some passive exercise, to sprinkle myself with sand, to walk around barefoot, in the bathroom, before plunging into the hot water, to pour wine over myself, to bathe without help and to give an Attic drachma to the bath attendant, in common to offer sacrifice to Asclepius, Epione and the

---

griech.-röm. Magie u. Religion, Diss. Berl., 1937, p. 49, n. 47) maintains the opposite view without giving any reasons for his opinion. The dogs are most probably the sacred animals kept in the temple; sacrifices to animals were not unusual. A. Furtwängler (*S. B. München,* 1897, p. 406) believed that these dogs were demons; Wilamowitz, *loc. cit.,* seems to agree with him, yet such a view is hardly correct because no demons of this kind are ever mentioned. Wilamowitz also suggests that the whole ritual echoes ceremonies originally dedicated to the hero from whom Asclepius took over the sanctuary. This is quite possible.

[10] Such a conclusion seems warranted by the development that took place in regard to the thank-offerings, cf. below, p. 188. Wherever the connection between Asclepius and Maleatas remained close throughout the centuries, as for instance in Epidaurus (cf. above, p. 99, n. 31), Maleatas may have continued to receive a share in the initial sacrifice.

[11] Cf. below, pp. 189 f.

Eleusinian goddesses, to take milk with honey " and several other things of a similar kind (T. 432). Even more marked and shocking is another combination that Aristides attests: " Thereupon purifications took place near the river by means of libations, and purifications at home by means of vomiting " (T. 518). In temple medicine sacrifices apparently were a matter of course, and great importance and healing power seem to have been attached to them.[12]

When the patient was healed immediately, or finally had regained his health, he was bound to offer a thanksgiving and to fulfill whatever vows he had made (T. 423, 5; 520). In early times, such a sacrifice again started out with the reverence paid to another god. In Erythrae at the beginning of the fourth century B. C. the person who wanted to sacrifice to Asclepius was expected first to march around the altar of Apollo three times and thrice to sing a paean in his honor (T. 521).[13] Herondas in the middle of the third century, when describing in great detail a thanksgiving in the temple of Cos (T. 482), no longer mentions Apollo. It was only Asclepius who now received all the remuneration for having aided his worshipper.[14] At the altar in the prophylaeum, where many divine statues were set up, the suppliants said a prayer of gratitude to the god (T. 482, vv. 1-18). Then the sacrificial animal was given to the sexton (v. 41).[15] It was for him to decide whether the offer was acceptable to the god, and, if so, he addressed him in another solemn prayer (vv. 79-85). After this, the priest received his share in the offering; a portion was put into the mouth of the snake as a small tribute to the sacred animal (v. 91).[16] Part of the dedicated cake was taken home and eaten

---

[12] The belief that such ceremonies would contribute towards recovery is on the same level with that in the greater efficacy of drugs taken near the sacred tripod (T. 411) or lamp (T. 411a).

[13] Cf. Wilamowitz, " Nordionische Steine," *Abh. Berl. Akad.*, 1909, pp. 37 ff.; Wünsch, Pauly-Wissowa, *s.v.* Hymnos, IX, p. 169, 12. The rules preserved concern any kind of sacrifice, whether offered after a healing through a dream or through a prayer.

[14] For a detailed interpretation of the Herondas passage, cf. R. Wünsch, *Arch. Rel. Wiss.*, V, 1904, pp. 95 ff.; cf. also S. Kalinka, *S. B. Wien*, CXCVII, no. 6, 1922, pp. 3 ff. The document is most important for ancient cult rites in general; cf. L. Ziehen, Pauly-Wissowa, *s.v.* Hiereis, VIII, p. 1422, 67: " wohl typisch für das gewöhnliche Privat-opfer." Whether the sacrifice was made after a healing through dream vision or prayer, cannot be ascertained. The god seems to have performed an immediate cure, cf. T. 482, *vv. 17 ff.*: " paying the price of cure from disease that thou didst wipe away, Lord, by laying on us thy gentle hands."

[15] The sacristan here acts in place of the priest, cf. below, p. 192.

[16] That the " offering to the snake " means a payment of money into a box on which an ornamental serpent was seated has been suggested by Herzog, *Arch. Rel. Wiss.*, X, 1907, pp. 205 ff.; such tills have been excavated. Farnell, *Hero Cults,* p. 257, speaks of *pelanoi* or money given to a serpent " who enjoyed an underground treasure chamber "; cf. *ibid.*, p. 262.

there. Last but not least, the worshippers took with them even some holy bread as a means of averting evil (*vv. 93-95*).[17]

This, it seems, was the general form of thank-offering prevailing in the Asclepius cult from Herondas' time down to the end of antiquity. To be sure, the details varied.[18] While in Cos the worshipper was permitted to take home his share in the god's meal (T. 482), in Epidaurus and Titane (T. 510), as well as at Athens (T. 510a), he had to consume his part within the precincts of the temple.[19] But everywhere, it seems, the devout feasted with the god. In the Asclepius cult the ancient concept of the sacrifice as a communion between god and man was upheld tenaciously.[20] Concerning the sacrificial animals there also were variations. In Cyrene goats could be offered (T. 532), in Epidaurus (T. 532) and Tithorea (T. 533) they were tabooed.[21] Generally speaking, however, all animals which the ancients were wont to give to their deities could also be offered to Asclepius.[22] The god evidently had few predilections.

Only in one respect did he show an extraordinary attitude: he was satisfied with small dedications. The Coan women who sacrificed a cock to him apologized for the cheapness of their offering; they were aware of

[17] Cf. Wünsch, *loc. cit.*, pp. 115 f.; his interpretation remains valid whether one attributes lines 93 ff. with Wünsch and Herzog to Cynno, or with Knox to the priest. Cf. also T. 503, where Libanius speaks of branches which were supposed to have an apotropaeic effect; in general, cf. above, pp. 182 f.

[18] In late Rome it was customary to give thanks before the assembled people (T. 438).

[19] Farnell, *Hero Cults*, p. 242, and Latte, *Gnomon*, VII, pp. 122 f., mention a *holokautesis* in honor of Asclepius, and Latte, *loc. cit.*, has explained this sacrifice as significant for the chthonian character of Asclepius. Apart from the fact that a burnt sacrifice does not prove a chthonian cult (cf. Nilsson, *Griech. Feste*, p. 428, n. 2; Farnell, *loc. cit.*; Wilamowitz, *Glaube*, I, p. 242), the inscription referred to (*I. G.*, IV², 1, no. 97) says only θεῷ (l. 23; 26). There is no certainty, then, that this ritual really concerns Asclepius; many other gods were revered in his temple.

[20] Cf. Ziehen, Pauly-Wissowa, XVIII, p. 622, 19 ff.; Nilsson, *Popular Religion*, pp. 74 f.

[21] Goat sacrifices were not very common in Greek cults, excepting that of Artemis; cf. Wilamowitz, *Glaube*, I, p. 290; Farnell, *Cults*, II, p. 449. Servius' explanation of the preference for goats (T. 535) seems fanciful; at any rate, it could equally well be used for an argument in favor of the rejection of goats: the god of health could not accept animals who are never free of fever. Or maybe in Epidaurus goats were forbidden, because in the opinion of the Epidaurians a goat had nursed the babe Asclepius (T. 7), unless this story itself is an explanation of the rule by which goat sacrifices were excluded. Farnell, *Hero Cults*, p. 242, says: "If this rule [*sc.* of tabooing goats] were aboriginal and universal, it would certainly strengthen our belief in his [*sc.* Asclepius'] original human-heroic nature." Yet the rule was not universal, though it seems to have been more common to refrain from goat sacrifices, cf. T. 534.

[22] For the animals usually sacrificed, cf. *e. g.* T. 536; in general, cf. Ziehen, *op. cit.*, p. 590, 36 ff. If he says (*ibid.*, p. 594, 49) that for the Asclepius cult male sacrificial animals alone seem to be attested, this is in contradiction with T. 691, where *gallinae* are mentioned; cf. also T. 535.

the fact that other clients would perhaps have offered an ox or a pig (T. 482, v. 15). Yet, the sacristan assured them that nobody could give to Asclepius a gift which would be more welcome to him (vv. 80-81). The cock was indeed the most common sacrifice; the phrase " a cock to Asclepius " was almost proverbial, not only on account of Socrates' offering.[23] The divine physician seems to have agreed with Hesiod's demand that everybody should give to the gods as much as was in his power,[24] and no more; in the case of the poor this meant a little; in the case of the rich of course it meant much. For if they gave on a small scale, they deprived the deity of the share that was rightly his.

Finally it should be mentioned that thank-offerings to Asclepius did not necessarily consist of animal sacrifices or of these alone. People in exchange for the fulfillment of their wishes could give and actually gave almost anything: money, frankincense, laurel, olive shoots, oak leaves, garlands, songs, branches, chaplets, pictures on which Asclepius was painted as well-doer (T. 539),[25] or brass rings (T. 542), candles (T. 544), offerings in gold and silver (T. 545; 484). Some patients even dedicated their sandals to the god; they had made a long trip in order to visit him, and thus it seemed fitting that they should give him their shoes.[26] Whatever it was, the god received it graciously: " The gods . . . do not strive after human goods. Wealth is not gratifying to them, nor any other gain. They are interested in beneficence alone. In this the gods find their riches " (T. 538). So the ancient sceptic himself said before he was asked to render his due to Asclepius.

---

[23] Cf. T. 523; 539 and 562 (for καλαῖς, cf. E. Fraenkel, Glotta, IV, 1913, pp. 33 f.). It has sometimes been doubted that cocks were used as a " Speiseopfer "; but the Herondas passage proves that this doubt is not warranted; cf. Ziehen, Pauly-Wissowa, XVIII, p. 590, 65 ff. In other worships the sacrifice of a cock is rarely attested, perhaps on account of the fact that the cock was brought to Greece only in post-Homeric centuries, cf. V. Hehn, " Homer," Die Antike, III, 1927, p. 74. Although it follows from Herondas and from other passages (e. g. T. 539) that the suppliants of Asclepius favored the cock sacrifice because it was cheap, it may also have been of importance that the cock was the herald of dawn—Asclepius was the god who allowed men to see the light of day. Or the belief that the cock was an apotropaeic bird may have contributed to the popularity of the sacrifice. Certainly the worshippers did not consider the cock a bird dedicated to chthonian deities, for in the cult of Eleusis the cock was tabooed, cf. Porphyry, De abstinentia, iv, 16, and for the significance of the cock in general, E. R. E., III, pp. 694 ff.

[24] Hesiod, Opera, 336. It is perhaps not by chance that Aristides refers to this passage in his speech on Asclepius (T. 317, 2).

[25] For such pictures, cf. below, p. 215. Incidentally, a special kind of garland seems to have been sacred to Asclepius (T. 835; cf. also 543).

[26] Cf. W. Amelung, Arch. Rel. Wiss., VIII, 1905, pp. 157 ff. (according to a monument). Mention should be made at least of those offerings which were made to Asclepius for help bestowed upon men in other than medical instances, cf. T. 546 ff.

3. TEMPLE SERVICE

Those who sought preservation of health and those who sought liberation from diseases tried to solicit Asclepius' help in their own individual concerns. The regular worship, however, which the devotees paid to their god was of a more disinterested nature; it was the god's unquestioned due, regardless of his beneficence. This ceremonial centered around the temple. When Asclepius rose to power, the Greek deities had long ceased to dwell outdoors and had taken up their residence in the magnificent palaces that men had built for them. It is true, a cave sacred to Asclepius is mentioned (T. 755), and so is an altar established in the open air (T. 757). But other deities, too, occasionally found worship at such places even in late centuries. Generally speaking, all ceremonies in honor of Asclepius were held within the precincts of the sacred temple.[1]

In ground plan and construction the Asclepius sanctuaries seem not to have differed from those of other divinities, with the exception of two features: the Asclepieia had buildings attached to them which were intended to house the patients and to provide the necessary means for their treatment; besides, there was a hall in which the patients slept. It was called *abaton* (T. 423, 2) or *adyton* (T. 516), names customarily given to the most sacred part of the temple; in the Asclepieia naturally this was the place of incubation, since it was here that the god made his appearance.[2] But apart from these features which resulted from the specific

[1] The fact remains, however, that in many places the worship of Asclepius originally was performed at an altar erected in the temple of Apollo. This was true of Epidaurus, for instance (cf. above, p. 99, n. 32), and of Cos (cf. *Kos*, ed. R. Herzog, I, *Asklepieion*, ed. P. Schazman, 1932, p. 72). Note that those sanctuaries which are referred to in the testimonies were dedicated to the god Asclepius; of the hero no shrines are attested. It is true, Aelian (T. 730) after having spoken in general terms of the respect of the Athenians for *heroa*, relates a story that refers to an Asclepius sanctuary. From this statement one can hardly conclude with certainty that a *heroon* of Asclepius existed in Athens; Aelian tells a number of stories illustrating the religious attitude of the Athenians in general. If his report is to be taken literally, one probably ought to connect this sanctuary with the worship of the Athenian physicians, cf. above, p. 184, n. 11.

[2] Cf. Pfister, Pauly-Wissowa, *s. v.* Kultus, XI, pp. 2140, 36 ff. It should be remembered, however, that in some places the incubants slept in the temple proper; cf. above, p. 150, n. 20. In Tricca, according to Isyllus (T. 516), Asclepius had a nether *adyton*. Farnell, *Hero Cults*, p. 239, thinks that "this subterranean structure may have prevailed elsewhere, accounting for the rise of a legend that such and such communities [Cynosura, Epidaurus, cf. above, p. 50] possessed the tomb of Asklepios." Whether his assumption is correct remains uncertain. Herzog, *Wunderheilungen*, p. 159, n. 34, believes that Isyllus' words are indicative of an "ansteigende Örtlichkeit"; yet καταβαίνειν εἰς ἄδυτον is the technical term for the descent into a nether sanctuary; cf. *e. g.* Pausanias, II, 2, 1. On the other hand, such a subterranean structure is not a chthonian feature (contrary to Farnell, *loc. cit.*). For nether *adyta* are known from many temples, even from that of Athena, cf. Pausanias, VII, 27, 2, and Daremberg-Saglio, *s. v.* Adytum. If a special significance can be attached at all to this underground part of the temple, it may be

character of the Asclepius religion the shrines were built according to the usual pattern, just as they enjoyed all the prerogatives commonly attached to holy places: whatever fell to the ground belonged to the god (T. 487); whoever took refuge in the temple was protected by the right of asylum against any persecution (T. 798).[3]

In these sanctuaries the god was honored with regular sacrifices. Their performance rested in the hands of the priests (cf. *e. g.* T. 494). It constituted in fact their main task, aside from the offering of special sacrifices made at a given time, either by individuals or by the city.[4] When executing their holy functions the priests usually were clad in white—in Pergamum they wore a purple robe (T. 492a) and Egyptian shoes, as Aristides says (T. 493)—, their hair was bound with a white fillet (T. 850, *v. 676*).[5] They were assisted by a sacristan. Great sanctuaries like that of Pergamum had even two sextons (T. 495), or the sacristan had

characteristic of Asclepius as an accessory of Apollo (cf. above, p. 99), for other minor deities also had nether *adyta, e. g.* Erechtheus (cf. Rhode, *Psyche,* I, p. 136), or Palaemon (cf. Pausanias, II, 2, 1).

[3] The various sanctuaries probably had sacred laws of their own (cf. Kern, *Religion,* I, p. 170; cf. also R. Herzog, *Heilige Gesetze von Kos, Abh. Berl. Akad.,* 1928, no. 6). In Epidaurus, birth and death could not take place within the temple limits (T. 488).

[4] For the participation of the sacristan in private sacrifices, cf. above, p. 188. Besides, the priests had other obligations: they supervised the buildings as well as the minor officials (T. 491). They administered the money offered to the god (in Epidaurus special officials, the *Hieromnemones* [T. 562], four in number, seem to have been in charge of the treasury, cf. Pauly-Wissowa, VIII, p. 1491, 39 ff.). The priests probably also kept an account of the other presents (T. 489) and took charge of the sale of offerings which became necessary from time to time, or of the removal of less valuable ones when space was lacking. Finally, they were responsible for the purification of the sanctuary which took place at stated intervals, a ceremony that attracted many people. In white garments they stood in the gateway during the performance of the holy ritual (T. 486). Concerning the duties of the priests in general, cf. L. Ziehen, Pauly-Wissowa, *s. v.* Hiereis, VIII, p. 1420, 57 ff.

[5] For the garments of priests in general, cf. Stengel, *Kultusaltertümer,* p. 47. The attire of the priests of Asclepius is described in more detail in T. 492. From T. 687 it seems to follow that in some places they wore Greek sandals. To add a few details concerning the election and remuneration of the priests: they were elected by the people, chosen by lot, or they would inherit, or even buy their office (T. 491 b-e). In short, all the procedures common in the installation of such officials were employed also in the Asclepius cult. Cf. L. Ziehen, Pauly-Wissowa, *s. v.* Hiereis, VIII, pp. 1413, 12 ff.; cf. also *E. R. E.,* X, pp. 305-306. In Pergamum, where the priesthood was hereditary, even women officiated as priests; cf. O. Deubner, *Das Asklepieion von Pergamon,* 1938, p. 11. For the chronology of the Athenian priests, cf. W. S. Ferguson, *The Priests of Asklepios, Univ. of Cal. Publications in Class. Phil.,* I, 1906, pp. 131 ff.; W. K. Pritchett and B. D. Meritt, *The Chronology of Hellenistic Athens,* 1940, pp. 74 ff. The remuneration which the priest received consisted mainly of a share in the sacrifices (T. 490). Wherever his office lasted only one year, he probably found himself adequately rewarded by the honor of his position. When in office for life, he may have received a salary. During his tenure the priest had certain public prerogatives. In Pergamum (T. 491) and in Athens (T. 491a) a special seat was reserved for him in the theater. After he had served his term and had given a satisfactory report to the authorities, public thanks and honorary inscriptions were decreed for him (T. 494).

subordinates who may also have officiated in the ceremonies.[6] As for the sacrifices made, it suffices to point out that there is no indication of a chthonian ritual. While offerings to the chthonian deities were received by pits, those made to Asclepius were placed upon altars.[7] Finally, it should be remembered that the word spoken was no less agreeable to the deity than was the gift presented. Nay, prayer rather than sacrifice formed the climax of the ritual. Hymns were sung and the temple resounded with the sacred responses, the " Hail " (T. 598), or " Hail Paean " (T. 592), or " Great is Asclepius " (T. 602).[8]

While it remains uncertain how often such regular observances were held in early times, whether monthly or twice a month or more rarely, in late centuries daily temple services seem to have been customary. At that time Asclepius, at least twice every day, in the morning and at dusk, received the veneration of the pious. Special morning songs are attested for Epidaurus (T. 598; cf. T. 466); Aristides, in reference to the Pergamene temple, alludes to its being opened again at night (T. 485).[9]

[6] Concerning the office of the sacristan (νεωκόρος), cf. Pauly-Wissowa, XVI, pp. 2422 ff. The procedure of his selection is not known; in Asia Minor, at least in late centuries, it was an honorary title (cf. Pauly-Wissowa, ibid., p. 2423, 66). In Athens, the sacristan was called ζάκορος (T. 500); his assistant was the ὑποζακορεύων (T. 498a). For the participation of the sacristan in sacrifices, cf. above, p. 188. For Eretria special hieropoioi are attested (T. 787), cf. Pauly-Wissowa, VIII, p. 1585, 60.

[7] The altar is usually called βωμός (T. 483), sometimes also τρίβωμος (T. 438). Walton, Cult of Asklepios, p. 43, understands the latter as "triangular in shape," but altars of such form are otherwise unknown (cf. Pauly-Wissowa, s.v. Altar, I, p. 1677, 20). Wissowa, Religion und Kultus, p. 308, n. 7, explains τρίβωμος as an altar dedicated to three gods, Apollo, Asclepius, Hygieia. Offering tables have been excavated at Cos; cf. R. Herzog, Arch. Rel. Wiss., X, 1907, p. 204. One of the Asclepius monuments probably shows an ἐσχάρα (cf. Farnell, Hero Cults, p. 240); ἐσχάρα and βωμός are to be found in chthonian and Olympian cults alike (cf. Pauly-Wissowa, loc. cit., p. 1664, 45 ff.; Wide, op. cit., p. 233). The excavations of the Athenian temple have unearthed a pit within the sanctuary of Asclepius (cf. U. Köhler, Ath. Mitt., II, 1877, p. 254). This does not necessarily mean that chthonian sacrifices were dedicated to Asclepius, they may have been given to other deities, as was the case in Pergamum (T. 504). Köhler thinks that the pit was used in connection with the Heroa, a sacrifice for the dead. But cf. above, p. 184, n. 11. The ὀμφαλός which appears on some of the Athenian reliefs (cf. Pauly-Wissowa, loc. cit., p. 1665, 25 ff.) denotes Asclepius as a hero. It seems that in Athens traces of a hero cult survived.

[8] Wilamowitz, Glaube, I, p. 301, characterizes the ancient cult ceremonial thus: "Freilich das Wort kam zu kurz, auch wenn Gebetsformeln, einzeln auch Litaneien, und die rituellen Rufe der Gemeinde . . . (hinzu) traten." This statement does not seem justified. For the connection between prayers and sacrifices, cf. Ziehen, Pauly-Wissowa, XVIII, p. 604, 47 ff. For the ὕμνος παραβώμιος, cf. Plato, Laws, VII, 800 C-D, and in general, Schmid-Stählin, Gesch. d. griech. Literatur, I, 1, p. 340.

[9] There was a special officer, called the κλειδοῦχος (T. 498, n. 1); cf. Pauly-Wissowa, XI, p. 595, 38 ff. Sometimes the sacristan was in charge of the keys of the shrine (T. 485). Herondas mentions the opening of the cella in the morning (T. 482, vv. 55 f.); for the rules in regard to the closing of the cella in various cults, cf. Wide, op. cit., p. 234. But it should be stressed that people could come to the sanctuary whenever they

Whereof these ordinary services consisted is not known in detail. One may, however, venture to say that everything was regulated minutely. Different hymns were recited at different hours. In many cases they were certainly sung by the congregation, but regular choirs also existed, sometimes perhaps made up of slaves.[10] In addition to singing, offerings were made. In view of the fact that they were presented so often, it is likely that they commonly consisted of libations and frankincense rather than of animal sacrifices.[11] Finally, at night, illuminations played an important part. Aristides speaks of the evening service as the time when the sacred lights are kindled (T. 485). Asclepius it seems was fond of candles.[12]

The average number of people who attended the regular observances, and especially the daily ceremonies, is hard to guess. In the great cult centers, in Epidaurus, Cos, or Pergamum the temple must have been crowded with the devout; Herondas mentions the rush of people in the early morning hours (T. 482, *v. 54*). In small towns or villages, the priest and the choir may have found themselves in the company of a small congregation, comprising mainly old men and women. Yet, it seems permissible to surmise that in the routine of daily life Asclepius was less likely to be forgotten than many another deity. The preoccupation of the ancients with their bodily well-being, their fear of disease which they felt to be a constant threat, must have caused them to be particularly strict in their adherence to the god of medicine.[13]

---

liked to offer prayer or sacrifice; for there were statues and altars in the propylaeum (T. 482) which could be approached at any time.

[10] For hymns destined to be sung at the first and the third hour, cf. Maas, *Epidaurische Hymnen*, p. 160, 3; 16. For an interpretation of divine poetry, cf. below, pp. 199 ff. Choirs are mentioned *e. g.* in T. 561; 562. With the requirement of so much singing, the temples must have found it necessary to hire a choir, and I think it permissible to assume that the slaves owned by the temples (*e. g.* T. 491, *26*; cf. O. Kern, *Hermes*, XLVI, 1911, p. 302) were used for this task also, in addition to their duty to perform all the heavy manual labor, such as cleaning the temple, or to perform the many odd jobs to be done in institutions where patients were taken care of, such as attendance in the bathhouse (T. 432).

[11] Latte, *Gnomon*, VII, 1931, pp. 133 f., has tried to reconstruct the daily ritual of the Asclepius temple in Epidaurus according to *I. G.*, IV², 1, no. 742. But Latte himself calls the document a "hoffnungslos verstümmelten Rest eines späten Kultrituals," and he admits that it is uncertain, whether the inscription in every detail really refers to Asclepius. For this reason I have not made use of this testimony. But from Latte I have borrowed the suggestion that bloodless sacrifices were the rule in these ceremonies. I should mention that daily services are attested already on an inscription of the second century B. C. (T. 553).

[12] For the λυχναψία, cf. Latte, *Gnomon*, VII, p. 133. There were also sacred lamps (T. 411a); in regard to them, cf. Nilsson, *Griech. Feste*, p. 345, n. 5; S. Eitrem, *Opferritus und Voropfer*, 1915, pp. 152 f. In this connection the office of the fire bearer (T. 423, *5*) should also be mentioned. Smoking censers are referred to in T. 432.

[13] This book was already in print when an article was published by M. P. Nilsson on "Divine Service in Late Antiquity," *The Harvard Theological Review*, XXXVIII, 1945, pp. 63 ff. Nilsson outlines the history of the daily ceremonies and shows that they

4. FESTIVALS

In addition to the regular services, Asclepius, like all the other gods, was worshipped with celebrations that recurred at certain intervals. To these festivals, at least to the more important ones, people came from far and near, often in troupes of pilgrims (T. 402).[1] Pergamum was one of the places " where they all run " to celebrate feast days (T. 569). Such observances were a treat for men no less than for the deity; they were intended to be not only religious ceremonies but also popular entertainments.

The names of the festivals and their dates varied a good deal. Many were called *Asclepieia* (e. g. T. 575);[2] but at Athens it was the *Epidauria* that were sacred to Asclepius (T. 564). Some of the ceremonies had no special designation and were distinguished from those devoted to other deities simply by the addition of Asclepius' name. A procession of Asclepius (T. 568) and a night festival of Asclepius (T. 553) are mentioned. As regards the time of year at which the various fêtes took place and as regards their frequency, sometimes they fell in the spring (T. 570), sometimes in the fall (T. 565); one city had an annual festival (T. 554), another commemorated the god twice a year (T. 572). As far as the day was concerned, the 8th seems to have enjoyed a certain preference. At least in Athens, Asclepius was honored on the 8th (T. 566) and on the 18th (T. 565), that is the 8th of the second decade of the month. This custom may also have prevailed elsewhere.[3]

deeply influenced the formation of the Christian service, that the pagan reverence of old, which accentuated the sacrifice of animals, gradually was replaced by a more spiritual worship in which songs and bloodless offerings became increasingly popular. Nilsson also stresses the importance which the cult of Asclepius assumed in this development (p. 67). For other details of the Asclepius worship reminiscent of the Christian service, cf. below, p. 206, n. 28. Incidentally Nilsson states (p. 64) that the epigram of Theocritus (T. 501, cf. above, p. 184) is one of the earliest testimonies mentioning a daily house cult.

[1] For pilgrimages, cf. M. P. Nilsson, *Jahrb. d. Dtsch. Arch. Inst.*, XXXI, 1916, p. 310. A θίασος of patients is mentioned on one of the inscriptions of Lebena, cf. M. Guarducci, *Inscriptiones Creticae*, I, 1935, XVII, 11, 6.

[2] Festivals with the name Asclepieia, as far as they were known from inscriptions found before 1896, are collected in Pauly-Wissowa, II, p. 1683. Since then, considerable new material has been found; yet, it is not my intention to investigate the regional extension of the cult.

[3] For the date of the Athenian *Epidauria*, cf. Deubner, *Attische Feste*, p. 72. The belief that the 8th day had a special significance for Asclepius rests mainly on the testimony concerning Athens referred to above, and on that concerning Pergamum. Here, certain ceremonies in honor of the priest of Asclepius were held on the 8th. In Elaea the people staged a procession to the Asclepius temple and sacrificed to the god and to Attalus on the 8th, cf. W. Schmidt, *Geburtstag im Altertum, R. G. V. V.*, VII, 1, 1908, p. 104; O. Deubner, *op. cit.*, p. 18. Schmidt, *loc. cit.*, therefore calls the 8th day the birthday of the god. Such a " birthday " in ancient terminology was either the day of the first

At such occasions the worshippers used to meet in the center of the city, and, raising the paean in honor of the god, they went in procession from there to the temple which often was situated at the outskirts of the town. In later centuries it was customary for the pious to carry the statue of the deity with them.[4] In Athens the procession was supervised by the archon (T. 567); in other places other officials were in charge of it. In some cases, it seems, everybody took part in these parades, men and women, and small children, even those under seven years (T. 787).[5] In Epidaurus, the communities which were on friendly terms with the host city had the right to send legations which in an official capacity participated in the pageant, a cause of pride to the smaller neighboring towns. The animals which they intended to sacrifice were taken along with those dedicated by the city of Epidaurus (T. 563). On special occasions, however, the procession was limited to certain groups of citizens, men selected from each tribe, the best of the city; they appeared in white garments and with flowing hair, holding in their hands garlands of laurel for Apollo and branches of olive for Asclepius (T. 296).[6] Or the procession had to be formed by young people, carrying in their hands suppliant boughs, bright offshoots of the olive (T. 593). In Cos, at the yearly festive assembly of the national god, the " children of Asclepius," the physicians, staged a very costly procession of their own (T. 568).[7]

epiphany of the deity, or the day of his main festival, chosen for some unknown reason and later understood as the birthday of the god. But considering the variety of ancient cult rites, it is not certain that what was true of Athens and Pergamum also applied to other temples. In Rome, the 11th of September (T. 855a) and the 1st of January (T. 855) were celebrated as *dies natalis* of Asclepius. These days, however, mark the respective foundings of the two Roman Asclepieia, dates which the Romans generally considered the "birthday" of a god, cf. Schmidt, *op. cit.*, pp. 116 ff. and 124.

[4] For processions in general, cf. Nilsson, *loc. cit.*, pp. 311 ff.; Kern, *Religion*, I, pp. 168 f.; Wilamowitz, *Glaube*, II, pp. 350 ff. Processions were very popular. They could also be held as independent ceremonies. For the carrying of the divine statues in processions, cf. Nilsson, *op. cit.*, p. 317; Wilamowitz, *Glaube*, II, p. 351. Only one testimony mentions a statue of Asclepius together with those of other gods in a Roman procession (T. 258). For the paean, cf. below, pp. 199 ff.

[5] Their names were included in the records of the festival. Nilsson, *Griech. Feste*, p. 412, thinks it probable that this custom was connected with the cult of the babe Asclepius; cf. above, p. 71, n. 14, and below, p. 224.

[6] Nilsson, *op. cit.*, p. 410, believes that the procession, for which Isyllus wrote his paean, formed part of the annual festival in honor of Asclepius. But it seems to me that the specific purpose of Isyllus, as outlined by Wilamowitz (*Isyllos*, pp. 38 ff.), intimates that this was an independent ceremony, not integrated into the main festival; cf. also below, p. 201. Other processions, such as that of the *ephebi* at Cos, are discussed by Nilsson, *op. cit.*, p. 412.

[7] Nilsson, *op. cit.*, p. 411, rightly claims that the phrase ἣν ἔθος ἀνάγειν τοῖς τῷ θεῷ προσήκουσιν presupposes " dass das Asklepiadengeschlecht beim Fest eine besondere Rolle spielte." The Coan physicians certainly had an old-established worship of Asclepius even before the great temple was founded (cf. below, p. 243) and the god became the main

When the worshippers had arrived at the temple, sacrifices were offered to Asclepius. The god himself was believed to come and take part in the celebration. From time immemorial the deity, in the opinion of the faithful, feasted with them. In early centuries his presence was taken for granted, no symbolic representation of the divinity was needed; later, it became more and more customary to set up a couch and to place upon it a puppet embodying the god. In Athens, Asclepius was honored in such a way relatively early (T. 553), and *lectisternia* for Asclepius are attested also for other temples (T. 719).[8] The offerings differed in the different centuries and at different places. Epidaurus, at the annual festival of its saint, celebrated him together with the other gods and goddesses who shared his temple (T. 562).[9] Asclepius and the male deities received a bull, the female deities were given a cow. In addition, Asclepius was honored with a cock, with barley, wheat and wine. What sacrifices were offered in Cos and Pergamum, the other famous seats of Asclepius, is not known.[10]

While processions and sacrifices formed the main content of the celebrations held during the daytime, those which took place during the night were characterized by other and no less spectacular solemnities. Nights holy to Asclepius were probably observed in Pergamum (T. 571),[11] they are unambiguously attested for Athens (T. 553). During the nightly vigils people went out into the streets. They indulged in merriment. For such nights, fraught as they were with religious emotion, were

deity of the city. Their custom (ἔθος) of celebrating Asclepius is here referred to, just as it is said of the sacrifice of the Athenian physicians that it was customary (πάτριον, T. 552).

[8] Concerning the *lectisternia*, cf. Pauly-Wissowa, XII, pp. 1108 ff.; Kern, *Religion*, III, p. 178; Wilamowitz, *Glaube*, II, pp. 350 ff., has stressed their relatively late attestation by Greek inscriptions. The Athenian inscription concerning Asclepius goes back to the second century B. C.; Latte, *Gnomon*, VII, 1931, p. 118, suggests that the festival was adopted from the Eleusinian ritual. Wissowa, *Religion u. Kultus*, pp. 421 ff., claims that in Italy *lectisternia* were especially favored in those cults that were taken over from Greece and performed in the Greek fashion. Asclepius, then, should have received many such celebrations.

[9] Nilsson, *Griech. Feste*, p. 409, understands T. 562 as referring to an annual festival. T. 561 specifies the sacrifice to Apollo, made apparently on the same occasion.

[10] For Cos, cf. R. Herzog, *Heilige Gesetze von Kos*, Abh. Berl. Akad., 1928, no. 6, pp. 39 ff.; the inscriptions preserved are too badly mutilated to be considered here. One of the Coan festivals was called τῆς ῥάβδου ἡ ἀνάληψις (T. 568); according to Nilsson, *op. cit.*, p. 411, this ceremony was connected with the taking over of the sacred rod by the new priest of Asclepius. Nilsson also suggests that this festival was a " Vorgänger der grossen Asklepieien." For Pergamum, the new excavations may bring some more material; cf. the preliminary report of Th. Wiegand, *Abh. Berl. Akad.*, 1932, no. 5.

[11] Thraemer, Pauly-Wissowa, II, p. 1643, 48, has drawn attention to this testimony; but it cannot be decided with certainty whether θεός, as is usual with Aristides, refers to Asclepius, or whether it refers to Poseidon whom he has just mentioned.

also devoted to gaiety. Even Aristides, in spite of his melancholic and morose disposition that made him stay at home, felt obliged to pass that night amusing himself (T. 571). The Athenian festival may have been connected with a torchlight race.[12] If so, it provided a fascinating spectacle that could not fail to attract people. How great the contrast between these nightly sights, these merry crowds, and the nightly temple sleep of the sick to whom the god revealed himself.[13]

What the Asclepius festivals meant in the life of the city, it is rather difficult to ascertain in general terms. For there were no uniform regulations in regard to the participants in the ceremonies or to the Sunday-rest, if one may use that expression. In Lampsacus (T. 572), where the cost of the sacrifices was defrayed by the temple treasury, all business ceased when twice a year, in the months of *Lenaion* and *Leucathion*, the *Asclepieia* were staged. Children had a day off at school; slaves were granted a day of rest.[14] In Athens, at the celebrations in the *Elaphebolion*, no political assemblies were held. Demosthenes did an unheard-of thing when he brought in a decree that the prytanes call a meeting of the citizens for that day (T. 566).[15] In the beginning of Asclepius' reign the extent to which his holidays were observed by the cities, apart from

[12] The connection suggested here between *Pannychis* and *Lampadophoria* is based on Plato, *Republic*, 328 A. For the latter ceremony, cf. Pauly-Wissowa, *s.v.* Lampadedromia, XII, pp. 569 ff.; its aim was to take fire from one altar, on which it was burning, to another, on which it was to be started, or more generally to light the fire on certain altars. It is interesting that in Athens this rite originally was restricted to Prometheus, Hephaestus, and Athena, the deities of artists and craftsmen (cf. Pauly-Wissowa, *loc. cit.*, pp. 576 f.); the fire was taken from the Prometheus sanctuary. *Lampadophoriae* are attested only for Paros (T. 547a); *ibid.*, p. 570, 46. The ceremony may, of course, have taken place in many of the Asclepieia. Alexander the Great staged a *Lampadophoria* for the god (T. 547). The rite seems to have gone out of fashion later on; cf. Wilamowitz, *Glaube*, II, p. 353. For the general importance of lights in the Asclepius cult, cf. above, p. 194.

[13] Kern, *Religion*, I, pp. 167 f., is the only one to have mentioned the general importance of the night for the Asclepius cult. With the exception of the mystery religions or the veneration of certain demons, the day was the main time of worship.

[14] The consideration shown here for children seems characteristic of the Asclepius festivals, cf. also T. 787.

[15] On the day of the *Epidauria* (T. 564), however, assemblies took place; Deubner, *Attische Feste*, pp. 72 f., explains this fact by the relatively small importance of this festival which was of rather late date. The *Epidauria* actually were a mystery rite rather than an Asclepius festival. The god supposedly had come to Athens on that day, and for this reason it was an important date for him as well as for the city. But the ceremonies performed in his honor served as introductory rites to the Great Mysteries, and they were much shorter than the rites preliminary to initiation had been in former times. Hence the legend arose that the Athenians initiated Asclepius, although he had arrived late for the mysteries (T. 565); for an interpretation of this testimony, cf. A. Mommsen, *Feste der Stadt Athen*, 1898, pp. 216 ff.; 219 ff. The text of T. 564 has been analyzed by Kern, Pauly-Wissowa, *s.v.* Epidauria, VI, p. 45, 25 ff.; for the date of the festival, cf. above, n. 3. For Asclepius as a *mystes*, cf. above, pp. 127 ff.

the ones that revered him as their principal deity, must have been relatively small. The more he gained in power, the more he became acknowledged as one of the foremost protectors of man, the greater must have been the citizen's concern with the festivals held in his honor.

5. HYMNS AND SERMONS

In all the ceremonies, be they daily exercises or solemn festivals, one feature recurs: the singing of hymns. They were recited in the temples morning and night; they were chanted during processions; they were accompaniment for sacrifices. The song characteristic of the Asclepius worship was the paean, a choral hymn with no music other than that of the cithara. Originally the paean had been sacred to Apollo; the fact that later it was connected especially with Asclepius indicates that in the belief of the people the son had usurped the place of the father in matters pertaining to medicine.[1] Although only a few of the hymns dedicated to Asclepius are still extant, one can evaluate, at least in broad outline, the characteristics of this divine poetry which finally came to include the composition of prose hymns, no less stately and elaborate than those written in verse. Of course, it is only the words of the songs that can still convey a message to the modern interpreter. The music is lost, and yet, for the Greeks, the tunes may have meant even more than the words. For as Plato says, more than anything else do rhythm and harmony find their way to the inmost soul and take strongest hold upon it, bringing with them and imparting grace if they are rightly chosen, or the contrary if they are chosen badly.[2]

The most famous of all Asclepius hymns and the oldest known is the one composed by Sophocles (T. 587).[3] The words that are still legible on the stone on which it is preserved are too few to convey an adequate impression of its religious fervor or artistic quality. The ancients' appreciation of the song is fairly expressed by the statement that the god himself had urged the poet "to write a paean and though 'famed for his skill' he does not disdain to hear it from you" (T. 590). Sophocles,

[1] In general, cf. A. Fairbanks, *A Study of the Greek Paean*, *Cornell Studies in Classical Philology*, XII, 1900, especially pp. 66 f.; Th. Reinach in Daremberg-Saglio, *s. v.* Paean, IV, pp. 265 ff. In later times, cithara and flutes were used as instruments.
[2] I am paraphrasing Plato's statement, *Republic*, III, 401 D. For the Greek concept of music, cf. E. Frank, *Plato und die sogenannten Pythagoreer*, 1923, pp. 1 ff., and for music in its relation to religious services, cf. G. Méautis, *Aspects ignorés de la religion grecque*, 1925.
[3] For the Homeric hymn (T. 31) which was also composed at the end of the fifth century B. C., cf. below, p. 210.

apparently in the dactylic metre, addresses Coronis, the mother of Asclepius, the warder-off of pain; he speaks of Apollo, the unshorn. His loud-voiced hymn calls on the helper of the sons of Cecrops.[4]

Not much later than Sophocles' paean the so-called Erythraean hymn (T. 592) must have originated: the inscriptions on which it is to be found can be dated between 380 and 360 B. C.[5] The dactylics first praise Asclepius' parent, Apollo "who begat great joy for man when he mingled in love with Coronis in the land of the Phlegyae." Then Asclepius' children by Epione are commended: Machaon and Podalirius, Iaso, fair-eyed Aegle, Panacea, Hygieia.[6] Finally the god himself is asked to come to the city and to grant that men "may see the sunlight in joy." Simple and formal though the words are, they nevertheless reflect the warmth of feeling and the confidence that Asclepius inspired in the hearts of the pious, the affection with which they worshipped him. Thrice he is hailed as the most renowned of demigods. Being himself a joy to men he imparts joy to them also through his children, above all through his daughter Hygieia, "the glorious," with whom he is invited to appear. The devout implore the god that he may make them "acceptable with bright Hygieia."[7]

The Erythraean hymn emphasizing health as the greatest gift of Asclepius is the perfect expression of the this-worldliness which throughout the centuries was so important an incentive in the veneration of the god. This motif remains dominant in the later hymns, but at the same time

[4] The latest reconstruction of the fragments is that given by J. H. Oliver, *Hesperia*, V, 1936, p. 112; for the metre, cf. Wilamowitz, *Griechische Verskunst*, 1921, p. 353. Oliver, *loc. cit.*, pp. 113 ff., suggests that this is not the famous paean to Asclepius, but rather one dedicated to Coronis and written "as an introductory hymn for a definite place in a familiar ceremony" (p. 121). This is not impossible, although it is difficult to ascertain, since so few words are preserved. That Sophocles should not have written a paean to Asclepius at all, and that another anonymous hymn should have been attributed to him at a later date on account of his well-known relationship to the god, seems, however, improbable to me. It is true, the paean to Asclepius is mentioned relatively late. But I think it was the famous paean that gave rise to the story concerning Sophocles and Asclepius, and not the reverse.

[5] Cf. Wilamowitz, *Abh. Berl. Akad.*, 1909, II, pp. 42 ff.; *Griechische Verskunst*, p. 353. I. U. Powell, *Collectanea Alexandrina*, 1925, pp. 136 ff. Powell thinks that the hymn originated in Thessaly. This assumption is rightly rejected by Wilamowitz, *Glaube*, II, p. 225, n. 3, who himself suggests Attic origin. Cf. also K. Keyssner, *Philologus*, XCII, 1937, pp. 281 f.

[6] Cf. above, p. 87.

[7] Wilamowitz, *Abh. Berl. Akad.*, 1909, II, p. 45, commenting on the word δοκίμους in the hymn, says: "[The Ptolemaic copy] bittet den Gott, um das Sonnenlicht freudig schauen zu lassen δοκίμους, acceptos als ἀνθρώπους εὐδοκίας. Da ist das Verhältnis zwischen dem Gläubigen und seinem Gotte in der Richtung entwickelt, die der Dünkel der Juden und Christen als ihre spezifische Frömmigkeit in Anspruch nimmt." Powell, *op. cit.*, pp. 137 f., has pointed out that even the Erythraean copy of the fourth century must have read δοκίμους.

they introduce other aspects of Asclepius' godhead. Thus Macedonius (T. 593),[8] in the first century A. D., hails Asclepius and noble, delightsome Hygieia; he addresses the deity that protects from disease. Yet a strong accent is laid on Asclepius' wisdom which was imparted to him by Chiron. It is in fact this wisdom that men glorify in the hymn, just as in Epidaurus they greeted Asclepius as the god of "lofty art" (T. 593a).[9] Again a late Epidaurian morning chant of six hexameters (T. 598), which in a charmingly naive way urges the god to disperse sleep from his eyes and to heed the prayers of his worshippers, propitiates Hygieia, his prime power. But twice it heralds Asclepius as the "gentleminded," recalling his name, his character, his way of healing.[10]

Isyllus' paean (T. 594) is of a slightly different character. Here, the praise of the god's origin fills most of the verses.[11] Such an exposition was common in hymnology; yet Isyllus had a special reason for devoting so much space to this feature: he wished to expound the new genealogy which he himself had devised. Those who sang his words or listened to them were probably deeply impressed by his story, although Isyllus was not a great poet and in spite of his personal devotion could hardly inspire enthusiasm in others through his art. The myth itself charmed the Greeks,[12] and Isyllus wrote as an Epidaurian for Epidaurians. It must have delighted them to hear that the god was their compatriot. The mention of Asclepius as the giver of health and the liberator from disease, which occurs at the end of the poem, sounds rather conventional and cold. Only in his final words does Isyllus find expression for one particular wish which the other songs fail to include: "send bright health to our hearts and bodies."

The Orphic hymn (T. 601) finally rounds out the circle of laudatory poetry in honor of Asclepius' achievements.[13] Himself young and mighty,

[8] Concerning Macedonius, a rather famous poet of the first century, cf. Wilamowitz, *Griechische Verskunst*, p. 133, n. 1. New readings are proposed by Keyssner, *op. cit.*, pp. 269 ff.

[9] Keyssner, *op. cit.*, p. 278, calls the hymn "dichterisch unerfreulich." That may well be true. But it seems exaggerated to claim (*ibid.*, p. 269, n. 1) that Macedonius slavishly follows the Erythraean hymn. The judgment of Bülow, *Xenia Bonnensia*, 1929, p. 39, n. 1, in my opinion is more adequate.

[10] For the metre of this hymn, cf. Maas, *Epidaurische Hymnen*, pp. 151 ff. I should add that in the older hymns the god is asked to protect their particular cities, whereas the Epidaurian hymn speaks of the worshippers of the god without any specification.

[11] For the metre of this hymn, cf. Wilamowitz, *Isyllos*, p. 19; also pp. 125 ff.; 158. The song was written for the procession introduced by Isyllus, cf. *ibid.*, p. 14. Nilsson, *Griech. Religion*, I, p. 508, n. 7, seems to consider this poem the official hymn of Epidaurus; this is hardly correct.

[12] For the rôle of the myth, cf. below, p. 205.

[13] That the Orphic hymns were composed and written down for Orphic congregations

"the enemy of diseases," he has as wife, "faultless Hygieia." He is the giver of gentle gifts, the softener of pains. But in these solemn hexameters it is also the terror of death that is felt: Asclepius is the one who stays the harsh fate of dying. In the same way Aristides, in one of the first lines which he dedicated to the god, had celebrated him: "From sharp-sighted death he rescued many who had advanced right to the gates of Hades whence none return" (T. 597), and thousands must have shared this experience. Moreover, the deity who is invoked by the Orphics, the savior, the blessed one, is also the one who grants "a good end to life." Death casts its dark shadow over man's heart, but death at the same time gives him a right understanding of life; it bestows upon him tranquility of mind and serene certainty concerning the Here and the Beyond; this-worldliness and other-worldliness are thus integrated.

These are the songs dedicated to Asclepius that have survived. Many more were written in the course of the centuries.[14] In some cases at least the name of the author is attested. Dionysius, the tyrant of Syracuse, composed a hymn to Asclepius (T. 603), and so did Apuleius (T. 608), and Isodemus (T. 588).[15] Aristides, the prophet of Asclepius, must have dedicated a great number of poems to the god. Allusions to these songs are frequent throughout his work (T. 604 f.). He wrote them in various metres, in hexameters and elegiacs, in anapaestics or iambic dimeters, and it seems that he was a poet of considerable talent.[16] Doubtless he saw to it that his compositions were brought before the members of the Asclepius congregation. He prides himself on having paid for no less than ten singings of the chorus (T. 605a); even in his dreams, he heard young boys sweetly intoning his verses (T. 597). Other hymns were probably written by others, for special occasions or for definite places.[17]

and were not merely literary products, has been shown by A. Dieterich, *Kleine Schriften*, 1911, p. 86; cf. also R. Wünsch, Pauly-Wissowa, IX, p. 171, 18 ff.; I. M. Linforth, *The Arts of Orpheus*, 1941, pp. 179 ff.

[14] Occasionally inscriptions record dedications to Asclepius in the form of hymns. Thus the priest Diophantus, saved by his god from podagra, uttered his thanks in a poem praising the might of the deity who had enabled him to come again to the sacred temple and to fulfill his duties (T. 428). The song is written in the favorite anapaestic iambic metre of late hymnology; cf. O. Crusius, Pauly-Wissowa, V, pp. 1050 f., who rightly rejects the assumption of Wilamowitz and Kaibel that the metre was invented "ad ipsum podagricorum usum"; for the interpretation of the poem, cf. also above, p. 125. Another example of such divine poetry is T. 596, attributed to Aristides by R. Herzog, *S. B. Berl.*, 1934, pp. 753 ff.

[15] Maas, *op. cit.*, p. 155, reads † Ἀλισόδημος; Proclus wrote a hymn to Asclepius Leontouchos (T. 826a).

[16] For the metrical form of Aristides' poems, cf. Wilamowitz, *Griechische Verskunst*, p. 134.

[17] Attention should be drawn to those hymns which are to be found at the beginning of medical books, or which were added to a prescription, invoking the god to come and

Whether there was one generally acknowledged cult hymn of Asclepius, it is difficult to decide. Although Epidaurus played so important a rôle in the dissemination of the Asclepius worship, the various temples remained independent in their rites, and the songs probably differed as greatly as did the rules of sacrifices.[18] On the other hand, it seems that some of the songs enjoyed a certain preference. The Erythraean hymn was performed in Athens as well as in Macedonian Dium and in Menschieh in Egypt; a few changes or additions were made to render the text suitable for the particular circumstances.[19] Moreover, as the dates of the various inscriptions show, this song, composed in the fourth century B. C., was still performed in the Christian era, just as Sophocles' paean was still heard in the Athenian temple around 200 A. D. (T. 589).[20] In every religion there is a tendency to conservatism and traditionalism. But new hymns were also written and were in use, as is proven by the number

impart his power to that which was written or that which was prepared for the recovery of the sick. The hymn that introduces the work of Serenus (T. 615), written in hexameters, is a perfect example of the "Du-Stil" (cf. E. Norden, *Agnostos Theos*, 1913, p. 174), with the predications of the god given in relative clauses: "you who know how to impart new lives, you who cherish Aegae, Pergamum, Epidaurus and the Tarpeian rocks." In another poem, in elegiac metre, preserved among the works of Galen (T. 595) and written by the elder Andromachus, a physician under Nero, the god is implored to enhance the efficacy of the drug to be given to the emperor: "Whether the Triccaean ridges hold you, O demigod, or Rhodes, or Cos and Epidaurus on the sea." For the disjunctive form in which the divine places are named here, cf. Norden, *op. cit.*, pp. 144 ff. T. 614 too may be mentioned here. It is taken from the poem found at the end of the work of Marcellus, a physician around 400 A. D. Riese, *ad locum*, is hardly right in rejecting the authorship of Marcellus, cf. M. Niedermann, *C. M. L.*, V, 1916, p. xxi. In hymnology, the epithets given to the god play an important part. A few of the cognomina are collected in T. 623 ff. For additional material, cf. C. F. H. Bruchmann, *Epitheta deorum*, in Roscher, *Lexikon*, Suppl., 1893; Walton, *Cult of Asklepios*, pp. 83 f. Another example of the use made of Asclepius' name should be mentioned at least: the swearing by Asclepius (T. 618 ff.; 266; 337). The testimonies do not indicate that it was any particular class of people to whom the oath was restricted. For the significance of the various oath formulas, cf. E. Ziebarth, *De iure iurando*, etc., Diss. Gött., 1892, pp. 10 ff.; Pauly-Wissowa, *s. v.* Eid, V, pp. 2076 f. In Galen's time it was apparently quite common to swear by Asclepius of Pergamum (T. 620). Late sources ascribe even to Socrates the oath by Asclepius (T. 622).

[18] Cf. above, p. 189.

[19] The changes made in the hymn of Dium (T. 592a) are summarized by Wilamowitz, *Griechische Verskunst*, p. 353, n. 3. The Egyptian copy, the first one found, was edited by J. Baillet, *Revue Archéologique*, XIII, 1889, pp. 70 ff. For its readings, cf. Powell, *op. cit.*, p. 138. Concerning the addition of a few lines, cf. Wilamowitz, *op. cit.*, p. 133. Following Powell's example only the oldest (T. 592) and the most complete version (T. 592a) of the hymn have been included in the collection of the testimonies. An Athenian copy is preserved, *I. G.*, II², 4509; cf. Oliver, *op. cit.*, p. 115. The Pergamene hymn (T. 607) began with a mention of Telephus, but failed to name Eurypylus, the Mysian hero who slew Machaon, cf. T. 512.

[20] The Egyptian hymn can be dated in 97 A. D. The Athenian inscription probably stems from the turn of the second to the third century, cf. Oliver, *op. cit.*, p. 115. The paean of Dium was inscribed in the second century, cf. G. P. Oikonomos, Ἐπιγραφαὶ τῆς Μακεδονίας, IV, 1915, pp. 8 ff.

of other songs preserved or mentioned. Especially in Hellenistic times novel expression was given to human feelings, and though the old hymns continued to be recited, new ones were favored.[21] Only the reactionary temper of the Roman period reverted to archaic compositions, at least in the official cult; yet even Lucian (T. 588) does not claim more than that the singing of old hymns is no less honor to the gods than the singing of those recently invented, and the example of Aristides is evidence of the fact that the production of hymns did not immediately cease in the first two Christian centuries. Later, Porphyry with apparent approval refers to Aeschylus who, when asked to write a new song to Apollo, answered that the old one would do better, " for new religious songs, like new religious statues, though probably more elaborate and refined, and consequently more admired, have less of the splendor of the divine about them." [22] After Porphyry, the general attitude certainly became even more conservative.

On the other hand, during the same period in which the performance of old hymns became more and more customary, the word sung in honor of the god was in ever increasing rivalry with the word spoken to the same end. The prose hymns or *encomia* may well be compared with sermons.[23] They were recited in the temple (T. 608), or in the theater sacred to the god (T. 806), whenever one of his great festivals took place; in white holiday attire the people sat and listened. The subject of the orations was the power and greatness of the god (T. 609; 388). In their arrangement the speeches usually followed a definite pattern, as it was evolved by ancient rhetoric.[24] To the modern mind, they may seem dry and artificial, bombastic and lacking in real fervor. Greeks and Romans were easily swayed by oratorical feats performed in rhetorically and stylistically perfect language. When distinguished orators like Apuleius (T. 608), or Aristides (T. 317), or Acacius (T. 388), celebrated the god in elaborate addresses, they must have stirred in their audience an aesthetic excitement that intensified the religious experience, and many a hearer's concept of the god was likely to be shaped by what he heard in these lectures during which the stage of Asclepius became a school house, as it were.[25]

---

[21] Cf. Wünsch, Pauly-Wissowa, *s. v.* Hymnos, IX, p. 164, 60 ff.

[22] Porphyry, *De abstinentia*, II, 18.   [23] Cf. E. Norden, *Agnostos Theos*, p. 165.

[24] Thus Aristides follows rhetorical theories (cf. Norden, *loc. cit.*). Weinreich, *Heilungswunder*, p. 7, n. 5, has rightly stressed the fact that some of the greater inscriptions, as for instance that of Apellas (T. 432), come close to the literary standards of a sermon.

[25] Such sermons were also published (T. 609) and consequently had a wide circulation among the educated. But their literary importance hardly equalled the influence of the spoken word.

Besides, in these sermons, even more than in the songs, two expectations of the worshippers must have found satisfaction: the one, characteristic of all Greek devotees, was to hear about their god's mythical deeds; the other, especially characteristic of the followers of Asclepius, was to be assured of their god's providence and care for the individual. As for the splendor of the myth, Aristides' speech in honor of the Asclepiads (T. 282) is a good example of the account that was expected. Here, the god and his sons are placed in their proper relationship to Zeus and Apollo, they are seen as outstanding members of the society of the Olympians, in which the Greeks never tired of reveling. But Machaon and Podalirius, and therewith Asclepius, were also connected with the world of the Homeric heroes. Superior to all of them they decided the fate of the Trojan war: the fall of Troy was entirely due to them. Afterwards the two took possession of the whole world; they themselves left Thessaly and went to Asia Minor, to the islands; their children, happier and better than the Heraclids, inherited the *oecumene*. Thus even the present moment was colored by the glory of the divine legend and the heroic myth of old. The believer in Asclepius, through kinship with his god, himself became part of the mythical world for which he longed.[26]

On the other hand, Aristides' speech on Asclepius (T. 317) shows the intensity with which the providence of the god was experienced, the help of him "who is looking after man." Aristides feels that through Asclepius he has lived not only once, but many times. Years and decades have been added to his span of life through the divine oracles. His body is a present which he has received from the god. Even more: everything that Aristides has accomplished he has achieved through Asclepius. Wherever he found success, it was Asclepius who caused men to applaud his faithful servant. In the care of Asclepius, Aristides, just like everybody else, is sheltered from all dangers and evils, he is secure and safe, whatever may befall him.[27] Granted that these encomiasts were given to

[26] For the importance of the myth in antiquity, cf. J. Burckhardt, *Griechische Kulturgeschichte*, I⁵, pp. 15 ff. He says (p. 29): ". . . der Mythus als eine gewaltige Macht beherrschte das griechische Leben und schwebte über demselben wie eine nahe, herrliche Erscheinung. Er leuchtete in die ganze griechische Gegenwart hinein, überall und bis in späte Zeiten, als wäre er eine noch gar nicht ferne Vergangenheit, während er im Grunde das Schauen und Tun der Nation selbst in höherem Abbilde darstellte." Pausanias, I, 35, 6, is a particularly instructive example for the mythical fantasy of the average people even in late centuries; cf. Burckhardt, *ibid.*, p. 31.

[27] For the god's concern with the individual, cf. above, pp. 111 f. The orator here becomes, so to speak, a witness of the deity. This, one must keep in mind if one wishes to understand correctly Aristides' *Sacred Speeches*. They are by no means merely egotistic representations of what happened to him. Others who did not want to speak in their own name took their text from the dedicatory inscriptions (T. 388). But in these cases, too, it was personal experiences that were recounted.

exaggeration and that they flattered the god and those who listened, a kernel of truth nevertheless remains: Asclepius was the god who took a deep interest in the individual, and it was proper and justifiable to praise him on that account.

If one looks back now on the vast mass of divine poetry and prose that is preserved or referred to in the testimony, one may safely state that there was an extremely rich store of literature dedicated to Asclepius.[28] This may be explained first of all by the fact that he was so popular a god, and by the circumstance that from the beginning of his healing activity he had among his clientele many poets, rhetors and philosophers who reciprocated his gifts with works of art. Aelian delights in pointing out that Asclepius was wont to take care of the educated (T. 456), and the testimonies, indeed, give the impression that the god was the special protector of this class of people.[29] Moreover, it would be erroneous to believe that they felt indebted to Asclepius only if he had healed their diseases. The god was the giver and preserver of health, and health is essential, if the spirit is to obey, as Aristides says (T. 323). Naturally, then, the intellectuals felt close to Asclepius, even when they were in good health. Nobody is more dependent on the obedience of the spirit than are literary men, who, as Celsus asserts somewhat sarcastically, are almost without exception fragile and liable to disorders, if they do not look after their well-being.[30]

There may be still another reason for the many poetic productions offered to the god: not only did Asclepius have so many devotees who were qualified to commemorate his power in words or songs; he was himself reputed to have inspired, or even requested literary gifts (T. 590). He was credited with giving the literati the ability to accomplish their task (T. 610), to write poetry and prose. But why should such a

[28] The frequency of divine songs addressed to Asclepius seems to indicate that in the worship of this god particular attention was paid to singing. If this is correct, the Asclepius cult would form an important link between the pagan and the Christian ritual; from the earliest times hymns were an integral part of the Christian ceremonial, cf. J. Kroll, *Die Antike*, II, 1926, pp. 258 ff.; H. Abert, *ibid.*, pp. 282 ff.

[29] This fact has been emphasized by Weinreich (*Heilungswunder*, pp. 4 f.; 7, n. 4; 113; 129 f.), who has also pointed out that Sarapis had to assume a similar rôle so as to be able to compete with Asclepius, *op. cit.*, p. 118. Among Asclepius' patients were Apollonius (T. 397), Aristarchus (T. 455), Theopompus (T. 456), the rhetor Aeschines (T. 404), Plutarch and Domninus (T. 427), Hermocrates (T. 434), Polemon (T. 433), Proclus, the philosopher (T. 445), and Aristides. Since the names of the god's patients are known only in relatively few cases, the list is rather impressive, although it must be admitted that names of famous patients were more likely to appear in literary documents than those of ordinary people.

[30] Celsus, *De Medicina*, I, 1, 2; cf. above, pp. 122 f.; 126.

belief have been current? Why should Asclepius have been especially fond of literary productions?

Libanius, the rhetor, thought that Asclepius as the son of Apollo had a natural affinity to literature and all the arts (T. 610).[81] Still the son must not necessarily have inherited the cultural interests or talents of his father, and therefore have been able to apportion the gift of genius to whomever he desired, as Libanius claims that he was. But was not the Asclepius cult, from an early date, closely connected with the names of great poets and with events memorable in their lives? Whether Sophocles actually gave shelter to the god in his own home, or whether the god appeared to him only in a vision—it was believed that he had written a paean in his honor (T. 587 ff.), inspired by the god himself. According to one tradition, it was Sophocles himself who introduced the Asclepius worship in Athens (T. 591).[82] The great poet then was known to have been the favorite of Asclepius. Was it not natural to conclude that the friendship which Asclepius had bestowed upon him would be extended and granted to Sophocles' fellow artists as well? In addition, in Athens the main festival of Asclepius was celebrated on the day preceding the Great Dionysia (T. 566), and it was on that day that poets and actors made their first appearance before the public.[83] What an occasion for all those interested in art! On this day that was dedicated to Asclepius, the whole city offered prayers and sacrifices to him. To be sure, the poets must have done likewise. Their thoughts must have turned to him for protection, for the fulfillment of their wishes and hopes. He who resided next to the great theater of Dionysus (T. 582) must have been near to their hearts. Finally one should remember that the word was considered a remedy of the soul (T. 617), that the *logos* was often named

[81] The opinion expressed by Libanius goes back to older tradition. Even Aristides acknowledges that he owes his skill in writing to Asclepius (T. 317, *12*). A physician whom Lucian quotes (T. 616) began his book by saying: "It is seemly for a doctor to write history, at least if Asclepius is the son of Apollo, and Apollo the leader of the Muses and the patron of all education." It is also noteworthy that as early as in the fourth century B.C. someone dedicated to Asclepius a prize won in a musical *agon* (T. 549).

[82] Sophocles' relation to Asclepius has been discussed in detail by F. R. Walton, *Harvard Stud. in Class. Phil.*, XLVI, 1935, pp. 170 ff.; cf. also above, p. 117. The singularity of Sophocles' connection with Asclepius has hardly been appreciated in the modern discussion. There are no other cults in which the paean of a poet of the fifth century B.C. played a rôle comparable to that of the Sophoclean paean in the Asclepius cult.

[83] Cf. E. Rohde, *Kleine Schriften*, II, 1901, pp. 381 ff.; Deubner, *Attische Feste*, p. 142. H. Usener, *Kleine Schriften*, III, 1914, p. 241, sees in the *proagon* of poets and actors only a procession and an offering of prayers to the god. Even if this assumption were correct, it is still true that the day itself, sacred to Asclepius, was of great importance for poets and actors.

a physician.[34] Thus one may readily understand how Asclepius came to be the inspirer of writers and their special protector, why they in turn felt indebted to him for their productions and offered their works to him with grateful devotion.

6. GAMES

One last feature of the cult remains to be considered: the games connected with many of the festivals dedicated to Asclepius. Not only in the main centers of worship, but also in smaller and less important communities, contests as well as prayers and sacrifices formed part of the ritual.[1]

Of all the games attested, those held in Epidaurus are the oldest known, although it is difficult to decide with exactness when they were celebrated for the first time. Pindar (T. 556) speaks of a victory in boxing and in the *pancratium* won at Epidaurus around 530 B. C.[2] In spite of the fact that he does not mention Asclepius, the ancient commentators did not hesitate to identify these *Epidauria* with the contests performed in honor of Asclepius (T. 557-59), the *Asclepieia*, as they were called later by Plato (T. 560). Such an identification seems plausible, yet it is by no means certain. At the end of the sixth century Asclepius was established as a god in Epidaurus, and the games may well have been founded at the same time. It is also possible that the contests mentioned by Pindar were much older and dedicated to another deity, that they were connected with Asclepius only after he had become the main god of the Epidaurians.[3]

---

[34] The identification of λόγος with φάρμακον or ιατρός, which first occurs in the writings of the Pre-Socratics, gained in importance in Hellenistic centuries. For the dependence of Aristotle's poetical theory on medical concepts, cf. E. Howald, *Hermes*, LIV, 1919, pp. 187 ff.

[1] A list of the games mentioned in literature and on inscriptions is to be found in Roscher, *Lexikon*, I, 1, p. 631, 10 ff.; Pauly-Wissowa, II, p. 1683; cf. also Nilsson, *Griech. Feste*, p. 413. Recent excavations have greatly increased the data available; for this material, cf. especially Th. Klee, *Zur Geschichte d. gymnischen Agone an griech. Festen*, 1918. From those games which were performed regularly others must be distinguished which were celebrated only once as thank-offerings or *soteria* (cf. *e. g.* T. 547); for a discussion of this custom, cf. Pauly-Wissowa, *s. v.* Agones, I, p. 842, 26 ff.

[2] Pindar refers to a victory of one called Themistius, the grandfather of the boy to whom his poem is addressed. *Nemean*, V usually is dated between 489 (L. R. Farnell, *The Works of Pindar, Critical Commentary*, 1932, p. 274) and 485/83 B. C. (Wilamowitz, *Pindaros*, 1922, p. 169). The year 530, therefore, would seem the earliest possible date for the victory of Themistius.

[3] Concerning Asclepius and Epidaurus, cf. above, p. 98. No games except those in honor of Asclepius are known from Epidaurus; cf. Farnell, *op. cit.*, p. 280. That Pindar does not speak of the *Asclepieia* may be due to his dislike of this god (T. 1). On the other hand, all the Epidaurian victories mentioned by Pindar occur in songs which are dedicated to citizens of Aegina (T. 556; 556a; 556b), a town that had one of the earliest temples of Asclepius (T. 733); cf. below, p. 245. This seems corroborative proof of the assumption that the games referred to by Pindar really were the *Asclepieia*. Yet,

However that may be, from the end of the fifth century the Epidaurian contests are unequivocally referred to as those of Asclepius, and from that time onward they enjoyed widespread fame. People came from far and near to vie with each other when every five years, probably in the spring of the fourth year of the Olympiad and nine days after the Isthmiae, the games sacred to Asclepius were held.[4] The judges of the contests were called *Hellanodikai* (T. 557a). This title implies that Epidaurus claimed Panhellenic importance for its games, and apparently it succeeded in substantiating this right.[5] The city that had given a new god to mankind commanded the respect of the whole ancient world.

The Epidaurian *Asclepieia* were celebrated throughout antiquity. So were the games of the god in other places, where Asclepius established his influence over the neighboring sections of the towns in which he resided and attracted visitors at least from the adjoining provinces.[6] His power remained great enough to protect his games against the competition of other contests that were introduced in Hellenistic and Roman centuries and in many instances superseded older festivals.[7]

as has already been pointed out, it is also possible that the Epidaurian games originally belonged to another deity, perhaps Maleatas, and were later taken over by Asclepius. Ancient tradition names the Asclepiads as the original founders and the Argives as the instigators of the penteteric games (T. 559). The text of the testimony, a Pindar scholion, is perhaps corrupt; cf. Drachmann, *ad. loc.* Yet the meaning of the statement seems clear: the Asclepiads were the mythical founders of the contest, while the Argives, the foremost political power, were the real founders in historical times; similar distinctions are to be found concerning other games. Penteteric games were prevalent in later centuries (cf. Pauly-Wissowa, I, p. 844, 7 ff.); many of the games were re-organized during the sixth century (cf. *ibid.*, pp. 853 f.). At any rate, if the Epidaurian games from the beginning were sacred to Asclepius, they cannot have been instituted long before 530 B. C.

[4] For the date of the festivals, cf. Klee, *op. cit.*, pp. 59; 70. Inscriptions concerning the victors in the games are extremely scarce (cf. Klee, *op. cit.*, pp. 57 f.). Pindar (T. 556) and Plato (T. 560) attest the fact that people from other parts of the country came to Epidaurus, some of whom are known as contestants at the other great festivals (Nemeae, Isthmiae, Panathenaeae). Wilamowitz, *S. B. Berl.*, 1909, p. 817, denies the sacred character of the Epidaurian games, because money was the prize. Yet T. 557 expressly says that the games were sacred to Asclepius.

[5] Latte, *Gnomon*, VII, 1931, pp. 117 f., has stressed the Panhellenic character of the *Epidauria*. He is probably right in assuming that the games were intended to connect the Epidaurian festivals with aristocratic tradition. His claim that the visitors came only from nearby places, however, seems exaggerated.

[6] To Cos, for instance, where contests were founded after 300 B. C., people came from all Asia Minor, cf. Klee, *op. cit.*, p. 118. The competitions were more or less identical with those of the Isthmiae, Nemeae, and Panathenaeae, yet prizes were given not only to the first, but also to the second contestant, in order to make the games more attractive. The epigraphical material from Cos yields considerable information concerning the names of the victors (cf. Klee, *op. cit.*, pp. 121 ff.).

[7] For the gradual decline of older games and the rise of new ones, cf. Pauly-Wissowa, I, pp. 858; 860. For games in general, cf. J. B. Bury, *The Nemean Odes of Pindar*, 1890, Appendix D, pp. 248 ff.; F. M. Cornford in J. E. Harrison, *Themis*, 1912, Ch. VII;

As regards the program of such games, they consisted mainly of gymnastic contests. All kinds of sports that the Greeks used to perform in honor of their gods were practiced also in honor of Asclepius. Why should the giver of health and the liberator from diseases not have enjoyed physical exercises and competitions?[8] Besides, musical contests were held. In Epidaurus they must have been introduced not long before the end of the fifth century B. C. For in the Platonic *Ion* (T. 560), Socrates expresses his astonishment that Ion, a rhapsodist, should have gone to Epidaurus to give a recital there. Apparently Socrates has never heard of singing contests at the Asclepius festivals, although he is familiar with the fact that games were held on that occasion. Ion has to tell him about " the contests of rhapsodies and of arts in general." At the end of the fifth century, then, the games which originally were of a merely gymnastic character must have been extended so as to include musical presentations, and such a development is in accord with changes in the programs of other games which are known to have occurred at that time.[9] It is not impossible that the Homeric hymn to Asclepius (T. 31), the oldest hymn to the god that is extant, was composed for these Epidaurian contests. For this poem, too, was written at the end of the fifth century B. C., and it is a prooemium intended to be sung at rhapsodical recitals.[10]

Not much is known about the repertories of the musical games. That epics were recited one may conclude from Ion's report (T. 560). The presentation of a dithyramb is attested only once, in a late Athenian inscription (T. 549a).[11] This kind of musical play originally was sacred

E. N. Gardiner, *Athletics of the Ancient World*, 1930, pp. 28 ff.; Pauly-Wissowa, *s. v.* Agones, I, pp. 836 ff. I need not dwell here on the origin and the meaning of such games. When they were adopted in the Asclepius worship, they were an acknowledged form of tribute to the gods, not restricted to any special group of deities.

[8] Some interpreters have expressed astonishment that the healer of diseases should have been celebrated with athletic contests. They forget that Asclepius was also the god of the healthy, and that sports traditionally formed part of divine festivals.

[9] Unfortunately the date of the Platonic dialogue, or rather the time when Socrates and Ion supposedly met, cannot be determined exactly. It is not even known who this Ion was, or when he lived. Concerning the increasing number of musical contests after 500, cf. Pauly-Wissowa, I, p. 854, 24 ff. Wilamowitz, *Platon*, II, 1919, p. 37, denies that any conclusions concerning the Epidaurian festival can be drawn from the *Ion*; his arguments are not convincing.

[10] All the shorter Homeric hymns are such prooemia, cf. Schmid-Stählin, *op. cit.*, I, 1, p. 232. The singer usually began his recital of epic poems by praising the god in whose honor the festival was held. The hexameters of the Homeric hymn to Asclepius are very formal and unemotional in their wording, naming the god by his usual cult epithets. For the date of T. 31, cf. above, p. 68 and *E. R. E.*, *s. v.* Hymns, VIII, p. 41; for the relationship of the Homeric hymnology to Delphi, cf. above, p. 68, n. 3.

[11] Cf. O. Crusius, Pauly-Wissowa, *s. v.* Dithyrambos, V, p. 1228, 31 ff., especially *ibid.*, p. 1208, 10. Deubner, *Attische Feste*, p. 142, n. 3, suggests that the dithyramb was

to Apollo, and from his cult it seems to have been taken over into the ritual of Asclepius. But its performance must have remained restricted to rare occasions.

Tragedies and comedies, on the other hand, were probably played regularly. Maybe the god appreciated such tragedies as that which Aristarchus wrote as a thanksgiving after he had been cured by Asclepius (T. 455).[12] Maybe the comic poet Theopompus dedicated a play to the god who had given him the power to continue his work (T. 456), and the god enjoyed its content.[13] But there is no certainty about any of the plays performed in Epidaurus, or Pergamum, or Cos. Literary sources and inscriptions are equally silent in this regard. Yet it seems permissible to suggest that in some of the works produced on the stage the god, his life and his deeds, were the subject which was elaborated. The Asclepius myth certainly lent itself to a representation in the true tragic fashion.[14] It may likewise be assumed that the god and his patients were ridiculed in the manner of the Aristophanic *Plutus* (T. 420 f.), or of the play that Plautus transposed into Latin verse (T. 429 f.).[15] The god and his worshippers were Greeks, they did not mind jokes, or fun poked at the divine. When they celebrated his festivals they could honor him with comedies no less than with tragedies.[16]

performed in honor of another god, while only the prize, a tripod, was dedicated to Asclepius. Telestes, the great dithyrambic poet of the fourth century B. C., wrote a poem 'Asclepius' (T. 613), of which a short fragment is preserved (for its stylistic analysis, cf. Crusius, *loc. cit.*, p. 1224, 41 ff.). Cf. also T. 73, a dithyramb of Cinesias.

[12] Concerning Aristarchus, cf. A. Dieterich, Pauly-Wissowa, II, p. 861 f.; he was an older contemporary of Euripides. No trace of his play is left; cf. Wilamowitz, *Glaube*, II, p. 232, n. 3.

[13] The text of T. 456 is corrupt and cannot be restored with certainty; cf. Wilamowitz, *Glaube*, II, p. 232, n. 4. For the interpretation of the testimony, cf. Weinreich, *Heilungswunder*, pp. 1 ff. Theopompus was a contemporary of Aristophanes. There were other comedies with the title 'Asclepius': one by Antiphanes (T. 611) whose first works were performed after 387 (cf. Pauly-Wissowa, I, pp. 2519, 28 ff.), and one by Philetaerus (T. 612), perhaps a son of Aristophanes, at any rate a contemporary of Hyperides (cf. Pauly-Wissowa, XIX, p. 2163, 65 ff.). The fragments are too short to allow a judgment concerning the content of these works.

[14] The Hippolytus story seems to have been a favorite topic of dramatists, as Staphylus attests (T. 69). If their work was to be played in the theater of the god, it can hardly have contained any allusion to the fact that Asclepius was enticed by gold to heal Hippolytus (*e. g.* T. 72). The same consideration excludes the possibility that any of the plays referred to by Plato (T. 99) was ever performed at Epidaurus or at any other of the cult centers. Wilamowitz, *Isyllos*, p. 83, n. 57, expresses the belief that Plato had contemporary tragedians in mind and that he cannot have meant Aristarchus. He adds that "die attische Tragödie die Asklepiossage beiseite gelassen hat, wenigstens diejenigen unter ihren Vertretern, welche die Kraft besassen, dem Sagenschatz der Vorwelt für alle Folgezeit den Stempel ihres Geistes aufzudrücken."

[15] E. Legrand, *Rev. des Études Anciennes*, VII, 1905, pp. 25 ff., suggests that the Greek original of Plautus was written for the Epidaurian theater.

[16] Cf. in general P. Friedländer, *Die Antike*, X, 1934, pp. 209 ff.; especially pp. 225 f.

Finally, in the Christian era the program of the games celebrated in honor of Asclepius was enriched by a singular innovation: in Ephesus, in the second century A. D., the doctors of that city held a medical contest at the festival of the god (T. 573).[17] What subjects they selected for their competitions cannot be determined exactly. The inscription on which the report about the event is preserved lists composition, a thesis, surgery and instruments. It seems that the competitors had to write an essay on a subject of their own choice and another one on a given problem. The treatise which was judged best carried off the prize. Subsequently operations were actually performed and the achievements classified, or more likely, the most successful and astounding cases of the past year were selected and given an award. Last, but not least, inventions or improvements of instruments were brought before the committee of judges, and whoever had done the most outstanding work received the crown. The jury consisted of the priest of Asclepius, of some especially appointed arbiter, and of the presiding officer of the association of physicians.

Whether such medical contests were restricted to Ephesus or not, it is too early to decide. New excavations may bring similar evidence from other cities. At any rate, physicians competed in the sanctuary of Asclepius, in addition to boxers, racers, singers, tragedians and comedians. There could hardly be a more fitting tribute to the god who was himself a physician.

This much can still be ferreted out concerning the outward ceremonial of the Asclepius cult. There are no features reminiscent of a chthonian worship. Asclepius was revered in the manner in which all celestial deities were venerated.[18] It seems appropriate to add that the devotees of the god did not succeed in creating new religious forms, but rather followed the old-established pattern of worship. In those centuries in which Asclepius rose to power, the productive capacity for innovations in matters of religious observance seems to have failed.[19] The new hymns which the god received, the new games that were introduced in his honor testify at least to the vitality with which the tradition inherited from the

[17] Cf. J. Keil, *Jahresh. d. Oest. Arch. Inst.*, VII, 1905, pp. 128 ff., whose interpretation of the inscription I follow.
[18] Farnell, *Hero Cults*, p. 242, admits that this was true of the later period of the Asclepius worship, and that the traits of chthonian ritual, which he thinks he can detect, are rarely to be found. But even such a claim, in my opinion, cannot be substantiated.
[19] Cf. Nilsson, *Griech. Feste*, p. 408.

past was handed on from one generation of the faithful to the next. And one may perhaps claim that the followers of Asclepius played a leading part in the development of the ancient ritual toward an intensification of worship, toward the introduction of daily ceremonies, toward greater emphasis on singing.

But one must be aware that in dealing with rites, processions, sacrifices and games, and even with hymns and sermons, one is concerned only with the framework of the cult. Behind the external procedures that are still recognizable there was a more inward, a more spiritual experience that the testimonies do not reveal. Nothing is known of the " holy play," and yet it is certain that such a performance took place on the days sacred to the god.[20] Moreover, although the god's healings were the true mystery into which he initiated men (T. 402), in late centuries there may also have been secret rites of worship which were not divulged to everybody, oral disclosures kept zealously from common men, sights that were shown only to the elect.[21] It was in such religious plays and initiations that the emotional life of the worshippers found its deepest satisfaction, and here the cult may have attained the goal of giving a new expression to a new belief.

[20] That every ancient cult included a religious play, a δρώμενον, portraying the divine myth, has been shown by H. Usener, Arch. Rel. Wiss., VII, 1904, pp. 281 ff.; cf. also Wilamowitz, Glaube, I, p. 301. By a curious chance of tradition the play performed in the cult of Asclepius Glycon is attested. On the first day, the childbed of Leto, the birth of Apollo, his marriage to Coronis, and the birth of Asclepius were represented (T. 29b), followed on the second day by an appearance of Glycon, and on the third by the union of Podalirius and the mother of Alexander (T. 210); cf. also Lucian, Alexander, 38-39. It is hardly possible to draw any conclusion from this report in regard to the content of the sacred play of the genuine Asclepius, although it is conceivable that his ceremony too lasted several days, and that on the first day the birth of Apollo, the marriage of Coronis and the birth of Asclepius were represented. Note that Aristides was asked by the god to sing of "the marriage of Coronis and of the birth of the god" (T. 606). For T. 29 b and 210, cf. also above, p. 160, n. 11.

[21] The office of the ἱεροφάντης is attested for Epidaurus (T. 498b), as it is for the cult of Asclepius Glycon (cf. Lucian, loc. cit.). For a characterization of the late τελεταί, in which supposedly the true name of the god was revealed to the devout, cf. Wilamowitz, Glaube, I, p. 301; II, pp. 475; 541. But cf. above, p. 129, n. 15.

V. IMAGE

Just as the ancients did not think it sufficient to venerate the gods by believing in them, but deemed it necessary to give visible evidence of their faith, so they were not satisfied with gods who reside in the human heart alone. They wished to have them before their eyes, to look at them in bodily shape. In spite of the objections raised by some philosophers, the statues of the deities were never banished from the temples. They were in fact considered the true embodiment of the divine. With time progressing, this conviction grew stronger rather than weaker. Even philosophical opposition turned into approval. While Stoics like Seneca had inveighed against the reverence paid by men to dead pictures of stone, Neo-Platonists like Plutarch could speak to the statue of Asclepius and receive an answer from it (T. 427).[1]

Natural though it may be for the modern to interpret the artistic representations of Asclepius solely in aesthetic categories, he must try to evaluate them also in terms of their religious significance. Did the Asclepius ideal, as it was evolved by the artists, agree with the concept of the god's nature, as it was held by the devout? What type of cult statue was the most common one and for what reason? What was the meaning of the attributes given to the god in his portrait? In regard to these questions the literary testimony assumes great importance. For it speaks clearly, while the monuments that are extant remain silent and are themselves in need of an interpretation which must be based on written information.[2]

[1] For the philosophical discussion of the question whether divine statues should be revered, cf. E. Zeller, *Die Philosophie d. Griechen*, III, 1, 1880, p. 313; 2, 1881, p. 697.
[2] Of course, the literary evidence is important also in other respects. The monuments that have survived are only a small fraction of those that once existed. The reports of ancient writers about images which they saw, often furnish data beyond the finds of excavations. Moreover, the authors refer to names of artists who made statues or other likenesses of the god; thus attributions largely depend upon the information that can be gathered from the written sources. Finally, just as the ancients themselves sometimes found it difficult to decide whether a statue was that of Prometheus or of Asclepius (T. 671), that of Hippolytus or of the god of medicine (T. 654), so has it been a matter of ardent debate among modern critics whether an image should be taken as that of Zeus or Asclepius, of Demeter or Hygieia. In such instances the literary reports may be decisive. For a discussion of the archaeological material, cf. B. Starck, *Die Epiphanien des Asklepios u. ihre Darstellung durch die Kunst*, Arch. Zeitschrift, 1851, pp. 341 ff.; Panofka, *Berl. Phil.-Hist. Abh.*, 1845, pp. 271 ff.; E. Loewe, *De Aesculapi figura*, Diss. Strassburg, 1887; Kjellberg, *Asklepios*, pp. 70 ff.; K. A. Neugebauer, *Asklepios*, 78. *Winckelmannsprogramm*, 1921; K. Sudhoff, *Handanlegung des Heilgotts auf attischen Weihtafeln*, Arch. f. Gesch. d. Medizin, XVIII, 1926, pp. 235 ff.

1. LOCATION—MATERIAL—ARTISTS

In the ancient world, wherever men turned their eyes, there were gods to behold, and Asclepius' likeness was one of those that were to be found everywhere. The houses of the rich probably sheltered his picture like that of other deities; the homes of physicians were in possession of Asclepius images for the private worship of their inhabitants (T. 501). But the god was also to be seen on the market place (T. 668a), or he graced a spring with his presence (T. 643). Nor did he fail to adorn the shrines of other gods and goddesses. He was painted together with Machaon and Podalirius in the Messenian temple of Triops (T. 657). Somewhere else he was represented together with his daughters, Hygieia, Aegle, Panacea, and Iaso (T. 665). A panel showing Asclepius and Hygieia decorated the door by which people entered one of the temples of Demeter and Kore (T. 639).[1] Statues of the god of medicine were erected in the sanctuaries of Athena (T. 653), or Juno (T. 651), or Concord (T. 661).

But no matter how frequently the god may have been represented outside his own temples, it was naturally within his sacred abodes that his image was prominent. The votive offerings on which he was painted as well-doer (T. 539, *36*) must have been innumerable. Made of cheap material, of wood or terra cotta, they were probably the favorite thanksgiving of the poor, hardly distinguished by any great aesthetic value; no ancient author takes the trouble of describing these mass products in detail.[2] But then there were the expensive marble reliefs which sometimes were of high artistic quality. The scene probably represented most frequently was a healing performed by the god; thus Aelian says of the offering of Theopompus: " The representation of his sufferings is very lively

---

[1] Mention should be made here of offerings to other deities on which Asclepius was represented, such as the table in the Heraion in Olympia (T. 641), or the statues dedicated by Micythus to the Olympian Zeus, among them Asclepius and Health (T. 642); cf. below, p. 245, n. 10.

[2] Such votive pictures are mentioned on the Epidaurian inscriptions (T. 423, *7*), and also by Herondas (T. 482, *v. 19*); for their material, cf. O. Crusius-R. Herzog, *Die Mimiamben des Herondas*², 1926, p. 110 (*ad* Herondas, IV, v. 19); for the custom in general, cf. R. Wünsch, *Arch. Rel. Wiss.*, VII, 1904, p. 107. These paintings apparently were so inexpensive that even the miserly father depicted by Ps. Libanius, T. 539, would gladly have paid for such a thank-offering. A. Koerte, *Ath. Mitt.*, XVIII, 1893, p. 236, distinguishes three kinds of votive offerings: (1) imitations of the diseased part of the body, (2) representations of the patient thanking the god for his aid, (3) portrayals of the scene of healing. The cheap tablets with the god as well-doer would fall into the third category. G. Karo, *Weihgeschenke in Epidaurus*, 1937, on the basis of the Epidaurian inscriptions (T. 423) has reconstructed the subjects of votives which were probably dedicated by the sick.

—even the sickbed itself being made of stone. On this lies his (*sc.* Theopompus'*) wasted figure, carved with artistic skill. Beside him stands the god, stretching forth his healing hand to him, and a young boy, and he too smiles a little " (T. 456).[8]

In addition to such votive offerings, and to statues that were scattered throughout the sacred district (T. 673), the temple itself was studded with images of the god. His statue adorned the vestibule (T. 427), the altar in the propylaeum, and the *cella* (T. 630). He was shown alone or together with his wife, Epione (T. 631), surrounded by his children (T. 636) or by other deities (T. 482).[4] In one case his likeness was placed between Isis and Sarapis (T. 672). The favorite grouping, however, seems to have been that of Asclepius and Hygieia (T. 638; 660), he seated and his daughter standing by his side (T. 663).[5]

The statues of the god, like those of other deities, were made of stone (T. 668), more particularly of white marble (T. 670), Pentelic (T. 671) or Parian (T. 673), sometimes of ivory (T. 674) or ivory and gold (T. 675), of wood (T. 677), or of bronze (T. 676). The selection of the material was determined by the varying tastes of the times, by the means available, or sometimes by ritualistic reasons: trees that were sacred to the god seemed an especially appropriate matter in which to carve his portrait (T. 677).[6] Again like other gods, Asclepius was represented either standing (T. 679, n. 2) or seated (T. 680, n. 2), sometimes on a throne. The latter position was apt to emphasize his superior rank, and it is not by chance that he was thus depicted in Epidaurus (T. 680), the size of the statue being half that of the Zeus of Phidias (T. 630), which was about seven times life-size.[7]

In only one respect did Asclepius' portrait differ considerably from

---

[8] In my interpretation of T. 456 I follow that of Weinreich, *Wunderheilungen*, pp. 1 ff. Even if the scene depicted were a "funeral repast," at which the god is present (cf. L. Ziehen, *Ath. Mitt.*, XVII, 1892, pp. 237 f.), the fact remains that many of the votives preserved resemble Aelian's description.

[4] The excavations at Cos have proved that Leto, Helius, Hemera, Hecate, Heracles, Aphrodite were depicted together with Asclepius; cf. Herzog, *Mimiamben, ad* v. 10.

[5] For the importance of Hygieia, cf. above, pp. 89 f.

[6] T. 677 attests that the Asclepius statue in Sparta was made of *agnus castus*. Since in the fabrication of cult statues wood was later commonly given up in favor of stone, Loewe, *De Aesculapi figura*, p. 7, concluded that this was one of the oldest images of the god. But this inference is not conclusive, as there were exceptions to this rule, cf. Pauly-Wissowa, *s. v.* Kultbild, Suppl. V, pp. 492 f. Kern, *Religion*, I, p. 86, seems right in claiming that *agnus castus* was sacred to Asclepius because of its medical importance. Note that the cult statue in the possession of the physician Nicias was made of cedar-wood (T. 659).

[7] For the Zeus of Phidias, cf. G. M. A. Richter, *The Sculpture and Sculptors of the Greeks*, 1930, p. 219, who adds that the statue occupied "the full height of the temple." For the various postures of the divine statue, cf. Pauly-Wissowa, Suppl. V, p. 500, 30 ff.

that of other deities, at any rate from that of the great gods: he always wore a chiton and in most cases also shoes. The garb of Asclepius and his sandals were proverbial (T. 686; 687).[8] The artistic tradition, then, was conscious of the fact that the god of medicine had originally been a human being, and that he never became an equal of the Olympians. For only the gods of old, the aboriginal gods, were represented as nudes; deified mortals were not pictured in this way; Asclepius, therefore, always appeared draped.[9] In general the chiton was probably made of the same matter as the statue. Sometimes it consisted of other material (T. 669), in one case the image is said to have been wrapped in a tunic of white wool and a cloak (T. 678).[10]

Some of the statues must have been masterpieces. The greatest names known in the history of Greek art are linked with the representation of Asclepius. Yet it is noteworthy that of the sculptors of the fifth century, only Alcamenes (T. 647), Colotes (T. 648), and Calamis (T. 649) are referred to.[11] Neither Myron, nor Phidias, nor Polyclitus is credited with the portrayal of the god of medicine. The ancients themselves, at least in late centuries, apparently took umbrage at the fact that none of these men who came to be recognized as the outstanding classical sculptors were reported to have created a likeness of Asclepius. It is probably for this reason that one testimony (T. 645) ascribes the Epidaurian cult statue, which was actually made by Thrasymedes in the fourth century B. C. (T. 630), to Phidias, or that Libanius (T. 646) pretends to know of an Asclepius image by Phidias, made in the likeness of Alcibiades. As a matter of fact, in the fifth century, the temples of the god did not yet attract the

---

[8] Note the expression "clad in the garb of Asclepius" (T. 686), and "When this mantle itself . . . and sandals . . . serve to flatter Asclepius" (T. 687); cf. also T. 492.

[9] In my interpretation of the Asclepius statue I have followed A. Furtwängler's analysis of the portrayal of Heracles who, at least in the classical period, was usually represented with a chiton (Roscher, *Lexikon, s. v.* Herakles, I, 2, p. 2147, 36 ff.). The literary evidence is in agreement with the archaeological findings: no nude Asclepius statue is known, cf. W. Wroth, *Numismatic Chronicle,* II, 3rd Ser., 1882, pp. 301 ff.; Kjellberg, *Asklepios,* p. 79; Pauly-Wissowa, II, p. 1690, 60 ff. Only a few coins of very late date show the god undraped; cf. O. Bernhard, *Griech. u. Röm. Münzbilder in ihrer Beziehung z. Gesch. d. Medizin, Veröffentl. d. Schweizer Ges. f. Gesch. d. Med. u. Naturw.,* V, 1926, p. 15. These deviations from the rule must be explained by the fact that at the end of antiquity Asclepius sometimes was equated with Zeus, cf. above, p. 107, n. 22. It was owing to the same development that in late centuries Heracles was portrayed nude, cf. Furtwängler, *loc. cit.*

[10] Farnell, *Hero Cults,* pp. 249 f., surmises that statues covered with a fabric chiton must have been extremely ancient. This is not correct. The wrapping of the image seems to have been due to a special ritual or mode of preservation, cf. Pauly-Wissowa, Suppl. V, p. 499, 37 ff.

[11] It has proved impossible to identify any of these statues among the extant monuments. Concerning Alcamenes, cf. L. Curtius, *Antike Kunst,* II, 1926 ff. (*Handb. d. Kunstwiss.*), p. 265; for Colotes, *ibid.,* p. 267; for Calamis, cf. Richter, *op. cit.,* p. 204.

best among the famous artists, or could not pay for their services.[12] The situation changed from the fourth century onward. Statues of the god by the sons of Praxiteles (T. 651; 482), if not by Praxiteles himself,[13] by Scopas (T. 652 f.), by Timotheus (T. 654), by Thrasymedes (T. 630), by Bryaxis (T. 655 f.) are attested. In later centuries, too, the most prominent sculptors are mentioned as having fashioned images of Asclepius, men like Phyromachus (T. 658), Damophon (T. 660), and others,[14] and occasionally it is said that their fame in part rested on their Asclepius representations (T. 661). The sanctuaries of the medical god, which in the fifth century B. C. were still relatively unimportant, in the course of time became so well established that they were able to commission those who were most skillful in the arts. Sculptors and painters, on the other hand, began to recognize that to work for Asclepius was as much of an honor as it had been for their ancestors to adorn the sanctuaries of Apollo, or Athena, or even of Zeus.

2. THE ASCLEPIUS IDEAL

None of the numerous Asclepius statues that are extant can be identified with certainty as one of the sacred cult statues of the god.[1] The literary testimonies, on the other hand, repeatedly refer to temple statues of Asclepius. But in all these references it is the external data alone that are given, the place where the image was erected, the material of which it was made, the posture of the god, and the like. Even those descriptions

[12] Archaeologists have tried to trace some of the extant statues to Myron, Phidias, Polyclitus or their pupils, cf. in general Pauly-Wissowa, II, p. 1691, 46 ff.; A. Furtwängler, *Meisterwerke*, 1893, pp. 394; 84, n. 4; 489, n. 2. None of these attributions have been generally accepted. In this connection it should also be noted that no likeness of the god is older than the fifth century B. C. No Asclepius representation occurs on early vases, cf. Welcker, *Götterlehre*, II, p. 742. Thraemer, Pauly-Wissowa, II, p. 1691, 18 ff., rightly rejects Kern's attempt (*Eph. Arch.*, 1890, pp. 131; 142) to interpret one Boeotian vase as picturing Asclepius.

[13] The sons of Praxiteles, Timarchus and Cephisodotus, worked for the Coan temple. Concerning their works and the fragments found, cf. Herzog, *Mimiamben, ad* IV, vv. 5; 23. Their father is credited only with a statue of Trophonius "which also resembles Asclepius" (T. 650).

[14] Phyromachus' statue was at Pergamum (T. 546); Damophon was especially famous for his attempt to revive the art of Phidias in the spirit of the Hellenistic period, cf. F. Winter, *Griech. Kunst*, Gercke-Norden, II[8], 1922, p. 182. Other artists named are Eëtion (T. 659), Niceratus (T. 661), Timocles and Timarchides (T. 662), Xenophilus and Strato (T. 663), Boëthus (T. 599, 600). Of painters are mentioned Omphalion (T. 657), Socrates (T. 665), and a woman, Aristarete (T. 666).

[1] A. Schober, *Oesterr. Jahresh.*, XXIII, 1926, p. 12, calls the Asclepius of Melos a cult statue; R. Carpenter, *The Greeks in Spain*, 1925, pp. 105 ff., does the same in regard to the Asclepius of Emporion; but such assumptions remain hypothetical, cf. Pauly-Wissowa, Suppl. V, p. 494, 63 ff.

which give a detailed account of the attributes with which the god was adorned fail to consider the expression of the face or the countenance of the figure. Only one feature is almost unfailingly mentioned that has some bearing on the type of representation: the presence or absence of a beard; in other words, it is noted whether Asclepius is portrayed as a mature man or as a youth (T. 682; 681).

Now, as is well known, bearded and unbearded representations of the same divine figure often varied in accordance with the changing taste of the centuries or existed simultaneously, though with regional variations, according to popular likes and dislikes.[2] At first glance, therefore, one is inclined to believe that in regard to Asclepius, too, the preference for one type or the other depended primarily on aesthetic ideals, and the data known concerning the early portraiture of the god seem to confirm such an opinion. The oldest cult statue attested, that by Calamis (T. 649), showed Asclepius as a youth. In Calamis' time youthful pictures of the gods were prevalent.[3] The young Asclepius type, in the fourth century B. C., survived in the images made by Scopas (T. 652) and Timotheus (T. 654). But Thrasymedes depicted the god as a mature man;[4] the fourth century B. C. began to prefer bearded statues of the deities. After 350 B. C., however, the situation changed: despite fluctuations in the aesthetic evaluation, the mature Asclepius apparently became and remained the favorite type of representation. The god of Pergamum and the god of Cos were bearded, like that of Epidaurus.[5] Moreover, the testimony insists on the fact that it was characteristic of Asclepius to have a beard (T. 685), just as it was characteristic of Apollo to have

[2] Furtwängler, Roscher, *Lexikon*, I, 2, pp. 2151 f.; 2162; 2178, has outlined the change which took place in the representation of Heracles as bearded or unbearded in accordance with the criteria referred to above.

[3] Farnell, *Hero Cults*, p. 277, calls the youthful Asclepius an "eccentric" work. This seems hardly appropriate. Even in the time of Calamis bearded images of Asclepius existed (cf. the relief in the Piraeus Museum, Curtius, *Antike Kunst*, II, p. 236, fig. 409). But there certainly was no general agreement concerning the representation of a god who was as yet so unimportant. Nor do I agree with Farnell's suggestion (*ibid.*) that Asclepius was represented beardless, because he was the son of Apollo. Nothing is known about Calamis' intention, and Farnell's argument sounds as if it were deduced from the objections of the ancients to the bearded representation of Asclepius (T. 683; 684). It seems more natural to explain the manner in which Calamis portrayed the god by the trend of the time.

[4] That the Epidaurian Asclepius was bearded is not mentioned by Pausanias (T. 630), but it is evident from T. 683; for the interpretation of this testimony, cf. also below, p. 249, n. 30. Cf. also T. 850, *vv. 654 ff.*, and the Epidaurian coins, Richter, *op. cit.*, p. 282 and fig. 733; Bernhard, *op. cit.*, table II, no. 28.

[5] For Cos and Pergamum, cf. Bernhard, *ibid.*, nos. 36; 29. It is noteworthy that even the coins of Sicyon and Gortys, where the beardless statues of Calamis (T. 649) and Scopas (T. 652) are attested, show the god with a beard, cf. Pauly-Wissowa, II, p. 1691, 50 f.

none, and the ancients were not slow in making mocking remarks about this curious divergence between father and son. Dionysius ordered the removal of the golden beard of the Asclepius statue at Syracuse, since he did not think it fitting for the son to wear a beard, while his father in all his temples appeared beardless (T. 683). Lucian, in his satirical vein, encourages Apollo to speak boldly and not to be sensitive about talking without a beard, although he has such a long-bearded, hairy-faced son in Asclepius (T. 684). Arnobius, comparing Apollo and Asclepius, refers to the great weight and the philosophic thickness of the latter's beard which makes it uncertain which of the two gods is the father, which the son (T. 683c). For the people at large, Asclepius was the god who was seen with a " flowing beard " (T. 850, *v. 656*). It is evident, then, that the bearded type was more commonly found than the beardless one, which seems to have been preserved and treasured as a relic of older times, but after the fourth century was hardly copied any more. And it is a fair assumption that this was so because the mature figure was held to be the more appropriate representation of the deified physician.[6] He who had become a god after having healed men during his life on this earth could be better imagined as advanced in years than as a young person, and aesthetic reasoning could not prevail against the claims of religious fantasy.

But nobody will believe that Greek artists were satisfied with indicating the nature of Asclepius' godhead only by the rather superficial trait of age. While in the fifth century B. C. they hardly individualized the images of the various deities except by their respective attributes, from the fourth century on sculptors commonly characterized the several gods more clearly by their facial expression and their general appearance. As regards the portrayal of Asclepius, there is at least one testimony that pretends to outline a likeness of this god with reference to the ideal which it is intended to express, to its " character " rather than to its external appearance: Callistratus' description of a statue of Paean (T. 627). Though the rhetor of the fourth century A. D., being more interested in words than in art, is not the best authority that can be wished for, his report requires careful interpretation if for no other reason than for its uniqueness.[7]

[6] Farnell who most strongly insists that the bearded Asclepius was "the canonical type" says, *Hero Cults*, p. 277: "It was felt that the full imaginative presentation of the divine physician demanded maturer forms, bearing the imprint of the thought and experience that comes with long studious years."

[7] That Paean here is identical with Asclepius must be concluded from the fact that in the words of Callistratus "Asclepius infused his own power into the statue." The description therefore has rightly been used by H. Brunn, *Griech. Götterideale*, 1893, p. 107, in his analysis of the Asclepius ideal. Concerning Callistratus, cf. A. Fairbanks

Callistratus begins by analyzing the countenance of the statue. It enthralls the senses, he claims, not because it is "fashioned to an adventitious beauty," but rather because it radiates dignity mixed with compassion.[8] This impression of an emotion of indescribable profundity apparently is strengthened by the peculiarity of the eye which is turned upward, looking saintly and benign. Next, Callistratus describes the hair. Curly locks fall on the back and over the forehead down to the eyebrows. Yet it is not the fullness of the hair that is most astonishing; the locks are curled "as if stirred by life and kept moist of themselves." Finally the statue as a whole is commented upon: "although all things that are born are wont to die, yet the form of the statue, as though carrying within itself the essence of health, flourishes in the possession of indestructible youth."

Thus the account of Callistratus. Do the features accentuated by him reflect a certain ideal? Do they become the god of medicine more than any other deity?

Dignity and compassion: to the ancients, the physician was the ruler of his patients, their king. He commanded and they had to obey. That is why dignity is ascribed to Asclepius. Yet the physician did not assume leadership in his own interest. His orders were given solely for the sake of others. He became master only through his desire to help; he laid down laws only out of regard for his fellow men. That is why the dignity of Asclepius is tempered with compassion.[9]

The eye turned upward: unlike Zeus, Asclepius does not face the spectator; unlike Zeus, the god of medicine does not simply nod approval or disapproval, he does not simply grant or reject the wishes of his suppliants. The help which he gives is not so much the consequence of his power as it is the result of experience, intelligence, wisdom. Asclepius ponders over the case before him. That is why his eye is not fixed on a definite point; it is vaguely directed into the distance, while his thought tries to visualize what he should do or advise.[10]

in his edition of *Philostratus and Callistratus*, 1931, pp. 369 ff. [Loeb]; concerning the history of the *Ekphrasis* in general, cf. P. Friedländer, *Johannes von Gaza und Paulus Silentarius*, 1912, Intr.

[8] The word αἰδώς, which I have translated as dignity, is rendered by Fairbanks as modesty. But αἰδώς at best is moderation, and most commonly it is respect for others, especially for the helpless. Farnell, *Hero Cults*, p. 278, also understands the words in question as "dignity mingled with compassion."

[9] The ancient concept of the physician as the ruler over his patients (cf. above, pp. 111 f.) accounts for the similarity of the Asclepius ideal with that of Zeus, a similarity which Brunn, *op. cit.*, p. 107, has rightly emphasized, but in my opinion has failed to explain fully.

[10] I have borrowed my interpretation from Brunn, *op. cit.*, pp. 99; 103, who in this way distinguished the eye of the Zeus statue from that of the Asclepius portrait.

The expression of the eye chaste and propitious: Asclepius is a physician, integrity and love of mankind are the two qualities which the divine healer embodies in agreement with the demands made upon his human followers.[11]

The locks curl as if stirred by life and kept moist of themselves: Asclepius is the giver of life; he is the one who postpones the dryness of death and restores to the body its natural moisture, the sap of life. The embodiment of indestructible youth: Asclepius gives health, he grants it to men to see forever the light of the sun.[12]

Certainly there is not one feature in Callistratus' account that cannot be understood as a symbol of those qualities which are characteristic of the ancient concept of Asclepius. Yet is it likely that a statue as Callistratus pictures it ever existed? He remains rather vague in his description; he does not name the artist of the portrait, nor the place where it is to be seen.[13] Twice he stresses the fact that the material does not render obedience to the law of art: the impossible, it seems, is here accomplished, and Asclepius is said to have infused into the statue his own power, giving it " purposeful intelligence " and thus making it " a partner with himself." Moreover, there is a certain contradiction in Callistratus' description. For the last words indicate that the statue is that of a person in the spring of life, the embodiment of youth, as it were. And yet, can dignity and compassion, which are mentioned in the beginning, be imagined in a youth? Are they not rather the virtues of a mature man? Are pity and sanctity and wisdom and experience the characteristics of a youthful person, or are they not rather the accomplishment of a man of ripe age, versed in the vicissitudes and sufferings that life has in store for men? Callistratus' portrait apparently is the outgrowth of the rhetor's own fantasy, an imaginary picture rather than the semblance of an artistic reality.

Still, it would be rash to conclude that the various features which Callistratus integrates into his imaginary likeness of the god of medicine were his own inventions. The youthful Asclepius, of whom the testimonies speak, certainly was the embodiment of youth, of health and life. The Asclepius who could be taken for Hippolytus (T. 654) must have been young and beautiful, and Libanius' claim that Phidias made an image of Asclepius in the shape of the young Alcibiades (T. 646) also points to a likeness of the beautiful youth.[14] On the other hand, the pictures of

---

[11] Cf. above, p. 113. [12] Cf. above, pp. 122 ff.

[13] In other instances Callistratus is more specific, cf. *e. g. Imagines*, 1; 6; 11.

[14] A. Kalkmann, *Arch. Zeitung*, XLI, 1883, p. 39, has claimed that Hippolytus was represented bearded (cf. also Eitrem, Pauly-Wissowa, *s. v.* Hippolytus, VIII, p. 1870,

Asclepius, as one Hellenistic writer attests (T. 448), were wont to represent him as gentle and calm, and these two qualities were equally characteristic of his posture as of the expression of his face. This statement, referring to the usual portrayal of the god, must concern the bearded type that was dominant.[15] The features which it describes as essential for the impression conveyed by the divine likeness, are not different from those traits which in addition to youthfulness and vigor of life Callistratus ascribes to his statue: compassion, benignity, thoughtfulness. In short, the rhetor, in his analysis of the Asclepius ideal, has combined the two intentions which were to determine the artistic representation of Asclepius. The youthful god, one must judge, in the sprightliness that pervades the crisp form of his body, came to symbolize the dispenser of health.[16] The mature god, in his serene dignity and composed energy, was the physician who heals men through treatment or advice. Not only was he a mature man, in his entire attitude he expressed the concept of the physician.

To create the likeness of Asclepius was a difficult task indeed. No poetical description could guide the fantasy of the artist, as it did in the creation of the ideal of those deities of whom Homer had sung. In the Epic, Asclepius was but a name. Nor did the divine myth recount any events which could serve to impart inspiration. The legend was devoid of action and passions.[17] Many an attempt may have been made in the beginning by various sculptors to translate into stone the concept of Asclepius' godhead, or it may have been one happy thought that immediately found general acclaim. Whatever the historical development, uniformity was reached only when the artist endeavored to express in his medium the quality of mildness which the name of the god implied for the ancients, when he succeeded in representing the benign and friendly character of the god, when he depicted that power of giving and pre-

---

66 ff.) simply on account of the assumption that otherwise the image could not have been taken for the likeness of Asclepius. But there were unbearded statues of the god of medicine. Moreover, Hippolytus at Troizen was venerated as the chaste young hero, the symbol of youth; the Troizenian maidens on the eve of their wedding used to sacrifice their hairlocks to him. F. Wieseler, *G. G. N.*, 1888, p. 146, therefore rightly rejected Kalkmann's opinion. For Phidias' portrayal of Asclepius, cf. above, p. 217.

[15] Archaeologists have always stressed the mildness expressed on the face of the bearded Asclepius. Thus Curtius, *Antike Kunst*, II, pp. 324 f., says of the oldest Asclepius relief preserved: " So hingebend milde war bisher kein griechischer Gott gewesen. Das ganze Relief lebt nur von seiner Gebärde."

[16] I hesitate to claim that the youthful Asclepius from the beginning had this symbolic meaning. For as has been shown above, p. 219, in Calamis' time youthful representations of the gods were quite common. Moreover the work of Calamis certainly had other implications, cf. below, p. 226.

[17] Cf. above, pp. 74; 85.

serving life which endeared Asclepius to his followers.[18] Nor is it astonishing that the bearded type was the one which occurred in the cult statues almost to the exclusion of the unbearded one, that the youthful figure was cast into oblivion by the mature god. Those who came to his temples needed the help of the physician even more than that of the dispenser of health, for they were sick, and to be liberated from disease was their main concern.

So strong was the hold of the Asclepius ideal over his worshippers that no other manner of portrayal could compete with this: Asclepius in the likeness of a child never became popular. In one of the Arcadian sanctuaries dedicated to Asclepius the Child (T. 773), there was a statue, a cubit high, showing the god as a boy (T. 679). Such a representation may have been connected with the saga of Asclepius' childhood in Arcadia, and it may have occurred also in other regions where the god was supposed to have grown up.[19] Or perhaps it was meant as an image of Asclepius, the patron of children. The god was fond of the young and showed great love for them.[20] Whatever the significance of such statues, they are rarely mentioned; apparently they were not commonly approved. This fact is the more remarkable since in Hellenistic and later centuries scenes from the youthful age of the gods became a favorite subject of artistic representation, as is evidenced by the images of the Dionysus child or the Hermes child.

Yet, not even that type of representation was accepted by the worshippers which is attested for a work of Boëthus, an artist of the second century B. C. He pictured Asclepius as a child " who has just been borne by his mother " (T. 599), " a young child such as the goddesses of childbirth brought forth from the maiden, daughter of Phlegyas, for unshorn Phoebus " (T. 600).[21] Whether the infant was depicted alone or together

[18] Cf. above, pp. 68 ff. Wilamowitz, *Glaube*, II, p. 232, points to the fact that the artistic representation reflects the ancient etymology of the name of Asclepius. But he is hardly correct in contrasting the ideal of the artist with that of the common worshipper. In this connection I should also mention the theory according to which the representation of Christ, in early centuries, was dependent on the Asclepius ideal; cf. A. v. Harnack, *Medicinisches aus der ältesten Kirchengeschichte*, 1892, p. 106; C. Bonner, *Harv. Theol. Rev.*, XXX, 1937, p. 123, n. 12. Modern histories of Christian art usually reject such a view, cf. *e. g.* C. M. Kaufmann, *Handbuch d. christl. Archäologie*, 1922, p. 368. However, considering the character of the Asclepius statue and the many resemblances that existed between Christ and Asclepius, the hypothesis is not entirely unfounded.

[19] For the worship of Asclepius the Child, and for the various legends concerning Asclepius' infancy, cf. above, p. 71, n. 14.

[20] Cf. above, p. 198. That the representation of Asclepius as a child points to his patronage over children has been suggested by Farnell, *Hero Cults*, pp. 276 f.

[21] That T. 599 and 600 actually refer to a work of the famous Boëthus of the second century B. C. is now generally agreed upon; cf. C. Robert, Pauly-Wissowa, *s. v.* Boëthos, III, p. 605, 21 ff.; Richter, *op. cit.*, p. 297.

with his mother, does not become entirely clear from the description. Reading it one can hardly fail, however, to be reminded of the new born child so often portrayed with his mother, the savior child, in whom the ancients put their trust.[22] It is impossible to tell whether Boëthus when creating his statue had this wonder-child in mind. But when Nicomedes, the physician of the third century A. D., dedicated the work of Boëthus in the Roman temple of Asclepius[23] and in his verses recounted the story of mother and infant (T. 599; 600), he must certainly have thought of Isis and her child, or even of the Christ Child. He must have been instigated by the wish to celebrate his savior child, to elevate him above that of the Egyptians and that of the Christians.[24] Yet, the followers of Asclepius refused to listen to the prophecies of other religions. Their savior was not a child, an infant, he was a mature man who through his divine wisdom preserved their health or liberated them from sickness.

3. ATTRIBUTES

Divine statues almost always showed the god adorned with attributes that symbolized his power or his rank. Even after the artistic representation of the various deities had taken on individual features, symbols continued to be added. The Asclepius statues formed no exception to this rule. Few if any images portrayed the god simply in a chiton and with sandals on his feet. At least a crown of laurel was wound around his head (T. 691), or other insignia were put into his hands or placed beside him.[1]

[22] The most famous document attesting this belief is Virgil's fourth *Eclogue*. For the interpretation of this poem and its religious background, cf. E. Norden, *Die Geburt des Kindes, Studien d. Bibliothek Warburg*, III, 1931.

[23] Cf. Robert, *loc. cit.*

[24] Note the words ἀρτίτοκος (cf. Lucian, *Dialogi Deorum*, VII, 1) and κόρη. Nicomedes dedicated the statue to Asclepius in gratitude for "recovery from illness" (T. 599) and for "avoidance of sickness" (T. 600). Certainly, then, it is the physician Asclepius, the savior, rather than the patron of children, who is addressed by his devotee. Any identification or equation of Asclepius with Harpocrates was facilitated by the fact that Asclepius was the son of Apollo, of Helius, as the Neo-Platonists said (cf. above, p. 108), just as Harpocrates in late Egyptian theology was regarded as son of Sarapis-Helius. Again like Asclepius, Harpocrates, the child god in the Egyptian trinity, was the one nearest to men. Influence of Egyptian ideas on the late Asclepius cult is indicated by the fact that in Epidaurus, Apollo and Asclepius were revered with the surname "Egyptians" (T. 739, 6). The development of the idea of the Savior Child has been analyzed by Norden, *op. cit.*; for the late Egyptian concept of the triad, cf. J. G. Milne, *E. R. E.*, *s. v.* Graeco-Egyptian Religion, VI, pp. 376 ff. Representations of the Madonna with child occur from the middle of the second century A. D., especially in the Roman catacombs; cf. J. Reil, *Die altchristlichen Bildzyklen des Lebens Jesu, Studien über christliche Denkmäler*, X, 1910, p. 2.

[1] For a survey of the attributes of Asclepius statues, cf. Pauly-Wissowa, II, pp. 1680 ff. Bowl, diadem, scepter, head-cloth, crowns of pine, laurel, medicinal plants are attested. A. Furtwängler, *Bonner Jahrb.*, LXXXX, 1891, p. 50, has drawn attention to the fact

The literary testimonies describing Asclepius' likeness discuss his attributes only occasionally. Most of the authors being of a relatively late date do not seem to have thought it important to speak about the symbols which had become stereotyped and were quite familiar to everyone. At any rate, accounts of attributes that were actually seen on monuments are less frequent than theoretical explanations of their significance, given without reference to any particular image.

It is fortunate that at least those symbols are attested which adorned the youthful Asclepius of Calamis and the mature god of Thrasymedes. Calamis, as Pausanias relates, represented Asclepius " holding in one hand a scepter, in the other a cone of the cultivated pine " (T. 649). Pausanias has nothing to say about the meaning of these attributes, nor are they ever mentioned by any other writer in connection with Asclepius. The scepter could indicate the god's power, as it does in the cases of Zeus and Hera, and probably in those of Poseidon and Hestia.[2] But in Calamis' time Asclepius was not yet one of the most highly respected and powerful deities. It is therefore more likely that his scepter, like that of Hermes, was the staff which brings sleep to men and wakes them up again, an appropriate symbol for the god of dreams.[3] The pine cone, on the other hand, is a conventional symbol of moisture, that is, of vitality and fertility; it is indicative of the sap of life. This attribute, usually connected with Dionysus or Poseidon, seems not unfitting for the god who restored to life and who granted and preserved health.[4]

Entirely different from the symbols which Calamis chose for his Asclepius were those with which Thrasymedes embellished the Epidaurian god. He depicted him " seated on a throne, grasping a staff, while holding the other hand over the head of the serpent, and a dog lying by his side " (T. 630). From the Epidaurian coins it can be concluded that

that the bowl, for instance, is to be found with many deities, even at a time when artists differentiated the divine figures by their facial expression rather than by their attributes.

[2] Cf. Gruppe, *Griechische Mythologie*, pp. 662, n. 6; 1132, n. 2; 1162, n. 5; 1404, n. 6.

[3] Cf. *Iliad*, XXIV, 343; *Odyssey*, V, 47; XXIV, 2. Homer speaks of the ῥάβδος of Hermes, but this magical staff is also called scepter, cf. Roscher, *Lexikon*, *s. v.* Zauberstab, V, p. 544, and in general *ibid.*, pp. 547 ff. In this connection it should be remembered that in Munychia around 400 B. C. Hermes was revered together with Asclepius, cf. above, p. 186.

[4] For the pine cone and its relation to Dionysus and Poseidon, cf. W. Otto, *Dionysos*, *Frankf. Stud. z. Rel. u. Kultur d. Antike*, IV, 1933, pp. 146-47. Gruppe, *op. cit.*, p. 1449, n. 5, has rightly stressed the resemblance between Calamis' representation of Asclepius and that of Dionysus. Cf. also Kjellberg, *Asklepios*, p. 71, n. 7; J. Murr, *Die Pflanzenwelt in d. griech. Mythologie*, 1890, p. 120 (remedy?). It is not known which were the attributes usually given to the youthful Asclepius, but one may assume that other artists followed Calamis not only in the type of the figure, but also in the choice of symbols.

it was the left hand that held the long staff, whereas the right hand was turned to the snake.[5]

The problem of the dog's selection as an attribute of Asclepius is not discussed by ancient authors. Yet it seems certain that Asclepius inherited this attribute from Maleatas, with whom he was associated in Epidaurus. Maleatas was a hunter and was fond of dogs. His sanctuary was situated on the hill *Kynortion* and itself was called *Kyon*. When Asclepius usurped the place of the deity whose accessory he had once been, the dog became sacred to him because of old it had been the friend of Maleatas.[6] It is also possible that in Thrasymedes' time the dog was regarded as the animal which had nursed the babe Asclepius after he was exposed by his mother. This story is reported not only by Tarquitius (T. 9) and Festus (T. 691) but also by Apollodorus (T. 5), who adds that the dog was suckling the infant when it was found by hunters. In the second century B. C., then, it was the dog who played nurse to the child, not the goat as the later saga had it (T. 7), and this older version of the Asclepius birth myth may well go back to the fourth century and may have been in Thrasymedes' mind when he chose to represent the dog lying by the side of the god. Later the story was forgotten and the dog ceased to be represented together with Asclepius.[7]

Staff and serpent, which, in addition to the dog, appeared on the Epidaurian Asclepius, were explained by the ancients in various ways. One

[5] For the Epidaurian coins, cf. Bernhard, *op. cit.*, no. 59. No copies of Thrasymedes' statue have been traced, cf. Gruppe, *op. cit.*, p. 1457, n. 6.

[6] Cf. Nilsson, *Griech. Feste*, p. 409, n. 7: "Dass der Hund ursprünglich diesem Gotte eignete [*sc.* Maleatas] ist ein sicherer Schluss Blinkenbergs (*Asklepios og hans Fraender*, Kopenhagen, 1893, S. 22 ff.) aus *I. G.*, II, 1651 [= T. 515; the sacrifice to dogs in Munychia] und dem Umstand, dass der Hund als Begleiter des Asklepios zuerst und am häufigsten in Epidauros auftritt." For Epidaurian coins (between 323 and 240 B. C.) showing Asclepius with the dog, and for monuments of this type from early Athens and Epidaurus, cf. F. R. Walton, *Harvard Studies*, XLVI, 1935, p. 178. For the place of Maleatas' sanctuary, cf. Pausanias, II, 27, 7; *I. G.*, IV², 1, no. 109, i, 128; the sanctuary of Asclepius was located at the foot of the *Kynortion*. Cf. also Wilamowitz, *Glaube*, II, pp. 226 f.

[7] T. 5, 9, and 691 are commonly understood as mistaken versions of the Epidaurian legend, as told in T. 7; cf. *e. g.* Pauly-Wissowa, II, p. 1682, 55 ff. But such an interpretation is not convincing considering the fact that T. 5 can be traced to Apollodorus. Rather does the version, according to which the dog was the nurse of Asclepius, seem to form an important link in the development of the divine myth which underwent so many changes, cf. above, pp. 72 f. When later on shepherds replaced the hunters, when the goat took the place of the dog, when the birth story of the peaceful god assumed a more Arcadian coloring, the dog was referred to only as the watchdog of the herd. On the assumption that the dog story is old, one also understands better why in Munychia a sacrifice was offered to hunters (T. 515), cf. above, p. 186. I should mention that Xenophon states that Asclepius learned hunting from Chiron (T. 56). The rôle played by the dog in Asclepius' life may partly account for the custom of keeping dogs in his temple, as Festus maintains (T. 691). But cf. also below, n. 10.

faction saw in the wand a sign of relief and support which the god granted (T. 706). Others found in it an indication of the medical inventions which prevent men from falling into sickness (T. 705). Still others took the stick, especially the knotted staff, as a symbol of the difficulty of the art which the god practiced (T. 691).[8] As for the serpent, some said, the god had reared this animal on Pelion and later had made it his servant (T. 697 f.). Others claimed that the serpent had been acquired by Asclepius after he had healed Glaucus (T. 704a).[9] Apart from such historical accounts there were symbolical interpretations. The serpent was taken as sign of the rejuvenation which Asclepius brought about; for the animal restores itself by shedding its skin (T. 701). Again, the serpent was understood as a symbol of sharp-sightedness which the physician needs (T. 706), or of vigilance and guardianship which are also his duties (T. 702), or of the healing power as such, for the snake was used as a remedy (T. 699). Finally, the belief was current that the reptile, being itself mild and friendly, indicated the mildness of the god of medicine (T. 700).[10]

Of all the attributes staff and serpent, to the ancients no less than to the moderns, are the typical symbols of the god, they are truly "Asclepiadic." Cornutus, when analyzing the significance of the concept of Asclepius, concentrates on these two attributes and does not even mention any others (T. 705). The same holds good of many other commentators (*e. g.* T. 706). Moreover, most of the statues preserved show the god with staff and serpent. What, then, was the real meaning of these symbols? What were they intended to indicate concerning the nature of the god?

As regards the staff, all the ancient interpretations may contain a kernel of truth: it may symbolize support or relief, inventiveness or profundity of knowledge. At least it may have done so in the opinion of later generations. Yet, it is hard to believe that from the beginning the

---

[8] This explanation is probably inspired by the famous first aphorism of Hippocrates: ὁ βίος βραχύς, ἡ δὲ τέχνη μακρή (IV, p. 458 L.).

[9] Thraemer, Pauly-Wissowa, II, p. 1681, 43 ff., claims that the Pelion story is more original than the Glaucus version (for Glaucus, cf. above, p. 51, n. 100). I do not think that the priority of either can be affirmed.

[10] Since serpents were sacred to Asclepius, they were kept in his temples (T. 696). But there were also other animals, geese (T. 423, *43*), dogs (T. 500), cocks (T. 466), sparrows (T. 730). These, too, were sacred to the god, or dedicated to him, or they served as guardians of his temple. The fact that the god sheltered animals is often taken as proof of the superstitious character of his worship, or of the revival of primitive concepts, cf. *e. g.* Wilamowitz, *Glaube*, I, pp. 25, n. 2; 154, n. 3. But Wilamowitz himself admits (*ibid.*, p. 149, n. 2) that animals were kept in other sanctuaries, without charging that this custom was based on superstition. Besides, it should be remembered that Asclepius was also a healer of animals (T. 378).

rod was of merely symbolic importance. The staff, particularly the long
one which reached up to the armpit, was the walking stick which people
used when on a journey. Asclepius, the physician, had to do extensive
travelling, not only while he was a human being, but also later when he
had become a god. He went from Troizen to Epidaurus and back again
(T. 423, 23). He wandered all over the earth to help everybody everywhere. The staff, therefore, may at first have been the natural attribute
of the travelling physician, and with this significance it may have been
retained on many images of the god even in late centuries. When the
divine power of Asclepius had become more and more a matter of course,
the wand lost its original meaning; it was now understood in a meta-
phorical sense, it became the symbol of Asclepius' helpfulness or wisdom.[11]

To determine the meaning of the serpent in connection with Asclepius
is very difficult. That the snake indicated the rejuvenation which the
god perfected, or his shrewdness, or his medical knowledge cannot be
denied entirely. But it seems more plausible that the animal symbolized
the mildness and goodness of Asclepius, his guardianship over men.
Friendly daemons were often revered in the shape of snakes. The serpents
of Asclepius were famous for being tame with men (T. 692),[12] and it was
known that their bite, if they struck at all, was not dangerous (T. 700).
The snakes that were sacred to the god of medicine were often kept as
house-snakes (T. 694); the ancients were fond of having serpents in
their homes, as guardians over men and possessions.[13] For whom, then,
should the serpent have been more appropriate as an attribute than for
Asclepius who embodied mildness and philanthropy?

One fact, however, must be emphasized: although to the modern the
serpent seems inseparably connected with the god of medicine, and although the ancients themselves felt that it was the particular servant of
Asclepius, the animal in fact was sacred to many more deities, and to the

---

[11] Thraemer, Pauly-Wissowa, II, p. 1681, 30 ff., suggests that Asclepius' staff originally had a religious meaning, and in confirmation of this opinion he refers to the Coan
festival of the ἀνάληψις ῥάβδου (T. 568); cf. above, p. 196. In general, cf. F. J. M.
de Waele, *The Magic Staff or Rod in Graeco-Italian Antiquity*, 1927, pp. 91 ff.
Clement, *Paedagogus*, I, Ch. VII (*Opera*, ed. O. Stählin, I, 1936, p. 126, 4 ff.) speaks
of the ῥάβδος παιδευτική which heals in those cases in which words and threats fail,
and he connects this symbol with Christ. But to apply such an interpretation to
Asclepius seems hardly warranted.

[12] It was only in Titane that people were afraid to approach the sacred serpents
(T. 700a). Sometimes the enormous size of the reptiles is mentioned (T. 421, *v. 734*;
448).

[13] The relation of the snake to the ancient house cult has been most strongly emphasized by Nilsson, *Popular Religion*, p. 71; concerning the Good Daemon, cf. *ibid.*,
p. 73. The serpent and its importance for Greek religion in general has been discussed
by E. Küster, *Die Schlange in der griech. Kunst u. Religion*, R. G. V. V., XIII, 2, 1913.

most divergent ones at that. Zeus, Sabazius, Helius, Demeter, Kore, Hecate, and all the Heroes were adorned with snakes (T. 703).[14] A statue of Trophonius was difficult to distinguish from that of Asclepius because the serpents on the image could be explained as sacred attributes either of the one or of the other (T. 690). It is impossible, therefore, from the serpent attribute to draw any conclusions regarding the original nature of the god, or if this is done, he may be called with equally good reason an Olympian, a chthonian god, or a hero.[15]

All the symbols with which the god was depicted were relatively late inventions. The hero of the Hesiodic poem, the patron of doctors, had no such attributes. When Asclepius became a god, when his statue was set up in his temples, he also received insignia of his power as they were customarily given to ancient deities. Tradition does not tell which attributes were connected with Asclepius in the beginning, nor does it indicate where the various symbols first occurred. Staff and serpent seem characteristic of the mature cult statue for which they are attested as early as the work of Thrasymedes. The staff cannot be traced to any local tradition, whereas the serpent seems to have been the typically Epidaurian attribute of the god. The specific kind of serpent which is sacred to Asclepius is to be found in Epidaurus only, says Pausanias (T. 692). And it was in the shape of the serpent that Asclepius came from Epidaurus to other places, to Athens (T. 720), Halieis (T. 423, *33*), Sicyon (T. 695), Epidaurus Limera (T. 757), Rome (T. 846).[16]

[14] Snakes are also known from the cults of Athena and Dionysus; cf. Gruppe, *Griech. Mythologie*, pp. 92; 1203, n. 2. For Zeus and the snake, cf. Nilsson, *op. cit.*, pp. 67 ff.; A. B. Cook, *Zeus*, II, 2, 1925, pp. 1091 ff. In the modern discussion, the fact that Asclepius was not the only god to whom snakes were sacred has been neglected, or at least not properly evaluated.

[15] Küster, *op. cit.*, p. 120, and Nilsson, *op. cit.*, pp. 71 ff., have shown that the snake is not indicative of the chthonian origin of the deity with whom it is connected, as had been generally supposed. The theory that Asclepius was a chthonian god largely rests on this obsolete interpretation of the snake as a chthonian symbol; cf. above, p. 66, n. 8.

[16] The fact that Asclepius appeared in the form of a snake is often taken as indicating that in the Asclepius religion old superstitions survived which had long been rejected by truly religious people; cf. Wilamowitz, *Glaube*, II, p. 231. But theriomorphic gods continued to be revered throughout antiquity; even Homer spoke of the metamorphosis of gods into animal form. When Ovid, describing the appearance of the serpent in the Epidaurian temple, makes the priest say: " The god! Behold the god " (T. 850, *v. 677*), he is giving expression to a belief almost universally shared: it was felt that the deity was present in his sacred animal. Moreover, in the dreams of his patients Asclepius appeared in human form. This is sufficient proof that the belief in his theriomorphic appearance cannot have been an experience that distinguished the Asclepius cult from those of other Greek gods. Finally, one should not forget that the god was supposed to be embodied in the snake mostly in those instances in which his worship was carried to other places. All translations of a cult, it seems, presupposed not only that the rites attached to the deity had to be learned (T. 848), but also that some symbol of the deity had to be acquired and installed at the new sanctuary, be it a statue or a stone or an animal or the like.

Wherever the god went, the animal followed him (T. 448), or it was twined around his staff (T. 706). Poets could allude to " the Epidaurian snake " as an expression familiar to everyone (T. 693; 693a; 289). It is most likely that the Epidaurians who created the god also gave him that attribute which was destined to remain his own forever.

Whatever the origin of the main attributes of Asclepius, this much is certain: they were suggested by his medical activity and by the specific character of his godhead, just as the image of the god was shaped in accordance with the Greek ideal of the physician.

VI. TEMPLES

Temple medicine, ritualistic rules, and artistic representations of Asclepius in the foregoing chapters have been dealt with in order to catch a glimpse of a reality that has passed away. The discussion of these subjects has also put to proof the hypothesis proposed in regard to the character of Asclepius' godhead. A last test remains to be applied which bears mainly on the presuppositions made concerning the time and place of Asclepius' deification and the position of his cult in ancient religious life. Does the history of the temples confirm the assumption that Asclepius was a late god, that Epidaurus was the city of his origin, that his worship, having spread throughout the *oecumene*, continued to exercise its power up to the end of antiquity? Moreover, which were the particular factors that in certain regions retarded or quickened the ascendancy of Asclepius? For one will not be satisfied with knowing that Asclepius became generally accepted because he was a true deity, because he was the preserver of health which the ancients cherished more than any other of the human goods.[1]

Of course it will never be possible to reconstruct the process of Asclepius' recognition in all its details. Although hundreds of sanctuaries are referred to in literary sources or have been accounted for by inscriptions, a great number of the shrines that once existed will be unknown forever. The judgment of the modern historian, therefore, to a certain extent must remain conjectural. On the other hand, it may seem that this is not yet the proper moment to attempt an evaluation of the data. Places of veneration not mentioned in the written evidence have been unearthed by archaeologists in the past, and their work is still progressing. New discoveries are bound to be made. But the more important sanctuaries, with the exception of Tricca, have been excavated. The regional expansion of the temples and their chronology have in the main become clear. The outlines of the history of the sanctuaries, I think, can be established now without too much fear that the conclusions drawn from the material available will be controverted by future findings.

1. GENERAL DESCRIPTION

Before turning to the discussion of historical problems proper, it seems worth while to ask whether the shrines of Asclepius were truly divine abodes. For the modern, it is difficult to realize that the Asclepieia should

[1] Cf. above, pp. 111 ff.; 119 ff.

have been places of worship on a level with those dedicated to Zeus or Apollo. Again and again he is tempted to imagine them as hospitals, religious hospitals to be sure, in which prayers and rites were as indispensable as dietetic treatment and drugs, but hospitals nevertheless. When he reads of Plutarch's claim that the Greeks selected clean and elevated places outside the city for the Asclepieia (T. 708), when he hears Vitruvius enjoin the rule that the shrines of the god of medicine should be built in healthful surroundings (T. 707), he feels confirmed in his belief that the temples of Asclepius were something unique in ancient religious life. But Plutarch's statement is in contradiction of the facts.[1] Vitruvius' theories are indicative of a sceptical rationalism toward the divine healing power which he supposes to be in need of a favorable climate and good air in order to be effective. In reality, the sanctuaries of Asclepius, like those of the other gods and goddesses, were situated in all places which pious reverence considered sacred, either on account of old tradition or because they seemed to have something of the divine about them.[2] Asclepieia were established in valleys (T. 739) and on the tops of mountains (T. 762), outside towns (T. 759), and within the walls of cities (T. 725). They were especially to be found in cities where human activities were concentrated. Just as every community had its Zeus or Apollo, it had its Asclepius. When people passed the Asclepieion, they entered to pray or to attend the ceremonies; here they met other people and could be found by those who wanted to meet them (T. 712).[3]

Moreover, the Asclepius of the town, the local god, acquired all the power of attraction that accrued to ancient deities through their presence in the community, through their connection with the daily pleasures and sorrows of one generation of men after another. In early centuries, it had been necessary for most people who sought Asclepius' help to make a lengthy journey, for his shrines were still scarce. Later, the invalids could go to the temples near their homes. For the town folk, this was the sanctuary of their own city; for the peasants, it was the shrine in their provincial capital. Themistius neatly expresses the common attitude when, referring to an Asclepieion that in his time lay in ruins, he says:

[1] Cf. above, p. 158. The sources of Plutarch's statement are not known; the thesis may well be his own (cf. H. J. Rose, *The Roman Questions of Plutarch*, 1924, p. 207).

[2] Thus the Epidaurian temple was built, where Apollo Maleatas had been revered; other Asclepieia were established over old Apollo temples; cf. below, p. 246, n. 16. If the shrine was situated outside the city, it was placed there not for hygienic, but for religious reasons. The solitude of mountains and valleys, for the Greeks, was awe-inspiring and fraught with divine presence, cf. Kern, *Religion*, I, p. 98. Naturally, in new cities where a site for the temple could be chosen wherever it seemed most advisable, people in some cases may have selected a spot for its healthfulness.

[3] For the ceremonies that took place in the temple, cf. above, pp. 191 ff.; 195 ff.

would you go to Tricca, if Asclepius were still here in his sanctuary? (T. 385). It was not only human laziness, it was not only the expenses and dangers of travelling, the inconveniences of a long trip, especially hard on the sick, it was mainly the preference for the ancestral god and the familiar temple that tended to give to the local Asclepius cult a predominant rôle in the life of his devotees.

On the other hand, in the worship of Asclepius as in that of all ancient gods, a strange spell was cast over distant sanctuaries. The places which the god himself seemed to like best were viewed as those in which his power was greatest, where he did especially good work. If the disease was dangerous and the local Asclepius failed to take care of it, if the patient was in a condition to travel and to spend money, or even if the pious simply wished to express his gratitude and his reverence for the deity in whom he trusted, pilgrimages to one of the great sanctuaries were made. Sometimes these were the most famous shrines of a certain province. Aegae was visited by all Cilicians (T. 817), Lebena by all Cretans and Libyans (T. 792). Cos was approached from far and near. All of Asia came to Pergamum (T. 792). And then there was Epidaurus, the city where Asclepius was born. Such places, above all Cos, Pergamum and Epidaurus, were the Olympia of Asclepius, his Delphi and Delos.

In these centers of worship to which everybody came who wished to pay special homage to Asclepius, his temples must have been magnificent indeed, comparable to those shrines in which Zeus or Apollo or Athena dwelt. It is very difficult to imagine their splendor and beauty. Even the most ingenious scholarly reconstruction in plaster or on paper gives but a faint semblance of the sanctuaries that now lie in ruins. If one happens to come across a passage in which an ancient writer describes such a temple which he saw still intact, which he viewed in the spirit of those for whom it was built, it is as if for a moment the past came to life again. Among the testimonies mentioning the shrines of Asclepius, there are two which have the peculiar charm of giving such an illusion to the reader: Herondas' fourth mimiamb, the scene of which is laid in Cos, and Pausanias' report of his visit to the temple of Epidaurus. In a certain sense the two descriptions are complementary. Herondas speaks mainly of the interior of the temple, while Pausanias pictures the sacred district in general; the poet dwells more emphatically on the artistic sights, the professional traveller more on the historical facts and spatial data. Although the two testimonies are separated by an interval of almost five centuries, although they deal with different sanctuaries, it seems permissible to consider them together so as to give a picture of an Asclepius temple as it was seen in antiquity.

In the open propylaeum, in front of the *cella* where Asclepius resides, the women of Herondas' poem (T. 482) stand before the altar on which the sacrifices are burned. They are face to face with statues made by the sons of Praxiteles, representing Apollo, Coronis, Epione, Hygieia, Panacea, Iaso, Podalirius, Machaon—Asclepius' whole family, his ancestors and his descendants.[4] Besides these divine images, however, the women see many more monuments, votive statues of extraordinary beauty: a girl looking up at an apple; an old man; a boy strangling a goose;[5] a portrait of a woman whom they know, a speaking likeness, so lively that one may look at this image " and be satisfied without the woman herself " (*v. 38*); and finally simpler gifts, placed somewhere near the divine statues, just as the women themselves put their painted terra cottas to the right of Hygieia (*vv. 19-20*).[6]

Captivating as these sights are, they are by far surpassed by the marvels that are to be found in the *cella* itself (*vv. 56 ff.*): here stands the sacred cult statue of the god; here the temple implements are kept.[7] But no word is uttered about them. Does awe of the divine keep the spectators silent?[8] Instead, they admire profusely the paintings adorning the walls of the *cella*. One of these pictures shows a procession of men and animals, drawn by Apelles so vividly that in gazing at it one almost forgets that the scenes depicted are not reality: " If I did not think it would be unbecoming for a woman, I should have screamed for fear the ox would do me a hurt: he is looking so sideways at me with one eye " (*vv. 69-71*). Where the artists were so great that in time they " will be able even to put life into stones " (*vv. 33 f.*), where works that had " the look of light and life " (*v. 68*) adorned the sanctuaries, the gods indeed were present in their statues. The visitors, on the other hand, even the common people—for the women are simple folk—, apparently were able to appreciate the beauty of the temples in which they worshipped (*vv. 20 f.; 39 f.*). They felt the realistic appeal of the world that surrounded them.[9] Moreover their aesthetic understanding bordered very closely on

---

[4] For the other gods who, according to the excavations, were represented in addition, cf. above, p. 216, n. 4. For an archaeological interpretation of the remains, cf. M. Bieber, *Arch. Jahrb.*, XXXVIII-IX, 1923-4, pp. 242 ff.

[5] This monument is known through copies, cf. O. Crusius-R. Herzog, *Die Mimiamben des Herondas*, 1926, IV, v. 30 and *ibid.*, Table IX.

[6] Cf. above, p. 215, n. 2.

[7] Herzog, *Mimiamben*, IV, vv. 56 ff., has rightly pointed out that at this moment the scene of the mime shifts from the outer temple to the *cella*.

[8] Pausanias, in his description of Epidaurus (T. 739, 2), gives a detailed report on the sacred statue, cf. above, p. 226, referring even to the reliefs carved on the throne on which the god is seated and to the myths there depicted, cf. also T. 632.

[9] Herzog, *op. cit.*, p. 19, has pointed out that the direct reference to Apelles and the praise of his achievements (*vv. 72 ff.*) are probably instigated by a contemporary debate

their religious experience. They believed that beauty could not be achieved without divine inspiration. So wonderful are the master-pieces that " you would say [they] were chiselled by Athene herself " (vv. 57 f.). Finally, how familiar everything was to these visitors, with what pride they looked at the treasures of their temple, and yet how little disturbed they were by the magnificence of the great works. With perfect naïveté they placed their simple offerings next to the statues of Timarchus and Cephisodotus; for the temple, in spite of all its grandeur, remained their property and part of their own life.

The women of Herondas' poem stand in the Asclepius temple proper. When the imaginary scene of this reconstruction is now changed from Cos to Epidaurus (T. 739), those who leave the temple with its golden roof and marble pavement through the splendid doors and descend the polished steps,[10] look toward the building where the incubations take place; this, with the exception of the *cella*, is the most important part of the sanctuary. Nearby, another structure catches the eye, the *tholos*.[11] It is adorned with the most beautiful paintings: Eros " who has thrown away his bow and arrows, and has picked up a lyre instead," and Drunkenness " drinking out of a crystal goblet; and in the picture you can see the crystal goblet and through it the woman's face." [12]

Next, Pausanias is attracted by the tablets recording the healings of patients. In his time, only six were left on which the names of patients who had been healed, together with the history of their disease, were engraved in the Doric dialect.[13] One other ancient tablet stands apart: the dedication of Hippolytus. Pausanias singles it out for a more detailed discussion, not only because he delights in telling all that he knows about

concerning the merit of the artist's work in which Herondas wishes to take sides in favor of Apelles. But this does not mean that the women's appreciation of the monuments is beyond their comprehension. Herondas would have been a bad poet indeed had he ascribed to his characters thoughts and judgments which they could not actually pronounce. Emphasis must also be laid on the fact that the description given by Herondas is fashioned according to older models (*e. g.* Euripides, *Ion*, vv. 184 ff.). Such aesthetic appraisals occur often and reflect a common attitude toward temples and monuments, cf. Herzog, *ibid.*, p. 20.

[10] These few details are taken from Ovid (T. 850, *vv. 671 f.*; *685*). Some of the shrines were roofless, cf. *e. g.* T. 761.

[11] Concerning the Epidaurian monuments in general, cf. K. A. Neugebauer, *Arch. Jahrb.*, XLI, 1926, pp. 82 ff.; for the *tholos* in particular, cf. F. Noack, *Arch. Jahrb.*, XLII, 1927, pp. 75 ff.

[12] It is interesting that Pausanias was impressed by the realism of the drawings in much the same way as the women of Herondas. He wrote for the average traveller and gave those data which people in general wanted to know. His judgment certainly did not go beyond the common aesthetic understanding. This circumstance makes his report the more important. For a more detailed analysis of Pausanias' description of Epidaurus, cf. C. Robert, *Pausanias als Schriftsteller*, 1909, pp. 111 f.

[13] These are the inscriptions still preserved (T. 423).

this hero, but also because stories that testify to the participation in mythical deeds of the past were important for the various sanctuaries, for the respect and veneration which they enjoyed.

The incubation hall, the *tholos*, the tablets are found near the temple. But the sacred district extends much farther; it includes many more buildings, older and newer temples mingled with one another. There is a shrine of Artemis, of Aphrodite and Themis, and, built in Pausanias' time by Antoninus, a sanctuary of the gods Bountiful, of Health and Asclepius and Apollo, the Egyptians.

Finally there is a theater, " in my opinion most especially worth seeing. It is true, in splendor the Roman theaters far transcend all the theaters in the world, and in size the theater at Megalopolis in Arcadia is superior; but for symmetry and beauty, what architect could vie with Polyclitus? For it was Polyclitus who made this theater and the round building also." [14] There is a stadium, a fountain " worth seeing for its roof and general splendor." There is a colonnade, restored by Antoninus, and last but not least, there is a bath of Asclepius.[15]

What a wealth of buildings and monuments was crowded within the bounds of the sacred grove of the god of medicine! Of course it would be a mistake to imagine that all the sanctuaries of Asclepius were built on as lavish a scale as Cos and Epidaurus and Pergamum, the main seats of the deity. But many places could compete with them, if not in the spaciousness of their layout, at least in the perfection of their design and the beauty of their adornment.[16] To be sure, the artistic genius of the ancients made even the shrine of the least saint delightful to the eye. But such heights as it reached in the great Asclepieia it attained only in the temples of the foremost gods.

[14] The god was much concerned with the way in which his patients passed their leisure. In Epidaurus, gaming tables have been found (cf. Chr. Blinkenberg, *Ath. Mitt.*, XXIII, 1898, pp. 1 ff.). In Pergamum, there was a library for the visitors (cf. O. Deubner, *op. cit.*, pp. 40 ff.).

[15] Sacred springs seem not to have been lacking in any of the sanctuaries of Asclepius; for Pergamum, cf. T. 804; 805, and in general Herzog, *Wunderheilungen*, pp. 155 ff. Whether this resulted from the fact that incubations were preceded by a bath, or from the old-established belief that water and mantic were closely connected, cannot be decided with certainty; cf. also M. Ninck, *Die Bedeutung des Wassers im Kult und Leben der Alten, Philol. Suppl.*, XIV, 2, 1921, pp. 47 ff.; Kern, *Religion*, I, p. 89. Colonnades are known from Pergamum too, cf. Deubner, *op. cit.*, p. 23. Incidentally, the colonnade in Epidaurus must have been a two-story building, for Apellas is told to take his walk in the upper portico (T. 432).

[16] Corinth and Ephesus, for instance, seem to have been of extraordinary beauty. Note that even the simpler sanctuaries in small towns or villages were sometimes in possession of artistic masterpieces. Cyllene, "a village of moderate size, has the Asclepius made by Colotes, an ivory image, wonderful to behold" (T. 648).

2. AFFILIATION

Of all the sanctuaries of Asclepius that arose in the course of time, that of Epidaurus, as has been suggested, was the one in which Asclepius was elevated to godhead. If this is correct, one would expect that the other places of worship did not spring up independently, that they had some connection one with another. The ancients thought indeed that most of the temples, the greatest and best known at any rate, were settlements of one center, Epidaurus.

Pausanias, discussing the question of Asclepius' birthplace, declares himself in favor of the Epidaurian claims on the god (T. 709). One reason for his acceptance of the Epidaurian legend is the fact that so many of the temples of Asclepius originated from Epidaurus. In this connection he names Athens, Pergamum, and Balagrae, besides Smyrna and Lebena which were established by Pergamum and Balagrae respectively; in certain other passages he connects Epidaurus with Cos and Epidaurus Limera (T. 757), with Sicyon (T. 748), and with Naupactus (T. 444). Julian agrees with Pausanias' statements: according to him (T. 307), Asclepius appeared first in Epidaurus; from there he went to Pergamum, to all the Ionian temples, and to Cos in particular, to Aegae and to the Italian shrines, Tarentum and Rome. The evidence on which Pausanias and Julian base their assertions is not known.[1] But the reliability of their contentions can be confirmed in some instances: Athens and Rome freely admitted their dependence on Epidaurus (T. 564 f.; 850), a fact that is in favor of the correctness even of those data which cannot be checked. Moreover, in no case can the claims made by Pausanias and Julian be disproved; it is possible even to add more names to their lists: Halieis was founded by Epidaurus (T. 423, *33*), Erythrae, Dium, and Ptolemais almost certainly were under Epidaurian influence, if they were not actually settled by Epidaurus.[2]

The impressiveness of the Epidaurian claim to being the metropolis of the Asclepius cult becomes abundantly clear if the testimonies concerning the Epidaurian origin of certain sanctuaries are compared with those attesting the dependence on other cult centers. One faction of

---

[1] Wilamowitz, *Isyllos*, p. 84, n. 61, thinks that Pausanias' source is Ister; for other theories, cf. Pauly-Wissowa, *s. v.* Mythographie, XVI, p. 1366 f. But all such " Quellenforschung " does not help to decide, whether the statement itself is correct or not.

[2] Cf. Wilamowitz, *Glaube*, II, pp. 225 f.: " Ob die Kulte von Erythrai, Dion und dem oberägyptischen Ptolemais, die dasselbe Kultlied hatten, direkt aus Epidauros stammen, ist fraglich, aber da sie die epidaurischen Göttinnen, sogar die Aigla, aufführen, stehen sie unter epidaurischem Einfluss."

ancient critics asserts that Tricca, not Epidaurus, had the oldest shrine of Asclepius (T. 714). Consequently there should have been at least some influence of Tricca on the foundation of later temples. Yet, the proponents of Tricca are silent concerning any such data. According to the evidence preserved, Tricca is directly related to other places of worship only in two instances: Gerenia in Messenia, Strabo says, had a temple of the Triccaean Asclepius (T. 715), and Herondas mentions that Asclepius came to Cos from Tricca (T. 714b).

Now, as for Gerenia the assertion made can be neither proved nor disproved.[3] Concerning Cos, the question arises what Herondas means by saying that Asclepius arrived from Thessaly (T. 714b). Does the poet refer to the hero of physicians, or to the god of medicine? It is most likely that Herondas is thinking of Asclepius, the father of Podalirius from whom the Coan physicians claimed to be descended. For he has just mentioned Heracles and his son, Thessalus, who were Coan heroes, too, and in the Coan legend Asclepius was the son-in-law of Heracles.[4] That this Asclepius, in the opinion of the Coans, came from Tricca, is quite probable. But the heroic saga does not permit any conclusion concerning the derivation of the Coan temple or of the divine cult.[5] Nor is it possible to prove the Triccaean origin of the Coan sanctuary simply by referring to the fact that in the prayer of the Coan women Herondas names Tricca first, before Cos and Epidaurus (T. 482, *vv. 1-2*).[6] For this, if it means anything at all, is counterbalanced by Herondas' claim that Asclepius inhabits Cos and Epidaurus, whereas he rules over Tricca. Certainly, to call a town the dwelling place of a god is a far greater compliment than to put it under his tutelage, and by such arguments one might be induced to quote Herondas as further witness for the Epidaurian

[3] Cf. below, pp. 239 f.

[4] For Asclepius, the forefather of Coan physicians, cf. above, p. 20; for Heracles, cf. Herzog, *Mimiamben*, II, vv. 94 ff. Cf. also above, p. 86, n. 42.

[5] Kjellberg, *Asklepios*, pp. 19 f., claims that the passage in question reflects the opinion of the Alexandrian scholars. It is true that in Herondas' time the scholarly tradition had established Tricca as the birthplace of Asclepius, cf. above, p. 71, but it is unlikely that the figure of the Mimiamb, a brothel-keeper, should advocate the latest results of scholarly research instead of relying on the popular Coan saga. Farnell, *Hero Cults*, pp. 257 f., accepts the Thessalian origin of the Coan temple on the evidence of T. 714b; still he maintains an interdependence of the Coan and Epidaurian cults. T. 757 presupposes that the Epidaurians gave the Coans their advice concerning the god; this may best be understood as referring to the transformation of the worship of the hero of physicians into the cult of the god. The prehistoric connection between Thessaly and Epidaurus, which Farnell, *op. cit.*, p. 120, assumes, is a hypothesis which is not very convincing in view of Herodotus' express statement (VII, 99) that the Coan population hailed from Epidaurus.

[6] Contrary to Herzog, *op. cit.*, IV, v. 2; cf. also *Kos*, p. x.

origin of Cos which is directly attested by other sources. However that may be, Herondas does not expressly say that Cos was settled by Tricca.[7]

The evidence for the Triccaean influence on the establishment of other Asclepius shrines actually consists, then, of one testimony: Strabo's report concerning Gerenia. Maybe some temples in Thessaly were also settled by Tricca; one or the other of the Asclepieia outside that region may have been connected with Tricca, although the sources do not mention this. Generally speaking, it is fair to state that Tricca's importance was locally restricted, that it did in no way equal that of Epidaurus.[8]

Epidaurus, then, was the metropolis of the Asclepius cult. Only in the Peloponnese, and only at a certain time was its position as the stronghold of Asclepius challenged. Messenia had an ancient title to Asclepius through Machaon, and after the restoration of its independence it is likely to have pressed its rights fervently. In the same way, the increasing importance that Arcadia assumed in the political life of the fourth century probably served as an incentive for the Arcadians to insist on their contention that Asclepius was a native of Arcadia.[9] Although the Delphic Oracle did not waver in its allegiance to Epidaurus and rejected Messenia's pretensions directly and those of Arcadia indirectly, people for quite some time must have been at a loss what to believe. The result of this uncertainty seems to have been a cult like that at Titane. This temple traced its origin to Alexanor, the son of Machaon, the Messenian hero (T. 749), but it had a statue of the Arcadian Asclepius (T. 668c) and agreed with Epidaurus in the acceptance of Coronis as Asclepius' mother (T. 555), and in certain ritualistic rules (T. 510).[10] On the other hand, if, of all places in Messenia, Gerenia which had the tomb of

---

[7] Wilamowitz, *Glaube*, II, p. 226, who states that the passage in question does not prove " einen älteren thessalischen Asklepios, der nach Kos gekommen wäre," explains the mention of Tricca before Cos and Epidaurus as a tribute to Homer.

[8] Farnell, *Hero Cults*, p. 247, thinks that the temples in Thessaly were settled by Tricca. This is not impossible. If Paros and Thera revered the Asclepius of Thessalian Hypata (cf. Farnell, *ibid.*, p. 247a), this would prove the influence of a specific Thessalian cult similar to the influence exercised by other individual temples, cf. below, p. 249, n. 30. That Tricca had only local significance has been pointed out by Wilamowitz, *Glaube*, II, p. 228.

[9] Concerning Machaon and Messenia, cf. above, p. 21. For Arcadia, cf. above, pp. 68 f. For the rise of nationalism in that province during the fourth century B. C., cf. Bury, *A History of Greece*, pp. 591 f. For the Messenian temples, cf. T. 768 ff.

[10] Farnell, *Hero Cults*, pp. 249 f., believes that Titane's worship was very old, his only argument being that the cult statue (T. 678) was covered with a chiton. Apart from the fact that this conclusion is not warranted, another statue of the god draped in a chiton is known from Achaia (T. 669, cf. above, p. 217), and this section of Greece certainly had no ancient Asclepius cult, cf. Pauly-Wissowa, II, pp. 1666 f. Besides, Messenian influence is impossible before 370 B. C., and the connection with Epidaurus also speaks against a very early temple at Titane. A peculiarity of the cult at Titane was the wild serpent (T. 700a). For the Arcadian temples, cf. T. 772 ff.

Machaon (T. 186; 190) built a sanctuary of the Triccaean Asclepius (T. 767), this disloyalty to the Messenian tradition may have been intended to contribute toward an adjustment of the differences that had arisen. Around 300 B. C. Tricca's claims to Asclepius had found support in the scholarly tradition; even Isyllus, at the very moment when with the approval of Delphi he set up in stone the Epidaurian birth legend of Asclepius, deigned to mention Tricca and its cult.[11] Such accommodations were natural as long as Messenia, Arcadia, Epidaurus, and Tricca were still contending for their respective titles to being the birthplace of Asclepius.

But even after the argument had been settled by Delphi in favor of Epidaurus, Arcadia and Messenia extended their influence abroad. Laconia seems to have accepted the Arcadian tradition, and this dependence on Arcadia would in itself suggest that the Laconian temples were of a rather late date. Moreover, the true Spartans of old had hardly much liking for the god of medicine who rose to power because in the opinion of the majority of the Greeks health had replaced glory as the supreme good.[12] But why did the Lacedaemonians follow Arcadia, not Epidaurus? Why did Titane have a cult statue of the Asclepius of Gortys (T. 668c) thus acknowledging its connection with the Arcadian cult, or at least with that of Gortys? Why were certain Argive sanctuaries reputed to have been founded by the children of Machaon (T. 187; 188)? Why did even Argos' most famous Asclepieion, with a cult statue of the second century B. C., trace its origin to Sphyrus, a brother of Alexanor and a son of Machaon, the Messenian (T. 663; 753)?[13] Why did Epidaurus not succeed in introducing its cult into the adjoining provinces; why was it scorned even by cities that were inhabited by men of the same race? And all this shortly after Epidaurus had won one of its greatest triumphs: Rome had accepted Asclepius from this city in 292 B. C. (T. 846).[14]

It seems that it was the political stand taken by the Epidaurians in the time after Alexander's death that made them unpopular with their neigh-

---

[11] Cf. above, pp. 97; 99, n. 32.

[12] Thraemer, Pauly-Wissowa, II, p. 1669, 3 ff., has suggested that the Laconian temples were dependent on Messenia. But a Messenian influence on Sparta is *a priori* unlikely. Moreover, Arsinoë whom the Lacedaemonians revered as mother of Asclepius must not be the Messenian Arsinoë, as Thraemer says, but may well be the Arcadian Arsinoë, cf. above, p. 69. Finally Machaon was not revered in Sparta, as Thraemer himself admits, and this is quite incomprehensible under the presupposition that Sparta was connected with the Messenian cult of Asclepius. For the Lacedaemonian temples, cf. T. 754-766; for Sparta in particular, cf. Ziehen, Pauly-Wissowa, *s. v.* Sparta, VI (2), p. 1471, 18 ff.

[13] The artists were Xenophilus and Strato, cf. above, p. 218, n. 14.

[14] Cf. below, p. 252.

bors. Isyllus, to be sure, had praised Asclepius as the savior of Sparta and had claimed that the Lacedaemonians agreed with him and paid thanks to the god as their savior from destruction. He had tried to introduce the Spartan political system into the state of Epidaurus.[15] Yet, in the third century B. C. the Epidaurians became staunch supporters of Macedonia, and in the second century B. C. they allied themselves with the Romans; in short, they were always on the side of the enemies of Greece, just as was the Delphic Oracle which favored Epidaurus. Other Greek states still fought on, among them Sparta and Arcadia.[16] That is why dedications of Spartan kings are conspicuously absent from Epidaurus in the third century B. C.; that is why during the second century B. C. Epidaurus lost influence in the Argolid, even though the internal strifes among the Greeks did not impair its world reputation as the birthplace of the son of Apollo and Coronis, as which the city remained famous to the end of antiquity.

But the highest title that Epidaurus could lay on Asclepius did not rest on her merit to have created the god and to have propagated his worship throughout the *oecumene*. It was based on the spiritual heritage that Asclepius had received from his home town, on the pattern which she had established for paying tribute to the god of medicine. Epidaurian ideals remained valid even in countries which were far remote from the center of Greek life. For as in Epidaurus the early temple inscription announced to the pious that in the god's opinion purity consisted in the thinking of holy thoughts (T. 318), so in provincial Africa a late inscription on one of the god's sanctuaries still proclaimed: " Enter a good man, leave a better one " (T. 319).

3. EARLY HISTORY

Definite and outspoken as the testimonies are regarding the dependence of most Asclepius sanctuaries on Epidaurus, they have almost nothing to tell about the time in which the dissemination of the cult took place. From the ancients' views concerning the date of Asclepius' life and that

---

[15] Cf. above, p. 103.

[16] For the history of Epidaurus in these centuries, cf. Latte, *Gnomon*, VII, 1931, pp. 123 ff. The Romans, after their conquest of Greece, granted a *foedus* to Epidaurus. Concerning the attitude of the Delphic Oracle toward Macedonia, cf. H. W. Parke, *A History of the Delphic Oracle*, 1939, pp. 244 ff. Up to the time of the Macedonian hegemony Sparta was the dominant power in the Peloponnese, Arcadia her enemy; later, their differences were forgotten in the fight against the common enemy. The earlier enmity between Sparta and Arcadia is a further confirmation of the proposed date of the Laconian cult.

of his deification one can at best surmise that in their opinion the Asclepieia were established shortly before the Trojan War or some time
afterwards, generalizations which in their vagueness and incompatability
are not of too much help.[1]

When modern scholars first attempted to write the history of the
spread of the sanctuaries, they usually assumed—in agreement with their
belief that Asclepius was an aboriginal god,—that his temples were as
ancient as all the other divine abodes. It was not everywhere, to be sure,
that Asclepius had been venerated in the distant past. But Thessaly, and
above all Tricca, presumably had very old temples. With the migration of
the Greek tribes the god travelled through Greece down to the Peloponnese
and even to Asia Minor, where Cos was supposed to have been his earliest
place of worship.[2]

The revolutionary finds of the excavations of the past decades have
changed this picture greatly. It has been discovered that the god Asclepius was not revered in Cos until the middle of the fourth century B. C.
The temple itself can be traced only to the third century.[3] Though other
sanctuaries in the colonies were older, the belief that the divine worship
in Asia Minor goes back to the time of the Epic or even beyond it, must
be given up. As far as Tricca is concerned, one cannot yet pronounce
judgment. Up to now it has proved impossible to find the sanctuary.[4]
However, after it has become clear that generally speaking the Asclepius
cult is of much later date than had been believed, that the importance of
Tricca for the history of the worship is relatively insignificant, one will at
least feel less apprehension about the fact that the Triccaean shrine cannot
be dated exactly. That it was very ancient, that it was founded long
before the other sanctuaries, is a mere hypothesis, and considering the
evidence as a whole not even a probable one.

Epidaurus, too, used to be considered an ancient seat of Asclepius.
Recent excavations have established the fact that it was only at the end of
the sixth century B. C. that an altar and a sacred building for Asclepius
were erected. The first temple, the famous sanctuary for which Thrasymedes made his cult statue, was built in the beginning of the fourth
century B. C.[5] If this is so, and if Epidaurus, not Tricca, was the place

[1] For the ancient testimony concerning the deification of Asclepius, cf. above, pp. 91 f.
[2] Cf. *e. g.* Wilamowitz, *Isyllos*, pp. 101 ff.
[3] Cf. Herzog, *Heilige Gesetze von Kos*, *Abh. Berl. Akad.*, 1928, no. 6, p. 48. Cf. also
Kos, ed. R. Herzog, I, *Asklepieion*, ed. P. Schazmann, 1932. Asclepius was first worshipped in the grove of Apollo, *ibid.*, p. 72.
[4] Cf. F. Stählin, *Das hellenische Thessalien*, 1924, p. 119; Wilamowitz, *Glaube*, II,
p. 228; Herzog, *Wunderheilungen*, p. 159.
[5] Cf. G. Karo, *Weihgeschenke in Epidaurus*, 1937, p. 1, n. 4; Kern, *Religion*, III,

where Asclepius became a god, then no shrine of Asclepius ought to antedate the sixth century. The question arises whether the testimonies preserved agree with this dating, and whether the expansion of the cult can be understood under the assumption that it originated that late.

Now with the exception of Epidaurus there are certainly no indications of any divine worship earlier than the end of the sixth century B. C. Even for the fifth century the evidence for the cult is of the scantiest.[6] To begin with Greece proper and more specifically, with the Peloponnese: Sicyon in the Argolid had a temple of Asclepius with a cult statue which was ascribed to Calamis (T. 649). Consequently, the sanctuary should be dated around the middle of the fifth century B. C., a least if it is the older Calamis who is referred to. If it was the younger artist of this name who made the statue, the temple would have been founded around 400 B. C.[7] Beyond the Argolid, in Mantinea in Arcadia, Pausanias saw a temple, half of which was dedicated to Leto and her children, the other half to Asclepius. The statue of the latter he describes as the "art of Alcamenes" (T. 647), while the pictures of Leto and her children, he adds, were made by Praxiteles, three generations later. According to Pausanias, then, this Asclepius image should be traced to the middle of the fifth century B. C.; according to the modern dating of Alcamenes it would be rather the end of the fifth century B. C.[8] At any rate, the

p. 155; Wilamowitz, *Glaube*, II, p. 227. The Epidaurian inscription which records the expenses for the building of the temple and which is of great importance for a study of the time and circumstances of its construction, has been discussed in part by Richter, *op. cit.*, pp. 274 ff.

[6] I do not wish to minimize the uncertainties and difficulties involved in this survey. Only in a few instances can the founding of shrines be deduced from direct evidence, and the means of filling the gaps of tradition are rather limited. Sometimes the date of one sanctuary can be determined by its dependence on another, whose time of establishment is known. Moreover, those cult statues which can be traced to certain artists indicate the period in which the sanctuaries themselves which they adorned were founded. Especially as regards a relatively late cult, it is a fair assumption that the creation of the divine image coincided with the building of the temple. Thraemer, Farnell and Wilamowitz, therefore, have rightly made use of such artistic data for the history of the Asclepius shrines. It is true, an argument based on such evidence is not absolutely cogent; it is possible that a new temple or a new statue replaced older monuments. Yet while the conclusions that have been drawn do not carry much weight so long as they are isolated or contradictory, they gain force if taken together and found to be in agreement with one another.

[7] Wilamowitz, *Glaube*, II, p. 224, says simply: "[Sicyon] einige Jahrzehnte früher (*sc.*, than Athens; cf. below, p. 245), die Zeit ergibt sich daraus, dass Kalamis das Kultbild machte." He does not consider the fact that two artists are known by that name; cf. Richter, *op. cit.*, pp. 201 ff., esp. p. 202, n. 64. That the temple at Sicyon was founded after 500 B. C. is certain; for its Epidaurian dependence is claimed (T. 748); cf. Herzog, *Wunderheilungen*, pp. 36 f. Thraemer, Pauly-Wissowa, II, p. 1665, 34 ff., is not justified in doubting this tradition. Hiller, *I. G.*, IV², 1, *Prolegomena*, p. xv a, ll. 46 ff., dates the temple after 480 B. C. For Sicyon cf. also T. 747.

[8] For Alcamenes, cf. Richter, *op. cit.*, p. 236. It is hard to imagine that in a sanctuary

Mantinean temple was not established before the second half of the fifth century. Moreover, it is noteworthy that Asclepius here was still worshipped together with other deities, rather than independently.[9] In Elis, Cyllene had a temple of the god which contained a statue by Colotes (T. 779; 648). If the artist was the elder Colotes, the temple would have been founded in the middle of the fifth century B. C.; if he was the younger artist of the same name, the cult statue would have been created considerably later. It is impossible to decide with certainty between these two possibilities.[10]

Outside the Peloponnese, in the Saronian gulf, Aegina possessed a building or a temple sacred to Asclepius as early as 422 B. C. Here, the god took care of his patients in their sleep, as can be gathered from the *Wasps* of Aristophanes (T. 733).[11] In 420 the god came to Athens (T.

that was divided into two sections only by a partition, the cult statues in each part should have been made at different times, and it is a fair conclusion that the Praxiteles referred to here was actually the elder Praxiteles, who seems to have been a contemporary of Alcamenes. For the elder Praxiteles, cf. Richter, *ibid.*, p. 266.

[9] A *terminus post quem* for the temple in Mantinea is afforded by the fact that that city was founded in the second half of the sixth century, before the Persian wars (cf. K. J. Beloch, *Griech. Gesch.*, I, 1, 1912, p. 335). The Argives took part in the settlement of Mantinea; this makes Epidaurian influence quite possible. Interestingly enough, Tegea, founded at the same time as Mantinea (cf. Beloch, *loc. cit.*), had strong trade-relations with Argos; one of the Epidaurian miracles (no. 47) deals with a man bringing fish from Argos to Tegea—a famous trade, as is clear from Aristotle, *Rhetoric*, 1365 a 27; cf. Herzog, *Wunderheilungen*, pp. 136 ff. The only Asclepius temple of uncertain date that is attested for Tegea, lies on the road to Argos (T. 772). The Tegean reputation of Asclepius, then, may well have been influenced also by Epidaurus, and it seems justifiable to stress the fact that Mantinea and Tegea were relatively recent foundations through συνοικισμός and therefore certainly more inclined to adopt new gods than were cities of old standing.

[10] Archaeologists seem to have no doubt that the Colotes referred to by Strabo is the elder artist (cf. Pauly-Wissowa, XI, p. 1123, 22 ff.). It should be noted, however, that the statue was made of ivory alone, not of ivory and gold, a technique for which the elder Colotes was famous (in Pauly-Wissowa, *ibid.*, the statue is listed among the "Goldelfenbeinwerke," contrary to Strabo, the only source). That the sacred table in the Heraeum at Olympia, which showed Asclepius and Hygieia (T. 641) and which was likewise made by Colotes, was not the work of the elder but rather of the younger artist is attested by Pausanias who, as he says, relies on other art critics. I see no reason to question his attribution (cf. however Pauly-Wissowa, *ibid.*, p. 1123, 7 ff.; 67 ff.). At any rate, from this monument nothing can be concluded about the date of the Asclepius in Elis (contrary to Thraemer, Pauly-Wissowa, II, p. 1667, 18 ff.). Among the dedications of Micythus at Olympia (T. 642) there were Asclepius and Hygieia. These monuments by the artist Dionysius were erected after Micythus had come to Tegea, that is, some time after 467; cf. Pauly-Wissowa, *s. v.* Dionysios, V, p. 999, 53 ff. That Asclepius was represented here as patron of the games, as Thraemer has claimed (Pauly-Wissowa, II, p. 1667, 22), is unlikely, since the occasion for the offering was the disease of Micythus' son.

[11] The *Wasps*, where the incubation in Aegina is mentioned, was performed in 422. The sanctuary may of course have been older; but the fact that Aegina was once in the possession of Epidaurus, from which it became independent at an unknown date (cf. Beloch, *op. cit.*, I, 1, p. 330, n. 1), suggests that the cult in Aegina was no more ancient than that of Epidaurus, and probably connected with the latter; cf. above, p. 208, n. 3.

720 ff.). He resided at first in the sanctuary of the Eleusinian goddesses, then probably in the shrine of another healing hero, until finally, in the year 420-19 B. C., he could take up residence in his own temple. In 408 another sanctuary had been established for him in the harbor of the city.[12] In Middle Greece in the fifth century B. C. the god was honored with a sacred district in Delphi. The Oracle accepted the son of Apollo within the holy city of his father.[13] In Northern Greece, Larissa in Thessaly may have had a sanctuary of Asclepius in the second half of the fifth century. Whether other places in this region had a cult of the god at that time, cannot be ascertained.[14]

In the fourth century Asclepius increased his hold over the Argolid and the Peloponnese. Halieis was founded (T. 423, *33*). Troizen was established, its cult statue being the work of Timotheus (T. 654); for Megara, Bryaxis created a likeness of Asclepius (T. 656).[15] The sanctuary of Corinth (T. 745) was founded. Here, the son of Apollo took over a temple where previously his father had been venerated.[16] When in 369

[12] Cf. Wilamowitz, *Glaube*, II, p. 223; his previous assumption (*Isyllos*, p. 83) that the Athenian cult was introduced around 460 was refuted by inscriptions that were found later. Concerning the Asclepius cult in Athens and its history, cf. Kutsch, *Attische Heilgötter*, pp. 16 ff.; also Walton, *Harvard Studies in Class. Phil.*, XLVI, 1935, pp. 172 ff. Eleusis, too, seems to have had an Asclepius sanctuary at the end of the fifth century; cf. Kern, *Religion*, II, pp. 314 f.

[13] Concerning Delphi and the excavations through which the holy district of Asclepius was discovered, cf. Pomtow, Pauly-Wissowa, Suppl. IV, *s. v.* Delphoi, pp. 1361, 18 ff.; especially p. 1362, 60 ff.

[14] Farnell, *op. cit.*, p. 247 a. Coins from Larissa between 450 and 400 B. C. show Asclepius feeding a serpent. Farnell, *ibid.*, p. 247, says: " Other communities in Thessaly had adopted him (*sc.*, Asclepius), at least as early as the fifth century." The only evidence adduced, however, is the coin from Larissa, *Brit. Mus. Cat.*, *Thessaly*, p. 28, pl. V, 9.

[15] Halieis is mentioned on the Epidaurian tablets (T. 423, *33*). The episode related here may have occurred earlier, yet the temple at Halieis was certainly later than that of Epidaurus by which it was founded; cf. also Herzog, *Wunderheilungen*, p. 36. For the date of Timotheus, cf. Richter, *op. cit.*, pp. 274 f. Troizen is also mentioned in T. 423, *23*; since the god of Epidaurus took care of patients in Troizen, the latter place seems to have been connected with Epidaurus. For the date of Bryaxis, cf. Richter, *op. cit.*, pp. 279 f. For other Argive temples, cf. T. 750 ff.

[16] The Corinthian temple has been excavated, and F. J. de Waele, *Am. Journ. Arch.*, XXXVII, 1933, p. 421, has claimed that " beyond any doubt there was a small temple dedicated to Asklepios and Hygieia in the second half of the sixth century B. C." But as de Waele himself says, in the earliest deposit of potteries on which the proposed dating depends, no votive offerings of limbs were found; the oldest dedication is one to Apollo, *ibid.*, p. 420. That even the archaic temple was sacred to Asclepius is concluded only from the fact that an Asclepius sanctuary was built here in the fourth century (pp. 424 ff.), yet it happened very often that Asclepius replaced Apollo, cf. *e. g.* Cos, above, p. 243, n. 3. De Waele adds that the Asclepius cult had migrated " possibly in the Greek Middle Ages " to the Peloponnese (p. 421), and that Titane also had an old temple. Such an argument, however, begs the question; it is not known that such a migration took place, nor had Titane an archaic sanctuary (cf. above, p. 240). Influence of an old Thracian Asclepius of which de Waele speaks elsewhere (Pauly-Wissowa,

Messenia regained her independence, Asclepius and his sons were pictured among the heroes (T. 657) whom the city of Messene called back to their abode after its restoration.[17] In Arcadia, the famous temple of Gortys was built at about the same time; here, the statue of the god was made by Scopas (T. 774; 652).[18]

As for other sections of Greece, in the first part of the fourth century B. C. Euboea had a temple of the god of medicine and honored him with a procession (T. 787). Around 300 B. C. the Asclepius cult reached Locris: the shrine at Naupactus was founded at that time (T. 717).[19] Not much later Asclepius may have been accepted even in Epirus. Ambracia had a famous temple of the god; when the Romans captured Ambracia (T. 716), monuments from that sanctuary were taken to Rome.[20] Finally, in Thessaly, Tricca in the fourth century must have been famous for its Asclepieion, and other Thessalian sanctuaries may have flourished at that time.[21]

This much can be ferreted out about the Asclepieia in Greece proper during the classical period. Granted that the information available at times is ambiguous, that the conclusions drawn in some instances may be open to objection, it seems certain that none of the shrines, with the exception of the altar and *abaton* at Epidaurus, can be traced beyond 500 B. C. Even for the fifth century, the earliest for which an Asclepius worship outside of Epidaurus can be proved at all, the data are few that

Suppl. VI, *s. v.* Korinthos, p. 195, 58), is again a merely hypothetical supposition. Farnell, *Hero Cults*, p. 249, had assumed an old temple at Corinth because of a bronze statuette dedicated to Asclepius and inscribed " in letters that point to one or the other of those two communities (*sc.* Corinth or Megara), and which may be dated near 500 B. C." This monument is certainly not sufficient proof for a temple, either in Megara or Corinth.

[17] The picture was painted by Omphalion who lived at the end of the fourth century B. C., cf. Pauly-Wissowa, XVIII, p. 398, 55 ff. For the Asclepieia at Messene, cf. Pauly-Wissowa, XV, p. 1241, 18 ff.

[18] Concerning Messenia and Arcadia in general and their influence around 300, cf. above, pp. 240 ff. For Scopas, cf. Richter, *op. cit.*, pp. 267 ff.

[19] Cf. Oldfather, Pauly-Wissowa, *s. v.* Naupaktos, XVI, p. 1999, 41 ff. The date follows from the lifetime of the founder of the temple and that of the other persons involved in the story told by Pausanias (T. 444). Oldfather rightly rejects the claim that the cult must have been older, adding that it is only natural " dass in eine ziemlich entlegene Stadt der ohnehin recht neue Kult des Asklepios aus Epidaurus relativ spät . . . eindringt." Herzog, *Wunderheilungen*, p. 38, puts the foundation of the temple around 320 B. C.

[20] Ambracia had a period of great prosperity under Pyrrhus who adorned the city with many monuments (cf. Pauly-Wissowa, *s. v.* Ambracia, I, p. 1806, 35 ff.). Here, as in the case of Naupactus, it can hardly be assumed that the Asclepius cult came to a city on the outskirts of the motherland long before Pyrrhus' reign.

[21] Triccaean coins of the fourth century show the head of Asclepius; cf. Pauly-Wissowa, II, p. 1662, 54. Around 300 Isyllus refers to the worship at Tricca (T. 516). For the Thessalian temples in general, cf. Stählin, *Das hellen. Thessalien, passim.*

attest a shrine of the god of medicine. This scarcity of material is the more noteworthy since for the same period innumerable temples of other deities are attested.

The expansion of the Asclepius cult was most extensive in the Peloponnese. Thessaly, too, seems to have had a number of sanctuaries. In both regions claims were made on Asclepius as one of their citizens. In Middle Greece, the cult made only small inroads. Testimonies from Boeotia are entirely missing. Small wonder! Boeotia had its own hero, Trophonius, so similar to Asclepius, and at the border of Attica and Boeotia there was Amphiaraus, who was also famous for his medical wisdom.[22] People are wont to adhere to their gods and do not like to give up allegiance to those in whom they trust in favor of a stranger. In Phocis, too, despite Delphi's acceptance of Asclepius, the god seems to have made but slow progress. The fact that it was here that Asclepius had been revered as the "leader of medicine" may have impeded the success of the Epidaurian god.[23] Generally speaking, the spread of the cult in the motherland confirms the assumption that it was from one place, Epidaurus, and at a rather late date that Asclepius began to conquer the Greek territory.[24]

To turn now to a consideration of the world outside the Greek mainland: here too, one can say with assurance, no Asclepieion is known before the fifth century B. C. Balagrae in Cyrene (T. 709; 831) used to be dated at the end of the sixth century. But it is more likely that it was founded in the fifth century; at any rate, it was a settlement of Epidaurus.[25] Nor are any temples of Asclepius attested in Asia Minor that

[22] Farnell, *Hero Cults*, p. 247, maintains that "at some indefinitely earlier date" (that is, before the fifth century) the worship "was carried into Boeotia." Yet he admits that "none of the records of an Asklepios-cult are ancient enough to serve as telling evidence" for his thesis. For Boeotia, cf. also Pauly-Wissowa, II, p. 1663, 54 ff.

[23] Concerning Asclepius *Archegetes*, cf. above, p. 96. The temple at Tithorea (T. 719) cannot be dated. The fact that no goats were sacrificed there, just as at Epidaurus (T. 532 f.), speaks for an Epidaurian affiliation; it is possible that the Tithoreans changed their ancient worship when at the end of the fourth century they rebuilt their town which had been destroyed during the Holy War. To explain the ritual through a birth legend similar to that of Epidaurus (Pauly-Wissowa, II, p. 1664, 18) seems unwarranted, since nothing is known about an Asclepius who was supposed to have been born in Phocis. Other temples in this region may have been of even later origin. The cult statue at Elateia was made by Timocles and Timarchides (T. 662; 718), artists of the second century B. C.; cf. Richter, *op. cit.*, p. 295.

[24] In this survey not all sections of Greece have been considered, partly because in certain regions no Asclepieia are known, partly because they were of a much later date. For Acarnania, cf. Pauly-Wissowa, II, p. 1664, 35 ff.; Illyria, *ibid.*, 50 ff.; Achaia, *op. cit.*, p. 1666, 64 ff., and T. 782 ff.

[25] The temple at Balagrae has not yet been found. Cf. Wilamowitz, *Kyrene*, 1928, p. 18, n. 1. Farnell, *Hero Cults*, p. 263, assumed that it was founded before 500, because

precede the one at Epidaurus. Even Cos, contrary to the assumptions formerly made by scholars, was established at a much later date.[26] It was in fact only in the first half of the fourth century that the god was admitted to the colonies. Erythrae received him at that time (T. 521; 592);[27] Pergamum was founded not long afterwards. It must have been around 350 that Archias brought the cult to that city (T. 801), and very shortly thereafter Ephesus (T. 573) built a sanctuary of Asclepius which remained famous throughout antiquity.[28] With the beginning of the third century the influence of the god over Asia Minor became consolidated through the dedication of the temple at Cos (T. 794 ff.), which replaced an altar erected to Asclepius in the grove of Apollo around 350 B. C. The worship paid to the hero Asclepius by the famous physicians of Cos probably retarded the recognition of the god.

Finally, as regards the Greek possessions in the Mediterranean, it was only at the turn of the fifth to the fourth century that they were reached by the divine cult,[29] which from then on seems to have spread rapidly in all directions. Far to the North, Thasos possessed a sanctuary around 380. Its priest is mentioned in an inscription (T. 786). On the Cyclades, the Asclepieion of Delos must have been founded in the fourth century, for a long and detailed document of the year 279 refers to the repair work done on that temple (T. 788). An inscription of Paros mentions sacrifices to Asclepius (T. 789). In the West, the cult took root in Sicily. The Syracusans had a temple adorned with a statue of the Epidaurian god (T. 683).[30] From the third century on the isles in the Aegean sea, small and large, were studded with temples of Asclepius. Moreover, in Crete a sanctuary was built that was to become a main center of the worship: Lebena (T. 791) rose under the influence of Cyrene.[31]

in his opinion Lebena, a foundation of Balagrae, was established in the fifth century. Excavations at Lebena have proved, however, that that temple belongs in the third century; cf. below, n. 31.

[26] Cf. above, p. 243.

[27] Cf. Wilamowitz, *Nordionische Steine*, *Abh. Berl. Akad.*, 1909, No. 2, p. 43; Farnell, *Hero Cults*, pp. 263 f. For the influence of Epidaurus on Erythrae, cf. above, p. 238.

[28] For Pergamum, cf. Wilamowitz, *Glaube*, II, p. 225, n. 2; Herzog, *Wunderheilungen*, p. 38; for Ephesus, J. Keil, *Jahresh. d. Oest. Arch. Inst.*, XXIII, 1926, pp. 263 f.

[29] Cf. Farnell, *Hero Cults*, p. 263.

[30] Cf. Pauly-Wissowa, II, p. 1679, 31 ff. Thraemer rightly concludes that Syracuse was dependent on Epidaurus because it had a statue of the Epidaurian Asclepius; in the same way, expressions such as the Gortynian Asclepius (T. 668 c), or the Triccaean Asclepius (T. 767), point to an influence of these respective cities on other temples. Thraemer, *ibid.*, p. 1676, 63 ff., by mistake speaks of an Asclepieion at Agrigentum in 491 B. C.; the year referred to in T. 839 is 491 *ab urbe condita* (263 B. C.). For Acragas, cf. P. Marconi, *Agrigento*, 1929, p. 87.

[31] Cf. Pauly-Wissowa, Suppl., VII, p. 369, 16 ff., where Kirsten rightly rejects the earlier assumption that Lebena was founded in the fifth century. The first buildings

The progress of Asclepius outside Greece, then, like his progress on the mainland, was a slow conquering of single positions. Perhaps the god in the overseas possessions faced fewer difficulties and proceeded more quickly. This would not be surprising. The religious life of Asia Minor certainly was not as conservative and hostile to innovations as was that of the cities of the old country.[32] That the isles were especially rich in temples one can well understand too. The worship of Asclepius seems soon to have become popular with seafaring people; to a certain extent the spread of the Asclepius cult followed the ancient trade routes. Many of the early temples were situated in harbors or in coastal towns, at the point of departure or the port of destination.[33] Separated as the sailors were from their homes and their gods, they were all the more inclined to put themselves under the tutelage of a new deity who seemed anxious and able to protect them. Moreover, among those engaged in navigation the need for divine medical help was great. Overseas commerce was a dangerous undertaking, threatened by disease no less than by piracy and the perils of the elements. Sickness, when it befell men on the high seas, was even more dreadful than at home, for the sailor's safety depended on his being well and fully in command of his physical strength. And who could help, if not a god, since their ships hardly carried along sufficient medical supplies and certainly had no doctor on board?[34] Thus one will readily believe that seamen starting on their voyage prayed to Asclepius as well as to Poseidon, and that they thanked both deities after they had landed safely. The islands were stop-overs on long trips, and Asclepius, it seems, was brought there by the sailors who had adopted him as their patron. But whatever the forces instrumental in the expansion of the cult beyond Greece, this much is certain: here, too, no worship can be detected that antedates that of Epidaurus. The early history of the temples does not contradict, it rather confirms the assumption that Asclepius was elevated to godhead by the Epidaurians.

---

actually dedicated to Asclepius can be traced only to the third century B. C. The older buildings may have been dedicated to Apollo; Asclepius often superseded his father, cf. above, p. 243, n. 3. For the islands, cf. T. 790; 793, and Pauly-Wissowa, II, p. 1670, 9 ff.

[32] In this connection it should be remembered that Tegea and Mantinea, which had early temples of Asclepius, both were new settlements, cf. above, p. 245, n. 9.

[33] Temples located in harbors were those of the Piraeus (T. 722), Cenchreae (T. 746); Cyllene, where a relatively old Asclepieion is attested (T. 779), was an important trade center, just as Aegina, which also had received the god in the fifth century (T. 733), was a large trading place connected with Cyllene.

[34] Homer, Odyssey, XIV, 250 ff., especially 255, expresses the characteristic attitude of the ancients toward sailing and its risks. Cf. also Hippocrates, On Regimen, III, 68 (VI, p. 594 L.). For Asclepius as helper in distress at sea, cf. above, p. 104, n. 13.

4. HELLENISTIC AND ROMAN TIMES

With the Hellenistic era began the period of the ascendancy of the Asclepius cult. The god became universally recognized. All major cities now had temples of the god of medicine; the country towns worshipped the famous son of Apollo. Kings and commoners, philosophers and business men alike invoked Asclepius. In the early Roman Empire, the Asclepieia were hardly less numerous than the sanctuaries of the other great Greek gods. Hundreds of temples are still known.[1] Until the end of the second century A. D. the influence of Asclepius increased steadily. Within the framework of this survey it is impossible to deal in detail with the development that took place during these five hundred years. Nor would such an analysis be of advantage for the evaluation of the cult as a whole.[2] There are, however, certain facts of general interest that stand out and deserve consideration.

First, it should be pointed out that Alexander the Great was a devotee of Asclepius, and of the Epidaurian god in particular.[3] The king celebrated Asclepius with offerings (T. 548) and with festivals (T. 547), he made dedications at Epidaurus (T. 821). It must have impressed the people at large that the greatest hero, the conqueror of the world, had sought Asclepius' advice and help. Many a patient may have consoled himself with the knowledge that even Alexander did not secure from the deity everything for which he had asked. And the private citizen too, even if he felt neglected by the god, may have made pious offerings, as Alexander had chosen to do; for it was only untrustworthy authors who asserted that the king in revenge of the god's failure to help, had destroyed the sacred temple (T. 821 ; 713). More important, to his successors, who were so intent on imitating his deeds, Alexander's example must have been a stimulus to further the cult of Asclepius. The early foundation of the sanctuary at the Egyptian Ptolemais (T. 513a) is ample proof

---

[1] Cf. Pauly-Wissowa, II, pp. 1662-1677, where Thraemer lists 186 sanctuaries. In E. R. E., VI, p. 550, he speaks of 410. A survey of the temples is also given by Walton, *Cult of Asklepios*, pp. 95 ff., yet this enumeration is not absolutely reliable (cf. Pauly-Wissowa, *ibid.*, p. 1677, 21), and is now of course antiquated.

[2] The evidence available for the various parts of the ancient world has been considered in books on the cults of certain sections of Greece or of the several provinces of the Roman Empire, cf. *e. g.* W. Immerwahr, *Die Kulte u. Mythen Arkadiens*, I, 1891; S. Wide, *Lakonische Kulte*, 1893; cf. also below, n. 10.

[3] Kern, *Religion*, III, pp. 47 f., seems the only one to have noted Alexander's relation to Asclepius. Whether the data given in the testimonies are reliable in every detail, it is difficult to tell. Yet one cannot fail to attribute a certain significance to the fact that Alexander was connected with Asclepius by the tradition.

of this attitude on the part of the Diadochi.[4] Especially in the eastern part of the world, the fame of Asclepius doubtless owed much to the support of Alexander.

The rise of the Asclepius worship in the East, however, was not limited to the acceptance of the Greek cult. The identification of the god with Oriental deities was another form of recognition paid to him. In Egypt, the son of Coronis became Imouthes or Imhotep; his sanctuaries in Memphis and Thebes were especially famous (T. 331; 827 ff.).[5] In Phoenicia, he became Esmoun, and in the guise of this god he even acquired a new genealogy (T. 826). In fact, the Phoenician Asclepius later was so highly respected that a man from Sidon could dispute with Pausanias about the true nature of the god, and claim that the Phoenicians had more adequate notions concerning him than had the Greeks (T. 297).[6] When this process of assimilation set in, can hardly be determined with exactness. The fact that Asclepius was identified with the main deity of the Carthaginians (T. 832 ff.) indicates at least that it did not take long before he was received even by the Semitic people.[7]

For the ascendancy of the god in the West, on the other hand, it was decisive that Asclepius was among the first foreign gods to be admitted in Rome. His temple was dedicated as early as 291 B. C. (T. 855). In its importance this event may be compared only with the entry of Asclepius into Athens.[8] It set a pattern for the attitude of Roman citizens and

[4] For the Ptolemaic shrine, cf. W. Otto, *Priester und Tempel im Hellenistischen Ägypten*, I, 1905, p. 395. For Alexandria, cf. E. Visser, *Götter u. Kulte im ptolemaeischen Alexandria*, 1938, pp. 39 f.

[5] Cf. Pauly-Wissowa, II, p. 1680, 13 ff.; above, p. 129, n. 15.

[6] As Esmoun, Asclepius was revered in Berytus (T. 825 ff.); between that city and Sidon he had a sacred grove (T. 823). Another Phoenician Asclepius was the one of Ascalon, called Leontouchos (T. 826a); cf. Pauly-Wissowa, XII, p. 2057, 20 ff. On Phoenician coins, Asclepius is represented " im Typus griechischer Kunst "; cf. Pauly-Wissowa, II, p. 1679, 65.

[7] In general, cf. F. Cumont, Pauly-Wissowa, *s. v.* Eshmun, VI, pp. 677. For Asclepius in New Carthage, cf. T. 838. The Orientalization of the god was reflected even in the late Greek concept of Asclepius. Certain features were due to foreign influence: the Egyptian shoes of the priest in Pergamum (T. 493), and more important, the temple in Epidaurus that was dedicated to Hygieia, Apollo and Asclepius, surnamed the Egyptians (T. 739, 6). But despite such Oriental touches, the Asclepius cult remained essentially Greek, and particularly Epidaurian; cf. above, p. 242.

[8] For Rome, cf. T. 845 ff. The Romans went to Epidaurus at the instigation of the Delphic Oracle (T. 850, *vv. 631 ff.*). The question of the transfer of the cult to Rome is complicated by the fact that two Roman sanctuaries seem to be referred to in T. 860, and it has therefore been assumed that the city had a shrine even before 291 B. C.; cf. Wissowa, *Kultus u. Religion*, p. 307, n. 7. Latte, *Gnomon*, VII, 1931, p. 121, n. 2, has pointed out that the Roman transliteration of the name Asclepius presupposes a spelling no longer used in the Epidaurus of the third century; cf. also Deubner, *Die Römer*, p. 462. Yet, this may be explained perhaps by the fact that the god was known in Rome even before his temple was founded. Concerning the sanctuary on the Tiber Island,

officials of later centuries. The Asclepius whom they found in Greece or in the Orient was one of the gods whom their city had made its own, and consequently he was certain to enjoy their favor. The emperors continued this policy.[9] Yet, most of all it was the Roman soldiers who contributed to the extension of Asclepius' reign. They took him to all the regions that came under Roman domination, to the farthest corners of the empire, to the ends of the inhabited world. The *Aesculapius castrorum* was the one who was received by the barbarian countries; the mild god, the friend of men, was brought to them by the conquering legions.[10]

Asclepius' world-wide reputation, however, should not obscure the fact that his old Greek sanctuaries still remained his strongholds, the places on which his fame rested most securely. They, too, profited by the steadily increasing reverence paid to the god. Epidaurus from the beginning had been intent on making its shrine a fane of Greece.[11] When the magnificent temple of the fourth century had been finished (T. 735 ff.) patients gathered from everywhere; there never was a dearth of sacrifices for Asclepius. Because the altar of the god was continually stained with blood, Epidaurus was called *Haimera* (T. 483). In spite of its unpopular political attitude, Epidaurus did not lose worshippers even in the third and second centuries, as is attested by the inscriptions. Aemilius Paulus, when visiting Epidaurus in 167 B. C., found it rich in offerings (T. 383). And although the Roman notables of the Republic did not show much interest in Epidaurus, although its monuments were carried away by Sulla (T. 743), although the sanctuary was plundered by pirates (T. 744), and although only a few benefactors of the shrine are known between 100 B. C. and 100 A. D., Strabo expressly says that the temple was always

cf. M. Besnier, *L'Ile Tibérine dans l'Antiquité*, 1902; Kern, *Religion*, III, p. 156. The late temple in the Thermae of Diocletian has been discussed by H. Jordan, *Commentationes Philologae in honorem Th. Mommseni*, 1877, p. 356. The Roman inscriptions that are extant are printed as T. 438. Julian claims that Tarentum was founded before Rome (T. 843), an assumption that can neither be proved nor disproved. A temple in Antium is attested for the second century B. C. (T. 844), another one in Croton is referred to in T. 842.

[9] Cf. below, p. 254.

[10] Cf. in general A. v. Domaszewski, *Die Religion des römischen Heeres, Westd. Zeitschr. f. Gesch. u. Kunst.*, XIV, 1895; Wissowa, *Religion u. Kultus*, p. 309. The religious movement in the late centuries has not yet been studied in its entirety. I list a few works which give information on Asclepius in the various provinces of the Roman empire: *Dacia*, L. W. Jones, *Univ. of Calif. Public.*, IX, 1929, pp. 245 ff.; 268 ff.; N. Igna, *Cultu lui Esculap i al Higei*, 1935; *Germania*, F. Drexel, *Die Götterverehrung im römischen Germanien, 4. Bericht der röm.-germ. Kommission*, 1922, pp. 1-68; *Helvetia*, O. Stähelin, *Die Schweiz in römischer Zeit*, 1927; *Illyria*, Pauly-Wissowa, Suppl. V, p. 337, 27 ff.; *Thracia*, Pauly-Wissowa, VI (2), p. 495, 62 ff. (*Moesia, ibid.*, p. 500, 23 ff.).

[11] Concerning the history of Epidaurus, cf. Hiller v. Gaertringen, *I. G.*, IV², 1, Prolegomena, pp. IX ff.; Latte, *op. cit.*, pp. 113 ff.; Kern, *Religion*, III, pp. 155 f.

filled with patients even in his own time (T. 382).[12] In the second century A. D. the Emperor Hadrian seems to have favored Epidaurus more than any other place. The senator Antoninus rebuilt the sanctuary and adorned it with magnificent monuments (T. 739, 6).[13]

The importance of Epidaurus at the end of the third century B. C. was challenged by Cos. The newly-founded temple rose quickly to a place of prominence. Even a Syrian king deigned to make a dedication to the Coan Asclepius (T. 796). Caesar upheld its sanctity (T. 797), a fact which certainly was of great consequence. The Coans themselves had shown that they respected the holiness of the place by not handing over to Mithridates the Romans who had sought refuge in the sanctuary (T. 798). Strabo calls the shrine very famous (T. 794). Tiberius recognized anew its right of asylum (T. 798). Claudius granted special honors to Cos (T. 799). When in the second century an earthquake damaged the sanctuary, help was given by Antoninus Pius for rebuilding the shrine.[14]

While Cos suffered severely from the damage inflicted upon it by the blind forces of nature, Pergamum had reached the pinnacle of its fame. Founded in the fourth century B. C., it added new buildings in the second. But during the same period, the temple was ravaged by Prusias (T. 802). The anti-Roman stand taken by the city in the first century— the Pergamenes failed to protect the Romans who had fled to the sanctuary before Mithridates (T. 809)—probably did much harm to the reputation of the shrine. However, it recovered quickly. Under the reign of Antoninus Pius the new temple of Asclepius Zeus was built by Rufinus (T. 803).[15]

[12] Cf. Latte, op. cit., pp. 127 ff.; he fails to mention Strabo, but concludes from the scarcity of inscriptions between 100 B. C. and 100 A. D. that the temple must have lost in importance. Livy says that Epidaurus is "now rich in traces of broken votives" (T. 383).

[13] Hadrian dedicated a bust of Epictetus in the temple; cf. I. G. IV^2, 1, no. 683, and Chr. Blinkenberg, Nord. Tidsskrift for filologi, III, 1894-5, p. 157. Concerning Antoninus, cf. Hiller v. Gaertringen, Hermes, LXIV, 1929, pp. 63 ff.

[14] For the history of Cos, cf. W. R. Paton and E. L. Hicks, The Inscriptions of Cos, 1891, pp. ix ff.; R. Herzog and P. Schazmann, Kos, I, Asklepieion, 1932, pp. IX ff. The inscriptions of Cos have not yet been published in full, but cf. Paton and Hicks, loc. cit.; A. Maiuri, Nuova Silloge epigrafica di Rodi e Cos, 1925; R. Herzog, Heilige Gesetze von Kos, Abh. Berl. Akad., 1926, no. 6. Claudius' help for the Coan sanctuary, which was instigated by his personal relation to Xenophon of Cos, his physician, was also motivated by the emperor's recognition of Cos as the birthplace of Hippocrates; cf. Herzog, Hist. Zeitschr., CXXV, 1921, pp. 216 ff.

[15] In general, cf. Th. Wiegand, Zweiter Bericht über die Ausgrabungen in Pergamon, 1928-32, Abh. Berl. Akad., 1932, n. 5; Wilamowitz, Glaube, II, pp. 506 ff.; Kern, Religion, III, pp. 157 ff.; O. Deubner, Das Asklepieion von Pergamon, 1938. A selection of the inscriptions from Pergamum is to be found in M. Fränkel, Die Inschriften von Pergamon, I-II, 1890-95. On the temple of Rufinus, cf. H. Hepding, Philologus, LXXXVIII, 1933, pp. 90 ff.; 241 ff.

Pergamum, in the opinion of Aristides, was the hearth of the god (T. 402). The whole island of Cos was sacred to him (T. 795). Epidaurus was his holy city (T. 400). Throughout the centuries the Greek sanctuaries remained the focus of the worship, even after Asclepius had conquered Orient and Occident.

5. THE CHRISTIAN ERA

The authority of the Asclepius temples was at its zenith when in the second century Christianity had grown from a small and unimportant sect that was anxious to live undisturbed by others into a proselytizing mass movement threatening to destroy the foundations of ancient religious life. Not only did the Christians face in Asclepius the god who had so much in common with their Savior;[1] the actual power of the god of medicine was highly dangerous to the new religion. To be sure, all the Greek gods were still worshipped. But, as Aristides testifies, in his time even in Rome it was Asclepius who had gained the most in his hold over men; of all Greek deities his fame alone had risen in equal measure with that of the Egyptian gods, who now came more and more to the fore, although they had not yet found the same legal recognition which was accorded to other foreign cults adopted by the state.[2] Small wonder that the Apologists denounced Asclepius so bitterly! However, the Asclepieia still proved impregnable to attack.

The situation changed somewhat during the third century A. D. In certain sections at least Christianity began to make inroads or Asclepius himself lost in power. Pergamum, destroyed by an earthquake between 253 and 260 A. D., was not rebuilt. Were no means available for a reconstruction, did nobody care what happened to the shrine? Soon the site was used as a graveyard.[3] But the other temples stood fast. In Epidaurus, new altars rose; Eleusinian priests administered the priesthood of Asclepius.[4] On the whole, the god held his own. Nay, the

---

[1] Cf. above, pp. 132 ff.

[2] Aristides (*Oratio*, XXVI, 105, p. 122, 23 f. K.) in reviewing the significance which the various gods had in the life of the Romans of his day says: αἱ δ' Ἀσκληπιοῦ χάριτες καὶ τῶν κατ' Αἴγυπτον θεῶν νῦν πλεῖστον εἰς ἀνθρώπους ἐπιδεδώκασιν. For the position of the Egyptian gods in the second century A. D., cf. Wissowa, *Religion u. Kultus*, p. 355.

[3] Cf. O. Deubner, *op. cit.*, p. 20.

[4] Cf. *I. G.*, IV², 1, no. 415; Latte, *Gnomon*, VII, 1931, p. 129. Latte, *ibid.*, p. 134, has drawn attention to the fact that no inscriptions can be traced to the period between 260-290 A. D. If this is indicative of a certain decline of the worship, the situation soon improved, as Latte himself has pointed out. Cos is not heard of after the second century A. D. However, it seems to have been in existence until 554, when it was demolished by an earthquake, cf. Herzog, *Kos*, p. XII.

Christian legend admits that Diocletian ordered the erection of new Asclepius temples, both in Pannonia and in Rome, and it tells how the Christian artists who refused to create an image of the god suffered martyrdom and death.[5]

The fourth century brought both reverses and successes. Constantine destroyed the sanctuary at Aegae. Holding the true Savior a jealous god, he commanded that this temple, too, be razed to its foundations: "At one nod it was stretched out on the ground . . . and with it (fell) the one lurking there, not a demon nor a god, but a kind of deceiver of souls, who had practiced his deceit for a very long time" (T. 818).[6] The last inscription preserved from Epidaurus and dating from the year 355 A. D. was dedicated to the Asclepius of Aegae. It commiserates the destruction of his temple and raises a protest against it.[7] Moreover, a general change is noticeable: the pagan religion was now adhered to above all by the educated, while it began to decline in popular appeal. Speaking of Aegae, Eusebius says: "The celebrated marvel of the noble philosophers [was] overthrown by the hand of the soldier" (T. 818). The Asclepius worship became esoteric. After Julian had assumed the purple, there seemed new hope even of winning back the masses. If the change in leadership was gratifying to all heathens, it must have been especially welcome to the followers of Asclepius, for the new emperor showed a marked preference for their god.[8] When Julian passed through Tarsus, the priest of the Asclepieion in Aegae asked him to restore the pillars which the high priest of the Christians had carried off from the shrine to build with them a temple for his own people. "And the Apostate straightway commanded that this be done at the expense of the bishop." But the efforts of the heathens were to no avail. "With great toil and expenditure, then, the Greeks barely managed to take down one of the columns and bring it to the lintel of the doorway of the church by means of mechanical devices; yet in a long period of time they were unable to drag it further. Therefore they left it and went away. When

---

[5] Cf. the so-called *Passio Sanctorum Quatuor Coronatorum*, ed. W. Wattenbach, *S. B. Berl.*, 1896, pp. 1281 ff.

[6] For the temple at Aegae, cf. T. 816 ff. and J. Geffcken, *Der Ausgang des griechisch-römischen Heidentums*, 1920, p. 279, n. 42; Latte, *op. cit.*, p. 121, n. 1. In Beroea in Syria, at approximately the same time, an Asclepius statue in the form of the beautiful son of Cleinias attracted so much attention among the pagans that the Christians destroyed the lovely monument, although it could not be proved that illegal services to the god were performed (T. 646).

[7] *I. G.*, IV[2], 1, no. 438; Latte, *op. cit.*, p. 134.

[8] Cf. G. Mau, *Die Religionsphilosophie Kaiser Julians*, 1908, pp. 66; 122.

Julian died, the bishop again easily lifted it up and restored it to its own place " (T. 820).[9]

The fate of Aegae was symbolic. To an ever increasing extent the sanctuaries of the pagans became quarries for the shrines of Christ. In Syria in the fifth century, the Asclepius cult seems to have been exterminated by the authorities. Nevertheless Theodoretus was afraid that in secret corners men continued to honor the god with libations and sacrifices (T. 5). In other sections, the faithful, though on the defensive, dared to express their belief openly. The Asclepius of Ascalon was worshipped even at that late date. The philosopher Proclus wrote a hymn in his honor (T. 826a; cf. 163). The same Proclus ascended to the Athenian Asclepieion to offer prayers and to invoke the help of the god (T. 582). At a time when the sanctuary of Athena Parthenos had already become a church, the temple of Asclepius was still frequented by the pious, by the intellectuals who clung to this god the longest.[10]

The temples of Asclepius, then, held out well nigh into the sixth century, the time at which even the last vestiges of paganism were finally stamped out all over the ancient world. The hero of physicians, who had become a god at the beginning of the classical period, proved to be as strong as those deities who were revered from time immemorial, if he was not even more powerful than they turned out to be. The temples of the Oriental gods withstood the onslaught of Christianity no longer than did the Asclepieia. The god of medicine who cured the sick had shown charity toward the poor, philanthropy toward all; he had been satisfied with small gifts in exchange for the greatest boon, health and freedom from disease; he had been mild and helpful, as he appeared to men in their dreams, and as he stood before their eyes in his statues. His deeds and his merits had endeared the son of Coronis to the ancients. That is why his divine abodes were among the last to fall.

---

[9] It is noteworthy that a writer like Firmicus Maternus mentions Asclepius only casually (T. 114), but inveighs against the Oriental deities. In the fourth century, these foreign gods had the greatest attraction for the people at large, and that is why the main attack of the Christians now turned against the cults of Isis or Mithras or Sarapis.

[10] Cf. Geffcken, *op. cit.*, p. 188. In this connection one must remember the importance which Asclepius assumed in late Neo-Platonic philosophy, cf. above, pp. 107 f.

LIST OF ABBREVIATIONS

Arch. Rel. Wiss.: *Archiv für Religionswissenschaft*

C. M. G.: *Corpus Medicorum Graecorum*

C. M. L.: *Corpus Medicorum Latinorum*

L. Deubner, *Attische Feste*: *Attische Feste*, 1932

———— *Die Römer*: *Die Römer, Lehrbuch der Religionsgeschichte*, ed. A. Bertholet and E. Lehmann, II, 1925, pp. 418 ff.

L. Edelstein, *Problemata*, IV: Περὶ ἀέρων *und die Sammlung der Hippokratischen Schriften, Problemata*, Heft IV, 1931

E. R. E.: *Encyclopaedia of Religion and Ethics*, ed. J. Hastings, I-XII, 1908-1922

L. R. Farnell, *Cults*: *The Cults of the Greek States*, I-V, 1896-1909

———— *Hero Cults*: *Greek Hero Cults and Ideas of Immortality*, 1921

O. Gruppe, *Griechische Mythologie*: *Griechische Mythologie und Religionsgeschichte, Handbuch der Klassischen Altertums-Wissenschaft*, 1906

R. Herzog, *Wunderheilungen*: *Die Wunderheilungen von Epidauros, Philologus*, Supplementband, XXII, Heft III, 1931

O. Kern, *Religion*: *Die Religion der Griechen*, I-III, 1926-1938

L. Kjellberg, *Asklepios*: *Asklepios, Upsala Universitets Årsskrift*, 1897. Filosofi, Sprakvetenskap och Historiska Vetenskaper, III, pp. 12 ff.; 70 ff.

F. Kutsch, *Attische Heilgötter*: *Attische Heilgötter und Heilheroen, Religionsgeschichtliche Versuche und Vorarbeiten*, XII, Heft 3, 1913

P. Maas, *Epidaurische Hymnen*: *Epidaurische Hymnen, Schriften der Königsberger Gelehrten Gesellschaft*, Geisteswissenschaftliche Klasse, IX, Heft 5, 1933

M. P. Nilsson, *Die Griechen*: *Die Griechen, Lehrbuch der Religionsgeschichte*, ed. A. Bertholet and E. Lehmann, II, 1925, pp. 280 ff.

———— *Griech. Feste*: *Griechische Feste von religiöser Bedeutung (mit Ausschluss der attischen)*, 1906

———— *Griechische Religion*, I: *Geschichte der Griechischen Religion*, I, *Handbuch der Altertumswissenschaft*, 1941

———— *Popular Religion*: *Greek Popular Religion*, 1940

Pauly-Wissowa: Pauly-Wissowa, *Real-Encyclopädie der Classischen Altertumswissenschaft*

F. Pfister, *Bursian*, 1930: *Die Religion der Griechen und Römer, Jahresberichte über die Fortschritte der Klassischen Altertumswissenschaft*, Band 229, 1930

L. Preller-C. Robert, *Mythologie*: *Griechische Mythologie*[4], I-II, 1894-1926

R. G. V. V.: *Religionsgeschichtliche Versuche und Vorarbeiten*

E. Rohde, *Psyche*: *Psyche, Seelencult und Unsterblichkeitsglaube der Griechen*[7-8], I-II, 1921

Roscher, *Lexikon*: *Ausführliches Lexikon der Griechischen und Römischen Mythologie*, ed. W. H. Roscher, I-VI, 1884-1937

H. Usener, *Götternamen*: *Götternamen. Versuch einer Lehre von der religiösen Begriffsbildung*, 1896

A. Walton, *Cult of Asklepios*: *The Cult of Asklepios, Cornell Studies in Classical Philology*, III, 1894

O. Weinreich, *Heilungswunder*: *Antike Heilungswunder, Religionsgeschichtliche Versuche und Vorarbeiten*, VIII, Heft 1, 1909

F. G. Welcker, *Götterlehre*: *Griechische Götterlehre*, I-III, 1857-1862

———— *Kleine Schriften*, III: *Kleine Schriften zu den Alterthümern der Heilkunde bei den Griechen*, etc., 1850

G. Wissowa, *Religion u. Kultus*: *Religion und Kultus der Römer, Handbuch der Klassichen Altertums-Wissenschaft*, 1912

U. v. Wilamowitz-Moellendorff, *Glaube*: *Der Glaube der Hellenen*, I-II, 1931-32

———— *Isyllos*: *Isyllos von Epidauros, Philologische Untersuchungen*, IX, 1886

INDEX OF TESTIMONIES *

* I am greatly indebted to Miss Janet Brock for her kind assistance in preparing the indices.

261

INDEX OF NAMES

Milton Keynes UK
Ingram Content Group UK Ltd.
UKHW021156130724
445466UK00010B/499